Oilfield Chemistry and Its Environmental Impact

Oilfield Chemistry and Its Environmental Impact

Henry A. Craddock
HC Oilfield and Chemical Consulting
Angus, UK

Registered Office(s)
John Wiley & Sons, Inc., 111 River Street, Hoboken, NJ 07030, USA
John Wiley & Sons Ltd, The Atrium, Southern Gate, Chichester, West Sussex, PO19 8SQ, UK

Editorial Office
9600 Garsington Road, Oxford, OX4 2DQ, UK

For details of our global editorial offices, customer services, and more information about Wiley products visit us at www.wiley.com.

Wiley also publishes its books in a variety of electronic formats and by print-on-demand. Some content that appears in standard print versions of this book may not be available in other formats.

Library of Congress Cataloging-in-Publication Data
Names: Craddock, Henry A., 1956- author.
Title: Oilfield chemistry and its environmental impact / Henry A. Craddock
  (HC Oilfield and Chemical Consulting, Angus, United Kingdom).
Description: Hoboken : Wiley, 2018. | Includes bibliographical references and
  index. | Description based on print version record and CIP data provided
  by publisher; resource not viewed.
Identifiers: LCCN 2018000479 (print) | LCCN 2018000920 (ebook) | ISBN
  9781119244264 (pdf) | ISBN 9781119244271 (epub) | ISBN 9781119244257
  (cloth)
Subjects: LCSH: Petroleum engineering. | Petroleum chemicals. | Gas
  engineering. | Green chemistry. | Environmental protection. |
  Demulsification. | Oil well drilling–Environmental aspects. | Drilling
  muds. | Oil field brines–Environmental aspects.
Classification: LCC TN870 (ebook) | LCC TN870 .C73 2018 (print) | DDC
  622/.3381–dc23
LC record available at https://lccn.loc.gov/2018000479

Cover Design: Wiley
Cover Image: © pan demin/Shutterstock

Set in 10/12pt WarnockPro by SPi Global, Chennai, India
Printed in Singapore by C.O.S. Printers Pte Ltd

10 9 8 7 6 5 4 3 2 1

To my dearest wife, Hazel, without whose tolerance, patience, support and love, this book would not have been possible.

Contents

Preface

In writing this book I have designed it primarily as a reference book for chemists and environmentalists practicing in the upstream oil and gas industry. The oilfield presents a large number of technical challenges to the chemist and many of these are set against a background of increasing regulation and control of environmental impacts. This book focuses on the key chemistries used across the oilfield as defined by various upstream oil and gas exploration, drilling, development, production, processing and transportation. It is an attempt to be as comprehensive as reasonable, however other sources should be consulted particular in reference to particular chemistries or issues of interest. The work includes chemistries used in cementing, completion, work-over and stimulation, and further covers a number of chemistries involved in enhanced oil recovery, however these areas are not as comprehensively covered as others such as drilling and production.

A number of other books and reference works have examined the subject matter but all from the viewpoint of function and then chemistry applied. This book is an attempt to look at chemistry types and their use and potential use alongside their actual and possible environmental impacts. In the final two chapters I have attempted to focus on issues concerned with the environmental impact and fate of chemicals used in the oilfield sector. An outline of the regulatory conditions in a number of regions of the globe is included as well as also a critique of these in terms of overall environmental protection and sustainability.

The organisation of the book is laid out to examine specific chemistry types (Chapters 2–8) with Chapter 1 laying out the historical context of the subject, from its early application as a fairly piecemeal approach to solving certain oilfield problems, to its development into a fully-fledged chemical discipline. Chapter 8 also includes the key area of formulation as this is a critical part of the application of chemicals to the oilfield as rarely can neat products be added or used directly. Chapters 9 and 10 are concerned with environmental issues which in the last 20 years have been a critical part of the discipline and a growing requirement for any oilfield chemist to understand. Chapter 10 is particularly focused on issues of sustainability which at the time of writing are becoming of growing importance and I believe will become vital in the continued use of chemicals and the application of chemistry to the oilfield.

I would like to acknowledge the works of Professors Johannes Fink and Malcolm Kelland whose books on 'Oilfield Chemicals and Fluids' and 'Production Chemicals' respectively were an invaluable source of information and guided me to numerous primary references. Laurent Schramm's seminal work on oilfield surfactants was of particular use in Chapter 3, and Malcolm Stevens third edition in Polymer Chemistry kept me on the straight and narrow in Chapter 2.

Both the Royal Society of Chemistry and Society of Petroleum Engineers library databases were extremely useful in affording primary sources of chemicals and chemistry applied in the oilfield context.

Finally, I would like to thank my daughter Dr Emma Craddock for her invaluable sense check on the critique and argument in Chapter 9.

Oilfield chemistry is a fascinating subject and continues to excite and stimulate and is sure to do so into the future. It is hoped this book will aid chemists and others to enjoy it further.

October 2017

Henry A. Craddock
Kirriemuir, Angus, UK

1

Introduction and History

This book is designed as a reference book for chemists and environmentalist examining the chemistry used in the oilfield and the environmental issues it can present. It is an attempt to be as comprehensive as reasonable and focuses on the key chemistries used across the oilfield as defined by various upstream oil and gas exploration, drilling, development, production, processing and transportation. It includes chemistries used in cementing, completion, workover and stimulation and further covers a number of chemistries involved in enhanced oil recovery (EOR). A number of other books and reference works have examined the subject matter but all from the viewpoint of function and then chemistry applied. This book is an attempt to look at chemistry types and their use and potential use alongside their actual and possible environmental impacts. The final two chapters are particularly concerned with the environmental impact and fate of chemical additives and chemical derivatives used in the oilfield sector, this contains an outline of the regulatory conditions in a number of regions of the globe and also gives a critique of these in terms of overall environmental protection and sustainability.

Oil production, as we know it today, originated in the mid-nineteenth century; however it was not until the twentieth century that the application of chemical inhibitors, particularly in the application of drilling additives such as lubricants and additives to separate oil and water and in corrosion protection, came into being.

As early as 1929, the physical chemistry around oil production was being investigated [1], and the limitations of experimental design in trying to reproduce the conditions in an oil well were becoming apparent.

In the early 1930s, there was a fundamental shift to a more scientific approach to the design and development of drilling fluids [2]. Until then drilling muds were considered as coarse suspensions and were formulated based on empirical knowledge as to their properties. However, the amount of field experience gained and theoretical knowledge to that point set the stage for an efficient programme of laboratory study and also the well economics, which can be effected by improved mud techniques. In particular laboratory work was undertaken to understand the functions of mud fluids and the properties necessary for their effective operation, such as weight, plasticity, viscosity, suspension efficiency or settling rate, permeability, non-gas cutting, maturity, gelling and gel strength and hydrostatic efficiency stability. It was also found that certain difficulties frequently occurred when drilling in limestone reservoirs.

Oilfield Chemistry and Its Environmental Impact, First Edition. Henry A. Craddock.
© 2018 John Wiley & Sons Ltd. Published 2018 by John Wiley & Sons Ltd.

1.1 Demulsifiers

One of the earliest production and processing problems encountered by the oilfield pioneers was the separation of crude oil from production water. There is anecdotal evidence of soaps and other detergent-based materials being used in oil wells in the late nineteenth century in the United States. This has a certain logic as it was already known that such materials or more correctly the surfactants they contain do act upon oils and fats, allowing their separation in water particularly if they have formed emulsions. These early soap products were at best able to aid in resolution at high concentrations added; however, the resulting water quality is likely to have been poor alongside the crude needing further treatment and/or settling. Over the years, however, better products and processes have resulted in a range of polymer- and surfactant-based demulsifiers and also a reduction in the demulsifier concentration required. This has come about due to a greater understanding of how surfactants work (see Section 3.1) and improvements in demulsification technology, which have been concentrated in the following areas:

- Chemical synthesis of new demulsifiers
- Development of laboratory and field techniques for testing demulsifier combinations
- Improvements in the engineering design of surface treating facilities, e.g. more favourable conditions for chemical emulsion destabilisation, forced coalescence in pipelines and treating units, faster and more efficient settling and cleaner phase separation

In modern crude oil production, efficient separation, dehydration and desalting of the crude oil, as well as treatment of effluent water to an environmentally acceptable level, are critical. In order to achieve this in a time-efficient manner and to ensure continuous and smooth oil production operations, demulsifiers have become a critical part of oilfield operations. Chemical demulsifiers are specially tailored to act where they are needed – at the oil/water interface, and their high efficiency at low dosages makes their use a very attractive economic way to separate oil and water [3].

In the 1940s the industrial availability of ethylene oxide (EO) allowed the production of fatty acid, fatty alcohol and alkylphenol ethoxylate nonionic surfactants, and this resulted in the first use of nonionic surfactants as demulsifiers in crude oil production [4].

With the creation of ethylene oxide/propylene oxide (EO/PO) block copolymers, the first 'genuine' demulsifiers were available. Addition of EO and/or PO to linear or cyclic (acid- or base-catalysed) alkylphenolformaldehyde resins and to diamines or higher functional amines yields classes of modified polymers that perform quite well at relatively low concentrations. Furthermore, these demulsifier bases were converted to high molecular weight products by reaction of one or more with difunctional compounds such as diacids, diepoxides, di-isocyanates and aldehydes, delivering a host of potential emulsion breakers. Over the decades many other classes of demulsifier were developed alongside better understanding of testing in 'live' oil situations. In recent years demulsifier development has focussed on the development of products to be environmentally acceptable [5], achieve good separation at low temperatures [6] and be effective in heavy waxy or asphaltenic crudes [7].

Significantly chemical demulsifiers have not only aided to the economics of modern oilfield production processing but also, by ensuring clean phase separation to give 'clean' water, reduced potential hazardous discharges of highly contaminated produced water.

Alongside the development of demulsifiers, the behaviour and stability of oilfield emulsion had been studied extensively, and a number of factors were isolated as being responsible for their stabilisation, primarily by film-forming asphaltenes and resins containing organic acids and bases. The effect of pH was also established [8]. In crude oil–brine systems, an optimum pH range over which the adsorbed film exhibits minimum contracted film properties. In this range, interfacial tension is high – frequently near its maximum value – indicating the absence of highly surface-active species; crude oil–brine emulsions generally show minimum stability and surfactant requirements for breaking these emulsions are significantly reduced – sometimes no surfactant is needed.

1.2 Corrosion Inhibitors

There is a long history of the application of chemical inhibitors for the protection of physical assets in the oilfield from the corrosive components of oil and gas production [9]. Chemical corrosion inhibitors used in the oilfield can be grouped into several common types or mechanistic classes: passivating, vapour phase, cathodic, anodic, film forming, neutralising and reactive. The common material of construction in oil and gas production is carbon or low-alloy steel, so the primary aim is inhibition of the corrosive effects of fluids and gases on steel.

Inorganic inhibitors, such as sodium arsenite (Na_2HAsO_3) and sodium ferrocyanide, were used up to the early twentieth century to inhibit carbon dioxide (CO_2) corrosion in oil wells, but the treatment frequency and effectiveness was relatively poor [8].

The development of many organic chemical formulations that frequently incorporated film-forming amines and their salts began in the early 1920s, as with demulsifiers investigating the use of detergents and soaps. Significantly in the mid-1940s, long-chain polar compounds (surfactants) were shown to have inhibitive properties [10]. This discovery dramatically altered the practice of inhibitor application on primary production oil wells and gas wells. It permitted operation of wells that, because of the corrosivity and volume of water produced along with the hydrocarbons, would not have been produced due to economic considerations [8]. Perhaps entire reservoirs would have been abandoned because of the high cost of corrosion. Inhibitors also allowed the injection and production of high volumes of corrosive water resulting from the secondary recovery concept of waterflooding. Tertiary recovery floods, such as CO_2, steam, polymer and in situ combustion floods, would usually be uneconomical without the application of corrosion inhibitors.

Over the years there have been many improvements in inhibitor technology, mainly in formulation and application methods [11–13]. Until the turn of the century, it seemed a very settled question that the primary chemistries involved, in particular with regard to filming surfactants, could be selected from the following, either individually or in formulated combinations. These include primary amines, quaternary salts of amines and imidazolines [12, 14]. These filming protection agents act either by filming on the metal surface or by interacting and creating a bond with a variety of scales, e.g. calcium carbonate deposited on these surfaces [15].

At the present time, the filming protection mechanism is the most widely used agent in oil and gas processing and transportation due to necessity of good performance within a highly dynamic environment [16]. The mixture of produced fluids can give rise to a highly aggressive corrosive medium, which is then in direct contact with carbon steel, and to further challenge this environment to place these mixtures under conditions of high flow, creating a number of shear stress conditions [17]. Until recently chemical corrosion inhibitors have been sought, which have the ability to film within fluid mixtures at the interface between the liquid and surface, i.e. surfactant-like materials [18] (see Chapter 3), and also are persistent under conditions of high flow, i.e. have a certain amount of persistence to removal within these conditions [19]. These materials are, in general, excellent corrosion inhibitors under a variety of field conditions [20]; however they have certain limitations to performance, such as conditions of high temperature [21], and can be hazardous to the environment [22], often having certain properties relating to marine toxicity and biodegradation, which make them less acceptable for use in the highly regulated offshore environments. In attempting to produce more environmentally acceptable corrosion inhibitors, a dilemma was observed in that the very technical effects that are being designed, for example, persistent materials, are not usually readily biodegradable. Similarly, a number of the chemistries are nitrogen based and possess inherent toxic properties [23]. These chemistries and many similar types are extensively explored in Chapter 3, alongside attempts to provide efficacious but environmentally acceptable products.

1.3 Drilling Fluids and Additives

Drilling fluids or 'muds' are probably among the oldest applied 'chemical' in the oilfield. The documented use of mud-ladened water for rotary drilling dates from the early twentieth century [24], but undoubtedly such products were already in common usage in such a way as to make them helpful to the driller, the operator or the engineer in solving his/her own special drilling problems. Alongside the use of these early water-based drilling fluids, the engineering particularly the tools and apparatus of the oilfield developed.

The materials encountered in wells drilled were varied, comprising unconsolidated sands, gravels and clays, in which thin layers of sandstones, shell conglomerates and shales would also be present. In these early days offshore drilling as we now come to think of it was not known; however many wells were drilled in tidal waters. All of this made it necessary for the oil exploration and production industry to start to employ specialists such as particular engineers, geologists and eventually oilfield chemists.

Drilling fluids are designed to exhibit thixotropic properties. Thixotropy refers to the property possessed by many suspensions of setting to form a jelly when allowed to stand undisturbed; this jelly can be broken down merely by shaking whereby the suspension reverts to the fluid condition. The change from jelly to liquid and liquid to jelly can be brought about in an indefinite number of times.

Thixotropy in drilling muds, particularly those to which reagents have been added, is largely due to the charges carried by particles in suspension and partly due to the size and shape of the particles. It may be that thixotropy is an intermediate stage between perfect stabilisation and complete coagulation of a suspension.

Over the years these muds have been refined and tailored to specific conditions and applications, and by the mid-twentieth century, three types of mud were established [25] as follows:

1) Clay freshwater-based muds
2) Organic colloids (both fresh-and saltwater based)
3) Water-based and oil-based blends (emulsions)

From field evidence and laboratory data, it was becoming possible to have a greater understanding of the drilling fluid on well performance. Over the coming decades until the present time, water-based drilling fluids have been a mainstay of drilling practices and well completion operations. However, as techniques have developed, chemical additives have been utilised to improve the rheological profile of the drilling fluid, aid in fluid loss control and ameliorate formation damage.

In the late 1930s, oil-based muds were being developed for drilling application. An oil-based mud is a drilling fluid in which oil has been substituted for water as the principal liquid ingredient. The composition of these muds was documented in the early 1940s and attempted to give through analytical techniques an understanding of the mud's characteristics and physical properties [26]. Over decades this work has also been developed to relate the design of the mud and its composition to its desired physical properties.

In an oil-based mud, the oil acts as the continuous phase and water as the dispersed phase in conjunction with emulsifiers, wetting agents and gelling agents, which are mainly surfactant and/or polymer based. The oil base can be diesel, kerosene, fuel oil, selected crude oil or mineral oil.

Emulsifiers are important to oil-based mud due to the likelihood of contamination. The water phase of oil-based mud can be freshwater or a solution of sodium or calcium chloride. The external phase is oil and does not allow the water to contact the formation.

Oil-based muds are more expensive but are worth the cost when drilling through

1) Troublesome shales that would otherwise swell and disperse in water-based mud
2) To drill deep, high-temperature holes that dehydrate water-based mud
3) To drill water-soluble zones
4) To drill producing zones, particularly deviated or horizontal completions

The disadvantages of using oil-based muds are as follows:

1) Inability to analyse oil shows in cuttings, because the oil-based mud has fluorescence confusing with the original oil formation
2) Contamination samples of cuttings, cores and sidewall cores for geochemical analysis of TOC (Total Organic Carbon) masks the determination of the real API gravity
3) Contaminate areas of freshwater aquifers causing environmental damage
4) Disposal of cuttings in an appropriate place to isolate possible environmental contamination

In recent years, the environmental concerns from the disposal of oil-based muds and contaminated drill cuttings have been prominent in a number of regions, particularly in offshore drilling and completion, and this is discussed in greater detail in Section 8.3.3.

However, the development of oil-based muds has allowed a great extension to the range of temperatures and pressures that drilling operations can undertake and also the complexity of such operations. Undoubtedly the shale gas developments of the last few decades would not have occurred without the development of horizontal drilling techniques, and in turn these would not have occurred without the greater use and understanding of oil-based muds [27].

1.4 Cementing

It was becoming common practice to cement line wells by the early twentieth century, and cement plugging for the exclusion of water was documented in 1919 [28]. The objective in this early work was twofold:

1) To prevent the oil sand from becoming flooded
2) To plug off bottom water, thereby preserving the individual well and reclaiming production

Over the next decade or so, the use of cement in oil wells increased dramatically, leading to a large degree of variation in the types and specification used, which was recognised as inadequate as it was primarily based on typical construction cement. A uniform specification for oil well cements was adopted in the late 1930s [29]. This led to the definition of specific test work and desirable properties and an attempt to standardise this specification across the entire industrial sector. To date this has largely applied with chemical additives to cements being allowed within specific parameters to achieve desirable properties pertinent to specific applications. Many chemistries have been used in achieving useful additives, and a number of examples are given throughout the following chapters. An additive of particular importance was the use of organic fluid loss additives [30]. These were developed for neat and gel cement slurries for use in squeeze operations to provide controlled dehydration of the slurry and, thus, fewer job failures. These materials were compatible with bentonite materials and all Portland cements, and did not adversely affect the physical characteristics of the set cement. They are effective over a wide range of well conditions and provide improved control of cementing treatments.

Fluid loss additives are now considered an essential component of any well cement. The fluid loss behaviour of a cement slurry basically occurs in two stages: (i) a dynamic one corresponding to the placement and then (ii) a static one, the waiting on the cement setting.

In general, these additives are based on polymers and in particular cellulose-type polymers; see Section 2.2.2. These materials are considered environmentally acceptable; indeed, they are PLONOR (pose little or no risk) listed materials [31], which means they do not have to be subjected to testing protocols; see Section 9.1.3. However, cellulose-type materials have several limitations: their fluid loss control properties decrease at temperatures above 200 °F, their salt tolerance is limited and over-retardation problems can occur at low temperature. Furthermore, the slurry viscosity to fluid loss ratio is unfavourable compared with that of synthetic fluid loss additives. When these cellulose-type materials are combined with other additives to help improve their performance, the environmental advantages are compromised and the resulting substance can no longer be listed as PLONOR. Modified cellulose materials have been developed to overcome this [32].

Additives of this type and others are considered across the following chapters of this book.

1.5   Well Stimulation and Improving Recovery

In the 1920s, the mechanisms that drive oil recovery were better understood [33], and the use of certain processes such as waterflooding was considered but was not deemed commercially viable.

In the 1930s, acid treatments on wells in carbonate reservoirs were being performed [34], which led to the study of acid treatment. This involved an analysis of the formation to be treated, the selection of a suitable type and amounts of acid, the time of treatments with respect to date of well completion and an economic analysis. The conclusion of the study was that it is considered good practice to use a properly inhibited acid, which practically eliminates the reaction of acid with steel and yet does not appreciably retard the action of the acid with limestone. These results and this objective form the basis of modern matrix stimulation practices.

Over the next 20 years or so, the understanding of what happens in acid stimulation developed and is discussed more fully in Section 5.6.2. Of particular importance has been the recognition of iron and iron salts being released during the stimulation process under certain conditions and need for its control [35]. Iron sulphide salts and free sulphur can be released, causing severe production impairment. Early treatments consisted of (i) making the plugging material water-wet with a non-ionic detergent, (ii) acidising to remove the iron salts, (iii) making the remaining free sulphur oil-wet with a surfactant, (iv) dissolving the free sulphur with a suitable solvent and (v) flushing all fluids into the formation. Standard practice is now to include a suitable sequestering agent to prevent the deposition of iron compounds following any acid treatment [36].

However, it is important to understand the acid treatment and design the chemical additive package appropriately. Many chemical additives for acid, including iron-control agents can be misused and overused with damaging results. Some agents precipitate if the expected downhole sources of iron are not present. In some cases, the iron actually keeps the sequestering agent in solution in spent acid. Thus, the effective use of iron sequestering agents depends upon the chemical conditions existing downhole during acid reaction. Since it is obviously impossible to know exactly what conditions will be encountered during an acid treatment, it is doubly important that care be used in selecting acid additives based on the best information available. The use of additives in acid stimulation is primarily discussed in Chapter 6.

1.5.1 Waterflooding

In the early 1930s, the use of waterflooding as a method of secondary recovery was being established indeed until that time there was only one place in the world where this method of rejuvenation was used on a large scale, that is, in the Bradford and Allegany fields of northwestern Pennsylvania and southwestern New York [37]. Here waterflooding was responsible for the complete revival of the Bradford field, which had been producing for sixty-two years. Although recoveries from flooding operations had been small, compared with the recoveries obtained from natural production in other producing districts of the world, they were nevertheless profitable.

Now the use of waterflooding as a means of secondary recovery is widespread and common practice. Indeed, the largest field in the world, the Ghawar, discovered in 1948 in the Eastern Province of Saudi Arabia, has a natural water drive support; therefore peripheral water injection was initiated to provide full pressure maintenance in 1968. Initially, water injection was conducted by gravity water injection. This was replaced by power water injection to provide flexibility in controlling the waterflood front propagation. The field has over 1500 wells, including producers and injectors both conventional and horizontal.

This use of water in maintaining reservoir pressure and production support, as well as a game changer in the economics of oil production, has had a significant effect on the complexion of oilfield chemistry. A large number of chemical additives used in the production of oil are really involved in water treatment, and these chemicals are described throughout the following chapters of this book, as well as some of the significant types such as scale inhibitors and biocides; see Section 1.6 by function being briefly detailed in this introductory chapter.

1.5.2 Enhanced Oil Recovery (EOR)

The previously described use of waterflooding has led on to more complex methods of improving and enhancing oil recovery. It is generally accepted that there are three primary techniques for EOR:

1) Thermal recovery
2) Gas injection
3) Chemical injection

Using EOR, 30–60%, or more, of the reservoir's original oil can be extracted [38], compared with 20–40% using primary and secondary recovery techniques [39]. Obviously, the techniques surrounding chemical injection have had the greatest bearing on oilfield chemistry and in particular the use of polymers and surfactants in EOR. These chemistries and their environmental consequences are substantially covered in Chapters 2 and 3, respectively.

1.6 Water Treatments

As stated in the previous section, the use of water has had a great influence on the chemicals used in oilfield activities; however, as equally important there has been a greater understanding of the chemistry of the water involved and occurring in the oilfield. This can be various types of water:

- *Connate water* that is naturally occurring in the reservoir and has been trapped in the pores of a rock during formation of the rock. The chemistry of connate water can change in composition throughout the history of the rock. Connate water can be dense and saline compared with seawater.
- *Formation water* or interstitial water in contrast to connate water is simply water found in the pore spaces of a rock and might not have been present when the rock was formed.
- *Produced water* is a term used in the oil industry to describe water that is produced as a byproduct along with the oil and gas. Oil and gas reservoirs often have water as well as hydrocarbons, sometimes in a zone that lies under the hydrocarbons and sometimes in the same zone with the oil and gas.
- *Injection water* is the water used in secondary recovery, and its sources are usually aquifer or other freshwater sources (e.g. river), seawater or produced water (this is water being reinjected from oil production operations).

A full understanding of the chemistry of these waters is essential for an understanding of the oilfield chemistry that may be utilised in oilfield operations. These waters can be used for the make-up of drilling fluids, used in workover or enhanced recovery operations and utilised in waterflooding. They will require treatment to improve their quality usually by addition of a biocide and often with additional oxygen removal to aid in corrosion prevention. The chemistries involved here are given in a number of chapters of this book and are varied as they employ a number of chemistries from phosphorous products (Chapter 4) to simple organic molecules (Chapter 6) and complex polymers (Chapter 2). The use of water as a make-up and carrier solvent in oilfield chemical formulation is described in Section 8.2.1.

Although the use of water and its production was commonplace in the oilfield, it was not until the 1960s that analytical chemistry was applied to oilfield waters to give a full picture of the potential interactions that could occur [40]. Until that time water was seen as both a useful instrument in secondary recovery and a nuisance requiring disposal.

At this time, the importance of analysis of water chemistry was becoming important in geophysics in the understanding of oil and gas in sedimentary rocks [41]. These analyses were utilised by a number of oilfield disciplines, and from the water chemistry the design of chemical treatments was developed and improved as the water chemistry could be replicated in the laboratory and placed under simulated reservoir and production conditions.

1.6.1 Scale Inhibitors

Scale deposition was recognised as a problem in oilfield operations in the late 1930s, and particularly in waterfloods scaling compounds frequently were deposited away from the wellbore in flow channels of waterflood wells [42]. Mineral deposition in formation flow channels and on tubing, casing and producing equipment has been and continues to be a major problem in the upstream oil and gas industry. Compounds such as calcium carbonate, calcium sulphate and barium sulphate carried in the produced water can crystallise or precipitate as a result of a pressure drop, a temperature change, or exceeding the solubility of the product. This scaling reduces and sometimes even stops oil production by plugging the formation, perforations or producing equipment.

Acid, salts and phosphates were suggested as possible remedial treating chemicals. However, it was not until much later that the practical application of chemical scale inhibitors was developed and utilised. In the first half of the twentieth century, the regular remediation was extraction of equipment from the well, and the deposits were removed mechanically with scrapers and other such devices. Hydrochloric acid was also used when the plugging material was basically carbonate in nature [43].

In the late nineteenth and early twentieth century, natural products such as tannins were being used in boiler water treatments partly to prevent scale formation interestingly because of their environmental acceptability such products have returned to application, mainly as a flocculant for water clean-up; see Section 6.1 [44]. Throughout the 1950s and 1960s, a variety of polymer- (Chapter 2) and phosphorous (Chapter 4)-based scale inhibitors were applied to a number of scaling issues often with mixed success, especially where deposition was occurring at the near wellbore region. A critical advance was the controlled placement of chemical in the near wellbore to allow controlled release of scale inhibitor [45]. This was achieved using the extremely slow solubility di-metallic polyphosphates injected into producing zones in conjunction with fracturing. As a result of this type of treatment, scale deposition in the formation in addition to the wellbore and producing equipment was observed to be absent for a prolonged period of time and in some cases for over a year.

Over the next three or four decades, the use of placement technologies in the design of scale inhibitor squeezes has been highly successful in preventing scale deposition and the corresponding productivity declines. Alongside this, the understanding of how the chemistry of the scale inhibitor interacts with the formation has improved further the design and application of this chemistry [44, 46].

Scale inhibitors function at concentrations significantly below the levels required to sequester or chelate the scaling cations. The molar ratio of precipitate held in solution to inhibitor is typically on the order of 10 000 : 1. It has been postulated that scale inhibitors prevent, slow or distort crystal growth by blocking growth sites. It is also believed that these inhibitors prevent the adhesion of scale to metal surfaces in some unknown manner. Regardless of the inhibition mechanism, scale inhibitors must be present during scale nucleation in order to function effectively.

Numerous chemicals are effective inhibitors, but only four classes of compounds have been widely applied in the oilfield: polyphosphates, phosphonates, phosphate esters and polyacrylates/polyacrylamides. These are substantially described in Chapters 2 and 4.

An equally significant breakthrough was in the understanding of barium sulphate scaling [47], its study in the oilfield and application of suitable treatments, particularly of chemical inhibitors. This is discussed in more detail in Section 5.3.

In the late 1960s, it was established that the usual methods of studying oilfield mineral scale deposition in the laboratory do not work for barium sulphate because only small non-adhering crystals are formed. On the other hand, barium sulphate scale found in downhole or surface equipment is strongly adhering and may contain very large crystals. The difference derives from the extremely low solubility of barium sulphate. The firm adherence of scale and the consistent development of oriented crystals of 100μ and larger, suggest a relationship between scale adherence and crystal growth.

In modern oilfield chemistry, much is now understood about scale deposition, and inhibition mechanisms [44] under various circumstances, from to the injection of seawater, causing barium sulphate deposition to the formation of the so-called exotic scales such as lead sulphide. Throughout the following chapters of this book, the reader will be introduced to a wide variety of chemistries that are used to treat, control, dissolve and inhibit the deposition of such inorganic scales.

1.6.2 Biocides and Bacterial Control

The organic origin of hydrocarbons has long been an accepted fact with nearly all geologists, with the weight of evidence favouring an organic origin. Crude oils themselves do not take long to be generated from appropriate organic matter. Most petroleum geologists believe crude oils form mostly from plant material, such as diatoms (single-celled marine and freshwater photosynthetic organisms) and beds of coal (huge fossilised masses of plant debris). It is somewhat obvious that given the right conditions bacteria will also be able to exist on such organic material. In introducing water or disturbing the amount of natural water present, it is possible for bacteria to become established and grow. However, it was not until the late 1930s that an evidential link relating bacteria in the oilfield to production problems was established, in this case the role of anaerobic bacteria in corrosion [48].

In 1950 the role that bacteria play in waterflood operations had been definitively established [49, 50], and numerous observations had been made that indicated that bacteria are very effective in plugging the sand faces in water-input wells. A number of treatments had been developed including batch treatment with sodium hypochlorite (see Section 5.1.1), which is very effective in cleaning out bacteria-infested lines and water-injection wells; but most often the effect was not sufficiently permanent, and follow-up treatments were necessary. Chemical compounds belonging to the broad classification of alkyl and aryl high molecular weight amine salts and certain condensed EO–amines were applied as potential bactericide/corrosion inhibitors [50]. These products have strong surface-active properties and therefore have additional cleaning properties in the injector wells and also increase the permeability of the sand to water.

As secondary recovery techniques employing water injection became more economic, more numerous injection problems became more apparent. One of the most important is the tendency of porous rock around the wellbore to become partially plugged. It has been shown that this plugging is the result of both the deposition of inorganic materials such as ferric hydroxide and ferrous and ferric sulphide and the formation of bacterial deposits. Bacterial plugging occurs not only on the face of the formation but also within the matrix of the rock [51].

Studies have shown that most injection systems contain microorganisms of some type. Of the various microorganisms commonly found in oilfield injection waters, almost all have been shown to produce some degree of formation damage. Bacteria-laden waters often are responsible for much of the reduction in injection rate observed in input wells.

Oxidising agents such as calcium and sodium hypochlorite have been used in treating injection wells damaged by bacteria. The beneficial effect of these materials has been attributed to the oxidation of living or dead bacterial cells or other easily oxidised organic matter produced as the result of bacterial action [52].

In modern oilfield practice, the role of bacterial control is well established, and the main biocides and their application proven and usually highly effective if well managed and maintained. This is the case for managing bacteria from added water [53]. In the reservoir things may be more complicated [54], where recent work would indicate that biocides may only cause cell injury and not completely 'kill', particularly if the dose rates are sublethal. In alternative strategies, the role of biostatic agents particularly in competitive 'feeding' so that bacteria are unable to produce harmful waste products has gained some traction [55–57]. These products are mainly based on nitrite and nitrate salts and are more fully considered in a number of sections in Chapter 5, primarily in Section 5.9.4, but also in Chapter 6, Section 6.7, where anthraquinone is described as a biostatic agent.

It would appear that the use of biocides and the chemical types available in the oilfield is settled particularly in light of increasing regulatory controls in many parts of the world; see Chapter 9. However, the understanding of the microbial population in the oilfield is far from complete, and new technologies in analysing and observing this population have opened a new door for further study [58].

1.7 Crude Oil Treatments

In the history of oilfield chemistry, the direct treatment of crude oil by deliberately designing chemical additives to change its properties, primarily its flow characteristics, is a relatively recent development. By contrast among the earliest crudes produced in what is recognised as the modern era, from the mid-nineteenth century onwards, are heavy crude oils [59]. The nature of paraffinic crudes was realised and, for the era, quite accurately determined in the 1920s where paraffin wax was isolated and its properties analysed, including the important recognition of the crude oil itself to aid in the solubility of the paraffin component [60].

Throughout the decades, the characterisation of heavy crude oils and the understanding of viscosity in crude oils developed. However, the main aids to producing heavier crude oils were by introduction of heat into the production process to keep the paraffin wax material from depositing and/or the addition of diluents materials such as aromatic solvents [61, 62]. Nevertheless deposition would often be problematic and be displaced from the flow lines to other equipment.

It was also noted that these 'waxes' also contained gums, resins, asphaltic material, crude oil, sand silt and in some instances water in addition to the wax crystals. In addition, the wax found in paraffin may range from the lowest to the highest melting point waxes present in the oil. The paraffin accumulations in wells in the same field will differ in percentages of wax of given melting points, since the conditions governing precipitation vary with each individual well [61].

1.7.1 Wax Inhibitors and Pour Point Depressants

It was not until the 1960s that direct chemical interaction to prevent wax depositing was proposed and developed, and at that time was known as crystal poisons [63]. Until then the main methods of paraffin deposition control were by regular mechanical removal, hot oiling and the maintenance of heat within the production system. Although this strategy was acceptable to an extent with onshore production, the advent of extremely deep production, offshore drilling and ocean floor completions, the application of such remedial measures becomes prohibitive economically. As a result, use of chemical additives as paraffin deposition inhibitors became more prevalent [64]. Since no one additive had proved to be universally effective, the selection of an efficient additive for a specific application becomes important, and as a consequence a better understanding of the mechanism of inhibition is also necessary. This has developed alongside the application of chemical additives.

Although pour point depressing additives were known in application in lubricating oils from the 1950s [65] and in refinery applications in the 1960s [66], it was not until over two decades later that their use in the oilfield was being reported [67, 68]. The majority of the chemistries developed for these applications are polymer based and are discussed throughout Chapter 2.

The mechanism of wax deposition has been studied extensively [69, 70]. Waxes are solids made up of long-chain ($>C_{16}$) normal or branched alkanes. These are naturally present in crude oils and some condensates. It has been conclusively established that normal alkanes, i.e. *n*-paraffin waxes, are predominantly responsible for deposition problems in pipelines; however wax deposition from paraffinic crudes can be both, a topside and downhole problem, blocking the flow of hydrocarbons as they are cooled during extraction and production [71].

The waxes in crude oils are much harder to control than those in condensates as they are composed of longer-chain alkanes. When the paraffinic wax is composed of alkanes of C_{16}–C_{25}, soft malleable waxes are observed. At higher molecular weights, C_{26}–C_{50} and greater, hard crystalline waxes are seen. The melting point of the wax is directly related to its molecular size and complexity; generally the higher the molecular size, the higher the melting point and the more difficult it is to keep the paraffin wax from forming deposits [72].

In the reservoir at high temperatures and pressures, any paraffinic waxes within the crude oil will be in solution. As the crude oil is produced, the temperature (and pressure) drops and wax will begin to precipitate from the crude oil if present in sufficient quantities and not solubilised by other components of the crude. Usually such precipitation or crystallisation occurs as the wax forming needles or plates [69].

As the pressure decreases, there is usually a loss of the lighter hydrocarbons to the gas phase. This reduces the solubility of the waxes in the crude oil.

In examining wax deposition, the most important measurement is the wax appearance temperature (WAT) or cloud point temperature. This is the temperature at which the first wax crystals begin to precipitate from the crude oil. It is not the same as the pour point. Typically, wax deposits when the pipe wall or other system surface temperature is below the WAT and below the temperature of the oil. The pour point temperature is generally reached when over 2% of the wax has deposited, whereas the WAT is observed when 0.05% or less of the wax has formed [69].

Wax deposition is considered to occur by two primary mechanisms:

1) If the surface (e.g. pipe wall) is colder than the WAT, then wax can form and deposit at this surface. This can occur even if the bulk fluid is above the WAT. This is known as the 'molecular diffusion mechanism'.
2) Already precipitated wax close to the surface will move to a region of lower velocity at the surface and deposit. This is known as 'shear dispersion'.

Therefore, wax deposition can occur by the first mechanism above and below the WAT, whereas the second mechanism only operates below the WAT. There are a number of other mechanisms and approaches, all of which contribute to the overall deposition effect; however by far the most significant effect has been shown to be the molecular diffusion mechanism [73].

Wax deposition and the WAT are both affected by the amount and type of asphaltenes present in the crude oil. In general, a significant amount of wax deposition is reduced in crudes containing high asphaltenic content [74].

In addition to deposition effects, waxy crudes exhibit problems related to increased viscosity and even gelation of the oil due to high amounts of wax precipitation. Crudes with a high paraffinic wax content are usually more prone to this problem. On cooling the waxes crystallise as plate-like arrays, which form a lattice structure, trapping the remaining liquid oil. This results in increased oil viscosity, decreased oil flow and reduced pressure in the pipeline [75].

1.7.2 Asphaltene Inhibitors, Dispersants and Dissolvers

Although there was some understanding of the composition of asphaltenes from as early as the early twentieth century, it was related to their constituency in bitumen and related materials and not as an overall component of crude oils. In the 1950s reports on the composition of asphaltenes began to appear [76]. In particular the understanding of the relationship between resins and asphaltenes began [77]. In addition understanding of the molecular structures and their properties in relation to a mineral oil residue was also reported [78]. From the early 1960s onwards, there has been an exponential growth in the study of asphaltenes and other components of crude oils with high molecular weight, and in recent times asphaltenes are probably the most studied, and still little understood, of oilfield chemical phenomena [79].

Asphaltenes are a component of all crude oils but are more prevalent in those of a low API gravity particularly of less than 10. The colloidal behaviour of asphaltenes in crude oil, the lack of knowledge about their structure and the complexity of the aggregation, flocculation, precipitation or deposition processes make their diagnosis and characterisation complex.

Asphaltenes comprise the heaviest and most polar fraction of crude oils. They are insoluble in low molecular weight alkanes, such as *n*-hexane, *n*-heptane, etc. (which can be used as an indicative test of their presence). The consensus of opinion is that asphaltenes exist in the form of a colloidal dispersions and are stabilised in solution by resins and aromatics, which act as peptising agents. Resins in particular are believed to stabilise asphaltenes by bridging between polar asphaltene particles and the non-polar oil surrounding them [80].

Asphaltenes are organic molecules with polycyclic structures of high molecular weights, 1000–10 000, usually having heteroatoms (N, S, O) possibly involved in metallic bonding and long aliphatic (alkane-like side chains), having an affinity for oil phase. The associated resins are also polycyclic molecules with associated aliphatic side chains; however these materials are soluble in *n*-heptane [81].

Together asphaltenes and resins form the asphalt particle, which carries a charge and can agglomerate under the right conditions; usually they are in equilibrium.

If conditions are right and equilibrium is disturbed such as in the presence of an aliphatic solvent and changes in pressure, temperature and composition during oilfield operations, such as operating below the bubble point, under such conditions the related resins desorb from the surface of the asphaltene colloidal particles, resulting in flocculation and precipitation.

In addition, mechanical shear forces, CO_2 in the crude and acidic conditions especially the presence of iron can drive destabilisation towards aggregation, flocculation and deposition.

The critical operational condition is the bubble point:

- Reducing towards the bubble point, the light fractions of the crude increase in volume as the pressure reduces – minimum stability.
- Above the bubble point, the light ends are released and solubility again increases.

However, in going below the bubble point, if asphaltenes are destabilised and at sufficient concentration, they will flocculate and deposit. Once this happens they cannot return into solution, even when operations are again above the bubble point. Asphaltene deposition is not reversible [80].

Until recently asphaltene deposition was treated remedially by mechanical removal or by dissolution in an aromatic solvent or similar chemical agent [82]. Chemical inhibitors and dispersants based on polymeric and surfactant chemistry are now available and are described in Chapters 2 and 3. Given an accurate diagnosis of the risk of asphaltene deposition, these can be an effective preventive measure against it.

These are the main oil treatment areas, namely, paraffin wax and asphaltene deposition. In the upstream oil and gas sector, the other major product line in this area is the application of ultra-high molecular weight polymers and related products as oil drag reducing agents. These are described in Section 2.1.1, mainly vinyl polymer-based materials.

1.8 Other Chemical Products

There are a wide range of other chemical products that have developed over the last 5 or 6 decades in the oil and gas industry such as hydrogen sulphide scavengers, deoilers, kinetic hydrate inhibitors, sand control agents, naphthenate control agents, flocculants, foamers and defoamers.

These and many more are covered in the range of chemistries described in the following chapters of this book. Many are small organic molecules (Chapter 6), but there are also a number of inorganic materials and metals and their salts used (Chapter 5). Phosphorous and silicon chemistries (Chapters 4 and 7) have also important functions and uses in the oilfield sector.

All of these materials have to a greater or lesser degree an environmental impact, and throughout this book an attempt is made to show the effort and strategies adopted in reducing these impacts.

1.9   Oilfield Chemistry

Throughout this chapter, there has been an attempt to place into context the development and application of oilfield chemicals with the evolution of a discipline, which today is known as oilfield chemistry. A few decades ago, an oil and gas company may have had very few chemists, and it was usually a drilling engineer or petroleum technologist that was charged with examining the application of chemical solutions to upstream oil and gas problems. The chemical industry also would have had very few specialists in oilfield chemistry but would have had a deep and fundamental understanding of how their chemical products functioned and their potential application to oil and gas drilling development and production problems.

In 1972, Vetter published what the author considers a seminal paper to probably christen the discipline of oilfield chemistry [83]. He argued that other industrial sectors were applying chemistry and sophisticated chemical technologies to better understand their processes and products while at the time the industry did not know why a scale inhibitor works or even, in most cases, why the scale is formed. He asked a number of fundamental questions:

- *How can we determine the stability of chemicals under oilfield conditions?*
- *What makes a dispersion of solids in water or oil break?*
- *How do different chemicals react with each other or with the formation when they are injected simultaneously?*

This lack of basic understanding was set against a background of millions of dollars being spent annually to inject chemicals into wells and reservoirs and to treat reservoirs, wells and well effluents with techniques and chemicals that are sometimes obscure and uncertain.

The results were at best unpredictable with millions of dollars wasted on the wrong chemicals because the reactions observed in a simple beaker or bottle test did not occur in the reservoir.

He concluded rightly that most of these failures were due to lack of knowledge about the specific chemistry involved and further concluded that some of the chemistry applied in the field would require the additional employment of a chemist. Additionally, he made a critical recommendation that the industry across all its component parts in the supply chain, oilfield operators, service companies, the chemical industry and various research organisations combine research efforts.

At that time oilfield chemistry was in a 'grey' area between the different sciences and technologies: the subject seemed too 'oilfield related' for the professional chemical societies and too 'chemistry related' for the Society of Petroleum Engineers (SPE). Things have changed, and over the next four decades, oilfield chemistry developed to such an extent that not only do many professional bodies recognise its importance with specific seminars directed at chemistry and its application to the oil and gas industry, but the industry itself has invested in fundamental research either in-house in oil operators and service companies or in funding academic research and other institutes to undertake fundamental examinations of chemistry both within the oilfield and applied to it. The chemical industry also has invested in much research and development and has created many chemical products to be applied to specific oilfield problems, and most large chemical manufacturers have specific oilfield chemical divisions or business units. The service companies also have grown almost to be as large as the oil operators, and within this across drilling, development and production operations in the upstream sector, a multibillion-dollar international business has grown.

Oilfield chemistry, too, has changed; in a number of academic institutions, it is now a taught subject, and within the service sector, an oilfield chemist is now a rare designation, usually being a specialist with either drilling, cement, completion, stimulation and workover, EOR, production chemistry, etc. Indeed, these specialisms have been further designated to areas such as corrosion, scale, emulsions, etc.

A critical co-development alongside the general understanding of oilfield chemistry has been the use of analytical chemistry. A few decades ago, this meant running iron and manganese counts to determine corrosivity in a system or measuring a phosphorous residual to determine a phosphonate scale-squeeze chemical-flowback profile. It was recognised from the 1970s onwards that analysis of returned treating fluids from the reservoir provides valuable information on the performance of chemical treatments [84].

Today, the techniques used to perform these analyses can be found in most, if not all, technical service laboratories or even can be implemented directly in the field. In the past two decades, analytics in the oilfield has grown to be a major discipline, integrally supporting the application of many different types of oilfield chemicals and becoming viewed by some as a technology differentiator.

Also within the last two decades, the importance of the environmental impact from oilfield chemicals has become in many areas preeminent, and regulation has striven to control and moderate this while also controlling the use of the chemicals themselves. Much of this is discussed in Chapters 9 and 10, with Chapter 10 concerned about the sustainability of the actions of both the oil and gas sector and the chemical activities within it. It is in this final chapter that the expectations of the industry are discussed while trying to reconcile its place in an oil and gas sector, which is dramatically changing to face the consequences of a more sustainable future.

Oilfield chemistry is certainly now a recognised disciple within the oil and gas sector although it still has a certain mysteriousness within an industry sector dominated by engineers. Vetter's 'grey area' still exists, but is perhaps a shade or two lighter; it is hoped that this volume of study adds to the luminosity.

References

1 Tickell, F.G. (1929). Capillary phenomena as related to oil production. *Transactions of the AIME* **82** (01): 343–361, SPE 929343.

2 Strong, M.W. (1933). Mud fluids, with special reference to their use in limestone fields. 1st World Petroleum Congress, London, UK (18–24 July 1933), WPC-1094.

3 Kokal, S.L. (2005). Crude oil emulsions: a state-of-the-art review. SPE Production & Facilities, Vol. 20 (01), SPE 77487.

4 Staiss, F., Bohm, R., and Kupfer, R. (1991). Improved demulsifier chemistry: a novel approach in the dehydration of crude oil. *SPE Production Engineering* **6** (03): 334–338, SPE 18481.

5 Newman, S.P., Hahn, C. and McClain, R.D. (2014). Environmentally friendly demulsifiers for crude oil emulsions. US Patent 8,802,740.

6 Lin, C., He, G., Li, X. et al. (2007). Freeze/thaw induced demulsification of water-in-oil emulsions with loosely packed droplets. *Separation and Purification Technology* **56** (2): 175–183.

7 Silva, E.B., Santos, D., Alves, D.R.M. et al. (2013). Demulsification of heavy crude oil emulsions using ionic liquids. *Energy Fuels* **27** (10): 6311–6315.

8 Strassner, J.E. (1968). Effect of pH on interfacial films and stability of crude oil–water emulsions. *Journal of Petroleum Technology* **20** (03): 303–312, SPE 1939.

9 Nathan, C.C. ed. (1973). *Corrosion Inhibitors*. Houston, TX: National Association of Corrosion Engineers.

10 Malik, M.A., Hashim, M.A., Nabi, F. et al. (2011). Anti-corrosion ability of surfactants: a review. *International Journal of Electrochemical Science* **6**: 1927–1948.

11 Dougherty, J.A. (1998). Controlling CO_2 Corrosion with Inhibitors. Paper No. 15. CORROSION-98 NACE International, Houston, TX.

12 Simon-Thomas, M.J.J. (2000). Corrosion inhibitor selection - feedback from the field. CORROSION 2000, Orlando, FL (26–31 March 2000), NACE 00056.

13 Gregg, M.R. and Ramachandran, S. (2004). Review of corrosion inhibitor develop and testing for offshore oil and gas production systems. Corrosion 2004, New Orleans, LA (28 March-1 April 2004), NACE 04422.

14 Raman, A. and Labine, P. ed. (1993). *Reviews on Corrosion Inhibitor Science and Technology*, vol. **1**. Houston, TX: National Association of Corrosion Engineers (NACE).

15 Hodgkiess, T. (2004). Inter-relationships between corrosion and mineral-scale deposition in aqueous systems. *Water Science and Technology* **49** (2): 121–128.

16 Jayaraman, A. and Saxena, R.C. (1996). Corrosion inhibitors in hydrocarbon systems. CORROSION 96, Denver, CO (24–29 March 1996), NACE 96221.

17 Efird, K.D., Wright, E.J., Boros, J.A., and Hailey, T.G. (1993). Correlation of steel corrosion in pipe flow with jet impingement and rotating cylinder tests. *CORROSION* **49** (12): 992–1003.

18 DeBerry, D.W. and Viehbeck, A. (1988). Inhibition of pitting corrosion of AISI 304L stainless steel by surface active compounds. *CORROSION* **44** (5): 299–230.

19 De Marco, R., Durnie, W., Jefferson, A. et al. (2002). Persistence of carbon dioxide corrosion inhibitors. *CORROSION* **58** (4): 354–363.

20 Roberge, P.R. (2002). *Handbook of Corrosion EngineeringCorrosion Inhibitors*, 2e, Section 10e. McGraw-Hill.

21 Chen, H.J., Hong, T. and Jepson, W.P. (2000). High temperature corrosion inhibitor performance of imdidazoline and amide. CORROSION 2000, Paper 00035, Orlando, FL (26–31 March 2000).

22 Garcia, M.T., Ribosa, I., Guindulain, T. et al. (2001). Fate and effect of monoalkyl quaternary ammonium surfactants in the aquatic environment. *Environmental Pollution* **111** (1): 169–175.

23 Taj, S., Papavinasam, S. and Revie, R.W. (2006). Development of green inhibitors for oil and gas applications. CORROSION 2006, San Diego, CA (12–16 March 2006), NACE 06656.

24 Knapp, I.N. (1916). The use of mud-ladened water in drilling wells. *Transactions of the AIME* **52** (01): 571–586, SPE 916571.

25 Kelly, W.R., Ham, T.F., and Dooley, A.B. (1946). *Review of Special Water-Base Mud Developments*, Drilling and Production Practice. New York: American Petroleum Institute, API – 46- 051.

26 Hindry, H.W. (1941). Characteristics and application of an oil-base mud. *Transactions of AIME* **142** (01): 70–75, SPE 941070.

27 Middleton, R.S., Gupta, R., Hyman, J.D., and Viswanathan, H.S. (2017). The shale gas revolution: barriers, sustainability, and emerging opportunities. *Applied Energy* **199**: 88–95.

28 Shidel, H.R. (1919). Cement plugging for exclusion of bottom water in the Augusta field, Kansas. *Transactions of the AIME* **61** (01): 598–610, SPE 919598.

29 Forbes, R.J. (1937). Specifications for oil-well cement. 2nd World Petroleum Congress, Paris, France (14–19 June 1937), WPC- 2120

30 Stout, C.M. and Wahl, W.W. (1960). A new organic fluid-loss-control additive for oilwell cements|. *Journal of Petroleum Technology* **12** (09): 20–24, SPE 1455.

31 www.ospar.org/documents?d=32652.

32 Bach, D. and Vijn, P. (2002). Environmentally acceptable cement fluid loss additive. SPE International Conference on Health, Safety and Environment in Oil and Gas Exploration and Production, Kuala Lumpur, Malaysia (20–22 March 2002), SPE 74988.

33 Clarke, H.C.O. and Lowe, H.J. (1926). Increasing recovery and its economic effects. *Transactions of the AIME* **G-26** (01): 241–247, SPE 926241.

34 Covel, K.A. (1934). *Acid Treatment of Michigan Oil Wells*, Drilling and Production Practice. New York: American Petroleum Institute, API-34-056.

35 Barnard, P. Jr. (1959). A new method of restoring water injection capacity to wells plugged with iron sulfide and free sulfur. *Journal of Petroleum Technology* **11** (09): 12–14, SPE 1299.

36 Smith, C.F., Crowe, C.W., and Nolan, T.J. (1969). Secondary deposition of iron compounds following acidizing treatments. *Journal of Petroleum Technology* **21** (09): 1121–1129, SPE 2358.

37 Nowels, K.B. (1933). Rejuvenation of oilfields by natural and artificial water flooding. 1st World Petroleum Congress, London, UK (18–24 July 1933), WPC-1075.

38 https://energy.gov/fe/science-innovation/oil-gas-research/enhanced-oil-recovery.

39 Amarnath, A. (1999). Enhanced Oil Recovery Scoping Study. Final Report *TR-113836*. Palo Alto, CA: Electric Power Research Institute.

40 Martin, W.C. (1967). Applying water chemistry to recovery. SPE Permian Basin Oil Recovery Conference, Midland, TX (8–9 May 1967), SPE 1789.

41 Overton, H.L. (1973). Water chemistry analysis in sedimentary basins. SPWLA 14th Annual Logging Symposium, Lafayette, LA (6–9 May 1973), SPWLA-1973.

42 Strong, M.W. (1937). Micropetrographic methods as an aid to the statigraphy of chemical deposits. 2nd World Petroleum Congress, Paris, France (14–19 June 1937), WPC-2036.

43 Morris, M.W. (1937). *Chemical Clean-Out of Oil Wells in California*, Drilling and Production Practice. New York: American Petroleum Institute, API-37-220.

44 Featherston, A.B., Mirham, R.G., and Waters, A.B. (1959). Minimization of scale deposits in oil wells by placement of phosphates in producing zones. *Journal of Petroleum Technology* **11** (03): 29–32, SPE 1128.

45 Frenier, W.W. and Ziauddin, M. (2008). *Formation, Removal and Inhibition of Inorganic Scale in The Oilfield Environment*. Richardson, TX: Society of Petroleum Engineers.

46 Meyers, K.O. and Skillman, H.L. (1985). The chemistry and design of scale inhibitor squeeze treatments. SPE Oilfield and Geothermal Chemistry Symposium, Phoenix, AZ (9–11 March 1985), SPE 13550.

47 Weintritt, D.J. and Cowan, J.C. (1967). Unique characteristics of barium sulfate scale deposition. *Journal of Petroleum Technology* **19** (10): 1381–1394, SPE 1523.

48 Bunker, M.H.J. (1937) The microbiological aspect of anaerobic corrosion. 2nd World Petroleum Congress, Paris, France (14–19 June 1937).

49 Plummer, F.B., Merkt, E.E. Jr., Power, H.H. et al. (1944). Effect of certain micro-organisms on the injection of water into sand. *Petroleum Technology* **7** (01): 1–13, SPE 944014.

50 Breston, J.N. (1949). *New Chemical Treatment of Flood Water for Bacteria and Corrosion Control*, Drilling and Production Practice. New York: American Petroleum Institute, API-49-334.

51 Ollivier, B. and Magot, M. ed. (2005). *Petroleum Microbiology*. Washington, DC: ASM Press.

52 Crow, C.W. (1968). New treating technique to remove bacterial residues from water-injection wells. *Journal of Petroleum Technology* **20** (05): 475–478, SPE 2132.

53 Maxwell, S., Devine, C. and Rooney, F. (2004). Monitoring and control of bacterial biofilms in oilfield water handling systems. CORROSION 2004, New Orleans, LA (28 March-1 April 2004), NACE 04752.

54 Campbell, S., Duggleby, A. and Johnson, A. (2011). Conventional application of biocides may lead to bacterial cell injury rather than bacterial kill within a biofilm. CORROSION 2011, Houston, TX (13–17 March 2011), NACE 11234.

55 Dennis, D.M. and Hitzman, D.O. (2007). Advanced nitrate-based technology for sulfide control and improved oil recovery. International Symposium on Oilfield Chemistry, Houston, TX (28 February-2 March 2007), SPE 106154.

56 Stott, J.F.D. (2005) Modern concepts of chemical treatment for the control of microbially induced corrosion in oilfield water systems. Chemistry in the Oil Industry IX (31 October - 2 November 2005). Manchester, UK: Royal Society of Chemistry, p. 107.

57 Burger, E.D., Crewe, A.B. and Ikerd, H.W. III (2001). Inhibition of sulphate reducing bacteria by anthraquinone in a laboratory biofilm column under dynamic conditions. Paper 01274, NACE Corrosion Conference.

58 Whitby, C. and Skovhus, T.L. ed. (2009). Applied microbiology and molecular biology in oilfield systems. Proceedings from the International Symposium on Applied Microbiology and Molecular Biology in Oil Systems (ISMOS-2), 2009. Heidelberg, London, NewYork: Springer Dordrecht, 2009.

59 Garfias, V.R. (1923). Present conditions in Mexican oil fields and an outlook into the future. *Transactions of the AIME* **68** (01): 989–1003, SPE 923989.

60 Wood, F.E., Young, H.W., and Buell, A.W. (1927). Handling congealing oils and paraffin in salt creek field, Wyoming. *Transactions of the AIME* **77** (01): 262–268, SPE 927262.

61 Reistle, C.E. Jr. *Paraffin Production Problems*, Drilling and Production Practice. New York: American Petroleum Institute, API-35-072.

62 Brown, W.Y. (1940). *Prevention and Removal of Paraffin Accumulations*, Drilling and Production Practice. New York: American Petroleum Institute, API-40-085.

63 Knox, J., Waters, A.B. and Arnold, B.B. (1962). Checking paraffin deposition by crystal growth inhibition. Fall Meeting of the Society of Petroleum Engineers of AIME, Los Angeles, CA (7–8 October 1962), SPE 443.

64 Mendell, J.L. and Jessen, F.W. (1970). Mechanism of inhibition of paraffin deposition in crude oil systems. SPE Production Techniques Symposium, Wichita Falls, TX (14–15 May 1970), SPE 2868.

65 Ruehrwein, R.A. (1951). Specificity of pour point depressants in lubricating oils. 3rd World Petroleum Congress, The Hague, The Netherlands, (28 May-6 June 1951), WPC-4632.

66 Tiedje, J.L. (1963). The use of pour depressants in middle distillates. 6th World Petroleum Congress, Frankfurt am Main, Germany (19–26 June 1963), WPC-10518.

67 Slater, G. and Davis, A. (1986). Pipeline transportation of high pour point New Zealand crude using pour point depressants. SPE Annual Technical Conference and Exhibition, New Orleans, LA (5–8 October 1986), SPE 15656.

68 Fielder, M. and Johnson, R.W. (1986). The use of pour-point depressant additive in the beatrice field. European Petroleum Conference, London, UK (20–22 October 1986), SPE 15888.

69 Becker, J.R. (1997). *Crude Oil Waxes, Emulsions and Asphaltenes*. Tulsa, OK: PennWell Books.

70 Frenier, W.W., Ziauddin, M., and Vekatesan, R. (2008). *Organic Deposits in Oil and Gas Production*. Richardson, TX: Society of Petroleum Engineers.

71 Misra, S., Baruah, S., and Singh, K. (1995). Paraffin problems in crude oil production and transportation: a review. *SPE Production and Facilities* **10** (01): 50–54, SPE 28181.

72 McCain, W.D. Jr. (1990). *The Properties of Petroleum Fluids*, 2ee. Penwell Publishing, Tulsa, OK.

73 Azevedo, L.F.A. and Teixeira, A.M. (2003). A critical review of the modeling of wax deposition mechanisms. *Journal of Petroleum Science and Technology* **21** (3–4): 393–408.

74 Ganeeva, Y.M., Yusupova, T.N., and Romanov, G.V. (2016). Waxes in asphaltenes of crude oils and wax deposits. *Petroleum Science* **13** (4): 737–745.

75 Bern, P.A., Withers, V.R. and Cairns, R.J.A. (1980). Wax deposition in crude oil pipelines. European Offshore Technology Conference and Exhibition, London, UK (21–24 October 1980), SPE 206-1980.

76 Mariane, E. (1951). Researches on the constitution of natural asphalts. 3rd World Petroleum Congress, The Hague, The Netherlands (28 May-6 June 1951), WPC- 4511.

77 Serguienko, S.R., Davydov, B.E., Delonfi, I.O. and Teterina, M.P. (1955). Composition and properties of high molecular petroleum compounds. 4th World Petroleum Congress, Rome, Italy (6–15 June 1955), WPC- 6428.

78 Padovani, C., Berti, V. and Prinetti, A. (1959). Properties and structures of asphaltenes separated from mineral oil residua. 5th World Petroleum Congress, New York (30 May-5 June 1959), WPC-8491.

79 Mullins, O.C., Sheu, E.T., Hammami, A., and Marshall, A.G. ed. (2007). *Asphaltenes, Heavy Oils and Petroleomics*. New York: Springer.

80 Akbarzadeh, K., Hammami, A., Kharrat, A. et al. (2007, Summer 2007). Asphaltenes— problematic but rich in potential. *Oilfield Review* **19** (2): 23–48.

81 Bunger, J.W. and Li, N.C. ed. (1981). *Chemistry of Asphaltenes*, vol. **195**. American Chemical Society.

82 Samuelson, M.L. (1992). Alternatives to aromatics for solvency of organic deposits. SPE Formation Damage Control Symposium, Lafayette, LA (26–27 February1992), SPE 23816.

83 Vetter, O.J. (1972). Oilfield chemistry a challenge for the industry. *Journal of Petroleum Technology* **24** (08): 994–995, SPE 3922.

84 Maddin, C.M. and Lopp, V.R. (1973). Analytical chemistry of oil well treating chemicals. SPE Oilfield Chemistry Symposium, Denver, CO (24–25 May 1973), SPE 4352.

2

Polymer Chemistry

In the oilfield a wide variety of polymers and polymeric chemistry are used to provide chemical solutions to a wide variety of issues. It is interesting to note that most, if not all, of these polymers and derivatives have historically been obtained from the use of petrochemical feedstocks such as ethylene, which in turn is obtained from the hydrolytic or steam 'cracking' of hydrocarbon feedstocks, usually naphtha or gas oil. Ethylene, the simplest of olefins, is used as a base product for many syntheses in the petrochemical industry: plastics, solvents, cosmetics, pneumatics, paints, packaging, etc. It is probably the single most important chemical material in our modern world and is certainly produced in the largest tonnage of any basic organic chemical. Today, the demand for ethylene is over 140 million tonnes per year with a growth rate of 3.5% per year. The average capacity of production plants, known as steam crackers, has risen from 300 kilotonnes per annum in the 1980s to over 1000 kilotonnes per annum today.

A polymer in its simplest form can be regarded as comprising molecules of closely related composition of molecular weight at least 2000, although in many cases typical properties do not become obvious until the mean molecular weight is about 5000. There is virtually no upper end to the molecular weight range of polymers since giant three-dimensional networks may produce cross-linked polymers of a molecular weight of many millions. Polymers (macromolecules) are built up from basic units known as monomers. These units can be extremely simple, as in addition polymerisation, where a simple molecule adds on to itself or other simple molecules by methods that will be indicated subsequently. Thus ethylene $CH_2:CH_2$ can be converted into polyethylene, of which the repeating unit is $—CH_2:CH_2—$, often written as $(—CH_2:CH_2—)n$, where n is the number of repeating units. The major alternative type of polymer is formed by condensation polymerisation in which a simple molecule is eliminated when two other molecules condense. In most cases the simple molecule is water, but alternatives include ammonia, an alcohol and a variety of simple substances. There are a number of variations on the construction of these polymers; however as to their application in the oil and gas industry, the other important method of note is copolymerisation. The term 'copolymer' is sometimes confined to a polymer formed from two monomers only. In a more general sense, it can be used to cover polymers formed from a larger number of monomers. The term 'terpolymer' is sometimes used when three monomers have been copolymerised. Across the upstream oil and gas exploration and production (E&P) industrial sector, a wide range of polymers and copolymers are used, either directly or as formulations to give a wide variety of effects.

This chapter will not only examine many of these products but also their environmental impact versus their necessity alongside the use of 'green' alternatives. It will show that a number of green products have been used historically and either continue to be used or have been reintroduced as they provide a better environmental impact. This is particularly the case where natural polymers such as guar gum have historically been used as thickeners and have found increased use in applications such as shale gas recovery [1].

Oilfield Chemistry and Its Environmental Impact, First Edition. Henry A. Craddock.
© 2018 John Wiley & Sons Ltd. Published 2018 by John Wiley & Sons Ltd.

In understanding polymers and their use in oil and gas applications, it is necessary to recognise that polymers exhibit certain properties, especially macroscopic ones that are very different from those of lower molecular weight materials used in oilfield applications. This is particularly true at the macroscopic level, and these differences will affect both the inherent environmental impact of the polymer application and its environmental fate.

The Swedish chemist Berzelius in 1833 [2] was the first recorded person to use the word polymer; however in this early chemistry there was no clear understanding of the molecular structure of these macromolecules. However, this did not mean that some fundamental and critical products were not developed, many of which are still in everyday use. In the oil and gas industry, poly(ethylene glycol) (PEG) (Figure 2.1) is widely used in a variety of applications; its synthesis was first developed in the 1860s [3]. Indeed this and other synthesis although not utilising petroleum-based monomer feedstocks became the basis of much of later polymer synthesis in the twentieth century.

Also in the latter half of the nineteenth century, isoprene (Figure 2.2) was isolated as a degradation product of natural rubber [4].

As shall be apparent later in this chapter, these fundamental monomer units and their ultimate biodegradation are important in the selection and design of green polymer chemistry as used in the oil and gas industry.

Two key discoveries are important in general polymer chemistry and its use in the oil and gas sector or indeed any industrial application. First was the recognition of the work of Staudinger who elucidated that the remarkable properties of polymers were due to the intermolecular forces between molecules of high molecular weight [5], and second was the work of Wallace Carothers [6] who put Staudinger's theories into a firm experimental and commercial basis, leading to the development of synthetic rubber polymers such as polyesters and polyamides, the latter leading famously to the invention of nylon.

In this chapter a wide variety on monomers and their polymers shall be detailed along with their use and potential uses in the upstream oil and gas industry. Of particular note will be their green credentials and their biodegradation. Many natural and biopolymers are used both for environmental acceptability and economy, and this relationship will be looked at in some detail. One of the greatest properties of synthetic polymers is their durability, but this is the environmental chemist's greatest dilemma as the rates of biodegradation, if any, can be very low. Recycling in other industries is gathering pace, albeit it is still low compared with other materials such as paper; however in the oil and gas industry, many polymers not used in hardware developments cannot be recycled. This chapter will also examine new developments in making polymers more degradable without compromising their macromolecular properties.

In the oil and gas industry polymers are used in material provision such as rubbers and plastics. These robust and durable materials, as in other industrial sectors, are coming under pressure to be recycled or reused. The main area of implementation of waste hierarchical regimens is in the UK sector of the North Sea. The UK continental shelf has over 470 installations, 10 000 km of pipelines, 5000 wells and 15 onshore terminals where over the next 25 years or so a large number of facilities and infrastructure will require to be decommissioned [7]. The oil

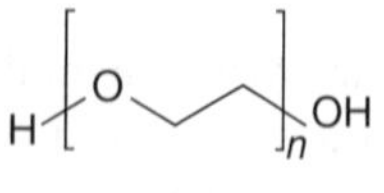

Figure 2.1 Poly(ethylene glycol).

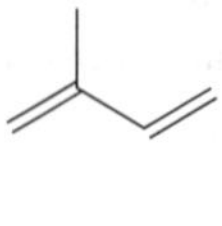

Figure 2.2 Isoprene.

and gas sector has set out an ambitious plan to reuse or recycle up to 97% of all materials by weight. Material polymers and plastics will be included in this target. As decommissioning progresses, it is anticipated that smarter end-of-life approaches such as reuse, remanufacture and design for reparability could help to reduce some of this expenditure for a sector under pressure to cut the environmental impacts of its operations [8]. The amount of reuse that currently takes place during the decommissioning process is very small; some observers estimate the figure to be just 1% by weight. An overview of materials used in the oil and gas industry is given in Section 2.4.

In terms of polymer chemistry, this material use although important is only part of the overall polymer chemical use in oil and gas industry. Many polymers, and in large volumes, are used as effect chemicals in drilling, production, stimulation, secondary recovery and other operations, and it is these chemistries that this chapter is primarily concerned. The challenge on reduction on environmental impact from these materials is great, and their opportunities for reuse and recycling, although potentially available, are still undergoing primary research and feasibility. The main thrust of environmental impact reduction is in the design and use of chemistries, which are less harmful to the environment than those currently used, and as we shall see throughout this book, this is a general principle, perhaps it could be argued that it is not the correct one, of green chemistry in the oil and gas industry. In this vain, in terms of green chemistry related to polymers, the natural polymers, particularly the polysaccharides, based on the repeating unit of D-glucose (Figure 2.3) are used in a variety of applications [9].

In the following sections of this chapter, a number of types of polymer and their derivatives will be explored as to their chemistry and application in the oilfield.

2.1 General Organic Polymers: Synthetic Polymers

In this section all of the major polymer classes used in the oilfield will come under consideration. This will include the following polymer types [10], the vinyl polymers such as polyethylene, polyacrylic acid, polystyrene and others and the non-vinyl polymers such as the polyethers, polyesters and polyamides. Polymer chemists generally classify polymers into these two main groups. Vinyl polymers are those derived by the chain reaction of alkenyl monomers and all others, while non-vinyl polymers are derived by all other monomer types. This latter group is much more varied and complex due to the larger number of possible repeating units in the polymer. However, the three main commercial types mentioned are the ones most applied in the oil and gas industry and will be the primary concern of this section.

In addition to these classical polymers or homopolymers being derived from a single monomer, the oil and gas industry also uses a selection of copolymers and other composite polymers, particularly terpolymers, in a variety of applications. Copolymers are polymers derived from more than one species of polymer, and terpolymer is used to denote a copolymer derived from three different monomers.

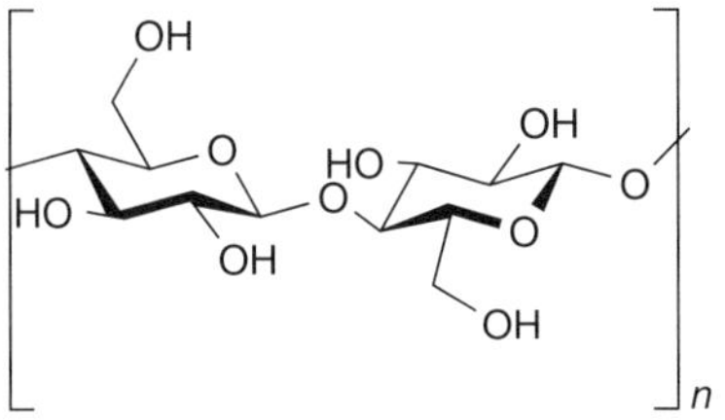

Figure 2.3 D-Glucose-based polysaccharide, cellulose.

In general, the properties that make these polymers stable make them non-biodegradable; however they are usually non-toxic to aquatic organisms being non-bioavailable.

2.1.1 Vinyl Polymers and Copolymers

The vinyl polymers and their copolymeric derivatives are used across a wide range of applications in the upstream oil and gas industry particularly in the drilling and production operations; however they are not as widespread in their use and applicability as the non-vinyl polymers. Nonetheless they are important in a number of oilfield applications. The vinyl polymers in oilfield application are usually addition polymers formed from diene monomers, such as styrene, vinyl chloride, acrylic acid, etc., where addition can be at (1,2 addition) or across (1,4 addition) the double bound.

2.1.1.1 Polyethylenes, Polyisobutylenes and Related Polymers

Drag-reducing agents (DRAs) are an important additive in the efficient and economic transport of both produced fluids and processed crude oil. One of the first uses, which continues to the present day, is the application of these polymeric materials to the *trans*-Alaskan pipeline [11]. Polyethylene [12] and copolymers of α-olefins [13] have been used in these applications. Of particular importance in enhancing the flow of liquid hydrocarbons by drag reduction is the use of ultra-high molecular weight polyethylene [14, 15]. Such materials are continuously injected to enhance the flow and allow greater capacities to be transported. At present none of the material used in the process is specifically recovered for reuse; however it can be argued that since the material is transported with the crude oil, it is ultimately recycled through the refining process and will return again into the petrochemical feedstock supply.

Homo- and copolymers of α-olefins are also used widely as drag reducers [16]. Of particular importance are polyisobutylenes (PIBs) [17] (see Figure 2.4), which are the basis of many oil-soluble DRAs.

It has been found that introducing copolymers based on larger monomers such as decene and tetradecene (see Figure 2.5) significantly improves the DRA performance [18].

Among vinyl polymers as DRAs, polystyrenes have been examined [19] but do not appear to be commercialised. Early work in this area concluded that drag reduction performance and degradation of the polymer through shear forces in the flow fluids are related to molecular weight distribution [20].

In conclusion what has been observed is that long polyolefin chains (e.g. PIB with 30–50 million Da) are the best performing DRAs. The greater the chain length, the better the DRA performance. However, these polymers are susceptible to degradation in turbulent flow, especially shearing in pumps. Putting larger side chains on the carbon backbone helps resist degradation and improves the DRA effect for a given backbone chain length. C10–C14 α-olefins are, for example, good for this, such as 1-tetradecene shown previously in Figure 2.5. However, polymerisation becomes more difficult due to steric hindrance, as the size of the olefin monomers

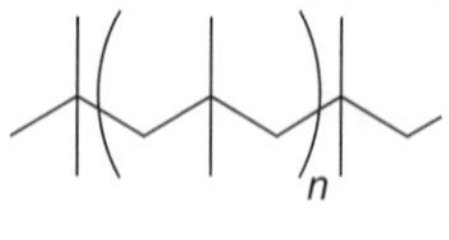

Figure 2.4 Polyisobutylene.

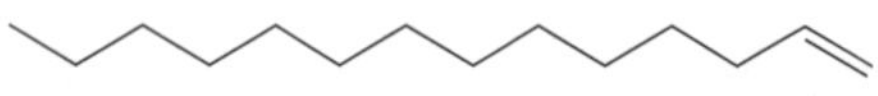

Figure 2.5 1-Tetradecene.

increases, such that it becomes difficult to get to the ultra-high molecular weights required for ultimate performance. Therefore there is a trade-off between better performance and C_{10}–C_{14} α-olefins, but not the top performance as the molecular weight is lower than would be expected.

The environmental dilemma for the oilfield chemist in designing efficient DRAs is that polymers, which are the most efficient in performance and activity and provide good stability to shear degradation, are also the least biodegradable. Polystyrenes are well known to have very low rates of biodegradation [21], whereas some polyethylenes, especially low-density polyethylenes, have remarkably good rates of biodegradation [22]. This is a property that, in terms of other polyethylene applications, could be better exploited; however these low molecular weight polymers are not suitable for DRA applications. This is a common occurrence, which as will become apparent is encountered many times in attempting to design products through a substitution approach using a similar but more biodegradable molecule.

Ethylene polymer and copolymers, which are involved in wax control, are, in the main, ethyl vinyl acetate derived and are described in a separate section on polyvinyl acetate (PVAc) polymers. In corrosion control polyethylene coatings can be used to prevent contact with corrosive materials and solutions, and these can also be impregnated with antioxidants to give good resistance to thermal oxidation [23]. They have been used as external coatings for corrosion protection across a wide range of temperatures between −30 and 120 °C [24].

Such polymers are, by their very design, durable and robust having to undergo a variety of aggressive and harsh conditions. This makes their greening in terms of degradation counterproductive to their required use. In such circumstances the best environmental outcome is to ensure that, where required to be used, such materials can be reused and/or recycled.

2.1.1.2 Poly(acrylic Acid) Products, Methacrylates, Acrylamides and Related Polymers

A versatile class of vinyl polymers is that derived from acrylic acid monomers. Poly(acrylic acid) (Figure 2.6) has a wide range of oilfield applications including demulsifiers, scale inhibitors, fluid loss additives, gelling agents and water shut-off applications.

The biodegradability (mineralisation to carbon dioxide) of acrylic acid oligomers and polymers has been studied, and mixtures of low molecular weight ($M_w < 8000$) poly(acrylic acid)s have shown to have reasonable degradation but not complete degradation to carbon dioxide [25]. Nonetheless this class of vinyl polymer is more acceptable in its ecotoxicity profile than many others. Indeed the ester derivatives of poly(acrylic acid) and the polyacrylamides (PAM) have enhanced biodegradation due to the side groups on the vinyl polymer backbone. This allows them to be used widely in oilfield chemical applications with applications in wax and asphaltene control as hydrate inhibitors, drag reducers and flocculants in addition to the ones previously mentioned. A number of these applications are now examined.

Polyacrylic acid or its sodium salt has been used in the construction of invert emulsion-based drilling fluids [26], where poly(acrylic acid) gives stability to the overall emulsion system. In drilling practices today the term has become an outdated distinction between two types of oil muds. In the past, invert emulsion oil muds were those with more than 5 vol.% emulsified water, and oil-based muds (OBM) were those with less than 5 vol.% water. Today, this distinction is not pertinent because the general term oil mud covers all water concentrations [27].

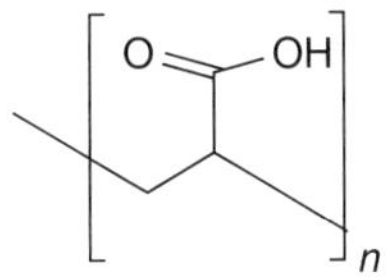

Figure 2.6 Poly(acrylic acid).

Fluid loss additives are a critical component of many drilling fluid systems and other completion and fracturing fluids. They are also used in cement slurries; they are used to prevent or retard leakage of the liquid phase of drilling fluid, slurry or treatment fluid containing solid particles into the reservoir formation. The resulting build-up of solid material or filter cake may be undesirable, as may the penetration of filtrate through the formation. Fluid loss additives are used to control the process and avoid potential reservoir damage. The loss of expensive fluid is also not tolerable from an economic viewpoint and may result in more drilling fluid being required and drilling work processes being lengthened in time. Finally there is obviously a potential undesirable environmental impact from fluid loss during drilling and related operations, which, in best practice, should be avoided [28]. Many types of polymer, as will be illustrated throughout this chapter, are involved in fluid loss additives including the polymers of acrylic acid [29] and acrylamide [30]. In a subsequent section on polyvinyl alcohols (PVAs) and acetates, it will be seen that these are more commonly used as fluid loss additives.

Polymers of acrylic acid and methacrylic acid have been examined as gelling agents, in particular in conjunction with surfactants to give a gel for application in water shut-off scenarios [31].

The main use, however, of poly(acrylic acid) products and related polymers is in scale inhibition. These materials have a wide range of action against common oilfield scales and therefore a broad range of application. They do however have a significant limitation in that they are intolerant to calcium ions at concentrations above 2000 mg l^{-1}. They are also, in general, not very biodegradable. However, a polymer based on maleic acid, the polymaleate (see Figure 2.7), has been found to be significantly biodegradable [32].

It has been found that most polymeric scale inhibitors have optimum performance in molecular weight range of 1000–30 000 [33] and also that in most cases the number of active repeating units in the polymer needs to be at least 15. For polymers based on acrylic acid, this means a molecular weight of at least 1000 Da; however the optimum biodegradation is at molecular weights of below 700 Da. As has already been observed with other polymers in other applications, there is a trade-off between biodegradability and performance. Performance improvement is claimed for branched homologues over linear polymers [34] and also for relevant copolymers such as acrylic acid/isoprene and for sulphonated copolymers [35].

Copolymers of acrylic acid and maleic acid are very common as polymeric scale inhibitors, and this is partly because maleic acid does not easily polymerise, and therefore to increase polymeric molecular weight and thus performance, it is usually copolymerised [36]. It is claimed that these types of polymers and copolymers have superior performance against barium and strontium sulphate scales. Methacrylic acid is the only other commercial copolymer in this class.

As has been stated earlier, a serous limitation to the application of these polymers is their relatively poor calcium ion tolerance, and it is observed that adding amide or hydroxyl functionality can improve calcium tolerance [37]. This is particularly the case of copolymers of maleic acid with maleamide groups and acrylic acid polymers with acrylamide, methacrylamide or hydroxypropylacrylate monomers.

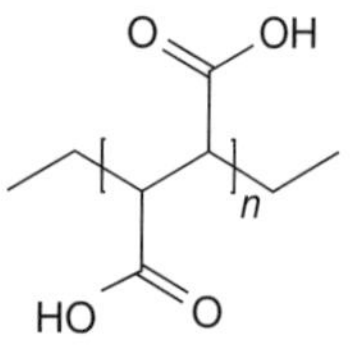

Figure 2.7 Polymaleic acid.

Serious improvements in biodegradation and general ecotoxicity without severe losses in performance can be achieved with other classes of polymeric materials for application as scale inhibitors. Polymers in this class, such as polyaspartate, are examined in Section 2.2.

Acrylate and (meth)acrylate polymers, especially their esters, are widely used as paraffin wax control additives [38], and their action as pour point depressants (PPDs) has been known for some decades [39]. These are also used in combination as a monomer in copolymers, with maleic anhydride [40]. As in scale inhibition previously described, these latter copolymers have a number of similarities but also some significant differences, and these can be challenging for the environmental chemist looking to provide a green chemical solution.

Acrylate ester copolymers (Figure 2.8), and in particular (meth)acrylate ester polymers, are a specific class of comb polymer. In the structure below, R can be H for an acrylate ester and can be methyl for a (meth)acrylate ester. Ri is usually a long linear alkyl chain.

These are more expensive than the ethylvinylacetate (EVA) copolymers, and the most effective on the problematic crudes are the most expensive variants. They are however biodegradable, although slowly, through oxidation of the ester functionality, which can then promote further degradation of the more stable backbone [41]. Biodegradation and other forms of polymer decomposition are discussed in Section 2.6.

Long-chain ester groups of greater than C20 are observed to be best in class across a wide range of crudes and condensates [42]. It has also been observed that maximum PPD effectiveness can be achieved with around 60% of the side chains at C18 or greater, with the remaining side groups being methyl esters [43]. An alternative method for achieving these long side chains has been developed using stearyl acrylate copolymerised with a small amount of hydroxyethyl acrylate and then esterifying the hydroxyl groups with stearic acid chloride [44]. This gives long-chain side groups of C20 plus without using expensive alcohols. Grafting of alkyl (meth) acrylate chains onto polyvinyl backbones has also been developed [45]. These methods and the improving performance of the paraffin inhibitor also enhance the biodegradation of the polymer molecule [46], which will be further discussed later in this chapter on polymer degradation and biodegradation.

Copolymers of (meth)acrylic acid ester of C16 plus alcohols have been synthesised with small amounts of vinyl pyridine and vinyl pyrrolidone and appear to have improved PPD and wax inhibitor characteristics [47]. Terpolymers containing (meth)acrylate esters, where the third monomer is a vinyl grouping such as 2- or 4-vinylpyridine, have good wax inhibitor properties [48].

There are a number of other developments in manipulation of these co- and terpolymers related to blending polymers and using varying side chain lengths to match the alkane chain lengths in the wax crystals [49].

Earlier in this chapter we described the use of polyalkenes, in particular PIBs as DRAs or flow improvers. The properties that make vinyl polymers with long side chains, such as polyacrylates, methacrylates and alkyl styrenes, good wax inhibitors and PPDs also make them good DRAs, particularly poly(meth)acrylate esters [50]. These polymers usually contain pendant groups of six to eight carbon atoms and can exhibit superior shear stability in comparison with the PIBs described earlier [51]. However, in terms of performance, these polymers do not

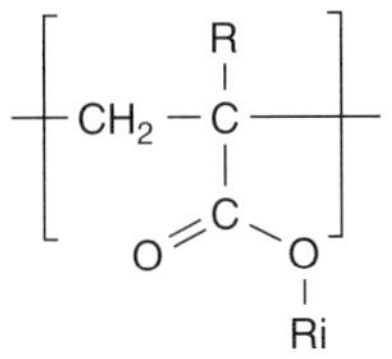

Figure 2.8 Acrylate ester copolymer.

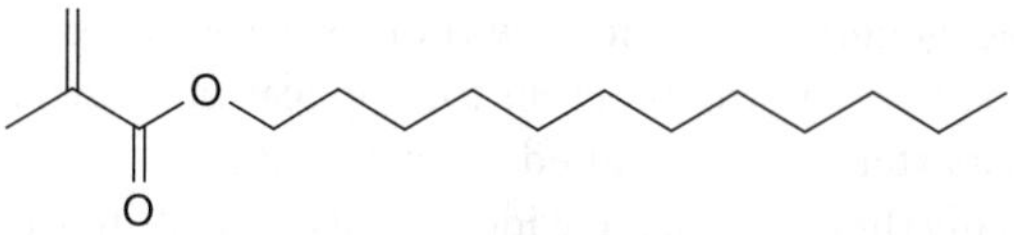

Figure 2.9 Lauryl methacrylate.

exhibit the same drag reduction properties, and this is primarily due to their lower molecular weight. It is difficult to synthesise ultra-high molecular weight polyalkyl (meth)acrylates compared with polyolefins. Nonetheless methods of synthesising improved molecular weights have been developed [52], and molecules such as polymers based on lauryl methacrylate (see Figure 2.9) have been shown to perform as well as polyolefin DRAs.

Polymers of such monomers are however not biodegradable, being very insoluble in water, although they are non-toxic aquatic organisms [53].

A number of copolymers of alkyl(meth) acrylates have also been reported as superior DRAs [54]; again however they are poorly biodegradable.

Finally, in the section of vinyl polymers based on acrylic acid, related derivatives and maleic acid/anhydride are an important class of demulsifiers. In general, when synthesising these molecules, both hydrophilic and hydrophobic component parts are required, and this is discussed more fully in Chapter 3. Polymers from vinyl monomers such as (meth)acrylic acid and maleic anhydride are subsequently ethoxylated or propoxylated (sometimes butoxylated) under basic and acidic conditions to create a variety of esters of various solubilities and surfactant properties.

These vinyl monomers can also be further reacted with polyalkoxylates (see Section 2.1.2) such as block copolymers of ethylene oxide (EO); ethylene oxide (EO)/propylene oxide (PO) and/or alkylphenol-formaldehyde resin alkoxylates to give an interesting polymeric demulsifier class [55]. A number of demulsifier types have been made using appropriate vinyl monomers. Indeed copolymers with variation on up to four different vinyl monomers with aromatic, oleophobic, ionisable and hydrophilic groups have been constructed and claimed as demulsifiers. Usually variations are around oleophilic and hydrophobic groups; however, in general these types of product are not as common and widely applied as those in the non-vinyl class of synthetic polymers.

Interestingly the use of esterification of these polymers to enhance their technical properties also gives a greater degree of biodegradation although the ultimate environmental fate may be the same.

2.1.1.3 Polyacrylamides

In contrast to polyacrylates and their esters, the 'sister' class of polymers, the PAM (see Figure 2.10), is water soluble and reasonably biodegradable.

These polymer types are in general synthesised from acrylamide, which in turn is derived from acrylic acid. The homogeneity of the molecular structure (see Figure 2.6) of polyacrylic acid is striking, and yet the biodegradation profile is markedly different. This is in part due to the polymers' high water solubility. This is in marked contrast to the monomer units where, although hydrogen bonding may enhance the water solubility of amides relative to hydrocarbons (alkanes, alkenes, alkynes and aromatic compounds), amides typically are regarded as

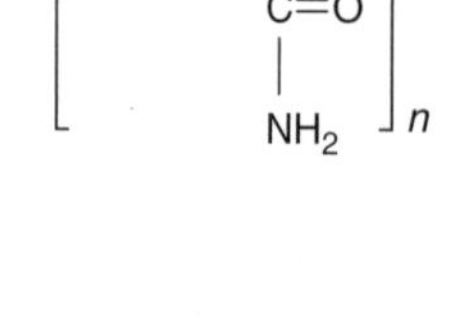

Figure 2.10 Polyacrylamide (PAM).

compounds with low water solubility. They are significantly less water soluble than comparable acids or alcohols due to their non-ionic character, the presence of non-polar hydrocarbon functionality, and the inability of tertiary amides to donate hydrogen bonds to water (they can only be hydrogen bond acceptors). Thus amides have water solubilities roughly comparable with esters, and typically amides are less soluble than comparable amines and carboxylic acids since these compounds can both donate and accept hydrogen bonds and can ionise at appropriate pH to further enhance solubility [56].

It has been found that PAM has a monopolar surface nature, which results in this type of polymer having good dissolution in water. The solubility of PAM in water is generally much greater than in the other organic solvents [57].

PAM is most often used to increase the viscosity of water (creating a thicker solution) or to encourage flocculation of particles present in water. PAM can be tailored to fit a broad range of applications.

One of the largest uses of PAM is to flocculate or coagulate solids in a liquid, such as in the wastewater treatment industry. The introduction of PAM as a chemical flocculating agent to wastewater causes suspended particles in wastewater to aggregate, forming what is known as a flocc [58, 59]. In addition, PAM products react with water to form insoluble hydroxides, which, upon precipitating, link together to form meshes, physically trapping small particles into a larger flocc. In simple terms, PAM makes the fine solids in treated water adhere to one another until they become big enough to settle out or be captured by filters to make sewage sludge. In either case, the flocc can be filtered or removed more easily. The same principles are applied in the oil and gas industry in its application as a flocculant for treating production water streams prior to discharge and/or reinjection. Here PAM is used where a non-ionic polymer is required; however, if the oily particles requiring flocculation carry a positive charge, then partially hydrolysed PAM is normally used in treatment [60].

Another common and major use of PAM is in subsurface applications such as enhanced oil recovery, where high viscosity aqueous solutions can be injected to improve the economics of conventional water flooding. For oil extraction applications, PAM is used to increase the viscosity of water to improve the effectiveness of the water flooding process. In other words, the PAM-injected solution assists in sweeping (or pushing) oil locked in a reservoir towards the producing well(s). The result is improved volumetric sweep efficiency – more oil is produced for a given volume of water injected into the well [61].

Historically, PAM oilfield projects primarily used PAM supplied as a solid powder. This powder is then hydrated to form an aqueous solution. More conveniently, most PAM oilfield projects now use PAM supplied in emulsion form as liquids, offering a more efficient and faster way to get the PAM into the aqueous solution. In dilute aqueous solution, such as is commonly used for EOR applications, PAM polymers are susceptible to chemical, thermal and mechanical degradation. Chemical degradation occurs when the labile amine moiety hydrolyses at elevated temperature or pH, resulting in the evolution of ammonia and a remaining carboxyl group [62]. Thermal degradation of the vinyl backbone can occur through several possible radical mechanisms, including the autoxidation of small amounts of iron and reactions between oxygen and residual impurities from polymerisation at elevated temperature [63]. Mechanical degradation can also be an issue at the high shear rates experienced in the near wellbore region. However, cross-linked variants of PAM have shown greater resistance to all of these methods of degradation and have proved to be much more stable [64]. This enhancement in stability of course comes at a price in terms of biodegradability. Also the cross-linking agents can be highly detrimental to the environmental profile of the polymer. Chromium that is commonly used is not thought to be particularly environmentally toxic, particularly chromium(III) [65]. However, information is lacking on the biological activities of water-soluble chromium(III) compounds, organochromium compounds and their ionic states. It is recognised that chromium(VI) is more toxic to aquatic species. In the marine environment, chromium(VI) toxicity has found to be strongly dependent on mean salinity due to the competitive role of sulphate [66]. The environmental properties of metals and metal salts are more fully discussed in Chapter 5.

Nonetheless PAM, as has been stated earlier, are reasonably biodegradable [60, 67], and their use in other oilfield applications, where performance and stability is less demanding, has been somewhat underdeveloped. In comparison with their use in enhanced oil recovery and flocculation applications, PAM have been used for the synthesis of polymer gel systems for sand control and are claimed to be more reliable than the more traditional resin polymer treatments [68].

There are a number of related applications that are of a similar nature in terms of the application chemistry and the use of the polymeric gel-like properties of PAM, e.g. acid diverters [69], foam diverters [70], cement additives [71], fluid loss additives [72] and water shut-off applications [73].

PAM has been utilised, unsurprisingly given the applications in enhanced oil recovery, in relative permeability modification [74]; this is alongside a wide variety of other vinyl polymers. It has also been used in fracturing fluids but usually in high-temperature applications or as a friction reducer and as part of a non-volatile phosphorous hydrocarbon gelling agent [75].

The properties of PAM are also suitable for application in drag reduction exhibiting specific viscoelastic properties under specific injection conditions. Indeed, under certain conditions, drag reductions of almost 50% have been achieved at low dose rates [76, 77].

Finally, two other applications have been reported. PAM have been utilised in the synthesis of synergists for kinetic hydrate inhibitors (KHIs) [78] and most interestingly in a reasonably biodegradable corrosion inhibitor application where it has been reacted with hydroxylamine [79]. This latter application potentially shows the possibilities of further investigating this class of polymers as a useful source of environmentally acceptable polymers for a wider application in the upstream oil and gas industry. The unique water-soluble properties as described earlier in the section have not been fully utilised in particular with regard to production chemical applications.

2.1.1.4 Poly(vinyl Acetate) and Related Polymers and Copolymers

In this class of polymers, as well as the PVAc polymers, other similar types will be considered, although their oilfield use is probably less. This will include polyvinyl alcohols (PVA), polyvinyl chloride (PVC) and polyacrylonitriles.

The most commonly used of these vinyl polymers in the upstream oil and gas industry are copolymers of ethylene and vinyl acetate. These EVA copolymers are used as wax control agents and primarily as PPDs [80]. These are supplied as random copolymers of low molecular weight [81] (see Figure 2.11).

The vinyl acetate groups disrupt the wax crystallisation, lowering the wax appearance temperature (WAT) and pour point. Critical for the performance in these polymers is the amount of vinyl acetate and hence the amount of side-chain branching. It is generally accepted that

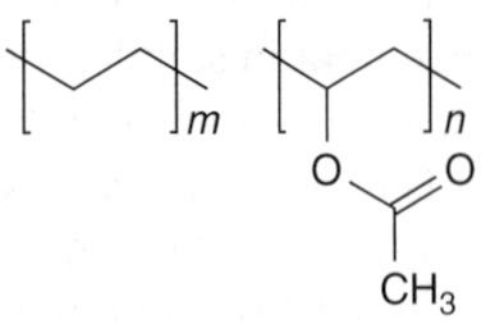

Figure 2.11 Polyvinyl acetate copolymers.

25–30% is the optimum vinyl acetate content [82]. Interestingly, EVAs have been shown to be wax nucleating agents as well as growth inhibitors [79, 83].

EVA copolymers are, in general, not as good at pour point depression as comb polymers, such as the (meth)acrylate ester polymers described earlier; however, they are considerably cheaper. It has been claimed, however, that partial hydrolysis of the EVA enhances performance [81]. This hydrolysis capability also makes them, in general, more biodegradable [84].

When dealing with crudes and condensates, especially those with WAT (over 40 °C), it appears to be beneficial to formulate these materials with para- and meta-cresols at around 10–15% of the formulation and usually an amine synergist such as cyclohexylamine (H.A. Craddock, unpublished results). This amine salt effect only manifests itself when the wax crystals have already been reduced by the EVA copolymer.

In general, either EVA copolymers in an aromatic solvent are supplied (this is certainly the case in the pipeline additive), or blends of EVA copolymers and acrylates, again in an aromatic solvent, are supplied.

A similar and cheap monomer to copolymerise with ethylene is acrylonitrile (see Figure 2.12). 10–20% acrylonitrile is the preferred copolymer ratio for optimum performance.

The lack of effectiveness of these polymers as a PPD tends to mean that they are not usually used for prevention of gelation in a pipeline situation. Also they are less biodegradable than ethyl vinyl acetate polymers and copolymers [85, 86].

Below 50 °C PVAcs are not fully water soluble but are water swellable and therefore have uses in drilling and wellbore cementing applications particularly as fluid loss additives [87, 88]. This behaviour also lends itself to the prevention of swelling in clays and shales [89] although PAM tend to be more favoured. In cementing applications the PVAc and PVA derivatives work by reducing the permeability of the filter cake. This occurs due to the microgel particles, which are formed when the polymer is hydrated, coalescing to form a polymer film. This is, however, a temperature-dependent behaviour, and film formation is not possible above 38 °C without cross-linking of the polymer to create additional thermal stability [90]. Again the environmental chemist is in the dilemma of the suitability of the application versus the stability and therefore biodegradation profile of the polymer.

The related PVAs in particular have good biodegradation profiles [91] and are one of the very few vinyl polymers soluble in water, which are also susceptible of ultimate biodegradation in the presence of suitable microorganisms. Accordingly, increasing attention is being devoted to the preparation of environmentally compatible PVA-based materials for a wide range of industrial and other applications; however, the efficacy and application in the oil and gas industry has been, to date, somewhat limited. PVAs have found uses in gel applications such as water shut-off [92] and as drilling lubricants applied as linear or cross-linked polymers [93].

PVAs have been used as DRAs [94]; however they are not as efficient as the ultra-high molecular weight PAM described earlier in this section. They have also been used as flocculants [95] but again have poorer efficacy than the more commonly applied polyelectrolytes and PAM previously described.

As will be described in more detail in Chapter 3, PVAs can be used in combination with certain surfactants to modify the viscosity of the water layer and improve the foam stability [96].

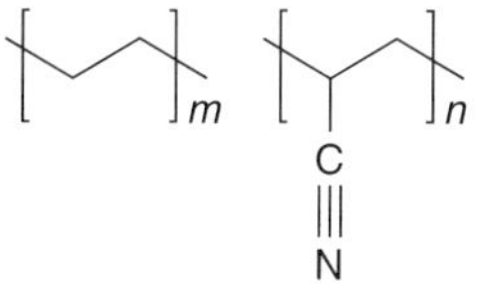

Figure 2.12 Polyvinyl acrylonitrile copolymer.

One of the main industrial vinyl polymers is PVC, which, although ubiquitous as a solid plastic material in the oil and gas industry, as in many other industrial and domestic areas, is absent in any direct form in application as a chemical agent. PVC has been studied as to its biodegradation characteristics, and although capable in soil to biodegrade in the right conditions over time [97], it has been found to be poorly degradable and biodegradable in the marine and other aquatic environments [98]. Also weathering degradation of plastics such as PVC results in their surface embrittlement and micro-cracking, yielding microparticles that are carried into water by wind or wave action. These microparticles become concentrated with persistent organic pollutants (POPs), which potentially can be ingested by marine biota. This and related topics are discussed further in Section 2.6.

2.1.1.5 Polystyrene and Other Unclassified Vinyl Polymers

There are a few, not easy to classify, vinyl polymers that have been found to be very useful in the oilfield area, which are discussed in this section.

Polystyrene (Figure 2.13) has found little use in the upstream oil and gas area with a report on its use as a cement additive [99] being one of the relatively few chemical additive applications. Although it has been assumed that polystyrene, along with other vinyl polymers, has a low rate of biodegradation [21] and, indeed, the styrene monomer is known to be toxic and is reasonably anticipated to be a human carcinogen, recent work has shown that it can be considered, under certain soil conditions, to be partly biodegradable [100]. It has also been shown that when combined with starch-based materials, it can be used as a composting material [101].

Polystyrene is being revisited especially where it has been adapted to allow partial degradation to occur to substances other than the styrene monomer. Also the degradation of the styrene monomer in soil remediation has been examined [102], and the polymer manufactured with a starch substrate in the processing should be reasonably biodegradable and produce less toxic degradation products [103, 104].

Two other vinyl polymer classes are particularly useful in the production of oil and gas, namely, the polyvinyl sulphonates and other related polysulphonates and the polyvinylcaprolactams (PVCap). Representative structures are shown in Figures 2.14 and 2.15, respectively.

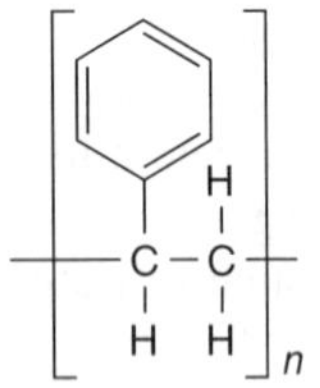

Figure 2.13 Polystyrene.

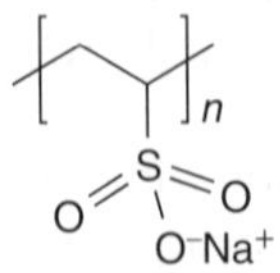

Figure 2.14 Poylvinyl sulphonate.

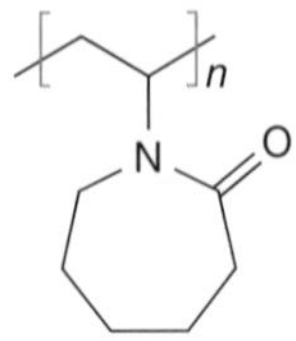

Figure 2.15 Polyvinylcaprolactam.

Polymers of vinyl sulphonate are usually manufactured as their metal salts, particularly sodium, such as illustrated in Figure 2.14, and potassium. They have a highly effective role in scale inhibition especially in controlling the formation and deposition of barium sulphate scales. This is particularly the case for copolymers with maleic acid or maleic anhydride [105]. These polyvinyl sulphonates are useful when working under harsh conditions such as high temperature, lower pH and higher than usual concentrations of calcium and magnesium ions than other types of scale inhibitor [106].

Polyvinyl sulphonates also have a use in well cementing technology as a dispersant [107]. Such dispersants improve the rheological properties of the cement, which allows blending at higher densities and lower water concentrations. They also allow the cement slurry to be pumped faster and therefore to produce a better bond between the well casing and the rock formation.

The environmental profile of these vinyl sulphonate polymers is under discussion and review. Many of the polymers and copolymers are known to be not sufficiently biodegradable to allow their use in the environmentally regulated areas of oil and gas production. However, many products and formulations have been developed, which do meet these regulatory requirements. This is an important issue especially in the North Sea region where a number of oil and gas fields are particularly prone to barium sulphate scale deposition and operate under harsh scaling conditions.

PVCap are the most widely applied and effective low dose KHIs found to date. Gas hydrates are a serious issue in oil and gas production and are formed from water and small hydrocarbon molecules, particularly methane at elevated pressures and low temperatures [108]. Blockages caused by gas hydrate clathrate during drilling operations and production can have serious consequences such as in the recent BP Macondo well blowout, which undoubtedly had a gas hydrate formation factor. The usual course of prevention and remediation is to use sufficient, usually large, volumes of thermodynamic inhibitor, methanol or ethylene glycol to give an anti-freeze effect, changing the equilibrium for gas hydrate formation to lower temperatures and/or higher pressures. The volumes used as mentioned are large, usually two volumes of hydrate inhibitor to one volume of water. This leads to some serious handling, logistics and safety (methanol is a highly flammable liquid) and environmental concerns. Low dose hydrate inhibitors can obviously help in alleviating some, if not all, of these issues, and to date, kinetic low dose inhibitors based on PVCap polymers and copolymers are the types mostly applied.

As with other vinyl polymers PVCap and other polyvinyllactams are poorly biodegradable; however, some products have been developed to show sufficient biodegradation [109] for application in regions such as the North Sea where regulatory compliance is a key requirement of application.

This section has illustrated a wide range of synthetic vinyl polymers, which are a highly important class of products in aiding the development and production of oil and gas. The challenges to using these products in oil and gas production are great, as, in the main, they exhibit very poor rates of biodegradation and have normally highly toxic monomer profiles. However, as has been illustrated, this will be further described in Section 2.1.3 and will be consolidated in Section 2.7. These challenges are being tackled by suitable chemistry modification in order to ensure, at least in the prescribed testing, reasonable rates of biodegradation are achieved. As will be discussed in the closing chapter of this book is this the right approach?

2.1.2 Non-vinyl Polymers and Copolymers

In general the non-vinyl polymers are not as ubiquitous in oilfield applications as the vinyl polymers. Both addition polymers and condensation polymers are manufactured in this class using a large variety of monomer units. Commercially, however there are three main classes of non-vinyl polymers: polyethers, polyesters and polyamides. Additionally polyamines and polyimines have important oilfield uses.

2.1.2.1 Polyethers and Polyols

Polyether compounds can be used as DRAs or friction modifiers in OBM formulations [110]. Polyethylene oxide (PEO) is synthesised by the metal-catalysed polymerisation of EO. This results in polymeric molecules of very high molecular weight, up to 8 million Da, which have a linear non-branched construction [111] (see Figure 2.16). It is, however, less often used in oil and gas operations as it exhibits shear degradation under standard chemical injection conditions and also under turbulent flow.

PEGs are formed from the base-catalysed polymerisation of EO. These materials have molecular weights of less than 100 000 Da. They also have a wide range of applications in the oil and gas sector. The structure of PEG was illustrated earlier in this chapter (Figure 2.1).

There are a number of polymer thickeners, and these include PEG [112]. Conversely PEG has been used in drilling lubricant formulations, particularly alongside polyether phosphate ester surfactants [113]. PEGs have a number of uses both as a carrier medium in formulated products and as a base polymer for further derivatisation. In the latter case, glycide demulsifier derivatives [114] can be prepared by the acid- or base-catalysed reaction with di-epoxides. These products are particularly efficacious in resolving emulsions formed in the extraction of bitumen like tar sands.

PEG is a highly hydrophilic material and is resistant to oxidative degradation. However, very little information on the decomposition of PEG under biological oxidative stress can be found in the literature. Pseudo-polypeptides [115] have attracted some attention as biomaterials and as an alternative to PEG with an altered stability against oxidative degradation. However, these are supposedly non-biodegradable, which could be seen as one of their main disadvantages. Recent evidence that PEG and some pseudo-polypeptides are degradable by oxidative degradation under biologically relevant conditions has been published [116] and mid- and long-term biodegradability in vivo appears feasible.

Related to these products and widely used in demulsifier chemistry are the polyols, such as glycerol, a simple polyol (see structure in Figure 2.17).

These polyols are used in artificial sweeteners, particularly sorbitol (see Figure 2.18) and are therefore not toxic, also being derived from sugars they are highly biodegradable. In demulsi-

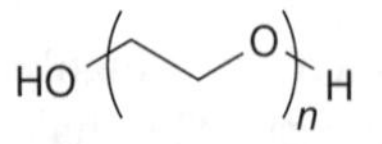

Figure 2.16 Polyethylene oxide.

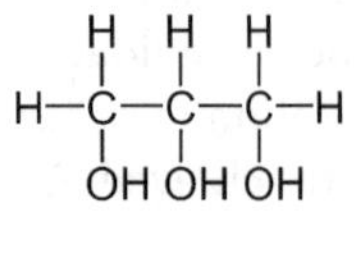

Figure 2.17 Glycerol or glycerine.

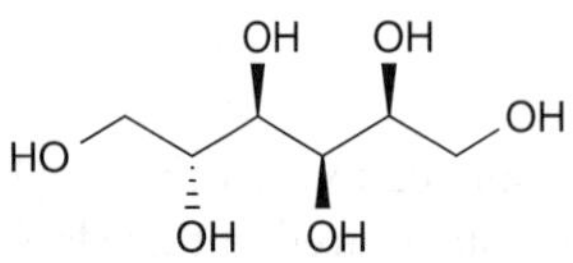

Figure 2.18 Sorbitol.

fier use, they are functionalised, copolymerised and cross-linked with, for example, acrylic acid [117].

Using polyols as the backbone of the demulsifier and/or functionalised with EO and PO, as a block copolymer structure, has led to a range of biodegradable demulsifiers. Such products, as with the polyethylene glycols mentioned previously, can also form glycidyl ethers and related polyethers, again forming yet another class of demulsifier products, which are highly biodegradable and have different and useful functionalities. In general it is the polyglycerols (see structure in Figure 2.19) that find the most use in the oil and gas industry primarily as demulsifiers.

In drilling operations, polyols have limited but specific use. They are found as an additive in clay stabilisation [118] and can be used as a delay or retardation agent when using borate in a hydraulic fracturing fluid containing galactomannan gum [119]. This later use has good temperature stability of up to 150 °C.

More general organic clay stabilisation and fluid loss agents can be based on polyether amines [120] such as the structures shown in Figure 2.20.

These materials are found to have shale swelling inhibition properties that are superior to potassium chloride, which is the conventional inhibitor, and they can be improved with a decrease of pH value. Another drawback of salts such as potassium chloride is that although they reduce clay swelling, they can also flocculate the clays, leading to high fluid losses and loss of thixotropic rheology. Attempting to counteract this by increasing the salinity often leads to further loss of functional characteristics [121].

The polyether amines have found application as synergists to the PVCap polymers used as KHIs [122, 123].

A combination of polyol and polyether chemistry can produce the polyether cyclic polyols. These materials are prepared by thermally condensing a polyol to its oligomers and cyclic ethers [124]. They have multifunctional properties particularly suited for enhanced drilling fluids in that they can inhibit hydrate formation, prevent shale dispersion and clay swell, reduce fluid loss and generally improve wellbore stability. They are useful substitutes for OBM, which have many associated environmental challenges for their use.

Polyethers have industrial uses as defoamers and are also applied in a growing number of areas where either other defoamers, usually silicone-based materials (see Chapter 7), cannot be used or not effective, as in aqueous drilling fluids. These defoamers are based on higher alcohols and polyols, which are suitably derivatised, such as by ethoxylation and/or propoxylation.

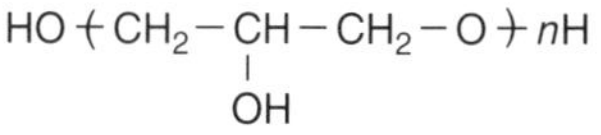

$$HO + CH_2 - CH - CH_2 - O + nH$$
$$\quad\quad\quad\quad |$$
$$\quad\quad\quad OH$$

Figure 2.19 Polyglycerol.

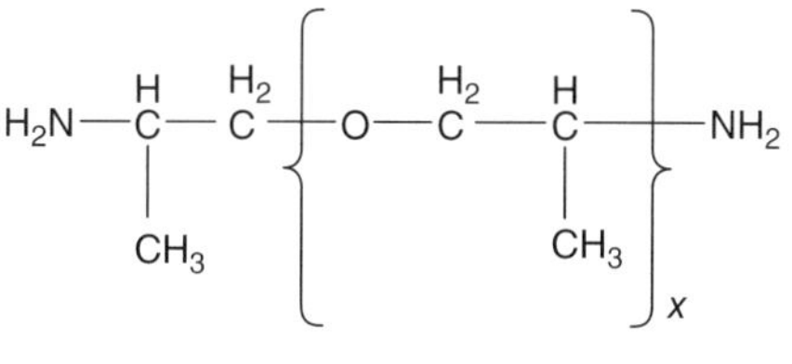

Figure 2.20 Polyether amines.

Environmentally acceptable aqueous active defoamers have been produced from fatty acid esters of polyols such as sorbitan monooleate [125]or sorbitan monolaurate with diethylene glycol monobutyl ether as a co-solvent [126].

Partially etherified cellulose polysaccharides have been used as fluid loss additives in mineral- and clay-based drilling muds [127]. The chemistry of polysaccharides is more fully detailed in Section 2.2.

Polyethers are classified into biodegradable synthetic polymers [128]. As has been described, PEGs are among the most widely used chemicals in various industrial fields including the oil and gas E&P sector. Consequently, biodegradability studies have been primarily focused on PEGs. PEGs are aerobically or anaerobically metabolised by various bacteria. Biochemical routes have also been studied for aerobic degradation of the compound. Bacterial degradation of polypropylene glycol has been observed and its metabolic pathways described. Another polyether, polytetramethylene glycol, can also be utilised by aerobic bacteria as the sole carbon and energy source.

Similarly polyols are amongst the most biodegradable polymers especially when suitably functionalised as will be described in Section 2.1.3.

2.1.2.2 Polyesters

Polyesters are a category of polymer that contains the ester functional group in their chain. As a specific material, it most commonly refers to a type called polyethylene terephthalate (PET). Polyesters include naturally occurring chemicals, as well as synthetics such as polybutyrate (see Figure 2.21). Natural polyesters are discussed in more detail in Section 2.2; these and a few synthetic ones are biodegradable, but most synthetic polyesters are not.

The primary use of polyester polymers is as fibres and resins for clothing and a wide variety of industrial uses. In the oil and gas industry, they are used in construction and plastics. Their use as chemical additives however is limited or relatively unexplored.

Polyesters, particularly polyorthoesters (see Figure 2.22) and aliphatic polyesters such as polylactides (see Figure 2.23), have properties that aid fluid loss control in hydraulic fracturing [129]. This is related to the hydrolytic degradation of these polymers in that the slow rate of degradation provides fluid loss control in these fracturing operations during proppant placement.

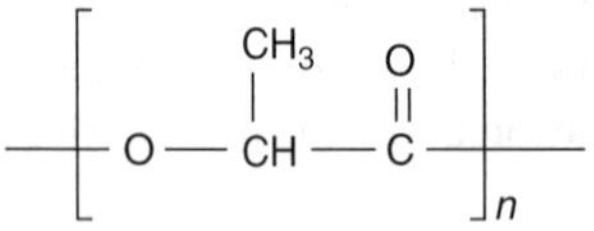

Figure 2.21 Polybutyrate.

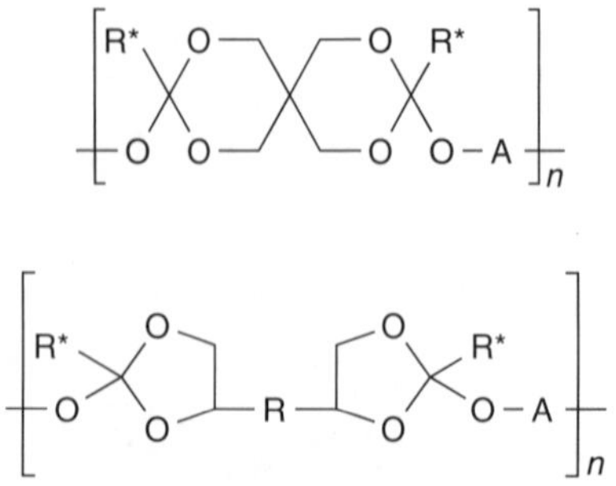

Figure 2.22 Polyorthoesters.

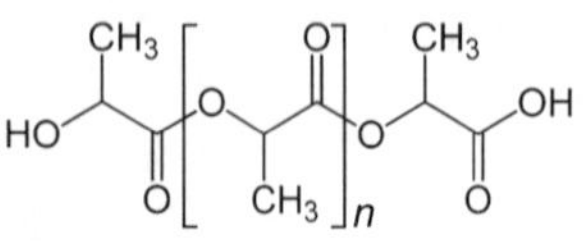

Figure 2.23 Polylactides.

During acidisation workover of wells and in scale inhibitor squeeze treatment diverters can be applied to temporarily block the zone or zones of highest permeability, allowing the main treatment to react with less permeable and more damaged zones within the near wellbore area. Within the categories of diverters used are polymer gels. Indeed recent work has developed solid-free diverters in which polyesters have been primarily used [130]. This is a considerable advantage over other methods within which the removal of the diverter can be problematic and its degradation leads to solids that cause further formation damage.

Asphaltene dispersants based on esters of polyhydric alcohols have been claimed in the patent literature [60]. However, polyester chemistry is commercially applied as asphaltene inhibitors, the ester functionality usually being derived from acrylic or maleic anhydride monomers. The structure of these polymers and copolymers is more akin to the vinyl polymers referred to earlier in Section 2.1.1.2. A typical example is the styrene/maleate ester shown in Figure 2.24.

The other application of polyester and related polymers is in gas hydrate inhibition where a number of products as low dose hydrate inhibitors have been developed. These are KHIs and provide alternative technology to the previously described polymer of vinylcaprolactam, being the preferred low dose hydrate inhibitor in use. In particular two types of polyester chemistry have been developed: the hyperbranched polyesteramides [131] (see Figure 2.25) and ester and polyester of pyroglutamate polymers [132] (see Figure 2.26), with the latter claiming to be highly biodegradable and affording superior performance at low subcooling.

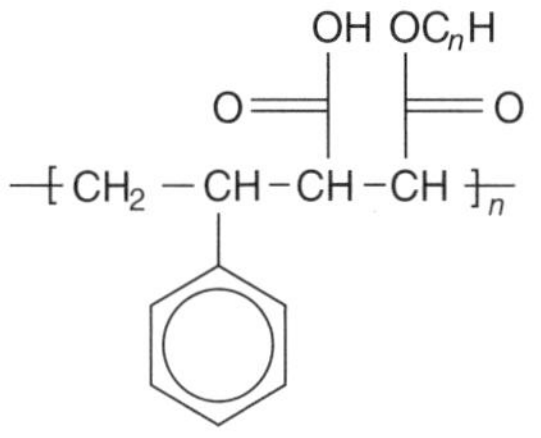

Figure 2.24 Styrene/maleate ester.

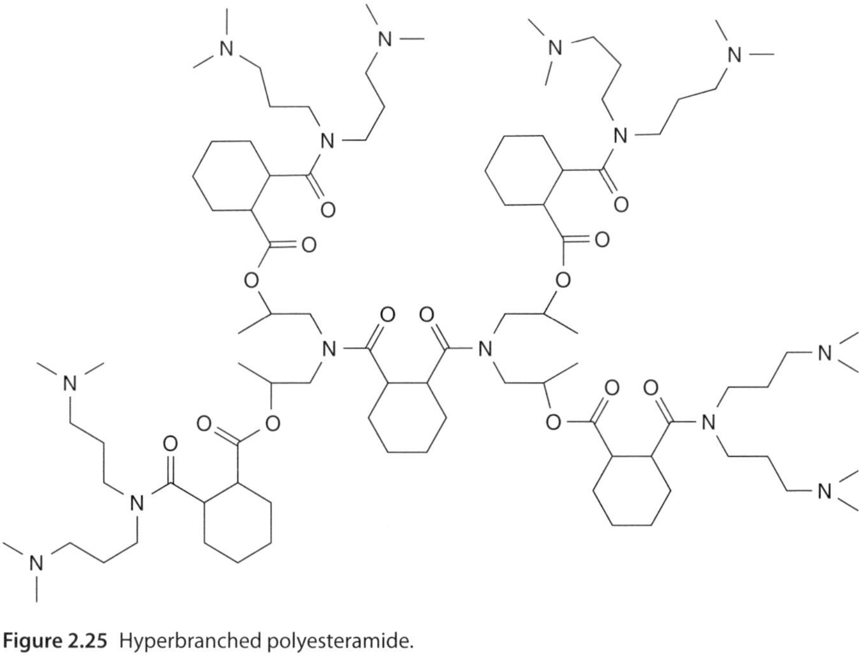

Figure 2.25 Hyperbranched polyesteramide.

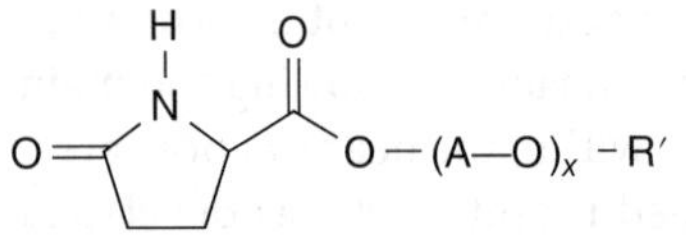

Figure 2.26 Pyroglutamic acid ester.

In the structure shown in Figure 2.26, A is C2–C4 alkylene group, x is from 1 to 100, and R′ is an aliphatic, cycloaliphatic or aromatic group as they are based on a natural amino acid with a polyester 'backbone' they are highly biodegradable. As will be apparent this tactic of using and/or reassembling natural products has become a major strategy in the design of environmentally acceptable products for application in the industry, particularly with respect to chemical additives.

2.1.2.3 Polyamines and Polyamides

There are a number of different amine products used in the upstream oil and gas industry, many of which will be described in later chapters, e.g. fatty acid amine and amide surfactants (Chapter 3) and low molecular weight organics (Chapter 6). These will be discussed as to application and chemistry in these chapters. In effect there are again relatively few true polyamines (and polyamides) applied as chemical additives.

The adducts of polyamides, such as polyethylene amine and formaldehyde, have been found suitable for use as gas hydrate inhibitors, particularly in drilling applications [133]. Many amines are derived from other polymers, for example, polyether amines (see Figure 2.27) as described previously have been used as synergist [122, 123] in formulations with PVCap KHIs. In the last decade this chemistry has been further developed and more widely applied [134]. Quaternised polyether amines have been shown to have anti-agglomerate behaviour [135], and related to this chemistry polypropoxylated polyamines also show interesting anti-agglomerate behaviour [136].

Polyether amines and other polyamines have also displayed good efficacy as corrosion inhibitors [137] although here the surfactant properties of the molecule are as important as its polymeric structure.

Polyethylene amines are also claimed to be good corrosion inhibitor, in particular polymethylene polyamine dipropionamides [138] (see Figure 2.28), and have been effectively used to mitigate CO_2 corrosion. It is also claimed that they have low toxicity to marine species, particularly marine algae.

In this class of products the boundary of classification between materials employed as polymers and those utilised as surfactants becomes indistinct. There are materials such as polyamines, polyamides and polyimines that are also surfactants. Indeed one of the main uses of these materials is as demulsifiers and phase separation agents. Polyamine-based demulsifiers are a main class of product [139]. One of the main types of polymer in this area is the polyamine

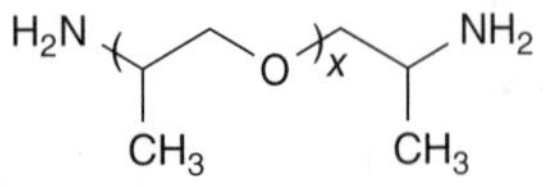

Figure 2.27 Polyether amines.

$$H_2NCO - (CH_2)_2 - NH - [-(CH_2)n - NH-]m - (CH_2)_2 - CONH_2$$

Figure 2.28 Polymethylene polyamine dipropionamides.

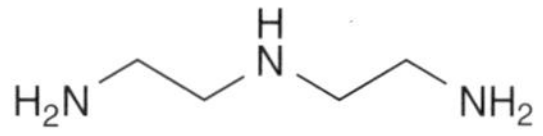

Figure 2.29 Diethylenetriamine.

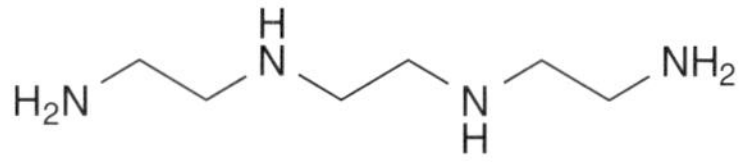

Figure 2.30 Triethylenetetramine.

polyalkoxylates and related cationic polymers, which are used for oil and water separation and resolution. As has just been described, they are also used as corrosion inhibitors where they exhibit film-forming surfactant properties.

These polyamine polyalkoxylates are synthesised from readily commercially available amines such as diethylenetriamine (DETA) (Figure 2.29) and triethylenetetramine (TETA) (Figure 2.30). These basic amines are important both as additives in their own right and as part of synthetic building block for a variety of oilfield chemicals.

The polyoxyalkylenated DETA demulsifier can be derivatised with various PEO and polypropylene oxide (PPO) copolymers to destabilise a stable water-in-oil emulsion [140]. It was found that the destabilisation of an emulsion is closely correlated with the PO and EO numbers. When the PO number in a molecule is much greater than the EO number, the surfactant gives a very low oil resolution rate, requires high dosages and produces a stable middle phase. When a surfactant/polymer contains more EO than PO, it gives a high oil resolution rate at a low dosage, but it is easily overdosed, and some surfactants of this type (high molecular weight) also produce a stable middle emulsion phase. When the PO and EO numbers in the surfactant are close to equal, the surfactant breaks the emulsion rapidly at a very low dosage, does not show overdosing at very high dosage, and does not produce a stable middle phase. Therefore, surfactants with balanced PO and EO numbers give optimal performance.

The related polyethyleneimines are also useful demulsifiers after being derivatised by ethylene, propylene or butylene oxide [141, 142]. The polyethyleneimines themselves are also useful as synergists in wax inhibition alongside the previously described vinyl polymers of acrylic acid, in particular (meth)acrylate ester polymers (H.A. Craddock, unpublished results).

Polyamines, polyamides and polyimides are all claimed to be useful as asphaltene inhibitors [60] and are similar in structure to the polyester types previously described.

Polyamides have uses as emulsifiers in the construction of emulsion and invert emulsion-based drilling fluids [143]. These emulsifiers are primarily for oil-based drilling fluids and muds comprising an emulsifier based on the polyamides derived from fatty acid/carboxylic acid and optionally alkoxylated polyamines.

Again these non-vinyl polymers compete with a large number of vinyl polymer alternatives such as the PAM previously described.

Interestingly polyamines, polyamides and related products have different metabolic and degradation pathways to the other non-vinyl polymers such as polyols and polyesters [144] as they in some ways mimic the nucleic acids found in living cells. This can increase their toxicity to marine species but conversely improve the rates of biodegradation.

2.1.2.4 Other Non-vinyl Polymers

In this short section, polymers, which are somewhat unclassified, shall be examined.

The alkylphenol resin polymers are a very useful class on non-vinyl polymers in the upstream oil and gas industry. These are usually synthesised from an alkylphenol with formaldehyde to give a number of oligomers of 2–12 alkylphenol groups connected by methylene bridges as shown in Figure 2.31 of a butyl phenol-formaldehyde resin polymer.

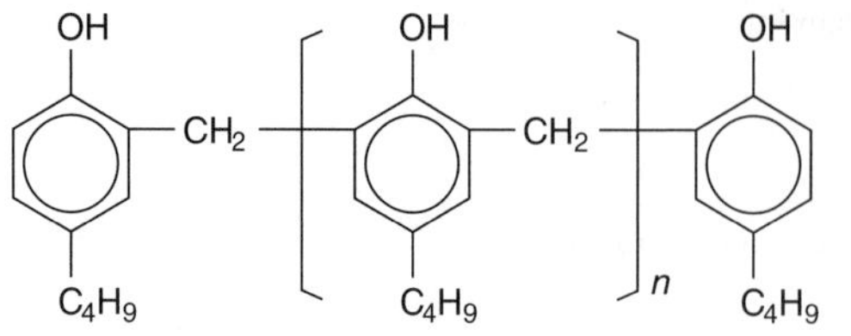

Figure 2.31 Butyl phenol-formaldehyde resin polymer.

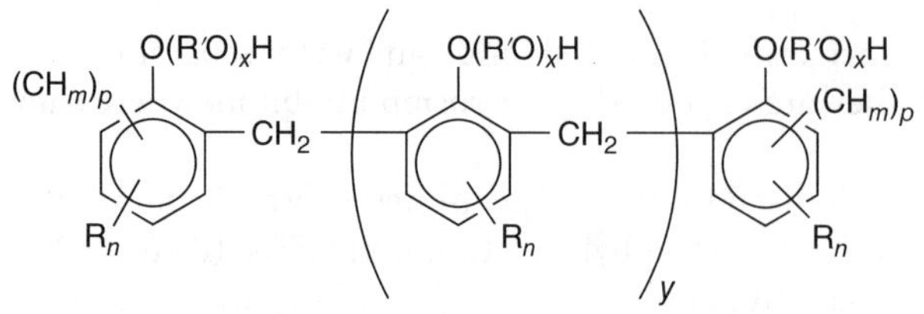

Figure 2.32 Alkylphenol alkoxylate polymer.

These products in their own right provide useful properties and are claimed to be asphaltene inhibitors, which can be used in squeeze applications [145]; however by far their biggest use is when derivatised to their alkoxylates, particularly with EO and/or PO, and used in demulsifier formulations [146]. These resin alkoxylates can have complex structures (see an example in Figure 2.32), and the most common are the so-called EO/PO block polymers. Indeed these are likely to be the most common type of demulsifiers used in oilfield applications.

A type of polymer, the linear polymer, is formed from an acid-catalysed condensation process; if the process is base catalysed, the interesting calixarenes [147] are formed (see Figure 2.33), which shows the *t*-butyl phenol calixarene. These macrocyclic molecules have different demulsification properties, particularly with respect to ensuring low salt concentrations in crude oil [146, 148].

Again in this area of polymers there is considerable overlap with having them classified as surfactants; indeed in their use as phase separation agents or demulsifiers, this is where they will be more fully described (see Chapter 3).

The alkylphenols have been classed xenoestrogens or endocrine disruptors, and most scientists consider this to have a serious environmental impact as they mimic natural oestrogens, giving rise to hormone disruptive effects in humans and wildlife. The European Union has therefore implemented sales and use restrictions on certain applications in which, in particular, nonylphenols (see Figure 2.34) are used because of their alleged 'toxicity, persistence, and the liability to bioaccumulate'; however, the US Environmental Protection Agency (EPA) has taken a slower approach to make sure that action is based on 'sound science'. This fundamental difference in regulatory regimens is further discussed in Chapter 9.

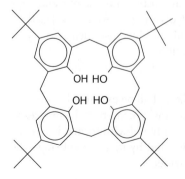

Figure 2.33 *t*-Butyl phenol calixarene.

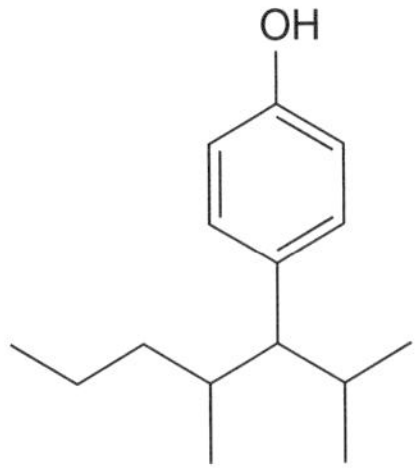

Figure 2.34 Nonylphenol.

This action on alkylphenols, particularly in the European area, impacts the oilfield, as the degradation products of ethoxylated alkylphenols are known to be the relevant alkylphenols. However this may not necessarily be the case for the polymeric alkylphenol-formaldehyde polymers, whose degradation pathways are not, as far as is known, back to the alkylphenol monomer [149].

Finally there are copolymers of non-vinyl monomers such as co-polyesters and co-polyamides in which their use in oilfield applications is undetermined and at best limited.

The environmental impact, biodegradation and environmental fate of these vinyl and non-vinyl and synthetic polymers have been briefly detailed as individual examples have been described. This will be further discussed in Section 2.6. However, in the next section some of the synthetic constructions and derivations employed by the oilfield chemist to achieve a better environmental profile of these synthetic polymers are discussed.

2.1.3 Environmental Derivations of Synthetic Polymers

As has been alluded to in the previous two sections of this chapter, most synthetic organic polymers are not environmentally benign. In most cases, but not all, they poorly and slowly biodegrade, and their ultimate fate to their monomers can lead to toxic degradation products. Indeed in the case of vinyl polymers, their monomers are often known to be toxic to aquatic species. Given the breadth of application of these products, the environmental and oilfield chemist in looking to design better more environmentally acceptable products has focused on three methods of modification:

1) Derivatising any functionality to increase the biodegradation profile, such as esterifying alcohol side chains.
2) Grafting or copolymerising more environmentally acceptable polymeric species.
3) Replacing the polymer backbone with one that is more biodegradable.

All of these methodologies are well tried and tested in oilfield applications.

A serious advantage in working with polymers and their variants is that the regulatory frameworks on toxicology and other impacts including environmental impacts are based on monomers. Therefore, as long as the monomers are registered under the appropriate regulatory requirements, the variations in polymer chains and functionality do not require further study work and registration. This allows the environmental chemist and the synthetic chemist to be able to examine a wide range of variations, safe in the knowledge that these can be commercially available, or at least not be unavailable, due to additional regulatory and registration costs.

As has been described, polymers are an important class of chemical additives in the field, and the challenge to the environmental chemist is to, wherever possible, maintain performance and efficacy and design more environmentally acceptable products. The three methodologies above have been widely adopted in this regard; however, inevitably in conducting these 'iterative' processes, there is a fundamental compromise between improving rates of biodegradation and maintaining product performance. This is simply due to two factors. Firstly, in moving to a more biodegradable product, the lifetime of the actual un-degraded product being available is shortened, sometimes dramatically. Secondly, in designing these products, the structural fit of the product, especially threshold inhibitors such as used in scale and wax control, is changed in such a way that the product is not as efficient in stopping the deposition it is trying to prevent. Both of these factors usually mean that increased dose rates of products are required to give an equivalent performance of the original polymer-based product. This means that the economics of application of such green products has become less favourable. In addition, the manufacture of these 'greener' products involves additional manufacturing costs, which means that these products are usually more expensive. Combined, this means that a green choice of a replacement product is challenging, unless it is necessary to comply with a regulatory requirement, and in most oilfield areas, environmental controls and regulation, as is further discussed in Chapter 9, 'Current Regulatory Frameworks for Chemical Use and Discharge', are not a prerequisite requirement or are not enforceable.

In the main, it is in the offshore environment, particularly in Europe and North America, that criteria are laid down for product requirements to minimise their impact on the marine environment.

Nonetheless it is a worthy exercise to attempt to reduce the overall environmental impact of the chemical additive or the chemicals used in a particular application, and some examples relating to polymer chemistry will now be illustrated.

2.1.3.1 Side-Chain Modification

As was described earlier, it is fortuitous that the alkylphenol-formaldehyde resins have enhanced surfactant properties if the alcohol groups are further functionalised such as by ethoxylation and propoxylation, as this also enhances their biodegradation characteristics [146, 148]. This general principle of using functionality in polymer side chains is a simple and relatively cheap means to enhance biodegradability, and it is founded on the principle that most enzymatic and other metabolic pathways employed by bacteria and other species in the degradation of chemical additives involves some form of hydrolysis. This is illustrated in the diagram (Figure 2.35) where sucrose is the substrate and importantly enzyme hydrolysis releases

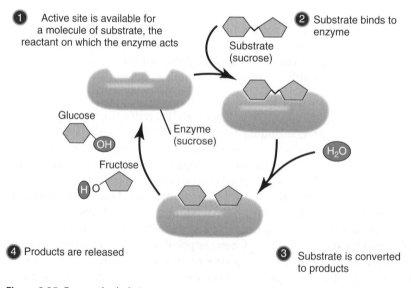

Figure 2.35 Enzyme hydrolysis.

products available to be further processed or consumed and release energy. Hydrolysis is simply the reaction of chemical species with water, and in metabolic pathways, this is usually catalysed by enzymes. In the overall process of hydrolysis, a bond in an organic polymer molecule is broken, and an O—H bond in a water molecule also breaks. Then, from the water molecule, an O—H group adds to one part from the organic molecule, and an H atom to the other.

Selecting functionalities, which are easily hydrolysed, should not only improve the overall environmental profile of the polymer but also the rate of biodegradation. In the main this means utilising available side-chain functionality to produce alkyl ethers, esters and amides as these produce weaker C—O bonds that are more readily, or a more available to be, hydrolysed [150].

This enhancement of functionality, as can be seen in other non-polymer examples, is widely used to influence the biodegradation test results of the molecule being applied, and this is discussed further.

Also in enhancing this biodegradation capability, particularly in polymers, it is often the case that the water-soluble characteristics of the polymer are also enhanced. This is further discussed in Section 2.6.

2.1.3.2 Copolymerisation and Grafting

As described earlier, when a polymer is made by linking only one type of small molecule, or monomer, together, it is called a homopolymer. When two different types of monomers are joined in the same polymer chain, the polymer is called a copolymer. Copolymers can be classified based on how these units are arranged along the chain [151]. These include:

- Alternating copolymers with regular alternating A and B units.
- Periodic copolymers with A and B units arranged in a repeating sequence (e.g. $(A-B-A-B-B-A-A-A-A-B-B-B)_n$).
- Statistical copolymers are copolymers in which the sequence of monomer residues follows a statistical rule. If the probability of finding a given monomer residue at a particular point in the chain is equal to the mole fraction of that monomer residue in the chain, then the polymer may be referred to as a truly random copolymer [10].
- Block copolymers are a special kind of polymer and comprise two or more homopolymer subunits linked by covalent bonds. As has been described earlier, these copolymers are of special significance in phase separation in the oil and gas industry, and some more detail on these polymers is given in the following subsection.

Copolymers may also be described in terms of the existence of or arrangement of branches in the polymer structure. Linear copolymers consist of a single main chain, whereas branched copolymers consist of a single main chain with one or more polymeric side chains.

Other special types of branched copolymers include star copolymers, brush copolymers and comb copolymers. A terpolymer is a copolymer consisting of three distinct monomers.

All of these techniques in making copolymers are used to manufacture specifically required polymers for application in the upstream oil and gas industry, and a number of examples have been detailed earlier in this chapter. In the main the technique is focused in providing the polymer with specific functionality and has been exemplified earlier; this has been applied to the design of molecules for a number of applications especially with regard to copolymers from vinyl monomers. However, the use of the technique for specifically including monomer units or groups of monomer units for the design of environmentally acceptable polymers has been hardly used. A few examples have been described such as the use of polymaleate (Figure 2.7) in polymeric scale inhibitor design [32] and the design of environmentally acceptable variants of the vinylcaprolactam polymeric KHIs [152].

The main reason for the lack of development in this area is twofold: the overall imperative for green polymers in the global oilfield and the cost of manufacture of these materials. As has been alluded to, the drive for the use of green products in the oil and gas industry although growing has been driven by regulatory requirements in certain geographic areas. Therefore, notably in the use of copolymer, variation and related design techniques to the development of specific green polymers has been, in the main, focused on product types such as scale inhibitors and hydrate inhibitors, which are coincidental for use in key highly regulated regions such as the North Sea and the US Gulf of Mexico. As will be illustrated, on more than one occasion, the need for a green product alternative is primarily driven by regulatory controls outweighing economic concerns.

Graft polymers or more correctly graft copolymers are a branched copolymer where the components of the side chain are structurally different than that of the main chain. Graft copolymers containing a larger quantity of side chains are capable of wormlike conformation, compact molecular dimension and notable chain end effects due to their confined and tight fit structures [153]. Generally, grafting methods for copolymer synthesis results in materials that are more thermostable than their homopolymer counterparts [154]. This is not a desirable property when trying to design a more environmentally friendly polymer, and hence grafting is not a technique widely employed for this purpose.

A related methodology has been employed in the design of a new class of scale inhibitor [155], namely, hybrid polymerisation where a synthetic polymer and a natural polymer have been manufactured to form a cross-linked hybrid polymer. An example of a hybrid scale inhibitor polymer in which a polyaminoamide and a polysaccharide have been 'hybridised' is shown in Figure 2.36.

2.1.3.3 Block Copolymers

As described earlier in section on other non-vinyl polymers, block polymers and copolymers are an important class of polymer in oilfield use. These copolymers are constructed of blocks of different polymerised monomers and can give a variety of functionalities. A variety of synthesis techniques are employed to make these block polymers and copolymers, and complex and exotic structures can be synthesised [156]. This is a dynamic area of polymer research, and many different types of block copolymer are being examined for a variety of industrial and other applications. As yet however only derivations such as ethoxylation or esterification of the

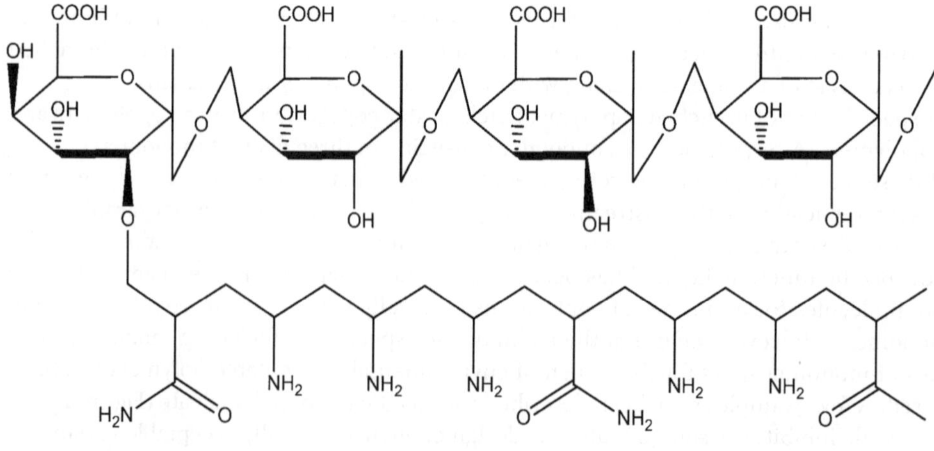

Figure 2.36 Hybrid polymer.

block copolymers have been examined in the oilfield that potentially enhance their environmental properties. Design of a specific block copolymer with functionality and environmental acceptability, at least for an oilfield application, has yet to be described. As has been previously described where environmental improvement has been seen, it has been alongside the targeted design changes in surfactant properties.

2.1.3.4 Selecting Polymer Backbone

This is probably the most fundamental of the three possible approaches to producing an environmentally more acceptable polymer product without losing key functionality. There appears to be few, if any, examples in published literature of this strategy being deliberately adopted.

There are instances where analogous polymers have been adopted and developed to provide a similar functionality to the existing but environmentally less benign product. For example, poly(2-alkyl-2-oxazoline)s, which in effect exist as poly(*N*-acylalkylene imine)s in polymeric form, in effect a PAM, have been examined as to their effectiveness as KHIs [157]. Figure 2.37 shows the ring-opening reaction of oxazoline to give the polymeric form.

Although some of these products showed comparable performance with commercial PVCap polymer products, none of the poly(2-alkyl-2-oxazoline)s exhibited acceptable biodegradation profiles. This is surprising as the amide-type moiety has introduced a weaker C—N bond into the polymer backbone; however the enzyme hydrolysis needs access to the amide/ester linkages, and if they are sterically crowded as in this type of KHI polymer, then the rate of biodegradation is low [158]. In a related and more successful approach, the introduction of other weak links has been developed for other functionalities, in particular demulsifiers [159]. Here block polymers, involving the ortho-ester derivation of PEG and polypropylene glycol, have led to the production of best performing demulsifiers with a high degree of biodegradability, over 60%, in the accepted marine test protocol [160].

As has been described earlier, the vinyl monomers tend to produce more stable and less degradable polymers, which are less water soluble in the main. The exceptions in vinyl polymers are polyacrylates, PAM and PVAs although it is possible in all three classes to synthesis polymeric forms, which are not water soluble. The non-vinyl polymers are generally water soluble. This is an important factor in the design of environmentally friendly polymers as the main pathway to biodegradation is through enzymatic hydrolysis, and therefore polymers, which are not available in solution, are likely to degrade and biodegrade more slowly than those that are soluble in water.

A good example of a polymeric form, which in the authors' view has been somewhat overlooked as a versatile and yet potentially highly biodegradable polymer source for iteration and variation, is the polyglycols. For some time these products have been used as demulsifiers [161] and have a variety of other oilfield uses, particularly the PEG [112–116].

Finally a recent development in polymer application to the oilfield has been described [162].

This work has involved the development of polyamines and polyquaternaries with hydrolysable bonds within the backbone of the polymers and/or at the terminal ends as derived from long fatty alkyl groups. An example of a polyamine type is shown in Figure 2.38; however, these molecules are again at the boundary of definition between polymers and surfactants. Indeed they are polymeric surfactants.

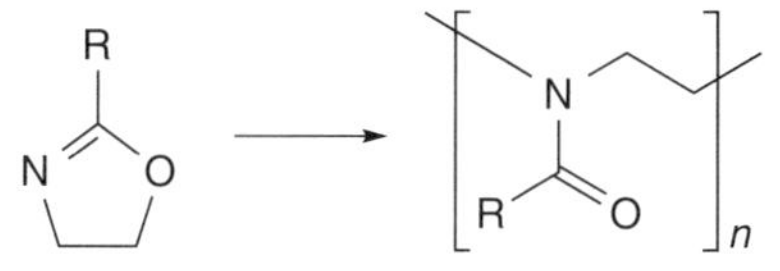

Figure 2.37 Formation of poly(2-alkyl-2-oxazolines).

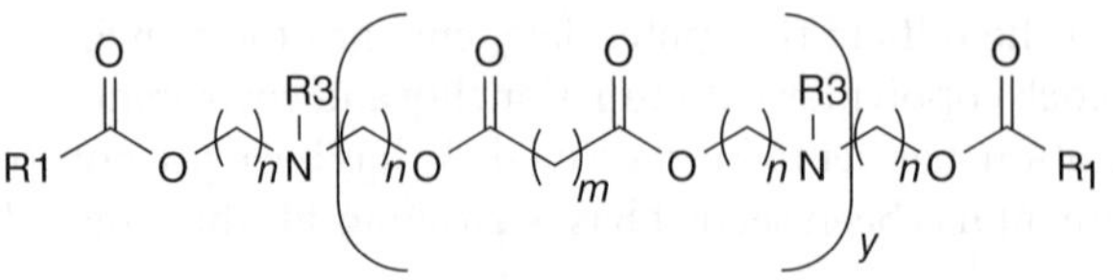

Figure 2.38 Polyamine type of corrosion inhibitor with hydrolysable backbone.

In concluding this section it can be seen that a wide variety of synthetic polymers have been applied in developing chemical additives to meet the requirements of mitigating a diverse range of problems found in the drilling, completion, development, production and other areas of the oil and gas E&P industry. This has led, primarily where required by regulation, to the development of a number of greener alternative polymers. There is little doubt that this area of chemical additive development will remain important, as polymers themselves are probably one of the largest classes, perhaps the largest class, of chemical additives used in the oilfield sector. Polymers also offer a unique advantage to the oilfield and environmental chemist in that they can be manipulated in a variety of ways to offer both performance functionality and environmental acceptability.

Therefore synthetic polymers are likely to remain at the forefront of chemical additive technology; however they will also see competition from natural derivatives as these can offer greater environmental performance, as will be illustrated in the next section of this chapter. The challenge is to be able to utilise the manipulation afforded in synthetic polymers to their natural equivalents to gain functional performance.

2.2 Natural Polymers and Related Materials

As is implied by the nomenclature, natural polymers are polymers that occur in nature. Most of the structures of living things are comprised of natural polymers, of which there are three main types:

1) Polynucleotides, which are chains of nucleotides
2) Polyamides, which are chains of proteins
3) Polysaccharides, which are chains of sugars

Most readers will be familiar with at least one naturally occurring polymer: deoxyribonucleic acid (DNA). The DNA molecule is made up of monomers called nucleotides. The monomers are linked by a condensation reaction so that many nucleotides are linked in a chain to make the DNA polymer molecule or polynucleotides (see Figure 2.39).

Another naturally occurring polymer is keratin found in nails and hair. It is among the most abundant proteins in humans. Proteins are condensation polymers made from amino acid monomers. One amino acid's —NH2 functional group reacts with the —COOH functional group of another amino acid, forming a peptide bond, —CO—NH—. The peptide bond is also called an amide bond; therefore proteins are also known as polyamides or polypeptides.

The final group of natural polymers is the polysaccharides, again familiar to most readers through the consumption of foodstuffs where sugars and starches are readily found.

By a wider definition, natural polymers can be man-made out of raw materials that are found in nature. Although natural polymers still amount to less than 1% of the 300 million tonnes of synthetic polymers and plastics produced per year, their production is steadily rising.

In the oil and gas industry, as in many other industrial sectors, polynucleotides are not considered for use in applications involving chemical additives. They do have a specialist applica-

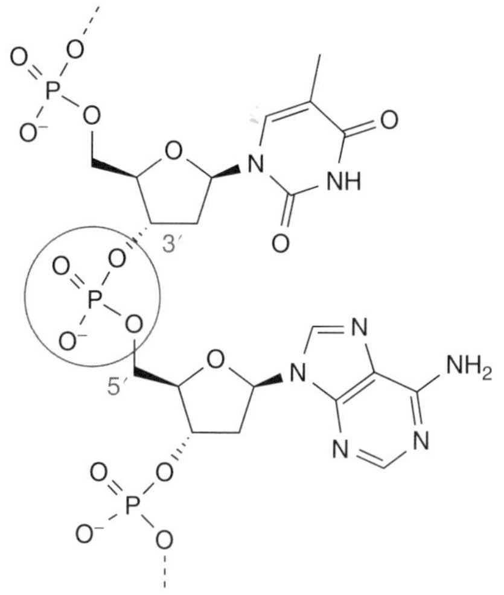

Figure 2.39 Segment of DNA molecule, a polynucleotide.

tion with their involvement in detection and assay of microbial species in oilfield samples to assay the effectiveness of control strategies [163, 164], including the use of chemical biocides.

2.2.1 Biopolymers

Simply defined biopolymers are polymers that are biodegradable. The input materials for the production of these polymers may be either renewable (based on agricultural plant or animal products) or synthetic [165] as has been described in a number of examples in Section 2.1. There are four main types of biopolymer based respectively on substrates used:

1) Polysaccharides
    a) Starch
    b) Sugar
    c) Cellulose
2) Synthetic materials

   Currently either renewable or synthetic starting materials may be used to produce biodegradable polymers. Two main strategies may be followed in synthesising a polymer. One is to build up the polymer structure from a monomer by the process of chemical polymerisation. The alternative is to take a naturally occurring polymer and chemically modify it to give it the desired properties. A disadvantage of chemical modification is that the biodegradability of the polymer may be adversely affected. Therefore it is often necessary, as has been described in a number of derived examples of synthetic polymers in Section 2.1.3, to seek a compromise between the desired material properties and biodegradability.

   This chapter will primarily focus on the two main industrially applied natural polymers and their derivation to improve functionality and/or environmental characteristics, namely, the polypeptides and the polysaccharides. The chapter will address other natural and biological polymers such as rubber, lignin, etc. and examine some other related biopolymers.

2.2.2 The Polypeptides

The polypeptides are the naturally derived class of polyamides, and these have a variety of uses in a number of sectors including the pharmaceutical, cosmetic and food industries. Their use in the upstream oil and gas industry has to date been fairly minor and specialised.

In the natural world the polypeptides are a ubiquitous and highly important class of chemical compounds known as proteins; these polyamides have α-amino acids as their monomer unit. Proteins are the largest and most varied class of biological molecules, and they show the greatest variety of structures. Many have intricate three-dimensional folding patterns that result in a compact form, but others do not fold up at all ('natively unstructured proteins') and exist in random conformations. The function of proteins depends on their structure, and defining the structure of individual proteins is a large part of modern biochemistry and molecular biology.

The term protein and polypeptide are interchangeable although those involved in the study of proteins reserve the term polypeptide exclusively for poly(α-amino acid)s of molecular weights of less than 10000, and other workers consider that this should apply to molecules containing less than 50 amino acids. In practice most polymer chemists use both polypeptide and protein but mainly apply the latter to the complex biological molecules. In this chapter these polymers will in the main be called polypeptides.

They are all constructed from amino acids polymerising through a condensation process, that is, the loss of a molecule of water to form the peptide bond as illustrated in Figure 2.40. Here two molecules of glycine have reacted to form a peptide with the loss of a water molecule.

There are 20 common amino acids, many of which can be manufactured biologically and provide a cheap substrate for polymer manipulation. Unlike polysaccharides, which are detailed in the next section, because of the large number of amino acid monomers available, a limitless number of sequential arrangements can form. Another key difference from the other main class of natural polymers is that polypeptides are monodisperse, that is, they are characterised by particles of uniform size in a dispersed phase. This uniformity in polymer size can have advantageous in the application of polymers to a variety of industries as, for example, it can allow greater 'packing' of the polymer to increase strength [166] and also aid in biodegradation [167].

In the upstream oil and gas industry, however, outside of the polyamides derived synthetically and discussed previously in this chapter, there are relatively few examples of the polypeptides derived from amino acids being used in chemical additives. The application of polyaspartate corrosion inhibitors is a good example of a reasonable recent development and application of polypeptide chemistry. It had been noted some time ago that aspartic acid (see Figure 2.41) has some reasonable corrosion inhibitor properties [168], and it was later found that polypeptide derivatives gave good CO_2 corrosion protection however at that time not as good as other cheaper materials.

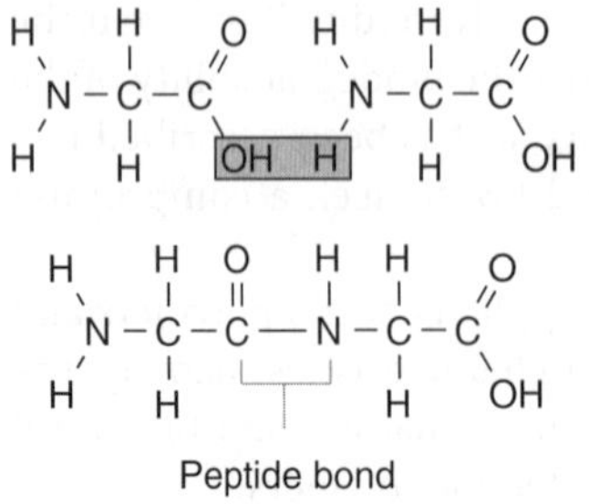

Figure 2.40 Peptide bond formation.

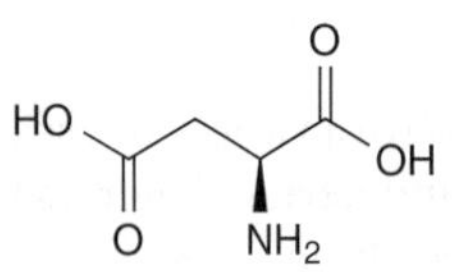

Peptide bond

Figure 2.41 Aspartic acid.

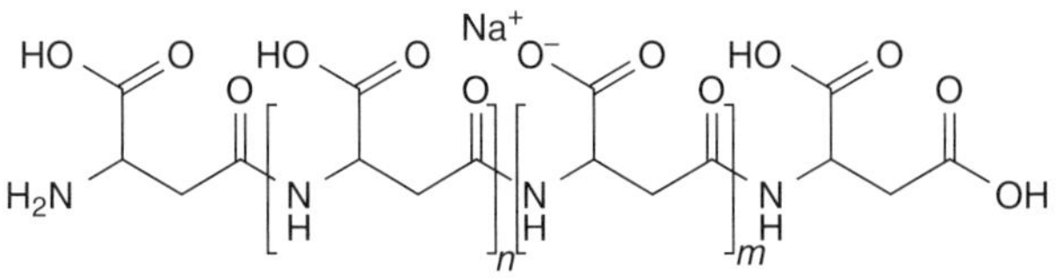

Figure 2.42 Sodium salt of polyaspartic acid.

These polyaspartate materials (see Figure 2.42) have now been re-examined in light of their excellent biodegradation and other non-toxic properties [169]. Used alone these materials are really only effective in low chloride and low pH conditions, which are not common in the oil and gas industry. However they do show improved activity if formulated with amino thiols, and indeed they have good activity across a range of conditions when formulated with the amino acids cysteine and cystine (formed from two molecules of cysteine) or their decarboxylated products cysteamine and cystamine [170]. Due to their high rate of biodegradation, non-bioaccumulation and low toxicity, they are much favoured in the some regions such as the North Sea basin. However, dose rates can be considerably higher than the more traditional products usually by a factor of 3 or more.

It has been found that polyaspartates can be formulated with alkyl polyglucosides to work synergistically, providing a formulated green corrosion inhibitor with a wider range of application conditions and better dose rates [171, 172]. The alkyl polyglucosides, in their own right, display interesting corrosion inhibitor properties and other functionality and are again mentioned later in this section on natural polymers under section 2.6, however, due to their surfactant nature, they are discussed in detail in Chapter 3.

Polyamino acids have been found in nature to interact in calcium carbonate formation and dissolution [173]. Polyaspartates also exhibit scale inhibition properties [166] and have also been formulated and applied to provide scale control ion the oilfield, particularly where environmental regulations require high rates of biodegradation. In terms of other scale control applications, these are found in cooling water systems, desalination process plants and wastewater treatment operations where scaling conditions are not usually as severe as is found in the oilfield. As with other highly biodegradable products, there is something of a compromise between required dose rates, economics of treatment and the rate of biodegradation. They have been derivatised to improve their performance such as in *N*-2-hydoxyalkylaspartamide [174]. The polyaspartates and their derivatives do have an advantage over some synthetic polymer chemistry in that they can exhibit superior calcium ion compatibility under certain pH conditions [175].

Interestingly there is the opportunity to deploy a single chemistry to control both scale and corrosion issues, and this can be advantageous in a number of oilfield situations [176]. This can be particularly the case in offshore locations where the use of a single product can facilitate a reduction in the number of umbilical lines to be used to provide flow assurance on the subsea flow lines.

The polyaspartates rather than being synthesised from a peptide condensation are commercially derived from polysuccinimides [177], which are produced from the thermal condensation of aspartic acid (see Figure 2.43). Polyaspartic acid can also be synthesised by polymerisation of maleic anhydride in the presence of ammonium hydroxide [178].

The intermediate polysuccinimides can be utilised to give other related materials such as polyaspartamides (see Figure 2.44), which shows the synthesis of polyaspartamides suitable as low dose KHIs [179]. Indeed these products display properties as scale inhibitors and low dose hydrate inhibitors [173].

Evidence that these products are suitably biodegradable in the marine environment has been published [180, 181] and suggests good biodegradation rates at 57–60% [182].

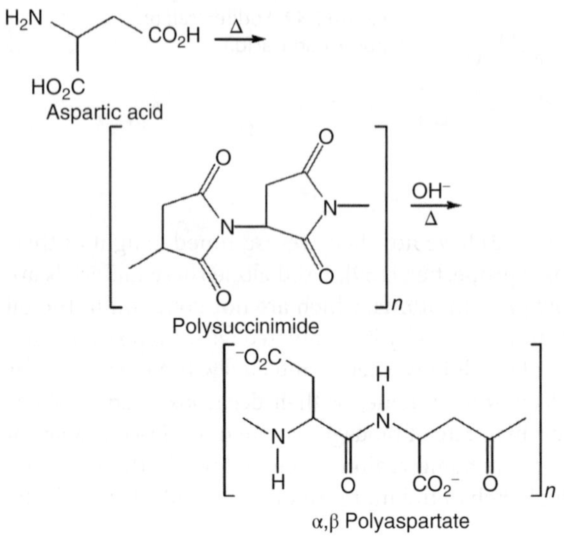

Figure 2.43 Commercial synthesis of polyaspartate.

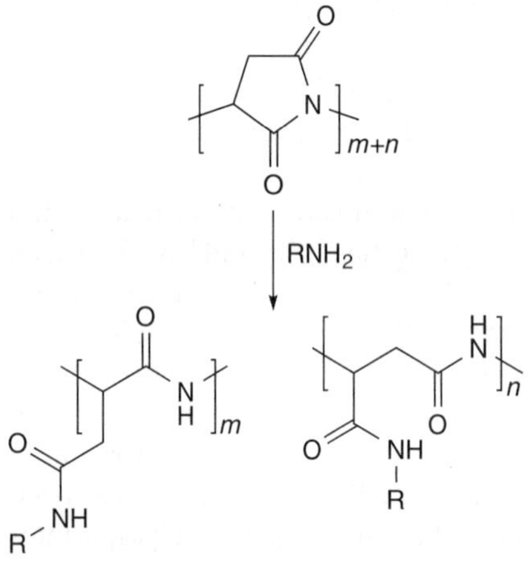

Aspartic acid

Polysuccinimide

α,β Polyaspartate

Figure 2.44 Synthesis of polyaspartamides.

Polyglutamic acid, unlike polyaspartic acid, can be biosynthetically manufactured to give polymers of very high molecular weights, up to 2 000 000 Da [183]. Commercially poly(γ-glutamic acid) (see Figure 2.45) is derived from the fermentation of *Bacillus subtilis* [184], and its application as a biodegradable anionic flocculant has been examined [185]. However there is no evidence of it being applied in the oil and gas industry.

Outside of these two main applications of polyaspartic acid derivatives, there are very few applications of polypeptides in the upstream oil and gas industry. It would appear that despite

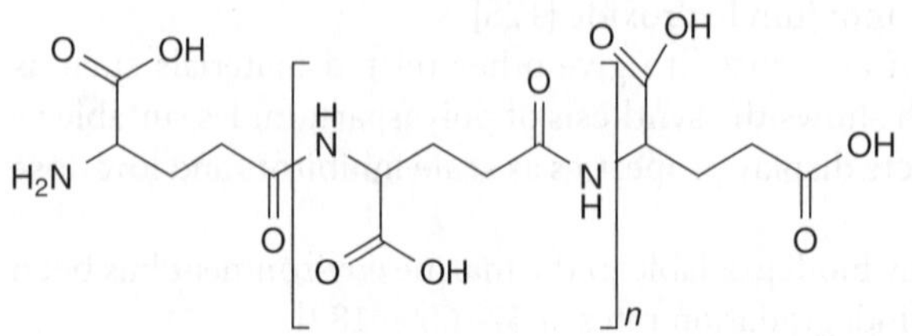

Figure 2.45 Poly(γ-glutamic acid).

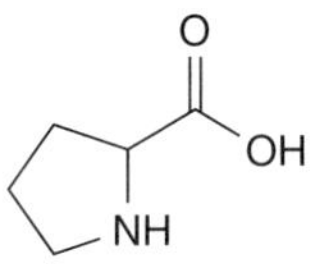

Figure 2.46 Proline.

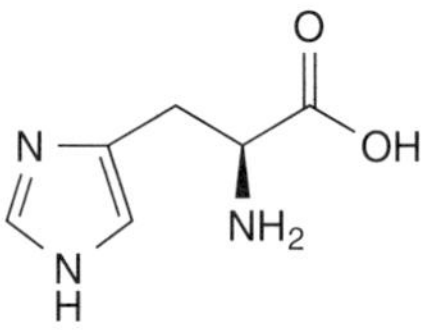

Figure 2.47 Histidine.

the availability of other products, such as polyglutamic acid, cost is a significant factor in their application.

There has been investigation of other derivatives of the polyaspartates such as copolymers of aspartic acid and the amino acids proline (Figure 2.46) and histidine (Figure 2.47), which are claimed as sand control agents [186]. These are usually as mixtures with other ingredients, namely, poly(diallylammonium salts).

A rare example of a chemical application of a protein is with collagen. Collagen is a structural protein from the repeating amino acids glycine (gly), proline (pro) and hydroxyproline (hyp), shown in Figure 2.48. Collagen will be familiar to readers as it is part of the connective tissue that in the skin helps in firmness, suppleness and constant renewal of skin cells. Collagen is vital for elasticity of the skin. For its chemical application, it has been modified to enrich the carboxyl group functionality, and these modified collagens have been shown to perform as scale inhibitors particularly on calcium carbonate [187].

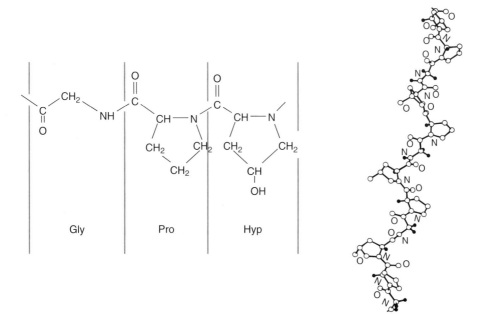

Figure 2.48 Collagen.

Environmentally friendly chemical additive inhibitors based on polypeptides and their derivatives are rare and are in the main focused on polyaspartic acid, polysuccinimide and the related polyaspartamides. These have been shown to be highly effective corrosion inhibitors in particular across a wide range of conditions including several highly sour oilfield brines and brine/hydrocarbon media [170, 171, 188]. The amide derivatives with C12–C18 alkyl groups have been proved to reduce the corrosion rate and localised attack in a way comparable with examples of more traditional inhibitors that were not environmentally benign [186].

Given the performance of the polyaspartates and related polymers, it is surprising that other polypeptides have been little examined, and as alluded to previously, it is thought that costs associated with their production make application in the oilfield sector prohibitive. However, it can be assumed that as biotechnologies associated with production become more efficient and able to cope with higher throughputs, these costs will reduce. This is a discussion that will be resumed later in this chapter.

2.2.3 The Polysaccharides

Unlike polypeptide, polysaccharides have much wider area of application across the various sectors of the upstream oil and gas industry.

Polysaccharides are polymeric carbohydrate molecules composed of long chains of monosaccharide units bound together by glycosidic linkages. On hydrolysis they give the constituent monosaccharides or oligosaccharides. They range in structure (see some examples in Figure 2.49), from linear to highly branched, and include products that will be familiar to the reader, such as the storage polysaccharides starch and glycogen and structural polysaccharides such as cellulose and chitin. Cellulose is used in the cell walls of plants and other organisms and is thought to be one of the most abundant organic molecules on Earth [6]. It has many uses such as a significant role in the paper and textile industries and is used as a feedstock for the production of a number of synthetic polymers, such as cellulose acetate, celluloid and nitrocellulose. Chitin has a similar structure, but has nitrogen-containing side branches, increasing its strength. It is found in the exoskeletons of arthropods (insects and crustacea) and in the cell walls of some fungi. As will be illustrated all of these polysaccharides have oilfield uses; however starches and cellulosic polymers are probably the most commonly used.

Polysaccharides have a general formula of $C_x(H_2O)_y$ where x is usually a large number between 200 and 2500. Considering that the repeating units in the polymer backbone are often six-carbon monosaccharides, the general formula can also be represented as $(C_6H_{10}O_5)_n$ where $40 \leq n \leq 3000$. It is accepted convention that polysaccharides contain more than ten monosaccharide units.

Monosaccharides, which also occur widely in nature, are simple carbohydrates with the general structures as the examples shown in Figure 2.50. They are usually colourless, water-soluble crystalline solids, and some monosaccharides have a sweet taste. They include glucose (dextrose), fructose, galactose and ribose.

Importantly these monomer units are all biodegradable or readily bioavailable to be utilised in metabolic processes [189] and are non-toxic [190]. Indeed in the last few decades much work has been done in the production of bioethanol and other fuels from monosaccharide feedstocks [191].

In the oil and gas exploration industry, use primarily lies within the starch and cellulosic polysaccharides and a number of examples are now discussed.

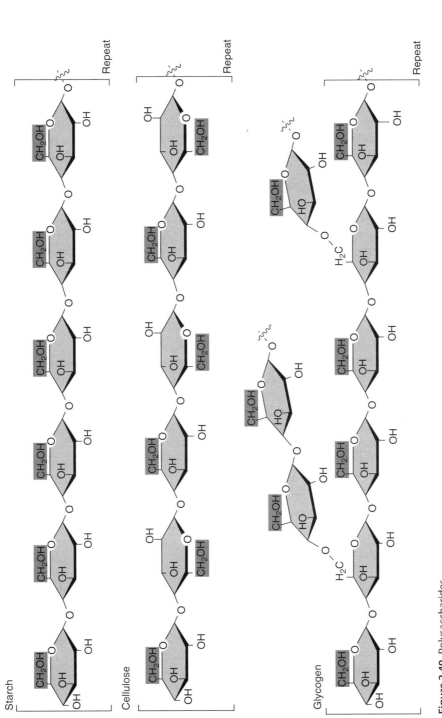

Figure 2.49 Polysaccharides.

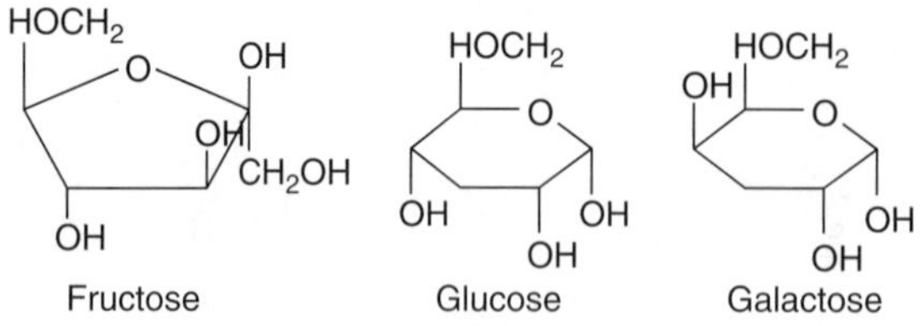

Figure 2.50 Examples of monosaccharides.

Fructose   Glucose   Galactose

2.2.3.1 Cellulosic Polymers

A number of derivatives of cellulose have a variety of applications in the upstream oil and gas industry, particularly in the areas of drilling, cementing and completion.

Fluid loss additives are required to be added to drilling fluids, especially water-based muds, and in this respect cellulose derivatives are among those of choice such as polyanionic cellulose, carboxymethyl starch, hydroxypropyl starch and in particular carboxymethyl cellulose (CMC). Figure 2.51 shows the relationship between cellulose and CMC. Water-soluble fluid loss additives can be categorised according to different effects on the negative electrical charge density of filter cake fines [192]. In the case of CMC the electrical charge density is increased where as with carboxymethyl starch it is unchanged. Pregelatinised starch decreases the electrical charge density.

Modified CMC in combination with a variety of other polymers, surfactants and related materials has been examined as a gelling agent on water-based systems to affect reservoir permeability and affect water shut-off [193]. The cellulosic polymers also have applications as additives in wellbore cementing in a related plugging-type function where modified methylcellulose is used in combination with monosodium phosphate [194].

However the main use of cellulosic polymers such as CMC, hydroxypropyl cellulose (HPC) and in particular guar gums and guar derivatives is in fracturing fluids. These materials are used as viscosifying agents to allow proppants such as sand particles to be suspended in the aqueous-based pumping fluids [1]. Indeed, the majority of fracturing treatments conducted to date have used fluids comprising guar gums or guar derivatives such as hydroxypropyl guar (HPG), HPC, carboxymethyl guar and carboxymethyl hydroxypropyl guar. All of the fracturing treatments to date in the United Kingdom have utilised HPG.

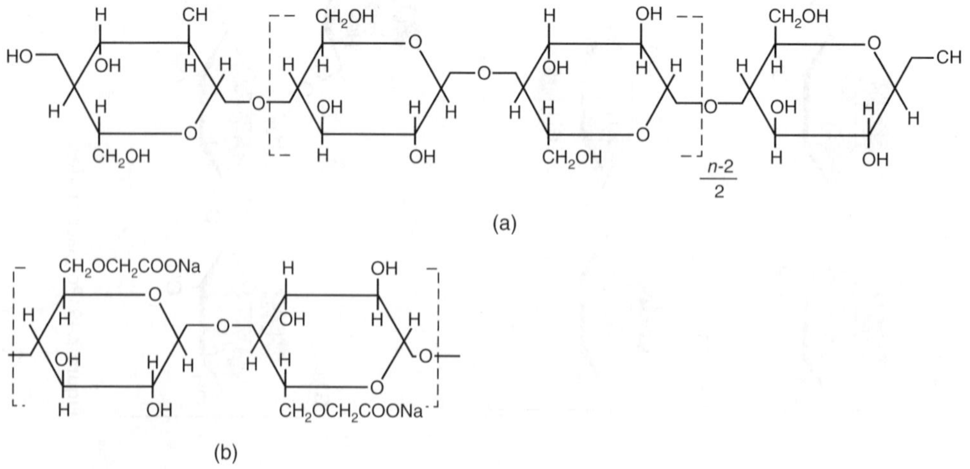

(a)

(b)

Figure 2.51 (a) Structure of cellulose. (b) Structure of CMC.

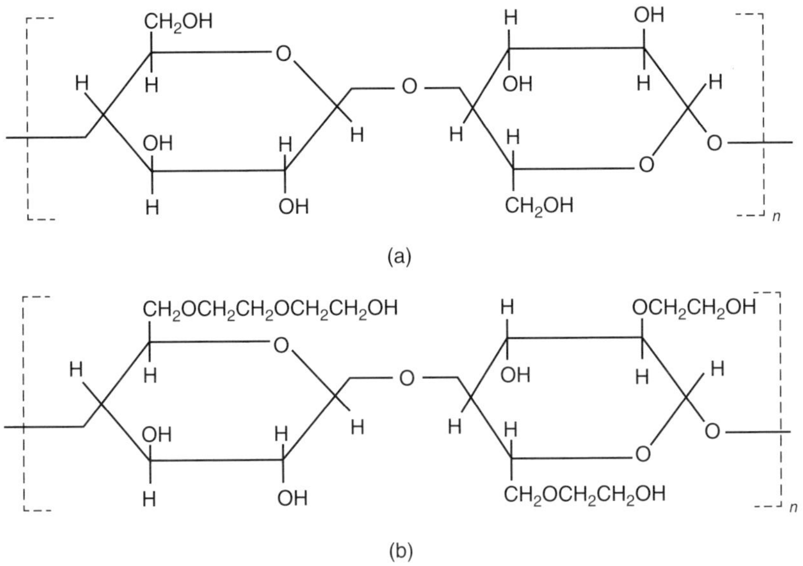

Figure 2.52 (a) Cellulose. (b) HEC.

Hydroxyethyl cellulose is an important drilling and cementing additive and is derived from cellulose through etherification (see Figure 2.52). This is an important property of polyhydroxy compounds as has been seen earlier in the synthetic polymers. In many of those examples, modifications were being utilised to imbue and enhance environmental characteristics, particularly biodegradation. In this case and other related derivations of polysaccharides, there is an attempt to gain chemical functionality.

The way that the monomer units in the polysaccharide are linked can have a significant effect on the physical properties of the polymer. A good example of this can be seen in two polymers of glucose, cellulose as represented in Figure 2.52 and amylose as seen in Figure 2.53.

As is stated in Figure 2.53, amylose has α-linkages, whereas cellulose has β-linkages. Although this difference may appear minor, it is for this reason that amylose is water soluble, whereas cellulose, thankfully, is not, as it is the major structural constituent of most plant life. Cellulose when chemically modified [195] can be made water soluble as in the cellulose ethers previously described.

2.2.3.2 Guar and Its Derivatives

Guar gum is a branched polysaccharide from the guar plant, *Cyamopsis tetragonolobus*. It is comprised of the sugars mannose and galactose in the ratio 2 : 1. The backbone is a linear chain

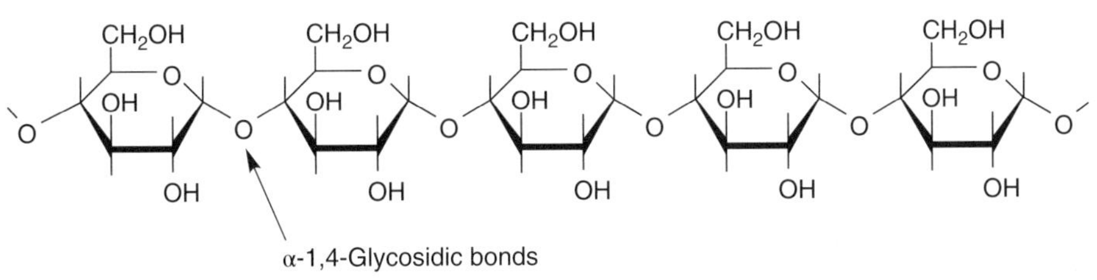

Figure 2.53 Amylose.

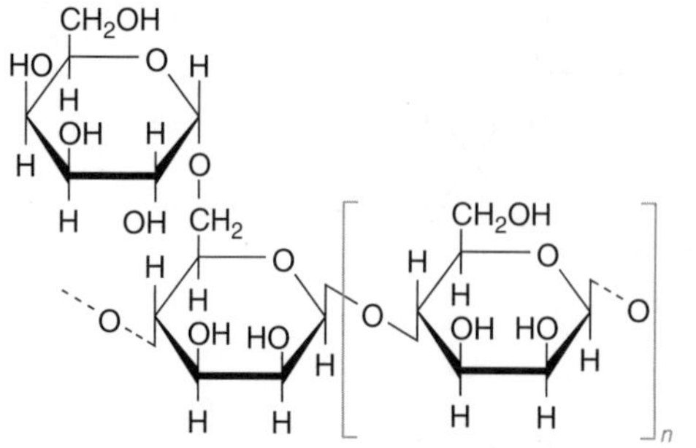

Figure 2.54 Guar gum.

of β-1,4-linked mannose residues to which galactose residues are 1,6-linked at every second mannose, forming short side branches as in Figure 2.54.

Polysaccharides, and in particular guar derivatives, having this structure are often referred to as galactomannans. Guar is more soluble than other similar gums and is a better stabiliser. It is not self-gelling; however, a cross-linking agent added to the guar gum will cause it to gel in water. It is not affected by ionic strength or pH, but will degrade at pH extremes at elevated temperature (e.g. pH 3 at 50 °C) [9]. It remains stable in solution over pH range 5–7. Strong acids cause hydrolysis and loss of viscosity, and alkalis in strong concentration also tend to reduce viscosity. It is insoluble in most hydrocarbon solvents.

Guar gum shows high low shear viscosity but is strongly shear thinning. It is very thixotropic above 1% concentration, but below 0.3% thixotropy is slight. It has much greater low shear viscosity than that of many other gums and also generally greater than that of other hydrocolloids. It is economical because it has almost eight times the water-thickening potency of similar materials, and therefore only a very small quantity is needed for producing sufficient viscosity. These properties make it ideal for use in fracturing fluid formulations as it acts as a viscosifier (when cross-linked) helping to gel the fluid and/or as a stabiliser because it helps to prevent solid particles from settling.

Guar gum is a direct food additive and it is registered on the PLONOR (Poses Little Or NO Risk to the Environment) List [196] as regulated by OSPAR (Oslo Paris Treaty Convention Organisation) and therefore is accepted for use and discharge in the North East Atlantic including the North Sea. It is highly biodegradable and is recognised as possessing little environmental or toxicological problems.

This is an excellent example of the key type of material that the environmental oilfield chemist requires; a polymer, a natural polymer, fits the requirements of a strict regulatory regimen and also has the necessary performance criteria to be more than fit for purpose in carrying out a specific chemical requirement. As has been seen these examples are relatively rare.

In the last few years guar and its derivatives have become important because of their viscosifying capability and suspension of proppants in the extraction of shale gas and other unconventional hydrocarbons. This has led to a resurgence of developments in chemical modification [1, 197] and examination of other 'greener' products [198].

Polyglycosides have been alkylated to give potentially biodegradable demulsifier bases [199] and as is described later in this section have similar structures (Figure 2.55) to the alkyl polyglucosides.

2.2.3.3 Xanthan, Other Gums and Related Products

Xanthan is produced by the bacterium *Xanthomonas campestris*. It is widely used as a food additive, in particular as a thickening agent such as in salad dressings. In the oil and gas industry

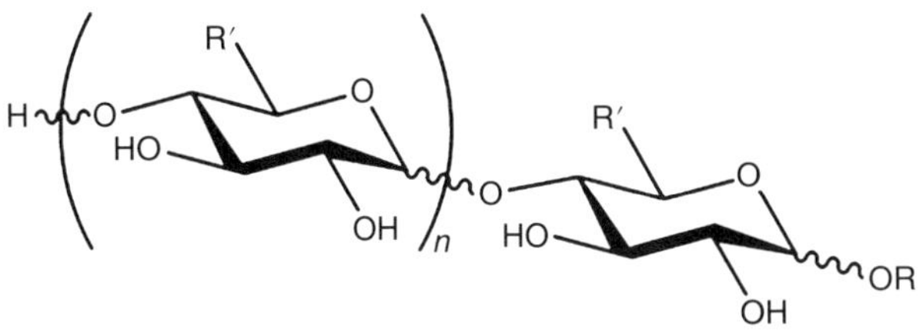

Figure 2.55 Alkyl polyglycoside.

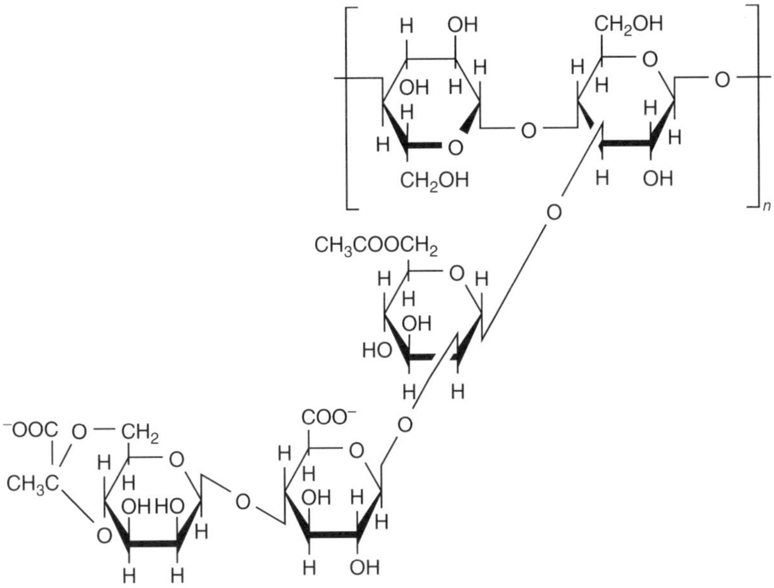

Figure 2.56 Xanthan repeating unit.

it has a particular use as a rheology modifier in drilling activity. It is composed of pentasaccharide repeating units, comprising glucose, mannose and glucuronic acid in the molar ratio 2 : 2 : 1 (see Figure 2.56). It is produced [200] by a fermentation process from which the polysaccharide is precipitated, dried and ground into a fine powder. Later, it is added to a liquid medium to form the gum.

Xanthans are water-soluble polysaccharides due to the configuration of the saccharide monomer linkages.

Xanthan gum has been deployed in enhanced oil recovery processes as the main polymer flooding agent [201] and examined under laboratory conditions for synergistic activity with other agents in alkaline surfactant flooding [202].

Gellan gum is a linear anionic polysaccharide produced by the bacterium *Sphingomonas elodea* (formerly *Pseudomonas elodea*) and consists of glucose, glucuronic acid and rhamnose in a basic structure shown in Figure 2.57.

Welan gum is identical in backbone to gellan gum but has side chains of L-mannose or L-rhamnose. Both of these products are used as fluid loss additives and other similar applications [203].

A number of polysaccharides of the guar gum type or other cellulose-based materials, such as HEC, have been considered as water-soluble DRAs for multiphase flow [204]; however, synthetic water-soluble polymers generally perform better due to the higher molecular weights

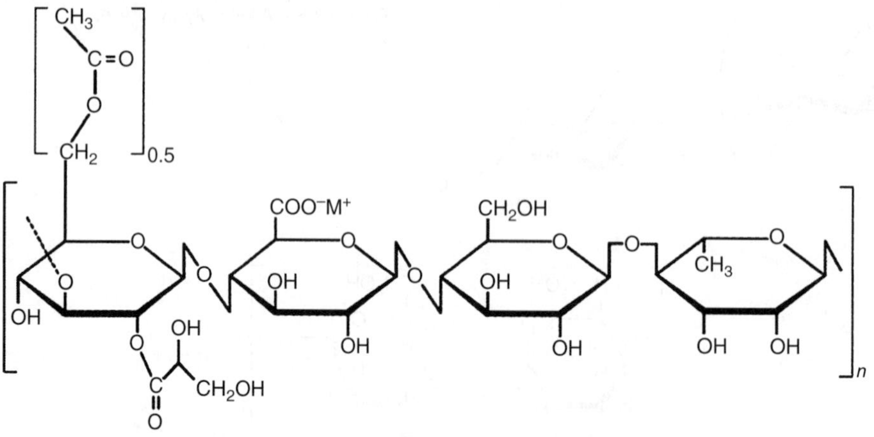

Figure 2.57 Gellan gum repeating unit.

and greater stability in the polymer, resulting in less shear problems [205]. Synergism has been reported between polysaccharides and PAM [206], and as has been illustrated, guar gum and its derivatives are used extensively in fracturing operations where they can also act as DRAs [207]. In a rare synthetic modification of these polymers, grafting of acrylamide moieties onto guar and related polysaccharides has been carried out to make them more resistant to shear, which also consequently makes them less biodegradable [208].

2.2.3.4 Starch and Its Derivatives

A water-based drilling fluid typically consists of sized calcium carbonate or salt particles, xanthan, starch, biocide, potassium chloride, potassium hydroxide, sodium sulphite, defoamer and lubricant. This complex mixture is required to perform many functions throughout the drilling process.

Starch (see Figure 2.49) and in particular colloidal starch and its derivatives are widely used in water-based drilling fluids [209, 210]. This is because of the ability of starch to reduce rapidly the permeability of the drilling mud filter cake, thereby reducing the invasion of the filtrate containing damaging water-soluble polymers (such as xanthan gum or scleroglucan), bridging agents and drilled solids into the formation. However, starch and other filter cake components can cause significant impairment of well performance. Formation damage treatments need to be performed to re-establish the native permeability of the critical wellbore area.

As described earlier and illustrated in Figure 2.36, hybrid polymers with starch-type polysaccharides have been developed and exhibit useful scale inhibitor properties as well as good rates of biodegradation. This area of work has been further developed, and copolymers have been modified with oxidised starch [211], and also graft copolymerisation has been adopted [212]. The aim of these synthetic methods is to produce an enhanced environmentally acceptable scale inhibitor based on the natural and inherent biodegradation of starch.

2.2.3.5 Chitosan

As mentioned earlier chitosan is an abundant structural polysaccharide and has the general structural formula as illustrated in Figure 2.58. It is composed of randomly distributed β-(1-4)-linked D-glucosamine (deacetylated unit) and *N*-acetyl-D-glucosamine(acetylated unit). It is commercially manufactured by treating shrimp and other crustacean shells with sodium hydroxide.

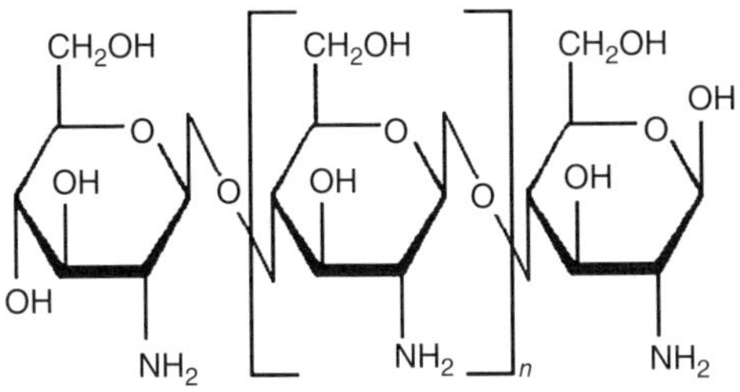

Figure 2.58 Chitosan.

There are a few uses for this interesting natural polymer in the oil and gas industry; however, as with much of these polysaccharides, their full potential seems under exploited.

Chitosan is used in pseudoplastic shear thinning fluids and enhances the thermal stability of such fluids [213]. Such fluids are useful in the make-up of particularly drilling fluids, drill-in fluids, workover fluids, completion fluids, perforating fluids, filter cake removal fluids, etc.

The pendant amine functionality of chitosan can be quaternised to generate a cationic polymer and incorporating *N*-(3-chloro-2-hydroxypropyl)trimethylammonium chloride is claimed to give a useful flocculant [214]. Also hydrophobically modified derivatives of chitosan are similarly claimed to act as improved flocculants in oil-in-water emulsions [215].

An issue of note with chitosan-type polymers and related biopolymers is that they are not readily biodegradable. This in many ways is part of their natural design as they are robust structural materials used by many animals in their exoskeletons. They do biodegrade over a period of time, and their ultimate biodegradation appears as good as if not better than other polysaccharides [216].

Very little work, if any, appears to have been investigated in relation to modification to improving the rate of biodegradation. There is however evidence of improved rates of biodegradation of polyethylene/chitin (PE/chitin) and polyethylene/chitosan (PE/chitosan) films, containing 10% by weight chitin or chitosan, by pure microbial cultures and in a soil environment. Both PE/chitin and PE/chitosan films degraded at a higher rate than the commercial starch-based film in a soil environment, indicating the potential use of chitin-based films for the manufacturing of biodegradable packaging material [217].

Modification of polysaccharides to aid functional properties has been carried out to oxidise the hydroxy groups to give carboxylic acid functionality. Such products when suitably degraded have been used as scale inhibitors [218].

2.2.4 Other Natural Polymers

There are a number of other interesting, and less often used in oil and gas industry applications, natural polymers. These include the well-known rubber and coal to others such as lignin, humus, kerogen, shellac and amber. This section will focus on those with relevance and application to the oil and gas industry. However in a number of ways these materials are an untapped resource of interesting and potentially 'green' chemistry.

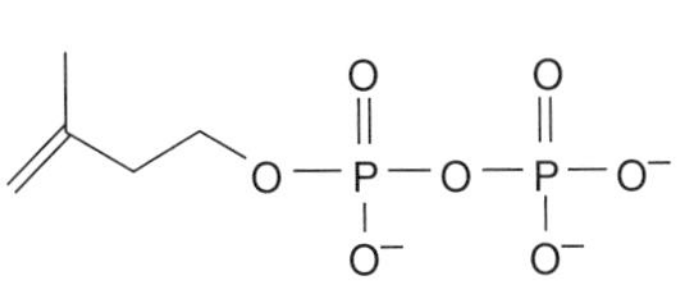

Figure 2.59 Isopentyl pyrophosphate.

2.2.4.1 Rubber

Rubber [219] is one of the most important natural polymers from an industrial use viewpoint. It is a polyterpene that is synthesised naturally by the enzymatic polymerisation of isopentyl pyrophosphate (see Figure 2.59). Readers will note that this has the inclusion of isoprene (see Figure 2.2) in its structure; indeed as stated before isoprene is the degradation product of rubber, and the polymers repeating unit is 1,4-polyisoprene.

The use and properties of the diene-based polymers, which are mainly synthetically derived from petrochemical feedstocks, have been discussed at some length in Section 2.1, and here the focus will be on latex-derived products.

Most natural rubber is harvested as a latex consisting of around 30–35% rubber and 5% other compounds including proteins, which give natural properties distinct from the synthetically derived *cis*-1,4-polyisoprene. Some of these proteins found in rubber derived from the hevea tree (*Hevea brasiliensis*) are thought to be responsible for allergic responses in humans to latex products, some of which have been life threatening. Another plant, the guayule shrub, also produces natural rubber but without these allergenic proteins. Most latex is coagulated and, through the vulcanisation process, used in a large array of products from chewing gum to car tyres.

Rubber is of very high molecular weight, around 1.5 million, and as such, as a latex form, has been used a drag reducer in the oil and gas E&P sector. This is applied as a colloidal dispersion in water and is usually formulated with additional surfactants (H.A. Craddock, unpublished results) [220].

Latex-based polymers both synthetically derived and originating from natural rubber are included in drilling fluids [221] and cement compositions [222] to reduce fluid loss and improve seal permeability. These polymer latexes can be incorporated in both water-based fluids [217] and oil-based drilling fluids [223].

This sealant property has been exploited for enhanced oil recovery operations particularly in reservoirs at high temperatures in forming a plug for oil and gas well diversion [224]. In a related application it is possible to create an all-oil reversible gelling system in which the component parts are oil soluble after the gel is formed, and therefore any residues are also oil soluble and do not cause formation damage [225].

Rubber biodegradation is a slow process, and the growth of bacteria-utilising rubber as a sole carbon source is also slow. Therefore, incubation periods extending over weeks or even months are required to obtain enough cell mass or degradation products of the polymers for further analysis. For over 100 years there have been efforts to investigate microbial rubber degradation; however, only recently have the first proteins involved in this process been identified and characterised. Analyses of the degradation products of rubber isolated from various bacterial cultures indicated without exception that there was oxidative cleavage of the double bond in the polymer backbone. This degradation of polymers will be further detailed in Section 2.6, alongside potential methods and derivations to improve biodegradation rates. As with other natural polymers, in comparison with synthetic polymers, little of this type of work has been explored and evaluated.

2.2.4.2 Lignin

Lignin [226] is nature's cement in the plant kingdom and provides wood with much of its dimensional stability. Wood is primarily composed of cellulose, hemicellulose and lignin. The structure of lignin (see Figure 2.60) is complex, and its molecular weight is thought to be very high; however, it is impossible to define accurately as in its separation from cellulose it is inevitably degraded.

Lignosulphonates, which are extracted from the wood-pulping process, have been used as drilling mud additives for many decades. Drilling muds [227] made from these products have good viscosity control, gel strength and fluid loss properties. Additionally they are temperature

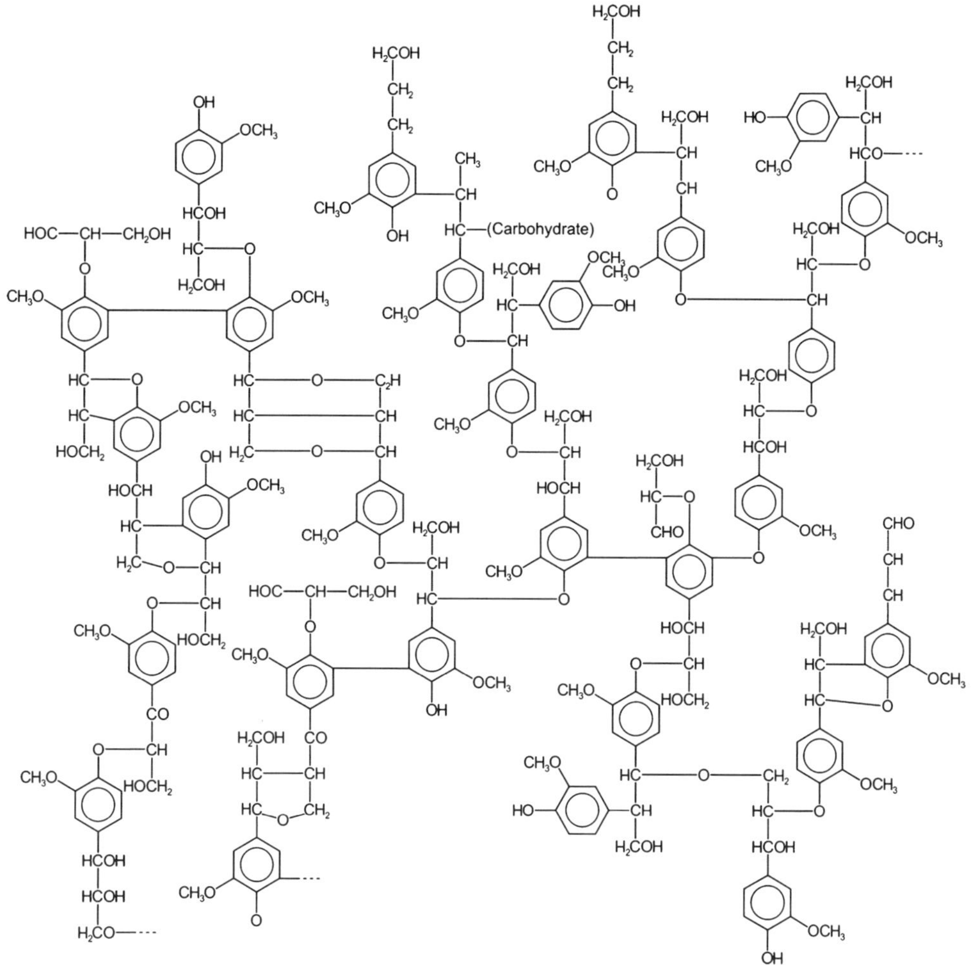

Figure 2.60 Structure of lignin.

resistant to 230 °C. In water-based muds they are also tolerant to high salt concentration and extreme hardness of the water. Lignin has also been modified to give other properties particularly by amination [228] or grafting synthetic co-monomers [29].

Lignosulphonate polymers are claimed for use as asphaltene inhibitors by application in squeeze treatments [229]. These products (see Figure 2.61) have a similar structure to the known

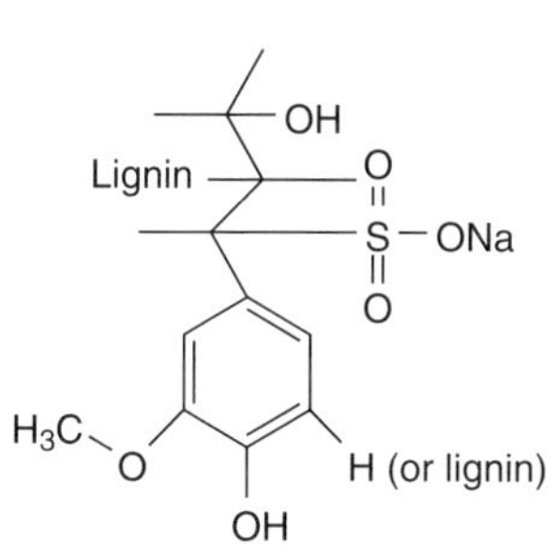

Figure 2.61 Partial structure of lignosulphonate polymer.

asphaltene inhibitors, the alkylaryl sulphonic acids, with a polar head group containing phe-nolic and sulphonic groups but now with other functional groups and a more complex poly-meric structure of phenolic groups derived from lignin.

An amount of uncertainty is attributed to the biodegradability of lignosulphonate polymers and similar derivatives. Lignin itself is insoluble, chemically complex and lacking in hydrolys-able linkages; it is therefore a difficult substrate for enzymatic depolymerisation [230]. Certain fungi, mostly basidiomycetes, are the only organisms able to extensively biodegrade it; white-rot fungi can completely mineralise lignin, whereas brown-rot fungi merely modifies lignin while removing the carbohydrates in wood.

Lignosulphonates and other derivatives, as can be seen from their composition, are unlikely to be environmentally acceptable. They are assumed to very slowly biodegradable with the products of biodegradation being toxic aromatic and polyaromatic compounds from the large number of benzene rings found in the lignin structure. However, there is evidence that they can biodegrade more readily under certain environmental conditions where synergistic actions of fungi were encouraged [231] and under conditions where particular fungal enzymes were pre-sent [232].

Although it is unlikely that good rates of biodegradation can be achieved in modification and that the biodegradation products are likely to be toxic, other than the few examples of deriva-tion illustrated, little work has examined this fascinating natural polymer as to its modification to a more biodegradable and perhaps more useful polymeric additive.

Related to lignin but not polymeric in nature are the tannins, which are mixtures of phenolic compounds and are further discussed in Chapter 6.

2.2.4.3 Humic Acid

Humus is the base component of soil and is relatively resistant to biodegradation processes. Humus products are classified as to their solubility below:

- Water soluble – Fulvic acid
- Water insoluble but base soluble – Humic acid
- Base insoluble – Humin

Humic materials are of great interest environmentally because of the role they play in soil drainage, water and nutrient motility and their metal scavenging and extraction proper-ties [233]. However, only humic acid (see Figure 2.62) has had any application in the upstream oil and gas industry.

Although humic acid is water insoluble, it can be treated with bases such as sodium hydrox-ide to make it water soluble and in such forms has been used as part of a fluid loss system in well cementing [234].

As can be seen, in comparing the structures of humic acid and lignin, there is great similarity; indeed humus products, such as coal, shale oil (kerogen) [235] and crude oil are derived from lignin-based plant materials or have a part to play in their derivation alongside other plant- and animal-based polymers. This gives humic acid good filtration characteristics [236], and alter-natively coal-derived humic acids have found uses as drilling fluid dispersants and viscosity control agents [237].

2.2.4.4 Tall Oil-Derived Polymers

Tall oil is obtained from dry wood, mainly pine trees, as a by-product of wood pulping. It is not polymeric but has been adapted for polymer synthesis as a potential renewable feedstock [238]. It is made up of two distinct components, rosin and tall oil fatty acids. Rosin, which is com-posed of a mixture of complex fused-ring monocarboxylic acids such as abietic acid (see

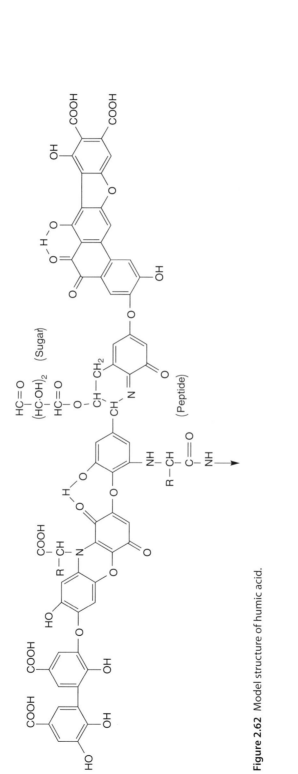

Figure 2.62 Model structure of humic acid.

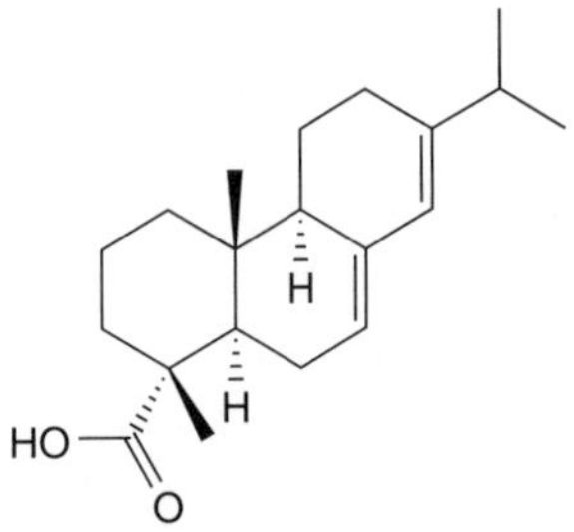

Figure 2.63 Abietic acid.

Figure 2.63), has no known, at least to the author, oilfield use. The tall oil fatty acids, which are separated from the rosin by distillation, have a number of oilfield applications. They are mainly 18-carbon atom organic acids and have significant surfactant properties. They are therefore discussed in some detail in Chapter 3.

As has been illustrated, in this section, there are a number of natural polymeric materials that, in the author's opinion, have been underutilised and relatively unexplored for potential derivation and application to the upstream oil and gas industry. In a number of cases they possess inherent qualities of good rates of biodegradation and are non-toxic. This is a theme that will be returned to in Section 2.6.

2.2.5 Other Biopolymers

As stated earlier in the chapter, biopolymers are simply defined as biodegradable polymers, although most practitioners of chemistry and biochemistry in this area of study would also expect them to have some natural product derivation, particular with respect to their monomer-based repeating units. As has been previously described in Section 2.1 'Synthetic Polymers', polyols, polyethers and polyesters have examples of biodegradable polymers whose monomer units are derived from pseudo-natural products, which in the context of this study means products that although found in nature are most commonly available to the polymer and synthetic chemist from petrochemical feedstocks. Both these types of products and those derived from natural feedstocks but however not already cover in the previous sections are exemplified and discussed here.

2.2.5.1 Glucan and Glycans

Curdlan is a microbial carbohydrate of the structure shown in Figure 2.64. It is of note that the polymer only has β-1,3-linkages. It forms elastic gels when heated and has been examined as to the possibility of using this material as a gelled biopolymer for permeability modification in reservoir rocks [239]. Work has continued with this system to improve placement within the reservoir and specificity of treatments [240].

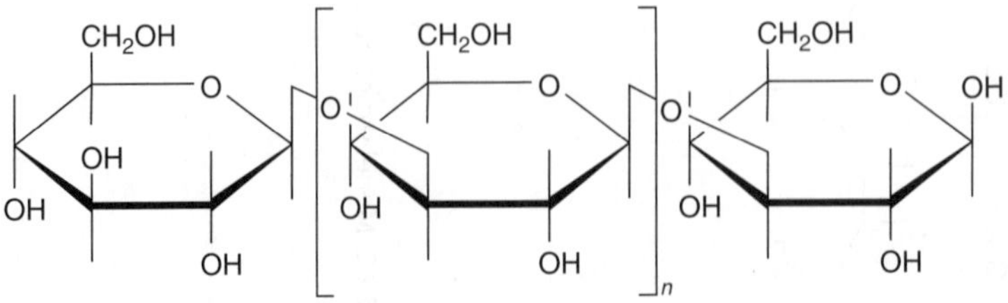

Figure 2.64 Curdlan.

Curdlan is part of the glucan family of polysaccharides being composed solely of glucose monomers with glycoside linkages. Glycan is a more general term and is synonymous with the polysaccharide classification.

Succinoglycan is a microbially produced polysaccharide with an eight-sugar repeating unit and has similar properties to xanthan. It has been successfully used in well completion fluids in the North Sea [241], where a unique property of partially reversible viscosity collapse, at a pre-determined temperature, as controlled by the brine composition, was considered advantageous. Succinoglycans are part of the microbial exopolysaccharide grouping and have complex structures and interesting and relatively unexplored properties.

An exopolysaccharide, Pseudozan, produced by *Pseudomonas* species has a high viscosity at low concentrations when added to oilfield brines [242]. It forms stable solutions over a wide pH range and is not degraded under shear conditions, and this polymer system has been proposed for use in enhanced oil recovery.

Cellulose produced by bacteria often has an intertwined reticulated structure. This material is unlikely conventional cellulose and has unique properties and functionality [243]. It has been shown to improve rheology in drilling fluids [244] and cements and can improve particle suspension in fracturing fluid composition over wide range of conditions [245].

2.2.5.2 Glycopeptides and Glycoproteins

Glycoproteins are proteins that 'contain' oligosaccharide chains (glycans) covalently attached to polypeptide side chains. The carbohydrate is attached to the protein in a co-translational or post-translational modification in a process known as glycosylation. Secreted extracellular proteins are often glycosylated. These materials are found in a number of plants, animals and microorganisms, particularly in the polar regions, where they act as antifreeze agents [246]. It has been shown that glycopeptides and glycoproteins act as antifreeze agents by interfering with the crystal growth of ice [247]. This discovery has directly led to the development of low dose hydrate inhibitors, previously described with both synthetic polymer examples such as the synthetic vinylcaprolactam polymers and the more naturally derived polyaspartamides.

Until relatively recently, this has been the only somewhat tenuous relationship of oilfield use of products derived from the glycoproteins and related peptides. The prospect of oil production from the Arctic region however has stimulated some interest in these materials. In arctic deepwater projects temperatures are near freezing. Combined with the high pressures present at these depths, oil congealment and ice-like hydrate formation is a commonly occurring phenomenon, which results into flow restriction. To counter these problems, soya slurry has been proposed as an anti-oil congealer [248].

The soybean slurry consists of glycoproteins and unsaturated fats, which act as an antifreeze and suppress ice nucleation effect. Preventing congealing by increasing its thermal hysteresis and acting as an ice nucleation barrier, the slurry also reduces the viscosity of the wellbore fluid substantially, resulting in an easy and smooth flow through the pipeline with a reduced amount of friction experienced between the fluid and tubing. Of course the material is environmentally friendly and produced sustainably.

Glycoprotein also has surfactant properties and will be explored and commented on in Section 3.7.

2.2.5.3 Non-polysaccharide Biopolymers

As has been illustrated in Section 2.1, there are a number of synthetically derived biopolymers; additional to this are a number of biopolymers that are derived from 'natural' monomers such

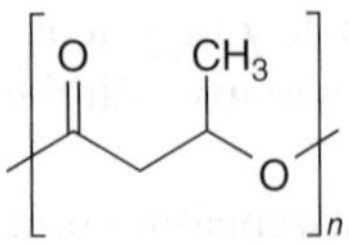

Figure 2.65 Poly-3-hydroxybutyrate (PHB).

as organic acids, glycerol, etc., which are not polysaccharide in nature, and these are explored as to their oilfield use in this following section.

Polyhydroxyacetic acid and the compounds formed with other carboxylic acids exhibit fluid loss behaviour and have been proposed as additives to drilling fluids and cements. A related product, poly-3-hydroxybutyrate, has been shown to be an efficient plugging agent at least in laboratory studies [249]. This is usually used as living cells of *Alcaligenes eutrophus*, which can be 70% of the cell weight. Other polymers such as polylactide and polyglycolic acid (PGA) have been used in high-temperature reservoirs as a source of acid stimulation [250].

These types of polymer have been well documented for use in plastics [251], particularly poly-3-hydroxbutyrate (see Figure 2.65) and polylactide acid. As mentioned earlier in this chapter, these materials are outside the scope of this current book; however Section 2.4, will briefly review their use in the upstream oil and gas industry. The key factor in the 'greenness' of these products is in the backbone where the polymer has heteroatoms included in the carbon backbone, making them more susceptible to enzymatic hydrolysis and therefore more likely to be biodegradable.

Other than the very few examples stated, there appears to be little work in utilising and derivatising these polymers for oilfield chemical additive use. This is also true of the highly biodegradable and related poly(lactic acid) [252]; however this has been found to have a few specialised uses in fracturing and stimulation applications [253, 254].

Polyglycerols (see Figure 2.19) exemplify this use of making the polymer backbone more biodegradable and have been previously described in Section 2.1.2. These 'natural' polymers have been manipulated by ethoxylation and other methods to produced biodegradable demulsifiers [255].

2.2.5.4 Lipids

As has been stated earlier in this section, although there are three main groups of polymers found in nature, there are four major types of macromolecules in living systems. These macromolecules and their monomeric building blocks are as follows:

1) Polysaccharides (complex carbohydrates) – monosaccharides (or simple sugars)
2) Proteins (polypeptides) – amino acids
3) Nucleic acids – nucleotides
4) Lipids

The lipids are a diverse group of molecules that are characterised by being water insoluble or hydrophobic in nature. This group does not neatly fit the polymer/monomer model. It contains a number of naturally occurring molecules that include fats, waxes, sterols, fat-soluble vitamins (such as vitamins A, D, E and K), monoglycerides, diglycerides, triglycerides, phospholipids and others. The triglycerides (see example in Figure 2.66) can be considered to fit the model as these are composed of fatty acids and glycerol.Triglycerides among other materials, such as vegetable oils, have been examined as potential base oil drilling lubricants [256] with a better environmental profile than the commonly used ester-based lubricants. They are however more susceptible to hydrolysis, especially in invert emulsion drilling fluids, and this lack of stability has restricted their use. Other non-polymeric, but nonetheless biodegradable materials, such as esters of long-chain carboxylic acids have been favoured. These will be referenced in Chapter 6.

Figure 2.66 A triglyceride.

2.2.6 Functional Derivations of Natural Polymers

As has been stated throughout this section, there are only a relatively few derivations of natural polymers and inherent biopolymers to shape their functionality, as compared with synthetic polymers being derived to make them more biodegradable and/or less environmentally toxic. There are two classes of products that do however show some utility in this respect: the carboxyinnulins being derived from polysaccharides and the alkyl polyglucosides having a polymer structure derived from glucose. This latter class has some powerful surfactant properties and again is in that crossover classification between polymers and surfactants. They have been mentioned in this chapter in relation to synergism with polyaspartates. Their own functional chemistry will be described in more detail in Chapter 3.

2.2.6.1 Carboxyinnulins

Inulins [257] are a group of naturally occurring polysaccharides produced by many types of plants, industrially most often extracted from chicory. They belong to a class of dietary fibre products known as fructans. They are used by some plants as a means of storing energy and are typically found in their root systems. Most plants that synthesise and store inulin do not store other forms of carbohydrate such as starch. Of note in the inulin structure (see Figure 2.67) is the predominance of the five-membered pentose structures.

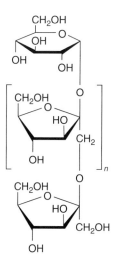
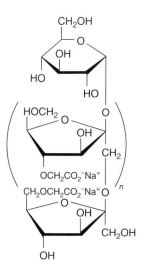

Figure 2.67 Structure of inulin and carboxymethyl inulin.

These products have been preliminarily investigated as to their potential for corrosion inhibition and show some capability of protection from corrosion under mild conditions [258]. To the author's knowledge the work to date is not extensive. They do however possess excellent ecotoxicological properties [259] being derived from sugars and are worthy of further investigation, especially in formulated products and with known synergists. They are used as scale inhibitors in the oilfield sector, especially in the Norwegian sector of the North Sea. In particular, carboxymethyl inulin (CMI) (see Figure 2.67) is a threshold scale inhibitor [260] for carbonate and sulphate scaling with calcium ion tolerance, high water solubility and low viscosity. It is considered as an alternative for traditional polymer and phosphorus-based scale inhibitors.

Other polysaccharides have been derivatised particularly to aid or enhance functionality as scale inhibitors [261]. Also as has been stated earlier, it is possible to derive a hybrid polymer of both natural and synthetic polymers [262] (see copolymerisation and grafting in Section 2.1.3).

In Section 2.1.1, relating to synthetic polymers, a wide range of different monomer chemistries and polymeric combinations were illustrated and discussed in Section 2.1.3. These are built on the fundamentals of polymer chemistry established by the brilliance of some great scientists. The invention shown by oilfield and environmental chemists in meeting the challenges of regulatory requirements and reducing environmental impact continues this tradition of scientific achievement. However, in examining polymer chemistry as found in nature, one has to conclude that our knowledge of polymer synthesis is merely at the beginning of discovery; indeed an entire field of scientific endeavour in biochemistry is very much involved with natural polymers and how they operate. It is therefore even more surprising that we have hardly exploited this resource in commercial terms and in particular in applications for the oil and gas industry.

A key tenant in the use of natural polymers, from the environmental chemist's viewpoint, is that unlike most synthetic polymers, the monomers, also natural products, are relatively non-toxic and highly biodegradable. This, along with their sustainability in manufacture and supply, makes them a credible and important resource of polymer and other chemistry for the future. This is a theme that will be returned to and further developed in the closing chapter of this book.

2.3 Dendritic Polymers and Other Unclassified Polymers

2.3.1 Dendritic and Hyperbranched Polymers

Dendritic polymers belong to a special class of macromolecules called 'dendrimers'. They are similar to linear polymers, in that they are composed of a large number of monomer units that were chemically linked together. Due to their unique physical and chemical properties, dendrimers have wide ranges of potential applications. These include adhesives and coatings, chemical sensors, medical diagnostics, drug delivery systems, high-performance polymers, catalysts, building blocks of supermolecules, separation agents and many more [263]. The name dendrimer is derived from Greek words *dendron* (meaning 'tree') and *meros* (meaning 'part'). A major difference between linear polymers and dendrimers is that a linear polymer consists of long chains of molecules, like coils, crisscrossing each other. A dendrimer consists of molecular chains that branched out from a common centre, and there is no entanglement between each dendrimer molecules. The first synthesis of these macromolecules is credited to Fritz Vogtle and co-workers in 1978 [264]. It consisted of Michael addition of acrylonitrile to primary amine groups. Each successive step involved reductions of the nitrile groups followed by additions of acrylonitrile (see Figure 2.68).

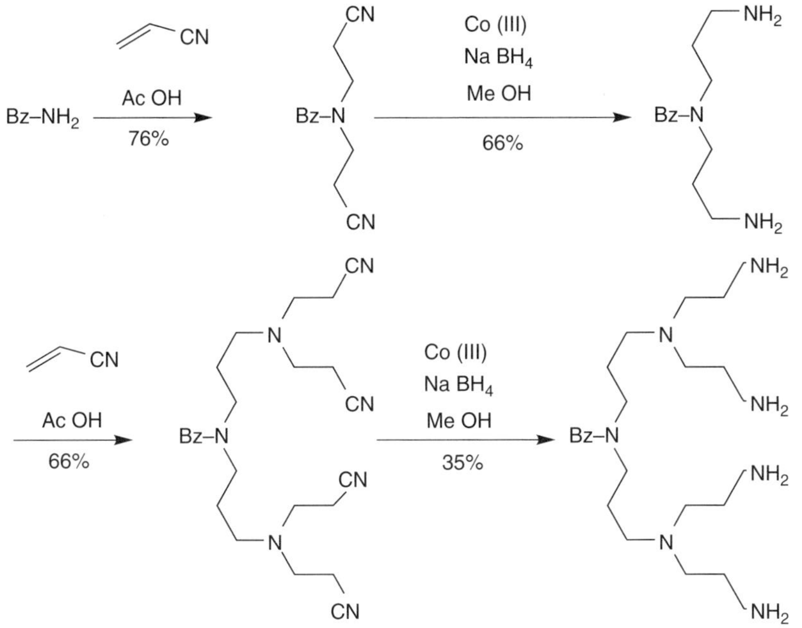

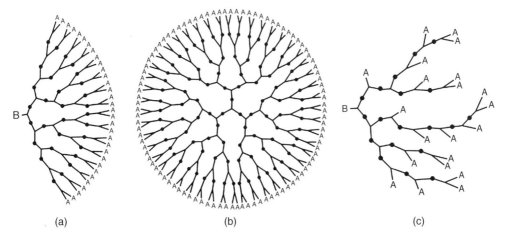

Figure 2.68 'Cascade' synthesis of poly(propylenimine) by Vögtle and co-workers [264].

Dendritic polymers have been classified into three main types (see Figure 2.69): hyperbranched polymers, dendrons and dendrimers. It is also possible to graft dendrons to themselves (dendrigrafts) and to traditional liner polymers (dendronised polymers).

The synthesis of dendrimers can be difficult and expensive [265], and dendrimer research and application is still at an early stage particularly with respect to the upstream oil and gas industry [266]. However, the application of dendrimers and dendritic polymers could revolutionise the fluid properties of the produced oil and associated hydrocarbon liquids due to their unique physical, chemical and biological characteristics of these materials. The ability to synthesise

(a)   (b)   (c)

Figure 2.69 Dendritic polymer types. (a) Dendron. (b) Dendrimer. (c) Hyperbranched polymer.

tailor-made dendrimer products with desirable functional behaviour also highlights the potential application of dendrimers in smart and intelligent fluid design for oil and gas field application.

It has been postulated that the internal cavities within the dendrimer structures can be used to store desirable chemicals, enzymes, surfactants, etc. to trigger appropriate interactions on demand at bottom-hole condition to negate, neutralise or reduce the unwanted changes in a variety of fluid applications such as drilling, drill-in, completion, cleaning, stimulation, fracturing, etc. Due to nanoscale dimension, dendrimers may provide effective external and internal inhibition to reactive shale surfaces, leading to long-term stabilisation of reactive shales. The tiny size and high surface area of dendritic materials will also provide superior fluid properties at a drastically reduced additive concentration. Also dendrimers and dendritic polymers with high thermal stability and affinity to acid gases such as H_2S and CO_2 could help to overcome the technical challenges of geothermal and sour gas drilling operations for a safe, risk-free and economic drilling operation.

The environmental pedigree of these polymers appears to be good, and the structural make-up allows for incorporation of weak hydrolysable bond, which can be subject to enzymatic hydrolysis and other biodegradation routes but also still give a good degree of thermal and material stability [261]. This is particularly the case where amide or ester linkages are contained in the dendritic or hyperbranched polymer.

The reader has already been introduced to hyperbranched polyesteramides in Section 2.1.2. These materials have hydrate inhibition properties [131, 267]. It has also been claimed that full dendrimer polymers also act as KHIs [268]. Hyperbranched polyesters can also be formed from a polyol core molecule by alkoxylation with EO or similar oxides to give potentially good demulsifier polymers [269]. This chemistry has been further developed and alkoxylated polyester dendrimers formed by reacting trimethylolpropane and 2,2-dimethylol (see Figure 2.70) and then suitably alkoxylated. These materials exhibit high performance and show good rates of biodegradation in the regulatory required test protocols [270].

In the oil and gas industry, applications of this dendrimer chemistry are relatively few, and there is much opportunity to examine many other applications. However much will depend on the economics of the polymers compared with other synthetic or natural product alternatives.

Recent work has been presented to show their application as potential paraffin wax control additives [271]. Importantly, and in contrast with other linear polymers, intrinsic viscosity of

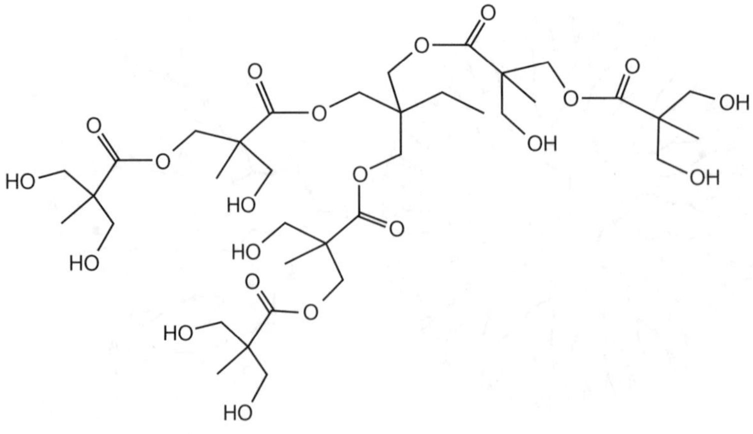

Figure 2.70 Dendritic polyester.

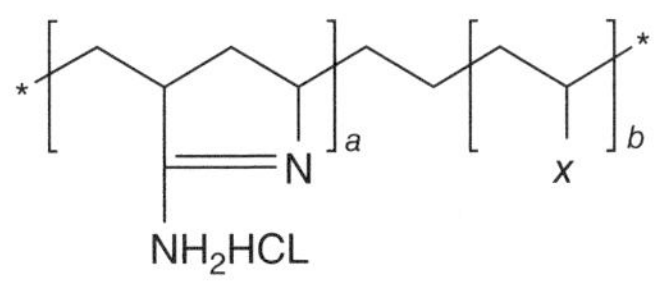

Figure 2.71 Polyamidine-based flocculant.

dendritic polymers does not increase with molecular mass but reaches a maximum at a certain dendrimer generation [272]. This means that there is the possibility of high activity and low viscosity inhibitors, and examples of dendritic polymers acting under laboratory conditions have been exemplified. However, evidence to date does not show these particular examples to have very good rates of biodegradation, which is perhaps, by necessity of application, due to their poor water solubility. They are however classified as polymers and are exempt from registration as new molecules under both US and European regulatory frameworks. This is discussed further in Chapter 9.

2.3.2   Other Polymers

There are a range of other polymers whose polymer make-up is not primarily dominated by a carbon–carbon molecular unit. A wide range of heterocyclic polymers can be derived synthetically from aromatic monomers such as pyrroles, furans and thiophenes; however their use in the oil and gas industry is not known to the author. A related 'hetero' polymer has been developed as a flocculant based on polyamidine [273] (see Figure 2.71). This has been commercialised as DIAFLOC KP7000 and has its main use in sludge treatment in a variety of industries including the oil and gas sector. It also displays a good environmental profile [269].

Polyphosphates and related polyphosphinates and polyphosphonates are used widely in applications particularly in scale and corrosion control, and these will be covered in Chapter 4. Similarly polymers of silicone and polysilicates will be covered in Chapter 7.

Polyaluminium chlorides (PAC) are synthetic polymers dissolved in water. They react to form insoluble aluminium polyhydroxides, which precipitate in big volumetric floccs. The floccs absorb suspended pollutants in the water, which are precipitated with the PAC and can together be easily removed [274]. In the upstream oil and gas industry, PAC can be used as a flocculant for all types of wastewater treatment. PAC will be considered in more detail alongside other aluminium salts in Chapter 5. PAC is manufactured in both liquid and powder form.

2.4   Plastics, Fibres, Elastomers and Coatings

The petroleum industry uses and consumes a vast array of oligomer/polymer products, many of which are used in chemical additives and as polymers in solution, as has been described in the previous sections of this chapter. The other main type of polymer used is a solid-state polymer, such as engineering materials. Engineering materials include those classified as plastics, fibres, elastomers and coatings in general, for use on offshore platforms, construction of pipelines and floating structures, among others. This section gives a brief overview of their use and environmental acceptability and impact.

2.4.1 Plastics

In common with most industries, the upstream oil and gas industry has increased the use of plastics and related materials. There are good economic and environmental reasons for this in that these materials weigh less and are resistant to corrosion. In general the total volume of plastics consumed worldwide outstrips that of iron and steel combined. In the upstream oil and gas industry, this is not the case and iron and steel are still the main volume construction materials for drilling rigs/platforms and processing facilities. In line with general growth in use and consumption, the oil and gas industry is increasing the use of plastics.

Plastics are divided into two main categories:

- *Commodity plastics*: These are characterised as disposable items such as packaging but also has application in more durable items.
- *Engineering plastics*: These are characterised as more durable and have superior mechanical properties. These are increasing in use in the oil and gas industry competing with alloys, ceramics and glass.

Both of these categories of plastics find widespread use in a wide variety of applications across the upstream oil and gas industry [275]. Indeed the more durable and engineered polymers are the materials of choice in many applications especially in subsea developments over the last three decades. The use of engineering polymers allows for highly advantageous qualities in the subsea environment, including high dimensional stability in addition to being very lightweight, almost one-seventh the weight of steel. Engineering polymers are also highly resistant to wear and abrasion, in addition to having a high chemical and corrosion resistance, something that is very commonly found within the subsea environment. Furthermore, engineering polymers offer the quality of self-lubrication and a very low coefficient of friction, resulting in both smoother, longer-running components and applications along with reduced or eliminated maintenance needs and costs [276].

There is a great focus on the reuse and recycling of these materials particularly in decommissioning so as to minimise environmental impact [7, 8]. The durability of the engineering plastics is a considerable advantage over the commodity, more disposable plastics, particularly in the marine environment as there is now a large body of evidence to support serious levels of marine pollution due to microplastics and their relationship to POPS [98].

This degradation and biodegradation of these plastic materials is further discussed in Section 2.6.

2.4.2 Fibres

Fibres are used in relatively small quantities in the upstream oil and gas sector and are really used for the same purposes as other industries. They are polymers of high strength and stretchability. They have good thermal stability and are able to be converted into filaments for use in textiles and other materials. The principal natural product fibres are cotton and wool. Cotton is a plant material and is derived from cellulose, and wool is a polypeptide or protein. Another protein fibre is silk.

Carothers [6] famously invented nylon, not only a breakthrough in polymer synthesis but also the first synthetic fibre. Now over half of the world's fibre production is derived from synthetic polymers. In the oil and gas industry urea-formaldehyde-based polymer fibres are found in woven coatings, as are epoxy-type polymers [277]. These materials are particularly relevant in corrosion retardation and prevention [278].

By their nature and functionality, these materials are highly durable and do not easily degrade, so their environmental impact can be very small if reused and/or recycled.

2.4.3 Elastomers

Elastomers are polymers that exhibit resilience, or the ability to stretch and retract rapidly [279]. They are widely used in many industrial applications, and nowadays synthetic rubber has surpassed the use of natural rubber. The terms rubber and elastomer are often used interchangeably; however it is scientifically accepted that rubber refers to the natural product and elastomer to synthetic polymers.

Elastomers are used in the oil and gas industry in critical locations especially in pump seals. Such pumps can be controlling flow of hydrocarbon products or delivery of chemical for treatment or injection fluids for reservoir pressure support. It is critical that any chemical additives do not adversely react or affect the properties of these elastomer materials, particularly in remote locations such as in subsea submersible pumps. Failures in such elastomers can cause serious production losses, potential pollution and expensive repairs.

Extensive compatibility testing is now normally undertaken to examine the compatibility of elastomers with a variety of chemical additives and their carrier solvents. This can be particularly important with aggressive dissolution and remedial treatments, for example, pyridine can be a useful asphaltene deposition removal agent and dissolver; it is however highly toxic and incompatible with a wide range of commonly used elastomers (H.A. Craddock, unpublished results). There are a number of published procedures and recommended practices for carrying out such testing and comparative work [280–282].

Elastomers (and to some extent resins) have been used in water conformance control. This is particularly the case for phenolic and epoxide polymers and are particularly applied to near wellbore situations where they can seal fractures, channels and perforations [92]. A variety of resin materials have been synthesised including the expected polyepoxides, polyesters and phenol-formaldehyde composites; however some potentially greener derivatives have also been examined such as glycidyl ether-based resins, and these have reported rates of biodegradation of over 40% in the prescribed 28-day testing protocols [283].

2.4.4 Coatings

The science of coatings [284] (and adhesives [285]) has developed slowly over millennia, as the reader will know that paints and painting are as old as civilisation. Adhesives too have a long history; however, with the development of synthetic polymers, these materials rapidly developed. Today they encompass most types of polymer and have a wide range of domestic and industrial applications, including high-temperature resistance in the aerospace industry.

The coatings and adhesive industries are far more complex than just the polymer types used and include solvents, fillers, stabilisers, etc. to provide a fully formulated and ready-to-use material. In the oil and gas industry, these are not used in any particular speciality, except that coatings can often be used to line transportation pipelines to aid in corrosion retardation [286]. The recovery of coatings and adhesives is an area for concern in the potential recycling and/or reuse of such materials and is further explored in the closing chapter of the book, Chapter 10.

2.5 Polymer Application and Injection

As has been illustrated and exemplified, polymers are used in a wide variety of applications across the upstream oil and gas sector, and this utility and versatility makes them one of the largest contributors to the volume of chemical additives used in the sector. The selection and application is, however, an important element in the design and use of any chemical additive

[287]. It is the primary reason that, in all but a few circumstances, the chemicals are formulated to specific design criteria in order to maximise their effectiveness.

This has established an important second-tier supply and service sector between the chemical manufacturer and the main users such as drilling companies and oil and gas operators. This element of supply is discussed further in Chapter 8. With regard to polymers, this is an area where in some circumstances applications and design for use are maintained between the manufacturers and the primary users. This is due to two factors: (i) in areas such as enhanced oil recovery or water conformance control, large volumes may be required and the economics direct a shorter supply chain, and (ii) a number of specialist areas require at least the direct input of the manufacturer in tailoring the polymer requirements, including possible environmentally acceptable properties. A critical issue is the inherent stability of the polymer used to the conditions in the receiving environment, and as will be illustrated in Section 2.6, polymers can be particularly prone to both thermal and mechanical degradation, and in the case of the latter their injection methodology can be critical to a successful application.

2.5.1 Enhanced Oil Recovery and Polymer Application

It is not the author's intention to review the chemistry and detailed application of polymers in enhanced oil recovery, and there are many reviews [288, 289] and books [290] that already do this. It is however important to recognise the need for specialist design and application in this area, the effect of polymer and associated ingredients on the properties of formulated products, their environmental fate, their impact on environmental acceptability, and the growing need for further 'green' chemistry in this area.

For over three decades and more, the industry and governments have been examining the types of polymer application and their economic effectiveness in obtaining more hydrocarbon from oil-bearing rocks and reservoirs [291]. Polymer injection can be a very useful as a way to enhance the effectiveness of water flooding in enhanced oil recovery. However the use of polymers can face a number of technical challenges, such as development of gels that can easily flow for several hundred feet into a reservoir rock and then harden to block water flow. For polymers that follow natural fractures, the set-up time could be only a few days; however, for polymers that must flow through the tiny pores in the rock matrix, the time to reach the intended location and harden may be several months, even a year or more. As has been described in this chapter, from an environmental chemist's standpoint, polymers can offer biodegradable options for these challenges; however often biopolymers and natural alternatives do not have the required strength and resistance to other forms of degradation. Indeed having polymer stability for several months may render the polymer extremely stable to any biodegradation processes.

In order to impart the required properties and stability to polymers and to form appropriate gels, many are cross-linked in situ by the application of appropriate cross-linkers. A good example of the problems with gel-forming cross-linkers is illustrated by the strengthening of guar gum used extensively in hydraulic fracturing [1].

It is desirable to have delayed cross-linking and this can be achieved by a number of systems, as explained later. Retarding the rate of reaction of the cross-linking means the fluid can be pumped more easily. The most common fracturing fluid cross-linkers are borate-based systems and can be formed by boric acid, borax, an alkaline earth metal borate or an alkali metal alkaline earth metal borate. It is essential that the borate source has around 30% boric acid. The boric acid forms a complex with the hydroxyl units of the guar gum polysaccharide, cross-linking the polymer units. This process lowers the pH (hence the need for pH control) [119]. These fluids provide excellent rheological, fluid loss control and fracture conductivity properties in fluid temperatures up to 105 °C [292].

Organic titanium systems have been found to be useful cross-linkers [293], but are seldom if ever used due to their hazardous nature and poor environmental characteristics. Various zirconium systems are used to ensure delayed cross-linking. These complexes are formed initially with low molecular weight compounds, which are then intramolecularly exchanged with polysaccharides; this process causes the delayed cross-linking. The zirconium complexes that are used are usually initiated from compounds with diamine-based molecules such as hydroxyethyltris-(hydroxypropyl)ethylenediamine [294], or hydroxyl acids such as glycolic acid, lactic acid, citric acid, etc. or polyhydroxy compounds such as arabitol, glycerol, sorbitol, etc. These materials can then form suitable gels with polysaccharides.

Clearly the environmental fate and toxicology of the cross-linkers is more complex than food-grade polysaccharide, such as guar gum. This complication in the application of polymers is fraught with difficulty in convincing regulators and other stakeholders of the overall environmental effect and the bioavailability of the cross-linking materials. Indeed boric acid is already found in abundance in the environment and naturally occurs in air, water (surface water and groundwater), soil and plants, including food crops. It enters the environment through weathering of rocks, volatilisation from seawater and volcanic activity [295]. Most boron compounds convert to boric acid in the environment, and the relatively high water solubility of boric acid results in the chemical reaching aquatic environments. Boric acid is therefore the boron compound of environmental significance [296]. It is assumed that boric acid is adsorbed into soil particles, aluminium and iron minerals and that this adsorption can be either reversible or irreversible, depending on soil characteristics. It is known that boric acid is mobile in soil [297]. Boron, in particular boric acid, is found in many aquatic environments and is in surprisingly high concentrations in seawater [298]. At present boric acid and borates are regulated through the normal channels, and regulators do not see any significant adverse effects to the environment from the use of boric acid and its derivatives. However, with the growth of shale gas fracturing, it is likely that permitting to operate will include the monitoring of boric acid and borates, as there may be some issues around bio-magnification [299]. There of course is a raft of other political and public pressure on these areas, and this will be returned to in Chapter 10.

2.5.2 Polymers in Water Shut-Off

Most cross-linked polymers used in water shut-off treatments use polymer systems that become cross-linked and gel in the reservoir formation at elevated temperatures [300]. These include many of the polymer and biopolymer systems discussed earlier in this chapter. Also, as discussed earlier, the systems often include metal ions as well as organic materials as the cross-linking agents. PAM and partially hydrolysed PAM are by far the most used materials in this area of application.

2.5.3 Drag Reducer Polymer Injection

As described earlier, in Section 2.1.1, ultra-high molecular weight polymers, especially isobutylenes, can be very effective DRAs. These materials are particularly prone to mechanical shear degradation [301] and are completely destroyed through boosting pressure pumps in place throughout long export crude pipelines, requiring the installation of new injection skids at the downstream of each pump station [302]. These long-chain polymers have been shown to be sensitive to mechanical degradation occurring during the transport within the pipeline, a phenomenon that progressively reduces the overall DRA efficiency [303]. In order to partly compensate for these effects, emulsion formulations and special injection delivery systems are recommended particularly for ultra-high molecular weight polymers utilised in oil transport drag reduction [304].

2.6 Polymer Degradation and Biodegradation

In this concluding section, the various mechanisms of polymer degradation are examined and discussed with particular emphasis on how such degradation pathways, again especially biodegradation pathways, affect the environmental fate of the polymer additive. As been shown throughout this chapter, there is normally a conflict between maximising product efficiency and the biodegradation of the product to other materials less harmful to the receiving environment. Polymers are usually designed to be durable and to resist degradation; however inevitably they will degrade over time and under harsh conditions. There are six main types of degradation as listed as follows:

- Thermal degradation
- Oxidative degradation
- Radiative degradation
- Mechanical degradation
- Chemical degradation
- Biological degradation

Each is now discussed and its effects on potential environmental fate are examined, although there is considerably more emphasis placed on the final degradation mechanism, namely, biodegradation. It should be noted that these types of degradation are often not occurring in isolation and that often thermal, oxidative, mechanical and radiative degradations can occur simultaneously and consequently. From a polymer chemist's standpoint, the generally accepted premise is that degradation is not a good process and polymer stability is paramount. From an environmental chemist's view, this is almost the opposite, in that stability can make polymers non-biodegradable and more susceptible to causing harm or pollution in the receiving environment. In general, polymers because of their high molecular weight are not bioavailable [305]. Compromise, on a number of levels, is inevitable; it is the fundamental premise of how this is achieved that can be questionable, and this in general terms as well as specifics relating to oilfield chemistry is discussed further in Chapter 10. Some specifics relating to polymers are discussed here.

2.6.1 Thermal Degradation

This section will consider thermal degradation on synthetic and natural polymers discretely as they have somewhat different mechanisms, and therefore the degradation effects on the potential environmental impact can differ.

2.6.1.1 Synthetic Polymers
The thermal degradation process in most synthetic polymers is related to a scission mechanism generating a free radical monomer, which is highly volatile [306]. In general in synthetic polymers, vinyl polymers are more prone to thermal instability than the non-vinyl examples.

Scission thermal degradation has two main mechanisms: (i) non-chain scission and (ii) random chain scission. Non-chain scission occurs to pendant groups and does not break the

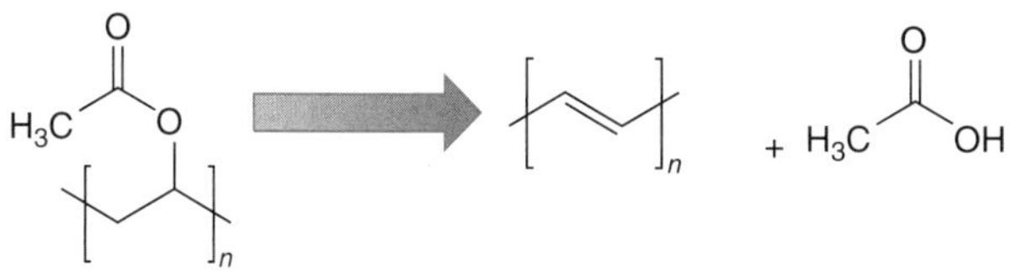

Figure 2.72 Elimination of acetic acid from poly(vinyl acetate).

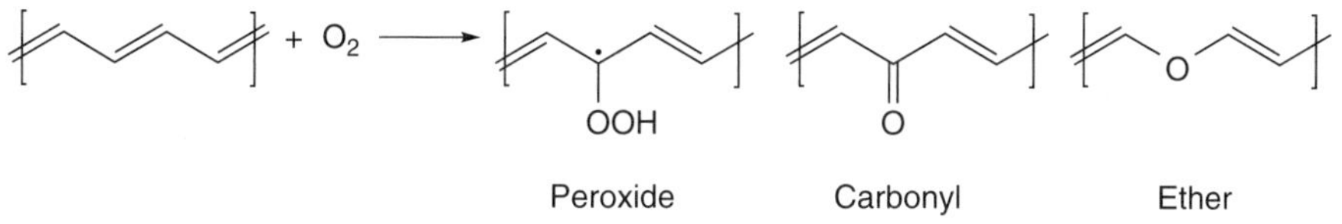

Peroxide   Carbonyl   Ether

Figure 2.73 Oxidation of polyacetylene.

polymer backbone, for example, the elimination of acetic acid from PVAc (see Figure 2.72) results in the formation of a polyalkene, polyacetylene.

It is worthy of note that this type of elimination reaction is not a useful method for synthesising polyacetylene.

The thermal degradation of PVAc shows that in examining environmental fate after use of such products, the impact of acetic acid should be considered alongside that of polyacetylene as well as the further degradation of polyacetylene where such products are in high-temperature usage such as in reservoir and wellhead conditions.

Such sophistication of environmental fate in determining overall environmental impact is seldom used, if at all, in the upstream oil and gas sector.

Continuing the thermal degradation of PVAc, it is polyacetylene's degradation that should be considered. Polyacetylene, although thermally stable [307] having many double bonds, is unstable in air. When polyacetylene is exposed to air, oxidation of the backbone occurs [308, 309] as shown in the schematic in Figure 2.73.

Therefore it is highly likely that the species that are likely to have environmental impact when apply and using polymers and copolymers, based on, or including, poly(vinyl acetate), are the thermal and other breakdown products, chemical and otherwise and not either the polymer itself or the constituent monomer. Yet as is shown in Chapter 9, this is rarely, if ever the case, with regulation driven by direct impact of the polymer and the possibility of residual monomer.

Random chain scission, the second of the main thermal degradation processes, occurs at weak points in the polymer chains or backbones and is the result of homolytic bond cleavage. This process occurs in most polymers and results in complex mixtures of degradation products including the monomer units. The complex nature of the degradation products is due to the reactions of the radical species formed in the bond cleavage and their recombination to new products as illustrated in Figure 2.74. This adds further to the complexity of what is potentially being discharged after product use into the receiving environment and related ecosystem as a variety of alkenes and polymeric alkenes can be derived.

$$-(CH_2CH_2CH_2CH_2)_n- \longrightarrow -(CH_2)_a\,CH_2CH_2\cdot + \cdot CH_2CH_2(CH_2)_b- \longrightarrow$$
$$-(CH_2)_a che = CH_2 + CH_3CH_2(CH_2)_b-$$

Figure 2.74 Thermolysis of polyethylene.

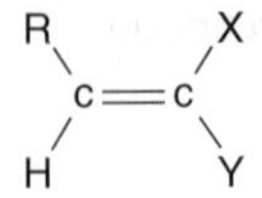

R, X   **Figure 2.75** Idealised 1,1-disubstituted monomer.

C=C

H   Y

Random chain scission processes occur in all vinyl polymers under thermal degradation. However this process occurs less with increasing substitution on the polymer backbone [310].

There is a third thermal degradation mechanism, namely, depropagation or 'unzipping' of the polymer to give the monomer units. This mainly occurs in synthetic polymers prepared from 1,1-disubstituted monomers such as the structure in Figure 2.75; therefore, this applies to a number of vinyl polymers.

In this third process a more expected result is observed in the formation of the starting monomers; however this process can occur at the chain ends or at random sites along the polymer backbone. With some polymers a mixture of depropagation and random chain scission occurs.

It of course should be emphasised that thermal degradation process in synthetic polymers, as applied to the upstream oil and gas industry, is not the norm and only occurs in circumstances where high temperatures prevail such as in the wellbore or the reservoir, or in processing at production facilities where again temperatures can be deliberately raised. It would seem sensible in considering environmental impacts from the use of polymer-based chemical additives, however, that suitable thermal degradation processes are taken into account, and this is even more the case with the natural polymers, discussed in the next section.

2.6.1.2 Natural Polymers

Many natural polymers are only thermally stable to temperatures of around 140 °C, particularly sugar and amino acid-based polymers. Readers will be familiar with the processes of thermal decomposition of many natural products and polymers through their experience of cooking and the flavours that it produces. Such flavours and colourisation are based on the Maillard reaction [311], named after the French chemist Louis Maillard, who first described it in 1912 while attempting to reproduce biological protein synthesis. It is the chemical reaction that gives browned food its desirable flavour. In the process, hundreds of different flavour compounds are created, and in the process these compounds, in turn, break down to form yet more new flavour compounds.

The cellulosic polymers also have a complex thermal degradation process and this occurs at lower temperatures, at least initially, than the synthetic polymers. The thermal decomposition involves at least four processes [312]. The first is the cross-linking of cellulose chains with the loss of water, dehydration. This occurs concurrently with the depropagation of the cellulose chain to form laevoglucosan (see Figure 2.76) alongside the formation of dehydrocellulose. This latter intermediate further decomposes to form a char and volatile products. Finally laevoglucosan can further decompose also to smaller volatiles and tars and eventually carbon monoxide

Of course some natural product-based polymers do display better thermal stability, and a good example is the alkyl polyglucosides mentioned earlier in Section 2.2.6. These materials

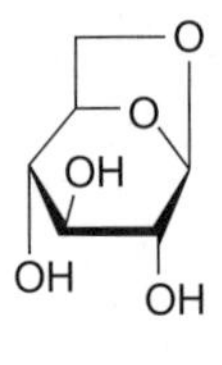

Figure 2.76 Laevoglucosan.

show thermal stability to over 170 °C [171] and still show excellent biodegradation characteristics [313].

Suffice to say that thermal degradation can affect the environmental profile of the product under question as the products of thermal degradation could be the ones that are entering the receiving environment. Indeed it is the author's belief that greater consideration should be given to this effect; as was stated earlier with the example of lignin, degradation products could be highly toxic to the receiving environment and seen as less harmful in the polymeric form.

2.6.2   Oxidative Degradation

Oxidative degradation can be defined as the cleavage of a carbon skeleton (often at a C=C double bond) with the introduction of new carbon–oxygen bonds and this can occur in polymers as well as other organic molecules.

The process of degradation by oxidation in polymers is related to thermal processes, in particular ones of combustion as they also involve oxygen. Needless to say it is not desirable for polymer-based additives to be involved in combustion processes, and they are not normally used or be in proximity to such environments. Therefore, thermally promoted oxidation processes, for the purposes of this current study, are not examined. However, in the oilfield, oxidation processes can readily occur as there is an abundance of oxidising agents used and also oxygen and other species capable of generating free radicals under the right conditions are abundant. As has been shown in the previous section, a number of polymer types are prone to autoxidation both in the absence and presence of additives capable of promoting free radicals. In some cases free radical inhibitors are added to polymer formulations to suppress these oxidative degradation effects [314].

Of course these free radical processes are also, in controlled circumstances, one of the main synthetic routes in forming many polymers. It is not within the scope of this book to examine the complex mechanisms and related kinetics of these processes. For the environmental chemist and the oilfield chemical practitioner, it is sufficient to be aware that oxidative degradation process can occur and may be very likely in the conditions in the oilfield, recognising that chemical additives, including polymer-based additives, could be oxidised. The inevitable conclusion is that the resulting oxidation products may be the materials to consider for environmental impact as opposed to the originally added products.

Polymers are susceptible to attack by atmospheric oxygen especially at elevated temperatures, and these can be encountered during processing to shape for plastics and in the oilfield for chemical additives as previously described. Oxidation tends to start at tertiary carbon atoms (terminal carbon atoms in polymers) because the free radicals formed are more stable and longer lasting, making them more susceptible to attack by oxygen. The carbonyl group formed can be further oxidised to break the chain, which weakens the material by lowering its molecular weight. In solid polymers such as polyethylene-based plastics, cracks appear and start to grow in the regions affected.

In general saturated polymers are degraded slowly by oxygen; however this process can be sped up, as has been shown in Section 2.6.1, by heat and by light (see Section 2.6.3), and the tertiary carbon atoms are the most susceptible to attack. This is reflected in the resistance to oxidative degradation by three common polymers, as shown in the following order:

Polyisobutylene > polyethylene > polypropylene

The reaction products, as has been shown in thermal degradation, (Section 2.6.1) are numerous.

Unsaturated polymers undergo oxidative degradation more rapidly by the complex free radical process previously described in Section 2.6.1 and can involve peroxide and hydroperoxide intermediates.

Antioxidants are often used to inhibit these degradation mechanisms [315] in commercially synthesised polymers.

Examining the potential oxidative processes in the degradation of polymers, which could happen to the chemical additive and other components of the production fluids, could be advantageous to the environmentalist and the production chemist as oxidation pathways are often some of the first stages in biodegradation. The formation of carbonyl groups by oxidative degradation can be sites for further chemical reaction and transformation. It has been known for some time that such biological oxidations can be used in the manipulation of microorganisms to convert potentially harmful agents to less harmful ones [316]. Indeed many industrial effluent systems use chemical oxidation alongside biological systems to treat some severely hazardous waste systems [317, 318].

As has been described in the previous section, natural product-based polymers can thermally degrade to their monomers or derivatives thereof. Where conditions are correct, oxidative degradation can occur in a similar manner and further degradation of the monosaccharide units can yield low molecular weight carboxylic acids and hydroxycarboxylic acids [319]. Again, it is argued by the author that a full recognition of the degradation of polymers, by a number of mechanisms acting in concert, is considered and that the substances formed should be recognised as important in the potential environmental impact of the added polymer substance and its ultimate environmental fate.

The oxidative degradation process alongside chemical hydrolysis (see Section 2.6.5), as the author would argue, should be considered in evaluating the environmental hazard of any chemical additive.

2.6.3 Radiative Degradation

Radiative degradation in polymers results usually in two types of reaction:

1) Chain scission, as has been described in a section on thermal degradation, gives low molecular weight moieties.
2) Cross-linking results in structures that are insoluble and infusible.

Also radiative degradation can be divided into two types:

1) Photolysis, usually from ultraviolet (UV) light. This process is due to the absorption of energy in discrete units by specific functional groups or chromophores.
2) Radiolysis, the absorption of ionising radiation such as X-rays. Here the reaction does not require specific chromophores as the energy imparted is much greater and transferred directly to any electron, which are in the path of the emitted photons.

In the normal operation of the oilfield, chemists and other practitioners will only encounter photolysis unless under the particular circumstance of using radioactive tracer materials. For the purpose of polymer degradation and polymer-based chemical additives, the discussion will centre on photolysis.

Photolysis or photo-oxidation is the degradation of a polymer surface in the presence of oxygen or ozone, although forms of photolysis can occur in the absence of oxygen. Photo-oxidation is a chemical change that reduces the polymer's molecular weight. The effect is facilitated by radiant energy such as UV or artificial light. This process is the most significant factor in weathering of polymeric materials such as plastics and is of concern as it aids in the generation of microplastics, which are responsible in the marine environment for the concentration of POPS [273]. With regard to polymeric chemical additive, these are not usually exposed to UV light, being mainly in reservoir or other enclosed conditions. At points of discharge then they may encountered light, and photolysis along with other degradation process may occur. Other degradations may have already commenced. Such a process may be useful to the overall degradation as well as biodegradation of the polymer species.

In offshore operations where produced water is discharged to the marine environment, photolysis despite misconceptions can occur to a depth of several metres, provided that the water is reasonably clear [320]. This is particularly so in shallow basins such as the Gulf of Mexico and the Arabian Gulf. The photolysis that is able to occur in these waters can be extensive and complex and involves surface layers of the water column and sediments [321] and is a topic worthy of more detailed discussion, which will be examined in Chapter 10.

2.6.4 Mechanical Degradation

As with many forms of degradation, mechanical degradation is not generally a desired effect with oilfield polymers and polymer-based additives. As has been discussed in a section on polymeric DRAs, in particular, the ultra-high molecular weight polymers can be highly susceptible to mechanical shear forces [20]. The reduction of drag in turbulent flow is very useful in the upstream oil and gas industry and includes applications in the long-distance transport of produced fluids and processed crude oil, well operations and the transport of suspended solids and slurries, such as in hydraulic fracturing. The drag reduction process, as has been illustrated, can be complicated by polymer degradation. It has been observed that for synthetic polymers (e.g. PAM and PIBs), the polymers degrade more in poor solvent systems at low Reynolds number, whereas the opposite effect is noted at high Reynolds numbers [322]. Somewhat contrary to expectations in aqueous systems, guar polysaccharides reduce friction drag tremendously even in turbulent flow and at small added amounts [323].

The undesirable effects of mechanical degradation mean that polymers are normally designed to be durable and to resist mechanical forces. This as we have seen has consequences on other degradation processes and usually means that most synthetic polymers, unless deliberately designed, are also resistant to other forms of degradation including biodegradation.

Not all mechanical degradation processes are detrimental to polymer integrity; indeed, in the production of natural rubber, it was realised in the early nineteenth century that the material could be improved by the process of mastication. It was later shown by Staudinger [324] that this was a result of lowering the molecular weight of the degraded rubber. These machining processes are essential in imparting rubber polymers with their viscoelastic properties [325].

2.6.5 Chemical Degradation

There is a large body of work on chemical degradation of polymers [326], and it is not the intention of this short section to examine this in great detail. Also, as there are a large number of chemicals that can attack polymers, the number requires to be restricted. Therefore, the section will focus primarily on chemical reaction and degradation of oilfield polymer-based additive as well as on some key material interactions with construction-type polymers. The chemicals considered in this context will also, as believed by the author, be those that are most important from a field experience point of view. Outside the application of polymers in the oilfield, agents such as organic pollutants, e.g. nitrogen dioxide and sulphur dioxide, would be probably among the most important agents in chemical attack [327]. However the most important is most probably oxygen, and this has been examined separately in the previous section on oxidative degradation.

In the oilfield and particularly for most polymer-based chemical additives, the predominant chemical attack is by hydrolysis. This is particularly the case with respect to polymers that are water soluble, and this was briefly discussed (Figure 2.10) in Section 2.1.1.

2.6.5.1 Water-Soluble Polymers

An alternative classification for polymers would have been to designate them as water soluble and sparingly soluble or insoluble. Water-soluble polymers can be categorised into the following three broad groupings:

- Synthetic, which are produced by the polymerisation of monomers synthesised from petroleum- or natural gas-derived raw materials.
- Semi-synthetic, which are manufactured by chemical derivatisation of natural organic materials, generally polysaccharides such as cellulose.
- Natural, including microbial-, plant- and animal-derived materials.

Synthetic water-soluble polymers are organic substances that dissolve, disperse or swell in water and thus modify the physical properties of aqueous systems undergoing gelation, thickening or emulsification/stabilisation. Semi-synthetic water-soluble polymers are derived by either chemical modification of natural polymers or from microbial sources. Most natural and semi-synthetic water-soluble polymers are polysaccharides that vary substantially in their basic sugar units, linkages and substituents. Derivatives are obtained by substitution, oxidation, cross-linking or partial hydrolysis. The products from animal sources are protein-based analogues of the more commonly used polysaccharide-based vegetable polymers.

Water-soluble polymer demand from US shale gas drilling, especially guar gum demand, has significantly increased in recent years. Guar gum, a guar bean derivative, is a vital component of the fluid used in hydraulic fracturing. During early/mid-2012, guar gum prices significantly increased. Accordingly, guar gum production in India and Pakistan has increased significantly. At the same time, water-soluble polymer alternatives to guar gum have been extensively explored. Thus, the US shale gas industry has had a global impact on the water-soluble polymer industry [1].

Water-soluble polymers, which perform various useful functions such as thickening, gelling, flocculating, rheology modifying and stabilising in any given application, are used for a wide variety of applications in the upstream oil and gas industry as has been illustrated throughout this chapter. These include water treatment, enhanced oil and natural gas recovery, oil separation and processing particularly in scale and corrosion control and many more. It is also worth noting that often these polymers perform more than one function in any given application. Much has been written across various industry sectors on the application of water-soluble polymers [328] and in particular on enhanced oil recovery in the upstream petroleum industry [329, 330]. Many of these polymers and their applications have been referenced and discussed throughout this chapter.

Water solubility is also a prerequisite characteristic to good biodegradation potential as has been briefly described in Section 2.1.3 (see Figure 2.35), and is further discussed in Section 2.6.6.

2.6.5.2 Hydrolysis

Hydrolysis is simply defined as chemical decomposition in which a compound is split into other compounds by reacting with water. In other words it is the reaction of water with another chemical compound, in this case a polymer molecule, to form two or more products, involving ionisation of the water molecule and usually splitting the polymer.

Hydrolysis can occur under neutral, acid or alkaline conditions. Neutral and acid hydrolysis are similar especially with respect to polymers, whereas alkaline hydrolysis can be quite different. This process can also occur enzymatically under conditions previously described and is one of the main pathways for biodegradation.

The hydrolysis of solid polymeric materials can be even more complex as their mechanism and rates of hydrolysis can be affected by a variety of factors such as film thickness, morphology, relative humidity, dielectric constant, etc. [331]. For chemical additives in a homogenous condition with the mainly aqueous fluids, the main factors affecting hydrolysis are pH and temperature [327]. Critically the other major factor is the type of polymer where the structure of the polymer can promote particular sites, such as functional pendant groups for preferential hydrolysis, as in polyesters , polyacrylates and PAM. Many of the so-called environmental derivations discussed in Section 2.1.3 are exploiting this property, particularly under acidic or neutral conditions.

Under alkaline conditions, the OH^- ion, which promotes hydrolytic conversion, may not be able to attack the polymer structure due to its dielectric constant [327, 332]. However, the alkaline hydrolysis of simple ester, for example, is offset by having a stronger nucleophile, OH^-, instead of H_2O.

In summary the hydrolysis of synthetic polymers and biopolymers involves the scission of susceptible molecular groups by reaction with water. This process can be acid or base (or enzyme) catalysed and is not surface-limited if water can penetrate the bulk structure, which is particularly relevant for water-soluble polymers. The main molecular and structural factors influencing this hydrolysis are as listed as follows:

- *Bond stability*: Introduction of hydrolysable bonds.
- *Hydrophobicity*: Increasing hydrophobicity reduces the potential for hydrolysis.
- *Molecular weight and polymer architecture*: The higher the molecular weight, the more difficult hydrolysis is.
- *Morphology*:
  - High crystallinity reduces the potential for hydrolysis.
  - High porosity increases the potential for hydrolysis.
- *Temperature*: The less mobility, i.e. the lower the temperature, then the slower the hydrolysis.
- pH, acid and base catalysis.

All of these factors are normally considered when designing a biopolymer.

There has been extensive study into the hydrolytic degradation of natural polymers, especially cellulose and proteins [333, 334], and primarily these degrade to the monomeric units of monosaccharides and amino acids. This is highly desirable in their credentials as biopolymers, as hydrolysis gives rise to bioavailable and useful substrates for microorganisms and other fauna, to utilise these materials in enzymatic hydrolysis.

The addition of functionality in the polymer can be used to allow hydrolysis to be a useful property. For example, bridging agents used in filter cake construction if made from a degradable material can enhance the removal of the filter cake. This is usually by hydrolysis over a period of time, allowing produced fluids to move more freely [335]. Various stereoisomers of polylactic acid (Figure 2.77), from D- and L-lactic acid (Figure 2.78), can degrade at different rates and can therefore provide both slow and rapid degradation through hydrolysis.

2.6.5.3 Other Chemical Degradation

There are a number of other chemical reactions that can lead to degradation of both solid polymers and their liquid additive equivalents. Contact with oxidising agents and highly corrosive

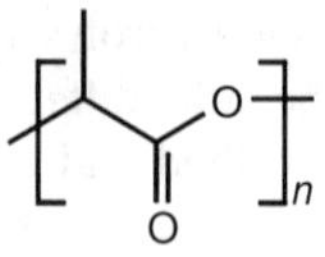

Figure 2.77 Polylactic acid.

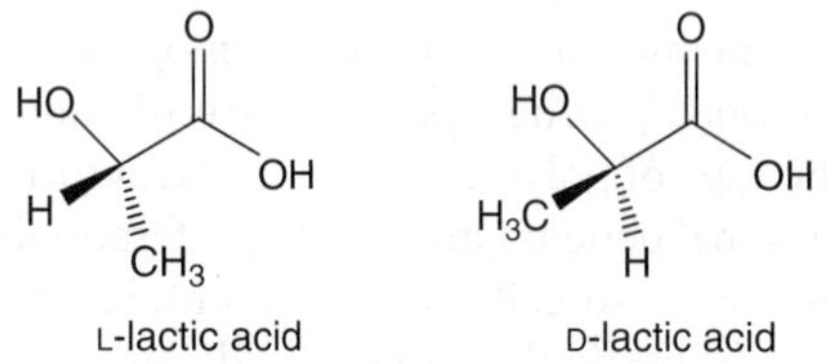

Figure 2.78 Lactic acid stereoisomers.

L-lactic acid   D-lactic acid

chemical agents should be avoided, if possible. However, in the oilfield the produced fluids can contain acidic gases and other corrosive species that will be detrimental to the integrity of any polymer as they contact both solid materials and liquid additives.

Polymer degradation by galvanic action was first described in the technical literature in 1990 [336]. This was the discovery that 'plastics can corrode', i.e. polymer degradation may occur through galvanic action similar to that of metals under certain conditions and has been referred to as the 'Faure effect'. Where plastics have been made stronger by impregnating them with carbon fibres, these carbon fibres can act as a noble metal similar to gold (Au) or platinum (Pt). In early 1990, it was reported that imide-linked resins in plastic composites degrade when bare composite is coupled with an active metal in saltwater environments. This is because corrosion not only occurs at the aluminium anode but also at the carbon fibre cathode in the form of a very strong base with a pH of about 13. This strong base reacts with the polymer chain structure, degrading the polymer. Degradation occurs in the form of dissolved resin and loose fibres. The hydroxyl ions generated at the graphite cathode attack the O—C—N bond in the polyimide structure.

Chlorine is a highly reactive gas that will attack susceptible polymers such as acetal resin and polybutylene found in pipework. There have been many examples of such pipes and acetal fittings failing as a result of chlorine-induced cracking, both in domestic use and in industrial use including the oil and gas industry. In essence, the gas attacks sensitive parts of the chain molecules (especially secondary, tertiary or allylic carbon atoms), oxidising the chains and ultimately causing chain cleavage. The root cause in the upstream oil and gas industry is traces of chlorine in the produced water, from seawater and other sources such as added in small amounts for its biocidal properties. Attack can occur even at parts per million traces of the dissolved gas.

2.6.5.4 Chemical Compatibility

Chemical degradation can be a complex problem and affect the polymeric nature of chemical additives (and polymeric materials), compromising their designed effects. It is now recognised as essential in the development, design and formulation that polymers are examined both for their resistance to chemical degradation and, if liquid additives, for their compatibility with other chemical agents and chemical conditions likely to be found in their application. To this end it is now seen as best practice to design a compatibility matrix for any chemical selection process to examine the chemical and material compatibility of the added substance. This is particularly important for polymers that can be water soluble and water immiscible as well as critical components in polymeric materials.

In general the compatibility of both additives and inhibitors against generic chemistries and materials encountered in drilling fluid or process systems is carried out to ensure that no detrimental effects will occur.

For example, compatibility tests for examining an additive in the process system are performed with the generic chemistries for scale inhibitor, deoiler, demulsifier and corrosion inhibitor wax inhibitor. Kerosene is normally used to simulate crude oil in all tests that are carried out. 100 ml samples of a 50 : 50 mix of kerosene and representative formation water are sequentially dosed with other chemicals present. All samples are mixed and heated to 65 °C and then monitored over a prescribed period for any incompatibilities.

Compatibility with the elastomers of nylon 11, teflon, nitrile, butyl and neoprene are usually evaluated. Samples of each material are submerged in the relevant chemical inhibitor for a period of 28 days at room temperature. The weight, hardness and dimensions of the elastomer samples are checked on a regular basis to determine whether any detrimental effects have occurred.

Although from an environmental standpoint hydrolysis of polymers and polymeric materials is normally a positive process affording substrates that can be further degraded or biodegraded, it needs to be recognised that such a chemical reaction can lead to toxic or otherwise harmful agents, and as such, in the authors' opinion, this should be considered in examining the environmental impact of the polymer use. For example, some polyacrylates, and in particular cyanoacrylates (used in dental adhesives), on hydrolysis generate formaldehyde [337], a biocidal agent with known carcinogenicity. Recognition of such a factor in examining the environmental fate of a polymer additive can be important, as such an agent could compromise any biodegradation study.

2.6.6   Biological Degradation: Biodegradation

Earlier in this chapter, polymer biodegradation was discussed in terms of the application of biopolymers and the manipulation of polymer chemistry to achieve enhanced environmental acceptability (see Section 2.1.3). Natural polymers and biopolymers were also discussed at some length (see Section 2.2). The intention of this section on polymer degradation is to further examine the biological processes involved in the degradation of polymers within the oilfield, and in the main this will be to the soil and marine environments, and to discuss the overall validity of this measure, i.e. the rate of biodegradation as applied to polymers. There will also be reference to the final chapter of this book, Chapter 10. It will also examine some chemistries that may provide other biodegradable polymers, which are not considered for application in the upstream oil and gas sector.

As has been illustrated throughout this chapter, environmental and oilfield chemists face a dilemma in designing materials and products with good biodegradability and retaining stability. The latter property can often, as has been previously illustrated, be inherent to the functional behaviour and efficacy of a polymer additive. Most polymers and polymer additives are very durable, and this is a property that has allowed solid polymers to be used as substitutes for metal and glass. As has been described, polymer waste [98] is significantly contributing to pollution. This has meant that in recent decades attention has been paid to making polymers that are degradable and most importantly biodegradable [338]. As has been illustrated in this section, polymers have various mechanisms of degradation, which from a deliberate enhancement perspective can be utilised to ensure the environmental impact is reduced; in particular light, oxygen and water (hydrolysis) play a major role in the designed degradation of polymers. These properties are also used in combination with the action of microorganisms, both soil and marine, to ensure suitable biodegradation of polymers.

Biodegradation is fundamentally chemical in nature; however the source of the 'chemical' attack is in the main from microorganisms, such as bacteria and fungi, and the main processes are catalytic in nature, based on enzymatic hydrolysis (see Figure 2.35) and other biological/metabolic processes. The susceptibility for a polymer to be subjected to microbial attack is generally dependent on the following:

- Enzyme availability
- A site on the polymer for enzymatic attack (this is a primary mechanism in water soluble polymers)
- The specificity of the enzyme for the polymer (this can be related to the available microfauna)
- Co-enzymes or other 'catalysts' if required

The process of biological degradation is happening all of the time. Microorganisms, in particular fungi and bacteria, exist in vast numbers in the biosphere. They thrive under particular growing conditions such as the oceans or the topsoil. Indeed, if fungi were not degrading leaf litter shed from deciduous forests at a constant rate, we would be overwhelmed by these materials and the planet would be a very different place. These microorganisms require a nutrient base, usually carbon, hydrogen, oxygen, nitrogen and a few other elements; consequently they can utilise all sorts of materials as substrates. A key ingredient for the process is usually water, and therefore materials such as water-soluble polymers have a distinct advantage in potentially being biodegradable.

Most, if not all, natural polymers such as cellulosics, other polysaccharides and polypeptides are biodegradable even if they are not water soluble. However, a substantial proportion are discarded as potential biopolymers, certainly with respect to oil and gas applications, as their rates of biodegradation are either slow or require specific conditions to ensure good rates of biodegradation. For example, the degradation of proteins can be very complex and necessarily so as specific enzymatic cleavages are used in the biological construction use and repair of these macromolecules [339]. This is a valuable resource that has not even begun to be explored and shall be discussed further in this section.

Most polymers and polymeric materials undergo a biodegradation process as generally outlined in Figure 2.79.

Of course this is driven by other conditions affecting the kinetics of hydrolysis, such as temperature and pH, and can occur over long periods of time and indeed in the crude oil itself over geological time periods. However, as we can see, the ultimate biodegradation is to small molecular gas molecules and water. Of ultimate environmental concern is the input of carbon dioxide (CO_2).

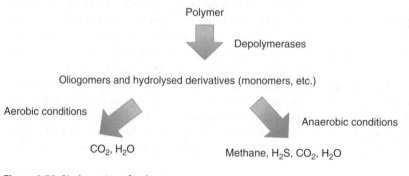

Figure 2.79 Biodegration of polymers.

Finally the emphasis that biodegradation profiles receive as a means of classifying the environmental acceptability or 'greenness' of a product requires to be challenged. As was outlined in Chapter 1, green chemistry is much more than achieving a good biodegradation profile in a specific test that is probably not that representative of the receiving environment. This is particularly the case in the upstream oil and gas industry where chemicals are constantly challenged as to their biodegradability, and although other factors such as toxicity are considered, this is the primary criterion for regulatory control. As has been illustrated throughout this chapter, most synthetic polymers are not very biodegradable; however the corollary is that they are often not bioavailable. This is an important point as polymers have large or very large molecular weights, cannot be absorbed into cells and cannot be bioaccumulated, at least by the normal cell transportation mechanisms [340]. As has been illustrated however, polymers can undergo other forms of degradation, primarily hydrolysis, and it is these 'fragments' that are primarily undergoing subsequent biodegradation. Making these degradation processes easier is the primary thrust of all modifications to synthetic polymers and the main *modus operandi* of the synthetic biopolymers.

This being stated a key driver in the provision of oilfield chemicals, especially in regions of the world where regulatory control is enforced, the industry is directed to provide products, including polymers, that are biodegradable. The design of biodegradable polymers, as has been illustrated earlier, is somewhat different than for other chemical additives in the upstream oil and gas industry, and the design for other chemical species will be explored in other relevant chapters. The following section considers some approaches that have not been previously detail, or not extensively so, earlier in this chapter.

2.6.6.1 Designing for Biodegradation

In Section 2.1.3, the three basic methodologies, which are adopted for application in the design of biodegradable synthetic polymers, were illustrated. An approach that has not been extensively utilised is to examine polymers that although synthetically derived mimic natural polymers.

For example, poly(2-oxazoline)s (see Figure 2.80) are known as pseudo-polypeptides [341] and are undergoing a resurgence of interest due to their versatility and diverse properties.

These polymers are made from the ring opening of 2-oxazolines from a 'living' polymerisation process, as illustrated in Figure 2.37, in Section 2.1.3. As described previously, these materials have been examined as to their application as KHIs; however other applications have not been examined, and the examples looked at did not give good biodegradation profiles.

The versatility, however, of this living polymerisation method allows copolymerisation of a variety of 2-oxazoline monomers to give a range of tunable polymer properties that enable, for example, hydrophilic, hydrophobic and hard and soft materials to be obtained. However, this class of polymers was almost forgotten in the 1980s and 1990s because of their long reaction times and limited application possibilities. In the new millennium, a revival of poly(2-oxazoline)s has arisen because of their potential use as biomaterials, among other functionality [342]. Their resemblance to polypeptides is all too apparent.

There are a number of other synthetic mimics of natural type polymers, and as mentioned earlier, the alkyl polyglucosides [171, 309] (see Sections 2.2.6 and 2.6.1) provide another example, this time mimicking a polysaccharide. They have the general structure shown in Figure 2.81.

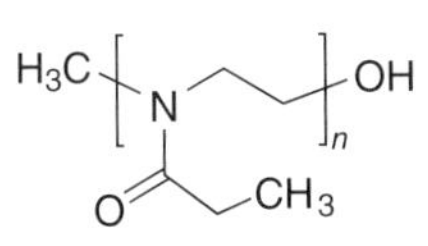

Figure 2.80 Poly(2-ethyl-2-oxazoline).

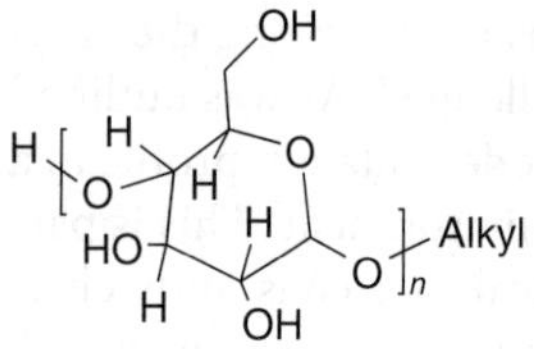

Figure 2.81 Alkyl polyglucosides.

These materials display surface-active properties and are considered in more detail in Chapter 3. Their use in the oil and gas industry has been limited to surface cleaner [343] and in application as a corrosion inhibitor [171]. However, there have been a number of other potential and actual applications explored [344] including drilling fluids, demulsifiers, flocculants and enhanced oil recovery. Given the excellent biodegradation profile couple to their temperature stability, it could be argued that a revisit of these interesting polymers is long overdue.

Polytartaric acid (see Figure 2.82) is a polymer with similarity to vinyl polymers such as polyacrylic acid but with a heteroatom, namely, oxygen, in the backbone. Although this is mimicking a synthetic polymer, it is derived from the monomeric units of natural product origin. Tartaric acid is industrially isolated from lees, a by-product of fermentation. This natural product origin coupled with the manipulation of the polymer backbone to one, which is more biodegradable, should enhance the environmental properties of the polymer.

These types of polymers have been described as scale inhibitors for calcium carbonate with particularly good calcium ion tolerance [345].

The previous examples merely illustrate the possibilities of this approach, particularly in combination with the previously described manipulations, in the design of more environmentally 'friendly' polymers.

2.6.6.2 The Unused Resource: Natural Polymers

The natural world is full of chemical products, which can provide an answer to almost every material and chemical problem in the process of ensuring a biological solution to plant and animal existence. It is a resource that mankind has only begun to understand and exploit. Earlier in the chapter, in Section 2.3, natural polymers were discussed, and numerous examples of use and potential use in the oilfield as chemical additives, with green credentials, described. However there is a huge amount of other products to examine and resources to consider. This will be particularly true if we are to consider alternative chemical feedstocks to our reliance on those derived from crude oil.

As was stated lignin is only slowly biodegradable under the specific conditions where white-rot fungi are present. The structure of lignin is complex and contains many aromatic and related units. It is recognised that this is a potential repository of higher value and useful products, some of which could be of value in the derivation of oilfield chemical additives, and work is ongoing to bio-convert lignin to other derivatives [346, 347].

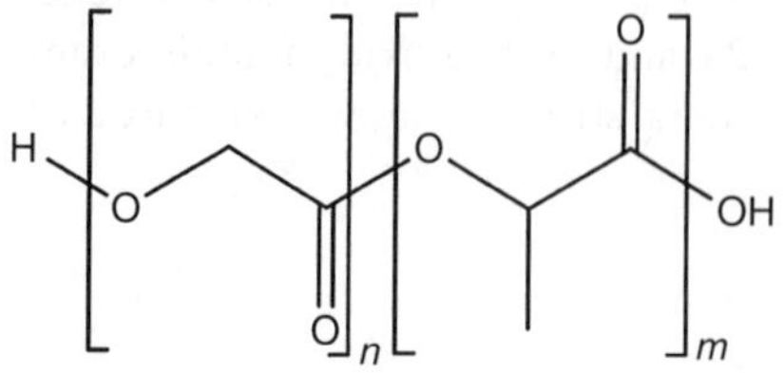

Figure 2.82 A polytartaric acid derivative.

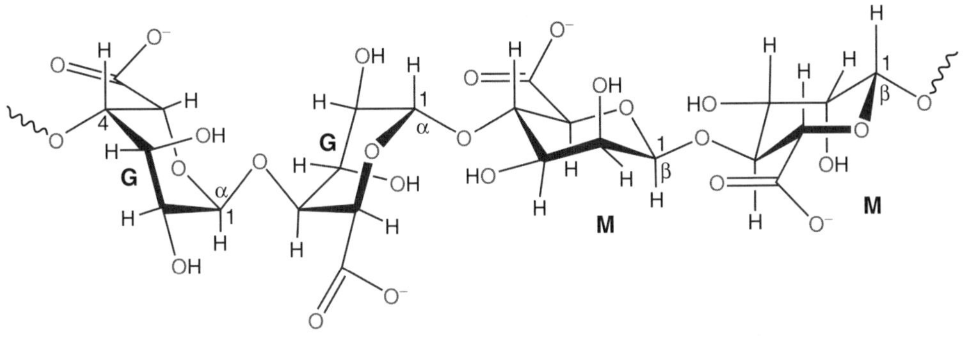

Figure 2.83 Alginate polysaccharide.

Another example would be alginate hydrocolloids (see Figure 2.83), which have been used as suspension agents in drilling fluids [348] but in few other oilfield applications. As can be seen, these have a complex polysaccharide structure and should be useful in a variety of applications. They are a common material in food thickeners and are available from many brown algae including seaweeds.

As was stated regularly in Section 2.3, the natural products are rarely manipulated as is more readily seen in the synthetic polymers to enhance functionality. The author would argue that this is an area where oilfield chemistry has hardly examined the possibilities and other industrial sectors are more advanced [349].

It has been shown that cellulose tristearate, a derivative of microcrystalline cellulose and the natural fatty acid stearic acid, has enhanced temperature stability [350], a useful property in may oilfield applications, and as has been previously described, many cellulosic polymers are employed in oilfield applications.

2.6.6.3 Reversing Biodegradation

In general the natural polymers exhibit good rates of biodegradation; however, this can counteract their efficiency, functionality and stability. This instability can compromise their usefulness in oilfield operations. A common method of affecting and reducing biodegradation in oilfield chemical additives is to formulate a biocidal product with the natural polymer. For example, biocides are often formulated in fracturing fluids especially where guar gum and its derivatives are used as a thickening agent or viscosity modifier. Of course here we have yet added another chemical ingredient to the complex mix of evaluating the overall environmental impact. Indeed we are reversing the natural process of biodegradation by eliminating, or at least reducing, the effectiveness of the microorganisms responsible.

Interestingly and somewhat counter-intuitively, recent work has been reported to reduce the biodegradation of PAM system by synthetic modification [351], as well as for application in enhanced oil recovery. A novel anti-biodegradable hydrophobic acrylamide copolymer has been prepared, and biodegradability testing indicated that unlike analogous copolymers this copolymer was not deemed to be readily biodegradable. Meanwhile the copolymer could significantly enhance the viscosity of the aqueous solution in comparison with partially hydrolysed PAM. This and other properties provide a good foundation for the application in enhanced oil recovery. In addition, the copolymer exhibited potential application for enhanced oil recovery in high-temperature and high-salinity reservoirs.

The question therefore lies in using highly biodegradable natural polymers: is it not also possible to enhance desired functionality and reduce the rate of biodegradation and in doing so open up a new realm of natural polymer chemistry targeted at not only at applications for the oil and gas industry but also at other industrial sectors?

2.6.6.4 Biodegradation of Polymeric Materials

The degradation, and particularly the biodegradation, of the many polymers, especially plastics, has been extensively studied [352, 353]. This has become particularly relevant in order to understand and overcome the increasing environmental problems associated with synthetic polymer and plastic waste. It is however not within the scope of this work to review and discuss this large area of study, which affects all industrial sectors. However a short précis of some of the relevant material to the upstream oil and gas sector is given.

Much focus is given to two main areas of study: the use of biopolymers, that is, biodegradable materials such as polyethers, polyesters, etc. as replacements for non-degradable materials, and understanding the microorganisms that can aid in the biodegradation of a variety of materials, especially plastics. Much has already been described and discussed on the use of biopolymer alternatives in the upstream oil and gas sector especially in regard to their application as chemical additives. The latter work is especially relevant to the oil and gas sector from not only the general impact of the use of polymers as with all other industrial sectors but also on the biodegradation of crude oil [354] in the event of a catastrophic event, leading to a major oil spill. As was illustrated in Figure 2.79, after an initial enzymatic hydrolysis step, particular microorganisms depending on environmental conditions and the polymer substrate degrade the oligomers and other hydrolysed derivatives.

Many polymeric materials such as plastics based on PVC degrade due to weathering. These results in their surface embrittlement and microcracking, yielding microparticles that are carried into water by wind or wave action [98]. Unlike inorganic fines present in seawater, microplastics concentrate POPs by partitioning. The relevant distribution coefficients for common POPs are several orders of magnitude in favour of the plastic medium. Consequently, microparticles laden with high levels of POPs can be ingested by marine biota. Bioavailability and the efficiency of transfer of the ingested POPs across trophic levels are not known, and the potential damage posed by these to the marine ecosystem has yet to be quantified, scientifically modelled and validated. However, given the increasing levels of plastic pollution in the oceans, it is important to better understand the impact of microplastics in the ocean food web.

This section of the chapter has examined the main degradation mechanisms of polymers and polymeric materials and looked at examples pertinent for application in the upstream oil and gas sector. The degradation of polymers is a complex process involving a number of individual mechanisms working simultaneously or in concert. These mechanisms however can be used to the advantage of the oilfield chemist as illustrated later.

To date there has only been one biopolymer considered strong enough for manufacturing downhole tools and related products, which offer degradation characteristics to aid in removal and disposal, that is, PGA (see Figure 2.84), which can degrade reasonably well in temperatures above 210 °F (above 99 °C). Consequently PGA materials are not considered for lower-temperature formations. PGA is a biodegradable, thermoplastic polymer and the simplest linear aliphatic polyester. It has been known since 1954 as a tough fibre-forming polymer.

The year 2015 saw the introduction of materials that offer degradation at lower temperatures [355]. These are degradable composite polymers that will degrade at ambient temperatures in

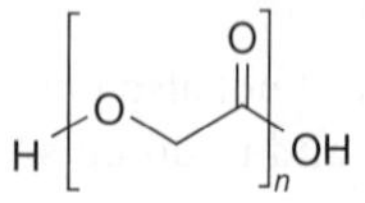

Figure 2.84 Polyglycolic acid.

fresh water. This effectively extends the concept of using a degradable tool that does not need to be retrieved or machined out.

As has been described, polymer degradation can be turned to the advantage of both the practising oilfield chemist and for the reduction of environmental impact.

2.7 Polymer Recycling and Reuse

In achieving a more sustainable approach to the extraction and exploitation of oil and gas resources, it is necessary to consider that the reuse and recycling of all materials and polymers are no exception. It is, however, not within the scope of this study to comprehensively examine the reuse and recycling of polymeric materials of construction, as used in the upstream oil and gas industry and as described earlier in Section 2.4. As was described, many synthetic polymers are used in the industry for a variety of construction and engineering functions [356], and it is in this regard that their durability is a positive factor. This durability, however, from an environmental viewpoint has a major impact on their waste characteristics, particularly for polymers such as plastics and elastomers, including rubber.

There are a number of ways in which the problem of polymer waste is being addressed across all industrial sectors and in domestic use. As has been discussed at length in this chapter, the substitution by biopolymers is a major strategy adopted in the oil and gas industry, in particular with respect to the use of polymer-based chemical additives. The recycling of these additives has also been examined especially with respect to the reuse of contaminated production waters. This has potentially large economic benefits for shale gas recovery through hydraulic fracturing where large volumes of water are used [357].

Similarly produced water is reinjected to provide reservoir support in conventional oil production particularly where conventional water supply is difficult and/or environmental concerns are of paramount [358], as well as having the added advantage of not being disposed of to land outside of the oil reservoir or in an offshore development [359] to the marine environment. Reinjection of produced water is now normal practice in the Norwegian sector of the North Sea [360]. Of course this reinjection is a form of use, and the chemical residuals contained therein could potentially be recycled through the production process and reused. This is a topic that will be returned to later in Chapter 10, where not only polymers are considered but all chemical additives. In this scenario, however, polymer-based materials that are durable have an advantage and are likely to still be present in useful forms. The reuse of chemicals here is challenging and high contentious but nonetheless could lead to a worthwhile economic and environmental benefit.

In considering the recycling and reuse of polymeric materials such as plastic, the processes are still being developed and established, and this is particularly true in the oilfield sector where large-scale decommissioning of installations is only arriving across the next few decades. It is the intention of the industry in the North Sea area to ensure a high recycling and reuse, over 97%, of polymeric materials during the decommissioning process.

In 1988 a coding system for plastic recycling was adopted by the Society of the Plastics Industry and has being widely accepted (see Figure 2.85). The American Society for Testing and Materials (ASTM) issued a more broad-based system, encompassing over 100 polymers and polymer blends, to identify industrial plastics for recycling and disposal. A third system is used exclusively in the automotive industry and is based on ISO abbreviations to identify plastics used in automotive parts so they can be used and recycled. The latter has been highly successful, and many automobiles now use over 95% recycled plastic components.

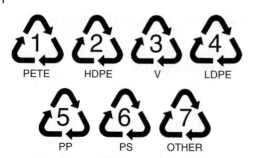

Figure 2.85 Plastic recycling symbols.

Another major strategy in the recycling of many plastics and related materials is to make them partly degradable and allow them to be compostable [361].

This chapter has shown the use of polymers and their diverse applications, particularly in chemical additives, across the upstream oil and gas industry. They will undoubtedly play an important role in the future of hydrocarbon exploitation and development. It has been shown that there is a move towards the use of green alternatives to synthetic polymers, particularly synthetic biopolymers and the use of natural polymers, and there has been an attempt to show there are still a number of areas, particular with regard to the derivation of natural polymers, to be explored. It is hoped that the reader will be challenged to look into these areas further and develop new, exciting and greener chemistries, with useful functionality.

2.8 Sustainable Polymers

A sustainable polymer is a plastic material or other polymer that addresses the needs of consumers or industry without damaging our environment, health and economy. The feedstocks for sustainable plastics are renewable sources such as plants. The production of sustainable polymers should use less net water and non-renewable energy, emit less greenhouse gases and have a smaller carbon footprint than their non-sustainable counterparts while still being economically viable. Other than the remarks on the reuse of produced water, little has been said about sustainability of polymers. However the main thrust of sustainable polymers is to derive them from natural sources. Therefore, in this sense, they have been extensively discussed in the section on natural polymers.

As has been described, traditional synthetic polymers are derived by the conversion of petroleum or natural gas into chemicals (monomers). These chemicals are made into useful polymers. Then after use these polymers can be disposed of, recycled or reused, and even today a significant and growing amount is still disposed of. Also the process is not sustainable as a non-renewable hydrocarbon resource is the primary feedstock. The process of applying sustainable concepts to the oil and gas industry is in its infancy; however the natural polymers (and the natural products) available, particularly from the plant world, offer a vehicle to introducing such a sustainable cycle. The increasing use of these materials will help to counteract modestly the imbalance of carbon footprint that the upstream oil and gas industry brings to bear on the planet's ecosystem.

An example of such a sustainable polymer would be the use of direct plant material from the aloe cactus. An aloe gel, which mainly comprises polysaccharides, has been solubilised in water and has been used a scale inhibitor [262] along with derivatives thereof. Indeed it is postulated that hydrolysis of the aloe gel favours the interaction with calcium and magnesium ions and may increase the scale inhibitor efficiency [362]. This is only one of many plants that could be considered as a more direct source of materials with limited processing and is highly sustainable.

References

1 Craddock, H. (2013). Shale gas: a new European Frontier, yesterday's news in North America – chemicals used, myths and reality. *Chemistry in the Oil Industry XIII* (4–6 November 2013). Manchester, UK: Royal Society of Chemistry.

2 Berzelius, J.J. (1833). *Jahresberichte* **12**: 63.

3 Lourenço, A.V. (1863). Recherches sur les composés polyatomiques. *Annales de Chimie Physique* **67**: 3, 273–279.

4 Williams, J.C.F. (1862). *Journal of the Chemical Society* **15**: 10.

5 Staudinger, H. (1924). *Chemische Berichte* **57**: 1203.

6 Hermes, M.E. (1996). *Enough for One Lifetime: Wallace Carothers, Inventor of Nylon.* American Chemical Society, Chemical Heritage Foundation. ISBN: 0-8412-3331-4.

7 Oil and Gas UK (2015). Economic Report 2015.

8 The RSA Great Recovery & Zero Waste Scotland Programme (2015). North Sea Oil and Gas Rig Decommissioning & Re-use Opportunity Report, Innovate UK, 2015.

9 Mathur, N.K. (2012). *Industrial Galactomannan Polysaccharides.* CRC Press.

10 Coleman, M.M. and Painter, P.C. (1997). *Fundamentals of Polymer Science*, 2nde. CRC Press.

11 Burger, E.D., Munk, W.R., and Wahi, H.A. (1982). Flow increase in the Trans Alaska pipeline through the use of a polymeric drag reducing additives, SPE 9419-PA. *Journal of Petroleum Technology* **34** (02): 377–386.

12 Aubanel, M.L. and Bailly, J.C. (1987). Amorphous high molecular weight copolymers of ethylene and alpha-olefins. EP Patent 243127, assigned to BP Chemicals Ltd.

13 Gessell, D.E. and Washecheck, P.H. (1990). Composition and method for friction loss reduction. US Patent 4,952,738, assigned Conoco Inc.

14 Smith, K.W., Haynes, L.V. and Massouda, D.F. (1995). Solvent free oil soluble drag reducing polymer suspension. US Patent 5,449,732A, assigned to Conoco Inc.

15 Dindi, A., Johnston, R.L., Lee, Y.N. and Massouda, D.F. (1996). Slurry drag reducer. US Patent 5,539,044, assigned to Conoco Inc.

16 Rossi, A., Chandler, J.E. and Barbour, R. (1993). Polymers and additive compositions. WO Patent 9319106, assigned to Exon Chemical Patents Inc.

17 Lescarboura, J.A., Culter, J.D., and Wahi, H.A. (1971). Drag reduction with a polymeric additive in crude oil pipelines, SPE 3087. *Society of Petroleum Engineers Journal* **11** (3).

18 Johnston, R.L. and Milligan, S.N. (2003). Polymer compositions useful as flow improvers in cold fluids. US Patent 6,596,832.

19 Milligan, S.N. and Smith, K.W. (2003). Drag-reducing polymers and drag-reducing polymer suspensions and solutions. US Patent 6,576,732.

20 Hunston, D.L. (1976). Effects of molecular weight distribution in drag reduction and shear degradation. *Journal of Polymer Science, Polymer Chemistry Edition* **14**: 713.

21 Kaplan, D.L., Hartenstein, R., and Sutter, J. (1979). Biodegradation of polystyrene, poly(methyl methacrylate), and phenol formaldehyde. *Applied and Environmental Microbiology* **38** (3): 551–553.

22 Otake, Y., Kobayashi, T., Asabe, H. et al. (1995). Biodegradation of low-density polyethylene, polystyrene, polyvinyl chloride, and urea formaldehyde resin buried under soil for over 32 years. *Journal of Applied Polymer Science* **56** (13): 1789–1796.

23 Myajima, Y., Kariyazono, Y., Funatsu, S., et al. (1994). Durability of polyethylene-coated steel pipe at elevated temperatures. Nippon Steel Technical Report No. 63, October 1994.

24 Arai, T. and Ohkita, M. (1989). Application of polypropylene coating system to pipeline for high temperature service. Internal and External Protection of Pipes – Proceedings of the 8th International Conference, Florence, Italy (24–26 October 1989), pp. 189–201.

25 Larson, R.J., Bookland, E.A., Williams, R.T. et al. (1997). Biodegradation of acrylic acid polymers and oligomers by mixed microbial communities in activated sludge. *Journal of Environmental Polymer Degradation* **5** (1): 41–48.

26 Monfreux, N., Perrin, P., Lafuma, F. and Sawdon, C. (2000. Invertible emulsions stabilised by amphiphilic polymers and applications to bore fluids. WO Patent 0031154.

27 http://www.glossary.oilfield.slb.com/en/Terms/i/invert_emulsion_oil_mud.aspx (accessed 7 December 2017).

28 Chin, W.C. (1995). *Formation Invasion: With Applications to Measurement while Drilling, Time lapse Analysis and Formation Damage.* Gulf Publishing Co.

29 Huddleston, D.A. and Williamson, C.D. (1990). Vinyl grafted lignite fluid loss additives. US Patent 4,938,803, assigned to Nalco Chemical Company.

30 Chueng, P.S.R. (1993). Fluid loss additives for cementing compositions. US Patent 5,217,531, assigned to The Western Company of North America.

31 Jones, T.G. J. and Tustin, G.J. (1998). Gelling composition for wellbore service fluids. US Patent 6,194,356, assigned to Schlumberger Technology Corporation.

32 Hogan, C., Davies, H., Robins, L., et al. (2013). Improved scale control, hydrothermal and environmental properties with new maleic polymer chemistry suitable for downhole squeeze and topside applications. *Chemistry in the Oil Industry XIII* (4–6 November 2013). Manchester, UK: Royal Society of Chemistry.

33 Jada, A., Ait Akbour, R., Jacquemet, C. et al. (2007). Effect of sodium polyacrylate molecular weight on the crystallogenesis of calcium carbonate. *Journal of Crystal Growth* **306** (2): 373–382.

34 LoSasso, J.E. (2001). Low molecular weight structured polymers. US Patent 6,322,708.

35 Gancet, C., Pirri, R., Boutevin, B., et al. (2005). Polyacrylates with improved biodegradability. US Patent 6,900,171, assigned to Arkema.

36 Rodrigues, K.A. (2010). Hydrophobically modified polymers. US Patent Application 20100012885, assigned to Akzo Nobel N.V.

37 Costello, C.A. and Matz, G.F. (1984). Use of a carboxylic functional polyampholyte to inhibit the precipitation and deposit of scale in aqueous systems. US Patent 4,466,477, assigned to Calgon Corporation.

38 Lindeman, O.E. and Allenson, S.J. (2005). Theoretical modeling of tertiary structure of paraffin inhibitors, SPE 93090. SPE International Symposium on Oilfield Chemistry, The Woodlands, TX (2–4 February 2005).

39 Van der Meij, P. and Buitelaar, A. (1971). Polyalkylmethacrylates as pour point depressants for lubricating oils. US Patent 3,598,736, assigned to Shell Oil Co.

40 Barthell, E., Capelle, A., Chmelir, M. and Danmen, K. (1987). Copolymers of n-alkyl acrylates and maleic anhydride and their use as crystallization inhibitors for paraffin-bearing crude oils. US Patent 4,663,491, assigned to Chemische Fabrik Stockhausen Gmbh.

41 Reich, L. and Stivala, S.S. (1971). *Elements of Polymer Degradation.* New York: McGraw-Hill.

42 Singhal, H.K., Sahai, G.C., Pundeer, G.S. and Chandra, L. (1991). Designing and selecting wax crystal modifiers for optimum performance based on crude oil composition, SPE 22784. Presented at SPE Annual Technical Conference and Exhibition, Dallas, TX (6–9 October 1991).

43 Duffy, D.M., Moon, C., Irwin, J.L., et al. (2003). Computer assisted design of additives to control deposition from oils. *Chemistry in the Oil Industry VIII* (3–5 November 2003). Manchester, UK: Royal Society of Chemistry.

44 Wirtz, H., Von Halasz, S.-P., Feustal, M. and Balzer, J. (1994). New copolymers, mixtures thereof with poly(meth)acrylate esters and the use thereof for improving the cold fluidity of crude oils. US Patent 5,349,019, assigned to Hoechst AG.

45 Meunier, G., Brouard, R., Damin, B. and Lopez, D. (1986). Grafted ethylene polymers usable more especially as additives for inhibiting the deposition of paraffins in crude oils and compositions containing the oils and said additives. US Patent 4,608,411, assigned to Elf Aquitaine.

46 Dumitriu, S. (2001). *Polymeric Biomaterials, Revised and Expanded*. CRC Press.

47 Brunelli, J.-F. and Fouquay, S. (2001). Acrylic copolymers as additives for inhibiting paraffin deposition in crude oils, and compositions containing same. US Patent 6,218,490, assigned to Ceca SA.

48 Shmadova-Lindeman, O.E. (2005). Paraffin inhibitors. US Patent Application 20050215437.

49 Mueller, M. and Gruenig, H. (1994). Method for improving the pour point of petroleum oils. US Patent 5,281,329, assigned to Rohm Gbmh.

50 Malik, S., Shintre, S.N. and Mashelkar, R.A. (1992). Process for the preparation of a new polymer useful for drag reduction in hydrocarbon fluids in exceptionally dilute polymer solutions. US Patent 5,080,121, assigned to Council of Scientific and Industrial Research.

51 Farley, D.E. (1975). Drag reduction in non-aqueous solutions: structure–property correlations for poly(isodecyl methacrylate). SPE 5308 SPE Oilfield Chemistry Symposium, Dallas, TX (13–14 January 1975).

52 Yu, P., Li, C., Zhang, C. et al. (2011). Drag reduction and shear resistance properties of ionomer and hydrogen bond systems based on lauryl methacrylate. *Petroleum Science* **8** (3): 357–364.

53 Russom, C.L., Drummond, R.A., and Huffman, A.D. (1988). Acute toxicity and behavioural effects of acrylates and methacrylates to juvenile fathead minnows. *Bulletin of Environmental Contamination and Toxicology* **41** (4): 589–596.

54 Schultz, D.N., Kitano, K., Burkhardt, T.J. and Langer, A.W. (1985). Drag reduction agent for hydrocarbon liquid. US Patent 4,518,757, assigned to Exxon Research and Engineering Co.

55 Taylor, G.N. (1997). Demulsifier for water-in-oil emulsions, and method of use. US Patent 5,609,794, assigned to Exxon Chemical Patents Inc.

56 Ralston, A.W., Hoerer, C.W., and Poll, W.O. (1943). Solubilities of some normal aliphatic amides, anilides and N,N-diphenylamides. *The Journal of Organic Chemistry* **8** (5): 473–488.

57 Wu, S. and Shanks, R.A. (2004). Solubility study of polyacrylamide in polar solvents. *Journal of Applied Polymer Science* **93** (3): 1493–1499.

58 Pearsce, M.J., Weir, S., Adkins, S.J., and Moody, G.M. (2001). Advances in mineral flocculation. *Minerals Engineering* **14** (11): 1505–1511.

59 Mortimer, D.A. (1991). Synthetic polyelectrolytes – a review. *Polymer International* **25** (1): 29–42.

60 Kelland, M.A. (2014). *Production Chemicals for the Oil and Gas Industry*, 2nde. CRC Press.

61 Putz, A.G., Bazin, B. and Pedron, B.M. (1994). Commercial polymer injection in the courtenay field, 1994 update, SPE 28601. SPE Annual Technical Conference and Exhibition, New Orleans, Louisiana (25–28 September 1994).

62 Gao, J.P., Lin, T., Wang, W. et al. (1999). Accelerated chemical degradation of polyacrylamide. *Macromolecular Symposia* **144**: 179–185.

63 Reich, L. and Stivala, S.S. (1971). *Elements of Polymer Degradation*. McGraw-Hill, Inc.

64 Al-Muntasheri, G.A., Nasr-El-Din, H.A., and Zitha, P.L.J. (2008). Gelation kinetics and performance evaluation of an organically crosslinked gel at high temperature and pressure, SPE 104071. *SPE Journal* **13** (3).

65 Fisler, R. (1986). Chromium hazards to fish, wildlife and invertebrates: a synoptic review. Biological Report 85(1.6) Contaminant Hazard Reviews, Report No. 6, US Fish and Wildlife Service.

66 Puddu, A., Pettine, M., La Noce, T. et al. (1988). Factors affecting thallium and chromium toxicity to marine algae. *Science of the Total Environment* **71** (3): 572.

67 Smith, E.A., Prues, S.L., and Oehm, F.W. (1997). Environmental degradation of polyacrylamides II, effects of environmental (outdoor) exposure. *Ecotoxicology and Environmental Safety* **37** (1): 76–91.

68 James, S.G., Nelson, G.B. and Guinot, F.J. (2002). Sand consolidation with flexible gel system. US Patent 6,450,260, assigned to Schlumberger Technology Corporation.

69 Tate, J. (1973). Secondary recovery process. US Patent 3,749,169, assigned to Texaco Inc.

70 Thach, S. (1996). Method for altering flow profile of a subterranean formation during acid stimulation. US Patent 5,529,122, assigned to Atlantic Richfield Company.

71 Reddy, B.R. and Riley, W.D. (2004). High temperature viscosifying and fluid loss controlling additives for well cements, well cement compositions and methods. US Patent 6,770,604, assigned to Halliburton Energy Services Inc.

72 Hille, M., Wittkus, H., Tonhauser, J., et al. (1996). Water soluble co-polymers useful in drilling fluids. US Patent 5,510,436, assigned to Hoechst AG.

73 Amjad, Z. ed. (1998). *Water Soluble Polymers: Solution Properties and Applications*. Springer.

74 Tielong, C., Yong, Z., Kezong, P., and Wanfeng, P. (1996). A relative permeability modifier for water control of gas wells in a low-permeability reservoir, SPE 35617. *SPE Reservoir Engineering* **11** (3): doi: 10.2118/35617-PA.

75 Lukocs, B., Mesher, S., Wilson, T.P., et al. (2007). Non-volatile phosphorus hydrocarbon gelling agent. US Patent Application 20070173413, assigned to Clearwater International LLC.

76 Al-Sharkhi, A. and Hanratty, T.J. (2002). Effect of drag-reducing polymers on pseudo-slugs, interfacial drag and transition to slug flow. *International Journal of Multiphase Flow* **28**: 1911–1927.

77 Hoyt, J.W. and Sellin, R.H.J. (1991). Polymer "threads" and drag reduction. *Rheologica Acta* **30** (4): 307–315.

78 Rivers, G.T. and Crosby, G.L. (2004). Method for inhibiting hydrate formation. WO Patent Application WO2004022910, assigned to Baker Hughes Inc.

79 Fong, D. and Khambatta, B.S. (1994). Hydroxamic acid containing polymers used as corrosion inhibitor. US Patent 5,308,498, assigned to Nalco Chemical Company.

80 Machado, A.L.C., Lucus, E.F., and Gonzalez, G. (2001). Poly(ethylene-co-vinyl acetate) (EVA) as wax inhibitor of a Brazilian crude oil: oil viscosity, pour point and phase behavior of organic solutions. *Journal of Petroleum Science and Engineering* **32** (2-4): 159–165.

81 Dorer, C.J. and Hayashi, K. (1986). Hydrocarbyl substituted carboxylic acylating agent derivative containing combinations, and fuels containing same. US Patent 4,623,684, assigned to the Lubrizol Corporation.

82 McDougall, L.A., Rossie, A. and Wisotsky, M.J. (1972). Crude oil recovery method using a polymeric wax inhibitor. US Patent 3,693,720, assigned to Exxon Research Engineering Co.

83 Marie, E., Chevalier, Y., Eydoux, F. et al. (2005). Control of n-alkanes crystallisation by ethylene-vinyl acetate copolymers. *Journal of Colloid and Interface Science* **209** (2): 406–418.

84 Amman, M. and Minge, O. (2011). Biodegradability of poly(vinyl acetate) and related polymers. In: *Synthetic Biodegradable Polymers*, Advances in Polymer Science, vol. **245**, 137–172. Berlin: Springer-Verlag.

85 Potts, J.E., Clendinning, R.A., Ackart, W.B., and Niegisch, W.D. (1973). The biodegradability of synthetic polymers. In: *Polymers and Ecological Problems*, Polymer Science and Technology, vol. **3** (ed. J. Guillet), 61–79. Boston, MA: Springer.

86 Chellini, E. and Solaro, R. (1996). Biodegradable polymeric materials: a review. *Advanced Materials* **8** (4): 305–3132.

87 Audibert, A., Rousseau, L. and Kieffer, J. (1999). Novel high pressure/high temperature fluid loss reducer for water based applications, SPE 50724. SPE International Symposium on Oilfield Chemistry, Houston, TX (16–17 February 1999).

88 Moran, L.K. and Thomas, T.R. (1991). Well cement fluid loss additive and method. US Patent 5,009,269, assigned to Conoco Inc.

89 Alford, S.E. (1991). North sea application of an environmentally responsible water-base shale stabilizing system, SPE 21936. SPE/AIDC Drilling Conference, Amsterdam (11–14 March 1991).

90 Plank, J., Recalde, N., Dugoniic-Bilic, F., and Sadasivan, D. (2009). Comparative study of the working mechanisms of different cement fluid loss polymers, SPE 121542. SPE International Symposium on Oilfield Chemistry, The Woodlands, TX (20–22 April 2009).

91 Chiellini, E., Corti, A., D'Antonne, S., and Solaro, R. (2003). Biodegradation of poly(vinyl alcohol) based materials. *Progress in Polymer Science* **28** (6): 963–1014.

92 Kabir, A.H. (2001). Chemical water and gas shutoff technology – an overview, SPE 72119. SPE Asia Pacific Improved Oil recovery Conference, Kuala Lumpur, Malaysia (6–9 October 2001).

93 Audebert, R., Janca, J., Maroy, P. and Hendriks, H. (1996). New, chemically crosslinked polyvinyl alcohol (PVA), process for synthesizing same and its applications as fluid loss control agent in oil fluids. CA Patent 2,118,070, assigned to Schlumberger Ca Ltd.

94 Oh-Kil, K. and Ling Sui, C. (1996). *Drag Reducing Polymers in The Polymeric Materials Encyclopedia* (ed. J.C. Salome). CRC Press.

95 Fernandez, R.S., Gonzalez, G., and Lucas, E.F. (2005). Assessment of polymeric flocculants in oily water systems. *Colloid and Polymer Science* **283** (4): 375–382.

96 Growcock, F.B. and Simon, G.A. (2006). Stabilised colloidal and colloidal like systems. US Patent 7,037,881, assigned to Authors.

97 Kaczmarek, H. and Bajer, K. (2007). Biodegradation of plasticized poly(vinyl chloride) containing cellulose. *Journal of Polymer Science Part B: Polymer Physics* **45** (8): 903–919.

98 Andrady, A.L. (2011). Microplastics in the marine environment. *Marine Pollution Bulletin* **62** (8): 1596–1605.

99 Boles, J.L. and Boles, J.B. (1998). Cementing compositions and methods using recycled expanded polystyrene. US Patent 5,736,594, assigned to BJ Services Co.

100 Yang, Y., Yang, J., Wu, W.M. et al. (2015). Biodegradation and mineralization of polystyrene by plastic-eating mealworms: Part 2. role of gut microorganisms. *Environmental Science & Technology* **49** (20): 12087–12093.

101 Pushnadass, H.A., Weber, R.W., Dumais, J.J., and Hanna, M.A. (2010). Biodegradation characteristics of starch–polystyrene loose-fill foams in a composting medium. *Bioresource Technology* **10** (19): 7258–7264.

102 Mooney, A., Ward, P.G., and O'Connor, K.E. (2006). Microbial degradation of styrene: biochemistry, molecular genetics, and perspectives for biotechnological applications. *Applied Microbiology and Biotechnology* **72**: 1–10.

103 Kiatkamjornwong, S., Sonsuk, M., Wiitayapichet, S. et al. (1999). Degradation of styrene-g-cassava starch filled polystyrene plastics. *Polymer Degradation and Stability* **66** (3): 323–335.

104 Burback, B.L. and Perry, J.J. (1993). Biodegradation and biotransformation of groundwater pollutant mixtures by *Mycobacterium vaccae*. *Applied and Environmental Microbiology* **59** (4): 1025–1029.

105 Pirri, R., Hurtevent, C. and Leconte, P. (2000). New scale inhibitor for harsh field conditions, SPE 60218. International Symposium on Oilfield Scale, Aberdeen, UK (26–27 January 2000).

106 Wat, R., Hauge, L.-E. Solbakken, K., et al. (2007). Squeeze chemical for HT applications – Have we discarded promising products by performing unrepresentative thermal aging tests? SPE 105505. International Symposium on Oilfield Chemistry, Houston, TX (28 February––2 March 2007).

107 Moran, L.K. and Moran, L.L. (1998). Composition and method to control cement slurry loss and viscosity. US Patent 5,850,880, assigned to Conoco Inc.

108 Sloan, D. ed. (2011). *Natural Gas Hydrates in Flow Assurance*. Elsevier.

109 Luvicap® Bio is the first biodegradable KHI (OECD 306 >50% in 28 days) Y1 registered in Norway. http://www.oilfield-solutions.basf.com/ev/internet/oilfield-solutions/en_GB/applications/production/low-dose-gas (accessed 7 December 2017)

110 Malchow, G.A. Jr. (1997). Friction modifier for water-based well drilling fluids and methods of using the same. US Patent 5,593,954, assigned to Lubrizol Corporation.

111 Little, R.C. and Weigard, M. (1970). Drag reduction and structural turbulence in flowing polyox solutions. *Journal of Applied Polymer Science* **14** (2): 409.

112 Lundan, A.O., Anaas, P.-H.V. and Lahteenmaki, M.J. (1996). Stable CMC slurry. US Patent 5,487,777, assigned to Metsa-Serla Chemicals Oy.

113 Dixon, J. (2009). Drilling fluids. US Patent 7,614,462, assigned to Croda International Plc.

114 McCoy, D.R., McEntire, E.E. and Gipson, R.M. (1987). Demulsification of bitumen emulsions. CA Patent 1,225,003, assigned to Texaco Development Corp.

115 Schlaad, H. and Hoogenboom, R. ed. (2012). Special Issue: Poly(2-oxazoline)s and related pseudo-polypeptides. *Macromolecular Rapid Communications* **33** (19): 1599.

116 Ulbricht, J., Jordan, R., and Luxenhofer, R. (2014). On biodegradability of polyethylene glycol, polypeptides and poly(2-oxazoline)s. *Biomaterials* **35** (17): 4848–4861.

117 Toenjes, A.A., Williams, M.R. and Goad, E.A. (1992). Demulsifier compositions and demulsifying use thereof. US Patent 5,102,580, assigned to Petrolite Corporation.

118 Hale, A.H. and van Oort, E. (1997). Efficiency of ethoxylated/propoxylated polyols with other additives to remove water from shale. US Patent 5,602,082, assigned to Shell Oil Company.

119 Ainley, B.R. and McConnell, S.B. (1993). Delayed borate crosslinked fracturing fluid. EP Patent 528461, assigned to Pumptech N.V. and Dowell Schlumberger S.A.

120 Zhong, H., Oiu, Z., Huang, W., and Cao, J. (2011). Shale inhibitive properties of polyether diamine in water-based drilling fluid. *Journal of Petroleum Science and Engineering* **78** (2): 510–515.

121 Patel, A.D., Stamatakis, E., Davis, E. and Friedheim, J. (2007). High performance water based drilling fluids and method of use. US Patent 7,250,390, assigned to MI LLC.

122 Pakulski, M. and Hurd, D. (2005). Uncovering a dual nature of polyether amines hydrate inhibitors. 5th International Conference on Gas Hydrates, Trondheim, Norway (13–16 June 2005), pp. 1401–1408.

123 Bakeev, K., Myers, R. and Graham, D.E. (2001) Blend for preventing or retarding the formation of gas hydrates. US Patent 6180699, assigned to ISP Investments Inc.

124 Blytas, G.C., Frank, H., Zuzich, A.H. and Holloway, E.L. (1992). Method of preparing polyethercycylicpolyols. EP Patent 505000, assigned to Shell International Research Maatschappij B.V.

125 Zychal, C. (1986). Defoamer and antifoamer composition and method for defoaming aqueous fluid systems. US Patent 4631145, assigned to Amoco Corporation.

126 Davidson, E. (1995). Defoamers. WO Patent 1995009900, assigned to ICI Plc.

127 Plank, J. (1993). Drilling mud composition and process for reducing the filtrate of metal hydroxide mixtures-containing drilling mud compositions WO Patent 1993012194, assigned to Sueddeutsche Kalkstickstoff.

128 Kawaki, F. (1990). Biodegration of polyethers. In: *Agricultural and Synthetic Polymers*, ACS Symposium Series, vol. **433**, Chapter 10, 110–123. American Chemical Society.

129 Todd, B.L., Slabaugh, B.F., Munoz, T. Jr. and Parker, M.A. (2006). Fluid loss control additives for use in fracturing subterranean formations. US Patent 7,096,947, assigned to Halliburton Energy Services Inc.

130 Reddy, B.R. and Liang, F. (2015). Solids-free diverting agents and methods related thereto. WO Patent 2015060813, assigned to Halliburton Energy Services Inc.

131 Kelland, M.A. (2006). History and development of low dosage hydrate inhibitors. *Energy and Fuels* **20** (3): 825–847.

132 Feustal, M. and Lienweber, D. (2009). Pyroglutamic acid esters with improved biodegradability. US Patent Application 20090124786.

133 Rivers, G.T. and Crosby, D.L. (2007). Gas hydrate inhibitors. US Patent 7,164,051, assigned to Baker Hughes Incorporated.

134 Pakulski, M.K. (2011). Development of superior hybrid gas hydrate inhibitors, OTC 21747. Offshore Technology Conference, Houston, TX (2–5 May 2011).

135 Pakulski, M.K. (2000) Quaternized polyether amines as gas hydrate inhibitors. US Patent 6,025,302, assigned to BJ Services Company.

136 Kelland, M.A., Svartaas, T.M., and Andersen, L.D. (2009). Gas hydrate anti-agglomerant properties of polypropoxylates and some other demulsifiers. *Journal of Petroleum Science and Engineering* **64** (1): 1–10.

137 Alsabagh, A.M., Migahed, M.A., and Awad, H.S. (2006). Reactivity of polyester aliphatic amine surfactants as corrosion inhibitors for carbon steel in formation water (deep well water). *Corrosion Science* **48** (4): 813–828.

138 Pou, T.E. and Fouquay, S. (2002). Polymethylenepolyamine dipropionamides as environmentally safe inhibitors of the carbon corrosion of iron. US Patent 6,365,100, assigned to Ceca S.A.

139 Treybig, D.S., Changand, K.-T. and Williams, D.A. (2009). Demulsifiers, their preparation and use in oil bearing formations. US Patent 7,504,438, assigned to Nalco Company.

140 Xu, Y., Wu, J., Dabros, T. et al. (2005). Optimizing the polyethylene oxide and polypropylene oxide contents in diethylenetriamine-based surfactants for destabilization of a water-in-oil emulsion. *Energy & Fuels* **19** (3): 916–992.

141 Eifers, G., Sager, W., Vogel, H.-H. and Oppenlaender, K. (1995). Oil-demulsifiers based on an alkoxylate and preparation of this alkoxylate. US Patent 5,401,439, assigned to BASF AG.

142 Wang, C., Fang, S., Duan, M. et al. (2015). Synthesis and evaluation of demulsifiers with polyethyleneimine as accepter for treating crude oil emulsions. *Polymers for Advanced Technologies* **26**: 442–448.

143 Yu, H., Steichen, D.S., James, A.D., et al. (2011). Polyamide emulsifier based on polyamines and fatty acid/carboxylic acid for oil based drilling fluid applications. US Patent Application 20110306523.

144 Bachrach, U. and Heimer, Y.M. (1989). *The Physiology of Polyamines*, vol. **I**. CRC Press.

145 Leonard, G.C., Rivers, G.T., Asomaning, S. and Breen, P.J. (2013) Asphaltene inhibitors for squeeze application. US Patent Application 20130186629.

146 Berger, P.D., Hsu, C., and Aredell, J.P. (1988). Designing and selecting demulsifiers for optimum filed performance on the basis of production fluid characteristics, SPE 16285. *SPE Production Engineering* **3** (6): 522.

147 Gutsche, C.D. (2008). *Calixarenes: An Introduction*, Monographs in Supramolecular Chemistry. RSC Publishing.

148 Stais, F., Bohm, R., and Kupfer, R. (1991). Improved demulsifier chemistry: a novel approach in the dehydration of crude oil, SPE 18481. *SPE Production Engineering* **6** (3): 334.

149 Jaques, P., Martin, I., Newbigging, C., and Wardell, T. (2002). Alkylphenol based demulsifier resins and their continued use in the offshore oil and gas industry. In: *Chemistry in the Oil Industry VII* (ed. T. Balson, H. Craddock, J. Dunlop, et al.), 56–66. The Royal Society of Chemistry.

150 March, J. (1977). *Advanced Organic Chemistry: Reactions, Mechanisms and Structure*, 2nde. McGraw-Hill.

151 Jenkins, A.D., Kratochvíl, P., Stepto, R.F.T., and Suter, U.W. (1996). Glossary of basic terms in polymer science (IUPAC Recommendations 1996). *Pure and Applied Chemistry* **68** (12): 2287–2311.

152 Frenzel, S., Assmann, A. and Reichenbach-Kliunke, R. (2009). Green polymers for the North Sea-biodegradability as key towards environmentally friendly chemistry for the oilfield industry. Chemistry in the Oil Industry XI, Manchester, UK (pp. 259–270).

153 Feng, C., Li, Y., Yang, D. et al. (2011). Well-defined graft copolymers: from controlled synthesis to multipurpose applications. *Chemical Society Reviews* **40** (3): 1282–1295.

154 Krul, L.P. (1986). Thermal analysis of polyethylene graft copolymers. *Thermochimica Acta* **97**: 357–361.

155 Holt, S. and Sanders, J. (2009). A technology platform for designing, high performance environmentally benign scale inhibitors for a range of application needs. Chemistry in the Oil Industry XI, Manchester, UK (2–4 November 2009), pp. 103–121.

156 Bellas, V. and Rehahn, M. (2009). Block copolymer synthesis via chemoselective stepwise coupling reactions. *Macromolecular Chemistry and Physics* **210** (5): 320–330.

157 Del Villano, L., Kommedal, R., Fijten, M.W.M. et al. (2009). A study of the kinetic hydrate inhibitor performance and seawater biodegradability of a series of poly(2-alkyl-2-oxazolidine) s. *Energy and Fuels* **23**: 3665–3673.

158 Kelland, M.A., Private Communication, 2016.

159 Hellberg, P.-E. (2007). Environmentally adapted demulsifiers containing weak links. Chemistry in the Oil Industry X, Manchester, UK (5–7 November 2007), pp. 215–228.

160 OECD (1992). OECD Guidelines for the Testing of Chemicals, Section 3: Degradation and Accumulation, Test No. 306: Biodegradability in Seawater.

161 Balson, T. (1998). The unique chemistry of polyglycols. In: *Chemistry in the Oil Industry – Recent Developments* (ed. L. Cookson and P.H. Ogden), 71–79. RSC Publishing.

162 Hellberg, P.-E. (2013). Polymeric corrosion inhibitors – a new class of versatile oilfield formulation bases. Chemistry in the Oil Industry XIII, Manchester, UK (4–6 November 2013), pp. 84-110.

163 Speicher, M.R. and Carter, N.P. (2005). The new cytogenetics: blurring the boundaries with molecular biology. *Nature Reviews Genetics* **6**: 782–792.

164 Price, A., Acuna Alvarez, L., Whitby, C., and Larsen, J. (2009). How many microorganisms present? Quantitative reverse transcription PCR (qtr.-PCR). In: *Proceedings from the International Symposium on Applied Microbiology and Molecular Biology in Oil Systems (IMOS-2)* (ed. C. Whitby and T.L. Skovhus). Springer.

165 Vroman, I. and Tighzert, L. (2009). Review: Biodegradable polymers. *Materials* **2**: 307–344.

166 Termonia, Y., Meakin, P., and Smith, P. (1985). Theoretical study of the influence of the molecular weight on the maximum tensile strength of polymer fibers. *Macromolecules* **18** (11): 2246–2252.

167 Xu, Q., Hashimoto, M., Dang, T.T. et al. (2009). Preparation of monodisperse biodegradable polymer microparticles using a microfluidic flow-focusing device for controlled drug delivery. *Small* **5**: 1575–1581.

168 Kalota, D.J. and Silverman, D.C. (February 1994). Behavior of aspartic acid as a corrosion inhibitor for steel. *Corrosion* **50** (2): 138–145.

169 Fan, L.-D.G., Fan, J.C. and Bain, D. (1999). Scale and corrosion inhibition by thermal polyaspartates, Paper 99120. NACE Corrosion 99, San Antonio, TX (25–30 April 1999).

170 Fan, J.C. and Fan, L.-D.G. (2001). Inhibition of metal corrosion. US Patent 6,277,302, assigned to Donlar Corporation.

171 Craddock, H.A., Caird, S., Wilkinson, H. and Guzzmann, M. (2006). A new class of 'green' corrosion inhibitors, development and application, SPE 104241. SPE International Oilfield corrosion Symposium, Aberdeen, UK (30 May 2006).

172 Craddock, H.A., Berry, P. and Wilkinson, H. (2007). New class of "green" corrosion inhibitors, further development and application. Transactions of the 18th International Oilfield Chemical Symposium, Geilo, Norway (25–28 March 2007).

173 Morse, J.W., Arvidson, R.S., and Lutte, A. (2007). *Chemical Reviews* **107**: 342.

174 Tang, J. and Davis, R.V. (1998). Use of biodegradable polymers in preventing scale build-up. US Patent 5,776,875, assigned to Nalco Chemical Company.

175 Fan, L.-D.G., Fan, J.C., Liu, Q.W., and Reyes, H. (2001). Thermal polyaspartates as dual function corrosion and mineral scale inhibitors. *Polymeric Materials Science and Engineering* **84**: 426–427.

176 Jordan, M.M., Feasey, N.D., Budge, M. and Robb, M. (2006). Development and deployment of improved performance "green" combined scale/corrosion inhibitor for subsea and topside application, North Sea basin, SPE 100355. SPE International Oilfield Corrosion Symposium, Aberdeen, UK (30 May 2006).

177 Low, K.C., Wheeler, A.P., and Koskan, L.P. (2009). Commercial poly(aspartic acid) and its uses. In: *Hydrophilic Polymers, Advances in Chemistry*, vol. **248**, Chapter 6 (ed. J.E. Glass), 99–111. American Chemical Society.

178 Boehmke, G. and Schmitz, G. (1995). Process for the preparation of polysuccinimide, polyaspartic acid and their salts. US Patent 5,468,838, assigned to Bayer AG.

179 Chua, P.C., Sæbø, M., Lunde, A. and Kelland, M.A. (2011). Dual kinetic hydrate and scale inhibition by polyaspartamides. Proceedings of the 7th International Conference on Gas Hydrates (ICGH 2011), Edinburgh, Scotland, UK (17–21 July 2011).

180 Craparo, E.F., Porsio, B., Bondì, M.L. et al. (September 2015). Evaluation of biodegradability on polyaspartamide-polylactic acid based nanoparticles by chemical hydrolysis studies. *Polymer Degradation and Stability* **119**: 56–56.

181 Giammona, G., Pitarresi, G., Cavallaro, G. et al. (1999). New biodegradable hydrogels based on a photocrosslinkable modified polyaspartamide: synthesis and characterization. *Biochimica et Biophysica Acta* **1428** (1): 29–38.

182 del Villano, L., Kommedal, R., and Kelland, M.A. (2008). Class of kinetic hydrate inhibitors with good biodegradability. *Energy & Fuels* **22** (5): 3143–3149.

183 Shih, I.-L. and Van, Y.-T. (2001). The production of poly(γ-glutamic acid) from microorganisms and its various applications. *Bioresource Technology* **79** (3): 207–225.

184 Kubota, H., Matsunobu, T., Uotani, K. et al. (1993). Production of poly(γ-glutamic acid)by Bacillus subtilis F-2-01. *Bioscience, Biotechnology, and Biochemistry* **57** (7): 1212–1213.

185 Yokoi, H., Arima, T., Hirose, J. et al. (1996). *Journal of Fermentation and Bioengineering* **82**: 84.

186 Kotlar, H.K. and Chen, P. (2009). Well treatment for sand containing formations. CA Patent 2, 569810, assigned to Statoil Asa.

187 Qiang, X., Sheng, Z., and Zhang, H. (2013). Study on scale inhibition performances and interaction mechanism of modified collagen. *Desalination* **309**: 237–242.

188 Schmitt, G. and Saleh, A.O. (2000). Evaluation of environmentally friendly corrosion inhibitors for sour service. NACE International, Paper No 00335 Corrosion 2000, Orlando, Florida (26–31 March).

189 Havakawa, C., Fujii, K., Funakawa, S., and Kosaki, T. (2011). Biodegradation kinetics of monosaccharides and their contribution to basal respiration in tropical forest soils. *Soil Science & Plant Nutrition* **57** (5): 663–673.

190 Omaye, S.T. (2004). *Food and Nutritional Toxicology*. CRC Press.

191 http://www.starch.dk/isi/bio/bioethanol.asp (accessed 8 December 2017).

192 Fink, J. (2013). *Petroleum Engineers Guide to Oilfield Chemicals and Fluids.* Elsevier.

193 Abramov, Y.D., Osipov, S.N., Ostryanskaya, G., et al. (1992). Gel forming plugging composition. SU Patent 1,776,766.

194 Tsytsymushkin, P.F., Khairullina, S.R., Tarnavskiy, A.P., et al. (1992). Cement slurry to isolate zones of absorption. SU Patent 1,740,627.

195 Bock, L.H. (1937). Water soluble cellulose ethers. *Industrial and Engineering Chemistry* **29** (9): 985–987.

196 OSPAR List of Substances Used and Discharged Offshore Which Are Considered to Pose Little or No Risk to the Environment (PLONOR), OSPAR Agreement 2012-06 (Replacing Agreement 2004-10), Revised February 2013 to correct footnote cross-references.

197 King, G.E. (2010). Thirty years of gas shale fracturing: What have we learned? SPE-133456. SPE Annual Technical Conference and Exhibition, Florence, Italy (19–22 September 2010).

198 Jung, H.B., Carrol, K.C., Kabilan, S. et al. (2015). Stimuli-responsive/rheoreversible hydraulic fracturing fluids as a greener alternative to support geothermal and fossil energy production. *Green Chemistry* **17**: 2799–2812.

199 Berkhof, R., Kwekkeboom, H., Balzer, D. and Ripke, N. (1992). Demulsifiers for breaking petroleum emulsions. US Patent 5,164,116, assigned to Huels AG.

200 Garcia-Ochoa, F., Santos Mazorra, V.E., Casas, J.A., and Gomez, E. (2000). Xanthan gum: production, recovery, and properties. *Biotechnology Advances* **18** (7): 549–579.

201 Guo, X.H., Li, W.D., Tian, J., and Liu, Y.Z. (1999). Pilot test of xanthan gum flooding in Shengli oilfield, SPE 57294. SPE Asia Pacific Improved Oil Recovery Conference, Kuala Lumpur, Malaysia (25–26 October 1999).

202 Solomon, U., Oluwaseun, T. and Olalekan, O. (2015). Alkaline-surfactant-polymer flooding for heavy oil recovery from strongly water wet cores using sodium hydroxide, lauryl sulphate, shell enordet 0242, gum arabic and xanthan gum, SPE 178366. SPE Nigeria Annual International Conference and Exhibition, Lagos, Nigeria (4–6 August 2015).

203 Navarette, R.C., Dearing, H.L., Constein, V.G., et al. (2000). Experiments in fluid loss and formation damage with xanthan-based fluids while drilling. IADC/SPE Asia Pacific Drilling Technology, Kuala Lumpur, Malaysia (11–13 September 2000).

204 Hoyt, J.W. (1985). Drag reduction in polysaccharide solutions. *Trends in Biotechnology* **3** (1): 17–21.

205 Interthal, W. and Wilski, H. (1985). Drag reduction experiments with very large pipes. *Colloid & Polymer Science* **263** (3): 217–229.

206 Malhotra, J.P., Chaturvedi, P.N., and Singh, R.P. (1988). Drag reduction by polymer–polymer mixtures. *Journal of Applied Polymer Science* **36**: 837–858.

207 Kuar, H., Singh, G. and Jafar, A. (2013). Study of drag reduction ability of naturally produced polymers from a local plant source, IPTC 17207. International Petroleum Technology Conference, Beijing, China (26–28 March 2013).

208 Singh, R.P. (1995). Advanced turbulent drag reducing and flocculating materials based on polysaccharides. In: *Polymers and Other Advanced Materials* (ed. P.N. Prasad, J.E. Mark and T.J. Fai), 227–249. Springer.

209 Thomas, D.C. (1982). Thermal stability of starch- and carboxymethyl cellulose-based polymers used in drilling fluids, SPE 8463. *Society of Petroleum Engineers Journal* **22** (2): doi: 10.2118/8463-PA.

210 Simonides, H., Schuringa, G., and Ghalambour, A. (2002). Role of starch in designing nondamaging completion and drilling fluids, SPE 73768. International Symposium and Exhibition on Formation Damage Control, Lafayette, Louisiana (20–21 February 2002).

211 Guo, X., Qui, F., Dong, K. et al. (2003). Scale inhibitor copolymer modified with oxidized starch: synthesis and performance on scale inhibition. *Polymer-Plastics Technology and Engineering* **52** (3): 261–267.

212 Liang, L.X. (2013). The synthesis of starch graft temperature scale inhibitor. *Applied Mechanics and Materials* **448–453**: 1412–1415.

213 House, R.F. and Cowan, J.C. (1998). Chitosan-containing well drilling and servicing fluids. US Patent 6,258,755, assigned to Venture Innovations Inc.

214 Ali, S.A., Sagar, P., and Singh, R.P. (2010). Flocculation performance of modified chitosan in an aqueous suspension. *Journal of Applied Polymer Science* **118** (5): 2592.

215 Bratskaya, S., Avramenko, V., Schwarz, S., and Philippova, I. (2006). Enhanced flocculation of oil-in-water emulsions by hydrophobically modified chitosan derivatives. *Colloids and Surfaces, A: Physiochemical and Engineering Aspects* **275** (1–3): 168–176.

216 Ratajska, M., Strobi, G., Wisniewska-Worna, M. et al. (2003). Studies on the biodegradation of chitosan in an aqueous medium. *Fibres & Textiles in Eastern Europe* **11** (3 (42), July/September).

217 Makarios-Laham, I. and Lee, T.-C. (1995). Biodegradability of chitin and chitosan containing films in soil environment. *Journal of Environmental Polymer Degradation* **3** (5): 31–36.

218 Baraka-Lokmane, S., Sorbie, K., Poisson, N., and Kohler, N. (2009). Can green scale inhibitors replace phosphonate scale inhibitors?: Carbonate coreflooding experiments. *Petroleum Science and Technology* **27** (4): 427–441.

219 Greve, H.-H. (2000). Rubber 2. Natural. In: *Ullmann's Encyclopedia of Industrial Chemistry*. Wiley-VCH.

220 Milligan, S.N., Harris, W.F., Smith, K.W., et al. (2008). Remote delivery of latex drag-reducing agent without introduction of immiscible low-viscosity flow facilitator. US Patent 7,361,628, assigned to ConocoPhillips Company.

221 Halliday, W.S., Schwertner, D., Xiang, T. and Clapper, D.K. (2008). Water-based drilling fluids using latex additives. US Patent 7,393,813, assigned to Baker Hughes Inc.

222 Reddy, B.R. and Palmer, A.V. (2009). Sealant compositions comprising colloidally stabilized latex and methods of using the same. US Patent 7,607,483, assigned to Halliburton Energy Services Inc.

223 Halliday, W.S., Schwertner, D., Xiang, T. and Clapper, D.K. (2007). Fluid loss control and sealing agent for drilling depleted sand formations. US Patent 7,271,131, assigned to Baker Hughes Inc.

224 V. L. Kuznetsov, G. Lyubitsk, E. Krayushkina, et al. (1992). Oilwell composition. SU Patent 1733624.

225 Ventresca, M.L., Fernandez, I., and Navarro-Perez, G. (2009). Reversible gelling system and method using same during well treatments. US Patent 7,638,476, assigned to Intevep S.A.

226 Pearl, I.A. (1967). *The Chemistry of Lignin*. New York: Marcel Dekker, published in Angewandte Chemie, 80(8), 328, 1968.

227 Azar, J.J. and Samuel, G.R. (2007). *Drilling Engineering*. PennWell Books.

228 Schilling, P. (1991). Aminated sulfonated or sulfomethylated lignins as cement fluid loss control additives. US Patent 4,990,191, assigned to Westvaco Corporation.

229 Bilden, D.M. and Jones, V.E. (2000). Asphaltene adsorption inhibition treatment. US Patent 6,051,535, assigned to BJ Services Company.

230 Reid, I.D. (1995). Biodegradation of lignin. *Canadian Journal of Botany* **73** (S1): 1011–1018.

231 Sundman, V. and Nase, L. (1972). The synergistic ability of some wood-degrading fungi to transform lignins and lignosulfonates on various media. *Archiv für Mikrobiologie* **86** (4): 339–348.

232 Cho, N.-S., Shin, W.-S., Jeong, S.-W., and Leonowicz, A. (2004). Degradation of lignosulfonate by fungal laccase with low molecular mediators. *Bulletin of the Korean Chemical Society* **25** (10): 1551–1554.

233 Wershaw, R.L. (1986). A new model for humic materials and their interactions with hydrophobic organic chemicals in soil-water or sediment-water systems. *Journal of Contaminant Hydrology* **1** (1–2): 29–45.

234 Lewis, S., Chatterji, J., King, B., and Brennies, D.C. (2009). Cement compositions comprising humic acid grafted fluid loss control additives. US Patent 7,576,040, assigned to Halliburton Energy Services Inc.

235 Goth, K., De Leeuw, J.W., Puttmann, W., and Tegelaar, E.W. (1988). Origin of messel oil shale kerogen. *Nature* **336**: 759–761.

236 Kelessidis, V.C., Tsamantaki, C., Michalakis, A. et al. (2007). Greek lignites as additives for controlling filtration properties of water-bentonite suspensions at high temperatures. *Fuel* **86**: 1112–1121.

237 Offshore Technology Report - OTO 1999 089, "Drilling Fluids Composition and Use within the UK Offshore Drilling Industry" Health and safety Executive, March 2000.

238 Maiti, S., Das, S., Mati, M., and Ray, A. (1983). Renewable resources from forest products for high temperature resistant polymers. In: *Polymer Applications of Renewable - Resource Materials*, Polymer Science and Technology, vol. **17** (ed. C.E. Carracher Jr. and L.H. Sperling), 129–147. Springer.

239 Stepp, A.K., Bailey, S.A., Bryant, R.S., and Evans, D.B. (1996). Alternative methods for permeability modification using biotechnology. SPE Annual Technical Conference and Exhibition, Denver, Colorado (6–9 October 1996).

240 Panthi, K., Mohanty, K.K. and Huh, C. (2015). Precision control of gel formation using superparamagnetic nanoparticle-based heating, SPE 175006. SPE Annual Technical Conference and Exhibition, Houston, TX (28–30 September 2015).

241 Clarke-Sturman, A.J., den Ottelander, D., and Sturla, P.L. (1989). Succinoglycan – a new biopolymer for the oil field. In: *Oil-Field Chemistry: Enhanced Recovery and Production Stimulation*, ACS Symposium Series, vol. **396** , Chapter 8 (ed. J.K. Borchardt and T.F. Ye), 157–168.

242 Lazar, I., Blank, L., and Voicu, A. (1993). Investigations on a new Romanian biopolymer(pseudozan) for use in enhanced oil recovery(EOR). *Biohydrometallurgical Technologies* **2**: 357–364.

243 Westland, J.A., Lenk, D.A. and Penny, G.S. (1993). Rheological characteristics of reticulated bacterial cellulose as a performance additive to fracturing and drilling fluids, SPE 25204. SPE International Symposium on Oilfield Chemistry, New Orleans, Louisiana (2–5 March 1993).

244 Cobianco, S., Bartosek, M., Lezzi, A., et al. (2001). New solids-free drill-in fluid for low permeability reservoirs, SPE 64979. SPE International Symposium on Oilfield Chemistry, Houston, TX (13–16 February 2001).

245 Zhao, H., Nasr-El-Din, H.A. and Al-Bagoury, M. (2015). A new fracturing fluid for HP/HT applications, SPE 164204. SPE European Formation Damage Conference and Exhibition, Budapest, Hungary (3–5 June 2015).

246 Devries, A.L. (1982). Biological antifreeze agents in coldwater fishes. *Comparative Biochemistry and Physiology – Part A* **73** (4): 627–640.

247 Klomp, U.C., Kruka, V.R., Reijnart, R. and Weisenborn, A.J. (1997). Method for inhibiting the plugging of conduits by gas hydrates. US Patent 5,648,575, assigned to Shell Oil Company.

248 Punase, A.D., Bihani, A.D., Patane, A.M., et al. Soybean slurry – a new effective, economical and environmental friendly solution for oil companies, SPE 142658. SPE Project and Facilities Challenges Conference at METS, Doha, Qatar (13–16 February 2011).

249 Li, Y., Yang, I.C.Y., Lee, K.-I., and Yen, T.F. (1993). Subsurface application of alcaligenes eutrophus for plugging of porous media. In: *Microbial Enhanced Recovery – Recent Advances* (ed. E.T. Premuzic and A. Woodhead), 65–77. Elsevier.

250 Braun, W., De Wolf, C. and Nasr-El-Din, H.A. (2012). Improved health, safety and environmental profile of a new field proven stimulation fluid (Russian), SPE 157467. SPE Russian Oil and Gas Exploration and Production Technical Conference and Exhibition, Moscow, Russia (16–18 October 2012).

251 Stevens, E.S. (2001). *Green Plastics: An Introduction to the New Science of Biodegradable Plastics*. Princeton, NJ: Princeton University Press.

252 Cardoso, J.J.F., Queiros, Y.G.C., Machado, K.J.A. et al. (2013). Synthesis, characterization, and in vitro degradation of poly(lactic acid) under petroleum production condition. *Brazilian Journal of Petroleum and Gas* **7** (2): 57–69.

253 Todd, B.L. and Powell, R.J. (2006). Compositions and methods for degrading filter cake. US Patent 7,080,688 assigned to Halliburton Energy Services Inc.

254 Nasr-El-Din, H.A., Keller, S.K., Still, J.W. and Lesko, T.M. (2007). Laboratory evaluation of an innovative system for fracture stimulation of high-temperature carbonate reservoirs, SPE 106054. International Symposium on Oilfield Chemistry, Houston, TX (28 February–2 March 2007).

255 Leinweber, D., Scherl, F., Wasmund, E. and Grunder, H. (2004). Alkoxylated polyglycerols and their use as demulsifiers. US Patent Application 20040072916.

256 Willey, T.F., Willey, R.J. and Willey, S.T. (2007). Rock bit grease composition. US Patent 7,312,185, assigned to Tomlin Scientific Inc.

257 Roberfroid, M.B. (2007). Inulin-type fructans: functional food ingredients. *The Journal of Nutrition* **137** (11): 2493S–2502S.

258 Verraest, D.L., Batelaan, J.G., Peters, J.A. and van Bekkam, H. (1998). Carboxymethyl inulin. US Patent 5,777,090, assigned to Akzo Nobel NV.

259 Johannsen, F.R. (2003). Toxicological profile of carboxymethyl inulin. *Food and Chemical Toxicology* **41**: 49–59.

260 Boels, L. and Witkamp, G.-J. (2011). Carboxymethyl inulin biopolymers: a green alternative for phosphonate calcium carbonate growth inhibitors. *Crystal Growth & Design* **11** (9): 4155–4165.

261 Decampo, F., Kesavan, S. and Woodward, G. (2008). Polysaccharide based scale inhibitor. International Patent Application, WO2008140729.

262 Holt, S.P.R., Sanders, J., Rodrigues, K.A. and Vanderhoof, M. (2009). Biodegradable alternatives for scale control in oilfield applications, SPE 121723. SPE International Symposium on Oilfield Chemistry, The Woodlands, TX (20–22 April 2009).

263 Vögtle, F., Gestermann, S., Hesse, R. et al. (2000). Functional dendrimers. *Progress in Polymer Science* **25** (7): 987–1041.

264 Buhlier, E., Wehner, W., and Vögtle, F. (1978). 'Cascade'- and 'nonskid-chain-like' syntheses of molecular cavity topologies. *Synthesis* **1978** (2): 155–158.

265 Frechet, J.M.J. and Tomalia, D.A. ed. (2001). *Dendrimers and Other Dendritic Polymers*. Wiley.

266 Amanullah, M. (2013). Dendrimers and dendritic polymers - application for superior and intelligent fluid development for oil and gas field applications, SPE 164162. SPE Middle East Oil and Gas Show and Conference, Manama, Bahrain (10–13 March 2013).

267 Klomp, U.C. (2005). Method for inhibiting the plugging of conduits by gas hydrates. US Patent 6,905,605, assigned to Shell Oil Company.

268 Rivers, G.T., Tian, J. and Trenery, J.B. (2009). Kinetic gas hydrate inhibitors in completion fluids. US Patent 7,638,465, assigned to Baker Hughes Incorporated.

269 Feustel, M., Grunder, H., Leinweber, D., and Wasmund, E. (2005). Alkoxylated dendrimers, and use thereof as biodegradable demulsifiers. WO Patent Application 2005003260.

270 Kaiser, A. (2013). Environmentally friendly emulsion breakers: vision or reality?, SPE 164073. SPE International Symposium on Oilfield Chemistry, The Woodlands, TX (8–10 April 2013).

271 Cole, R., Nordvik, T., Khandekar, S., et al. (2015). Dendrimers as paraffin control additives to combat wax deposition. Chemistry in the Oil Industry XIV (2–4 November 2015). Manchester, UK: Royal Society of Chemistry.

272 Bosman, A.W., Janssenan, H.M., and Meijer, E.W. (1999). About dendrimers: structure, physical properties and applications. *Chemical Reviews* **99** (7): 1665–1688.

273 http://www.mrc-flocculant.jp/english/product/polymer/diaflockp.html (accessed 9 December 2017).

274 Gebbie, P. (2001). Using polyaluminium coagulants in water treatment. 64th Annual Water Industry Engineers and Operators Conference, Bendigo, Victoria, Australia (5 and 6 September 2001).

275 Plunkett, J.W. ed. (2009). *Plunkett's Chemicals, Coatings & Plastics Industry Almanac: The Only Complete Guide to the Chemicals, Coatings and Plastics Industry*. Houston, TX: Plunkett Research Ltd.

276 Oilfield Engineering with Polymers: Institute of Electrical Engineers, Conference Proceedings, London, UK (3–4 November 2003).

277 Kirkpatrick, D., Aguirre, F. and Jacob, G. (2008). Review of epoxy polymer thermal aging behavior relevant to fusion bonded epoxy coatings, NACE 08037. CORROSION 2008, New Orleans, Louisiana (16–20 March 2008).

278 Hartley, R.A. (1971). Coatings and corrosion. Offshore Technology Conference, Houston, TX (19–21 April 1971).

279 Morton, M. ed. (1999). *Rubber Technology*, 3rde. Dordrecht, The Netherlands: Kluwer Academic Publishers.

280 Slay, J.B. and Ray, T.W. (2003). Fluid compatibility and selection of elastomers in oilfield completion brines, NACE 03140. CORROSION 2003, San Diego, CA (16–20 March 2003).

281 Reid, W.M. (2001). A proposed recommended practice to determine elastomer/oil mud compatibility, NACE 01113. CORROSION 2001, Houston, TX (11–16 March 2001).

282 Frostman, L.M., Gallagher, C.G., Ramachandran, S. and Weispfennig, K. (2001). Ensuring Systems Compatibility for Deepwater Chemicals. SPE International Symposium on Oilfield Chemistry, Houston, TX (13–16 February 2001).

283 Environment Canada and Health Canada (2010). Report on Screening Assessment for the Challenge Oxirane, (butoxymethyl)-(n-Butyl glycidyl ether). Chemical Abstracts Service Registry Number 2426-08-6 (March 2010).

284 Paul, S. ed. (1995). *Surface Coatings: Science and Technology*, 2nde. Wiley.

285 Lee, L.-H. ed. (1984). *Adhesive Chemistry: Developments and Trends*. New York: Springer.

286 Papavinasam, S. and Revie, R.W. (2006). Protective pipeline coating evaluation, NACE 06047. CORROSION 2006, San Diego, CA (12–16 March 2006).

287 Zhang, X. and Liu, H. (2001). Application of polymer flooding with high molecular weight and concentration in heterogeneous reservoirs, SPE 144251. SPE Enhanced Oil Recovery Conference, Kuala Lumpur, Malaysia (19–21 July 2001).

288 Taylor, K.C. and Nasr-El-Din, H.A. (1998). Water-soluble hydrophobically associating polymers for improved oil recovery: A literature review. *Journal of Petroleum Science and Engineering* **19** (3–4): 265–280.

289 Weaver, D.A.Z., Picchioni, F., and Broekhuis, A.A. (2011). Polymers for enhanced oil recovery: a paradigm for structure–property relationship in aqueous solution. *Progress in Polymer Science* **36** (11): 1558–1628.

290 Sheng, J.J. (2011). *Modern Chemical Enhanced Oil Recovery: Theory and Practice*. Oxford, UK: Gulf Professional Publishing and Elsevier.

291 Oil and Gas Journal Article (1996). DOE, Industry Aid Polymer Injection Studies. Society of Petroleum Engineers (15 July 1996).

292 Brannon, H.D. and Ault, M.G. (1991). New delayed borate-crosslinked fluid provides improved fracture conductivity in high-temperature applications, SPE 22838. SPE Annual Technical Conference and Exhibition, Dallas, TX (6–9 October 1991).

293 Putzig, D.E. and Smeltz, K.C. (1990). Organic titanium compositions useful as cross-linkers. US Patent 4,953,621, assigned to E.I. Du Pont De Nemours and Company.

294 Putzig, D.E. (2007). Zirconium-based cross-linker compositions and their use in high pH oil field applications. US Patent 8,236,739, assigned to Dork Ketal Speciality Catalysts LLC.

295 World Health Organization (1998). Boron, Environmental Health Criteria, 204, Geneva, Switzerland.

296 Eisler, R. (1990). Boron hazards to fish, wildlife, and invertebrates: a synoptic review. U.S. Department of the Interior, Fish and Wildlife Service, Biological Report, **82**, 1–32.

297 U.S. Environmental Protection Agency (1993). Reregistration eligibility decision document: boric acid and its sodium salts, EPA 738-R-93-017. Office of Pesticide Programs (September 1993), U.S. Government Printing Office: Washington, D.C..

298 Zeebe, R.E., Sanval, A., Ortiz, J.D., and Wolf-Gladrow, D.A. (2001). A theoretical study of the kinetics of the boric acid-borate equilibrium in seawater. *Marine Chemistry* **73** (2): 113–124.

299 Suedel, B.C., Boraczek, J.A., Peddicord, R.K. et al. (1994). Trophic transfer and biomagnification potential of contaminants in aquatic ecosystems. *Reviews of Environmental Contamination and Toxicology* **136**: 21–89.

300 Topguder, N. (2010). A review on utilization of crosslinked polymer gels for improving heavy oil recovery in Turkey (Russian), SPE 131267. SPE Russian Oil and Gas Conference and Exhibition, Moscow, Russia (26–28 October 2010).

301 Southwick, J.G. and Manke, C.W. (1988). Molecular degradation, injectivity and elastic properties of polymer solutions, SPE 15652. *SPE Reservoir Engineering* **3** (4): doi: 10.2118/15652-PA.

302 Berge, B.K. and Solsvik, O. (1996). Increased pipeline throughput using drag reducer additives (DRA): field experiences, SPE 36835. European Petroleum Conference, Milan, Italy (22–24 October 1996).

303 I. Henaut, P. Glenat, C. Cassar, et al. (2012). Mechanical degradation kinetics of polymeric DRAs. 8th North American Conference on Multiphase Technology, BHR Group, Banff, Alberta, Canada (20–22 June 2012).

304 Motier, J.F., Chou, L.-C. and Tong, C.L. (2003). Process for homogenizing polyolefin drag reducing agents. US Patent 6,894,088, assigned to Baker Hughes Incorporated.

305 Hamelink, J. ed. (1994). *Bioavailability: Physical, Chemical, and Biological Interactions*. Setac Special Publications Series.

306 David, C. (1975). Thermal degradation of polymers. In: *Comprehensive Chemical Kinetics*, vol. **14**, Chapter 1, 1–173. New York: Elsevier.

307 MacDiarmid, A.G. and Heeger, A.J. (1980). Organic metals and semiconductors: The chemistry of polyacetylene, $(CH)_x$, and its derivatives. *Synthetic Metals* **1** (2): 101–118. (Symposium on the structure and properties of highly conducting polymers and graphite.)

308 Will, F.G. and McKee, D.W. (1983). Thermal oxidation of polyacetylene. *Journal of Polymer Science, Polymer Chemistry Edition* **21**: 3479–3492.

309 Saxon, A.M., Liepins, F., and Aldissi, M. (1985). Polyacetylene: its synthesis, doping, and structure. *Progress in Polymer Science* **11**: 57.

310 Otsu, T., Matsumoto, A., Kubota, T., and Mori, S. (1990). Reactivity in radical polymerization of N-substituted maleimides and thermal stability of the resulting polymers. *Polymer Bulletin* **23** (1): 43–50.

311 Chichester, C.O. ed. (1986). Advances in food research. In: *Advances in Food and Nutrition Research*, vol. **30**. Orlando, FL: Academic Press Inc.

312 Madorsky, S.L., Hart, V.E., and Straus, S. (1958). Thermal degradation of cellulosic materials. *Journal of Research of the National Bureau of Standards* **60** (4), Research Paper 2853): 343–349.

313 Craddock, H.A., Simcox, P., Williams, G., and Lamb, J. (2011). Backward and forward in corrosion inhibitors in the North Sea, Paper III in an occasional series on the use of alkyl polyglucosides as corrosion inhibitors in the oil and gas industry. Chemistry in the Oil Industry XII (7–8 November 2011) Manchester, UK: Royal Society of Chemistry.

314 Braunecker, W.A. and Matyjaszewski, K. (2007). Controlled/living radical polymerization: Features, developments, and perspectives. *Progress in Polymer Science* **32** (1): 93–146.

315 Boersma, A. (2006). Predicting the efficiency of antioxidants in polymers. *Polymer Degradation and Stability* **91** (3): 472–478. (Special Issue on Degradation and Stabilisation of Polymers.)

316 Gibson, D.T., Koch, J.R., and Kallio, R.E. (1968). Oxidative degradation of aromatic hydrocarbons by microorganisms. I. Enzymatic formation of catechol from benzene. *Biochemistry* **7** (7): 2653–2662.

317 Horsch, P., Speck, A., and Himmel, F.H. (2003). Combined advanced oxidation and biodegradation of industrial effluents from the production of stilbene-based fluorescent whitening agents. *Water Research* **37** (11): 2748–2756.

318 Zimbron, J.A. and Reardon, K.F. (2011). Continuous combined Fenton's oxidation and biodegradation for the treatment of pentachlorophenol-contaminated water. *Water Research* **45** (17): 5705–5714.

319 Novotný, O., Cejpek, K., and Velšek, J. (2008). Formation of carboxylic acids during degradation of monosaccharides. *Czech Journal of Food Sciences* **26**: 117–131.

320 Calkins, J. ed. (1982). *The Role of Solar Ultraviolet Radiation in Marine Ecosystems*, NATO Conference Series 4: Marine Science 7. New York and London: Plenum Publishing Corporation.

321 Crosby, D.G. (1994). Photochemical aspects of bioavailability. In: *Bioavailability: Physical, Chemical, and Biological Interactions*, Setac Special Publications Series (ed. J. Hamelink).

322 Warholic, M.D., Massah, H., and Hanratty, T.J. (1999). Influence of drag reducing polymers on turbulence: effects of reynolds number, concentration and mixing. *Experiments in Fluids* **27** (5): 461–472.

323 Hong, C., Zhang, K., Choi, H., and Yoon, S. (2010). Mechanical degradation of polysaccharide guar gum under turbulent flow. *Journal of Industrial and Engineering Chemistry* **16** (2): 178–180.

324 Staudinger, H. and Heuer, W. (1934). *Chemische Berichte* **67**: 1159.

325 Darestani Farahani, T., Bakhshandeh, G.R., and Abtahi, M. (2006). Mechanical and viscoelastic properties of natural rubber/reclaimed rubber blends. *Polymer Bulletin* **56** (4): 495–505.

326 Comstock, M.J. ed. (2009). *Chemical Reactions on Polymers*, ACS Symposium Series. ACS Publications.

327 Davis, A. and Sims, D. (1983). *Weathering of Polymers*. New York: Elsevier.

328 Williams, P.A. ed. (2007). *Handbook of Industrial Water Soluble Polymers*. Wiley.

329 Chatterji, J. and Borchardt, J.K. (1981). Applications of water-soluble polymers in the oil field, SPE 9288. *Journal of Petroleum Technology* **33** (11): 2042–2056.

330 Stahl, G.A. and Schulz, D.N. (1986). *Water-Soluble Polymers for Petroleum Recovery: National Meeting Entitled "Polymers in Enhanced Oil Recovery and the Recovery of Other Natural Resources"*. American Chemical Society.

331 Grassie, N. and Scott, G. (1988). *Polymer Degradation and Stabilisation.* Cambridge University Press.

332 Rudakova, T.Y., Chalykh, A.Y., and Zaikov, G.E. (1972). Kinetics and mechanism of hydrolysis of poly (ethylene terephthalate) in aqueous potassium hydroxide solutions. *Polymer Science U.S.S.R.* **14** (2): 505–511.

333 Dussan, K.J., Silva, D.D.v., Moraes, E.J.C. et al. (2014). Dilute-acid hydrolysis of cellulose to glucose from sugarcane bagasse. *Chemical Engineering Transactions* **38**: 433–439.

334 Fountoulakis, M. and Lahm, H.-W. (1998). Hydrolysis and amino acid composition analysis of proteins : a review. *Journal of Chromatography A* **826**: 109–134.

335 Munoz Jr., T. and Eoff, L.S. (2010). Treatment fluids and methods of forming degradable filter cakes comprising aliphatic polyester and their use in subterranean formations. US Patent 7,674,753, assigned to Halliburton Energy Services Inc.

336 Faudree, M.C. (1991). Relationship of graphite/polyimide composites to galvanic processes. 36th International SAMPE Symposium (15–18 April 1991).

337 Wade, C.W.R. and Leonard, F. (1972). Degradation of poly(methyl 2-cyanoacrylates). *Journal of Biomedical Materials Research* **6** (3): 215–220.

338 Smith, R. (2005). *Biodegradable Polymers for Industrial Applications.* CRC Press.

339 Steinberg, D. and Mihalyi, E. (1957). The chemistry of proteins. *Annual Review of Biochemistry* **26**: 373–418.

340 American Petroleum Institute (1997). Bioaccumulation: how chemicals move from the water into fish and other aquatic organisms. Health and Environmental Sciences Department, Publication Number 4656 (May 1997).

341 Schlaad, H. and Hoogenboom, R. (2012). Special issue: poly(2-oxazoline)s and related pseudo-polypeptides. *Macromolecular Rapid Communications* **33** (19): 1593–1719.

342 Hoogenbloom, R. (2009). Poly(2-oxazoline)s: a polymer class with numerous potential applications. *Angewandte Chemie (International Ed. in English)* **48** (43): 7978–7994.

343 Knox, D. and McCosh, K. (2005). Displacement chemicals and environmental compliance – past present and future. Chemistry in the Oil Industry IX (31 October to 2 November 2005). Manchester, UK: Royal Society of Chemistry.

344 Balzer, D. and Luders, H. ed. (2000). *Nonionic Surfactants: Alkyl Polyglucosides*, Surfactant Science Series, vol. **91**. Marcel Dekker Inc.

345 Saeki, T., Nishibayashi, H., Hirata, T. and Yamaguchi, S. (1999). Polyalkylene glycol-polyglyoxylate block copolymer, its production process and use. US Patent 5,856,288, assigned to Nippon Shokubai Co. Ltd.

346 http://www.luxresearchinc.com/news-and-events/press-releases/read/first-higher-value-chemical-derived-lignin-hit-market-2021 (accessed 9 December 2017).

347 Upton, B.M. and Kasko, A.M. (2015). *Strategies for the Conversion of Lignin to High-Value Polymeric Materials: Review and Perspective.* Chemical Reviews. ACS Publications.

348 Kehoe, J.D. and Joyce, M.K. (1993). Water soluble liquid alginate dispersions. US Patent 5,246,490, assigned to Syn-Chem Inc.

349 Dufresne, A., John, M.J., and Thomas, S. (2012). Natural polymers. In: *Nanocomposites*, Green Chemistry Series, vol. **2**. RSC Publishing.

350 Huang, F.-Y. (2012). Thermal properties and thermal degradation of cellulose tri-stearate (CTs). *Polymer* **4**: 1012–1102.

351 Guo, S., He, Y., Zhou, L. et al. (2015). An anti-biodegradable hydrophobic sulphonate-based acrylamide copolymer containing 2,4-dichlorophenoxy for enhanced oil recovery. *New Journal of Chemistry* **39**: 9265–9274.

352 Hsiao, M. (2001). Review: biodegradation of plastics. *Current Opinion in Biotechnology* **1** (3): 242–247.

353 Premraj, R. and Doble, M. (2005). Biodegradation of polymers. *Indian Journal of Biotechnology* **4**: 186–193.

354 Connan, J. (1984). Biodegradation of crude oils in reservoirs. In: *Advances in Petroleum Geochemistry*, vol. **1** (ed. J. Brooks and D.H. Welte), 299–335. London: Academic Press.

355 http://www.bubbletightusa.com/ (accessed 9 December 2017).

356 Koutsos, V. (2009). Polymeric materials: an introduction. In: *Manual of Construction Materials*, Chapter 46. Institution of Civil Engineers.

357 Guerra, K., Dahm, K. and Dundorf, S. (2011). Oil and gas produced water management and beneficial use in the Western United States. Prepared for Reclamation Under Agreement No. A10-1541-8053-381-01-0-1, U.S. Department of the Interior Bureau of Reclamation Technical Service Center Water and Environmental Resources Division, Water Treatment Engineering Research Group, Denver, Colorado (September 2011).

358 Navarro, W. (2007). Produced water reinjection in mature field with high water cut, SPE 108050. Latin American & Caribbean Petroleum Engineering Conference, Buenos Aires, Argentina (15–18 April 2007).

359 Hjelmas, T.A., Bakke, S., Hilde, T., et al. (1996). Produced water reinjection: experiences from performance measurements on Ula in the North Sea, SPE 35874. SPE Health, Safety and Environment in Oil and Gas Exploration and Production Conference, New Orleans, Louisiana (9–12 June 1996).

360 Norwegian Ministry of the Environment (1996–1997). White Paper no. 58. Environmental Policy for a Sustainable Development – Joint Effort for the Future.

361 Rudnik, E. (2008). *Compostable Polymer Materials*. Oxford, UK: Elsevier.

362 Viloria, A., Castillo, L., Garcia, J.A. and Biomorgi, J. (2010). Aloe derived scale inhibitor. US Patent 7,645,722, assigned to Intevep. S.A.

3

Surfactants and Amphiphiles

The use of surfactant chemistry and related products is universal in the upstream oil and gas industry with just about every type of application having a surfactant chemical capable of fulfilling the desired function or contributing to the required effect. They are used in every area from drilling and completion to a variety of inhibitors in processing and production to stimulation and enhanced recovery operations. Alongside polymers, they account for the majority of oilfield chemical sales.

Commercially surfactants are usually classified according to their use. This form of classification is, however, not very useful because many surfactants have several uses. The most generally recognised and scientifically sound classification of surfactants is based on their dissociation in water [1]. This is the method that will be primarily adopted in this chapter. This classification broadly categorises surfactants as follows:

1) Anionic surfactants
2) Non-ionic surfactants
3) Cationic surfactants
4) Others

This chapter will also include specific sections on amphoteric (zwitterionic) surfactants and biosurfactants. There will also be discussion on polymeric surfactants, and although a number of examples have been described in Chapter 2, a number will be examined in more detail, as their surfactant behaviour is more relevant to oilfield application than their polymeric structure. There will also be specific detail on certain important types of surfactant or individual chemicals or chemical types.

Surfactants can broadly be defined as compounds that lower the surface tension (or interfacial tension (IFT)) between two liquids or between a liquid and a solid. Surfactants may act as detergents, wetting agents, emulsifiers, foaming agents and dispersants. In their action of reducing surface tension, surfactants increase their spreading and wetting properties.

The term amphiphile is derived from the Greek amphi, which means 'double', or 'from both sides', as in amphitheater, and philos, which expresses friendship or affinity, as in 'philanthropist' (the friend of man) or 'hydrophilic' (compatible with water).

An amphiphilic substance therefore exhibits a double affinity, which can be defined from the physico-chemical point of view as a polar–apolar duality. A typical amphiphilic molecule consists of two parts: on the one hand, a polar group that contains heteroatoms such as oxygen, sulphur, phosphorus or nitrogen included in functional groups such as alcohol, thiol, ether, ester, acid, sulphate, sulphonate, phosphate, amine, amide, etc. The molecule also has an essentially apolar group that is in general a hydrocarbon chain of the alkyl or alkylbenzene type, sometimes with halogen atoms and a few non-ionised oxygen atoms. The polar part exhibits a

Oilfield Chemistry and Its Environmental Impact, First Edition. Henry A. Craddock.
© 2018 John Wiley & Sons Ltd. Published 2018 by John Wiley & Sons Ltd.

strong affinity for polar solvents, particularly water, and it is often called hydrophilic part or hydrophile. The apolar part is called hydrophobe or lipophile, again from the Greek phobos (fear) and lipos (grease).

Because of its dual affinity, an amphiphilic molecule does not feel 'at ease' in any solvent, be it polar or non-polar, since there is always one of the groups that 'does not like' the solvent environment. This is why amphiphilic molecules exhibit a very strong tendency to migrate to interfaces or surfaces and to orientate so that the polar group lies in water and the apolar group is placed out of it, and eventually in oil. This is a particularly useful characteristic in the design of oil-treating chemicals as in oilfield operations where mixtures of hydrocarbon fluids and water are usually being extracted and processed. It is worthy of note that all amphiphilic molecules do not display such activity; in effect, only the amphiphiles with more or less equilibrated hydrophilic and lipophilic tendencies are likely to migrate to the surface or interface. This does not happen if the amphiphilic molecule is too hydrophilic or too hydrophobic, in which case it stays in one of the preferred phases. This behaviour is a crucial parameter when examining the design of surfactant-based oilfield chemicals and is known as the hydrophilic–lipophilic balance (HLB).

The terms surfactant and amphiphile are often used interchangeably although surfactant has more common usage particularly in oilfield application. Another common nomenclature is 'Tenside', which is referencing the IFT change properties of surfactants and is primarily the German use of surfactant. In the main, the term surfactant will be used throughout this chapter.

Surfactants of course are used in a number of industrial and domestic areas, and oilfield applications, although significant, do not account for more than 5% of the manufactured volume, with the personal care and detergent industries being the major users of these products. The surfactants industry is generally considered complex due to a variety of factors, some of which are listed as follows [2]:

- The broad-ranging definition of the term surfactant
- A large number of surfactant suppliers (more than 500 worldwide)
- A wide variety of product chemistries (more than 3500), intermediates and blends
- A combination of specialty and commodity products and business
- A wide range of applications and customer base

The global surfactants market size was worth over 29 billion US dollars in 2014 at an estimated production volume of around 15 000 kilotonnes and set to grow in the next 4–5 years to over 22 000 kilotonnes. The vast majority of this production is from anionic and non-ionic surfactants. Cationic and other classes of surfactants account for less than 5% of the volume manufactured. Interestingly, and as will be described further, cationic surfactants are highly important in oilfield use and offer a particular challenge to the environmental chemist.

Commercially produced surfactants are often not pure chemicals, and within each chemical type, there can be tremendous variation. Oilfield chemist and other practitioners who are not familiar with surfactants are frequently bewildered by the enormous variety of different products on the market and the vast body of literature, which exists on the composition and properties of surfactants. The selection of the best surfactant for any given use therefore can be a major problem. In this chapter the author will attempt to show the major types of surfactant used in the upstream oil and gas industry and their applications. It will not be a comprehensive review of the subject matter on the application of surfactants but a guide to further information and study. The chapter will cover the imperative requirement to provide greener surfactants and the need for sustainability.

Surfactants, which will be familiar to the reader, are soaps and detergents, which are also the highest volume application of surfactants at nearly half of the worldwide consumption [2].

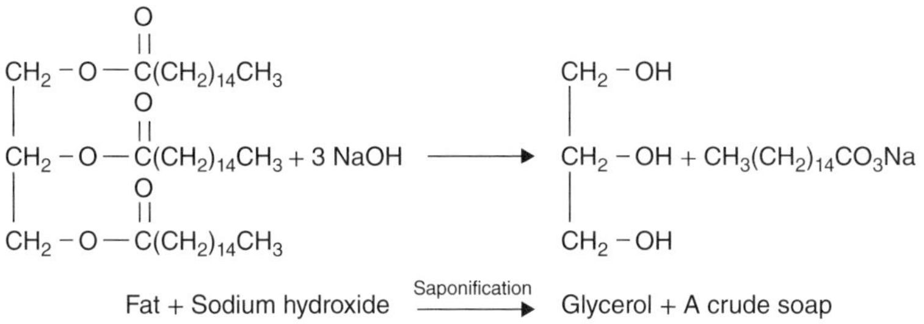

Figure 3.1 Saponification.

Soaps were the earliest surfactants and are obtained from a process known as saponification. In this process, particular fats known as glycerides are hydrolysed by heating with sodium hydroxide solution to form soaps, the sodium salts of the acids (the soap) and propane-1,2,3-triol(glycerol) (see example in Figure 3.1). The glycerides are esters formed from the reaction of glycerol, with long-chain carboxylic acids (fatty acids).

Soap making is known to have occurred in antiquity and goes back many thousands years. The most basic supplies for soap making were those taken from animal and nature; many people made soap by mixing animal fats with lye derived from wood ash. In the early beginnings of commercial soap making, it was an exclusive technique used by small groups of soap makers. The demand for soap was high, and it was expensive as there was a monopoly on soap production in many areas. The price of soap was significantly reduced in 1791 when Nicolas Leblanc discovered a chemical process to make soda ash, sodium carbonate. This allowed soap to be easily manufactured and be sold for significantly less money. As advances in chemistry developed, more was understood about the ingredients of soap, and in the mid-nineteenth century, soap for bathing became a separate commodity from laundry soap, with milder soaps being packaged, sold and made available for personal use.

As the chemistry of soap making is now well understood, this has led to the production of better technological products. Most of the cleaning agents we call soaps are really detergents. Detergents have the same characteristic property as soap, in that they let water attach to oil and grease molecules but are much less affected by hard water.

Synthetic surfactants are not very new, either. In the early nineteenth century, the first forerunner of today's surfactants was produced in the form of a sulphated castor oil, which was used in the textile industry. One of the first instances of industrially manufactured detergents happened during the First World War when Germany economy was strained and left without easy access to soap because of a shortage of animal and vegetable fats and oils. Additionally a substance that was resistant to hard water was needed to make cleaning more effective, and petroleum was found to be a plentiful source for the manufacture of these surfactants [3]. In the 1930s the commercially available routes for creating fatty alcohols have seen the development of not only the detergent industry but also the modern chemical industry.

In the 1950s, soap was almost completely displaced as a means of cleaning clothes in developed countries. Since then the use of detergents has grown exponentially, introducing many new ways of washing and reaching all four corners of the world. The last decade or so has seen the development of biodegradable and eco-friendly detergents. Today, detergent surfactants are made from a variety of petrochemicals (derived from petroleum) and/or oleochemicals (derived from fats and oils). This is an important dichotomy as it allows the chemist to derive both synthetic and natural product-based surfactant molecules with a variety of properties and environmental acceptability. However, as shall be further described later in this chapter, the very surface-active nature of these molecules makes them difficult to quantify their effects on the environment especially with respect to biodegradation and bioaccumulation.

The glycerides used to make surfactants contain saturated and unsaturated carboxylic acids, which have an even number of carbon atoms, generally within the range 12–20, for example, octadecanoic acid (stearic acid). Synthetic surfactants have one very important advantage over soaps. Soaps form insoluble calcium and magnesium salts with the calcium and magnesium ions in hard water and in the clays, which are present in dirt; much of the soap is wasted, forming an insoluble scum. However, this is avoided when using a synthetic surfactant. For example, in many anionic surfactants, a sulphonate or sulphate group as the hydrophilic component replaces the carboxylate group in soap. The corresponding calcium and magnesium salts are more soluble in water than the calcium and magnesium salts of carboxylic acids.

Synthetic/chemical-based surfactants still hold the largest share in the overall surfactants market as of 2014. However, due to stringent regulations, bio-based surfactants (biosurfactants) are expected to gain traction and grow at a higher rate in the coming years. Biosurfactants are priced higher than synthetic surfactants because of the expensive raw materials. However, considering the existing research and development taking place in the greater surfactants industry, prices are estimated to come down in the near future.

This increasing demand for eco-friendly (bio-based) surfactants, specifically from the European and North American regions, is driving the growth of the global surfactants market. Regulations and non-toxicity are among the main factors favouring the use of bio-based surfactants in these regions.

Biosurfactants are primarily of two types, namely, non-ionic or amphoteric. At present, the anionic surfactants segment captures the larger share of the surfactants market. However, due to higher growth of bio-based surfactants and eco-friendly and less toxic nature of non-ionic surfactants, the non-ionic surfactants segment is expected to witness higher growth in the near future.

Applications of biosurfactants in the oil and gas industry are also growing for the same reasons as that in the general market, and these applications are benefitting from the increasing research and development for other industrial applications.

3.1 How Surfactants Work

It is not the intention of this section to give a detailed or comprehensive review of how various surfactants work. This can be exceedingly complex and can involve many physico-chemical behaviours [4]. It is however necessary to equip the reader with an overview of the mechanistic application of surfactants and in particular how these substances work in the oilfield; in doing so the author has attempted to make this as descriptive and simple as possible so that the reader can grasp a fundamental overview of the actions of surfactants without having to have an in-depth knowledge of physical chemistry and its accompanying mathematics. As each class of surfactant is described, in subsequent sections of this chapter, then more detail will be given where appropriate.

Surfactants function as detergents by breaking down the interface between water and oils and/or dirt. They also hold these oils and dirt in suspension and so allow their removal. They are able to act in this way because they contain both a hydrophilic (water-loving) group, such as an acidic anion, e.g. a carboxylate or sulphate group, and a hydrophobic (water-hating) group, such as an alkyl chain. Molecules of water tend to congregate near the former and molecules of the water-insoluble material congregate near the latter as shown in Figure 3.2.

This head and tail structure of the surfactant, as represented in Figure 3.2 by the round head and sticklike tail, is of critical consequence for applications in the upstream oil and gas industry.

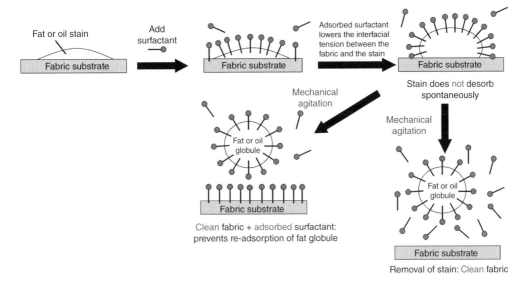

Figure 3.2 Detergent cleaning action of a surfactant.

It is also of note that in the classification of surfactants stated earlier, it is the nature of the head group upon which the classification is based:

Anionic surfactants: In these surfactants the hydrophilic group is negatively charged. They are the most widely used type of surfactants for laundering, dishwashing liquids and shampoos. They are particularly good at keeping the dirt, once dislodged, away from fabrics. These are the type of surfactants represented in Figure 3.2 and are widely used in detergent applications. Their use in the oilfield will be discussed in Section 3.2.

Non-ionic surfactants: These surfactants do not bear an electrical charge and are often used together with anionic surfactants. An advantage is that they do not interact with calcium and magnesium ions in hard water. Again their use in the oilfield will be discussed in Section 3.3.

Cationic surfactants: With these surfactants, the hydrophilic head is positively charged. Although they are produced in much smaller quantities than the anionics and non-ionics, there are several types, each used for a specific purpose. Specific oilfield use will be discussed in Section 3.4.

There are other types of surfactant, such as polymeric surfactants, that are also relevant to oilfield use, and these will be further discussed later in this chapter.

As stated earlier there are a number of critical physico-chemical properties of these surfactants, which are important in determining their functionality and therefore the application to specific chemical challenges such as cleaning fabric. In the oilfield these properties often determine the use and application of the surfactant; however they can also be manipulated to enhance a particular surfactant effect and therefore be more effective in particular or specific circumstances. This can be achieved not only by specific surfactant design but also by formulation of a combination of surfactants (and other chemicals, such as polymers).

3.1.1 The Hydrophobic Effect and Micelle Formation

In aqueous solution at dilute concentrations, surfactants, especially anionic and cationic surfactants, act as salts or electrolytes; however at higher concentrations different behaviours result. This is due to the organised self-assembly of the surfactant molecules into aggregates known as micelles (see Figure 3.3). In these micelles the lipophilic or hydrophobic parts of the

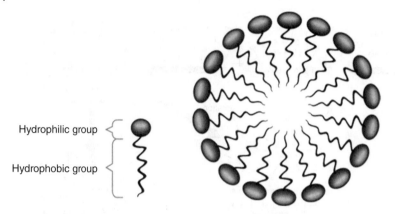

Hydrophilic group

Hydrophobic group

Figure 3.3 Micelle structure.

surfactant associate in the interior of the structure, and the hydrophilic parts face the aqueous medium.

The formation of micelles in aqueous solution is generally thought to be a compromise between the tendency of the hydrophobic alkyl groups to avoid unfavourable contacts with water and the drive of the hydrophilic polar parts to maintain contact with the aqueous environment [5]. As is well known, hydrocarbons, such as crude oil, and water are immiscible, and the limited solubility of hydrophobic species in water is known as the hydrophobic effect [6]. Although the micelle is assembled from 'free molecules', the non-aggregated units are sometimes referred to as unimers and can include single polymer units, which are, as has been described in Chapter 2, composed of two or more monomer units. The term unimer is somewhat ambiguous and not scientifically robust and so will only be referred to sparingly throughout this text.

The micellisation of surfactants is also an example of the hydrophobic effect [7]. In micellisation there are two opposing forces at work. The first is the hydrophobicity of the hydrocarbon tail, favouring the formation of micelles, and the second is the repulsion between the surfactant head groups. There exists today a considerable amount of information on the effect of temperature, pressure and addition of solutes on the strength of the hydrophobic interactions. Unfortunately there are many discrepancies between the results obtained by the different methods. Some of the models used in the study of the hydrophobic interactions are also relevant to the understanding of the micellisation process. The driving mechanism for micellisation is the transfer of hydrocarbon chains from water into the oil-like interior. The mere fact that micelles are formed from ionic surfactants is an indication of the fact that the hydrophobic driving force is large enough to overcome the electrostatic repulsion arising from the surfactant head groups.

As will also be described, this hydrophobic effect is involved in the surface adsorption of surfactants.

3.1.2 Surfactant Solubility, Critical Micelle Concentration (CMC) and Krafft Point

It is well established that the physico-chemical behaviour of surfactants varies dramatically above and below a specific surfactant concentration, the critical micelle concentration (CMC) value [5, 7]. Below the CMC value the physico-chemical properties of an ionic surfactant resemble those or a strong electrolyte, whereas above the CMC value, these properties change

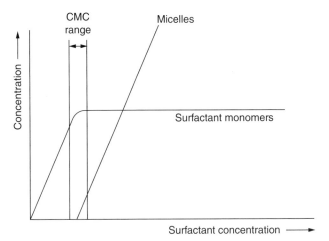

Figure 3.4 Relationship between surfactant concentration and CMC value.

dramatically, with the highly cooperative association process of micellisation occurring. Indeed almost all of the physico-chemical properties versus concentration plots for a given surfactant–solvent system will show an abrupt change in slope over a narrow concentration range, the CMC value, and was first illuminated in 1948 by Preston [8] in a groundbreaking piece of work in which he correlated a number of physico-chemical parameters for the surfactant sodium dodecyl sulphate.

The formation of micelles from the constituent monomers involves a rapid, dynamic association–dissociation equilibrium. Micelles are undetectable in dilute solutions of surfactant monomers but become detectable over a narrow range of concentrations as the total concentration of surfactant is increased, above which nearly all additional surfactant species form micelles. The concentration of free surfactant, counter-ions and micelles as a function of the overall surfactant concentration is represented in Figure 3.4. Above the CMC, the concentration of free surfactant is essentially constant, while the counter-ion concentration increases and the micellar concentration increases approximately linearly.

The value of the CMC is an important parameter in a wide variety of industrial applications, including oilfield applications, involving adsorption of surfactant molecules at interfaces, such as foams, froths, emulsions, suspensions and surface coatings. It is probably the simplest means of characterising the colloid and surface behaviour of a surfactant solute, which in turn determines its industrial usefulness. Many oilfield processes are also dynamic processes in that they involve a rapid increase in interfacial area, such as foaming, surface (rock) wetting, emulsification and emulsion breaking.

A large number of methods have been applied to the determination of the CMC of surface-active agents. Most physico-chemical property changes can be used to determine the CMC value, provided that the measurement can be carried out accurately. An extensive compilation of the CMCs of surfactants in aqueous and non-aqueous media has been compiled [9, 10]. Some of the most common methods for obtaining the CMC value of a surfactant are listed as follows:

- UV/Vis, infrared (IR) spectroscopy
- Fluorescence spectroscopy
- NMR spectroscopy
- Conductivity
- Calorimetry

- Light scattering techniques
- Surface tension

An issue particular to the oilfield in the determination of CMC values is that, in the main, they are determined under ambient laboratory conditions and it is assumed that these hold for conditions of elevated temperature and pressure [11, 12]. Also the aqueous fluids under question are usually of high salinity and high hardness. Furthermore, in the upstream industry, all of the main classes of surfactant are of interest, and for some of these it can be difficult to derive CMC values, particularly non-ionic and amphoteric surfactants. Also many surfactants applied to the oilfield are in mixtures or formulations alongside other chemical products, which can have complicating effects and depend on whether the added substance is soluble in the micelle or in the intermicellar solution [12].

In the oilfield, surfactants are used at high temperatures as additives for steam floods to generate foams for the improvement of sweep efficiency. In enhanced oil recovery (EOR), from high-temperature reservoirs, surfactants are part of chemical slugs. In these applications, the knowledge of the CMC value as well as the ability to calculate activity coefficients at various temperatures (e.g. for solubility, CMC and phase behaviour properties) is vital to their successful, or at least most economic, implementation.

Most surfactants must be above the CMC in order to have appropriate properties for forming foams or mobilising oil. The CMC of a surfactant increases dramatically at high temperatures. Thus, it is important to know the CMC value at the temperature of application for economical engineering design. However very little work has been conducted in this area. Also crucially very little work has been done in examining other additive effects on CMC values.

At moderate temperatures, the presence of brine and alcohol is known to lower the CMC of surfactants [13]. Aliphatic alcohols are known to partition into micellar aggregates to varying extents, depending on the alkyl chain lengths of the alcohol and the surfactant, the structure of the surfactant, temperature, micellar size and electrolyte concentration. It is believed that this partitioning is largely responsible for the change in the micellisation behaviour of surfactants in the presence of alcohol.

While cationic surfactants are seldom used as EOR surfactants, studies have shown that cationic and anionic surfactants with the same tail show similar trends in their behaviour [12].

Where there is a need to establish the CMC value for non-ionic and amphoteric surfactants at temperatures (and pressures) that may be realistically found in oilfield applications, most of the established methods are not feasible. Two methods have been found to be applicable:

1) The captive drop technique, which is measuring surface tension [14]
2) Dynamic foam stability measurement [15]

In general CMC values exhibit a weak dependence on temperature [16] and pressure [17]. The addition of electrolytes to an ionic surfactant solution results in a linear dependence on the concentration of the added electrolyte [18]. In non-ionic micelle situations the addition of an electrolyte has little or no effect on the CMC value. When non-electrolytes are added to a surfactant micellar solution, the effects are dependent on the nature of the additive. For polar additives such as alcohols, the CMC value decreases with increasing concentration of the alcohol [19]; however species such as urea have the opposite effect, increasing the CMC value, and may even inhibit micelle formation. Non-polar additives have no appreciable effect on the CMC value [20]. These relationships can be highly important in the selection of surfactants for specific oilfield applications and also in the design of formulations of combinations of surfactants and other additives.

The solubility of ionic micelle-forming surfactants shows a strong increase above a certain temperature, which is known as the Krafft point [21]. This is due to the fact that single surfactant molecules have a limited solubility due to their dual nature; however micelles are very soluble as they behave as either hydrophilic or hydrophobic entities. At the Krafft point a relatively large amount of surfactant can be dispersed as micelles, and solubility increases greatly. Above the Krafft point the maximum reduction in surface tension or IFT occurs at the CMC value because the CMC value determines the surfactant monomer concentration. Below the Krafft point the solubility of the surfactant is too low for micellisation to occur; therefore solubility alone determines the surfactant concentration.

Non-ionic surfactants do not exhibit Krafft points [21]; their solubility decreases with increasing temperature, and these surfactants lose their surface-active characteristics above a transition temperature known as the cloud point. Above this temperature the micelle-rich surfactant phase separates, and often a marked increase in turbidity is observed.

3.1.3 Surface Tension Effects

In oil (and gas) extraction and processing, two-phase dispersions are the norm, with a thin intermediate region, known as the interface, lying between the two phases. The physical properties of this interface layer can be very important with respect to oil recovery and processing operations as from the reservoir rock to surface processing there are large interfacial areas exposed to a large number of chemical reactions. Additionally and making the situation more complex, many of these recovery and processing operations involve colloidal dispersions such as foams and emulsions, again having large interfacial areas. In these interfaces there are large amounts of free energy present, and if in processing the crude oil (or gas) there is a need to interact and/or resolve this interface, a large amount of energy needs to be inputted. A convenient and efficient way to input this energy is to use surfactant chemistry, which lowers the interfacial free energy or interfacial (surface) tension. The addition of a very small quantity of surfactant, a few parts per million, can significantly lower the surface tension and therefore the amount of energy required for foam formation [22].

Surface tension may be visualised by considering the molecules of a liquid. The attractive intermolecular forces (van der Waals forces) are exerted equally between the molecules except at the surface or interfacial region. This imbalance pulls those molecules at the interface towards the interior bulk of the liquid. This contracting force at the surface is known as surface tension, and it acts to minimise the surface area. Therefore, in a gas, bubbles adopt a spherical shape to minimise the surface free energy, and in an emulsion of two immiscible liquids (such as crude oil and water), a similar situation applies to the droplets of one of the liquids. In the latter example, it may, however, not be obvious which liquid is forming, or will form, the droplets. In any case there will still be an imbalance, which results in IFT, and the interface will adopt a configuration, which minimises the interfacial free energy.

The surface tension of aqueous solutions of surface-active agents decreases very rapidly until the CMC is reached and then stays constant above the CMC. Above this concentration, the surface tension of the solution remains constant since only the monomeric form contributes to the reduction of the surface or IFT. For concentrations below but near the CMC, the slope of the curve is essentially constant, indicating that the surface concentration has reached a constant maximum value. In this range, the interface is saturated with surfactant molecules, and any continued reduction in the surface tension is mainly due to the increased activity of the surfactant in the bulk rather than at the interface.

There are many methods available to measure surface and IFT, and it is beyond the scope of this brief explanation to detail these [23]. Pendant and sessile drop methods are commonly used in oil and gas applications [24, 25]. For ultralow IFT measurements the spinning drop method is normally used [23, 24, 26].

3.1.4 Surface Adsorption Effects

When surfactant molecules adsorb at the interface, the IFT decreases, at least up to the CMC value; this is because the surfactant molecules provide an expanding force against the normal IFT. Thus the addition of surfactant tends to lower the IFT. This phenomenon is known as the Gibbs effect.

As has been discussed, surface tension effectively describes how difficult it is to extend the area of a surface (by stretching or distorting it). If surface tension is high, there is a large free energy required to increase the surface area, so the surface will tend to contract and hold together. The *composition* of the surface may be different from the bulk. For example, if water is mixed with a tiny amount of surfactant, the bulk water may be 99.9% water molecules and 0.1% surfactant molecules, but the topmost surface of the water may be 50% water molecules and 50% surfactant molecules. In this case, the surfactant has a large and positive 'surface excess'. In other examples, the surface excess may be negative: for example, if water is mixed with an inorganic electrolyte such as sodium chloride, the surface of the water is on average less salty and more pure than the bulk fluid.

Consider again the example of water with a small concentration of surfactant. Since the water surface needs to have higher concentration of surfactant than the bulk, whenever the water's surface area is increased, it is necessary to remove surfactant molecules from the bulk and add them to the new surface. If the concentration of surfactant is increased a small amount, the surfactant molecules are more readily available, so it is easier to 'pull' them from the bulk in order to create the new surface. Since it is easier to create new surface, the surface tension is lowered. This effect only persists until the surfactant equilibrium is re-established at the surface or interfacial boundary layer. For thick films and bulk liquids, this can occur quickly (seconds); however, for thin films, there may be insufficient surfactant in the interfacial region to establish the equilibrium quickly, which then requires diffusion from other parts of the film. The film restoration process is then one of surfactant moving along the interface from a region of low surface tension to one of high surface tension. This is an important mechanism in the design of a number of surfactant applications in the upstream oil and gas industry but particularly corrosion inhibitors, which adsorb at the interfacial surfaces between oil–water and metal.

In principle the same processes are being applied at all interfacial surfaces. Firstly, the available monomers adsorb onto the freshly created interface. Then, additional monomers must be provided by the break-up of micelles. Especially when the free monomer concentration (i.e. CMC value) is low, the micellar break-up time or diffusion of monomers to the newly created interface can be rate-limiting steps in the supply of monomers, which is the case for many non-ionic surfactant solutions [27].

As has been described, experimentally, the CMC value for any surfactant can be determined from the discontinuity or inflection point in the plot of a physical property of the solution as a function of surfactant concentration. These distinctive breaks of almost every measurable physical property in aqueous media are demonstrated by all types of surfactants, i.e. non-ionic, anionic, cationic and zwitterionic, and depend on the size and the number of particles in solution [8].

3.1.5 Detergency, Oil Displacement and Wettability

Detergency involves the property of surfactants to alter the interfacial effects such as tension and viscosity to the removal of a phase from a solid surface. This effect is most utilised in the application of surfactants as laundry detergents, in which the ability of a detergent to lift soil (dirt and grease) from a surface by displacing it with surfactants, which adhere more readily to the surface being cleaned than to the soil, is exploited. This set of properties has been applied to a number of oilfield applications, including but not limited to recovery processes, cleaning process and removal of unwanted or spent muds in drilling operations and oil spill dispersion.

These effects can be exceedingly complex and have been extensively studied over the preceeding decades [28–31]. Discussion on surfactant applications, the types of surfactant used and specific examples will be discussed in the following sections of this chapter. This section will concentrate on the general mechanistic aspects of inducing oil displacement, motility and dispersion.

When a drop of oil in water comes into contact with a solid surface, the oil can form a bead on the surface or spread to form a film. A liquid having a strong affinity with the surface will seek to maximise its contact with the surface and form a film, maximising its interfacial area. A liquid with a lower affinity may form a bead. The affinity is termed *wettability*. The wettability is measured by the contact angle and the IFT. The contact angle is the angle where a liquid–vapour interface meets a solid surface. It is conventionally measured through the liquid and quantifies the wettability of a solid surface by a liquid. Any given system of solid, liquid and vapour at a given temperature and pressure has a unique equilibrium contact angle. This is expressed mathematically through the Young equation (see Figure 3.5), which quantifies the wettability of a solid surface. In reality the situation is more complex than this simple model.

In oil recovery, primary production using the energy inherent in the reservoir will recover around 15% of the original oil in place. Secondary recovery techniques, usually flooding the reservoir with water, can achieve an additional 15% recovery. This means that 70% of the original oil in place remains trapped in the reservoir rock pores. In tertiary recovery (EOR), techniques are employed to alter the capillary forces, viscosity, IFT and wettability to drive the trapped oil out of the rock pores. This can be a complex engineering and economic challenge [14, 32] and is outside the scope of this book. This study will concentrate on the use of surfactants within this area.

Critically it has been found, over three decades ago, that to remove the oil (often termed residual oil) that remains as a discontinuous phase after a completed water flood (secondary recovery) is a function of the ratio delta P/L sigma, where delta P is the pressure drop across the distance L and sigma is pressure drop across the distance L and sigma is the IFT between the oil and water [33].

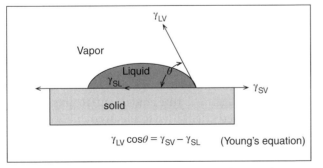

$$\gamma_{LV} \cos\theta = \gamma_{SV} - \gamma_{SL} \quad \text{(Young's equation)}$$

γ_{LV} = liquid–vapor interfacial tension or surface tension
γ_{SV} = solid–vapor interfacial tension, not true surface energy
γ_{SL} = solid–liquid interfacial tension
θ = contact angle (angle liquid makes with solid surface)

Figure 3.5 The Young equation and contact angle.

It has been established that no residual oil can be removed from a porous rock until a critical value of delta P/L sigma is exceeded. The critical ratio is P/L sigma and is a fundamental property of the reservoir rock. If this critical value is exceeded by applying more pressure or by reducing the oil–water interfacial pressure or by reducing the oil–water IFT (or both), some additional oil is invariably produced. The exact amount of additional oil is always a monotonically increasing function of the increase in the value of the ratio delta P/L sigma. To recover economically attractive amounts of oil, the critical delta P/L sigma value should be exceeded by at least one order of magnitude. By reducing the oil–water IFT utilising various combinations of surfactant and pressure gradients, excellent recoveries can be obtained in the laboratory. However, in translating the laboratory results to the reservoir, it becomes apparent that extremely low values of the oil–water IFT must be used to avoid excessively high water flooding pressures that would fracture the formation. Generally, to achieve this, a variety of surfactants are used in combinations with polymers and at alkaline pH.

It has been known at least since the 1950s [34, 35] that surfactants can significantly aid in extraction and EOR processes, particularly for heavy oils and bituminous tar sands. This is particularly the case for non-ionic surfactants under an alkaline condition or a pH of at least 12. Concentrations of about 0.1% are usually sufficient for the purpose of spontaneous emulsification, therefore increasing the recoverable yield of oil from the tar sands. Generally, water-soluble alkali metal halides, sulphates, carbonates, phosphates and the like are suitable for use for this purpose, the amount thereof in the solution ranging from substantially negligible amounts to as high as about 5% by weight, although still larger amounts may be sometimes desirable or even necessary for certain field conditions.

Importantly, although surfactants, surfactant combinations and alkali surfactants are used in EOR operations, they are also used in combination with polymers, in particular polyacrylamides (PAM) (see Chapter 2). Although one of the main uses of surfactants in EOR is to reduce IFT, they are also used to alter wettability, which is discussed in the next section.

3.1.6 Wettability Alteration

A consequence of surfactant adsorption onto a porous media, such as a reservoir rock, is that it may alter the surface wetting properties. This can be a considerable advantage in EOR where reservoir is of mixed wettability or is predominantly oil wet. In such cases surfactants are used to decrease the contact angle, making the reservoir more water wet.

More than half of the world's oil reserves are held in carbonate reservoirs, and many unfavourable factors contribute to low oil recovery in these reservoirs. The oil-wet nature of such reservoirs is among the leading factor for poor oil recovery, and therefore much research has focused on wettability alteration and reduction in IFT. Chemical EOR, particularly the use of surfactants, can lead to wettability alteration and IFT changes. Studies [36, 37] have shown that wettability alteration only plays important roles when IFT is high and is primarily effective in early field life application. IFT plays very important roles with or without wettability alteration and is effective during the entire process. The implication is that anionic surfactants are preferred to reduce IFT and cationic surfactants are preferentially used to alter wettability. Another observation is that in surfactant-induced wettability alteration with low IFT, gravity drive is a very important mechanism. Molecular diffusion of chemicals affects oil recovery rate in the early field application, but not ultimate oil recovery.

In fractured sandstone reservoirs, the efficiency of water flood is governed by spontaneous imbibition of water into oil-containing matrix blocks. When the matrix is oil wet or mixed wet, little oil can be recovered by imbibition. It has been shown that surfactants that can be added to the injection water that can induce imbibition into an originally mixed-wet, tight, fractured sandstone reservoir. It has also been shown [38] that the use of dilute (0.1 wt%) anionic surfactant solutions can alter the wettability from oil-wet towards more water-wet conditions. Incremental oil recovery as high as 68% original oil in place has been shown under laboratory conditions to be feasible.

Much research and development on the application of surfactants in this area continues particularly with regard to gas and gas condensate reservoirs [39, 40], especially with regard to liquid unblocking.

3.1.7 Surface Potential and Dispersions

When substances are brought into contact with a polar medium such as water, they acquire a surface electrical charge. In crude oil/aqueous mixtures, the charge can be due to the ionisation of surface acid functionalities; in gas/aqueous systems, the charge could be due to the adsorption of surfactant ions; in porous rock or solid suspensions, the charge could come from diffusion of counter-ions away from the mineral surface whose internal structure carries the opposite charge. In the field, the nature and degree of such surface charged systems is much more complicated. Surfactants through adsorption may cause this surface charge to increase, decrease or not significantly change.

The presence of a surface charge influences the distribution of nearby ions. Ions of opposite charge (counter-ions) are attracted to the surface, whereas those of the same charge are repelled. An electrical double layer (EDL) is therefore formed. This layer is likely to be diffuse in character due to mixing caused by thermal motion effects. This EDL can be viewed as having an inner layer that will include adsorbed ions and an outer or diffuse layer where ions are distributed according to electrical forces and thermal effects.

Any molecule covalently bonded to the surface moves with the particle when it diffuses or is induced to move as in an applied electric field. When wetting, dispersing or stabilising agents, such as surfactants, are strongly adsorbed onto the surface, they too move with the particle. Counter-ions very near the surface, perhaps within the first nanometre or two, also move with the particle. Finally, solvent molecules are sometimes also strongly bound to the surface. However, at some short distance from the surface, the less tightly bound species are more diffuse and do not move with the particle. This imaginary yet useful theoretical layer is defined as *the shear plane*. Everything inside the shear plane is considered to move with the particle; everything outside of the shear plane does not. In other words, as the particle moves, it shears the liquid at this plane.

It is possible to measure or quantify this surface potential through electrokinetic measurements, and this leads to a quantification of the moving particles known as the zeta potential. This is the potential difference across phase boundaries between solids and liquids and is a measure of the electrical charge of particles that are suspended in liquid. Therefore the zeta potential is defined to be the electrostatic potential difference between an average point on the shear plane and one out in the liquid away from any particles. As a consequence the zeta potential in aqueous suspensions is a function of two variables: charge at the shear plane and free salt ion concentration, where 'free' means not attached to the particle surface.

Colloidal suspensions are stabilised in one of two ways. Surface charge, naturally occurring or added, enhances electrostatic stability. Adsorption of non-polar surfactants or polymers enhances stability through static stabilisation.

The square of the zeta potential is proportional to the force of electrostatic repulsion between charged particles. Zeta potentials are, therefore, measures of stability. Increasing the absolute zeta potentials increases electrostatic stabilisation. As the zeta potential approaches zero, electrostatic repulsions become small compared with the ever-present van der Waals attraction.

Eventually, instability increases, which can result in aggregation followed by sedimentation and phase separation, an important mechanism in oil–water separation, for example.

The zeta potential is important because, for most real systems, as the surface potential cannot be measured, one cannot measure the zeta potential directly either; however, one can measure the electrostatic mobility of the particles and calculate zeta potential. Though strictly incorrect, it is common to hear the zeta potential spoken of as a substitute for the surface potential. The surface potential is a function of the surface charge density. The zeta potential is a function of the charge density at the shear plane. The magnitude of the zeta potential is almost always much smaller than the surface potential.

Understanding zeta potentials in oil recovery and oil processing scenarios can be critical to economic and efficient operations [41–44] as well as examining potential alteration by chemical surfactant addition [45–47].

Surfactants, particularly ionic surfactants, are involved in the stabilisation of colloidal dispersions by increasing the repulsive electrostatic forces between particles (dispersions), droplets (emulsions) or bubbles (foams). It is proposed that this action counterbalances the van der Waals attraction between molecules and stabilises films and dispersions. This classical concept of dispersion stability was formulated by Derjaguin, Landau, Verwey and Overbeek (DLVO) and is known as the DLVO theory [48]. More recent experimental data [49] has shown that considerable deviations from the conventional DLVO theory appear for short surface-to-surface distance (hydration repulsion) and in the presence of bivalent and multivalent counterions (ionic correlation force). Both effects can be interpreted as contributions to the double-layer interaction not accounted for in the DLVO theory.

There are a number of other forces such as oscillatory forces that can affect the stability of films and dispersions including surfactant micelles [50–52]; however this level of complexity is outside the scope of this current study, suffice to say that in terms of application in upstream oilfield chemistry, these forces appear to have minimal impact on the selection and use of surfactants.

When non-ionic surfactants are adsorbed at the film or particle surfaces, the formed polymer–surfactant complex can give rise to a steric interaction between the two surfaces [49, 53].

The surfactant molecules in adsorption monolayers or lamellar bilayers are involved in thermally exited motion, which brings about the appearance of fluctuation capillary waves. The latter also cause a steric interaction (though a short-range one) when two thermally corrugated interfaces approach each other.

Finally ionic contaminants can exist in many non-ionic surfactants and sometimes they are surface active; therefore these contaminants bring some negative charge to emulsion surfaces.

3.1.8 Surfactants in Emulsions and Hydrophilic–Lipophilic Balance (HLB)

At liquid–liquid interfaces, surfactants can, as described earlier, lower IFT, increase surface elasticity, increase the double-layer electric repulsion (particularly ionic surfactants) and potentially increase surface viscosity, all of which influence the stability of the emulsion. The addition of a surfactant to an emulsion system can determine the arrangement of the phases within the emulsion and which phase will be dispersed and which will form the continuous phase.

In oilfield emulsifiers and demulsifiers, mixtures of agents are often more effective than single components. It is thought that such mixtures have greater effects in lowering IFT and potentially forming strong interfacial films.

An empirical scale has been developed to characterise single-component or non-ionic surfactant mixtures used in this area, namely, the HLB. The HLB of a surfactant is a measure of the

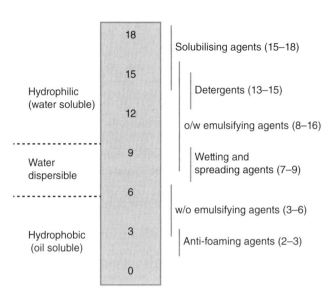

Figure 3.6 The HLB scale.

degree to which it is hydrophilic or lipophilic and is determined by calculating values for the different regions of the molecule [54–56].

The non-ionic surfactant HLB scale (Figure 3.6) varies from 0 to 20. A low HLB value (<9) refers to a lipophilic or oil-soluble surfactant, and a high HLB value (>11) to a hydrophilic or water-soluble surfactant. Most ionic surfactants have HLB values greater than 20 and are therefore mainly water soluble.

In general, natural emulsifiers that stabilise a water-in-oil emulsion exhibit an HLB value in the range of 3–8. Thus, demulsifiers with a high HLB value will destabilise these emulsions. Demulsifiers act by total or partial displacement of the indigenous stabilising interfacial film components (polar materials) around the water droplets [57]. This displacement also brings about a change in properties such as interfacial viscosity or elasticity of the protecting film, thus enhancing destabilisation. In some cases, demulsifiers act as a wetting agent and change the wettability of the stabilising particles, leading to a break-up of the emulsion film.

A limitation of the HLB system is that other factors are also important, for example, temperature. Also the HLB value of a surfactant is an indicator of its characteristics as an emulsifier or demulsifier but not its efficiency. For example, all emulsifiers with a high HLB value will tend to promote oil-in-water emulsions; there will be considerable variation in efficiency in which these systems are stabilised.

The HLB values of many non-ionic surfactants also vary with temperature; therefore a surfactant can stabilise not only an oil-in-water emulsion at a low temperature but also a water-in-oil emulsion at a higher temperature. The temperature at which the surfactant changes form stabilising oil-in-water emulsions to water-in-oil emulsions is known as the phase inversion temperature (PIT) [58]. At the PIT the hydrophilic and lipophilic natures of the non-ionic surfactant are essentially equivalent.

As can be seen the situation can be exceedingly complex and make the determination of suitable surfactants and surfactant mixtures for use in a variety of oilfield applications somewhat problematic. Nonetheless, from the understanding of some basic properties of a surfactant or surfactant mixture (the CMC, the Krafft point, adsorption characteristics, surface and IFT), it is possible to at least have an indicative view as to the performance of a surfactant in a given oil extraction or recovery process. However, this still requires some caution as in practice dynamic phenomena are in play. In the following sections the specific chemistries including examples in application will be explored.

In terms of examining 'green' characteristics of surfactants, this can also be very problematic. Many surfactants are derived from oleochemicals, which in turn are derived from plant and animal fats. They are analogous to petrochemicals derived from petroleum. The formation of basic oleochemicals such as fatty acids, fatty acid methyl esters, fatty alcohols, fatty amines and glycerols is by various chemical and enzymatic reactions. This pedigree would strongly suggest that they are inherently biodegradable and relatively non-toxic; however much debate centres around surfactants' potential to bioaccummulate [59] and this will be discussed further later in the chapter. In general anionic and non-ionic surfactants are less toxic than cationic surfactants; however, unlike other industrial sectors, the anionic surfactants are probably the least used in oilfield applications.

In the oilfield there are a number of applications across the range of surfactants and a great overlap between the surfactant classes described as products are applied for their effects on characteristics such as IFT. Nonetheless the author has attempted to show the main uses against the main recognised surfactant chemical classifications.

3.2 Anionic Surfactants

Anionic surfactants are dissociated in water in an amphiphilic anion, i.e. the polar group is negatively charged, and a cation, which is in general an alkali metal (e.g. sodium or potassium) or a quaternary ammonium species. They include alkylbenzene sulphonates (detergents), fatty acid derivatives (soaps), lauryl sulphate (foaming agent), dialkyl sulphosuccinate (wetting agent) and lignosulphonates (dispersants), to name but a few. They account for about 50% of the world production and are the most applied materials across a wide number of industry sectors. In the oil and gas sector, they are also widely used in a variety of functions. They are the most commonly used surfactants across all industries; however they are not the most commonly used in the upstream oil and gas sector.

In general anionic surfactants are manufactured from the reaction of a chemical agent such as sulphur trioxide, sulphuric acid or ethylene oxide (EO), which provides the hydrophilic head group, and hydrocarbons derived from petroleum or fats and oils (oleochemicals) to produce new acids similar to fatty acids. A second reaction adds an alkali metal to the new acids to produce one type of anionic surfactant.

In the upstream oil and gas industry, various anionic surfactants are used:

- Alkyl sulphates and alkyl ether sulphates
- Alkyl aryl sulphonates and other sulphonates
- Ethoxylated derivatives
- Other salts such as taurates and sarcosinates

3.2.1 Alkyl Sulphates and Related Derivatives

These are among the most commonly used surfactants, particularly the dodecyl (or lauryl) sulphate (Figure 3.7) as a sodium, ammonium or ethanolamine salt, which is the foaming agent found in shampoos, toothpaste and some detergents. They have a number of oilfield uses, as exemplified later in this section. They are generally prepared by neutralisation of the alkyl-ester-sulphuric acid by the appropriate base.

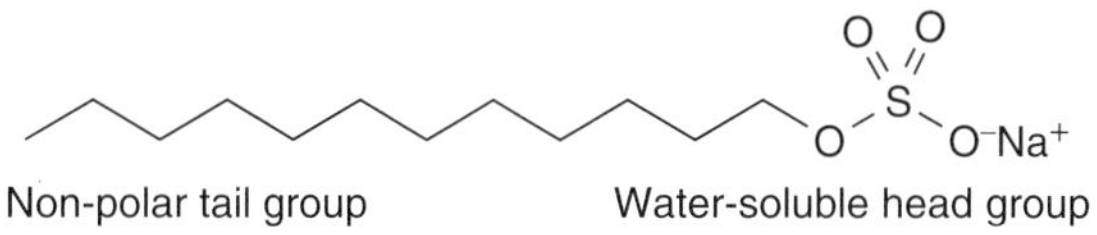

Figure 3.7 Sodium lauryl sulphate, an alkyl sulphate.

Non-polar tail group   Water-soluble head group

Sodium lauryl sulphate is an extremely hydrophilic surfactant. Lesser hydrophobicity can be achieved with a longer chain (up to C16) or by using a weaker base such as ammonia or ethanolamine.

The alcohols used in the non-polar portion of the surfactant are either produced from carboxylic acids obtained from oils, obtained naturally, for example, from palm kernel oil or coconut oil [60], or alternatively from long-chain alkenes, manufactured from ethylene and derived from petroleum cracking.

There are two main commercial processes for making alcohols from ethylene. The first commercial process is the Ziegler process [61], which uses an aluminium catalyst to oligomerise ethylene and allow the resulting alkyl group to be oxygenated. The usually targeted products are fatty alcohols, which are otherwise derived from natural fats and oils. The other main commercial process for making the alcohols from ethylene is known as Shell higher olefin process (SHOP) [62]. In the first stage, ethylene is passed, under pressure of about 100 atmospheres, into a solvent (usually a diol, such as butane-1,4-diol) containing a nickel salt at $400\,^{\circ}$K. It yields a mixture of α-alkenes, which are separated by fractional distillation of which about 30% are in the range C_{10}–C_{14}. These are subsequently reacted with carbon monoxide and hydrogen (hydroformylation) to yield straight-chain aldehydes, which on reduction give the required alcohols. It is possible to convert the other alkene fractions (C_4–C_{10} and C_{14}–C_{40}) into the more desirable C_{10}–C_{14} fraction. The fatty alcohols themselves are feedstocks for other surfactants as will described in subsequent sections of this chapter.

Sodium lauryl sulphate has been applied in a mixture with a polymer to form a coacervate gel [63]. Such gel systems are made with anionic or cationic polymers, a smaller amount of a surfactant having a charge opposite that of the polymer and a hydrophobic alcohol. The zeta potential of the gel is maintained at an absolute value of at least 20. It has been shown that a preferred gel comprises polydiallyldimethylammonium chloride, a lesser amount of sodium lauryl sulphonate and lauryl alcohol. Coacervate gels have excellent shear viscosities and other properties and are particularly useful in well drilling fluids and well fracturing fluids.

Alkyl sulphates, as well as other surfactants, have also used in the composition of foam cements [64]. Foamed cement forms a ductile and compressible medium that flexes and tolerates stresses that otherwise damage conventional cements. Inherent ductility of the foamed-cement sheath helps keep the casing and wellbore bond intact, eliminating the formation of a micro-annulus while providing greater resistance to stress cracking. These outcomes help prevent accumulation of pressure in the annulus.

A number of surfactant classes are capable of showing drag reduction capability, and anionic surfactants such as sodium dodecyl sulphate (sodium lauryl sulphate) have been shown to exhibit such properties [65].

In general however the simple alkyl sulphates are not commonly used in the upstream oil and gas sector; however the propoxylated and ethoxylated derivatives are used in a number of applications, particularly in EOR having the general structure as shown in the example in Figure 3.8. In such applications it is important that interfacial surface between oil and water is at least covered by a monolayer of surfactant molecules. This is so that a low enough IFT is achieved. The surfactant must also smoothly change from oil solubility to water solubility along the interphase. Finally it is often desired that such surfactants have a high tolerance to divalent and multivalent cations and alkyl propoxy–ethoxy sulphates, among others, satisfy these criteria.

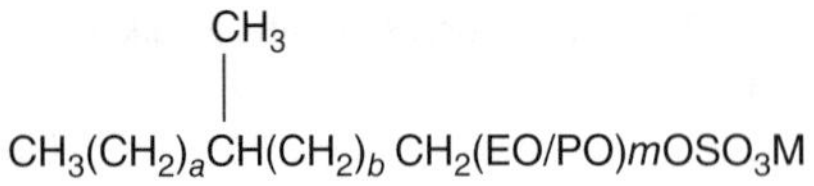

$$CH_3$$
$$|$$
$$CH_3(CH_2)_a CH(CH_2)_b\ CH_2(EO/PO)mOSO_3M$$

Figure 3.8 An alkyl propoxy ethoxy sulphates.

These products have been utilised alongside polymers as a relatively cheap and cost-effective method of enhanced water flooding [66].

Branched alkyl alcohol propoxylated sulphate surfactants have also been investigated as candidates for chemical EOR applications. Results show that these anionic surfactants may be preferred candidates for EOR, as they can be very effective at creating low IFT at dilute concentrations, without requiring an alkaline agent or co-surfactant. Also, some of the formulated products exhibit a low IFT at high salinity and hence may be suitable for use in more saline reservoirs [67].

Hydrolysis of sulphate surfactants may occur within the range of reservoir conditions at rates that can interfere with the performance of these surfactants. Studies have shown that the alkyl ethoxy sulphates have greater stability under many EOR process conditions [68].

As mentioned earlier there is an issue with adsorption of anionic surfactants on to undesired rock surfaces particularly where multivalent cations are present. This is particularly the case for simple alkyl sulphates [69]. The alkyl ether sulphates and the alkyl sulphonates and alkyl aryl sulphonates discussed in the following sections have more desirable characteristics in this respect.

3.2.2 The Alkyl Ether Sulphates (Alcohol Ethoxy Sulphates)

More widely used than simple alkyl sulphates are various types of alkyl ether sulphates; however their uses are still restricted mainly to EOR operations and some specialised drilling and completion applications.

In the manufacture of these products, the primary alkyl alcohol (from a synthetic or natural source) is typically a blend based on dodecanol. This is first ethoxylated with 1–3 molar equivalents of EO, and the reaction product is then sulphated using sulphur trioxide and neutralised with alkali to form the alkyl ether sulphate as represented in Figure 3.9.

As can be seen this is a very similar structure to that of Figure 3.7 except the hydrophilic (water-soluble head) group is much enlarged.

These products are preferred for a number of applications in the upstream oil and gas industry as they generate less foam, which is an advantage in the formulation and stability of many products both in drilling applications and in oil recovery and processing. Some, however, have good foaming characteristics such as the alcohol ethoxy sulphates and have the potential to aid foams to act as flow diverting agents in low permeability regions of a reservoir [70].

Like the alkyl sulphates, the alkyl ether sulphate surfactants have also been used in the composition of foam cements [64].

The alkyl ether sulphates have been applied in niche applications in drilling and completion scenarios, in particular in stabilised colloidal drilling fluids [71] where they work in combination with polyvinyl alcohols to modify the water viscosity to such an extent that an elastomeric

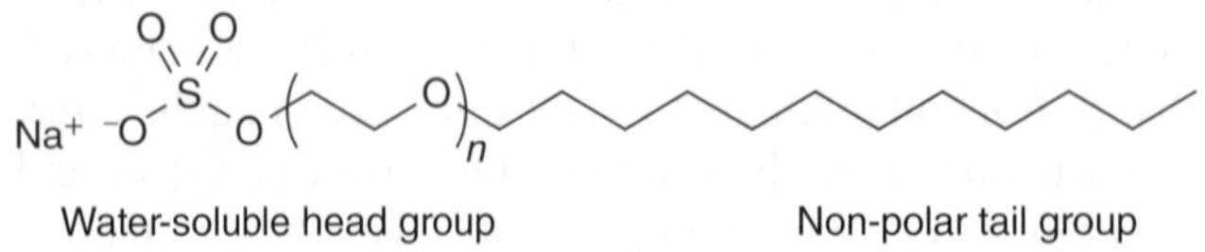

Figure 3.9 Sodium lauryl ether sulphate, an alkyl ether sulphate.

Water-soluble head group   Non-polar tail group

membrane is created to improve the stability of the colloid and create a better seal between production zones.

Where surfactants are used in drilling fluid mixtures, it is critical that they are highly compatible with the polymers present in the drilling fluid composition. Alkyl ether sulphates are one of the few anionic surfactants that are generally compatible for these applications [72].

From an environmental impact viewpoint, it has been established that the alkyl ether sulphate, and in particular sodium lauryl sulphate, are not carcinogenic [73]. Some products containing alkyl ether sulphates have been found to also contain traces of 1,4-dioxane. This is formed as a by-product during the ethoxylation step of their synthesis. Many regulatory authorities recognise 1,4-dioxane to be a probable human carcinogen and a known irritant; however it has no observed adverse effects at concentrations of $400\,\mathrm{mg\,m^{-3}}$, which is significantly higher than those found in commercial products [74].

3.2.3 Alkyl Aryl Sulphonates, Related Sulphonates and Sulphonic Acid Derivatives

This group is the most common of the synthetic anionic surfactants and is based on the straight-chain alkylbenzene sulphonates. Benzene, in slight excess, is mixed with an alkene or chloroalkane in the presence of an acid catalyst, usually a solid zeolite (ion exchange), aluminium chloride ($AlCl_3$) or hydrofluoric acid (HF), to produce an alkylbenzene. The alkylbenzene is then sulphonated using an air/sulphur trioxide mixture, and the resulting sulphonic acid is then neutralised with an aqueous solution of sodium hydroxide (often in situ), as in Figure 3.10.

The alkylbenzene can vary in average molecular mass depending upon the starting materials and catalyst used and is often a mixture in which the length of the alkyl side chain varies between 10 and 14 carbon atoms. Historically these included branches in the side chains with the result that they biodegrade very slowly, which in general industrial uses lead to pollution problems due to foaming in rivers and sewage plants. By law, in most countries today, the surfactant must have side chains, which are not branched so they degrade more rapidly.

Sodium dodecylbenzene sulphonates are a member of the linear alkylbenzene sulphonates (LAS), meaning that the dodecyl group ($C_{12}H_{25}$) is unbranched. This is the most widely used anionic surfactant across all industrial sectors including the upstream oil and gas sector, and billions of kilogrammes are produced annually. The dodecyl chain is attached at the 4-position of the benzene sulphonate group. Linear dodecyl-4-benzenesulphonate anions can exist in six geometric isomers, depending on the carbon of the dodecyl group that is attached to the benzene ring. The isomer shown in Figure 3.11 is 4-(5-dodecyl)benzenesulphonate (4 indicating the position of the benzene ring, 5 indicating the position on the dodecane chain). Branched isomers, e.g. those derived from tetramerised propylene, are also known Figure 3.12 but are not as widely used because they biodegrade too slowly.

Linear alkylbenzene products have been extensively evaluated due to concern about its effect on the environment and human health. European Council (EC) Regulation 1488/94 life cycle analysis considered the emissions and resulting environmental and human exposures.

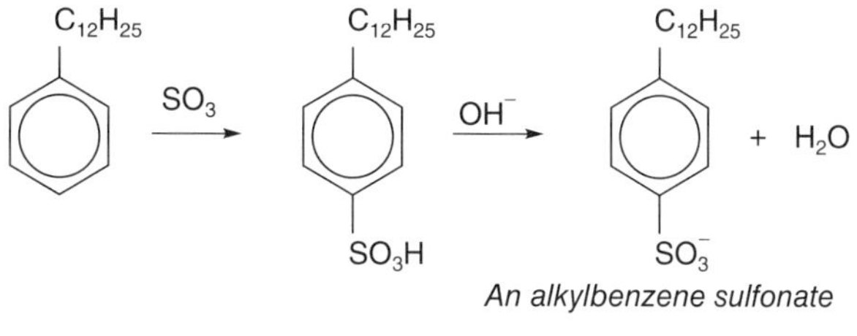

Figure 3.10 Manufacture of alkylbenzene sulphonate.

An alkylbenzene sulfonate

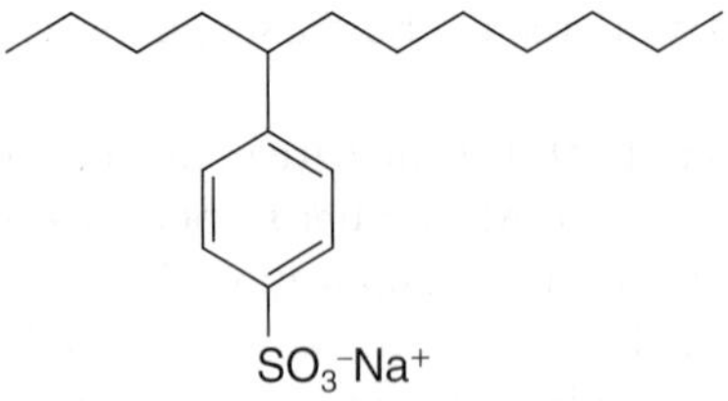

Figure 3.11 4-(5-Dodecyl)benzenesulfonate.

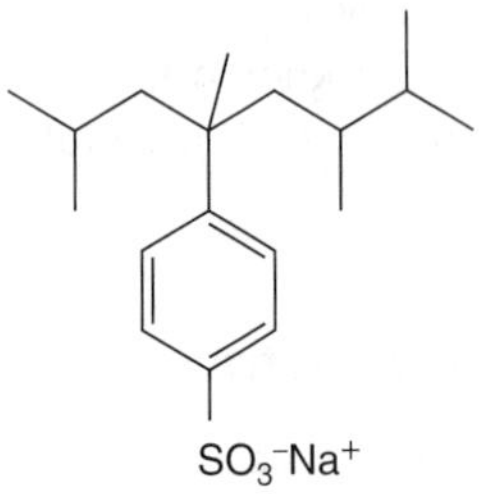

SO$_3^-$Na$^+$

Figure 3.12 A branched dodecylbenzenesulphonate.

SO$_3^-$Na$^+$

Following the exposure assessment, the environmental risk characterisation for each protection target in the aquatic, terrestrial and soil compartment was determined. For human health, the scenarios for occupational exposure, consumer exposure and human exposure indirectly via the environment have been examined, and the possible risks identified.

The biodegradability of LAS has been well studied [75] and is affected by, as previously stated, isomerisation (branching). The salt of the linear material has an LD$_{50}$ of 2.3 mg l^{-1} for fish, about 4× more toxic than the branched compound; however the linear compound biodegrades far more quickly, making it the safer choice over time. It is biodegraded rapidly under aerobic conditions with a half-life of approximately 1–3 weeks [76] and that oxidative degradation initiates at the alkyl chain. Under anaerobic conditions it degrades very slowly or not at all, causing it to exist in high concentrations in sewage sludge, but this is not thought to be a cause for concern, as it will rapidly degrade once returned to an oxygenated environment.

The report concluded that there are no concerns for the environment or human health and that there is no need for further testing or risk reduction measures beyond those currently practised [77].

The use of alkyl aryl sulphonate surfactants for oil recovery has been well studied for over 80 years. Water-soluble surfactants, such as polycyclic sulphonate and wood sulphate, were first described in 1931 by De Groote as an aid to improve oil recovery [78]. Reisberg and Doscher in 1959 [35], using a California crude and surfactant solutions containing NaOH, demonstrated in the lab that the addition of alkali produced interfacial activity related to certain components in the crude oil and that the addition of surfactant such as an alkylbenzene sulphonate could enhance this activity. In the following decades injection of a solution containing both surfactant and alkali for EOR has become widely applied. Such processes, described as alkaline surfactant flooding, have attracted and continue to attract considerable interest.

In the 1960s the surfactants used were made either by direct sulphonation of aromatic groups in refinery streams or crude oils or by organic synthesis of alkyl aryl sulphonates. Throughout the 1970s and early 1980s, extensive research, field testing and implementation were triggered by an expectation of high oil prices, especially in the United States, by a decline of overall oil production. Petroleum sulphonates (together with an alcohol co-solvent in most cases) gained popularity during this time. A series of systematic studies led to the recognition that the capillary number controlled the amount of residual oil remaining after flooding an oil-containing core. These studies revealed that at typical reservoir fluid velocities, the crude oil-brine IFT had to be reduced from 20 to $30\,\mathrm{mN\,m^{-1}}$ (or dyne/cm) to values in the range of 0.001–$0.01\,\mathrm{mN\,m^{-1}}$ to achieve low values of residual oil saturation. Gale and Sandvik [79] proposed four criteria for selecting a surfactant for a tertiary oil recovery process:

1) Low oil–water IFT
2) Low adsorption
3) Compatibility with reservoir fluids
4) Low cost

The alkyl aryl sulphonates and related products tend to meet all of the aforementioned criteria. An area where these products may not be as applicable is in alkaline surfactant EOR scenarios, and this is discussed in Section 3.9, which examines the thermal stability and degradation of surfactants.

In the last two decades there have been some significant developments in the use of derivatives of alkyl aryl sulphonates at very low concentrations, in particular to produce ultralow IFTs for sandstone and limestone formations [80–82]. In these products the sulphonate group is attached to the end of the alkyl chain as opposed to being directly attached to the aromatic ring, as illustrated in Figure 3.13.

For a number of years the alkyl aryl sulphonic acids and their base salts, particularly the magnesium salt, have been used as asphaltene dispersants [83]. These materials, mainly based on the 4-isomer of dodecylbenzene sulphonic acid (DDBSA) (Figure 3.14), are also used as asphaltene dispersants in downstream refinery applications [84].

The development of these sulphonic acid and related salts for application as asphaltene dispersants has taken two directions.

One area of development to improve dispersant performance has focused on the use of branched alkyl polyaromatic sulphonates with sulphonated alkyl naphthalenes being the preferred structures [85]. These synthetic dispersants greatly increase the solubility of asphaltenes in crude oils at low concentrations. They have one or more sulphonate head groups that complex with the polynuclear aromatic structures in asphaltenes and long paraffinic tails that

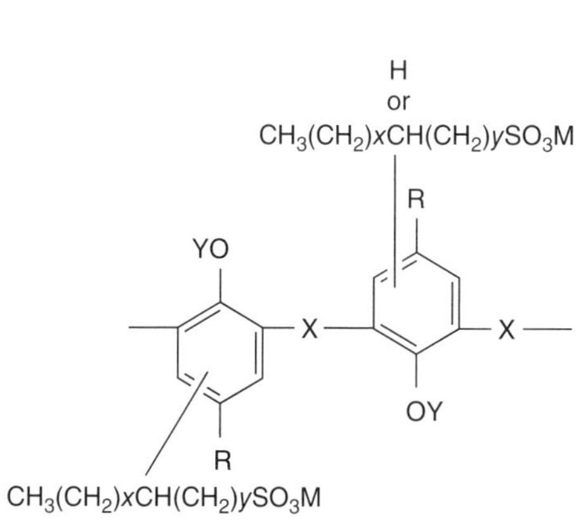

Figure 3.13 Sulphoalkylated phenol surfactants.

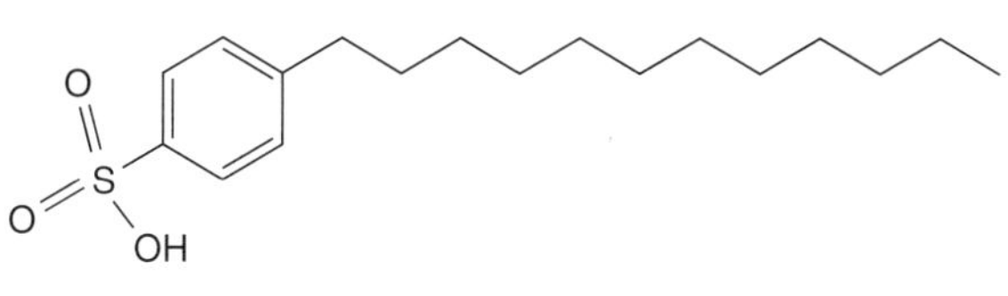

Figure 3.14 4-Isomer of dodecylbenzene sulphonic acid.

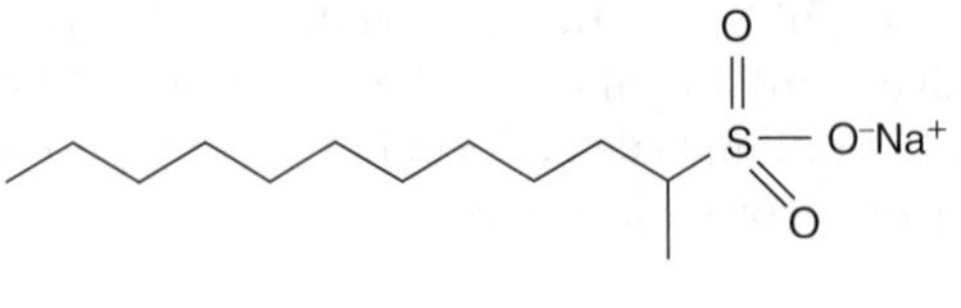

Figure 3.15 Secondary alkane sulphonate.

promote solubility in the rest of the oil. As a result, synthetic dispersants can be much more effective than the natural dispersants in the oil, namely the resins. At high concentrations, synthetic dispersants can even make all the asphaltenes soluble in *n*-heptane and therefore convert asphaltenes to resins. It has been determined that one sulphonic acid group was the most effective head attached to the two ring naphthalene aromatic structure. It has also been found that a straight-chain paraffinic tail is not effective above 16 carbons and that this is because of decreased solubility in the oil caused by crystallisation with other tails and with waxes in the oil. In addition, *n*-alkyl aromatic sulphonic acids lose their ability to disperse asphaltenes with time. Both of these issues are addressed by using two branched tails of varying length proportions between the two tails. As a result, the effectiveness of the dispersant increases with total tail length, well above 30 carbons, and it remains effective with time.

One group is based on the removal of the aryl group, leaving the sulphonate group directly bounded to aliphatic alkyl chains [86], as illustrated in Figure 3.15.These materials are analogous to the alkyl sulphates described in Section 3.2.1. Significantly it appears only secondary alkyl moieties are functional and are required to have chain lengths of between 8 and 22 carbon atoms. This is probably due to better overall solvent interaction in the formulated product as applied.

These latter materials are also analogous to α-olefin sulphonates, which are more fully described in Section 3.2.4.

3.2.4 α-Olefin Sulphonates

These surfactants, as illustrated in Figure 3.16, have good wetting and excellent, stable foaming properties and are used widely in the detergent industry [87], particularly in laundry and personal care as they exhibit good cleaning and high foaming properties in both hard and soft waters and are non-irritant with good skin mildness.

The stable foaming characteristics and the added advantage of rapid biodegradability have meant that these surfactants are used in a number of oilfield applications.

As with other sulphonates and in particular the alkyl aryl sulphonates, they have been applied in water flooding and secondary recovery processes. The α-olefin sulphonates have a particular efficacy in alcohol-free surfactant floods [88]. Surfactant flooding with an alcohol-free slug has considerable appeal in that it would potentially yield a high final oil recovery for a small amount of surfactant injected, thanks to the high solubilisation ratios produced compared with formulations with co-surfactant alcohols. The inclusion of ethoxylated groups, aromatic rings, branching of alkyl chains and in particular double bonds destabilise the viscous aggregates, allowing the micelles to accommodate water or oil, working like a 'built-in' co-surfactant.

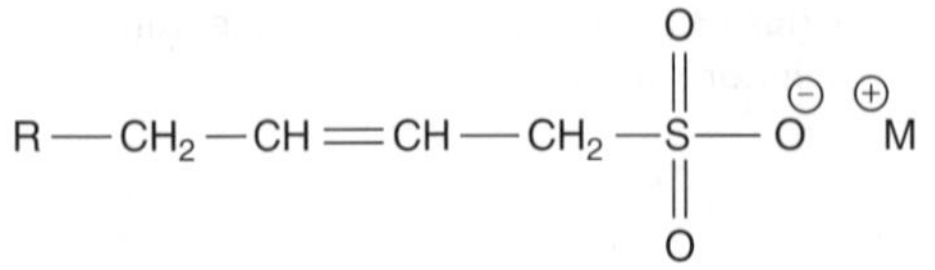

Figure 3.16 α-Olefin sulphonates.

Conversely, longer alkyl chains (more paraffinic), less branching and fewer aromatic groups all lead to lower IFTs. The α-olefin sulphonates satisfy both of these divergent criteria.

Secondary recovery using steam or carbon dioxide can be considerably improved when α-olefin sulphonates are used to both increase foam stability and reduce the permeability of the flood-swept zones [89]. α-Olefin sulphonates have also been proposed as suitable foamers for foam-based hydraulic fracturing applications [90].

A critical application for foaming surfactants is in gas shut-off and gas production control and has been found to be particularly useful in reservoirs having a high gas–oil ratio (GOR) [91]. This is because the foam is confined within the pore network in the reservoir rock matrix and therefore consists of thin liquid films spanning the pores, making the gas phase discontinuous. This drastically reduces the gas mobility without altering the liquid relative permeability. The foam is generated from nitrogen using a suitable surfactant, which can under suitable circumstances be an α-olefin sulphonate; however these have been found to be poor foam blockers in the presence of a high crude oil ratio [92]. This issue, however, has been successfully addressed with the use of mixed α-olefin sulphonate polymer formulations such as with PAM [93] and also when α-olefin sulphonates are combined with fluorosurfactants [94].

From the 1970s the fate of fluorinated chemicals in the environment has been regarded with considerable interest, as it emerged that chlorofluorocarbons were primarily responsible for the destruction of stratospheric ozone [95]. This consequently led to their use and manufacture being banned due to their extreme potential for destroying the ozone layer.

The incorporation of fluorine into organic molecules can give rise both positive and adverse effects. Although fluorine imparts positive and unique properties such as water and oil repellency and chemical stability, adverse effects often pervade members of this compound class. More recent concerns connected with fluorinated chemicals have arisen within the last two decades when the perfluorinated carboxylic acids and perfluorinated alkane sulphonates were found to be environmentally persistent [96, 97]. It has been shown, however, that biodegradation plays an important role in understanding how fluorinated substances reach the environment and, once they do, what their fate is. It appears that desulphonation of a highly fluorinated surfactants can be achieved if an α-situated H atom, in relation to the sulphonate group, is present, at least under sulphur-limiting conditions and molecules that are less heavily fluorinated can show very complex metabolic behaviour. There are, however, indications that fluorinated functional groups, such as the trifluoromethoxy group and the *p*-(trifluoromethyl) phenoxy group, may be useful derivations in novel environmentally benign fluorosurfactants [98]. This is detailed further in Section 3.3, where the direct use of fluorosurfactants in the upstream oil and gas industry is discussed.

3.2.5 Lignosulphonates and Other Petroleum Sulphonates

Lignosulphonates, and in particular the polymeric forms, were discussed in some detail in Section 2.2.4. Lignin, which is derived from the wood-pulping process, has been used as drilling mud additives for many decades. Drilling muds [99] using lignosulphonates have a number of advantageous properties, including good viscosity control, gel strength and fluid loss properties and high temperature resistance up to 230 °C. In water-based muds lignosulphonates are also tolerant of high salt concentration and extreme hardness of the water.

As was detailed previously, however lignin and lignosulphonates have very poor rates of biodegradation, and consequently under environmental scrutiny and regulatory pressure, drilling fluid compositions have largely excluded and/or replaced these products.

The petroleum sulphonates are a related class of complex aromatic products, and a representative example is illustrated in Figure 3.17.

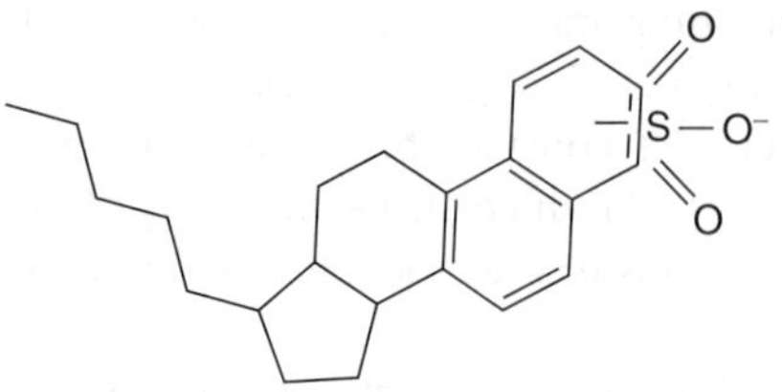

Figure 3.17 A petroleum sulphonate.

Thirty to forty years ago, the majority of surfactant systems considered for EOR included petroleum sulphonates as the primary component [100]. Previous and simultaneous work had shown a marked dependence of petroleum sulphonate performance upon its composition [101, 102]. Natural petroleum sulphonates are defined as those manufactured by sulphonation of crude oil, crude distillates or any portion of these distillates in which hydrocarbons present are not substantially different from their state in the original crude oil. These natural materials, then, are quite different from synthetic sulphonates, which are derived most commonly, as has been described in earlier sections of this chapter, from sulphonation of olefin polymers or alkyl aromatic hydrocarbons. In general, natural petroleum sulphonates are much more complex mixtures than synthetics. The major reason for this difference in complexity is that the natural materials contain condensed-ring, as well as single-ring, aromatics that permit multiple sulphonation to occur. These di- and polysulphonated materials cause the equivalent weight distributions of natural sulphonates to be much broader than those of monosulphonated synthetics. These petroleum sulphonate mixtures generate low IFTs; however, it has been noted that the alkyl groups pendant to the aromatic ring complexes largely govern their behaviour [103].

At first sight it would be reasonable to deduce that these complex organic petroleum surfactants were not easily biodegraded; however the picture is a complex one, and there is some evidence that petroleum sulphonates being surfactants act synergistically with the microbial fauna and aid the overall biodegradation of the crude oil [104]. This is an area that requires further study, as the prospect of using the naturally occurring petroleum feedstock could be advantageous economically and environmentally; indeed, in the last decade some studies have been conducted [105, 106].

3.2.6 Other Sulphonates

One of the most commonly used scale inhibitors in oilfield applications, particularly for barium sulphate control, are polysulphonates based on vinyl sulphonic acid monomers and the related styrene and allyl derivatives [107]. These are not surfactant in nature and have been detailed in Section 2.1.1 on unclassified vinyl polymers.

Alkylamphohydroxypropyl sulphonates are amphoteric surfactants that have been used in drilling fluids as clay stabilisers [108] and are discussed further in Section 3.5.

Calcium sulphonates are used as the basis of grease formulations and often provide superior carrier performance for controlled friction reduction in oilfield drilling and production threading operations. These materials can often form the basis of environmentally acceptable grease formulations, which would include vegetable oils and related fatty acids or polymers [109]. These materials also have corrosion-inhibiting properties [110].

The aryl sulphonates described in Section 3.2.3 can also form Gemini surfactants, which can be used in water flooding applications [81] in EOR. Gemini surfactants are described in more detail in Sections 3.2.11.5 and 3.6.6. Naphthyl sulphonates can also perform a similar function.

Propoxy ethoxy alkyl sulphonates, analogous to the sulphate derivatives described in Section 3.2.1, are used in a number of EOR operations [111].

Sodium asphalt sulphonate has historically been used as a cheap and reasonably effective drilling additive [112]. This has more or less been replaced by modern synthetic surfactants more tailored to specific modern drilling practices.

Methyl ester sulphonates (MES) have been commercially available since the mid-1960s; however until recently the only known oilfield uses of MES are as anti-sludging agent and as demulsifier in acidising stimulation work [113]. Recently these products have been found useful in forming gelled fluids.

Surfactant-gelled acid systems have been used for several decades in fracture acidising, matrix acidising, gravel packing, frac packing and hydraulic fracturing, and one of the perceived benefits of these fluids is that they are considered less damaging than polymer-based fluids. However, surfactant gels typically lose their viscosity in contact with hydrocarbons or by dilution with formation fluids. This often also results in an undesirable emulsion being formed. Another potential problem of some cationic surfactant gels is that they may adversely alter the wettability of sandstone formations. Until recently gel-forming surfactants were predominantly cationic (such as a quaternary amine) or amphoteric/zwitterionic (such as betaine), both of which types of surfactant are discussed later in this chapter in more detail. As described earlier, MES, an anionic surfactant, has been successfully developed for this application [114].

MES are prepared by the addition of sulphur trioxide to the α-carbon of a methyl ester and subsequently neutralised with a base. There are three advantages for use of MES:

1) It is less expensive than α-olefin sulphonate.
2) It is derived from renewable resources such as palm kernel oil.
3) It has a better environmental profile than many surfactant gels because it is biodegradable and exhibits low aquatic toxicity [115].

This 'new' type of sulphonate surfactant from non-edible vegetable oils has been further developed for general EOR applications, particularly the C16–18 fatty acid derivatives that exhibit good stability and IFT. The main mechanism involved in surfactant flooding is reducing IFT between trapped oil and water and increasing the capillary number, which finally results in improvement of microscopic displacement efficiency. This new class of surfactants offers a greener option in this respect [116].

Sulphonate vegetable oils, in particular soya and castor oils, can be used as lubricants in water- and oil-based systems in drilling fluid applications [117]. They do however exhibit significant foaming, particularly in water-based fluids, which can restrict their usefulness.

3.2.7 Sulphonic Acids

Not all sulphonic acids are surfactants, and those that have been primarily described in Section 3.2.3 are the alkyl aryl derivatives, which are powerful and versatile detergents and surfactants widely used in the oil and gas industry. A number of other sulphonic acids also have uses in the oilfield sector, and these are described in Chapter 6, where sulphonic acids are classed as a member of the organosulphur compounds.

3.2.8 Sulphosuccinates

Sulphosuccinates are part of a larger group of anionic surfactants the sulpho-carboxylic acids. These compounds display at least two hydrophilic groups: the sulphonate group and one (or two) carboxylic group(s) as carboxylate or ester. The monocarboxylic compounds are not widely used and have no applications in the oil and gas sector. The best-known product of this

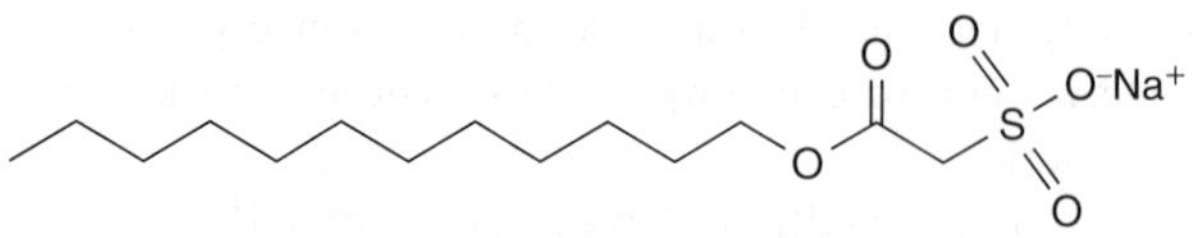

Figure 3.18 Sodium lauryl sulphonate.

category is the sodium lauryl sulphoacetate (see Figure 3.18), which is found in toothpastes, shampoos, cosmetics and slightly alkaline soaps.

C_{18} compounds are used in butter and margarine as anti-splattering agents, because they are able to fix the water in food emulsions so that the evaporation is not explosive when they are heated in a pan.

On the other hand, sulpho-dicarboxylic compounds, such as sulphosuccinates and sulpho-succinamates, are well known and used in many industrial and personal care applications.

Succinic acid (Figure 3.19) is a diacid that should be named 2-butene-1,4-dicarboxylic. The diester (succinate) is produced by direct reaction of maleic anhydride with an alcohol, followed by sulphonation.

Sulphosuccinates based on sulphosuccinic acid esters of alcohols, particularly dioctyl sulphosuccinate (see Figure 3.20), of which there are two main geometric isomers, exhibit low dynamic surface tensions in aqueous solutions and are therefore recommended as wetting agents and as dispersing agents for hydrophobic solids. These products have a number of upstream oil and gas applications; however their utility and relatively low environmental impact have not been as widely utilised as in other sectors, presumably due to cost compared with other anionic surfactants.

Dioctyl sulphosuccinates are available and are dissolved in a number of different carrier liquids. Concentrated grades are available at 70% or 60% in water and ethanol or glycol.

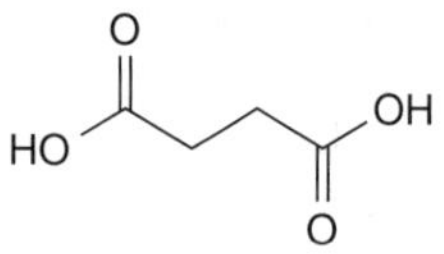

Figure 3.19 Succinic acid.

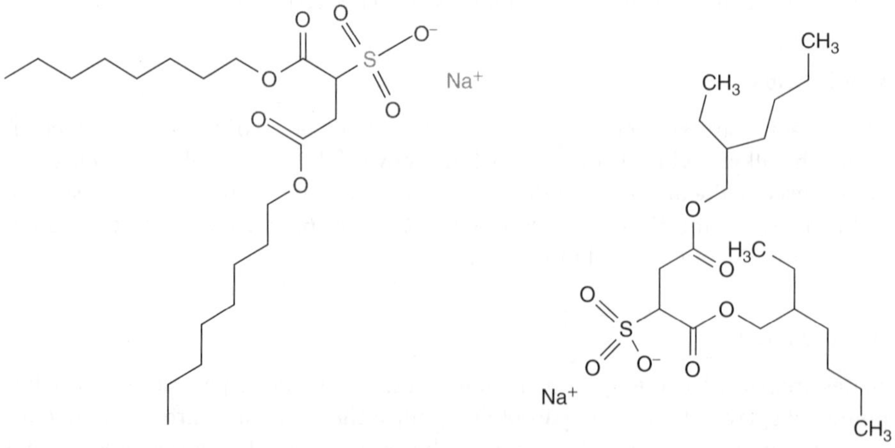

Figure 3.20 Sodium dioctyl sulphosuccinate.

The 60% active grades dissolve more rapidly in water. These products are soluble in paraffinic oil and non-polar solvents such as hydrocarbons and may be used as emulsifiers and as dispersants.

In general, sulphosuccinates are of low toxicity, biodegradable and stable in neutral solutions but may hydrolyse at the ester link when diluted in strong acid or alkaline solutions [118].

Mixtures of water-soluble polymers containing sulphosuccinate-derived surfactants are used as clay stabilisers when drilling in highly water-absorbent shales [119]. Dioctyl sodium sulphosuccinate is also a key constituent in the formulation of many oil spill dispersants [120, 121].

In recent years high strength acid fracturing treatments have been developed to release certain difficult heavy oils. Although the crude oil typically has a low concentration of asphaltenes, these oils can be very sensitive to acid- and/or iron-induced asphaltene precipitation. As the acid strength increases and ferric iron is dissolved into solution, it becomes increasingly difficult to chemically prevent the asphaltenes from precipitating. Acid blends designed to prevent asphaltene precipitation also tend to be very emulsifying with the crude oil; therefore a careful balance between anti-sludge additives and non-emulsifiers must be found. Sulphosuccinates are particularly useful in this respect [122].

In recent decades, particularly in North America, there has been a dramatic increase in the use of hydraulic fracturing in the exploitation of shale gas and oil trapped in tight reservoirs [123]. Fracturing fluid loss into the formation can potentially damage the hydrocarbon production in shale or other tight reservoirs. Well shut-ins are commonly used in the field as a way to dissipate the trapped water into the matrix near fracture faces; however surfactants, particularly sulphosuccinates, can be potentially used to reduce the impact of fracturing fluid loss on hydrocarbon permeability in the matrix as well by controlled reductions in IFT. The benefits of shut-ins and reduction in IFT by surfactants on hydrocarbon permeability for different initial reservoir conditions have been compared, and criteria that can be used to optimise the fracturing fluid additives and/or manage flowback operations to enhance hydrocarbon production from unconventional tight reservoir have been described [124].

3.2.9 Phosphate Esters

Phosphorus chemistry is particularly interesting and can be complex. The use of phosphorus products and their environmental effects as related to the oil and gas industry are discussed in some detail in Chapter 4. This section will consider the main class of organophosphorus surfactants used in the upstream oil and gas sector, namely, the phosphate esters and also some related surfactants. Phosphoric acid (Figure 3.21) can form mono- or diesters of fatty alcohols whose extra hydrogen atoms are neutralised by either and alkaline hydroxide or a low molecular weight amine to give an anionic surfactant.

The major use of phosphate ester anionic surfactants is as components of oilfield corrosion inhibitors, and phosphate esters are classified as film-forming corrosion inhibitors (FFCIs), and as will be further described in Sections 3.3 and 3.4, these are important products in oilfield in strategies to mitigate corrosion. These FFCIs also include various nitrogenous compounds,

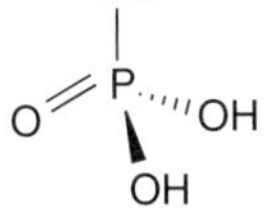

Figure 3.21 Phosphoric acid.

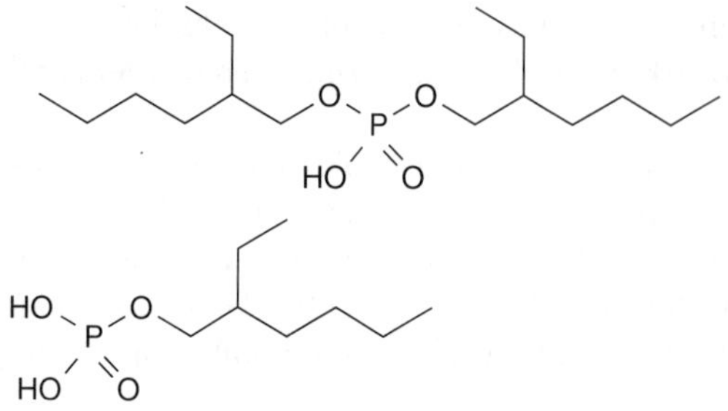

Figure 3.22 Mono- and di-2-ethylhexyl (isooctyl) esters of phosphoric acid.

many of which are surfactants, and sulphur compounds. It has been observed that the adsorption of these inhibitors depends on the physico-chemical properties of the functional groups and the electron density at the donor atom. The adsorption occurs due to the interaction of the lone pair and/or p-orbitals of inhibitor with d-orbitals of the metal surface atoms, leading to the formation of a corrosion protection film [125].

Phosphate ester-type corrosion inhibitors are manufactured by the reaction of suitable alkoxylated (ethoxylated, propoxylated or butoxylated) alcohols or phenols with a phosphating agent such as phosphorus pentoxide or orthophosphoric acid [126, 127].

Phosphate esters in general and ethoxylated phosphate esters in particular are used in different areas of corrosion inhibition, such as drilling fluids, well stimulation, oil and gas production and pipeline transportation. They are very effective especially at moderate temperatures or in the presence of trace amounts of oxygen [128–130]. Mono- and diphosphate esters (see Figure 3.22 for an example) have also been found useful in mitigation of under deposit corrosion [131]. The phosphate esters containing a hydrophobic nonylphenol group have been shown to be better FFCIs than linear or branched aliphatic phosphate esters, with the diesters being more effective than the monoester [129].

Phosphate esters can form insoluble salts with Fe^{2+} and Ca^{2+} ions, which are thought to form deposits on pipe walls and further hinder corrosion [132]. They have also been synthesised using poly-alkoxylated thiols, instead of alcohols, and these have been claimed to be especially useful in mitigation of pitting corrosion [133], particularly in highly oxygenated systems and also in deep gas wells [134].

These alkoxylated phosphate esters and related products have been shown to provide extremely effect corrosion inhibition when formulated with other surfactants such as quaternary ammonium salts and other materials, in particular didecyldimethyl quaternary ammonium chloride [135], various amines and ethoxylated fatty amines [136] and imidazolines [137]. Synergistic effects have also been noted with phosphate esters, in particular imidazolines in mixed formulations [138].

In addition to these properties and functions, water-soluble ethoxylated aliphatic phosphate esters are used effectively in an environmentally acceptable corrosion inhibitors as they pose low toxicity and are readily biodegradable [139].

Some mixtures of mono- and diphosphate esters derived from alkyl ethoxylates reacted with phosphorus pentoxide are claimed as low dose naphthenate inhibitors [140], where it is believed that the surfactant properties of these molecules cause them to align and concentrate at the oil–water interface and therefore prevent interactions between the naphthenic acids in the oil phase and the cations such as Ca^{2+} in the water phase.

Alkyl phosphate esters in conjunction with aluminium isopropoxide have been used as fluid loss additives in drilling fluid compositions [141]. It is important for drilling fluids to efficiently and quickly form a filter cake to minimise fluid loss and allow flow of fluids into the wellbore during production [112]. In the alkyl phosphate ester case, it is cross-linked with the aluminium compound to form a complex anionic polymer, which acts as a gelling agent to prevent fluid loss.

It has been found that the inclusion of polyether phosphate esters in aqueous drilling fluids can provide additional lubricating properties and as such enhance their suitability to a wider range of applications [142].

Phosphate esters have general utility as lubricants and have found numerous industrial applications due to their overall stability, especially in the presence of oxygen [143]. In oilfield completion operations, saline liquid is usually used and, increasingly so, when penetrating a pay zone. Brines are preferred because they have higher densities than fresh water and they lack solid particles that might damage producible formations. Classes of brines include chloride brines (calcium and sodium), bromides and formates. Lubricants based on phosphate esters can be added to these brines to aid in production penetration and also for ease of tool retrieval (H.A. Craddock, unpublished results).

Phosphoric monoesters such as isooctyl acid phosphate (see Figure 3.22), particularly when blended with dodecylbenzene sulphonic acid, give an effective asphaltene dispersant [144]. The phosphate esters however, unlike the alkyl aryl sulphonates, do not have aromatic groups that could give π—π bond interactions with asphaltenes; however, they do have highly acidic hydrogen ions that can hydrogen bond to amines and hydroxyl groups in the asphaltenes or with metal ions and therefore destabilise the asphaltene agglomeration mechanism. A similar behaviour is observed with the alkoxylated carboxylic acids (see Section 3.2.10).

The phosphate ester class of anionic surfactant is one of the most environmentally acceptable classes of surfactants and will be discussed further in Section 3.9.

3.2.10 Ethoxylated (or Ether) Carboxylic Acids

Alkyl ether carboxylic acids, such as laureth-3 carboxylic acid (see Figure 3.23), are alkyl ethoxylates with a terminating acetic acid group. They are used in a range of industries as emulsifiers, corrosion inhibitors, dispersing agents, wetting agents and detergents.

Surprisingly these anionic surfactants are under-represented in terms of oilfield applications. Like some phosphate esters, they are known, particularly when blended with sulphonic acids and/or phosphate esters, to be useful asphaltene dispersants [145].

Like phosphate esters they are highly biodegradable and non-toxic to aquatic species [146]. There have been some recent investigations in examining all oxygen-based acid corrosion inhibitors that are biodegradable, non-bioaccumulating and low in toxicity, which meet industry standards for protection of carbon steel oilfield tubing up to 120 °C (~250 °F) [147]. These molecules were proposed as replacements to quaternary amines, including pyridine and quinoline quats, in corrosion inhibitors used in matrix acidising packages, and included some ether carboxylic acids.

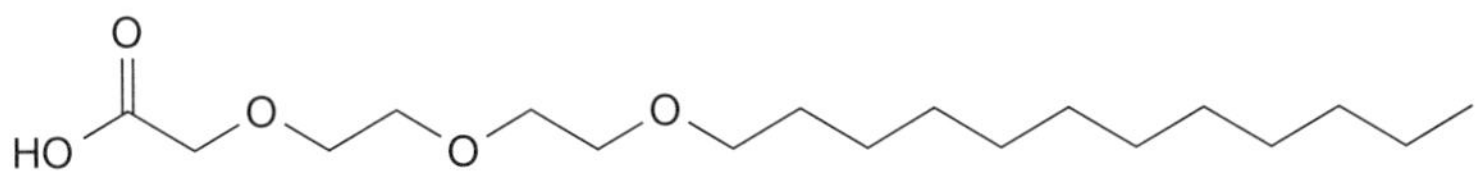

Figure 3.23 Laureth-3 carboxylic acid.

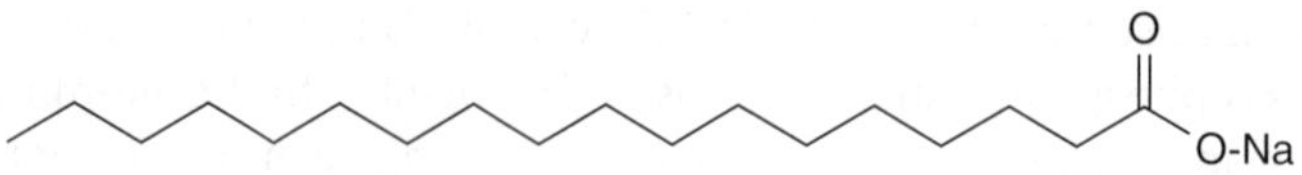

Figure 3.24 Sodium stearate.

3.2.11 Other Anionic Surfactants

3.2.11.1 Simple Soaps

Many simple soaps are anionic detergents and have the general structure as shown in Figure 3.24, in this case sodium stearate. As we have seen with other classes of anionic surfactant, they have a hydrophobic tail and hydrophilic head.

Although, in general, these molecules are readily biodegradable and non-toxic, they are seldom used in the upstream oil and gas industry. They are of course used in general industrial cleaners such as rig washes and have been examined as potential drag reducers in crude oil transportation pipelines [147].

3.2.11.2 Sarcosides and Alkyl Sarcosinates

This category of surfactant is derived from sarcosine, or methyl glycine, a cheap synthetic amino acid that has the structure illustrated in Figure 3.25.

The acid acylation is carried out with a fatty acid chloride, resulting in a surfactant that displays a fatty amide group as the lipophilic or hydrophobic tail. This reaction can take place with many different amino acids, particularly those that come from the hydrolysis of proteins. This results in the so-called sarcosides, whose structure is very similar to our biological tissues.

The most used synthetic product is lauroyl sarcosinate, which is both a strong bactericide and a blocking agent of hexokinase (putrefaction enzyme). Since it is not cationic, it is compatible with anionic surfactants; it is not widely used in the oil and gas sector but is in personal care and other domestic applications being a component in toothpaste, shaving foams and 'dry' shampoos for carpets and upholstery.

Notably since the nitrogen atom is in part of an amide linkage, the nitrogen is not pH active and is neutrally charged in all aqueous solutions regardless of pH. The carboxylate has a pK_a of about 3.6 and is therefore negatively charged in solutions of pH greater than about 5.5.

Alkyl sarcosinates having the structure shown in Figure 3.26 are classified as viscoelastic anionic surfactants. These products produce shear-thinning gels in the presence of cations and are therefore easily pumped into the rock matrix. Once in the formation the viscosity of the gel can increase by 100-fold and therefore restrict fluid flow, particularly water flow. Contact with hydrocarbons breaks the gel and reduces the viscosity. This means that in a formation treatment only pores with hydrocarbon saturation will flow freely, leaving them clear and water wet, whereas pores with high water saturation remain plugged with gel [148].

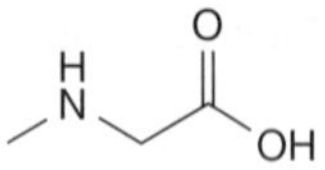

Figure 3.25 Sarcosine.

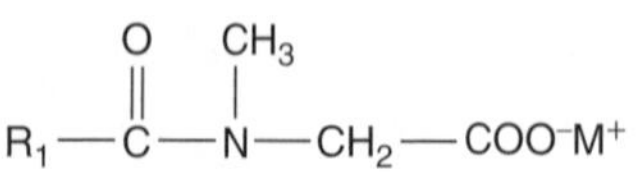

Figure 3.26 Alkyl sarcosinates.

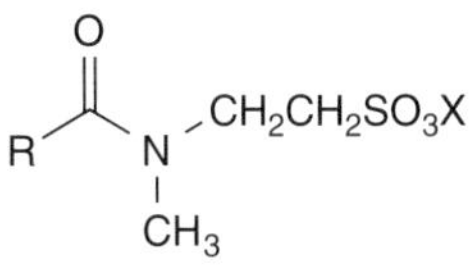

Figure 3.27 Alkyl taurate.

However, given their derivation from a semi-synthetic amino acid and their likely low environmental impact, they have been little developed for oilfield use, presumably on economic grounds.

3.2.11.3 Alkyl Taurates

In a related viscoelastic application, anionic surfactants based on alkyl taurates (Figure 3.27) have been developed [149]. These viscous surfactants are proposed for use as fluid diverting agents in acid stimulation applications [150].

Taurates are anionic acylamino alkane surfactants with a chemical structure very close to that of isothionates. They are used to increase the viscosity in a surfactant mix and are generally used as co-surfactants across a wide range of industrial and household applications. Historically they were used a lot in commercial shampoos and body washes but have now been replaced by the lauryl ether sulphonates. Many are derived from vegetable or related fatty acids and have good sustainability profiles as well as good rates of biodegradation.

Taurates are a surfactant of choice in thickening mixtures and are widely used in cosmetics as they are suitable for all skin types thanks to their gentle cleansing and moisturising, and because the pH of certain taurate combinations is neutral to alkaline, they are a primary choice for baby care products.

Again, despite their good environmental characteristics and low toxicity, they are not widely used in the upstream oil and gas sector.

As has been described a wide variety of anionic surfactants are utilised in the upstream sector of the oil and gas industry and have a wide variety of functions. They are however restricted in use, and the major volume uses tend to be in a few applications. This is primarily due to issues with compatibility and their foaming characteristics. The latter is not necessarily a welcome function in many applications where a surfactant would be a useful chemical type to apply in solving an oilfield processing problem. This being stated, anionic surfactants have lent themselves to ease in formulation and a certain amount of synthetic manipulation to enhance desired properties.

3.2.11.4 Alkyl o-Xylene Sulphonates

Products that are based on *ortho*-xylene and olefins such as alcohols from SHOP [62] can produce products such as the petroleum sulphonate illustrated in Figure 3.28.

These derivatives are more oil soluble than other alkyl aryl sulphonates described in Section 3.2.3 and have been successfully applied, over many decades, in EOR processes

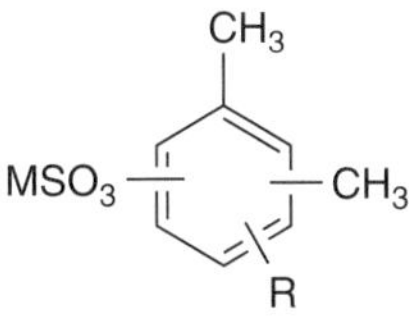

Figure 3.28 Alkyl o-xylene sulphonate.

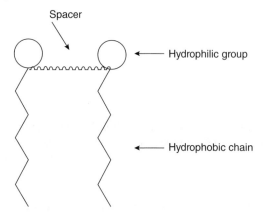

Spacer

Hydrophilic group

Hydrophobic chain

Figure 3.29 Schematic for Gemini surfactants.

[151, 152]. The use of modification techniques to target specific surfactant molecules could provide a number of solutions not only in anionic surfactants but also generally. This could not only enhance efficacy or certain desired properties but also lend itself to designing 'greener' molecules. This is discussed further in Section 3.6.

3.2.11.5 Gemini Surfactants

Finally, in the last three decades, anionic surfactants, particularly alkylbenzene sulphonates, have been utilised in the design of new Gemini surfactants [153]. A Gemini surfactant has in sequence a long hydrocarbon chain, an ionic group, a spacer, a second ionic group and another hydrocarbon tail. A schematic representation of Gemini is given in Figure 3.29, and related surfactants with more than two tails are also known. Gemini surfactants are considerably more surface active than conventional surfactants. Menger and Littau [153] assigned the name Gemini to bis-surfactants with rigid spacer (i.e. benzene, stilbene). The name was then extended to other bis- or double-tailed surfactants, irrespective of the nature of spacers.

The spacer type and number, as well as the length of the alkyl chains have a considerable effect on the CMC value and on Krafft temperatures, in turn leading to useful enhancement of surface tension properties [154, 155]. To date the application of these materials in the oilfield is somewhat embryonic although some work has been conducted on EOR application [81, 156] and examination of their viscoelastic properties in applications such as hydraulic fracturing fluids [157]. Relatively recently some new Gemini surfactants based on non-ionic imidazolines have been utilised as oilfield corrosion inhibitors [158] and are further discussed in Section 3.3.

As will be detailed in Section 3.6, Gemini surfactants offer a great opportunity in the design of biosurfactants and new possibilities of maintaining performance and having good environmental properties.

3.3 Non-ionic Surfactants

Non-ionic surfactants do not produce ions in aqueous solution. As a consequence, they are compatible with other types of surfactants and are excellent candidates for blending in complex mixtures. This makes them particularly useful in many industrial applications, particularly in the upstream oil and gas industry where products and additives used are predominantly blends of two, three or more ingredients. They are much less sensitive to electrolytes, particularly divalent cations such as calcium and barium, than ionic surfactants and can be used with high salinity or hard water. Again, this helps in their use in the oilfield sector. Having stated this they are surprisingly underutilised within the upstream oil and gas sector, possibly due to their expense but more likely due to their lack of historic application.

Non-ionic surfactants are good detergents, wetting agents and emulsifiers. Some of them have good foaming properties. Some categories exhibit a very low toxicity level and are used in pharmaceuticals, cosmetics and food products. These surfactants are found in a large variety of domestic and industrial products, such as powdered or liquid formulations. However the overall market is dominated by poly-ethoxylated products, i.e. those whose hydrophilic group is a polyethylene glycol chain produced by the polycondensation of EO on a hydroxyl or amine group.

Non-ionic surfactants account for about 40–45% of the world production and are the second most applied surfactant materials across a number of industry sectors. In the oil and gas sector, they are also used in a variety of functions however as stated earlier not as widely as may have been anticipated.

The major group in the class of surfactants is the ethoxylates, namely, the ethoxylated linear alcohols and ethoxylated alkylphenols. These are made by condensing long-chain alcohols, which can come from either a synthetic or natural source, with EO with to form ethers (see Figure 3.30).

These non-ionic surfactants can be reacted further with sulphur-containing acids to form some of anionic surfactants detailed in Section 3.2.

The use of EO as a commercial process is one of the most economical means of introducing hydrophilic ethoxy groups to achieve the desired balance with lipophiles. Although non-ionic surfactants do not contain an ionic group as their hydrophilic component, the presence of a number of oxygen atoms in one part of the molecule, which are capable of forming hydrogen bonds with molecules of water (see Figure 3.31), confers it with hydrophilic properties. As in other surfactants the long alkyl group forms the hydrophobic component of the surfactant.

Non-ionic materials having a variety of properties and uses are formed by condensing EO with alkylphenols, higher aliphatic alcohols, polyhydric alcohol partial esters, carboxylic acids, higher alkyl amides, alkyl mercaptans and polypropylene glycols. Condensation of EO with higher alkylamines yields cationic molecules, as is referenced in Section 3.4.

As the temperature of the surfactant solution is increased, the hydrogen bonds gradually break, causing the surfactant to come out of solution. This is commonly referred to as the cloud point and is characteristic for each non-ionic surfactant. Non-ionics are generally more surface-active and better emulsifiers than anionics at similar concentrations. They are less

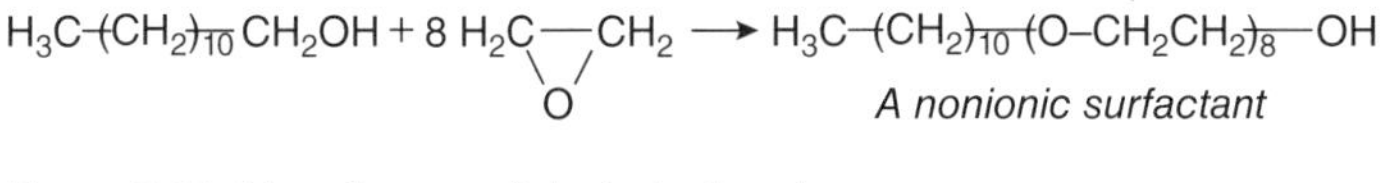

$$H_3C\text{-}(CH_2)_{\overline{10}}\,CH_2OH + 8\,H_2C\text{---}CH_2 \longrightarrow H_3C\text{-}(CH_2)_{\overline{10}}\,(O\text{--}CH_2CH_2)_{\overline{8}}\text{---}OH$$

A nonionic surfactant

Figure 3.30 Manufacture of alcohol ethoxylates.

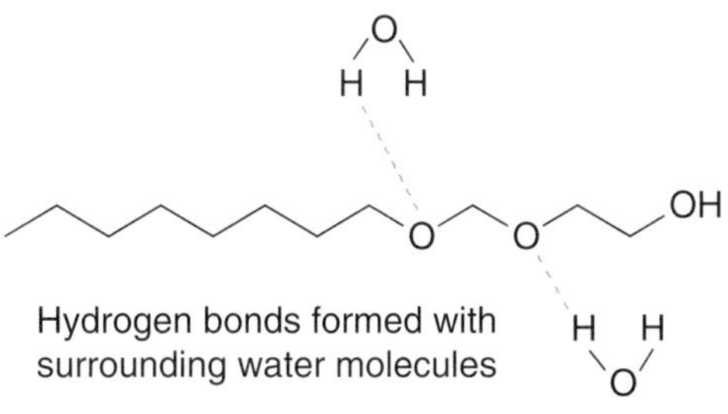

Figure 3.31 Hydrogen bonding of non-ionic surfactant with water.

Hydrogen bonds formed with surrounding water molecules

soluble than anionics in hot water and produce less foam. They also have more of a detergent effect than anionics in that they are more efficient in removing oily and organic dirt than anionics. Depending on the type of fibre, they can be active in cold solution and so are useful in countries that lack hot water supplies and in developed countries where there is a desire to lower the wash temperatures either to save energy or because of the type of fabric being washed. Non-ionics are used in fabric washing detergents (both powders and liquids), in hard surface cleaners and in many industrial processes such as emulsion polymerisation and agrochemical formulation.

They have a wide utility in the upstream oil and gas industry, as they are highly compatible in formulated products, with other additives and chemicals and with drilling and production fluids. In environmental terms they are less water soluble generally than ionic surfactants, as they do not ionise in aqueous solution, because their hydrophilic group is of a non-dissociable type, such as alcohol, phenol, ether, ester or amide. A large proportion of these non-ionic surfactants are made hydrophilic by the presence of a polyethylene glycol chain, obtained by the polycondensation of EO. They are called poly-ethoxylated non-ionics. The polycondensation of propylene oxide (PO) produces a polyether, which, in contrast to polyethylene oxide (PEO), is slightly hydrophobic. This polyether chain is used as the lipophilic group in the so-called poly-EOpolyPO block copolymers, which are most often included in a different class, the polymeric surfactants. These have been described in Chapter 2 and will be further detailed in Section 3.6.1, 'Polymeric Surfactants'. Also the use of EO to introduce improve water solubility will be further detailed in, 'Biosurfactants'.

Glucoside (sugar-based) head groups have been introduced in the market because of their low toxicity. As far as the lipophilic group is concerned, it is often of the alkyl or alkylbenzene type, the former coming from fatty acids of natural origin.

The following are the main categories of non-ionic surfactants, all of which are used in various applications in the upstream oil and gas industry:

- Ethoxylated linear alcohols (alcohol ethoxylates)
- Ethoxylated alkylphenols (alkylphenol ethoxylates (APE))
- Fatty acid esters
- Amines and amine derivatives
- Alkyl polyglucosides (APGs)
- EO and PO copolymers
- Polyalcohols and ethoxylated polyalcohols
- Thiols and derivatives.

As will be illustrated, the use of EO is extremely important in the synthesis and manufacture of this class of surfactants. EO was discovered by Würtz [159] in the mid-nineteenth century. However it is only after 1931 that it was prepared by direct oxidation of ethylene by air on a silver catalyst (300 °C, 10 atm.) [160]. It is a very unstable gas, very dangerous to manipulate and because of its triangular structure is submitted to extreme tension. Currently, EO is produced by direct oxidation of ethylene with air or oxygen; and the annual worldwide production capacity is over 20 million tonnes per annum, making it a highly important industrial chemical.

3.3.1 Alcohol Ethoxylates

The alcohol starting materials come from various origins, both synthetic and naturally derived; however, those with linear alkyl groups are generally more biodegradable and environmentally acceptable. They are in general primary alcohols with their —OH group at the end of the chain. They are usually prepared by moderate hydrogenation of fatty acids or fatty acid esters [161],

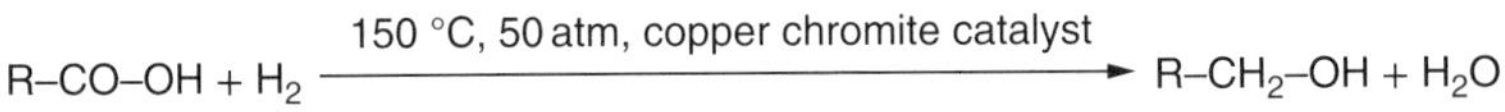

$$\text{R–CO–OH} + \text{H}_2 \xrightarrow{\quad\text{150 °C, 50 atm, copper chromite catalyst}\quad} \text{R–CH}_2\text{–OH} + \text{H}_2\text{O}$$

Figure 3.32 Hydrogenolysis of fatty acid to fatty alcohol.

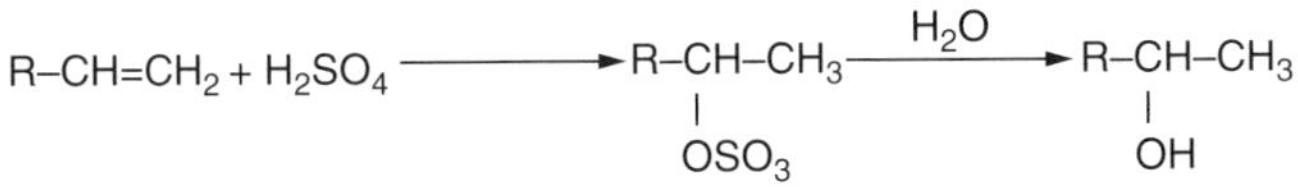

$$\text{R–CH=CH}_2 + \text{H}_2\text{SO}_4 \longrightarrow \underset{\underset{\text{OSO}_3}{|}}{\text{R–CH–CH}_3} \xrightarrow{\text{H}_2\text{O}} \underset{\underset{\text{OH}}{|}}{\text{R–CH–CH}_3}$$

Figure 3.33 Manufacture of secondary fatty alcohols.

many of which from natural sources such as palm oil, castor oil, etc., or by catalytic hydrogenolysis (see Figure 3.32).

They can also be prepared by Ziegler hydroformylation of olefins (OXO process) [61] or controlled oxidation of paraffins [62] as was described in Section 3.2.1.

Secondary alcohols, which have their hydroxyl group attached on the second carbon atom of the alkyl chain, are produced by hydration of α-olefins in sulphuric medium [162] as outlined in Figure 3.33.

These alcohols and others are reacted with EO to give alcohol ethoxylates. Ethoxylation is carried out with care in anhydrous conditions and in the absence of oxygen as the probability of reacting with alcohol already ethoxylated is the same as that for unreacted alcohol [163]. Consequently there are a large number of oligomers formed with different degrees of ethoxylation. As a consequence of its ethylene oxide number (EON) distribution [164], a commercial surfactant may contain substances with widely different properties. In the presence of both an oil and a water phase, this can result in an independent solution behaviour of each substance that may cause problems in certain applications and formulations. Therefore degree of ethoxylation can be very important in determining specific surfactant properties and also its water solubility and its susceptibility to hydrolysis, both of which, as has been described in Chapter 2, can be important factors in determining the biodegradability of the chemical compound and its environmental acceptability.

As stated earlier the use of alcohol ethoxylates is surprisingly small given their utility and ease of formulation. They are used as foamers especially in drilling applications where anionic surfactants cannot be used; such surfactants have a C_9–C_{11} alkyl chain and 8 moles of EO [4]. Alcohol ethoxylates also display inhibitive properties and have been used in a drill-in fluid designed to minimise formation damage in low permeability sandstones while drilling horizontal sections and to facilitate formation clean-up [165].

Alcohol ethoxylates along with other surfactants, primarily anionic surfactants of the previously described sulphates and sulphonates (see Section 3.2), have been examined as to their potential as foamers in flow diversion with respect to EOR operations [166].

In acid stimulation surfactants are included in the additive package to provide a variety of functions. Non-ionic surfactants are selected to provide low IFTs between the acid and the oil, and this includes alcohol ethoxylates [167], in particular the decyl alcohol capped with 8 moles of EO. This alcohol ethoxylate is a strong water-wetting agent as well as a good aqueous foaming agent. However, many are not very effective at higher temperatures as the surfactant can separate from the solution to form two immiscible phases, which can lead to plugging and/or loss of injectivity or productivity of the well [165].

Alcohol ethoxylates are among the most common surfactants formulated in well cleaners [168] and rig washes (H.A. Craddock, unpublished results) as they are relatively cheap but also, as stated earlier, are highly compatible with other surfactants, additives and fluids.

3.3.2 Alkylphenol Ethoxylates

Until recently APE were probably the most common non-ionic surfactants in use especially nonyl- and octylphenol ethoxylates. Concerns surrounding their lack of biodegradation and their potential for endocrine disruption have seen them being replaced by other, albeit less effective, surfactants, in particular the linear alcohol ethoxylates.

Ethoxylated alkylphenols are produced in two ways, depending on the available raw material. The first method consists in alkylating the phenol according to a classical Friedel–Crafts reaction [169] (see Figure 3.34).

The mixture of isomers can be separated by distillation; however the synthesis can be made exclusively for the *para* derivative if the R group on the alkyl chloride is sufficiently bulky to deter reaction at the *ortho* position due to steric hindrance [170]. However such groups capable of inducing steric hindrance are usually less biodegradable.

The second method consists in adding an α-olefin to phenol [171] as illustrated in Figure 3.35. This type of synthesis can result in nonyl-, dodecyl- and octylphenols, with branched and less biodegradable alkylates.

The common commercial products are the octyl-, nonyl- and dodecylphenol with a degree of ethoxylation ranging from 4 to 40. Octyl and nonylphenols with EON = 8–12 are used in detergents. With EON < 5 the attained products are anti-foaming agents or detergent in non-aqueous media. With EON ranging from 12 to 20, they are wetting agents and oil–water emulsifiers. Beyond EON = 20 they exhibit detergent properties at high temperature and high salinity.

The main use of alkylphenols was and still is as ingredients for domestic and industrial detergents, particularly for high electrolyte level such as in acid solution for metal cleaning, detergents for dairy plants, agrochemical emulsions, styrene polymerisation, etc.

Since branched alkylates are not readily biodegradable [172], the trend has been in the past decades to go into more linear products. However this comes at additional cost, and in the recent past another way to cut price and toxicity has been utilised, namely, the elimination of the benzene ring altogether, for example, by substitution with linear alcohol ethoxylates. The dilemma is that alcohol ethoxylates are not as good detergents as their counterpart phenol compounds, just as is the case with alkylbenzene sulphonates versus alkane or olefin sulphonates as seen in Section 3.2.

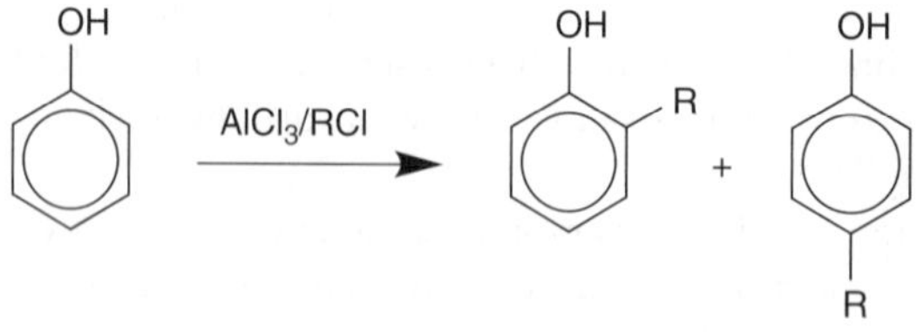

Figure 3.34 Friedel–Crafts alkylation of phenol.

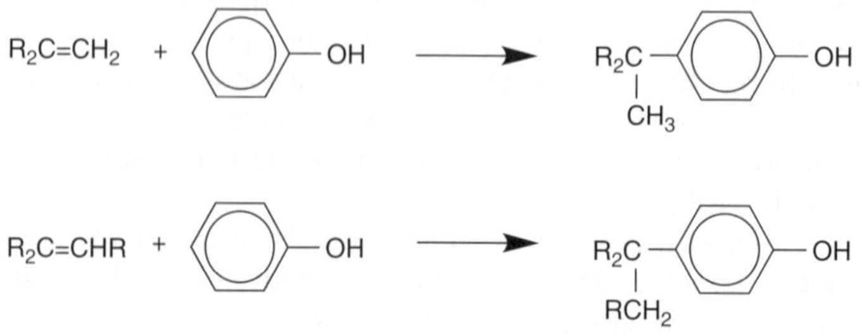

Figure 3.35 α-Olefin synthesis of alkylphenols.

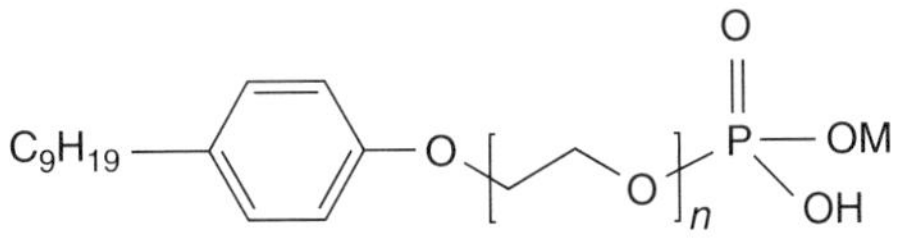

Figure 3.36 A phosphoric ester of an alkylphenol ethoxylate.

In the oilfield these products are under the same regulatory and environmental acceptability pressures as in other industries and perhaps more so. Consequently in a number of oilfield regions, the use of APE has vastly decreased if not been entirely eliminated. However, it is probably still the case that APEs are used as a key monomer in the formaldehyde condensation polymers, the so-called resin polymers, and Calixarenes. These products are still widely used and have been described in Section 2.1.2 and form the basis of many demulsifier formulations [173, 174].

The alkylphenols and their ethoxylates are however being used as asphaltene dispersants. The selection of an asphaltene dispersant is dependent on the properties of the crude oil, particularly its polarity and aromatic content [83]. It has also been reported that the phosphoric esters of APE (Figure 3.36) have performed well as asphaltene dispersants especially when blended with fatty acid diethanolamides [175].

Alkylphenols are well known as asphaltene dispersants [176, 177] and are widely used in refinery applications; however their use in many regions where oil and gas is exploited offshore is a concern along with their ethoxylates as they are known to be endocrine disrupters particularly in fish species [178, 179]. The subject of the environmental impact of potential and actual endocrine disrupters will be further examined in Chapter 10.

A number of studies have examined the stabilisation of asphaltenes with alkylphenols and their ethoxylates, and in general it has been concluded that the use of a surfactant with an aromatic ring performs better than using an aliphatic surfactant [180, 181]. The interfacial film properties with asphaltenes and non-ionic surfactants, in particular the APE, have been investigated, and those with an HLB value of 14.2 are declared to be the most efficient in preventing asphaltene absorption at the oil–water interface [182].

The range of HLB values available from APE has made them a widely used surfactant in the formulation of drilling fluids; however, concerns over their oestrogenic effects, as stated earlier, have largely led to their replacement [183].

In effect the use of APEs, which are not polymerised and therefore the alkylphenol is potentially available to the environment, is fairly well phased out from widespread use in the upstream oil and gas industry. Further uses of alkylphenols are discussed in Chapter 6.

3.3.3 Fatty Acid Esters and Related Organic Acid Esters

The esterification of a fatty acid by an alcohol functional group (—OH group) from a PEO chain tip or a polyalcohol generates an important family of non-ionic surfactants due to their low toxicity and compatibility with biological tissues, which make them suitable for uses in pharmaceuticals, cosmetics and foodstuffs. These materials have been rarely explored in the oil and gas industry, and both their functionality and potential environmental acceptability would make them suitable candidates for application, although fatty acids and other derivatives have been applied in many different areas in drilling, cementing and production scenarios.

There are three main types of fatty acid ester surfactant in common usage:

• Polyethoxy esters
• Glycerol esters
• Sugar base esters

RCO(OCH$_2$-CH$_2$)$_n$OH

Figure 3.37 Polyethoxy ester.

Polyethoxy esters (acid ethoxylated fatty acids) are formed by the condensation of EO on a carboxylic acid [184] and are the same type of reaction as previously described for alkylphenol ethoxylation. Polyethoxy esters (Figure 3.37) are produced, which are identical to those obtained by the esterification of the acid by polyethylene glycol.

Polyethoxy esters of fatty acids and other natural carboxylic acids are among the cheapest non-ionic surfactants; however they are seldom used in the upstream oil and gas industry. This could be due to their instability in alkaline pH, as they will readily hydrolyse. Despite this the esters of high molecular fatty acids and polyethylene glycols or polyethylene glycol mono-alkyl ethers have a variety of surface-active properties and are used as wetting agents, detergents and emulsifiers in other industrial sectors.

Triglycerides (see Figure 3.38) are found in most vegetal and animal oils and fats and are the fatty acid triesters of glycerol. They are not hydrophilic enough to be water soluble (see also Section 2.2.5). However, glycerol mono- and diesters and mono- and diglycerides can exhibit surfactant properties.

These lower esters can be synthesised by the reaction of glycerol with fatty acids, but the industrial method is by reacting a triglyceride with an excess of glycerol under alkaline conditions [185].

The size of the hydrophilic part can be increased by using a polyglycerol (Figure 3.39), which can be obtained by the dehydration of glycerol (see also Section 2.1.2).

Glycerol esters and derivatives are used in the conditioning and preservation of foodstuffs such as bread and dairy produce and are also used to produce emulsions and foams for beverages, ice creams, margarine, butter etc. They are also used in pharmaceuticals as emulsifiers, dispersants and solubilising agents. They are readily biodegradable and non-toxic; again they have been seldom used in the upstream oil and gas industry despite their good environmental characteristics. It has been found, however, that these partial glycerides are useful lubricants in both water- and oil-based drilling muds at low temperatures [117]. In water-based drilling fluid systems, these products are particularly useful as they are non-foaming.

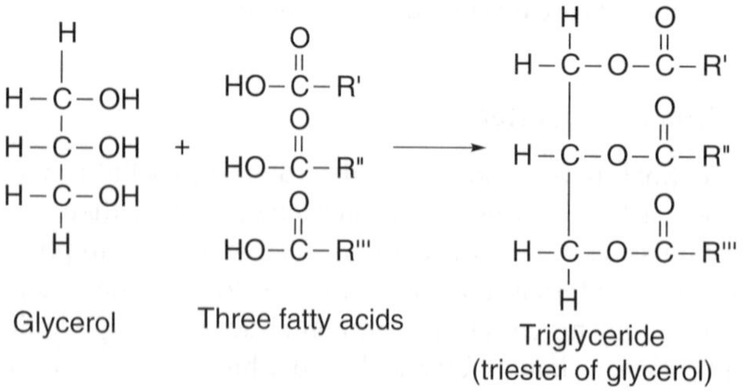

Glycerol   Three fatty acids   Triglyceride (triester of glycerol)

Figure 3.38 Synthesis of triglyceride.

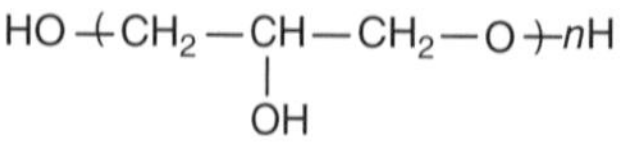

HO–(CH$_2$–CH–CH$_2$–O)$_n$H
|
OH

Figure 3.39 Polyglycerol.

CH₂–OH–CHOH–CHOH–CHOH–CHOH–CHO ⟶

CH₂OH–CHOH–CHOH–CHOH–CHOH–CH₂OH

Figure 3.40 The reduction of glucose to sorbitol.

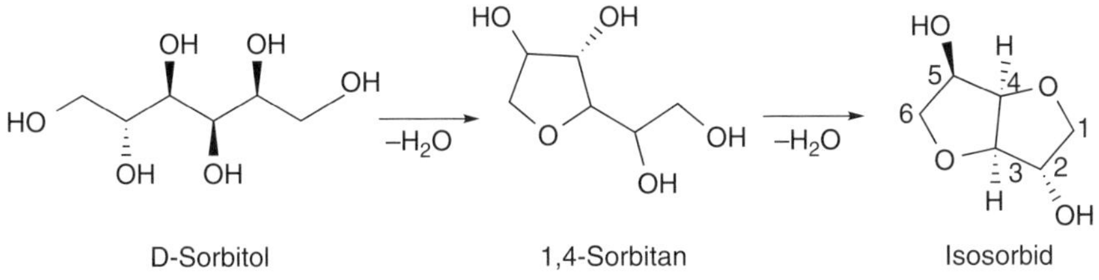

D-Sorbitol 1,4-Sorbitan Isosorbid

Figure 3.41 Synthesis of sorbitan and isosorbide.

The esters of the sugar-based molecules, the hexitols and related cyclic compounds, are the third type in this group of non-ionic surfactants. Hexitols are hexa-hydroxy-hexanes obtained by the reduction of hexoses or other monosaccharides. The most common is sorbitol, which is obtained by the reduction of D-glucose (Figure 3.40).

Monosaccharides can form a cycle or ether loop called hemi-acetaldehyde. The same happens to hexitols when they are heated at acid pH. Two hydroxyl groups merge to produce an ether link, resulting in a 5- or 6-atom cycle called hydrosorbitol or sorbitan. In some cases a two-cycle bi-anhydrosorbitol product, isosorbide, is produced. Therefore sorbitan is produced by the dehydration of sorbitol and is an intermediate in the conversion of sorbitol to isosorbide (see Figure 3.41). The dehydration reaction usually produces sorbitan as a mixture of five- and six-membered cyclic ethers (1,4-anhydrosorbitol, 1,5-anhydrosorbitol and 1,4,3,6-dianhydrosorbitol) with the five-membered 1,4-anhydrosorbitol form being the dominant product. The rate of formation of sorbitan is typically greater than that of isosorbide, which allows it to be produced selectively, providing that the reaction conditions are carefully controlled. Interestingly, the dehydration reaction has been shown to work even in the presence of excess water [186].

The sorbitan ring exhibits four hydroxyl groups, whereas isosorbide bicycle has only two. These hydroxyl groups can be reacted with fatty acids, to add to the molecule one or various lipophilic groups, or a PEO condensate, to increase hydrophilicity. Because of these two possibilities and the fine-tuning that can be adjusted for each of them, it is feasible to prepare a specifically tailored surfactant molecule.

Commercial sorbitan esters (SPAN brand or equivalent) and their ethoxylated counterparts (Tween brand or equivalent) can have a lipophilic group ranging from monolaurate (one C_{12}) to trioleate ($3C_{18}$). In the ethoxylated products, the EO groups (often as many as 20) are distributed on the different available hydroxyl groups before the esterification is carried out. The following figure indicates a likely formula for an isomer of sorbitan 20 EO monolaurate or polysorbate 20, sold as Tween 20 (see Figure 3.42), which obviously displays a hydrophilic part that is much bulkier than its lipophilic tail.

The number 20 following the 'polyoxyethylene' part refers to the total number of oxyethylene $-(CH_2CH_2O)-$ groups found in the molecule. The number following the 'polysorbate' part is related to the type of fatty acid associated with the polyoxyethylene sorbitan part of the molecule. Monolaurate is indicated by 20, monopalmitate by 40, monostearate by 60 and monooleate by 80.

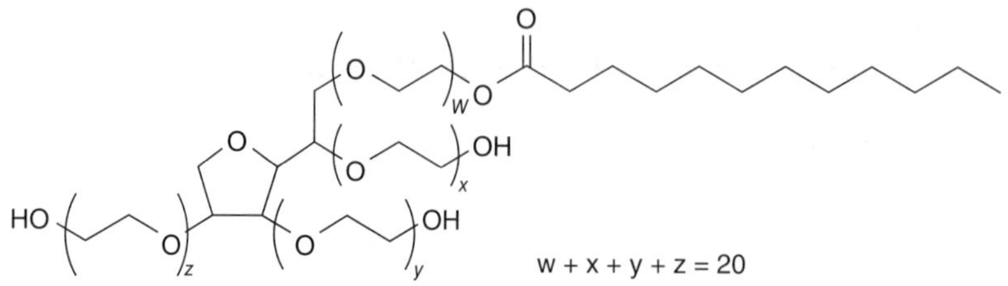

$$w + x + y + z = 20$$

Figure 3.42 Polysorbate 20(polyoxyethylene(20)sorbitan monolaurate).

These molecules appear very complex; however, they are quite easy to manufacture from commonly available natural raw materials, e.g. fat and sugar, which make them biologically compatible for food and pharmaceutical use and also non-toxic and environmentally accepta-ble, with SPANs being considered readily biodegradable and Tweens inherently biodegrada-ble [187].

It is relatively straightforward to adjust the hydrophilic character of these surfactants, either by manipulating the EO/fatty acid ratio within the molecule or by mixing different types in a formulation. It is worth noting that ethoxylation and esterification result in different types. Hence, the commercial product is always a mixture of different substances. These products produce excellent emulsifying agents and are widely employed in food conditioning (creams, margarine, butter, ice cream, mayonnaise) as well as in pharmaceuticals and cosmetics. They have been used in other applications, such as the preparation of microemulsions [188].

In the upstream oil and gas industries, the fatty acid esters of sorbitan such as the monooleate (Figure 3.43) or monolaurate are used as defoamers [189, 190]. They can be as effective as more toxic acetylenic alcohol-based products and are readily biodegradable.

The monolaurate has also been considered for applications in EOR by water flooding [191]. Other sorbitan esters have been used as emulsifiers in emulsion systems for in situ polymerisa-tion of acrylamide in water and gas shut-off scenarios [192].

A major application of these sorbitan-based fatty acid ester surfactants is in oil spill treat-ments, both in surface waters and for oiled shorelines. The oleates, mixtures of the triester and monoester, are used as emulsifiers in combination with other types of surfactants that act as dispersants such as sodium sulphosuccinate [193, 194].

Of course, esters are formed with other organic acids, some of which are useful in oilfield applications, in particular the esters of tall oil. Tall oil, which can be classed as a fatty acid, has been previously described in Section 2.2.4, and as was stated then many of its derivatives are surfactant in nature, in particular the amine and amide derivatives that are described in

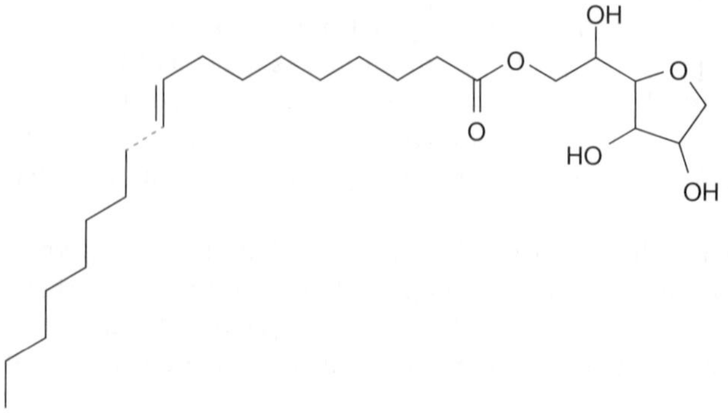

Figure 3.43 Sorbitan monooleate.

Section 3.3.4. Tall oil is also used in the manufacture of imidazoline-based corrosion inhibitors, which are described in Section 3.5.

As can be observed, the use of fatty acid esters and related products in the upstream oil and gas sector is rather meagre compared with their use in other industrial sectors despite their undoubted utility and flexibility. They are inexpensive and have a good environmental profile. The author can find no reason for these products not being explored other than similar derivatives that are either better performing with similar environmental characteristics are more economic or that the products perform poorly in oilfield conditions, and these results are unpublished.

3.3.4 Amine and Amine Derivatives

The majority of organic amines used in the oil and gas sector are amidoamines and imidazolines derived from naturally occurring fatty acids and polyethylene amines. A common feature of these filming amines is the presence of amide and/or amidine groups linking the polar amine and fatty tail portions of the molecule. The imidazolines are considered in Section 3.5. The other organic amines, particularly the fatty amines, have their corrosion inhibitor and film-forming characteristics enhanced by quaternisation, and these are then characterised as cationic surfactants and are examined in Section 3.4.

Fatty amine ethoxylates are non-ionic surfactants used as wetting and dispersing agents, stabilisers, sanitisers and defoaming agents in various industries like textile, paper, drilling, chemical, paint, metal, etc. [195]. In the upstream oil and gas industry, fatty amine ethoxylates are used as emulsifiers and can also be employed in the formulation of emulsifier blends and of drilling fluids and other drilling additives. They have also been found to have useful corrosion inhibition properties, particularly in acid chloride media [196], although these products in oilfield applications tend to be superseded by the amido analogues (see later in this section).

The amide derivatives of fatty acids and other related products have more use in the upstream oil and gas industry than the esters described in the previous section. These are primarily used as corrosion inhibitors [197, 198]; however a number of specific derivatives have properties that make them useful as asphaltene dispersants. Alkyl succinimides such as polyisobutylene succinimide (see Figure 3.44) can be considered as a low molecular weight amphiphile and has been shown to be a useful asphaltene dispersant [199, 200].

Polyisobutylenes were shown in Chapter 2 to be useful drag-reducing agents (DRAs), especially at high molecular weights. However, these products also pose a dilemma for the chemist in design and development, in that those that are the most efficient in performance and activity and provide good stability to shear degradation are also the least biodegradable.

It has been described that fatty acid diethanolamides such as the coco derivative shown in Figure 3.45, particularly in blends with phosphoric esters of APE, are useful asphaltene dispersants [175].

Figure 3.44 Polyisobutylene succinimide.

PIB = Polyisobutylene

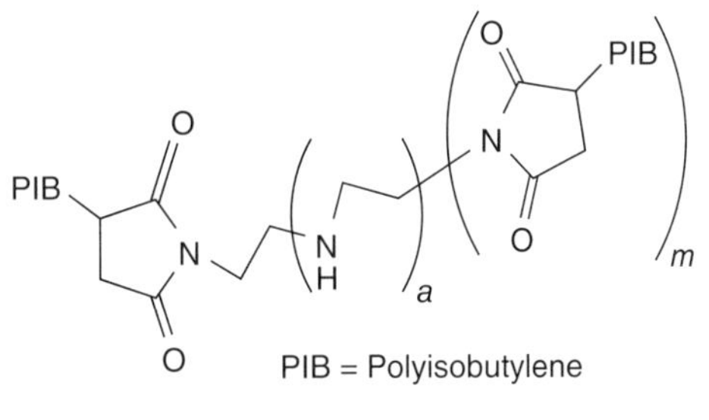

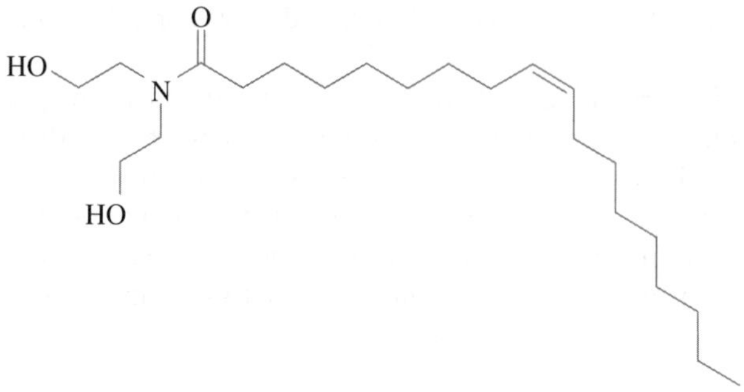

Figure 3.45 Coco-diethanolamide.

Such products have three functional groups capable of hydrogen bonding with the forming asphaltene complex, preventing agglomeration, as well as possessing a long lipophilic 'tail'. These products are also useful corrosion inhibitors [201] and have been used historically in film-forming applications, in particular in acidising stimulation. In the exploration and recovery of oil from underground fields, it is common to acidise both new and producing wells with aqueous solutions of strong acids. There are many inhibitors especially designed to prevent acid attack of the well casings, and very few provided satisfactory protection particularly at high temperatures, i.e. above 120 °C. The acetylenic alcohols, as will be described in Chapter 6, are the most widely used; however, the toxicity and environmental profile of these products is under increased pressure. These fatty amide derivatives, such as illustrated in Figure 3.45, have a reasonable environmental profile being at least water dispersible [202].

An amide-based surfactant, which results from the condensation of a linear N-alkyl polyamine and a cyclic anhydride, such as maleic anhydride, has also found application as an asphaltene dispersant [203]. Polyamines and polyamides and their surfactant-related products are covered further in Section 2.1.2.

The amide bond is prone to enzymatic hydrolysis by amidase enzymes [204]. These enzymes belong to the family of hydrolases, those acting on carbon–nitrogen bonds other than peptide bonds, specifically in linear amides. The hydrolysis of the amide bond will be further discussed in Section 3.9. Derivatives of these fatty amine and fatty amido products could be useful in providing environmentally acceptable additives, as they are often more water soluble and therefore more susceptible to hydrolysis and biodegradation.

As with other compounds in this group, the amine- and amide-based non-ionic surfactants can be ethoxylated. The first EO group must be added at acid pH, whereas the other ones (from the second group on) are added at alkaline pH, because RNH_2 is not acidic enough to release a proton at alkaline pH. During the first step, the first mole of EO is added at acid pH, so that the amine is transformed in an ammonium ion. Once the ethanolamine is produced, the EO polycondensation is carried out at alkaline pH. The first ethoxylation can be stopped when the monoethanol alkylamine is formed in order to avoid the polycondensation in more than one chain [205] (see reaction sequence in Figure 3.46).

$RNH_3^+ \longrightarrow RNH_2 + H^+$

$RNH_2 + EO \longrightarrow RNH\text{-}CH_2CH_2OH$ (mono-ethanol alkyl amine)

$RNH\text{-}CH_2CH_2OH + EO \longrightarrow RN(CH_2CH_2OH)_2$ (di-ethanol alkyl amine)

Figure 3.46 Reaction of EO with an amine.

This ethoxylation method is applied not only to amines but also to amides, such as alkyl amides and urea, as well as to imidazoles. Ethoxylated amines consisting of a fatty amine with one or two polyethylene glycol chains and with 2–4 EO groups behave as cationic surfactants at acid pH. They are used as corrosion inhibitors and emulsifiers with a better water solubility than most cationics. This solubility in water makes hydrolysis including enzymatic hydrolysis more likely, and hence these products tend to be more environmentally acceptable.

Fatty amine ethoxylates are made by ethoxylation of primary fatty amines and combine the wetting, emulsifying and dispersing properties of non-ionic and cationic surfactants. At neutral and acid conditions, they have a positive charge; at alkaline conditions they behave like non-ionic surfactants [184].

Ethoxylated alkyl amides are good foamers, which are used as additives. Because of their partial cationic character, they also provide anti-static and anti-corrosion effects. Ethoxylated and acylated urea also result in fabric softening substances. The same is obtained with an imide [184].

As can be observed, there are a wide variety of uses for these ethoxylated products; however, the uptake in the upstream oil and gas sector is mainly confined to corrosion inhibitor applications. Ethoxylated fatty amines and diamines (Figure 3.47) are commonly used in formulated FFCIs.

These ethoxylated amines have incorporated a biodegradable link such as the amide group previously described, and this can make them more environmentally acceptable [206]. As has been previously stated, ethoxylation makes the product more water soluble; however, there is an advantage in terms of corrosion inhibitor performance in that the additional oxygen atoms offer extra adsorption sites to the metal surface. This is therefore a rare example where making the product more environmental acceptable, performance is not reduced and perhaps enhanced.

There is another method with which to improve water solubility and increase oxygen functionality, and that is to incorporate hydroxyl groups into the amine and the deoxyglucityl derivatives of alkylamines [207] such as in Figure 3.48.

These products are surfactants that provide reduced dynamic and equilibrium surface tension, good solubility, moderate foaming and good cleaning performance. They are also claimed to be good FFCIs.

Imidazoles (see Figure 3.49) are related to amines, in that the amine is part of an aromatic ring system. They can be ethoxylated to produce useful non-ionic surfactant products in fabric

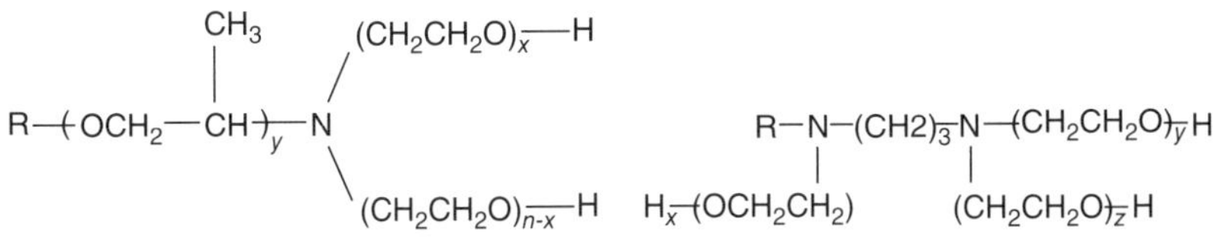

Figure 3.47 Ethoxylated fatty amine and diamine.

Figure 3.48 R1 and R2 are octyl or similar groups.

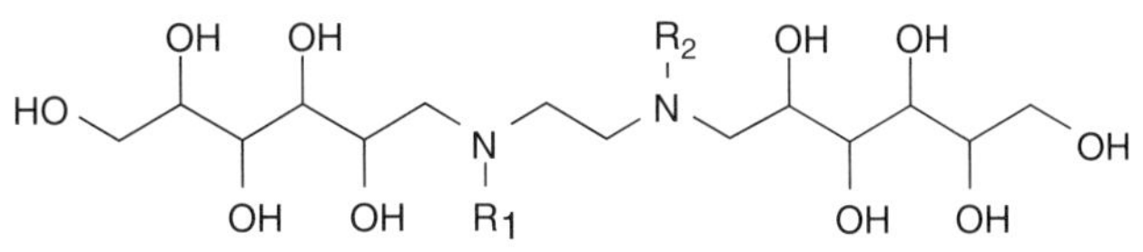

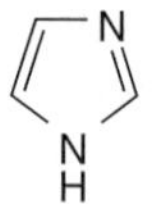

Figure 3.49 Imidazole.

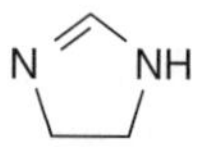

Figure 3.50 Imidazoline.

softeners for machine washing; such products also provide an anti-corrosion protection for the hardware. It is important to distinguish these separately from imidazoline products (see Figure 3.50), which are amphoteric surfactants and described in Section 3.5.

Imidazole is an aromatic heterocycle, classified as a diazole, and having non-adjacent nitrogen atoms. It is a white or colourless solid that is soluble in water, producing a mildly alkaline solution. Many natural products, especially alkaloids, contain the imidazole ring. These imidazoles share the $1,3\text{-}C_3N_2$ ring but feature varied substituents. This ring system is present in important biological building blocks, such as histidine, and the related hormone histamine. Many drugs contain an imidazole ring, such as the nitroimidazole series of antibiotics. When fused to a pyrimidine ring, it forms purine, which is the most widely occurring nitrogen-containing heterocycle in nature.

Imidazoline is a class of heterocycles formally derived from imidazoles by the reduction of one of the two double bonds, although as is described in Section 3.5, it is not manufactured in this fashion. Three isomers are known: 2-imidazolines, 3-imidazolines and 4- imidazolines. The 2- and 3-imidazolines contain an imine centre, whereas the 4-imidazolines contain an alkene group. The imidazoline derivatives used in oil and gas functional chemical additives will be discussed in Section 3.5. It is, however, important not to confuse the two classes of heterocycle, as can commonly be the case, as chemically they are quite different species, and this is reflected functionally in their properties and their environmental characteristics.

The imidazoles and ethoxylated imidazoles appear to be seldom used in upstream oil and gas applications. A graft copolymer containing an *N*-vinylimidazole monomer has been utilised as an asphaltene inhibitor [208].

Imidazole itself is moderately toxic and not thought to be particularly biodegradable [209]; this is in contrast to the imidazolines described later in Section 3.5. However, ethoxylated derivatives such as in Figure 3.51 are relatively non-toxic and readily biodegradable [210].

These products have been examined as ionic liquids and have a variety of functionalities and are used in a variety of industries as lubricants and surfactants. They have no reported use in the oilfield, although a brief examination of their structure would show the possibility of them being good corrosion inhibitors.

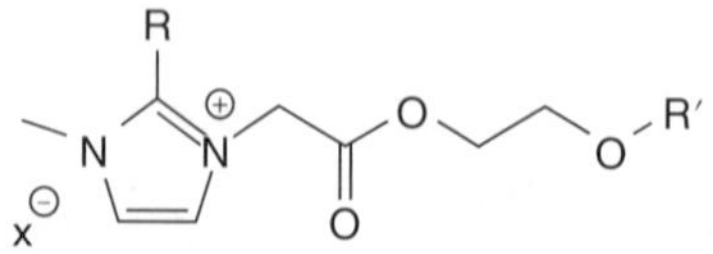

Figure 3.51 An imidazolium ester.

3.3.4.1 Amine Oxides

An amine oxide, also known as amine-*N*-oxide and *N*-oxide, is a molecule that contains the functional R_3N^+—O^-, an N—O bond with three additional hydrogen and/or hydrocarbon side chains attached to the nitrogen. Sometimes it is written as $R_3N \rightarrow O$ or, wrongly, as $R_3N=O$.

In the strict sense the term amine oxide applies only to oxides of tertiary amines; however, sometimes it is also used for the analogous derivatives of primary and secondary amines.

Amine oxide surfactants are classified as non-ionic as they act as such in neutral and alkaline pH. However at acidic pH amine oxides are similar to betaines and because of protonation they behave as cationic surfactants [211].

Amine oxides are good foaming agents and also act as foam booster and foam stabilisers for anionic surfactants. In combination with anionic surfactants, amine oxides can form a complex that has better surface activity than either the anionic or the amine oxide. Another important property of amine oxides is their resistance to oxidation.

Commercial amine oxide surfactants usually contain one long alkyl chain and two short chains, usually methyl alkyls, for instance, the oxide of lauryl dimethyl ammonium (Figure 3.52).

Lauryldimethylamine oxide, also known as dodecyldimethylamine oxide (DDAO) (also *N*, *N*-dimethyldodecylamine *N*-oxide) is one of the most frequently used surfactants of this type. Despite having only one polar atom that is able to interact with water – the oxygen atom (the quaternary nitrogen atom is hidden from intermolecular interactions) – DDAO is a strongly hydrophilic surfactant. This can be explained by the fact that it forms very strong hydrogen bonds with water [212].

Tertiary amine oxides are used as foam boosters. It is worth noting that the polarisation of the N—O bond in which the nitrogen atom provides both required electrons results in a negative oxygen atom, which is able to capture a proton in aqueous solution. As a consequence the actual form of an amine oxide foam booster is the cationic hydroxylamine.

Many of this class of surfactant products are biocompatibilising agents. They therefore exhibit a favourable effect on the skin tissues and are used as additives in hand dishwashing products and liquid soaps.

Some products include two amine oxide groups, with the amine hydrogen often replaced by ethanol groups. These are often incorporated as foam boosters used in bubble bath, hand dishwashing detergents and baby shampoos.

These properties make amine oxides non-toxic but relatively stable to hydrolytic biodegradation.

Amine oxides have a number of uses in the upstream oil and gas industry, however mainly in applications around diversion and gelling due to their good viscoelastic properties. Some of the more important ones are now described, along with other interesting potential applications.

Amine oxides can be used in viscoelastic surfactant (VES) applications, especially in controlling the viscosity of fluids such as fracturing brines as applied in the hydraulic fracturing of shale gas. Often these VES products are controlling the saponification of natural fatty acids from substrates such as canola oil or corn oil. The reaction can be controlled during mixing and pumping of the fracturing fluid or where the saponification is designed to occur within the reservoir shortly after the treatment is complete. Alternatively the amine oxide surfactant may

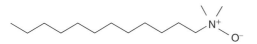

Figure 3.52 Lauryldimethylamine oxide.

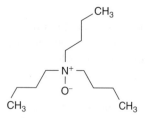

Figure 3.53 Tributylamine oxide.

be added as an external breaker to remove the VES-gelled fluids that are already in place downhole [213].

Dimethylaminopropyl tallow amide oxide is a related VES material that has been claimed as a useful diverting agent in matrix acidisation [214]. It is interesting to note that this type of product is likely to act as a cationic surfactant (see Section 3.4) due to the low pH of the medium. Dimethylcocoalkylamine oxide has been claimed to act with dodecyl sodium sulphate (an anionic surfactant) as a conjugate ion pair to prevent sludging in acid treatments without interfering with the added corrosion inhibitor [215]. A zwitterionic amine oxide (see Section 3.5 for further descriptions of this type of surfactant class) has been shown to exhibit good temperature stability, up to 125 °C, in self-diverting acid systems [216].

VES-based fluid systems that form plugs for water and gas shut-off applications can include a variety of surfactant types; however, amine oxide VESs are a preferred class as they can offer more gelling functionality per unit weight, making them more cost effective than alternatives.

Amine oxides have been claimed as potential hydrogen sulphide scavengers [217]; however, to the author's knowledge there is no infield use of these products for this application.

Various synergists have been proposed to boost the performance of the polyvinylcaprolactam-based kinetic hydrate inhibitors (see section 2.1.1). These include amine and polyamine oxides, for example, tributylamine oxide (see Figure 3.53) is known to be a good synergist for enhancing these polymers [218]. Polyamine oxides are also claimed to be useful kinetic hydrate inhibitors in their own right [219].

The aerobic and anaerobic biodegradability as well as the aquatic toxicity of fatty amine oxides and fatty amido amine oxide surfactants have been investigated [220]. All of the amine oxide-based surfactants tested were readily biodegradable under aerobic conditions, but only the alkyl amido amine oxide was found to be easily biodegradable under anaerobic conditions. Toxicity of these products was found to be relatively low with fatty amido amine oxides, showing the lowest aquatic toxicity.

3.3.5 Alkyl Polyglucosides (APGs)

APGs have briefly been referred to in Chapter 2 on polysaccharides and related natural polymers; this section will now examine these products, as well as chemistry and application in the upstream oil and gas sector in more detail.

APGs are polymers constructed from alkyl glucosides, which in turn are derived from the reaction of glucose and fatty alcohols (see Figure 3.54) by an acid-catalysed condensation process.

The industrial manufacture of APGs is by the Fischer glycosidation process and refers to the formation of a glycoside, again as in Figure 3.54, by the reaction of the glucose with an alcohol, in this case a fatty alcohol, in the presence of an acid catalyst. The reaction is named after the German chemist Emil Fischer who developed this method between 1893 and 1895 [221].

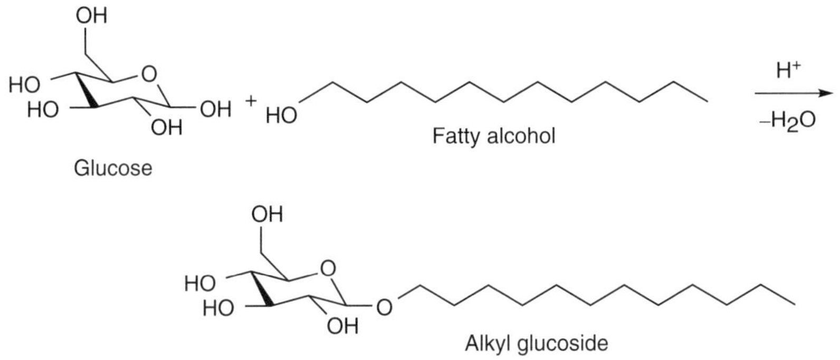

Figure 3.54 Synthesis of alkyl glucosides.

A significant development was the synthesis of a surfactant mixture with lauryl alcohol using sulphuric acid as the catalyst [222].

Commonly, the reaction is performed using a solution or suspension of the carbohydrate in the alcohol as the solvent. The carbohydrate is usually completely unprotected. The Fischer glycosidation reaction is an equilibrium process and can lead to a mixture of ring size isomers and anomers, plus in some cases small amounts of acyclic forms. With hexoses, short reaction times usually lead to furanose ring forms, and longer reaction times lead to pyranose forms. These and other technical problems on an industrial scale can make the manufacture of the APGs problematic; however modern practices have also led to a number of control techniques that have enabled specific and consistent manufacture of a narrow range of oligosaccharide variants [223].

A number of alkyl glucosides are non-ionic surfactants in their own right and consist of a hydrophobic alkyl group from the fatty alcohol and a hydrophilic saccharide structure derived from D-glucose. Alkyl glucosides with alkyl residues of C6–C18 carbon chain lengths show the most significant surfactant properties. The hydrophilic head group is constituted by one or several glycosidically linked D-glucose units. D-Glucose is widely found in nature in the form of sugars and can provide an inexhaustible and renewable raw material. Alkyl glucoside mixtures are often used technically as surfactants and commonly referred to as APGs. These mixtures will contain monosaccharides, disaccharides and other oligosaccharides.

Alkyl glucosides and APGs exhibit significant surfactant properties in aqueous solution. APGs due to their complex nature and their different isomerism can offer a wide range of surfactant behaviour, and selective synthesis can tailor this to give a high degree of specificity. Depending on the nature of the alkyl chain, different phase behaviours can be observed. This can be important in selecting an APG to meet certain conditions such as pH and temperature [224]. Also the chain length and the degree of glycosidation can affect the CMC value. Studies have shown that APGs have similar CMCs to the non-ionic surfactants, the alcohol ethoxylates; however, the behaviour of the APGs can be similar to anionic surfactants [225].

APGs are found to have, in common with other non-ionic surfactants, good solubilising properties [226]. In terms of flow behaviour, then the APGs have some unique characteristics, and they exhibit Newtonian behaviour in micellar solutions at concentrations as high as 50% [227] and some $C_{12}C_{14}$. APGs show this behaviour at concentrations as high as 60%. These same products exhibit shear-thinning behaviour even at low concentrations.

These characteristics make the APGs ideal for blending and formulating with other surfactants, particularly anionic surfactants. Synergistic effects are often displayed by these mixed surfactant systems. This can lead in turn to beneficial rheological effects and also environmental and toxicity benefits [228].

Since APGs are non-ionic surfactants with a certain amount of anionic character, their behaviour should fit somewhere between that of ionic surfactants and alkoxylates; however, they display microemulsion behaviour including one-phase microemulsions [229], which is stable and invariant to temperature and salinity [228].

This wide range and behaviour as well as the low toxicity and low environmental impact properties has meant that these materials are more and more used in a number of surfactant applications across a wide range of industries. Indeed, in the upstream oil and gas industry, APGs could be used in areas where other surfactants are currently used. They have a wide range of unique properties and much of this could be exploited advantageously. In summary and particularly important for application in the oilfield, APGs display a number of interesting and often unique properties, and these are detailed as follows along with some suggested application area:

- Thermal and thermotropic behaviour – They do not appear to follow normal melting processes and display 'double melting points' showing an initial melting point and then usually a much higher (up to 200 °C higher) clearing point. This interesting thermal phasic behaviour could lead to high-temperature and high-pressure well applications.
- Micellular formation behaviour – APGs have significantly differing micellulisation properties from other non-ionic surfactants, giving them unique surface tension and IFT properties, which may find serious application in EOR processes.
- Solubility – APGs are highly water soluble and paraffin insoluble, which could have applications in paraffin (wax) dispersants.
- Synergistic effects – Combining APGs with other surfactants results in the mixture displaying superior properties. This is particularly important in terms of surface and IFT activity. It is also useful in modifying the flow behaviour of the material itself. In the oil and gas industry, this is becoming of increasing importance to allow delivery across low subsea tie-back at low temperatures while still maintaining reasonable product activity. Work has been conducted to show their synergistic effect with aspartate polymers in the development of a new class of green corrosion inhibitor [230].
- Microemulsions – Adding medium-chain alcohols as a co-surfactant with APGs can result in the formation of microemulsions in oil and water systems. Such systems have found application in a number of difficult wellbore and downhole applications, in particular well completion in horizontal and deviated wells.
- Wetting properties – APGs possess good water wetting and adsorption properties, and this has recently been exploited in agrochemical applications. There are a number of downhole stimulation and EOR techniques that could exploit this ability.
- Foaming and foam stability – In comparison with anionic surfactants, non-ionic surfactants are considered to be low foaming. With respect to APGs this can be misleading; indeed the foaming of APGs differs from that of anionics and non-ionics and may behave synergistically either in foam boasting or in foam reduction. Most of this behaviour appears to be dictated by the hydrocarbon chain length, which in the case of APGs appears to be the converse of the general behaviour in that foaming, and foam stability decreases with increasing alkyl chain length. Although in the oil and gas industry foaming is often an undesirable characteristic in terms of gas–liquid unloading, it is absolutely necessary. This growing application requires good stable foams to be built in the liquid phase so additional gas can be produced from water-blocked reservoirs.

The aforementioned is a brief outline of the potential applications of APGs, which are a chemistry class not yet widely exploited in the upstream oil and gas industry; indeed use has been quite restricted mainly due to the high cost of the APGs compared with other surfactants. They are competing of course with some cheaper technologies, namely, fatty alcohol ethoxylates and polyoxyethylene alkyl ethers.

The work on the application of APGs as corrosion inhibitors has been further developed [231] and significantly shown to not only have a good environmental profile but also have good corrosion protection at temperatures above 120 °C [232]. This work also indicated the potential IFT effects of the APG, which has been further investigated as to the potential for a biodegradable demulsifier (H.A. Craddock, unpublished results). Of note in this later application is the use in a specific underbalanced drilling scenario where there is significant scale and solids production and stabilising of returned drilling fluid and initial production fluids as an emulsion. The 'formulated' APG is used to provide both a demulsifying effect and a dispersant/settling effect on the solids.

In the last few years, APGs have been considered as surfactants for EOR as they are non-toxic and non-ionic and their phase behaviour and IFT values are quasi-independent of temperature and salinity [233]. The products examined showed some application potential with solid adsorption of the APG surfactants being dependent on the APG alkyl chain length. A larger chain length led to more adsorption.

The alkyl glucoside monomers, preferentially the methyl glucoside, have been claimed as environmentally safe fluid loss additives in drilling fluids [234]; however, their actual use is thought to be minimal if at all.

The main and continuing use for the APGs is the exploitation of their detergent effects at low concentration levels [235]. The APGs are used domestically and industrially in powder formulations and in particular in liquid detergents, dishwashing liquids and shower gels. They possess a high foaming power and show a high foam-stabilising and stability- and viscosity-increasing effect in combination with anionic surfactants, such as alkylbenzene sulphonates. They are non-toxic, mild on the skin and readily biodegradable. In the upstream oil and gas sector, they have found application as rig washes and wellbore completion cleaners [236].

In well cleaning applications there are a number of properties that are important in constructing an efficient and acceptable well cleaner [237], and the following are the most important:

- *HLB:* A hydrophilic–lipophilic balance number in the range 10–15 is the best at 'dirt' removal (H.A. Craddock, unpublished results).
- *Cloud point:* Surfactants precipitate out of solution at a specific temperature, their cloud point. This is usually measured in monoethylene glycol. At just below this point, it is widely held that surfactants operate at their optimum. However, using solely the cloud point to design well cleaners is exceedingly difficult due to the wide variations in temperature that the cleaning formulation is likely to encounter. Nonetheless it is a good indicator within the range above as to those products likely to be effective. APGs have the desired cloud point profile [238].
- *Foaming characteristics:* It is desirable that excessive foaming is not present as this can cause considerable operational problems at the rig site.
- *Biodegradability:* In this respect, the main factors under consideration, as has been described before, which cause surfactant molecules to biodegrade, are their molecular size and the degree of side-chain branching. Small linear molecules tend to biodegrade more efficiently than those that are highly branched.

The properties mentioned require to be balanced as the HLB efficiency and biodegradability act against each other, and therefore commercial products are blends of chemicals, which take into account the interdependency of these properties.

The utility and versatility of the APGs in the upstream oil and gas industry have been rarely explored mainly due to their high unit cost; however it has been shown that they can be applied effectively in low concentrations. Their ready biodegradation profile and sustainable sourcing make them highly important candidates in green chemical application for the future.

3.3.6   Ethylene Oxide and Propylene Oxide Copolymers

The idea of using the hydrophilic nature of the PEO chain and the hydrophobic nature of the polypropylene oxide (PPO) chain to produce a surfactant is several decades old. In this respect it is worth noting that though the PEO chain is globally hydrophilic, each EO group contains 2 methylene ($-CH_2-$) units, which are hydrophobic. This duality becomes evident when it is known that the PPO chain, i.e. the three carbon atoms counterpart, is globally hydrophobic. It can be said that the hydrophilicity conferred by the oxygen atom is thus compensated by approximately 2.5 methylene groups.

This remark is quite important, because it clearly indicates that the PEO hydrophilic group is not an extremely hydrophilic group, a characteristic that explains why this kind of surfactant is soluble in organic solvents. Moreover, any change in formulation or temperature that affects the interaction between the PEO chain and the water–oil physico-chemical environment is likely to affect the behaviour of this kind of surfactant.

As described in Chapter 2 and earlier in this chapter, EO is an important constituent in introducing hydrophilicity into a surfactant-like molecule. Similar behaviour is observed with PO (see Figure 3.55). It is a colourless volatile liquid that is produced on a large scale industrially. The major application for PO is its use in the production of polyether polyols for making polyurethane plastics. In surfactant terms it gives the same functionality as EO; however mixed EO/PO surfactant polymers can have advantageous properties particularly in some oil and gas sector applications.

As described in Chapter 2, both vinyl and non-vinyl polymers and copolymers can be ethoxylated and propoxylated, giving a number of demulsifier candidates, particularly the block copolymers and the alkylphenol resin alkoxylates giving rise to a class of polymeric demulsifiers, the polyalkylene glycol adducts of EO [239]. This class of products is particularly useful in desalting applications [240].

Block copolymers consisting of PEO and PPO can self-assemble in water and water–oil mixtures (where water is a selective solvent for PEO and oil a selective solvent for PPO) to form thermodynamically stable spherical micelles. Important observations on the copolymer micellisation in water as affected by co-solutes and on the time dependency of the surface activity have been made, which supports their properties in oil and water separation and moving salts and ions to the aqueous phase [241].

It has also been observed that the polymer structure has effects on the micellular structure and solubilisation of polycyclic aromatic hydrocarbons [242], an important characteristic in their application as demulsifiers.

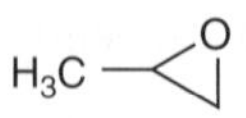

Figure 3.55 Propylene oxide.

Again as has been detailed earlier in this chapter, and in Chapter 2, ethoxylation (and propoxylation) enhances the rates of biodegradation of these polymeric materials; indeed these surfactants are recognised as environmentally acceptable and biodegradable products and used in a number of industrial and domestic sectors for their defoaming and wetting properties.

They are also closely related as non-ionic surfactants to the polyalcohols, or polyols, described in the next section.

3.3.7 Polyalcohols and Ethoxylated Polyalcohols

Polyalcohols and their related products have been extensively described in Chapter 2, and in this short section some of their behaviour as surfactants is detailed both in general and specifically in oilfield applications. Although the polyalcohols are classed as non-ionic surfactants, they behave as a hydrophilic solvent and work with other non-ionic surfactants to increase wettability [243]. This property is also widely exploited in pharmaceutical delivery applications [244] to provide a water-soluble film or coating with good muco-adhesive properties for ease of oral administration.

Ethoxylation of the polyalcohols can change their characteristics to being anionic in nature and behaviour; however they are also more biodegradable [245, 246].

In the oilfield it does not appear that both the surfactant and ecotoxicological properties are fully exploited.

Polyalcohols and their ethoxylated derivatives can be used as intermediates in the production of ester-based corrosion inhibitors, which are claimed to have good persistency as FFCIs [247]. The polyols have also been used in complex formation in fracturing fluids [248].

Finally, as has been previously described, polyalcohols are involved in a number of demulsifier applications, particularly when cross-linked with other polymers such as polyamines [249].

3.3.8 Ethoxylated Thiols

Thiols and related molecules and their use in the upstream oil and gas sector will be more fully described in Chapter 6. However the ethoxylated thiols show interesting surfactant behaviour and are classified as non-ionic surfactants.

In thiols, the oxygen atom of the alcohol structure is replaced by a sulphur atom as in Figure 3.56. Thiols can be ethoxylated just as alcohols or phenols. The corresponding products are excellent detergents and wetting agents, which are however only used in industrial applications as the possibility of releasing noxious smelling mercaptans banned them from domestic use.

tert-Dodecyl mercaptan ethoxylate exhibits a good solubility in both water and organic solvents. Moreover it is an excellent industrial detergent. It is use in raw wool treatment and agrochemical emulsions, in which its wetting ability enhances the cleansing action. It has a specialist application in cleansing metal surfaces as well as passivating them as a corrosion inhibitor [250].

As far as is known to the author, this class of non-ionic surfactants has had no application in the upstream oil and gas industry. They are, however, likely to have intriguing properties and depending on the degree of ethoxylation are biodegradable.

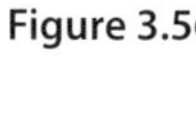

Figure 3.56 Ethanethiol.

3.4   Cationic Surfactants

Cationic surfactants are dissociated in water into an amphiphilic cation, i.e. the polar group or hydrophilic head, which is positively charged, and an anion, most often of the halogen type (e.g. chloride or bromide).

A very large proportion of this class corresponds to nitrogen-containing compounds such as fatty amine salts and quaternary ammoniums, with one or several long-chain alkyl groups, often coming from natural fatty acids. These surfactants are in general more expensive than anionics, because of their route of manufacture. As a consequence, they are only used in two cases in which there is no cheaper substitute, i.e. (i) as bactericide and (ii) as positively charged substance, which is able to adsorb on negatively charged substrates to produce anti-static and hydrophobant effects. The latter is often of significant commercial importance such as in corrosion inhibition, and until recently cationic surfactant-based corrosion inhibitors were among the dominant types of chemistry used in the formulations of corrosion inhibitors as applied in the upstream oil and gas industry [251]. Their decline in use has been primarily due to their aquatic toxicity to the marine environment, particularly alkyl quaternary ammonium products [252].

Cationic surfactants account for only less than 10% of the total surfactant production. However, they are extremely useful for some specific uses, because of their peculiar properties. They are not good detergents nor foaming agents, and they cannot be mixed in formulations that contain anionic surfactants, with the exception of non-quaternary nitrogenated compounds. The low foaming characteristic is another factor in their preference for use as oilfield corrosion inhibitors.

They exhibit two highly important characteristics:

1) Their positive charge allows them to adsorb on negatively charged substrates, as most solid surfaces are at neutral pH. This capacity gives to them an anti-static behaviour and a softening action for fabric and hair rinsing. The positive charge enables them to operate as floatation collectors, hydrophobating agents, corrosion inhibitors and solid particle dispersants. They are used as emulsifiers, as coatings in general, in inks, in wood pulp dispersions and as already stated as corrosion inhibitors in a variety of industrial applications.
2) Many cationic surfactants are bactericides. They are used to clean and aseptise surgery hardware to formulate heavy-duty disinfectants for domestic and hospital use and to sterilise food bottle or containers, particularly in the dairy and beverage industries.

Although they are produced in much smaller quantities than the anionics, there are several types, each used for a specific purpose. The following subsections of this section describe the more important cationic surfactants, in particular those that are used in the upstream oil and gas industry and those with interesting and acceptable environmental toxicity characteristics.

3.4.1   Alkyl Quaternary Systems and Related Products

The simplest quaternary system is the ammonium ion (Figure 3.57).

An alkyl quaternary nitrogen system has alkyl groups attached to the nitrogen atom. An example is shown in Figure 3.58.

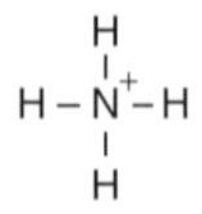

$$H - \overset{\overset{\textstyle H}{|}}{\underset{\underset{\textstyle H}{|}}{N^+}} - H$$

Figure 3.57 The ammonium ion.

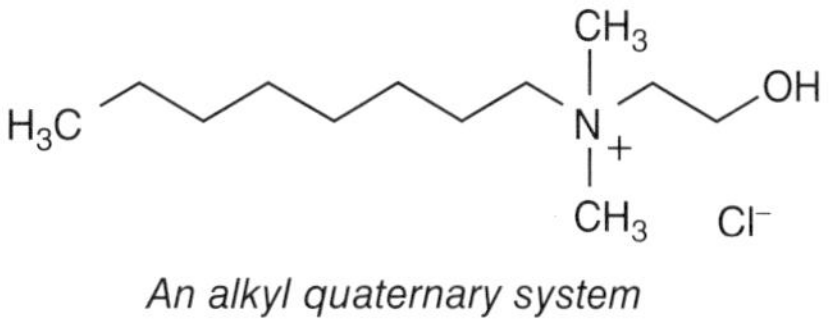

Figure 3.58 An alkyl quaternary ammonium chloride.

An alkyl quaternary system

Figure 3.59 Synthesis of alkyl quaternary surfactants.

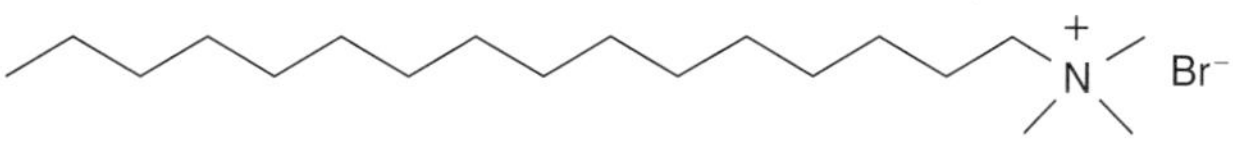

$$R_1R_2NH + 2\ CH_3Cl \longrightarrow R_1R_2N^+(CH_3)_2\ Cl^- + HCl \uparrow$$

Figure 3.60 CETAB.

They are used as fabric softeners with anionic surfactants, helping them to break down the interface between the dirt/stain and the water. In the oil and gas industry, many are used as corrosion inhibitors or as key components in corrosion inhibitor formulations [156].

These alkyl quaternaries are produced from primary and secondary amines, which are quaternised by exhaustive methylation with methyl chloride, with removal of produced HCl in order to displace the reaction as illustrated in Figure 3.59.

Another way is to react an alkyl bromide with a tertiary amine. This is the usual way to prepare cetyltrimethylammonium bromide (CETAB) (see Figure 3.60). It is one of the key components of the topical antiseptic cetrimide. The cetrimonium (hexadecyltrimethylammonium) cation is an effective antiseptic agent against bacteria and fungi. The closely related compounds cetrimonium chloride and cetrimonium stearate are also used as topical antiseptics and may be found in many household products such as shampoos and cosmetics. CETAB, due to its relatively high cost, is typically only used in cosmetics.

If a sulphate anion is required, the quaternisation of the tertiary amine is carried out with dimethyl or diethyl sulphate.

All these methods result in alkyl ammoniums displaying different alkyl groups.

Many quaternaries are derived from fatty amines, which in turn are derived from fatty acids. The overall reaction is referred to as the nitrile process [253] and begins with a reaction between the fatty acid and ammonia at high temperature (>250 °C) and in the presence of a metal oxide catalyst (e.g. alumina or zinc oxide) to give the fatty nitrile (see Figure 3.61).

The fatty amine is obtained from this by hydrogenation with any of a number of reagents/catalysts including Raney nickel. When conducted in the presence of excess ammonia, the hydrogenation affords the primary amines (Figure 3.62).

In the absence of ammonia, secondary and tertiary amines are produced, as in Figure 3.63 [254].

Alternatively, secondary and tertiary fatty amines can be generated directly from the reaction of fatty alcohols with alkylamines. These tertiary amines are precursors to quaternary ammonium salts used in a variety of applications.

$$RCOOH + NH_3 \rightarrow RC \equiv N + 2\ H_2O$$

Figure 3.61 Synthesis of fatty nitrile.

$$RCN + 2\ H_2 \rightarrow RCH_2NH_2$$

Figure 3.62 Synthesis of primary fatty amine.

$$2\ RCN + 4\ H_2 \rightarrow (RCH_2)_2NH + NH_3$$ **Figure 3.63** Synthesis of secondary and tertiary fatty amines.

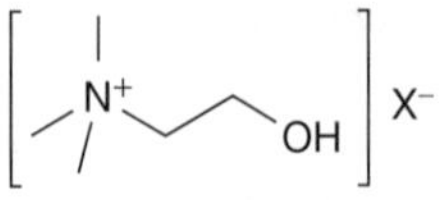

$$3\ RCN + 6\ H_2 \rightarrow (RCH_2)_3N + 2\ NH_3$$

Figure 3.64 Choline salt.

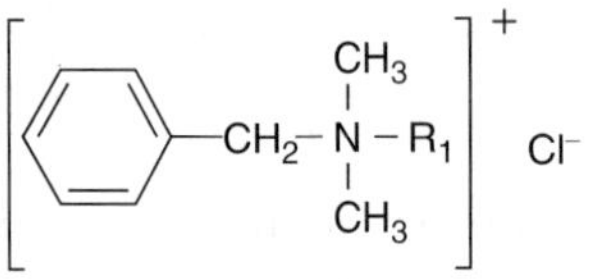

Figure 3.65 Alkyl dimethyl benzyl ammonium chlorides.

$R_1 = C_{12}H_{25} : C_{14}H_{29} : C_{16}H_{33} : C_{18}H_{37}$

In oilfield applications CETAB has been examined as a cationic VES for application in hydraulic fracturing [255]. This and other similar quaternary ammonium salts have also been examined as viscosity surfactants in EOR applications [256].

Quaternary ammonium salts such as choline salts have been used in underbalanced drilling applications as anti-swelling drilling fluid additives [257]. Choline (Figure 3.64) is quaternary ammonium salt containing the N,N,N-trimethylethanolammonium cation often as the chloride salt.

Other quaternary ammonium salts and in particular C10–C18 alkylbenzyldimethylammonium chloride are used as cement additives to increase adhesion properties and strengthen the cement bonding to the formation rock [258].

Quaternary ammonium salts such as alkyldimethylbenzylammonium chloride (Figure 3.65) have been used for the control of microbes in industrial wastewaters for a long time and are used in the upstream oil and gas industry both as biocides and for corrosion inhibitor applications.

Outside of the highly regulated areas for environmental compliance, such as the North Sea basin, the nitrogen quaternary salts, especially the common surfactants, for example, the benzalkonium chlorides and alkyl pyridine quaternaries [259], are still a major class of oilfield FFCI, particularly when formulated with other surfactants [251].

However, this type of cationic surfactant has a poor ecotoxicological profile, and in particular CETAB has been extensively studied [259]. They have high aquatic toxicities [260] and poor rates of biodegradation [261], which have led to their use being heavily reduced, as they are required to be substituted and only allowed where no suitable alternatives exist. Interesting and somewhat supportive of their toxicity is that their bacteriostatic efficiency decreases as their alkyl chain length increases, in particular where the counter-anion is derived from a fatty acid carboxylate [262].

A quaternary ammonium salt, which has an acceptable environmental profile, is $N,$ N-didecyl-N,N-dimethylammonium chloride; the structure is illustrated in Figure 3.66. It is claimed to be useful in downhole applications as an FFCI and exhibits biostatic and bactericidal properties [263].

A polymeric biodegradable quaternary ammonium salt that is claimed to function as a biocide and an FFCI is synthesised from a polymer reaction of epichlorohydrin and a tertiary amine [264].

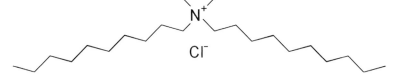

Figure 3.66 *N,N*-didecyl-*N,N*-dimethylammonium chloride.

These quaternary surfactants can be made more biodegradable by the introduction of weak linkages between the long-chain hydrocarbon groups and the quaternary nitrogen atom. The most common of these linkages are ester groups, and these products the ester quats are discussed in Section 3.4.2.

Quaternary ammonium salts are used in other applications such as anti-agglomerants (AAs) in mitigation of gas hydrate formation [265]. As expected these have poor toxicities and low rates of biodegradation unless modified [266]. The quaternary salts of polyether amines, usually prepared by reaction with alkyl bromides, are also effective gas hydrate inhibitors and have been found to prevent the growth of hydrate crystals already formed [267]. As has been described in Section 2.1, these polyethers offer a polymer backbone with potentially hydrolysable linkages and therefore potentially a better environmental profile.

The quaternary ammonium chlorides have some undesired effects, in that they can cause excessive foaming and can react antagonistically with anionic surfactants. In this respect they have been largely replaced by bis[tetrakis(hydroxymethyl)phosphonium] sulphate (THPS) particularly with regard to sulphate-reducing bacterial control [268]. This product and related phosphonium salts are discussed in more detail in Chapter 4.

3.4.2 Ester Quats

The directly quaternised fatty acid surfactants described earlier have been replaced in a number of domestic and household applications such as washing powders, fabric softeners, etc. [269] by more complicated structures in which there is an ester linkage between the alkyl chains and the quaternary head group. These are more biodegradable and less toxic and are known as ester quats. An example is shown in Figure 3.67: a diester quaternary of a distearic acid-based quat.

Despite their enhanced biodegradation characteristics, these materials have been seldom used or applied in the upstream oil and gas industry.

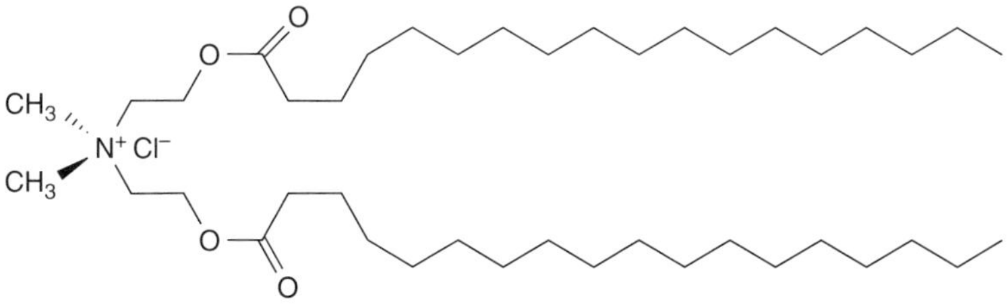

Figure 3.67 An ester quat.

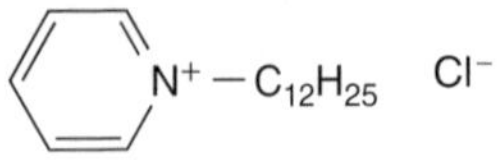

Figure 3.68 *N*-dodecylpyridinium chloride.

3.4.3 Heterocyclic Cationic Surfactants

An important class of cationic surfactants contains aromatic or saturated heterocycles includ-ing one or more nitrogen atoms. This is the case of the commonly used *N*-dodecylpyridinium chloride (Figure 3.68), which is prepared by reacting dodecyl chloride with pyridine. *N*-Dodecylpyridinium chloride is used as bactericide and fungicide. If a second hydrophilic group is added (amide, EO), the product is the both a detergent and a bactericide.

As stated in Section 3.4.1, these alkyl pyridinium quaternaries are also useful FFCIs. This dual functionality is of great use in the upstream oil and gas industry, particularly in deliver-ing a 'one-pot' treatment package such as in hydrotesting subsea process infrastructure [270]. In this package a biocide and oxygen scavenger are often required to be delivered simultane-ously, and the use of a non-oxidisable biocide or biostatic agent that also delvers corrosion protection is highly desirable. The alkyl pyridinium and other quaternary ammonium salts offer this facility. As has been already described, however, they do not have good environ-mental properties, have poor rates of biodegradation and high levels of aquatic toxicity. It is a significant challenge to formulate a product that can give both the technical and environ-mental properties required, and this will be further discussed in Chapter 8 on formulation practices.

3.4.4 Amide, Ester and Ether Amines

Reacting a fatty acid with ammonia can produce an amide. Another way to prepare an amide is to react a fatty acid with a short-chain alkylamine or alkyldiamine, e.g. ethylenediamine or diethylenetriamine (DETA) as in Figure 3.69.

The hydrophilicity of this substance (di-alkyl-amido triamine) can be enhanced by quaterni-sation or by adding a non-ionic moiety such as polyethylene glycol. They are primarily used as anti-static agents and bactericides in the textile industry.

Ester amines are prepared by reacting a fatty acid with an ethanolamine as in Figure 3.70.

Ether amines are prepared by condensing an alcohol on acrylonitrile double bond, followed by the hydrogenation of the nitrile (Figure 3.71).

$$2R-COOH + H_2N-CH_2CH_2NHCH_2CH_2-NH_2 \longrightarrow RCO-(NHCH_2CH_2)_2-NH-COR$$

Figure 3.69 Synthesis of fatty amides.

$$RCOOH + HO-CH_2CH_2N(CH_3)_2 \longrightarrow RCO-O-CH_2CH_2-N(CH_3)_2$$

Figure 3.70 Synthesis of ester-amines.

$$ROH + H_2C=CHCN \longrightarrow R-O-CH_2CH_2CN \longrightarrow R-O-CH_2CH_2CH_2NH_2$$

Figure 3.71 Synthesis of ether-amines.

These products are seldom used in the upstream oil and gas industry but have good environmental properties and offer interesting surfactant and detergent application characteristics [271].

The related alkanol amides are commonly used in a variety of industry as wetting and foaming agents; however their application in the upstream oil and gas industry is sparse but have been examined for use in EOR by chemical flooding [272].

3.4.5 Amine Oxides

Amine oxides have already been discussed in section on non-ionic surfactants in some detail. They are generally prepared by reacting a peroxide or a peracid with a tertiary amine. The amine oxide possesses a semi-polar N→O bond in which the nitrogen atom provides the two electrons so that there is a strong electronic density on the oxygen atom. Amine oxides capture a proton from water to become quaternised cationic hydroxyamines at acid pH and remain non-ionic under neutral and alkaline conditions. They are among the best foam boosters available at neutral and alkaline pH, with additional corrosion inhibition properties at neutral pH.

Cationic surfactants in general are incompatible with anionic surfactants, since they react with one another to produce insoluble catanionic compounds. This is can be quite a practical problem since the most inexpensive surfactants are anionics of the sulphonate or sulphate ester type. It can be therefore desirable to some cationic substance for additional or enhanced effects, and this will be further discussed in Chapter 8.

Cationic surfactants can present the oilfield chemist with a few dilemmas in selection as they exhibit effective rather than efficient behaviour, as like anionic surfactants the charged hydrophile does not favour micelle formation at low concentrations [273], which is useful in selection for FFCIs. Also cationic surfactants do not relate to the HLB value scale as developed for non-ionic surfactants (see Section 3.1.8), and since most reported applications of cationic surfactants in emulsions involve surface modification, there are few reported HLB values. Importantly however there are a few reported partitioning coefficients of cationic surfactants [274], and this is an unusual property in surfactants particularly, as will be further described in Section 3.9, and in Chapter10, the determination of bioaccumulation potential of surfactants is fraught with difficulties [59].

One of the major difficulties facing the environmental chemist in examining the oilfield use of cationic surfactants is their acute and chronic aquatic toxicity [275]. Lower molecular weight cationic surfactants such as alkylamines of C_{18} and less are toxic and corrosive, whereas higher molecular weight analogues are classified as harmful [252]. Increasing their water solubility through ethoxylation only, in most cases, increases toxicity, which has been contrary to previous cases where such methodology is used to improve rates of biodegradation [261]. Nonetheless there have been modifications made to cationic surfactants to improve their environmental impact [276] although to date these newer surfactants have yet to be applied for oilfield use.

3.5 Amphoteric Surfactants

Amphoteric (or zwitterionic) surfactants are so called because the head group carries both a negative and positive charge. A range of methods is used to produce such materials, almost all of which contain a quaternary ammonium ion (a cation). The negatively charged group can be:

- Carboxylate, $-CO_2^-$
- Sulphate, $-OSO_3^-$
- Sulphonate, $-SO_3^-$

Depending on the substance and on the pH, they can exhibit anionic or cationic tendencies. In most cases it is the pH that determines which of the groups dominate, by favouring one or the other ionisation: anionic at alkaline pH and cationic at acid pH. The ionic nature of amphoteric surfactants therefore depends on the pH. At an intermediate pH, the strength of both positive and negative charge becomes equal, and the molecule will have positive as well as negative charge. At this point it is a zwitterion and the pH at which both charges are equal in strength is called the isoelectric point. The isoelectric point is not a sharp point but depends on the nature of the anionic and cationic groups. At the isoelectric point, amphoteric surfactants generally have their minimum solubility.

True amphoteric surfactants contain both the positive and negative groups in the same molecule, which leads to their unique and useful properties.

As with other surfactants there are a number of different types of amphoteric surfactants.

3.5.1 Betaines

The most common class in this grouping is the alkyl betaines, which have a carboxyl group. A long-chain carboxylic acid reacts with a diamine to form a tertiary amine, which on further reaction with sodium chloroethanoate forms a quaternary salt, as shown in Figure 3.72. Betaines are neutral compounds with a cationic and an anionic group, which are not adjacent to one another but are contained in the hydrophilic head group. The alkyl group provides the hydrophobic tail.

The betaines (Figure 3.72) have a single methylene group between the acid and the quaternary ammonium groups. The sulphobetaines can have more than one methylene group between the sulphonic acid grouping and the quaternary nitrogen atom as in Figure 3.73, *N*-dodecyl-*N,N*-dimethyl-3-ammonio-1-propanesulphonate.

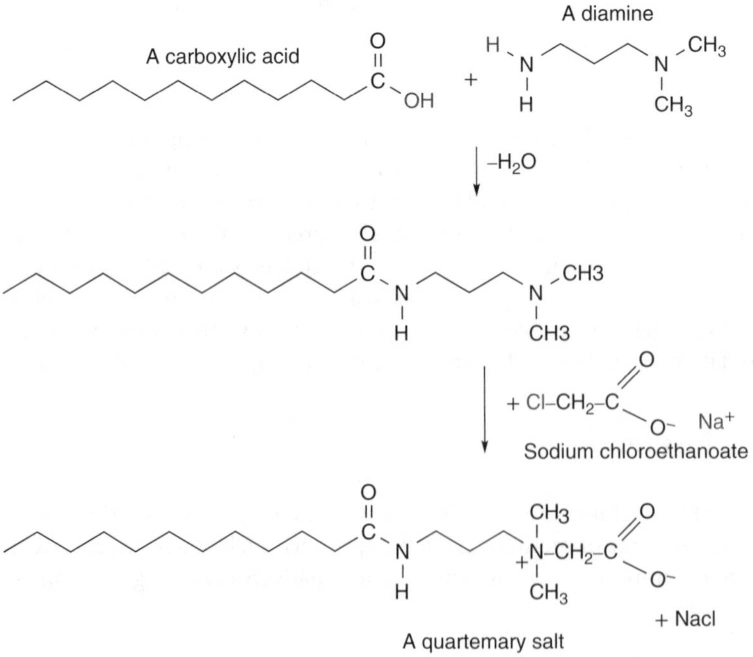

Figure 3.72 Alkyl betaine synthesis.

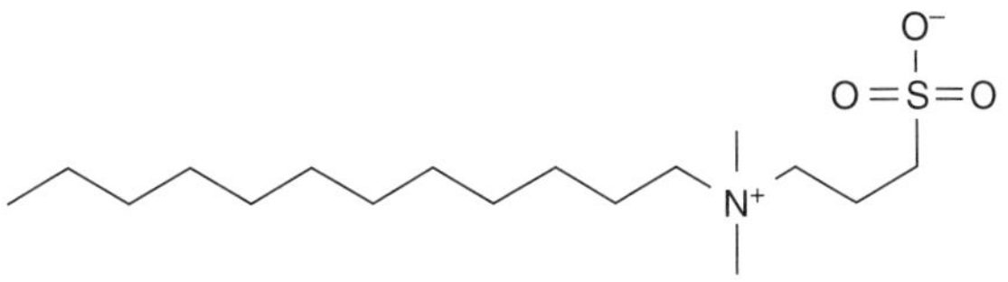

Figure 3.73 *N*-dodecyl-*N*,*N*-dimethyl-3-ammonio-1-propanesulphonate.

These surfactants, the betaines, are amphoteric at neutral and alkaline pH and cationic at acid pH (at which the carboxylic acid is not ionised). Since the nitrogen atom is quaternised, these surfactants always display a positive charge. They tolerate a high salinity, particularly divalent cations, e.g. calcium and magnesium.

They are widely used in softeners for textiles, hair rinse formulas and corrosion inhibition additives. They are good foam boosters because of their cationic characteristics. This class of amphoteric surfactants are described as being very mild in terms of cosmetic and personal care use is said to be pH balanced.

Various amphoteric surfactants, including alkyl betaines, have been found to impart viscoelastic properties when combined with organic or inorganic salts [277]. This is highly usefully when trying to maintain fracturing fluid viscosity in high brine concentrations where traditional cationic surfactants tend to lose viscosity [278]. These viscoelastic properties have found application in diversion techniques for gas and water during acid stimulation of petroleum reservoirs [279].

These viscoelastic properties along with their pH stability and wettability characteristics have made this type of surfactant a candidate for EOR by polymer/surfactant flooding without the need for additional alkali [280].

Similarly the sulphobetaines exhibit VES properties, and some specific examples such as the hydroxypropyl sulphobetaine display excellent stability in high-temperature and high-salinity reservoirs [281].

The alkyl betaines and other zwitterionic surfactants can act as FFCIs [282]. The most prominent of these are the imidazolines, which are discussed in Section 3.5.3. The alkyl betaines have also been used to retard the corrosive attack from 1.0 M hydrochloric acid [283]; however despite the better environmental profile of these products as referenced later, this work does not seem to have been developed to replace a number of toxic alternatives still used in acid corrosion retardation.

In general the betaines are much less toxic in the marine environment than their equivalent cationic surfactants but show much of the same characteristics and performance [284]. Indeed the long-chain alkyldiamine-based betaines reacted with acrylic acid are claimed to have low marine toxicities, certainly lower than the alkyldiamine [285]. Although these products show promising environmental toxicity profiles, they are not widely used in the upstream oil and gas industry, being usually outperformed by equivalent imidazolines.

3.5.2 Amino Acids

Amino acids can be regarded as amphoteric surfactants that contain both an acid and an amine group. The amino acid-based amphoteric surfactants are biocompatible and are widely used in pharmaceuticals and cosmetics. This type of surfactant will also be discussed in Section 3.7, and as will be developed there is an urgent and growing need for efficient surfactants that are both biocompatible and biodegradable.

Surfactant molecules from renewable raw materials that mimic natural lipoamino acids are one of the most promising lines of enquiry. Their natural and simple structure means they

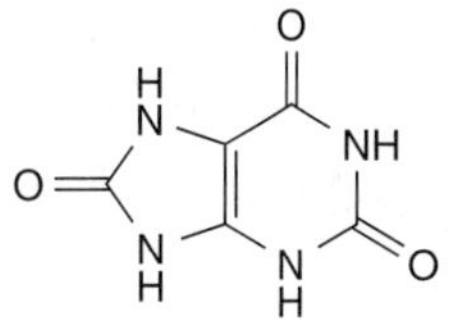

Figure 3.74 Uric acid.

exhibit low toxicity and good biodegradation rates. The combination of polar amino acids/peptides (hydrophilic moiety) and non-polar long-chain compounds (hydrophobic moiety) for building up the amphiphilic structure has produced molecules with high surface activity [286]. However, to date there has been little application of this class to the upstream oil and gas sector.

Recently a diacyl amino acid surfactant was prepared from lauryl chloride and l-lysine. Ultralow IFT could be achieved when the diacyl amino acid surfactant was mixed with sulphonate in the presence of polymer as well as absence of polymer, and its application in EOR is being explored [287].

A related class of amphoteric surfactants are the imido acids. An imido acid is an organic acid consisting of one or more acid radicals that are united with the imido group that contains replaceable acid hydrogen and plays the part of an acid; for example, uric acid (see Figure 3.74) is an imido acid.

The imido propionic acids are widely used as textile softeners, but to date the author knows of no upstream oil and gas applications. These amphoteric surfactants have a lower isoelectric point and therefore higher water solubility, which could be useful in the provision of environmental acceptability.

3.5.3 Imidazolines

One of the most important classes of chemicals used in the upstream oil and gas industry are the imidazolines. These products are classed as amphoteric surfactants particularly in relation to higher molecular weight fatty heterocyclic surfactants, which are highly effective in a wide variety of acidities and alkalinities [288]. They are able to provide a highly effective detergent action that will maintain a fine bubble foam over long periods of time in textile processing, without deterioration or loss of effective foam properties and without decreasing the volume of the foam regardless of the change in pH whether it be strongly acid or become strongly alkaline.

The high molecular weight fatty nitrogenous heterocyclic compounds, the imidazolines, are able to provide high corrosive inhibiting properties at low dilutions, which may be widely utilised in metal processing such as electroplating, radiator liquids and coolants for metal working processes and in the upstream oil and gas industry.

Imidazolines are the basis of possibly the most common type of general FFCI, and their mechanistic action has been well studied. In general without modification they exhibit a high degree of environmental toxicity.

There is a lot of published work regarding the interaction of imidazolines at solid–liquid surfaces, particularly with regard to the oleic imidazolines, which showed that the molecule is primarily bonding through the five-membered nitrogen ring that is lying in a planar orientation to the metal surface [289]. It was also concluded that the long carbon chain, although playing an important role in the mechanism of inhibition, was not the only critical component as varying this pendant chain did not affect inhibitor performance. Contrasting this other work using density functional theory and Monte Carlo simulations has indicated that imidazolines

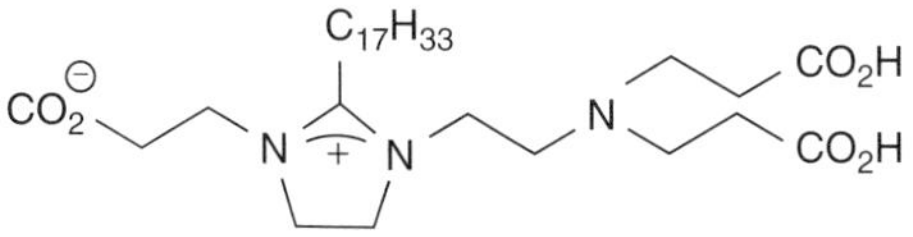

Figure 3.75 Betaine imidazoline.

favour perpendicular adsorption to the metal surface, while their protonated (or alkylated) species adsorb parallel to the metal surface [290]. Other studies have shown that the length of the hydrocarbon chain on the imidazoline is important. With a hydrophobic chain length of less than 12 carbon atoms, no corrosion inhibition is observed, and the optimum performance is a carbon chain length of 18 carbon atoms [291], which is the oleic moiety. This, however, has the poorest environmental properties.

Ethoxylation of the *N*-nitrogen in the imidazoline ring or the side-chain amine or alcohol can provide more water-soluble products with lower bioaccumulation potential and toxicity. Another way to make the imidazoline more water soluble and less toxic is to react a pendant alkylamine group of an imidazoline intermediate with stoichiometric amounts of acrylic acid [292], with, for example, three moles of acrylic acid, giving the structure in Figure 3.75.

Imidazolines are usually manufactured by condensing a polyamine containing a 1,2-diaminoethane functionality with a carboxylic acid. In terms of the upstream oil and gas industry, the source of carboxylic acid is usually tall oil fatty acid (TOFA). TOFA is principally oleic acid, linoleic acid and its isomer mixtures and forms a C_{17} alkyl group. *N*-Substituted 2-alkyl imidazolines are formed if the 1,2-diaminoethane is substituted on one of the nitrogens. In the oil and gas industry diethylene triamine (DETA) is by far the most commonly used polyamine (see the reaction scheme in Figure 3.76).

Pure oleic acid is the next most used fatty acid, which is the purified fatty acid with unsaturation at C_9–C_{10}. The double bond in the long alkyl chain is thought to impart some steric rigidity on the film formed and make the film more persistent. It is the author's experience that there is greater persistence at temperatures up to 120°C with imidazolines based on oleic acid (H.A. Craddock, unpublished results). There is usually a higher *trans* isomer content in this material; however, at higher temperatures, imidazoline-based corrosion inhibitors, in general, are not as persistent as the amines and amides [293, 294].

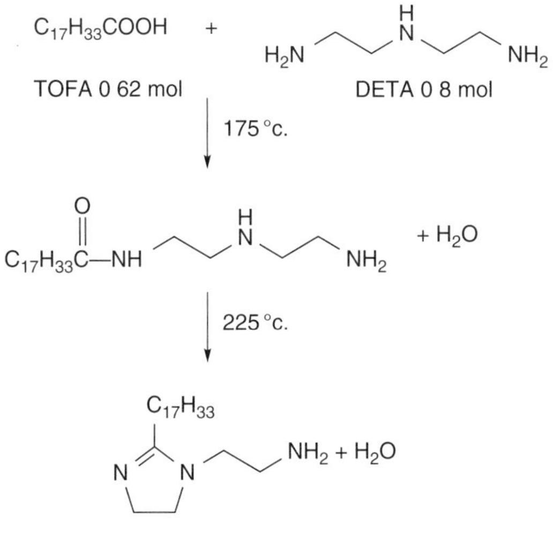

Figure 3.76 Amine imidazoline synthesis.

Stearic acid, which commercially usually contains some palmitic acids, is the fully saturated C_{18} fatty acid. It has been used to form imidazolines for corrosion inhibition [294]; however these homologues do not appear to be widely applied in the upstream oil and gas sector. A difficulty may be that they are waxy solids whereas TOFA is liquid.

Palm oil, whose main constituent is palmitic acid C_{16} and has a carboxylic acid spread between C_7 and C_{17}, is also used as a fatty acid starting material as it can be relatively cheap. The imidazolines formed are not as efficacious as corrosion inhibitors; however, they are widely used, as the raw material is cheap and readily available in Southeast Asia in particular.

Tallow fatty acids have been used to make imidazolines, as they are a mixture of stearic and oleic acids. Their use is uncommon, as their source, from animal fats, is more expensive than the plant-derived equivalents.

In general, fatty amines derived from fatty acids are used to make imidazolines; however, ether carboxylic acids have also been used, and these products are claimed to have greater water solubility and greater persistence [295].

Imidazoline-based amphoacetate surfactants are capable of being utilised to tailor foam characteristics, which can be useful in EOR applications and gas dewatering.

Other amphoteric materials are not widely used in the upstream oil and gas industry although there are a number of applications relating to polymeric amphoteric surfactants, which are described in Section 3.6. Zwitterionic surfactants have been described for use in clay swelling control during drilling operations [119].

3.6   Polymeric and Other Surfactant Types

3.6.1   Polymeric Surfactants

A polymeric macromolecule can have an amphiphilic structure and exhibit surfactant behaviour. Asphaltenes, which are natural components of crude oils, have polar and non-polar groups. However the location and segregation of these groups are often ill defined or at least less defined than in smaller molecules. Indeed there appears to be no clear definition of a polymeric surfactant. EO/PO copolymers (see Chapter 2) and silicon surfactants (see Chapter 7) could be looked upon as polymeric surfactants. In Chapter 2, polyvinyl alcohols and maleic acid demulsifiers can be considered as specific examples of polymeric surfactants.

Synthetic resins based on natural fatty acids such as tall oil alkyl resins, again described in Chapter 2, can be made oil soluble or partially water soluble and show some surfactant properties. Natural products such as alginates, pectins and protein-based products could all be looked upon as polymeric surfactants. Most of these products show only weak surfactant properties with respect to surface tension reduction in aqueous solution and the formation of micelles. The word polymeric would suggest high molecular weight with a large number of repeating molecular units. Most polymeric surfactants have only a few repeating units, and the phrase oligomeric would seem more descriptive. However the word polymeric has now been firmly established and so is used in this book.

In considering polymeric surfactants there are two main configurations, 'block' and 'graft', with respect to copolymer organisation. In such structures there are hydrophilic and lipophilic monomer units that are illustrated in Figure 3.77, where H and L represent hydrophilic and lipophilic monomer units.

In the first case hydrophilic monomer units H are linked together to form a hydrophilic group, and lipophilic units L just do the same to form a lipophilic group. The result is a polymeric surfactant with well-defined and separated hydrophilic and lipophilic parts, which is just

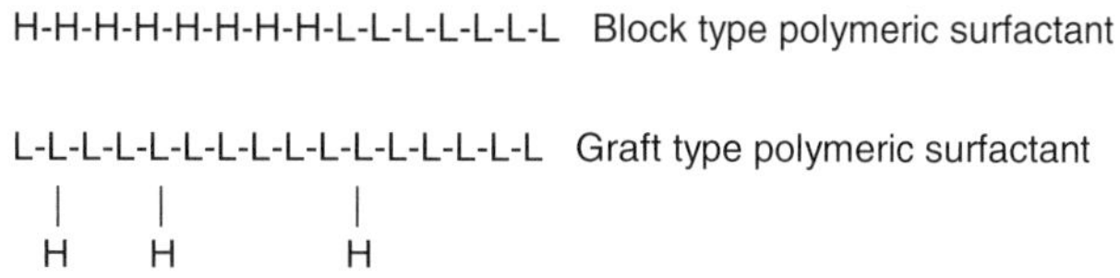

Figure 3.77 General structure of polymeric surfactants.

H-H-H-H-H-H-H-H-L-L-L-L-L-L-L Block type polymeric surfactant

L-L-L-L-L-L-L-L-L-L-L-L-L-L-L Graft type polymeric surfactant
   | | |
   H   H   H

much larger than a conventional surfactant molecule. The most used block polymer types are the copolymer of EO and PO, which have been extensively described in Section 2.1.2. Although the hydrophilic and lipophilic parts are quite separated, the polymer polarity segregation is not obvious since both groups are slightly polar, one (PEO) just slightly more polar than the other (PPO). These surfactants have many uses, such as colloid and nano-emulsion dispersants, wetting agents and detergents, particularly important to the oil and gas industry they are used to dehydrate crude oils [239].

Most polymeric surfactants, however, belong to the graft type of molecule, particularly synthetic products such as polyelectrolytes, which are not strictly surfactants or are not used for their surfactant properties. There exists a wide variety of graft-type polymeric surfactants, and the main way to prepare them is to produce a lipophilic polymer with functional group where a hydrophilic group can be attached later. A relevant example would be alkylphenol-formaldehyde resins, which can be turned hydrophilic by adding PEO or sulphate or ether sulphate groups. Again these products are widely in use as crude oil dehydration additives to break water-in-oil emulsions during production and processing. As was described in Chapter 2, many graft polymers exhibit non-surfactant effects, and in the oil and gas industry, many of these polymers are used for their dispersant and viscosity-enhancing properties, particularly the polyacrylic acid derivatives and sodium carboxymethyl cellulose (CMC) (see Figure 3.78).

Cellulose is insoluble in water; however sodium CMC is soluble in hot or cold water. This is a highly effective additive for improving product and processing characteristics in various fields, including food and pharmaceuticals. In the upstream oil and gas industry, CMC and its derivatives are widely used especially in drilling additives. These polymers and some of their functions have been previously described in Section 2.2,'Cellulosic Polymers'. Here the functionality of these materials and their surfactant effects are further detailed. It is worthy of note that most of these cellulosic materials have good environmental credentials, being non-toxic and highly biodegradable.

A major use of CMC is in drilling fluids formulations for rheological and fluid loss control, particularly in water-based muds [112, 296]. These functionalities and their applications have

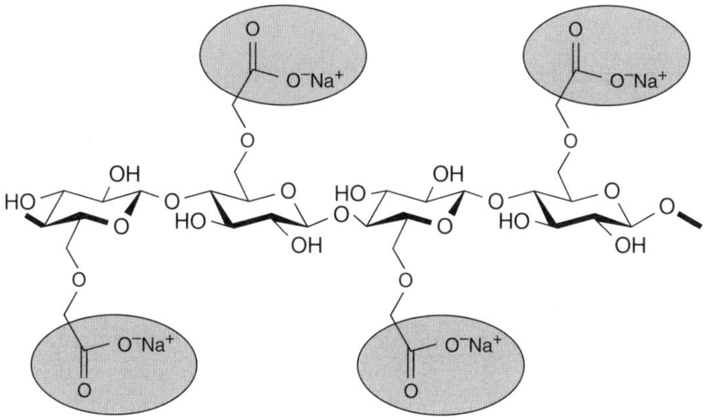

Figure 3.78 Sodium carboxymethyl cellulose.

been well studied over the last three decades [297]. Furthermore CMC has been used in a variety of other drilling directed applications such as clay stabilisation [298], as thickening and gelling agents in permeability modification [299] and in fracturing fluids for rheological control [300]. CMC and derivatives have also been investigated and claimed as potential 'green' scale inhibitors [301].

The related derivative surfactants, which are carboxymethylated and ethoxylated, have been utilised as suitable surfactants for EOR during water flooding operations [302].

Polyelectrolytes, in particular anionic polymers, are used in the upstream oil and gas industry as flocculants. These are most often copolymers of acrylic acid or partially hydrolysed PAM of high molecular weights. These anionic PAM can have good environmental properties and be useful in well clean-up operations [303]. Other anionic surfactant-like polymers would suggest themselves as useful environmentally acceptable flocculants, e.g. poly-γ-glutamic acid is a biodegradable anionic flocculant that does not appear to be used in the oil and gas industry [304].

Amphoteric polymers are also commercially available; however these do not appear to be much used in the oil and gas sector despite some potentially interesting characteristics and good environmental profiles. An amphoteric starch graft polymer has been reported and has exhibited good results in wastewater treatment [305].

In general the application of polymeric surfactants is an area worthy of further investigation by researchers and oilfield chemists and could yield a number of environmentally acceptable products.

3.6.2   Other Surfactants

There are a number of other surfactant types that do not fit the classifications so far used. These are now briefly described or referenced to other chapters and sections in this book.

3.6.3   Silicon Surfactants

Silicon surfactants are hydrophobic in character, particularly dimethylpolysiloxane. The introduction of an organosilicon group in a surfactant molecule tends to increase its hydrophobicity. Since silicon is a heavier atom than carbon, a similar hydrophobicity is obtained with less silicon atoms than carbon atoms. Essentially all surfactant types can be made with a silicon-based hydrophobic tail by replacing several carbon atoms by one silicon atom or one dimethyl-siloxane group. These materials will be explored further, and their use in the upstream oil and gas sector is detailed in Chapter 7.

3.6.4   Fluorinated Surfactants

Fluorinated surfactants are mainly produced by substitution of the hydrogen atoms of the surfactant hydrocarbon tail by halogens, particularly fluorine, to produce fluorinated hydrophobes, which exhibit properties similar to polymerised tetrafluoroethylene (PTFE), which will be known to the reader under the commercial brand name Teflon. This material exhibits high chemical inertia, mechanical and thermal resistance, low surface energy and therefore very high hydrophobicity. These products are primarily used in foam stabilisation and related uses in EOR applications [306]. They are much more expensive than other surfactants, and some fluorosurfactants have high persistence and toxicity, with a widespread occurrence in the blood of general populations [307]. As studies give evidence that longer fluorosurfactants are more bioaccumulative and toxic, there are also concerns about the potential impacts of fluorinated alternatives including short-chain poly- and perfluorinated alkyl substances on human health and the environment [308]. As a corollary to these properties, it has been discovered that the use of trifluoromethoxy groups enhances biodegradation when it appears in structures that are not normally biodegradable [98]. Interdisciplinary investigations on fluorinated surfactants are still very much needed and will need to continue over the coming years.

3.6.5 Acetylenic Surfactants

A class of acetylenic surfactants, which have the general structure as shown in Figure 3.79, has been developed to resolve or break water-in-oil emulsions [309]. The surfactants are of particular advantage in resolving crude oil emulsions of the type encountered in desalter and similar apparatus designed to extract brines from the crude as they partition to the aqueous phase. Also, the surfactants may be used to separate oil from oil sands and similar oil/solid matrices.

3.6.6 Gemini Surfactants

It is well known that most surfactants self-associate in solution to produce aggregates such as the micelles described in Section 3.1.2; however they can also form liquid crystals and microemulsions. Some surfactants, in particular double-tailed species, tend to associate as a bilayer (see Figure 3.80), which illustrates a bilayer (a) and micelle (b) formed in water (P) and oil (U).

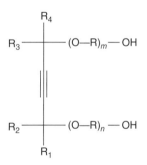

Figure 3.79 Acetylenic surfactants.

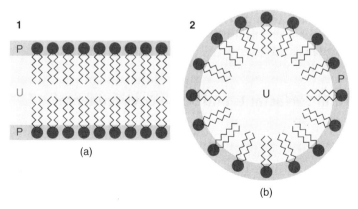

Figure 3.80 Bilayer and micelle structures.

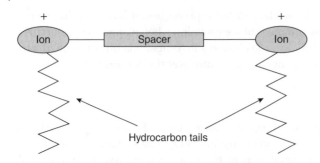

Figure 3.81 Schematic representation of a Gemini surfactant.

It is worth noting that these bilayers are the structural skeleton of many biological membranes in plants and animals, such as those produced by phospholipid association. These natural surfactants are further discussed in Section 3.7.

Gemini surfactants are a family of surfactant molecules possessing more than one hydrophobic tail and hydrophilic head group. These surfactants usually have better surface-active properties than corresponding conventional surfactants of equal chain length [153, 310].

Gemini surfactants are used in a variety of applications including industrial detergency, skin care and construction of high porosity materials. They have also been shown to exhibit antibacterial properties [311].

Where a conventional surfactant has a single hydrophobic tail connected to an ionic or polar head group, a Gemini has in sequence a long hydrocarbon chain, an ionic group, a spacer, a second ionic group and another hydrocarbon tail (see schematic representation in Figure 3.81). Related surfactants with more than two tails are also known. Gemini surfactants are considerably more surface active than conventional surfactants. The name Gemini was originally ascribed to bis-surfactants with rigid spacer (i.e. benzene, stilbene, etc.). The name was then extended to other bis- or double-tailed surfactants, irrespective of the nature of spacers.

To date the application of Gemini surfactants in the upstream oil and gas sector has been limited despite the technology being over two decades old. Presumably pricing has been a significant factor in the uptake of this technology. Some Gemini quaternary surfactants appear to show good corrosion inhibition properties particularly in acidising applications [312] and in EOR operations [313]. The unique properties of Gemini surfactants such as low CMC, good water solubility, unusual micelle structures and aggregation behaviour, high efficiency in reducing oil–water IFT and interesting rheological properties should be attractive for oilfield applications. Importantly rheological characterisation and determination of the IFT are two of the most important screening techniques for the evaluation and selection of chemicals for EOR.

Recently a new Gemini imidazoline has been developed through the reaction of oleic acid and triethylenetetramine and is claimed to have a greater performance and better environmental profile than conventional imidazolines [314].

3.7 Biosurfactants, Natural Surfactants and Some Environmental Considerations

Biosurfactants are diverse groups of surface-active molecules and chemical compounds, which are produced by microorganisms [315]. They are produced extracellularly or as part of the cell membrane by bacteria, yeasts and fungi. These are amphipathic molecules having both hydrophilic and hydrophobic domains that confer the ability to accumulate between fluid phases, thus reducing surface and IFTs at the surface and interface respectively [316]. Therefore, unlike

Type of organic compound	Biosurfactant type	Micro-organism source
Lipopeptides	Surfactin A and B	*Bacillus subtilis*
	Inturin A	*Bacillus subtilis*
	Pumilacidin	*Bacillus pumilus*
Glycolipids	Rhamnolipids	*Pseudomonas sp.*
	Sophorolipids	*Sacchromyces sp.*
	Trehalolipids	*Mycobacterium sp.*
Polymeric surfactant	Bioemulsan	*Acinobacet sp.*
Secondary Metabolite	Vicosinamide	*Pseudomonas fluorescens*

Figure 3.82 Surfactants sourced from microbes.

the biopolymers that are referenced as being classed as biodegradable, the biosurfactants are defined as being derived from a natural source.

Most biosurfactants are either anionic or neutral, and the hydrophilic moiety can be a carbohydrate, an amino acid, a phosphate group or some other compound. The hydrophobic moiety is mostly a long carbon chain fatty acid. These molecules reduce surface and IFTs in both aqueous solutions and hydrocarbon mixtures.

Examples of biosurfactants include the rhamnolipids produced from *Pseudomonas aeruginosa*, and *Bacillus subtilis* produces a lipopeptide called surfactin and the sophorolipids from *Candida* (formerly *Torulopsis*) *bombicola*, one of the few yeasts to produce biosurfactants [317]. The environmental applications of these biosurfactants for soil and water treatment have been investigated due to their biodegradability, low toxicity and effectiveness in enhancing biodegradation and solubilisation of low solubility compounds. The role of biosurfactants in natural attenuation processes has not been determined.

An anionic glycolipid biosurfactant has also been produced in a low cost process from the yeast *Candida lipolytica* using animal fat and corn steep liquor, which is hoped to have particular application in EOR applications in the oil and gas sector [318].

The table (Figure 3.82) outlines the major biosurfactants and their microbial source.

Glycolipids are the most investigated biosurfactants. They are conjugates of carbohydrates and fatty acids. The linkage is by means of either ether or an ester group. Among the glycolipids, the best known are rhamnolipids, trehalolipids and sophorolipids with most research, until now being performed with rhamnolipids.

Rhamnolipids have one or two molecules of rhamnose linked to one or two molecules of β-hydroxydecanoic acid. While the —OH group of one of the acids is involved in glycosidic linkage with the reducing end of the rhamnose disaccharide, the —OH group of the second acid is involved in ester formation (see Figure 3.83 for an example).

These and similar molecules have been investigated in oilfield applications for agents in EOR [319] and potential treatment for oil-based mud drill cuttings [320]. The other biosurfactants need to be investigated as they may have more promising properties, and when compared with synthetic surfactants, biosurfactants have several advantages including high biodegradability and low toxicity [321]. Other advantages of biosurfactants over synthetic ones include selectivity and specific activity at extreme temperatures, pH and salinity, all of which could be useful in oilfield applications.

Biosurfactants are biodegradable in nature, and as has been described previously, biodegradability is a critical issue concerning the potential environmental impact of chemical compounds and their possible pollution. Being able to be broken down by natural processes, by bacteria,

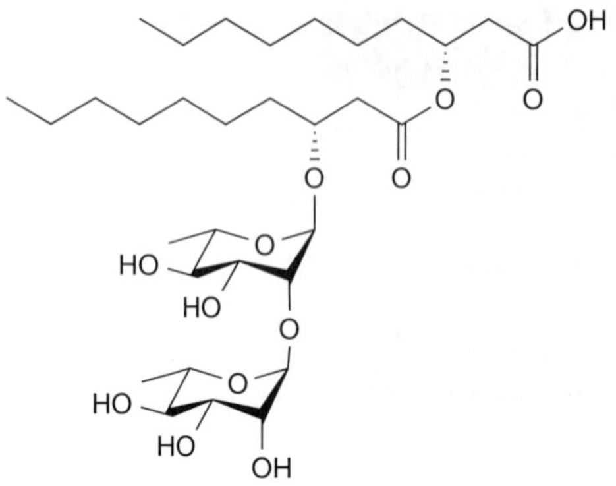

Figure 3.83 A rhamnolipid.

fungi or other simple organisms into more basic components, they have a much reduced impact on the environment compared with synthetic surfactants and are particularly suited for environmental applications such as bioremediation [322] and dispersion of oil spills [323]. The use of surfactants in oil spill treatment and dispersion is considered further in Section 3.8.

Very little data is available in the literature regarding the toxicity of biosurfactants. The available evidence suggests that they can be considered as low or non-toxic products and therefore appropriate for pharmaceutical, cosmetic and food uses. Biosurfactants do not appear to cause serious damage or harm to the biotic ecosystem since their toxicity level is low already being part of the normal ecology. Many synthetic chemical surfactants are toxic to various biota including marine aquatic species, which potentially makes their use more problematic particularly in regulated marine areas. When comparing the toxicity of six biosurfactants, four synthetic surfactants and two commercial dispersants, it was found that most biosurfactants degraded faster, except for a synthetic sucrose stearate that showed structure homology to glycolipids and was degraded more rapidly than the biogenic glycolipids. It was also reported that these biosurfactants showed higher EC50 (effective concentration to decrease 50% of test population) values than synthetic dispersants [324]. In another example a biosurfactant from *P. aeruginosa* was compared with a widely used commercial, synthetic surfactant in terms of toxicity and mutagenic properties. Both assays indicated a higher toxicity and mutagenic effect from the chemical-derived surfactant, whereas the biosurfactant was considered to be non-toxic and non-mutagenic [325].

A significant advantage in using biosurfactants is that they are biocompatible in nature, which means they are well tolerated by living organisms. These materials when interacting with living organisms do not change bioactivity of the organisms. This property allows their safe application in cosmetics and pharmaceuticals and as functional food additives [326].

The area of biosurfactant research and application is in its relative infancy particularly with respect to the purified and isolated molecules; however, the use of these molecules and their related properties, in application of the whole microorganism, particularly in EOR is well established [327]. Microbial EOR and in general a detailed examination of the use of surfactants in EOR applications would require a separate publication and are outside the scope of the current work.

A related biosurfactant class is the glycoproteins, some of whose properties have been described in Section 2.2.5. The glycoproteins are exploited in many biological systems for their surfactant behaviour such as the mucins, which are secreted in the mucus of the respiratory and digestive tracts of many animals, including humans. The application of these molecules outside of pharmaceutical and medicals areas is so far limited, and some oil applications have been discussed in Chapter 2. There has been some exploration of their potential application in EOR in the complex and unrefined form of gum arabic [328, 329] as well as the previously described soya bean slurry as an anti-oil congealer in oil hydrate formation [330].

Crude oil contains a number of organic molecules displaying surfactant behaviour including asphaltenes, resins and naphthenic acids, and because of the high shear characteristics at the wellhead and the mixing during transportation and processing, often some or all of the produced water will be emulsified with the oil phase, and as has been described, a number of polymers (Chapter 2) and surfactants are utilised to resolve this. These natural surfactants, however, can be utilised in hydrate prevention and management [331, 332].

A number of other natural surfactants are available but have been seldom used or modified for industrial use with the exception of the pharmaceutical and cosmetic industries where the higher costs can be adsorbed. This is, however, likely to change with the necessity to utilise and modify a wider range of green chemistries. For example, peptides and lipopeptides could offer a new range of green surfactant chemistry, but they are difficult to bio-produce at low cost and are not easily genetically engineered [333]. Designed peptide surfactants have also been examined, and these can self-assemble to give interesting structures with useful surfactant properties [334].

For the environmental chemist and oilfield applications specialist, the area of biosurfactants and related natural surfactants hold much promise in delivering biocompatible, low toxicity and environmentally acceptable products. This has become even more relevant in commercial terms with the announcement of the 'first' industrial-scale production of biosurfactants [335].

3.8 Oilfield Dispersants

Throughout this chapter, relevant surfactant applications, particularly those aligned to the upstream oil and gas industry, have been detailed and discussed. It should be noted that alongside the major applications of corrosion control, oil and water separation and EOR, oilfield dispersants require some further explanation. Many of these applications have been extensively reviewed in the literature [4, 112, 140], and the use of surfactants in EOR in particular is a continuously changing and developing area [336, 337].

Surfactants and surfactant mixtures have formed part of the clean-up and containment methods employed in managing oil spills since the 1980s. Oil spills and their resulting environmental pollution can cause major ecological damage, and the cost of clean-up can be monumental. Exxon states it spent over $2.1 billion on the clean-up of the Exxon Valdiz [338].

As has been described earlier in this chapter, surfactants have varying solubilities in water and various actions with respect to oil and water, and the HLB value (see Section 3.1.8) characterises the surfactants' oil and water solubility. In general a surfactant with an HLB value of 1–8 promotes the formation of water-in-oil emulsions, and one with a value of 12–20 promotes oil-in-water emulsions. A surfactant with an HLB value between 8 and 12 may promote either type of emulsion but more often promotes oil–water emulsions; dispersants are found to have HLB values in this range.

In treating oil spills, surfactants are used in three or four different ways. Dispersants are usually mixtures of surfactants and formulated to disperse oil slicks into the sea or other body of water. So-called beach cleaners are formulations designed to remove oil from surfaces such as sand and rocks. Emulsion breakers (demulsifiers) and emulsion inhibitors are designed to break water-in-oil emulsions or prevent their formation. Over 100 dispersants have been evaluated over the decades; however only a handful are approved for use and commercially available [339].

Developing effective oil spill dispersants is a major challenge to the chemical formulator as crude oil and refined oils are not consistent in their composition and have varying types and amounts of molecules with a variety of molecular sizes. What is effective for a polar alkane may not be effective for an aromatic component. The environmental conditions such as the sea state and temperature can also affect the performance of the dispersant.

In achieving an approval for using a spill dispersant or other treating agent, both effectiveness and toxicity are tested although the level of toxicity in the past has been wholly inadequate [340]. Much controversy has been generated over the use of oil dispersants, and this has been refuelled by their use during the Deepwater Horizon oil spill. Here over 7 million litres of dispersant was applied to very little overall effect. The major impact was that the dispersant used reduced the amount of floating oil, which reduced the risk for some organisms and environments, but increased the risk for others as the oil was dispersed into the water column. Also in hindsight, it was recognised that marine species are even more sensitive to oil than was previously thought, especially for some developmental stages of offshore fish such as tuna, which in taking the oil into the subsurface environment damaged the junvenile fish stock.

The comparative effectiveness of dispersants is relatively easy and straightforward to measure [342]; however the testing of effectiveness on real conditions is somewhat problematic since it is impossible to represent all the environmental conditions and oil compositional factors. Over 50 different tests have been developed to attempt to assess more realistically the dispersant performance [341, 342].

Toxicity testing has become more sophisticated, and now the dispersants in use are less toxic than the oils being dispersed. Toxicity of diesel and light crude oils is between 20 and 50 mg l^{-1} (LC50 on rainbow trout over 96 h), and all approved dispersants have an LC 50 of less than 20 mg l^{-1}. However the dispersion of the oil and chemical mixture can have serious toxic effects if applied in shallow waters on the marine biota, particular the fauna on the seabed.

Toxicity testing for the use of oil spill dispersants in regulated sea areas has become more defined [343], and these have become standards for the international use of oilfield dispersants.

A typical oilfield dispersant formulation consists of a pair of non-ionic surfactants in proportions to yield an HLB value of around 10 and some additional anionic surfactants. Studies have been conducted on this type of mixture to optimise the proportions and determine the best three ingredients [344]. In terms of chemistry the dispersants seem to be determined around the fatty acids particularly as the monooleate salt of certain molecules such as sorbitan and the sulphosuccinates. All of these materials are readily biodegradable.

To be effective the chemical dispersant must be applied as soon as possible after the oil spill has occurred. This is because if the oil ages, is weathered and/or thins out, then they will disperse poorly as the chemical dispersant is applied. There are a large number of other physical and environmental factors that affect the dispersant performance as well as how the dispersant is physically applied, including droplet size. Discussion of these is outside the scope of this book.

In the last decade much attention has been given to the possibility of oil spills under the Arctic Sea ice as the industry now explores and develops oil finds in the Arctic region. The consequential environmental damage from an oil spill could be ecologically catastrophic with perhaps chemical treatment being the only way to contain and remediate such a disaster.

Fundamentally no one wants to have any oil spill; however, unless the human race stops extracting, using, storing and transporting oil, the risk of a large oil spill is always present. The decision to use dispersants or not use dispersants will always be problematic and judgemental. Hopefully, as we progress the understanding of the physical process and progress green chemistry with similar advances in dispersion effectiveness, the use of dispersants will be recognised as a useful and environmentally responsible choice.

3.9 Degradation, Biodegradation and Environmental Fate of Surfactants

Many of the mechanistic aspects of degradation and biodegradation, particularly hydrolysis, have been described previously in Chapter 2 and will not be repeated in detail here. However, unlike other chemical additives, surfactants and amphiphiles have significant differences in response to physical, chemical and biological degradation pathways. Although chemical hydrolysis and biologically enzymatic hydrolysis are still the most important degradation pathways, it is often the case that surfactant molecules are not water soluble and therefore less susceptible to these mechanisms. Furthermore surfactants are a chemical group for which it is difficult to obtain reliable data pertaining to their bioavailability in particular data relating to partitioning (log P_{ow}) or bioconcentration factor (BCF). Such data is necessary for performing environmental risk assessments using the currently available assessment models.

The challenges in assessing the environmental fate of surfactants centre around the intrinsic property of surface-active substances to adsorb to surfaces and to accumulate at phase interfaces. Techniques for estimating bioaccumulation potential (e.g. OECD 107 Shake Flask and OECD 117 HPLC) are therefore unsuitable for determining a log P_{ow} for a surfactant [59].

In the OECD 107 shake-flask test, it is likely that the bulk concentrations of a surfactant would not be in equilibrium between the water and octanol phases, but in equilibrium with the octanol–water interface concentration. The experimental procedures used in this test will generally produce fine emulsions with large total surface area. In addition, partitioning of complex mixtures in this test requires substance-specific analysis of all the components in the mixture to obtain realistic results.

The OECD 117 (HPLC) test is only applicable for non-ionic, non-surface-active chemicals. Normally, retention times in the HPLC column are determined by a chemical's relative affinity for the mobile (usually methanol/water) phase and the immobile (lipophilic) phase. The affinity of surfactants for surfaces (mobile/immobile phase interfaces and substrate/immobile phase) will invalidate this method for this type of chemical.

The chronic and sublethal toxicities of surfactants are also differentiating particularly with regard to aquatic species as mainly the data is for freshwater toxicity and species. Toxicity data for marine species needs to be validated with both laboratory and field-derived chronic toxicity assessments [275].

These factors make the selection of surfactants to reduce environmental impact problematic and challenging.

3.9.1 Chemical and Physical Degradation

The main degradation of surfactants, as with other chemicals discharged into an aquatic environment, is through chemical hydrolysis, and the enzymatic hydrolysis of surfactants resulting biodegradation is discussed in Section 3.9.2. As with other entities the chemical hydrolysis of surfactant molecules has the rate of hydrolysis governed by temperature and pH [345], for example, in fatty amide corrosion inhibitors, hydrolysis of the amide and amidine groups is relatively easy under acidic conditions, which is probably the predominant chemical degradation mechanism of the conventional corrosion inhibitors under field use conditions [346]. It is also of note that the structure and functional groups available determine the ease of hydrolysis both chemically and biologically, although in the latter steric factors often play a major role in the kinetics [347].

Unlike polymers mechanical degradation plays only a minor role in overall surfactant degradation; however in certain applications such as corrosion inhibitors and DRAs, stability to shear forces is an important aspect of design. Shear stability can also explain, at least impart, foam and emulsion stability [348]. Thermal and photolytic degradation is more important in the overall stability of surfactants.

At extended exposure to temperatures above 250 °C, in the presence of oxygen, surfactants are likely to undergo decomposition, resulting in chain shortening, with a number of volatile products of decomposition arising such as acetaldehyde, formaldehyde, miscellaneous glycol ethers, etc., and the residual starting alcohol or other base molecule of the surfactant. For example, a study on the thermal degradation of alcohol ethoxylates, covering topics such as weight loss with time, calorimetry, the formation of aldehydes (using IR spectroscopy as an analysis technique) and also the degree of EO chain cleavage, found that surfactants are stable up to 300 °C in an inert nitrogen atmosphere and that oxidative degradation, concurrent with the formation of gases and chain cleavage, occurred in the presence of air [349]. Another study has examined the evaporation and decomposition of TRITON™ X-100, a commercially available alcohol ethoxylate surfactant under short-term (20 min) exposure at high temperatures (350 °C, max). It was found that in the absence of oxygen, TRITON X-100 surfactant evaporated with no associated production of thermal degradation products. In the presence of oxygen, unidentified carbonyl compounds were formed, corresponding to some oxidative degradation. However, most of the evaporated material was TRITON X-100 surfactant [350].

Mobility control is one of the main problems in steam flooding for EOR. Channelling through highly permeable zones and gravity override of steam leads to early steam breakthrough at the production wells. Accordingly, volumetric sweep efficiency and oil recovery are reduced. Surfactant-stabilised foam has been used as a permeability blocking agent in steam flooding to improve oil recovery. Surfactants have been investigated for their longevity under conditions typical of those found for steam injection oil recovery, namely, at 400 °F (205 °C) and 300–500 psi (20–34 bars), to test for thermal degradation, adsorption and phase partitioning. Surface tension, surfactant concentration, pH and electrical conductivity were measured with time. Some surfactants exhibited a rapid decrease in concentration and pH with heating time, while other surfactants showed better high temperature stability, with half-lives as long as several months at 400 °F [351].

It can be concluded that many surfactants are thermally stable; however this usually comes at a price in that these products are usually not as prone to hydrolysis and their rates of biodegradation are low. However this is not always the case, and many alkyl aryl sulphonates have good rates of biodegradation. They can also be thermally stable, particularly in alkaline media such as for alkali surfactant EOR. The rates of thermal degradation have been estimated at half-lives of hundreds of years if not longer, with the alkyl chain degrading by scission to a lower alkylbenzene sulphonate or sulphonic acid and desulphonation also occurring [352].

Photolysis is an important mechanism of degradation with surfactants; indeed surfactants are often used to enhance the photolytic reaction of other molecules, a particularly important effect in pesticide application [353]. This effect of enhancement of other properties and facilitation of specific reactions has been used synergistically as described earlier throughout the chapter; however it can also have an important positive effect on improving rates of biodegradation as will be described in the next section.

Photolytic pretreatment, with UV irradiation, of surfactants in anaerobic digestion have shown improvement in the biodegradation and detoxification of such molecules with respect to untreated materials where the surfactants were poorly biodegradable and toxic to methanogenic bacteria. This was the case for all main classes of surfactant anionic, non-ionic and cationic. This means that a combined system of photolytic and biological processes would be applicable to the treatment of wastewater containing toxic organics such as surfactants. It has also been indicated that the photolysis of aromatic surfactants is primarily a cleavage of the benzene ring [354].

3.9.2 Biodegradation

In Western Europe all surfactant components of domestic detergents must be biodegradable. This requirement has been from the fact that the original alkylbenzene sulphonate anionics were based on branched alkenes, and these proved resistant to degradation by bacteria at sewage treatment works, causing many rivers to suffer from foam. There was also a fear that surfactants could be 'recycled' into drinking water. Similar concerns were expressed about nonylphenol ethoxylates, which in addition could have toxic properties around endocrine disruption, and so in the last few decades, industry has moved to other more environmentally acceptable surfactants such as LAS and alcohol ethoxylates as the major ingredients of their detergent formulations.

Effective sewage treatment has ensured that detergent components, which are part of household effluent water, are not discharged untreated into rivers and watercourses. The oil and gas industry has followed this trend and has demanded that where possible surfactants as with other chemical additives are biodegradable. There is again a dilemma in evaluating the rate of biodegradation of surfactants in that the usual test OECD 306 can often be prone to substantial errors not taking into account the surface-active nature of the chemical. It is therefore usually recommended that a variant of this test protocol for surfactant biodegradation rate evaluation is adopted, namely, marine BODIS ISO TC 147/SC5/WG4 N141, which has been developed for water-insoluble materials.

As with polymers and other chemical additives, surfactant biodegradation is driven by oxidative aerobic pathways and enzymatic hydrolysis. The required amphiphilic properties of surfactants can be achieved by a seemingly endless variety of chemical structures. However, most of this diversity lies with the hydrophilic moieties and their means of linkage to the hydrophobe on which basis surfactants are broadly classified, as has been described throughout this chapter, according to the chemical nature of the polar group as being anionic, nonionic, cationic or zwitterionic; within these large groups there is further subdivision on the basis of the chemistry that has been detailed and illustrated. However, in practice a relatively small number of surfactants account for the majority of usage worldwide [355].

It is the general view that biodegradation of synthetic surfactants (and other chemical products) is a beneficial activity because it prevents undesirable pollution of the natural environment while still allowing the compounds to be used. From the microorganisms' viewpoint, surfactants represent a potential source of energy and reduced carbon for growth. Bacteria utilise essentially two strategies to access the carbon in a surfactant, the bulk of which (at least in ionic surfactants) is generally present in the hydrophobic moiety. The first strategy involves initial separation of the hydrophile from the hydrophobe, which is then oxidatively attacked. In the second mechanism, the alkyl chain is oxidised directly while it is still attached to the hydrophile. Both strategies lead to an immediate loss of surfactant activity in the molecule. For some surfactants, only one or the other route has so far been observed, whereas for others both of these general mechanisms have been found to operate.

Biodegradation of anionic surfactants and related molecules makes them one of the most common components in domestic and industrial waste streams. In man-made disposal plants, e.g. activated sludge and filter-bed systems, and natural water systems, e.g. rivers and lakes, bacteria are considered to be largely responsible for the biodegradation of surfactants although other microorganisms may also contribute [355].

Similar mechanisms exist for cationic and amphoteric surfactants, and it has been shown that a number of cationic and amphoteric surfactants can be readily biodegraded, with their degradation exceeding 94%. The relationship between structure and biodegradability is again important, and the nature and structure of hydrophobic group has a strong effect on the bio-degradability of these surfactants. Biodegradability has been observed to decrease with increasing chain length. Also the presence of hydrophilic groups affects the degradation rate of these surfactants, but not their ultimate biodegradability. Biodegradability is deterred, and degradation is slowed as steric hindrance increases and degradation rates increase markedly when hydrophilic groups containing an amide bond are present [356].

Biodegradation of cationic surfactants may be inhibited where they display biocidal activity as in the quaternary ammonium salts although the corollary is that the initial high concentration in the evaluation tests at the initial phase may contribute significantly to the biodegradation result [357]. It is generally accepted that the toxicity of quaternary ammonium salts increases with increasing alkyl and related chain lengths [358].

Non-ionic surfactants also have similar biodegradation mechanisms; however as has been described in Section 2.6, oxidative degradation of any polymer backbones is an important mechanism in overall biodegradation [359].

The ecotoxicology of a number of types of non-ionic surfactants such as fatty alcohol ethoxylates, nonylphenol polyethoxylate and APGs has been investigated. Also, the relation between metabolites and ecotoxicity during the biodegradation process has been evaluated. In these studies a solution of the surfactant, representing the sole carbon source for the microorganisms, was tested in a mineral medium, inoculated and incubated under aerobic conditions in the dark. The toxicity of the surfactants as expected was related to their molecular structure. For APGs, toxicity expressed as EC(50) was found to be related with the CMC, the HLB of the surfactant and the hydrophobic alkyl chain. The results indicated that toxicity increased as the CMC decreased and as the hydrophobicity increased and alkyl chain length increased. For fatty alcohol ethoxylates, the characteristic parameters studied were the HLB value, number of units of EO and the alkyl chain length. The conclusions found are in agreement with the fact that increasing the alkyl chain length leads to a lower EC(50), whereas increasing ethoxylation leads to a lower toxicity. An analysis of the behaviour of the toxicity and HLB again indicates that the toxicity was greater for surfactants with a smaller HLB. For all the non-ionic surfactants assayed, except for a nonylphenol polyethoxylate, a major decline was found in toxicity during the first days of the biodegradation assay and at all the concentrations tested [360].

The majority of work on contamination of the environment by surfactants has been conducted on detergent formulations, in particular the two of the major surfactant types, namely, the LAS and the APE. These pass into the sewage treatment plants where they are partially aerobically degraded and partially adsorbed to sewage sludge that is applied to land. Although the application of sewage sludge to soil can result in surfactant levels generally in a range of $0–3\,mg\,kg^{-1}$, in the aerobic soil environment, a surfactant can undergo further degradation so that the risk to the biota in soil is very small, with margins of safety that are often at least 100. In the case of APE, while the surfactants themselves show little toxicity, their breakdown products, principally nonyl- and octylphenols, adsorb readily to suspended solids and are known to exhibit oestrogen-like properties [183]. The conclusion of this work is that while there is little serious risk to the environment from commonly used anionic surfactants, cationic surfactants are known to be much more toxic, however overall the amount and quality of data on the degradation of cationics and other classes of surfactant and their fate in the environment is not as well established [361].

As has been stated previously, there is much controversy in the test methodologies used to evaluate the ecotoxicological profiles of surfactants. It is suggested that using 'read-across' data may help in obtaining data for risk and environmental impact evaluation particularly with regard to biodegradation. It has been recently shown that there is good correlation in the aerobic biodegradation pathway for primary and secondary fatty acid amides that are hydrolysed by the microorganisms of *P. aeruginosa* and *Aeromonas hydrophila* [362]. Read-across of previous reported ready biodegradability results of primary and secondary fatty acid amides is therefore justified based on the broad substrate specificity and the initial hydrolytic attack.

3.9.3 Developing 'Green' Surfactants

The development and design of 'green' surfactants is primarily focused on developing biodegradable surfactants.

Surfactants are relatively stable molecules from a physical viewpoint and, usefully from the environmental view, are prone to biological degradation. Therefore, many surfactants are already able to be classified as green from the point of view that they are readily biodegradable. In order to enhance biodegradation, the following strategies are considered:

1) Ensuring that the surfactant has the maximum water solubility, this may compromise its performance and have a detrimental effect on its bioavailability.
2) Changing the structure of the surfactant to make it more biodegradable such as in the development of hydrolysable surfactants by derivatising any functionality to increase the susceptibility to hydrolysis, e.g. by alkoxylating or esterifying any alcohol functional.
3) Replacing the hydrophobic grouping (tail) with one from a natural product source, which potentially is more biodegradable.
4) The use of biosurfactants (see Section 3.7).

However the concerted use of the aforementioned in designing and developing greener surfactants for application in the upstream oil and gas industry is at best in its infancy, with applications of microbial biosurfactants perhaps being the most developed [319].

In examining enhancement or targeting of water solubility, those with a lower alkyl chain length, i.e. shorter hydrophobe, are more water soluble. This however also affects the CMC and therefore the surfactant characteristics [363]. However examining water-soluble polymers with ampholytic properties is a possible route to ensure good surface-active performance, for example, the water-soluble ampholytic polysaccharides with carboxybetaine groups, which were prepared by grafting the zwitterionic monomer 2-(2-methacryloethyldimethylammonio) ethanoate onto hydroxyethyl cellulose (HEC), have been tested for their properties as a multifunctional drilling mud additive regarding clay hydration inhibition and mud rheological control [364].

It is well known that the branching of alkyl chains retards rates of biodegradation through steric hindrance [172, 365, 366], and therefore the use of linear alkyl derivatives has been widely promoted throughout the application of surfactants particularly with regard to detergents where linear alkyl derivatives, such as the linear alcohol ethoxylates, have largely replaced branched derivatives.

The use of hydrolysable bonds within the structure of the surfactant is in growing demand, i.e. surfactants that break down in a controlled fashion by either inducing a change in pH or a change in pH due to environmental pressures. Indeed environmental regulation and concern is the main driving force behind the interest in these surfactants; however they are also of interest in applications where surfactants are needed in one stage but later undesirable at another stage of a process. Surfactants that break down either on the acid or on the alkaline side have been developed, and it has been shown that the susceptibility to hydrolysis for many surfactants depends on whether or not the surfactant is in the form of micelles or as free molecules (or unimers) in solution. It has been observed that whereas non-ionic ester surfactants are more stable above the CMC, cationic ester surfactants break down more readily when aggregated than when present as free molecules or unimers [367].

A number of surfactant derivatives containing hydrolysable units are now available and are broadly classified as follows:

- Surfactants containing normal ester bonds [367]
- Ester quats (see Section 3.4.2)
- Betaine ester surfactants [368] (see Section 3.5.1)
- Surfactants containing carbonate bonds [367, 369]
- Surfactants containing amide bonds [367, 369]
- Surfactants containing cyclic acetals and ketals [367, 370]
- Ortho ester surfactants [371]

Applications in the oil and gas industry have been few despite these product types being developed over a decade ago. There is undoubtedly more scope for use of these products in the formulation of a number of oilfield additives.

As has been described in Section 3.2.8, phosphate ester surfactants have a number of uses in applications for the oilfield sector, in particular in corrosion inhibitor formulations. These products are naturally present in living organisms e.g. as phospholipids and nucleic acids and the phosphate ester bond is prone to enzymatic hydrolysis, which is catalysed by a variety of phosphatase enzymes, and these are also able to attack the ester linkages in these surfactants [139, 372].

Whereas most of the developments concerning the use of hydrolysable modifications have centred on the hydrophobic 'tail' of the surfactant, some recent work has interestingly examined possible hydrolysable counter-ions constituting in effect the hydrophilic 'head' group. This work has focused on a cationic 'Gemini' surfactant based on the didodecyldimethyammonium cation (see Figure 3.84) with a variety of hydrolysable cations such as phosphate oxalates and carbonates replacing the bromide ion illustrated [373].

In the upstream oil and gas industry, certain modified imidazolines are being used in corrosion inhibitor formulations. As has been previously described in Section 3.5.3, these imidazolines are usually derived by amine condensation reactions with tall oil, which imparts a C_{17} and/or C_{18} alkyl chain. The product is then subjected to selective ethoxylation, which enhances its water solubility and biodegradation. These modified products also meet required environmental regulatory criteria in appropriate areas.

Polysaccharides can be modified to be ampholytic; HEC has been grafted with 2-(2-methacryloethyldimethylammonio)ethanoate [374]. This derivative has been utilised as a highly biodegradable mud additive with shale inhibition and rheological control properties.

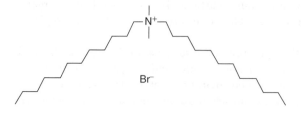

Figure 3.84 Didodecyldimethyl-ammonium bromide.

Other polysaccharides such as the alkyl polyglycosides have also been considered for oilfield surfactant applications. In particular the oxyalkylated polyglycosides have been used as low toxicity, biodegradable demulsifiers [375].

The use of oleochemicals, particularly the fatty alcohols and their derivatives, in the manufacture of surfactants is well documented [376]. Their use and application in providing hydrophobic moieties for surfactant molecules across the range of types of surfactant has been described throughout this chapter. These molecules are natural or naturally derived materials with good ecotoxicological profiles [377]. These have provided the bulk of the green surfactants to date.

The development of green surfactants is clearly a challenging and developing area; however the uptake in many industries including the oil and gas sector has been to date disappointing, primarily driven by the lack of clarity regarding whether their application can save costs across the supply chain. There is a clear need for affordable and viable sustainable surfactant technologies. At present 'green' surfactants are not expected to make a major change in the overall pattern of consumption, as products emphasising consumer value seem to be more appealing than the benefits of product sustainability. The question remains, however, as to how extensively green product features will affect customer behaviour. In some segments of the oilfield market, the 'green' label is promoting sales primarily driven by regulatory requirements.

More investments into oleochemicals and an increase in their supply are expected, especially in Asia. Most of the new investments in surfactants are taking place in the ASEAN region and in China and South America, and Asia is seen as the strategic growth area for most global surfactant manufacturers.

The use of biosurfactants and in particular those derived from marine microorganisms could provide interesting and useful molecules for further investigation given their good ecotoxicological profiles and their inherent sustainability [364].

3.9.4 Environmental Fate

Most of the discussion throughout this chapter regarding environmental impact of surfactants has been centred on their biodegradation and some of their toxic effects. It could be argued, however, that chronic and sublethal toxicity and their bioavailability are far more important to their overall environmental impact. In this closing section to the chapter, some of these aspects will now be considered alongside some strategies to alleviate these concerns as well as discussing some of the positive benefits the use of surfactants can and could bring to environmental damage and remediation of the effects of other chemical additives.

As was described earlier there are significant challenges in assessing the environmental fate of surfactants, which are centred around the intrinsic properties of surfactants and their desired property to adsorb to surfaces and to aggregate and self-assemble at phase interfaces. The most important physical and chemical properties for an environmental risk assessment are aqueous solubility, vapour pressure and the octanol–water partition coefficient or other partition coefficients to evaluate the bioaccumulation potential, for example, those between water and environmental matrices such as soil, sediment or sewage sludge. The readily available techniques for estimating bioaccumulation potential (e.g. OECD 107 Shake Flask and OECD 117 HPLC) are deemed by most regulatory authorities and the general scientific community to be unsuitable for determining the bioavailability of a surfactant [59]. This is not to say there are not similar techniques, which can give valuable information such as the 'slow-stirring' method [378]. This method is particularly appropriate for hydrophobic molecules and has good reproducibility and correlations with accepted data at log values of 4.5 or less; however this indicates materials that are not highly bioaccumulating, and therefore those potentially damaging materials at log values above 4.5 can give inconsistent and misleading results.

Many risk assessment models for environmental impact of chemicals require an input for bioaccumulation potential, usually based on the log of the partitioning ratio between octanol and water (log K_{ow}), and this is true for marine environmental risk assessment models also, where a predicted effect concentration (PEC) can be calculated and rationalised against a predicted no effect concentration (PNEC).

A more robust, but expensive and time consuming, methodology in obtaining meaningful data for surfactants bioaccumulation potential is to determine their bioconcentration factor (BCF). Bioconcentration is the result of uptake of a compound from the surrounding medium and elimination to the environment. Assuming that chemical uptake and elimination kinetics are first-order processes, the time course of the concentration of a chemical in an organism can be expressed mathematically and a quantitative measure of bioconcentration determines the BCF [379]. Two possible experimental approaches to determining the BCF are possible. In the first, the steady-state concentrations of a compound in the organism and water are measured and the BCF calculated as their ratio. The second approach is to measure the time course of a compound's concentration in water and/or organism during an uptake and elimination experiment [380].

Often to get around the lack of meaningful log K_{ow} data, the comparative assessment can be derived across a group of similar assessments where other robust data is known. For example, for alkyl ethoxy sulphates, all groups of homologues have sufficiently low volatility that the sensitivity of the risk assessment to the values of this parameter will be negligible. Therefore it is possible that a meaningful log K_{ow} value can be modelled from molecular structure. However, it is advised that all assessments based on partitioning coefficients that are not established experimentally but from log K_{ow} values determined by calculation and modelling should be considered only as a first and conservative estimate. For simple molecules of low molecular weight with one or two functional groups, a deviation of 0.1–0.3 log Pow units between the results of the different fragmentation methods and experimentally derived data can be expected. As the complexity of the molecule increases, the reliability of the various methods decreases due to the possibility of a variety of error factors, and the significant ones are listed as follows:

- The suitability of the various fragments used
- The analysts ability to recognise intramolecular interactions (e.g. hydrogen bonding)
- The correct use of the various correction factors (e.g. branching factors, proximity effects)

Clearly further work is required in determining the potential for bioconcentration and biomagnification from surfactants. It is established that the potential of many surfactants to bioaccumulate is primarily due to a general association of increasing alkyl chain length (i.e. increasing hydrophobicity) with an increase in BCF being noted [381]. Conversely, increasing the length of the hydrophilic section of a surfactant molecule (i.e. decreasing overall hydrophobicity) results in a reduction in BCF [382]. It has also been found that uptake rates and BCFs for alcohol ethoxylate surfactants increase when hydrophobicity is increased [383].

These apparent steric influences on surfactant toxicity and BCF appear to be consistent and may offer a means of predicting likely toxic effects of surfactants on marine organisms through a consideration of steric factors. If modifications to the molecular structure of surfactants can result in predictable influences on bioaccumulation and toxicity to aqueous organisms, then environmental effects of new surfactant formulations could be predicted at an early stage in product development.

There is normally a tendency for surfactant molecules to be retained on epithelial surfaces rather than to cross-cellular or epithelial membranes (resulting in uptake) and hence bioaccumulate. This may be a possible explanation for the longer-chain/lower toxicity observations. Surfactant molecules bound to an epithelial membrane surface may be expected to disrupt membrane integrity and interact with mucus. The number of binding sites on epithelial or cellular membranes is usually limited, and a critical number of (surfactant) molecules must occupy the available binding sites in order for lethal poisoning to occur; therefore surfactants that can more easily cross the membrane and bioaccumulate (as indicated by a higher BCF) are less likely to exhibit acute toxic effects [384].

In general, BCFs for surfactants are reported as being comparatively low and are generally below the conventional level for concern (i.e. log *P*ow value of 3–4) [59]. Biotransformation and biomagnification are processes that may occur once a chemical has entered an organism. Evidence for biotransformation of surfactants in aquatic organisms is limited and inconclusive. In order for biomagnification of a chemical to take place, the compound must be stable in the environment for significant periods of time. Compounds that degrade and or biodegrade relatively rapidly or that are readily metabolised will not be biomagnified within the food chain. While the bioaccumulation of a chemical can still present a problem where exposure levels and uptake rates are sufficiently high in relation to depuration and metabolism rates, a high bioaccumulation potential does not automatically imply the potential for biomagnification. Indeed, for some chemicals, which are readily taken up by organisms near the bottom of the food chain, a capacity for metabolism is more likely in successively higher trophic levels.

The available information indicates that most commonly used surfactants do not have the properties required to exhibit biomagnification, i.e. they have a tendency to be rapidly degraded and metabolised and are not highly hydrophobic.

BCFs in the aqueous phase are generally below the level of concern and (for some non-ionic surfactants at least) can be quantitatively related to the length of the hydrophobic and hydrophilic components. There is also evidence that overall molecular size may place constraints on biological uptake.

The fate of metabolites from surfactants and other chemicals discharged to the marine environment has not been comprehensively studied, and consequently there is a degree of uncertainty as to the fate and longer-term effects of some hydrophobic components (such as some alkylphenols) following partial metabolism [385, 386].

The overall toxicity of many surfactants is not of great concern; however some of the longer-term chronic effects are, in particular, those relating to endocrine disruption [183]. This is particularly the case for alkylphenol ethoxylates that have been used widely in the formulation of drilling fluids and demulsifiers. The popularity of these surfactants is based on their cost effectiveness, availability and the range of HLB values obtainable as previously described in Section 3.3.2.

The microbial degradation of products containing APE results in metabolites that are resistant to further biodegradation and are more toxic than their parent compounds. Nonylphenol has been demonstrated to be toxic to both marine and freshwater species to induce oestrogenic responses in male trout and may bioaccumulate in freshwater organisms. Concern over their possible environmental effects has led to the banning of nonylphenol ethoxylates from cleaning products in Europe. Norway has also banned their use in the offshore oil and gas sector. A voluntary ban has been in force for some time on their domestic use in the United Kingdom and the OSPARCOM Treaty signatures in the oil and gas sector [387].

Surfactants can also have positive effects in the environment, particularly in facilitating enzymatic hydrolysis on non-biodegradable or poorly biodegradable substances. It has been found that certain biosurfactants (sophorolipids, rhamnolipid, bacitracin) and other synthetic surfactants such as Tween 80 increased the rate of hydrolysis of cellulose derivatives as measured by as much as seven times and the hydrolysis of steam-exploded wood was increased by 67% in the presence of sophorolipids [388].

In examining the use and application of surfactants and particularly their use in the upstream oil and gas industry, this chapter has covered a large number of chemistries and their surfactant behaviour with particular emphasis on their environmental acceptability. It can be seen that in general surfactants are not environmentally damaging and have a significant and important role in the extraction and production of oil and gas.

A significant omission from this chapter is the analysis of surfactants, and other than some methods related to their environmental assessment, this topic has not been covered. The reader is directed towards the work of Schramm (Ref. [4]) and *The Analysis of Surfactants* by Schmitt [389].

References

1 Porter, M.R. (1991). *Handbook of Surfactants*. Springer.
2 IHS Markit (2015). Specialty Chemicals Update Program: Surfactants Global Market Report.
3 Noweck, K. and Grafahrend, W. (2002). Fatty alcohols. In: *Ullmans Encyclopedia of Industrial Chemistry*. Wiley-VCH.
4 Schramm, L.L. ed. (2000). *Surfactants: Fundamentals and Applications in the Petroleum Industry*. Cambridge University Press.
5 Israelachvili, J.N., Mitchell, D.J., and Ninham, B.W. (1976). Theory of self-assembly of hydrocarbon amphiphiles into micelles and bilayers. *Journal of the Chemical Society, Faraday Transactions II* **72**: 1525.
6 Ben-Naim, A.Y. (1982). Hydrophobic interactions, an overview. In: *Solution Behavior of Surfactants, Theoretical and Applied Aspects*, vol. **1** (ed. K.L. Mittal and E.J. Fendler), 27–40. Springer.
7 Kronberg, B., Costas, M., and Silveston, R. (1995). Thermodynamics of the hydrophobic effect in surfactant solutions-micellization and adsorption. *Pure and Applied Chemistry* **67** (6): 897–902.
8 Preston, W.C. (1948). Some correlating principles of detergent action. *The Journal of Physical Chemistry* **52** (1): 84–97.
9 Mukerjee, P. and Mysels, K.J. (1971). *Critical Micelle Concentrations of Aqueous Surfactant Systems, NSRDS-NBS 36*. Washington, DC: US Department of Commerce.
10 van Os, N.M., Haak, J.R., and Rupert, L.A.M. (1993). *Physico-Chemical Properties of Selected Anionic, Cationic and Nonionic Surfactants*. Amsterdam: Elsevier.
11 Evans, D.F. and Wightman, P.J. (1982). Micelle formation above 100 °C. *Journal of Colloid and Interface Science* **86** (2): 515–524.
12 L.A. Noll (1991). The effect of temperature, salinity, and alcohol on the critical Micelle concentration of surfactants, SPE 21032. SPE International Symposium on Oilfield Chemistry, Anaheim, CA (20–22 February 1991).
13 Mittal, K.L. and Fendler, E.J. ed. (1982). *Solution Behavior of Surfactants, Theoretical and Applied Aspects*, vol. **1**. Springer.
14 Yuan, Y. and Lee, T.R. (2013). *Surface Science Techniques*, Chapter 1, Springer Series in Surface Sciences 51 (ed. D.R. Karsa). Berlin and Heidelberg: Springer-Verlag Industrial applications of surfactants: the proceedings of a symposium organized by the North West Region of the Industrial Division of the Royal Society of Chemistry, University of Salford (15–17 April 1986).
15 Rosen, M.J. and Solash, J. (1969). Factors affecting initial foam height in the Ross-Miles foam test. *Journal of the American Oil Chemists' Society* **46** (8): 399–402.

16 van Os, N.M., Daane, G.J., and Bolsman, T.A.B.M. (1988). The effect of chemical structure upon the thermodynamics of micellization of model alkylarenesulfonates: II. Sodium *p*-(3-alkyl)benzenesulfonate homologs. *Journal of Colloid and Interface Science* **123** (1): 267–274.

17 Brun, T.S., Hoiland, H., and Vikingstad, E. (1978). Partial molal volumes and isentropic partial molal compressibilities of surface-active agents in aqueous solution. *Journal of Colloid and Interface Science* **63** (1): 89–96.

18 Corti, M. and Degiorgio, V. (1981). Quasi-elastic light scattering study of intermicellar interactions in aqueous sodium dodecyl sulfate solutions. *The Journal of Physical Chemistry* **85** (6): 711–717.

19 Abu-Hamdiyyah, M. and Kumari, K. (1990). Solubilization tendency of 1-alkanols and hydrophobic interaction in sodium lauryl sulfate in ordinary water, heavy water, and urea solutions. *The Journal of Physical Chemistry* **94** (16): 6445–6452.

20 Bostrom, S., Backlund, S., Blokhus, A.M., and Hoiland, H. (1989). *Journal of Colloid and Interface Science* **128** (1): 169–175.

21 Shinoda, K. (1963). *Colloidal Surfactants: Some Physicochemical Properties*. New York: Academic Press.

22 Rosen, M.J. and Hua, X.Y. (1988). Dynamic surface tension of aqueous surfactant solutions: 1. Basic parameters. *Journal of Colloid and Interface Science* **124** (2): 652–659.

23 Rusanov, A.I. and Prohorov, V.A. (1996). *Interfacial Tensiometry*, Studies Interface Science 3, Series Editors (ed. D. Mobius and R. Miller). Amsterdam: Elsevier.

24 Morrow, N.R. ed. (1990). *Interfacial Phenomena in Petroleum Recover*. CRC Press.

25 McCaffery, F.G. (1976). Interfacial tensions and aging behaviour of some crude oils against caustic solutions , PETSOC-76-03-09. *Journal of Canadian Petroleum Technology* **15** (3).

26 Drelich, J., Fang, C., and White, C.L. (2002). Measurement of interfacial tension in fluid-fluid systems. In: *Encyclopedia of Surface and Colloid Science*. Marcel Dekker, Inc.

27 Patist, A., Oh, S.G., Leung, R., and Shah, D.O. (2001). Kinetics of micellization: its significance to technological processes. *Colloids and Surfaces A* **176**: 3–16.

28 Taber, J.J. (1980). Research on enhanced oil recovery: past, present and future. *Pure and Applied Chemistry* **52**: 1323–1347.

29 G. J. Hirasaki, C. A. Miller, G. A. Pope and R. E. Jackson (2004). Surfactant Based Enhanced Oil Recovery and Foam Mobility Control. 1st Annual Technical Report, DE-FC26-03NT15406.

30 Alvarado, V. and Manrique, E. (2010). Enhanced oil recovery: an update review. *Energies* **3**: 1529–1575.

31 Iglauer, S., Wu, Y., Schuler, P. et al. (2010). New surfactant classes for enhanced oil recovery and tertiary oil recovery potential. *Journal of Petroleum Science and Engineering* **71**: 23–29.

32 Lake, L.W., Schmidt, R.L., and Venuto, P.B. (1992). A niche for enhanced oil recovery in the 1990s. *Oilfield Review* 55–61.

33 Taber, J.J. (1969). Dynamic and static forces required to remove a discontinuous oil phase from porous media containing both oil and water, SPE 2098. *Society of Petroleum Engineers Journal* **9** (1): 3–13.

34 Dunning, H.N., Hsiao, L., and Johanson, R.T. (1953). *Displacement of Petroleum from Sand by Detergent Solutions*. U.S. Dept. of the Interior Bureau of Mines, Technology and Engineering.

35 Doscher T.M. and Reisberg J. (1959). Recovery of oil from tar sands. US Patent 2,882,973, assigned to Shell Dev.

36 Sheng, J., Morel, D. and Gauer, P. (2010). Evaluation of the effect of wettability alteration on oil recovery in carbonate reservoirs. AAPG GEO 2010 Middle East Geoscience Conference & Exhibition, Manama, Bahrain (7–10 March 2010).

37 Standnes, D.C. and Austad, T. (2000). Wettability alteration in chalk: 2. Mechanism for wettability alteration from oil-wet to water-wet using surfactants. *Journal of Petroleum Science and Engineering* **28** (3): 123–143.

38 Kathel, P. and Mohanty, K.K. (2013). Wettability alteration in a tight oil reservoir. *Energy & Fuels* **27** (11): 6460–6468.

39 Tannich, J.D. (1975). Liquid removal from hydraulically fractured gas wells, SPE 5113. *Journal of Petroleum Technology* **27** (11).

40 Noh, M. and Firoozabadi, A. (2008). Wettability alteration in gas-condensate reservoirs to mitigate well deliverability loss by water blocking, SPE 98375. *SPE Reservoir Evaluation and Engineering* **11** (4): 676–685.

41 Street N. and Wang F.D. (1966). Surface potentials and rock strength. 1st ISRM (International society for Rock Mechanics) Congress, Lisbon, Portugal (25 September to 1 October 1966).

42 Alkafeef, S.F., Gochin, R.J. and Smith, A.L. (1995). Surface potential and permeability of rock cores under asphaltenic oil flow conditions, SPE 30539. SPE Annual Technical Conference and Exhibition, Dallas, TX (22–25 October 1995).

43 Johnson, J.D., Schoppa, D., Garza, J.L., Zamora, F., Kakadjian, S. and Fitzgerald, E. (2010). Enhancing gas and oil production with zeta potential altering system, SPE 128048. SPE International Symposium and Exhibition on Formation Damage Control, Lafayette, LA (10–12 February 2010).

44 Shehata, A.M. and Nasr-El-Din, H.A. (2015). Zeta potential measurements: impact of salinity on sandstone minerals, SPE 173763. SPE International Symposium on Oilfield Chemistry, The Woodlands, TX (13–15 April 2015).

45 Kanicky, J.R., Lopez-Montilla, J.-C., Pandey, S., and Shah, D.O. (2001). Surface chemistry in the petroleum industry. In: *Handbook of Applied Surface and Colloid Chemistry*, Chapter 11 (ed. K. Holmberg). John Wiley & Sons, Ltd.

46 Hirasaki, G. and Zhang, D.L. (2004). Surface chemistry of oil recovery from fractured, oil-wet, carbonate formations, SPE 88365. *SPE Journal* **9** (02): 151–162.

47 Somasundaran, P. and Zhang, L. (2006). Adsorption of surfactants on minerals for wettability control in improved oil recovery processes. *Journal of Petroleum Science and Engineering* **52**: 198–212.

48 Derjaguin, B.V. (1989). *Theory of Stability of Colloids and Thin Liquid Films*. New York: Plenum Press/Consultants Bureau.

49 Israelachvili, J.N. (2011). *Intermolecular and Surface Forces*, 3rde. London: Academic Press.

50 Bergeron, V. and Radke, C.J. (1992). Equilibrium measurements of oscillatory disjoining pressures in aqueous foam films. *Langmuir* **8**: 3020–3026.

51 Richetti, P. and Kékicheff, P. (1992). Direct measurement of depletion and structural forces in a micellar system. *Physical Review Letters* **68**: 1951.

52 Wasan, D.T., Nikolov, A.D., Kralchevsky, P.A., and Ivanov, I.B. (1992). Universality in film stratification due to colloid crystal formation. *Colloids and Surfaces* **67**: 139–145.

53 Russel, W.B., Saville, D.A., and Schowalter, W.R. (1989). *Colloidal Dispersions*. Cambridge: Cambridge University Press.

54 Griffin, W.C. (1949). Classification of surface-active agents by HLB. *Journal of the Society of Cosmetic Chemists* **1** (5): 311–326.

55 Griffin, W.C. (1954). Calculation of HLB values of nonionic surfactants. *Journal of the Society of Cosmetic Chemists* **5** (4): 249–256.

56 Davies, J.T. (1957). A quantitative kinetic theory of emulsion type, I. Physical chemistry of the emulsifying agent, gas/liquid and liquid/liquid interface. Proceedings of the International Congress of Surface Activity, pp. 426–438.

57 Rondon, M., Bourait, P., Lachaise, J., and Salager, J.-L. (2006). Breaking of water-in-crude oil emulsions. 1. Physicochemical phenomenology of demulsifier action. *Energy & Fuels* **20**: 1600–1604.

58 Lin, T.J., Kurihara, H., and Ohta, H. (1975). Effects of phase inversion and surfactant location on the formation of OIW emulsions. *Journal of the Society of Cosmetic Chemists* **26**: 121–139.

59 McWilliams, P. and Payne, G. (13–14 November 2001). Bioaccumulation potential of surfactants: a review. In: *Chemistry in the Oil Industry VII* (ed. T. Balson, H.A. Craddock, J. Dunlop, H. Frampton, G. Payne and P. Reid). Manchester, UK: Royal Society of Chemistry.

60 Zoller, U. and Sosis, P. ed. (2008). *Handbook of Detergents, Part F: Production*, Surfactant Science Series, vol. **142**. CRC Press.

61 Ziegler, K., Gellert, H.-G., Lehmkuhl, H. et al. (1960). Metallorganische verbindungen, XXVI aluminiumtrialkyle und dialkyl-aluminiumhydride aus olefinen, wasserstoff und aluminium. *Justus Liebigs Annalen der Chemie* **629**: 1–13.

62 Lutz, E.F. (1986). Shell higher olefins process. *Journal of Chemical Education* **63** (3): 202.

63 Scwartz, K.N., Smith, K.W. and Chen, S.-R.T. (2009). Polymeric gel system and use in hydrocarbon recovery. US Patent 7,575,057, assigned to Clearwater International L.L.C.

64 Chatterji, J., Brenneis, D.C., King, B.J., Cromwell, R.S. and Gray, D.W. (2007). Foamed cement compositions, additives, and associated methods. US Patent 7,255,170, assigned to Halliburton Energy Services Inc.

65 Inaba, H. and Haruki, N. (1998). Drag reduction and heat transfer characteristics of water solutions with surfactants in a straight pipe. *Heat Transfer-Japanese Research* **27** (1): 1–15.

66 Kalpakei, B., Arf, T.G., Barker, J.W., Krupa, A.S., Morgan, J.C. and Neira, R.D. (1990). The low-tension polymer flood approach to cost-effective chemical EOR, SPE 20220. SPE/DOE Enhanced Oil Recovery Symposium, Tulsa, OK (22–25 April 1990).

67 Wu, Y., Iglauer, S., Shuler, P. et al. (2010). Branched alkyl alcohol propoxylated sulfate surfactants for improved oil recovery. *Tenside, Surfactants, Detergents* **47** (3): 152–161.

68 Tally, L.D. (1988). Hydrolytic stability of alkylethoxy sulfates, SPE 14912. *SPE Reservoir Engineering* **3** (1).

69 Lawson, J.B. (1978). The adsorption of non-ionic and anionic surfactants on sandstone and carbonate, SPE 7052. SPE Symposium on Improved Methods of Oil Recovery, Tulsa, OK (16–17 April 1978).

70 Goodyear, S.G. and Jones, P.I.R. (1995). Assessment of foam for deep flow diversion. Proceedings of 8th EAPG Improved Oil Recovery Europe Symposium, Vienna, Austria (15–19 May 1995) pp. 174–182.

71 Growcock, F.B. and Simon, G.A. (2004). Stabilized colloidal and colloidal-like systems. US Patent 7,037,881, assigned to authors.

72 Brookey, T.F. (2001). Aphron-containing well drilling and servicing fluids. US Patent 6,716,797, assigned to Masi Technologies L.L.C.

73 Inventory Multi-Tiered Assessment and Prioritisation (IMAP). (2016). Human health tier II assessment for sodium and ammonium laureth sulfate. Australian Government, Department of Health, National Industrial Chemicals Notification and Assessment Scheme.

74 US Environmental Protection Agency. (2000). 1,4-Dioxane (74,4-Diethyleneoxide), Hazard Summary-created in April 1992; revised in January 2000.

75 Mackay, D., Di Guardo, A., Paterson, S. et al. (1996). Assessment of chemical fate in the environment using evaluative, regional and local-scale models: Illustrative application to chlorobenzene and linear alkylbenzene sulfonates. *Environmental Toxicology and Chemistry* **15** (9): 1638–1648.

76 Jensen, J. (1999). Fate and effects of linear alkylbenzene sulphonates (LAS) in the terrestrial environment. *Science of the Total Environment* **226** (2–3): 93–11.

77 Technical Guidance Document on Risk Assessment in support of Commission Directive 93/67/EEC on Risk Assessment for new notified substances and Commission Regulation (EC) No 1488/94 on Risk Assessment for existing substances, April 2003.

78 De Groote, M. (1931) Flooding process for recovering oil from subterranean oil bearing strata. US Patent 1,823,439, assigned to Tretolite Co.

79 Gale, W.W. and Sandvik, E.I. (1973). Tertiary surfactant flooding: petroleum sulfonate composition-efficacy studies, SPE 3804. *Society of Petroleum Engineers Journal* **13** (4): 191–200. doi: 10.2118/3804-PA.

80 Buckley, J.S. and Fan, T. (2005). Crude oil/brine interfacial tensions. International Symposium of the Society of Core Analysts held in Toronto, Canada (21–25 August 2005).

81 Berger, P.D. and Lee, C.H. (2002). Ultra-low concentration surfactants for sandstone and limestone floods, SPE 75186. SPE/DOE Improved Oil Recovery Symposium, Tulsa, OK (13–17 April 2002).

82 Berger, P.D. and Berger, C.H. (2004). Method of using alkylsulfonated phenol/aldehyde resins as adsorption reducing agents for chemical flooding. US Patent 6,736,211, assigned to Oil Chem Technologies.

83 Ghloum, E.F., Al-Qahtani, M., and Al-Rashid, A. (2010). Effect of inhibitors on asphaltene precipitation for Marrat Kuwaiti reservoirs. *Journal of Petroleum Science and Engineering* **70** (1–2): 99–106.

84 Dickakian, G.B. (1990). Antifoulant additive for light end hydrocarbons. US Patent 4,931,164, assigned to Exxon Chemical patents Inc.

85 Wieche, I. and Jermansen, T.G. (2003). Design of synthetic sispersants for asphaltenes. *Journal of Petroleum Science and Engineering* **21** (3–4): 527–536.

86 Miller, D., Vollmer, A. and Feustal, M. (1999). Use of alkanesulfonic acids as asphaltene-dispersing agents. US Patent 5,925,233, assigned to Clariant Gmbh.

87 Raney, K.H., Shpakoff, P.G., and Passwater, D.K. (1998). Use of high-active alpha olefin sulfonates in laundry powders. *Journal of Surfactants and Detergents* **1** (3): 361–369.

88 Sanz, C.A. and Pope, G.A. (1995). Alcohol-free chemical flooding: from surfactant screening to coreflood design, SPE 28956. SPE International Symposium on Oilfield Chemistry, San Antonio, TX (14–17 February 1995).

89 Osterloh, W.T. (1994). Long chain alcohol additives for surfactant foaming agents. US Patent 5,333,687 assigned to Texaco Inc.

90 Pauluiski, M.K. and Hlidek, B.T. (1992). Slurried polymer foam system and method for the use thereof. WO Patent 9214907, assigned to The Western Company of North America.

91 Ali, E., Bergren, F.E., DeMestre, P., et al. (2006). Effective gas shutoff treatments in a fractured carbonate field in Oman, SPE 102244. SPE Annual Technical Conference and Exhibition, San Antonio, TX (24–27 September 2006).

92 Dalland, M., Hanssen, J.E., and Kristiansen, T.S. (1994). Oil interaction with foams under static and flowing conditions in porous media. *Colloids and Surfaces A: Physicochemical and Engineering Aspects* **82** (2): 129–140.

93 Thach, S., Miller, K.C., Lai, Q.J., et al. (1996). Matrix gas shut-off in hydraulically fractured wells using polymer-foams, SPE 36616. SPE Annual Technical Conference and Exhibition, Denver, CO (6–9 October 1996).

94 van Houwelingen, J. (1999). Chemical gas shut-off treatments in Brunei, SPE 57268. SPE Asia Pacific Improved Oil Recovery Conference, Kuala Lumpur, Malaysia (25–26 October 1999).

95 Molina, M.J. and Rowland, F.S. (1974). Stratospheric sink for chlorofluoromethanes: chlorine atom-catalysed destruction of ozone. *Nature* **249**: 810–812.

96 Giesy, J.P. and Kannan, K. (2001). Global distribution of perfluorooctane sulfonate in wildlife. *Environmental Science & Technology* **35**: 1339–1342.

97 Hansen, K.J., Clemen, L.A., Ellefson, M.E., and Johnson, H.O. (2001). Compound-specific, quantitative characterization of organic fluorochemicals in biological matrices. *Environmental Science & Technology* **35**: 766–770.

98 Fromel, T. and Knepper, T.P. (2010). Biodegradation of fluorinated alkyl substances. *Reviews of Environmental Contamination and Toxicology* **208**: 161–177.

99 Azar, J.J. and Samuel, G.R. (2007). *Drilling Engineering*. PennWell Books.

100 Sandvik, E.I., Gale, W.W., and Denekas, M.O. (1977). Characterization of petroleum sulfonates, SPE 6210. *Society of Petroleum Engineers Journal* **17** (3).

101 Bae, J.H., Petrick, C.B. and Ehrilich, R.E. (1974). A comparative evaluation of microemulsions and aqueous surfactant systems, SPE 4749. SPE Improved Oil Recovery Symposium, Tulsa, OK (22–24 April 1974).

102 Bae, J.H. and Petrick, C.B. (1977). Adsorption/retention of petroleum sulfonates in Berea cores, SPE 58192. *Society of Petroleum Engineers Journal* **17** (5).

103 Cavias, J.L., Schechter, R.S., and Wade, W.H. (1977). The utilization of petroleum sulfonates for producing low interfacial tensions between hydrocarbons and water. *Journal of Colloid and Interface Science* **59** (1): 31–38.

104 Sundaram, N.S., Sawar, M., Bang, S.S., and Islam, M.R. (1994). Biodegradation of anionic surfactants in the presence of petroleum contaminants. *Chemosphere* **29** (6): 1253–1261.

105 Wang, H., Cao, X., Zhang, J., and Zhang, A. (2009). Development and application of dilute surfactant–polymer flooding system for Shengli oilfield. *Journal of Petroleum Science and Engineering* **65** (1-2): 45–50.

106 Olajire, A.A. (2014). Review of ASP EOR (alkaline surfactant polymer enhanced oil recovery) technology in the petroleum industry: prospects and challenges. *Energy* **77**: 963–982.

107 Salimi, M.H., Petty, K.C. and Emmett, C.L. (1993). Scale inhibition during oil production. US Patent 5,263,539, assigned to Petrolite Corp.

108 Alonso-de Bolt, M. and Jarret, M.A. (1995). Drilling fluid additive for water sensitive shales and clays, and method of drilling use the same. EP Patent 668339, assigned to Baker Hughes Inc.

109 Oldigies, D., McDonald, H. and Blake, T. (2007). Use of calcium sulfonate based threaded compounds in drilling operations and other severe industrial applications. US Patent 7,294,608, assigned to Jet-Lube Inc.

110 Olson, W.D., Muir, R.J., Eliades, T.I. and Steib, T. (1994). Sulfonate greases. US Patent 5,308,514, assigned to Witco Corporation.

111 Austad, T., Ekrann, S., Fjelde, I., and Taugbol, K. (1997). Chemical flooding of oil reservoirs Part 9. Dynamic adsorption of surfactant onto sandstone cores from injection water with and without polymer present. *Colloids and Surfaces, A: Physicochemical and Engineering Aspects* **27** (1–3): 69–82.

112 Fink, J.K. (2013). *Petroleum Engineer's Guide to Oilfield Chemicals and Fluids*. Elsevier.

113 Mokadam, A.R. (1998). Surfactant additive for oilfield acidizing. US Patent 5,797,456, assigned to Nalco Exxon Energy Chemicals LP.

114 Welton, T.D., Bryant, J. and Funkhouser, G.P. (2007). Anionic surfactant gel treatment fluid, SPE 105815. International Symposium on Oilfield Chemistry, Houston, TX (28 February to 2 March 2007).

115 Ghazali, R. and Ahmad, S. (2004). Biodegradability and ecotoxicity of palm stearin based methyl ester sulphonates. *Journal of Palm Oil Research* **16** (1): 39–44.

116 Majidaie, S., Muhaammad, M., Tan, I.M., et al. (2012). Non-petrochemical surfactant for enhanced oil recovery, SPE 153493. SPE EOR Conference at Oil and Gas West Asia, Muscat, Oman (16–18 April 2012).

117 Mueller, H., Herold, C.-P., Bongardt, F., et al. (2004). Lubricants for drilling fluids. US Patent 6,806,235, assigned to Cognis Deutschland Gmbh & Co. Kg.

118 Karsa, D. and Porter, M.R. (1995). *Biodegradability of Surfactants*. Blackie Academic and Professional.

119 Alonso-Debolt, M. (1994). New polymer/surfactant systems for stabilizing troublesome gumbo shale, SPE 28741. International Petroleum Conference and Exhibition of Mexico, Veracruz, Mexico (10–13 October 1994).

120 Steffy, D.A., Nichols, A.C., and Kiplagat, G. (2011). Investigating the effectiveness of the surfactant dioctyl sodium sulfosuccinate to disperse oil in a changing marine environment. *Ocean Science Journal* **46** (4): 299–305.

121 Nyankson, E., Ober, C.A., DeCuir, M.J., and Gupta, R.B. (2014). Comparison of the effectiveness of solid and solubilized dioctyl sodium sulfosuccinate (DOSS) on oil dispersion using the baffled flask test, for crude oil spill applications. *Industrial and Engineering Chemistry Research* **53** (29): 11862–11872.

122 Lalchan, C.A., O'Neil, B.J. and Maley, D.M. (2013). Prevention of acid induced asphaltene precipitation: a comparison of anionic vs. cationic surfactants, SPE 164087. SPE International Symposium on Oilfield Chemistry, The Woodlands, TX (8–10 April 2013).

123 US Energy Information Administration (2011). Review of emerging resources: U.S. shale gas and shale oil plays. US Department of Energy.

124 Liang, T., Longaria, R.A., Lu, J., et al. (2015). Enhancing hydrocarbon permeability after hydraulic fracturing: laboratory evaluations of shut-ins and surfactant additives, SPE 175101. SPE Annual Technical Conference and Exhibition, Houston, TX (28–30 September 2015).

125 Popova, A., Sokolova, E., Raicheva, S., and Christov, M. (2003). AC and DC study of the temperature effect on mild steel corrosion in acid media in the presence of benzimidazole derivatives. *Corrosion Science* **45** (1): 33–58.

126 Naraghi, A. and Grahmann, N. (1997). Corrosion inhibitor blends with phosphate esters. US Patent 5,611,992, assigned to Champion Technologies Inc.

127 Martin, R.L., Poelker, D.J. and Walker, M.L. (2004). High performance phosphorus-containing corrosion inhibitors for inhibiting corrosion by drilling system fluids. EP Patent 1076113, assigned to Baker Hughes Incorporated.

128 Dougherty, J.A. (1998). Controlling CO_2 corrosion with inhibitors, NACE 98015. Corrosion 98, San Diego, CA (22–27 March 1998).

129 Alink, B.A., Outlaw, B., Jovancicevic, V., et al. (1999). Mechanism of CO_2 corrosion inhibition by phosphate esters, NACE 99037. Corrosion 99, San Antonio, TX (25–30 April 1999).

130 Yepez, O., Obeyesekere, N. and Wylde, J. (2015). Development of novel phosphate based inhibitors effective for oxygen corrosion, SPE 173723. SPE International Symposium on Oilfield Chemistry, The Woodlands, TX (13–15 April 2015).

131 Brown, B., Saleh, A., and Moloney, J. (2015). Comparison of mono- to diphosphate ester ratio in inhibitor formulations for mitigation of under deposit corrosion. *Corrosion* **71** (12): 1500–1510.

132 Yu, H., Wu, J.H., Wang, H.R. et al. (2006). Corrosion inhibition of mild steel by polyhydric alcohol phosphate ester in natural sea water. *Corrosion Engineering, Science and Technology* **41**: 259–262.

133 T.J. Bellos (1982). Corrosion inhibitor for highly oxygenated systems. US Patent 4,311,662, assigned to Petrolite Corporation.

134 Outlaw, B.T., Alink, B.A., Kelley, J.A and Claywell, C.S. (1985). Corrosion inhibition in deep gas wells by phosphate esters of poly-oxyalkylated thiols. US Patent 4,511,480, assigned to Petrolite Corporation.

135 Martin, R.L., Brock, G.F. and Dobbs, J.B. (2001). Corrosion inhibitors and methods of use. US Patent 6,866,797, assigned to BJ Services Company.

136 Martin, R.L. (1988). Multifunctional corrosion inhibitors. US Patent 4,722,805, assigned to Petrolite Corporation.

137 Bassi, G. and Li, W. (2003). Corrosion inhibitors for the petroleum industry, EP Patent 1333108, assigned to Lamberti S.p.A.

138 McGregor, W. (2004). Novel synergistic water soluble corrosion inhibitors, SPE 87570. SPE International Symposium on Oilfield Corrosion, Aberdeen, UK (28 May 2004).

139 Saeger, V.W., Hicks, O., Kaley, R.G. et al. (1979). Environmental fate of selected phosphate esters. *Environmental Science & Technology* **13** (7): 840–844.

140 Kelland, M.A. (2014). *Production Chemicals for the Oil and Gas Industry*, 2nde. CRC Press.

141 Reid, A.L. and Grichuk, H.A. (1991). Polymer composition comprising phosphorous-containing gelling agent and process thereof. US Patent 5,034,139, assigned to Nalco Chemical Company.

142 Dixon, J. (2009). Drilling fluids. US Patent 7,614,462, assigned to Croda International Plc.

143 Johnson, D.W. and Hils, J.E. (2013). Review: phosphate esters, thiophosphate esters and metal thiophosphates as lubricant additives. *Lubricants* **1**: 132–148.

144 Miller, R.F. (1984). Method for minimizing fouling of heat exchangers. US Patent 4,425,223, assigned to Atlantic Richfield Company.

145 Miller, D., Volmer, A., Feustel, M. and Klug, P. (2001). Synergistic mixtures of phosphoric esters with carboxylic acids or carboxylic acid derivatives as asphaltene dispersants. US Patent 6,204,420, assigned to Clariant Gmbh.

146 Alameda, E.J., Fernandez-Serrano, M., Lechuga, M., and Rios, F. (2012). Environmental impact of ether carboxylic derivative surfactants. *Journal of Surfactants and Detergents* **15**: 1–7.

147 A.R. Mansour and T. Aldoss (1988). Drag reduction in pipes carrying crude oil using an industrial cleaner, SPE 179188. Society of Petroleum Engineers.

148 Di Lullo, G., bte. Ahmad, A., Rae, P., et al. (2001). Toward zero damage: new fluid points the way, SPE 69453. SPE Latin American and Caribbean Petroleum Engineering Conference, Buenos Aires, Argentina (25–28 March 2001).

149 Hartshorne, R., Hughes, T., Jones, T. and Tustin, G. (2005). Anionic viscoelastic surfactant. US Patent Application 20050124525, assigned to Schlumberger Technology Corporation.

150 Hartshorne, R., Hughes, T., Jones, T., et al. (2009). Viscoelastic compositions. US Patent Application 20090291864, assigned to Schlumberger Technology Corporation.

151 Gale, W.W. (1975). Oil recovery process utilizing aqueous solution of a mixture of alkyl xyene sulfonates. US Patent 3,861,466, assigned to Exxon Production Research Co.

152 Campbell, C. and Sinquin, G. (2006). Alkylxylene sulfonates for enhanced oil recovery processes. US Patent Application 20060014650, assigned to Chevron Oronite Company Llc.

153 Menger, F.M. and Littau, C.A. (1993). Gemini surfactants: a new class of self-assembling molecules. *Journal of the American Chemical Society* **115** (22): 10083–10090.

154 Renouf, P., DHerbault, J.-R.D., Mercier, J.-M. et al. (1999). Synthesis and surface-active properties of a series of new anionic gemini compounds. *Chemistry and Physics of Lipids* **99** (1): 21–32.

155 Du, X., Lu, Y., Li, L. et al. (2006). Synthesis and unusual properties of novel alkyl benzene sulfonate gemini surfactants. *Colloids and Surfaces, A: Physicochemical and Engineering Aspects* **290** (1-3): 132–137.

156 Gao, B. and Sharma, M.M. (2013). A new family of anionic surfactants for enhanced-oil-recovery applications, SPE 159700. *SPE Journal* **18** (5).

157 Yang, J., Guan, B., Lu, Y., Cui, W., Qiu, X., Yang, Z., Qin, W. Viscoelastic evaluation of gemini surfactant gel for hydraulic fracturing, SPE 165177. SPE European Formation Damage Conference and Exhibition, Noordwijk, The Netherlands (5–7 June 2013).

158 Yang, J., Gao, L., Liu, X., et al.(2015). A highly effective corrosion inhibitor by use of gemini imidazoline, SPE 173777. SPE International Symposium on Oilfield Chemistry, The Woodlands, TX (13–15 April 2015).

159 Wurtz, A. (1859). *Justus Liebigs Annalen der Chemie* **110**: 125–128.

160 Eliot, D.E. and McClements, W.J. (1958). Production of ethylene oxide. US Patent 2,831,870, assigned to Allied Chemical and Dye Corporation.

161 Adams, P.T., Selff, R.B., and Tolbert, B.M. (1952). Hydrogenation of fatty acids to alcohols. *Journal of the American Chemical Society* **74** (9): 2416–2417.

162 Rubinfeld, J., Bian, W. and Ouw, G. (1969). Alkene sulfonation process and products. US Patent 3,428,654, assigned to Colgate Palmolive Co.

163 Guilloty, H.R. (1980). Process for making ethoxylated fatty alcohols with narrow polyethoxy chain distribution. US Patent 4,223,163, assigned to Proctor and Gamble Company

164 Funasaki, N., Hada, S., and Neya, S. (1988). Monomer concentrations of nonionicsurfactants as deduced with gel filtration chromatography. *The Journal of Physical Chemistry* **92** (25): 7112–7116.

165 Jachnik, R.P. and Green, P. (1995). Horizontal well drill-in fluid utilising alcohol ethoxylate, SPE 28963. SPE International Symposium on Oilfield Chemistry, San Antonio, TX (14–17 February 1995).

166 Sevigny, W.J., Kuehne, D.L. and Cantor, J. (1994). Enhanced oil recovery method employing a high temperature brine tolerant foam-forming composition. US Patent 5,358,045, assigned to Chevron research and Technology Company.

167 R. Gdanski (1995). Molecular modeling gives insight into nonionic surfactants, SPE 28971. SPE International Symposium on Oilfield Chemistry, San Antonio, TX (14–17 February 1995).

168 Knox, D. and McCosh, K. (2005). Displacement chemicals and environmental compliance – past present and future. *Chemistry in the Oil Industry IX*, pp. 76–91. Royal Society of Chemistry, Manchester, UK (31 October to 2 November 2005).

169 Thomas, C.A. (1941). *Anhydrous Aluminium Chloride in Organic Chemistry*, 178–186. New York: American Chemical Society.

170 March, J. (1977). *Advanced Organic Chemistry: Reactions, Mechanism and Structure*, 2nde. Kogakusha, Japan: McGraw-Hill.

171 Oswald, A.A., Bhatia, R.N., Mozeleski, E.J., et al. (1988). Alkylphenols and derivatives thereof via phenol alkylation by cracked petroleum distillates. World Patent application WO 1988003133, assigned to Exxon Research and Engineering Company.

172 Fujiwara, Y., Takezono, T., Kyono, S. et al. (1968). Effect of alkyl chain branching on the biodegradability of alkylbenzene sulfonates. *Journal of Japan Oil Chemists' Society* **17** (7): 396–399.

173 Berger, P.D., Hsu, C., and Aredell, J.P. (1988). Designing and selecting demulsifiers for optimum filed performance on the basis of production fluid characteristics, SPE 16285. *SPE Production Engineering* **3** (6): 522.

174 Stais, F., Bohm, R., and Kupfer, R. (1991). Improved demuslifier chemistry: a novel approach in the dehydration of crude oil, SPE 18481. *SPE Production Engineering* **6** (3): 334.

175 von Tapavicza, S., Zoeliner, W., Herold, C.-P., et al. (2002). Use of selected inhibitors against the formation of solid organo-based incrustations from fluid hydrocarbon mixtures. US Patent 6,344,431, assigned to Authors.

176 Stephenson, W.K., Mercer, B.D. and Comer, D.G. (1992). Refinery anti-foulant – asphaltene dispersant. US Patent 5,143,594, assigned to Nalco Chemical Company.

177 Kraiwattanawong, K., Folger, H.S., Gharfeh, S.G. et al. (2009). Effect of asphaltene dispersants on aggregate size distribution and growth. *Energy & Fuels* **23** (3): 1575–1582.

178 Tyler, C.R., Jobling, S., and Sumpter, J.P. (1998). Endocrine disruption in wildlife: a critical review of the evidence. *Critical Reviews in Toxicology* **28** (4): 319–361.

179 Barber, L.B., Loyo-Rosales, J.E., Rice, C.P. et al. (2015). Endocrine disrupting alkylphenolic chemicals and other contaminants in wastewater treatment plant effluents, urban streams, and fish in the Great Lakes and Upper Mississippi River Regions. *Science of the Total Environment* **517**: 195–206.

180 Hu, Y.F. and Guo, T.M. (2005). Effect of the structures of ionic liquids and alkylbenzene derived amphiphiles on the inhibition of asphaltene precipitation from CO_2 injected reservoir oils. *Langmuir* **21** (18): 8168–8174.

181 Gonzalez, G. and Middea, A. (1991). Peptization of asphaltene by various oil soluble amphiphiles. *Colloids and Surfaces* **52**: 207.

182 Fan, Y., Simon, S., and Sjoblom, J. (2010). Influence of nonionic surfactants on the surface and interfacial film properties of asphaltenes investigated by Langmuir balance and Brewster angle microscopy. *Langmuir* **26**: 10497–10505.

183 Getliff, J.M. and James, S.G. (1996). The replacement of alkyl-phenol ethoxylates to improve the environmental acceptability of drilling fluid additives, SPE 35982. SPE Health, Safety and Environment in Oil and Gas Exploration and Production Conference, New Orleans, Louisiana (9–12 June 1996).

184 Fine, R.D. (1958). A review of ethylene oxide condensation with relation to surface-active agents. *Journal of the American Oil Chemists' Society* **35** (10): 542–547.

185 Mattil, K.F. and Sims, R.J. (1954). Monoglyceride synthesis. US Patent 2,691,664, assigned to Swift & Company.

186 Yamaguchi, A., Hiyoshi, N., Sato, O., and Shirai, M. Sorbitol dehydration in high temperature liquid water. *Green Chemistry* **13** (4): 873–888, 201.

187 OECD (2002). OECD guidelines for the testing of chemicals/OECD series on testing and assessment detailed review paper on biodegradability testing (12 December 2002). OECD Publishing.

188 Vasudevan, M. and Wiencek, J.M. (1996). Mechanism of the extraction of proteins into Tween 85 nonionic microemulsions. *Industrial and Engineering Chemistry Research* **35** (4): 1085–1089.

189 Zychal, C. (1986). Defoamer and antifoamer composition and method for defoaming aqueous fluid systems. US Patent 4,631,145, assigned to Amoco Corporation.

190 Davidson, E. (1995). Defoamers. WO Patent 9509900, assigned to Imperial Chemical Industries.

191 Shpakoff, P.G. and Raney, H.H. (2009). Method and composition for enhanced hydrocarbons recovery. US Patent 7,612,022, assigned to Shell Oil Company.

192 Leblanc, M.-C.P., Durrieu, J.A., Binon, J.-P.P., Provin, G.G. and Fery, J.-J. (1990). Process for treating an aqueous solution of acrylamide resin in order to enable it to gel slowly even at high temperature. US Patent 4,975,483, assigned to Total Compagnie Francaise Des Petroles Establissements Vasset.

193 Charlier, A.G.R. (1991). Dispersant compositions for treating oil slicks. US Patent 5,051,192, assigned to Labofina S.A.

194 Canevari, G.P., Fiocco, R.J., Becker, K.W. and Lessard, R.R. (1997). Chemical dispersant for oil spills. US Patent 5,618,468, assigned to Exxon Research and Engineering Company.

195 Maag, H. (1984). Fatty acid derivatives: Important surfactants for household, cosmetic and industrial purposes. *Journal of the American Oil Chemists' Society* **61** (2): 259–267.

196 Migahed, M.A., Abd-El-Raouf, M., Al-Sabagh, A.M., and Abd-El-Bary, H.M. (2006). Corrosion inhibition of carbon steel in acid chloride solution using ethoxylated fatty alkyl amine surfactants. *Journal of Applied Electrochemistry* **36** (4): 395–402.

197 Jovancicevic, V., Ramachandran, S., and Prince, P. (1999). Inhibition of carbon dioxide corrosion of mild steel by imidazolines and their precursors. *Corrosion* **55** (5): 449–455.

198 Miksic, B.M., Furman, A.Y. and Kharshan, M.A. (2009). Effectiveness of corrosion inhibitors for the petroleum industry under various flow conditions, NACE 095703. Corrosion 2009, Atlanta, Georgia (22–26 March 2009).

199 Chavez-Miryauchi, T.E., Zamudio, L.S., and Baba-Lopez, V. (2013). Aromatic polyisobutylene succinimides as viscosity reducers with asphaltene dispersion capability for heavy and extra-heavy crude oils. *Energy & Fuels* **27** (4): 1994–2001.

200 Firoozinia, H., Abad, K.F.H., and Varamesh, A. (2015). A comprehensive experimental evaluation of asphaltene dispersants for injection under reservoir conditions. *Petroleum Science* **13** (2): 280–291.

201 Monroe, R.F., Kucera, C.H., Oakes, B.D. and Johnston, N.G. (1963). Compositions for inhibiting corrosion. US Patent 3,077,454, assigned to Dow Chemical Co.

202 Stoufer, W.B. (1987). Cleaning composition of terpene hydrocarbon and a coconut oil fatty acid alkanolamide having water dispersed therein. US Patent 4,704,225, assigned to the inventor.

203 Bernasconi, C., Faure, A. and Thibonnet, B. (1986). Homogeneous and stable composition of asphaltenic liquid hydrocarbons and additive useful as industrial fuel. US Patent 4,622,047, assigned to Elf France.

204 Wei, B.Q., Mikkelsen, T.S., McKinney, M.K. et al. (2006). A second fatty acid amide hydrolase with variable distribution among placental mammals. *The Journal of Biological Chemistry* **281** (48): 36569–36578.

205 Folmer, B.M., Holmberg, K., Klingskog, E.G., and Bergström, K. (2001). Fatty amide ethoxylates: synthesis and self-assembly. *Journal of Surfactants and Detergents* **4** (2): 175–183.

206 Hellberg, P.-E. (2009). Structe-property relationships for novel low-alkoxylated corrosion inhibitors. Chemistry in the Oil Industry XI, Manchester, UK (2–4 November 2009).

207 Ford, M.E., Kretz, C.P., Lassila, K.R., et al. (2006). N,N′-dialkyl derivatives of polyhydroxyalkyl alkylenediamines. EP Patent 1637038, assigned to Air Products and Chemicals Inc.

208 Boden, F.J., Sauer, R.P., Goldblatt, I.L. and McHenry, M.E. (2004). Polar grafted polyolefins, methods for their manufacture, and lubricating oil compositions containing them. US Patent 6,686,321, assigned to Castrol Ltd.

209 OECD SIDS (2003). Imidazole. UNEP Publications 2, SIDS Initial Assessment Report For SIAM 17, Arona, Italy (11–14 November 2003).

210 Morrissey, S., Pegot, B., Coleman, D. et al. (2009). Biodegradable, non-bactericidal oxygen-functionalised imidazolium esters: a step towards 'greener' ionic liquids. *Green Chemistry* **11**: 475–483.

211 Friedli, F.E. (2001). *Detergency of Specialty Surfactants*. New York, NY: Dekker.

212 Kocherbitov, V., Verazov, V., and Soderman, O. (2007). Hydration of trimethylamine-*N*-oxide and of dimethyldodecylamine-*N*-oxide: an ab initio study. *Journal of Molecular Structure: THEOCHEM* **808** (1-3): 111–118.

213 Crews, J.B. (2010). Saponified fatty acids as breakers for viscoelastic surfactant-gelled fluids. US Patent 7,728,044, assigned to Baker Hughes Incorporated.

214 Fu, D., Panga, M., Kefi, S. and Garcia-Lopez de Victoria, M. (2007). Self diverting matrix acid. US Patent 7,237,608, assigned to Schlumberger Technology Corporation.

215 Cassidy, J.M., Kiser, C.E., and Lane, J.L. (2006). Methods and aqueous acid solutions for acidizing wells containing sludging and emulsifying oil. US Patent Application 2006004083, Halliburton Energy Services.

216 Nasr-El-Din, H.A., Chesson, J.B., Cawiezel, K.E. and De Vine, C.S. (2006). Lessons learned and guidelines for matrix acidizing with viscoelastic surfactant diversion in carbonate formations, SPE 102468. SPE Annual Technical Conference and Exhibition, San Antonio, TX (24–27 September 2006).

217 Collins, B.C., Mestesky, P.A. and Saviano, N.J. (1995). Method of removing sulfur compounds from sour crude oil and sour natural gas. US Patent 5,807,476, assigned to United Laboratories Inc.

218 Klug, P. and Kelland, M.A. (1998). Additives for inhibiting formation of gas hydrates. WO 1998023843, assigned to Clariant GmbH and Rogaland Research.

219 Kelland, M.A. (2013). Method of inhibiting the formation of gas hydrates using amine oxides. International Patent Application WO2013053770.

220 Garcia, M.T., Campos, E., and Ribosa, I. (2007). Biodegradability and ecotoxity of amine oxide based surfactants. *Chemosphere* **69** (10): 1574–1578.

221 Fischer, E. and Beensch, L. (1894). Ueber einige synthetische glucoside. *Berichte der Deutschen Chemischen Gesellschaft* **27** (2): 2478–2486.

222 Heinrich, B. and Gertrud, R. (1936). Process for the production of glucosides of higher aliphatic alcohols. US Patent 2,049,758, assigned to Firm H Th Boehme Ag.

223 Koeltzow, D.E. and Urfer, A.D. (1984). Preparation and properties of pure alkyl glucosides, maltosides and maltotriosides. *Journal of the American Oil Chemists' Society* **61**: 1651–1655.

224 Balzer, D. (1993). Cloud point phenomena in the phase behavior of alkyl polyglucosides in water. *Langmuir* **99** (12): 3375–3384.

225 Lange, H. and Schwuger, M.J. (1968). Colloids and macromolecules. *Kolloid Zeitschrift & Zeitschrift fur Polymere* **223**: 145.

226 Christian, S.D. and Scamehorn, J.F. ed. (1995). *Solubilization in Surfactant Aggregates*, Surfactant Science Series, vol. **55**. CRC Press.

227 Balzer, D. (1991). Alkylpolyglucosides, their physical properties and their uses. *Tenside, Surfactants, Detergents* **28**: 419–427.

228 Balzer, D. and Luders, H. ed. (2000). *Nonionic Surfactants: Alkyl Polyglucosides*, Surfactant Science Series, vol. **91**. New York: Marcel Dekker.

229 Kahlweit, M., Busse, G., and Faulhaber, B. (1995). Preparing microemulsions with alkyl monoglucosides and the role of n-alkanols. *Langmuir* **11** (9): 3382–3387.

230 Craddock, H.A., Caird, S., Wilkinson, H. and Guzzmann, M. (2006). A new class of 'green' corrosion inhibitors, development and application, SPE 104241. SPE International Oilfield Corrosion Symposium, Aberdeen, UK (30 May 2006).

231 Craddock, H.A., Berry, P. and Wilkinson, H. (2007). A new class of "green" corrosion inhibitors, further development and application. Transactions of the 18th International Oilfield Chemical Symposium, Geillo, Norway (25–28 March 2007).

232 Craddock, H.A., Simcox, P., Williams, G., and Lamb, J. (2011). Backward and forward in corrosion inhibitors in the North Sea paper III in an occasional series on the use of alkyl polyglucosides as corrosion inhibitors in the oil and gas industry. In: *Chemistry in the Oil Industry XII*, 184–196. Royal Society of Chemistry.

233 Iglauer, S., Wu, Y., Schuler, P. et al. (2010). New surfactant classes for enhanced oil recovery and their tertiary oil recovery potential. *Journal of Petroleum Science and Engineering* **71**: 23–29.

234 Walker, T.O. (1992). Environmentally safe drilling fluid. CA patent 2,152,483, assigned to Newpark Drilling Fluids Llc.

235 von Rybinski, W. and Hill, K. (1998). Alkyl polyglucosides-properties and applications of a new class of surfactants. *Angewandte Chemie International Edition* **37** (10): 1328–1345.

236 Berg, E., Sedberg, S., Kararigstad, H., et al. (2006). Displacement of drilling fluids and cased-hole cleaning: What is sufficient cleaning? SPE 99104. SPE/IADC Drilling Conference, Miami, FL (21–23 February 2006).

237 Knox, D. and McCosh, K. (2005). Displacement chemicals and environmental compliance-past present and future. Chemistry in the Oil Industry IX, Manchester, UK (31 October to 2 November 2005).

238 Balzer, D. (1993). Cloud point phenomena in the phase behaviour of alkyl polyglucosides in water. *Langmuir* **9**: 3375–3384.

239 Taylor, G.N. (1997). Demulsifier for water-in-oil emulsions, and method of use. US Patent 5,609,794, assigned to Exxon Chemical Patents Inc.

240 Merchant, P. Jr. and Lacy, S.M. (1988). Water based demulsifier formulation and process for its use in dewatering and desalting crude hydrocarbon oils. US Patent 4,737,265, assigned to Exxon research and Engineering Co.

241 Alexandridi, P. (1997). Poly(ethylene oxide)/poly(propylene oxide) block copolymer surfactants. *Current Opinion in Colloid & Interface Science* **2** (5): 478–489.

242 Hurter, P.N. and Hatton, T.A. (1992). Solubilization of polycyclic aromatic hydrocarbons. *Langmuir* **8** (5): 1291–1299.

243 Inada, M., Kabuki, K., Imajo, Y., et al. (1999). Cleaning compositions. US Patent 5,985,810, assigned to Toshiba Silicone Co. Ltd.

244 Zerbe, H.G., Guo, J.-H. and Serino, A. (2001). Water soluble film for oral administration with instant wettability. US Patent 6,177,096, assigned to Lts Lohmann Therapie-System Gmbh.

245 Sharvelle, S.R., Lattyak, R., and Banks, M.K. (2007). Evaluation of biodegradability and biodegradation kinetics for anionic, nonionic, and amphoteric surfactants. *Water, Air, and Soil Pollution* **183** (1–4).

246 Sharvelle, S.E., Garland, J., and Banks, M.K. (2008). Biodegradation of polyalcohol ethoxylate by a wastewater microbial consortium. *Biodegradation* **19** (2): 215–221.

247 Fischer, E.R., Boyd, P.G. and Alford, J.A. (1994). Acid-anhydride esters as oilfield corrosion inhibitors. GB Patent 2,268,487, assigned to Westvaco US.

248 Putzig, D.E. (1994). Zirconium chelates and their use in cross-linking. EP Patent 0278684, assigned to E.I. du Pont de Nemours and Company.

249 Baur, R., Barthold, K., Fischer, R., et al. (1993). Alkoxylated polyamimes containing amide groups and their us in breaking oil-in-water and water-in-oil emulsions. EP Patent 0264755, assigned to BASF Ag.

250 Rodzewich, E.A. (1997). Method of cleaning and passivating a metal surface with acidic system and ethoxylated tertiary dodecyl mercaptan. US Patent 5,614,028, assigned to Betzdearborn Inc.

251 Gregg, M. and Ramachandran, S. (2004). Review of corrosion inhibitor developments and testing for offshore oil and gas production systems, NACE 04422. Corrosion 2004, New Orleans, LA (28 March to 1 April 2004).

252 Garcia, M.T., Ribosa, I., Guindulain, T. et al. (2001). Fate and effect of monoalkyl quaternary ammonium surfactants in the aquatic environment. *Environmental Pollution* **111** (1): 169–175.

253 Foley, P., Kermanshahi, A., Beach, E.S., and Zimmerman, J.B. (2012). Critical review: derivation and synthesis of renewable surfactants. *Chemical Society Reviews* **41**: 1499–1518.

254 Barrault, J. and Pouilloux, Y. (1997). Synthesis of fatty amines. Selectivity control in presence of multifunctional catalysts. *Catalysis Today* **37** (2): 137–153.

255 Huang, T. and Clark, D.E. (2012). Advanced fluid technologies for tight gas reservoir stimulation, SPE 106844. SPE Saudi Arabia Section Technical Symposium and Exhibition, Al-Khobar, Saudi Arabia (8–11 April 2012).

256 Degre, G., Morvan, M., Beaumont, J., et al. (2012). Viscosifying surfactant technology for chemical EOR: a reservoir case, SPE 154675. SPE EOR Conference at Oil and Gas West Asia, Muscat, Oman (16–18 April 2012).

257 Kippie, D.P. and Gatlin, L.W. (2009). Shale inhibition additive for oil/gas down hole fluids and methods for making and using same. US Patent 7,566,686, assigned to Clearwater International Llc.

258 Tsytsymushkin, P.F., Kharjrullin, S.R., Tarnavskij, A.P., et al. (1991). Grouting mortars for fixing wells of salt deposits. SU Patent 170020, assigned to Volga Urals Hydrocarbon.

259 Shah, S.S., Fahey, W.F. and Oude Alink, B.A. (1991). Corrosion inhibition in highly acidic environments by use of pyridine salts in combination with certain cationic surfactants. US Patent 5,336,441, assigned to Petrolite Corporation.

260 Becker, L.C., Bergfeld, W.F., Belisto, D.V. et al. (2012). Safety assessment of trimoniums as used in cosmetics. *International Journal of Toxicology* **31**: 296S–341S.

261 Garcia, M.T., Campos, E., Sanchez-Leal, J., and Ribosa, I. (1999). Effect of the alkyl chain length on the anaerobic biodegradability and toxicity of quaternary ammonium based surfactants. *Chemosphere* **38** (15): 3473–3483.

262 Yan, H., Li, Q., Geng, T., and Jiang, Y. (2012). Properties of the quaternary ammonium salts with novel counter-ions. *Journal of Surfactants and Detergents* **15** (5): 593–599.

263 Martin, R.L., Brock, G.F. and Dobbs, J.B. (2005). Corrosion inhibitors and methods of use. US Patent 6,866,797 assigned to BJ Services Company.

264 Naraghi, A. and Obeyesekere, N. (2006). Polymeric quaternary ammonium salts useful as corrosion inhibitors and biocides. US Patent Application 20060062753.

265 Klomp, U.C., Kruka, V.R., Reijnhart, R. and Weisenborn, A.J. (1995). A method for inhibiting the plugging of conduits by gas hydrates. World Patent Application WO 199501757.

266 Buijs, A., Van Gurp, G., Nauta, T., et al. (2002). Process for preparing esterquats. US Patent 6,379,294, assigned to Akzo Nobel, NV.

267 Pakulski, M.K. (2000). Quaternized polyether amines as gas hydrate inhibitors. US Patent 6,025,302, assigned to BJ Services Company.

268 Jones, C.R., Talbot, R.E., Downward, B.L., Fidoe, S.D., et al. (2006). Keeping pace with the need for advanced, high performance biocide formulations of oil production applications, NACE 06663. Corrosion 2006, San Diego, CA (12–16 March).

269 Mishra, S. and Tyagi, V.K. (2007). Esterquats: the novel class of cationic fabric softeners. *Journal of Oleo Science* **56** (6): 269–276.

270 Darwin, A., Annadorai, K. and Heidersbach, K. (2010). Prevention of corrosion in carbon steel pipelines containing hydrotest water – an overview, NACE 10401. Corrosion 2010, San Antonio, TX (14–18 March).

271 Ying, G.-G. (2005). Fate, behavior and effects of surfactants and their degradation products in the environment. *Environment International* **32** (2006): 417–431.

272 Barnes, J.R., Smit, J., Smit, J., et al. (2008). Development of surfactants for chemical flooding at difficult reservoir conditions, SPE 113313. SPE Symposium on Improved Oil Recovery, Tulsa, OK (20–23 April 2008).

273 Rosen, M.J. (1972). The relationship of structure to properties in surfactants. *Journal of the American Oil Chemists' Society* **49** (5): 293–297.

274 Leo, A., Hansch, C., and Elkins, D. (1971). Partition coefficients and their uses. *Chemical Reviews* **71** (6): 525–616.

275 Lewis, M.A. (1991). Chronic and sub-lethal toxicities of surfactants to aquatic animals: a review and risk assessment. *Water Research* **25** (1): 101–113.

276 Yamane, M., Toyo, T., Inoue, K. et al. (2008). Aquatic toxicity and biodegradability of advanced cationic surfactant APA-22 compatible with the aquatic environment. *Journal of Oleo Science* **57** (10): 529–538.

277 Allen, T.L., Amin, J., Olson, A.K. and Pierce, R.G. (2008). Fracturing fluid containing amphoteric glycinate surfactant. US Patent 7,399,732, assigned to Calfrac Well Services ltd. and Chemergy Ltd.

278 Stournas, S. (1984). A novel class of surfactants with extreme brine resistance and its potential application in enhanced oil recovery, SPE 13029. SPE Annual Technical Conference and Exhibition, Houston, TX (16–19 September 1984).

279 Sultan, A.S., Balbuena, P.B., Hill, A.D. and Nasr-El-Din, H.A. (2009). Theoretical study on the properties of cationic, amidoamine oxide and betaine viscoelastic diverting surfactants in gas phase and water solution, SPE 121727. SPE International Symposium on Oilfield Chemistry, The Woodlands, TX (20–22 April 2009).

280 Hill, K.L., Sayed, M. and Al-Muntasheri, G.A. (2015). Recent advances in viscoelastic surfactants for improved production from hydrocarbon reservoirs, SPE 173776. SPE International Symposium on Oilfield Chemistry, The Woodlands, TX (13–15 April 2015).

281 Zhang, F., Ma, D., Wang, Q., et al. (2013). A novel hydroxylpropyl sulfobetaine surfactant for high temperature and high salinity reservoirs, IPTC 17022. International Petroleum Technology Conference, Beijing, China (26–28 March 2013).

282 Gough, M. and Bartos, M. (1999). Developments in high performance, environmentally friendly corrosion inhibitors for the oilfield, NACE 99104. Corrosion 99, San Antonio, TX (25–30 April 1999).

283 N. Hajjaji, I. Rico, A. Srhiri, A. Lattes, M. Soufiaoui and A. Ben Bachir, Effect of *N*-alkylbetaines on the corrosion of iron in 1 M HCl solution, NACE-93040326, *Corrosion*, Vol. **49** (4), 1993.

284 Chalmers, A., Winning, I.G., McNeil, S. and McNaughton, D. (2006). Laboratory development of a corrosion inhibitor for a North Sea main oil line offering enhanced environmental properties and weld corrosion protection, NACE 06487. Corrosion 2006, San Diego, CA (12–16 March 2006).

285 Clewlow, P.J., Haslegrave, J.A., Carruthers, N., et al. (1995). Amine adducts as corrosion inhibitors. US Patent 5,427,999, assigned to Exxon Chemical Patents Inc.

286 Infante, M.R., Perez, L., Pinazo, A. et al. (2004). Amino acid-based surfactants. *Comptes Rendus Chimie* 7 (6–7): 583–592.

287 Ren, H., Shi, C., Song, S. et al. (2016). Synthesis of diacyl amino acid surfactant and evaluation of its potential for surfactant–polymer flooding. *Applied Petrochemical Research* 6 (1): 59–63.

288 Katz, J. (1966). Imidazoline surfactant having amphoteric properties. US Patent 3,555,041.

289 Edwards, A., Osborne, C., Klenerman, D. et al. (1994). Mechanistic studies of the corrosion inhibitor oleic imidazoline. *Corrosion Science* 36 (2): 315–325.

290 Turcio-Ortega, D., Paniyan, T., Cruz, J., and Garcia-Ochoa, E. (2007). Interaction of imidazoline compounds with Fe_n ($n = 1$–4 atoms) as a model for corrosion inhibition: DFT and electrochemical studies. *Journal of Physical Chemistry C* **111**: 9853.

291 Duda, Y., Govea-Rueda, R., Galicia, M. et al. (2005). Corrosion inhibitors: design, performance and simulations. *Journal of Physical Chemistry B* **109** (47): 22674.

292 Meyer, G.R. (2003). Corrosion inhibitor compositions including quaternized compounds having a substituted diethylamino moiety. US Patent 6,599,445, assigned to Ondeo Nalco Energy Services.

293 Durnie, W.H. and Gough, M.A. (2003). Characterisation, isolation and performance characteristics of imidazolines: part II development of structure-activity relationships, Paper 03336, NACE. Corrosion 2003, San Diego, CA (March 2003).

294 Chen, H.J. and Jepson, W.P. (2000). High temperature corrosion inhibition of imidazoline and amide, Paper 00035, NACE. Corrosion 2000, Orlando, FL (26–31 March 2000).

295 Liu, X., Chen, S., Ma, H. et al. (2006). Protection of iron corrosion by stearic acid and stearic imidazoline self-assembled monolayers. *Applied Surface Science* **253** (2): 814–820.

296 Lunan, A.O., Anaas, P.-H.V. and Lahteenmaki, M.V. (1996). Stable CMC slurry. US Patent 5,487,777, assigned to Metsa-Serle Chemicals Oy.

297 Iscan, A.G. and Kok, M.V. (2007). Effects of polymers and CMC concentration on rheological and fluid loss parameters of water-based drilling fluids. *Energy Sources, Part A: Recovery, Utilization, and Environmental Effects* **29** (10): 939–949.

298 Palumbo, S., Giacca, D., Ferrari, M. and Pirovano, P. (1989). The development of potassium cellulosic polymers and their contribution to the inhibition of hydratable clays, SPE 18477. SPE International Symposium on Oilfield Chemistry, Houston, TX (8–10 February 1989).

299 Mumallah, N.A. (1989). Altering subterranean formation permeability. US Patent 4,917,186, assigned to Phillips Petroleum Company.

300 Welton, T.D., Todd, B.L. and McMechan, D. (2010). Methods for effecting controlled break in pH dependent foamed fracturing fluid. US Patent 7,662,756, assigned to Halliburton Energy Services Inc.

301 Decampo, F., Kesavan, S. and Woodward, G. (2010). Polysaccharide based scale inhibitor. EP Patent 2148908, assigned to Rhodia Inc.

302 Llave, F.M., Gall, B.L. and Noll, L.A. (1990). Mixed surfactant systems for enhanced oil recovery. National Institute of Petroleum and Energy Research, USA. *Technical report NIPER-497*.

303 Yunus, M.N.M., Procyk, A.D., Malbrel, C.A. and Ling, K.L.C. (1995). Environmental impact of a flocculant used to enhance solids transport during well bore clean-up operations, SPE 29737. SPE/EPA Exploration and Production Environmental Conference, Houston, TX (27–29 March 1995).

304 Shih, I.-L. and Van, Y.-T. (2001). The production of poly-(γ-glutamic acid) from microorganisms and its various applications. *Bioresource Technology* **79** (3): 207–225.

305 Song, H. (2011). Preparation of novel amphoteric flocculant and its application in oilfield water treatment, SPE 140965. SPE International Symposium on Oilfield Chemistry, The Woodlands, TX (11–13 April 2011).

306 Murphy, P.M. and Hewat, T. (2008). Fluorosurfactants in enhanced oil recovery. *The Open Petroleum Engineering Journal* **1**: 58–61.

307 Calafat, A.M., Wong, L.Y., Kuklenyik, Z. et al. (2007, 2007). Polyfluoroalkyl chemicals in the U.S. population: data from the National Health and Nutrition Examination Survey (NHANES) 2003–2004 and comparisons with NHANES 1999–2000. *Environmental Health Perspectives* **115** (11): 1596–1602.

308 Scheringer, M., Trier, X., Cousins, I.T. et al. (2014). Helsingør statement on poly- and perfluorinated alkyl substances (PFASs). *Chemosphere* **114**: 337–339.

309 Engel, Goliaszewski, A.E. and McDanield, C.R. (2010). Separatory and emulsion breaking processes. US Patent 7,771,588, assigned to General Electric Company.

310 Menger, F.M. and Littau, C.A. (1991). Gemini surfactants: synthesis and properties. *Journal of the American Chemical Society* **113** (4): 1451–1452.

311 Hait, S.K. and Moulik, S.P. (2002). Gemini surfactants: a distinct class of self-assembling molecules. *Current Science* **82** (9): 1101–1111.

312 Sharma, V., Borse, M., Jauhari, S. et al. (2005). New hydroxylated cationic gemini surfactants as effective corrosion inhibitors for mild steel in hydrochloric acid medium. *Tenside, Surfactants, Detergents* **42** (3): 163–167.

313 Kamal, M.S. (2015). A review of gemini surfactants: potential application in enhanced oil recovery. *Journal of Surfactants and Detergents* **19** (2).

314 Yang, J., Liu, X., Jia, S., et al. (2015). A highly effective corrosion inhibitor based on gemini imidazoline, SPE 173777. SPE International Symposium on Oilfield Chemistry, The Woodlands, TX (13–15 April 2015).

315 Desai, J.D. and Banat, I.M. (1997). Microbial production of surfactants and their commercial potential. *Microbiology and Molecular Biology Reviews* **61** (4): 47–64.

316 Karanth, N.G.K., Deo, P.G., and Veenanadig, N.K. (1999). Microbial production of biosurfactants and their importance. *Current Science* **77** (1): 116–126.

317 Mulligan, C.N. (2005). Environmental applications for biosurfactants. *Environmental Pollution* **133** (2): 183–198.

318 Santos, D.K.F., Rufino, R.D., Luna, J.M. et al. (2013). Synthesis and evaluation of biosurfactant produced by *Candida lipolytica* using animal fat and corn steep liquor. *Journal of Petroleum Science and Engineering* **105**: 43–50.

319 Fang, X., Wang, O., Bai, B., et al. (2007), Engineering rhamnolipid biosurfactants as agents for microbial enhanced oil recovery, SPE 106048. International Symposium on Oilfield Chemistry, Houston, TX (28 February to 2 March 2007).

320 Nwinee, S., Yates, K., Lin, P.K.T. and Cowie, E. (2010). A sustainable alternative for the treatment of oil based mud (OBM) drill cuttings. Chemistry in the Oil Industry XIV, Manchester, UK (2–4 November 2010).

321 O.P. Ward, " Microbial biosurfactants and biodegradation": In *Biosurfactants* Volume **672** of Advances in Experimental Medicine and Biology pp 65-74, Springer, 2010.

322 Bustamante, M., Durán, N., and Diez, M.C. (2012). Biosurfactants are useful tools for the bioremediation of contaminated soil: a review. *Journal of Soil Science and Plant Nutrition* **12** (4): 667–687.

323 de Cassia, R., Silva, F.S., Almeida, D.G. et al. (2014). Applications of biosurfactants in the petroleum industry and the remediation of oil spills. *International Journal of Molecular Sciences* **15** (7): 12523–12542.

324 Poremba, K., Gunkel, W., Lang, S., and Wagner, F. (1991). Marine biosurfactants, III. Toxicity testing with marine microorganisms and comparison with synthetic surfactants. *Zeitschrift für Naturforschung* **46c**: 210–216.

325 Flasz, A., Rocha, C.A., Mosquera, B., and Sajo, C. (1998). A comparative study of the toxicity of a synthetic surfactant and one produced by *Pseudomonas aeruginosa*. *Medical Science Research* **26** (3): 181–185.

326 Mulligan, C.N., Sharma, S.K., and Mudhoo, A. ed. (2014). *Biosurfactants: Research Trends and Applications*. CRC Press.

327 Zajic, J.E., Cooper, D.G., Jack, T.R., and Kosaric, N. ed. (1983). *Microbial Enhanced Oil recovery*. PennWell Books.

328 Onuoha, S.O. and Olafuyi, O.A. (2013). Alkali/surfactant/polymer flooding using gum arabic; a comparative analysis, SPE 167572. SPE Nigeria Annual International Conference and Exhibition, Lagos, Nigeria (5–7 August 2013).

329 Atsenuwa, J., Taiwo, O., Dala, J., et al. (2014). Effect of viscosity of heavy oil (class-A) on oil recovery in SP flooding using lauryl sulphate and gum arabic, SPE 172401. SPE Nigeria Annual International Conference and Exhibition, Lagos, Nigeria (5–7 August 2014).

330 Punase, A.D., Bihani, A.D., Patane, A.M., et al. (2011). Soybean slurry – a new effective, economical and environmental friendly solution for oil companies, SPE 142658. SPE Project and Facilities Challenges Conference at METS, Doha, Qatar (13–16 February 2011).

331 Sjoblom, J., Ovrevoll, B., Jentoft, G. et al. (2010). Investigation of the hydrate plugging and non-plugging properties of oils. *Journal of Dispersion Science and Technology* **31** (8): 1100–1119.

332 Camargo, R.M.T., Goncalves, M.A.L., Montesanti, J.R.T., et al. (2004). A perspective view of flow assurance in deepwater fields in Brazil, OTC 16687. Offshore Technology Conference, Houston, TX (3–6 May 2004).

333 Dexter, A.F. and Middelberg, A.P.J. (2008). Peptides as functional surfactants. *Industrial and Engineering Chemistry Research* **47** (17): 6391–6398.

334 Zhang, J.H., Zhao, Y.R., Han, S.Y. et al. (2014). Self-assembly of surfactant-like peptides and their applications. *Reviews Special Topic Biophysical Chemistry: Science China Chemistry* **57** (12): 1634–1645.

335 http://corporate.evonik.com/en/media/press_releases/pages/news-details.aspx (accessed 6 December 2017).

336 Hirasaki, G.J., Miller, C.A. and Puerto, M. (2008). Recent advances in surfactant EOR, SPE 115386. SPE Annual Technical Conference and Exhibition, Denver, CO (21–24 September 2008).

337 Lu, J., Liyanage, P.J., Solairaj, S., et al. (2013). Recent technology developments in surfactants and polymers for enhanced oil recovery, IPTC 16425. International Petroleum Technology Conference, Beijing, China (26–28 March 2013).

338 http://www.evostc.state.ak.us/%3FFA=facts.QA (accessed 6 December 2017).

339 US Environmental Protection Agency (2016). National Contingency Plan: product schedule.

340 Marine Board National research Council (1989). *Using Oil Spill Dispersants on the Sea.* Washington, DC: National Academic Press.

341 Fingas, M. (2002). A review of literature related to oil spill dispersants especially relevant to Alaska; for Prince William Sound Regional Citizens' Advisory Council (PWSRCAC) Anchorage, Alaska. Environmental Technology Centre, Environment Canada.

342 Venosa, A.D. *Laboratory-scale testing of dispersant effectiveness of 20 oils using the Baffled flask test.* U.S. Environmental Protection Agency, National Risk Management Research Laboratory, Cincinnati, OH. http://oilspilltaskforce.org/wp-content/uploads/2015/08/Venosa-and-Holder-baffled-flask.pdf (accessed 5 January 2018).

343 Kirby, M.F., Matthiessen, P. and Rycroft, R.J. (1996). Procedures for the Approval of Oil Spill Treatment Products. *Fisheries Research Tech. Rep. Number 102.* Directorate of Fisheries Research, UK.

344 Daling, P.S. and Indrebo, G. (1996). Recent improvements in optimizing use of dispersants as a cost-effective oil spill countermeasure technique, SPE 36072. SPE Health, Safety and Environment in Oil and Gas Exploration and Production Conference, New Orleans, Louisiana (9–12 June 1996).

345 Mori, A.L. and Schaleger, L.L. (1972). Kinetics and mechanism of epoxy ether hydrolysis. II. Mechanism of ring cleavage. *Journal of the American Chemical Society* **94** (14): 5039–5043.

346 Papir, Y.S., Schroeder, A.H. and Stone, P.J. (1989). New downhole filming amine corrosion inhibitor for sweet and sour production, SPE 18489. SPE International Symposium on Oilfield Chemistry, Houston, TX (8–10 February 1989).

347 Buchwald, P. and Bodor, N. (1999). Quantitative structure-metabolism relationships: steric and nonsteric effects in the enzymatic hydrolysis of non-congener carboxylic esters. *Journal of Medicinal Chemistry* **42** (25): 5160–5168.

348 Zell, Z.A., Newbahar, A., Mansard, V. et al. (2014). Surface shear inviscidity of soluble surfactants. *Proceedings of the National Academy of Sciences USA* **111** (10): 3677–3682.

349 Evetts, S., Kovalski, C., Levin, M., and Stafford, M. (1995). High-temperature stability of alcohol ethoxylates. *Journal of the American Oil Chemists' Society* **72** (7): 811–816.

350 Mitsuda, K., Kimura, H., and Murahashi, T. (1989). Evaporation and decomposition of TRITON X-100 under various gases and temperatures. *Journal of Materials Science* **24** (2): 413–419.

351 Al-Khafaji, A.A., Castanier, L.M. and Brigham, W.E. (1983). Effect of temperature on degradation, adsorption and phase partioning of surfactants used in steam injection for oil recovery. US Department of Energy, Contract No. DE-AC03-81SF-11564.

352 Shupe, R.D. and Baugh, T.D. (1991). Thermal stability and degradation mechanism of alkylbenzene sulfonates in alkaline media. *Journal of Colloid and Interface Science* **145** (1): 235–254.

353 Harrison, S.K. and Thomas, S.M. (1990). Interaction of surfactants and reaction media on photolysis of chlorimuron and metsulfuron. *Weed Science* **38** (6): 620–624.

354 Tanaka, S. and Ichikawa, T. Effects of photolytic pretreatment on biodegradation and detoxification of surfactants in anaerobic digestion. *Water Science and Technology* **28** (7): 103–110.

355 White, G.F. and Russell, N.J. (1994). Biodegradation of anionic surfactants. In: *Biochemistry and Microbial Degradation* (ed. C. Ratlegde), 143–177. Kluwer Academic Publishers.

356 Qin, Y., Zhang, G., Kang, B., and Zhao, Y. (2005). Primary aerobic biodegradation of cationic and amphoteric surfactants. *Journal of Surfactants and Detergents* **8** (1): 55–58.

357 Larson, R.J. (1983). Comparison of biodegradation rates in laboratory screening studies with rates in natural water. *Residue Reviews* **85**: 159–161.

358 Dean-Raymond, D. and Alexander, M. (1977). Bacterial metabolism of quaternary ammonium compounds. *Applied and Environmental Microbiology* **33** (5): 1037–1041.

359 Swisher, R.D. (1970). *Surfactant Biodegradation*. New York: Marcel Dekker.

360 Jurado, E., Fernandez-Serrano, M., Nunez-Olea, J. et al. (2009). Acute toxicity and relationship between metabolites and ecotoxicity during the biodegradation process of non-ionic surfactants: fatty-alcohol ethoxylates, nonylphenol polyethoxylate and alkylpolyglucosides. *Water Science and Technology* **59** (12): 2351–2358.

361 Scott, M.J. and Jones, M.N. (2000). The biodegradation of surfactants in the environment. *Biochimica et Biophsica Acta (BBA) – Biomembranes* **1508** (1-2): 235–251.

362 Geerts, R., Kuijer, P., van Ginkel, C.G., and Plugge, C.M. (2014). Microorganisms hydrolyse amide bonds; knowledge enabling read-across of biodegradability of fatty acid amides. *Biodegradation* **25** (4): 605–614.

363 Tadros, T.F. (2005). *Applied Surfactants: Principles and Applications*. Weinheim: Wiley-VCH.

364 Zhang, L.M., Tan, Y.B., and Li, Z.M. (2001). New water-soluble ampholytic polysaccharides for oilfield drilling treatment: a preliminary study. *Carbohydrate Polymers* **44** (3): 255–260.

365 Stjerndahl, M., van Ginkel, C.G., and Holmberg, K. (2003). Hydrolysis and biodegradation studies of surface-active ester. *Journal of Surfactants and Detergents* **6** (4): 319–324.

366 O'Lenick, T. and O'Lenick, K. (2007). Effect of branching on surfactant properties of sulfosuccinates. *Cosmetics & Toiletries* **22** (11): 81.

367 Lunberg, D., Stjerdahl, M., and Homberg, K. (2008). Surfactants containing hydrolyzable bonds. *Advances in Polymer Science* **218** (1): 57–82.

368 Tehrani-Bagha, A.R., Oskarsson, H., van Ginkel, C.G., and Holmberg, K. (2007). Cationic ester-containing gemini surfactants: chemical hydrolysis and biodegradation. *Journal of Colloid and Interface Science* **312** (2): 444–452.

369 Stjerndahl, M. and Holmberg, K. (2005). Hydrolyzable nonionic surfactants: stability and physicochemical properties of surfactants containing carbonate, ester, and amide bonds. *Journal of Colloid and Interface Science* **291** (2): 570–576.

370 Hellberg, P.-E., Bergstrom, K., and Holmberg, K. (2000). Cleavable surfactants. *Journal of Surfactants and Detergents* **3** (1): 81–91.

371 Hellberg, P.-E., Bergstrom, K., and Juberg, M. (2000). Nonionic cleavable ortho ester surfactants. *Journal of Surfactants and Detergents* **3** (3): 369–379.

372 Vincent, J.B., Crowder, M.W., and Averill, B.A. (1992). Hydrolysis of phosphate monoesters: a biological problem with multiple chemical solutions. *Trends in Biochemical Sciences* **17** (3): 105–110.

373 Liu, C.K. and Warr, G.G. (2015). Self-assembly of didodecyldimethylammonium surfactants modulated by multivalent, hydrolyzable counterions. *Langmuir* **31** (10): 2936–2945.

374 Zhang, L.M., Tan, Y.B., Huang, S.J., and Li, Z.M. (2000). Water-soluble ampholytic grafted polysaccharides. 1. Grafting of the zwitterionic monomer 2-(2-methacryloethyldimethylammonio) ethanoate onto hydroxyethyl cellulose. *Journal of Macromolecular Science, Part A* **37** (10): 1247–1260.

375 Berkof, R., Kwekkeboom, H., Balzer, D. and Ripke, N. (1992). Use of oxyalkylated polyglycosides as demulsifying agents for breaking mineral oil emulsions. EP Patent 468095, assigned to Huls AG.

376 Noweck, K. (2011). Production, technologies and applications of fatty alcohols. Lecture at the 4th Workshop on Fats and Oils as Renewable Feedstock for the Chemical Industry, Karlsruhe, Germany (20–22 March 2011).

377 Mudge, S.M. (2005). Fatty alcohols – a review of their natural synthesis and environmental distribution for SDA and ERASM. Soap and Detergent Association.

378 DeBruijn, J., Busser, F., Seinen, W., and Hermens, J. (1989). Determination of octanol/water partioning coefficients for hydrophobic organic chemicals with the "slow-stirring" method. *Environmental Toxicology and Chemistry* **8** (6): 499–512.

379 Smith, D.J., Gingerich, W.H., and Beconi-Barker, M.G. ed. (1999). *Xenobiotics in Fish*. Springer.

380 Fraser, G.C. (2009). Method for determining the bioconcentration factor of linear alcohol ethoxylates, SPE 123846, Offshore Europe, Aberdeen, UK (8–11 September 2009).

381 Tolls, J., Kloepper-Sams, P., and Sijm, D.T.H.M. (1994). Surfactant bioconcentration – a critical review. *Chemosphere* **29** (4): 693–717.

382 Staples, C.A., Weeks, J., Hall, J.F., and Naylor, C.G. (1998). Evaluation of aquatic toxicity and bioaccumulation of C8- and C9-alkylphenol ethoxylates. *Environmental Toxicology and Chemistry* **17** (12): 2470–2480.

383 Tolls, J., Haller, M., Labee, E. et al. (2000). Experimental determination of bioconcentration of the nonionic surfactant alcohol ethoxylate. *Environmental Toxicology and Chemistry* **19** (3): 646–653.

384 Stagg, R.M., Rankin, J.C., and Bolis, L. (1981). Effect of detergent on vascular responses to noradrenaline in isolated perfused gills of the eel, *Anguilla anguilla* L. *Environmental Pollution* **24**: 31–37.

385 Lewis, M.A. (1991). Chronic and sub-lethal toxicities of surfactants to aquatic animals: a review and risk assessment. *Water Research* **25** (1): 101–113.

386 The Environment Agency (2007). *Environmental risk evaluation report: para-C12-alkylphenols(dodecylphenol and tetrapropenylphenol)*. UK: The Environment Agency.

387 Jaques, P., Martin, I., Newbigging, C., and Wardell, T. (2002). Alkylphenol based demuslifier resins and their continued use in the offshore oil and gas industry. In: *Chemistry in the Oil Industry VII*, The Royal Society of Chemistry (ed. T. Balson, H. Craddock, H. Frampton, et al.), 56–66.

388 Helle, S.S., Duff, S.J.B., and Cooper, D.G. (1993). Effect of surfactants on cellulose hydrolysis. *Biotechnology and Bioengineering* **42**: 611–617.

389 Schmitt, T.M. (2001). *The Analysis of Surfactants*, 2nde. New York: Marcel Dekker.

390 Sitz, C., Frenier, W.W. and Vallejo, C.M. (2012). Acid corrosion inhibitors with improved environmental profiles, SPE 155966. SPE International Conference and Workshop on Oilfield Corrosion, Aberdeen, UK (28–29 May 2012).

4

Phosphorus Chemistry

The chemistry of phosphorus is complicated and its behaviour can be varied due to its capability to form a variety of complex compounds. This is because of the unusual valence states for a first row element of the periodic table. Phosphorus forms covalent bonds and much of its behaviour can be explained in terms of first row non-metallic covalent bonding; however, there are aspects of its chemistry that cannot be explained with analogy to first row elements [1]. In particular phosphorus can form pentacoordinated compounds such as phosphorus pentachloride (PCl_5) (see Figure 4.1). In non-metals this behaviour is exceedingly rare, and especially in such variety, as displayed by phosphorus.

The bonding of phosphorus compounds is highly interesting and complex. It is, however, outside the scope of this study to examine a detailed description of the bonding characteristics and behaviour. In terms of oilfield application, the key behaviour is that the double valency (3 and 5) leads to phosphorous acid that forms phosphite salts and phosphoric acid that in turn forms phosphates. Phosphorous acid is in an equilibrium and displays two tautomers as shown in Figure 4.2, the major dihydroxy tautomer being phosphonic acid that forms phosphonate salts. Only the reduced phosphorus compounds are spelled with an 'ous' ending.

Additionally phosphorus displays tetracoordination behaviour as exemplified by phosphoryl chloride. Like phosphate, phosphoryl chloride is tetrahedral in shape. It features three P—Cl bonds and one strong P═O double bond (see Figure 4.3).

The most important use of phosphorous acid (phosphonic acid) is the production of phosphates (and phosphonates) that are used in water treatment including produced water from oil and gas production. Phosphorous acid is also used for preparing phosphite salts, such as potassium phosphite. These salts, as well as aqueous solutions of pure phosphorous acid, are fungicides. Phosphites have also shown effectiveness in controlling a variety of plant diseases, and antimicrobial products (biocides) containing salts and complexes of phosphorous acid are marketed in a variety of industrial sectors including the upstream oil and gas sector; however, simple phosphite salts are rarely used. Phosphorous acid and its salts, unlike phosphoric acid, are somewhat toxic and should be handled carefully.

The main area of interest in this chapter will be the salts described earlier, namely, phosphates, phosphonates and related salts and complexes. In addition organophosphorus and phosphorous amides and amines will be described. It is of note that the phosphate ester surfactants have already been examined in Section 3.2.9, and the reader is directed to this section for a full description of their chemistry and application.

The environmental effects and impacts of phosphorus and its products as used in the upstream oil and gas industry are somewhat different from the other products described throughout this study and will be covered in Section 4.8.

Oilfield Chemistry and Its Environmental Impact, First Edition. Henry A. Craddock.
© 2018 John Wiley & Sons Ltd. Published 2018 by John Wiley & Sons Ltd.

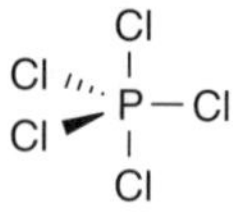

Figure 4.1 Phosphorus pentachloride.

Figure 4.2 Phosphonic and phosphorous acid.

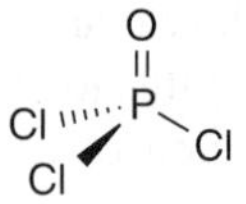

Figure 4.3 Phosphoryl chloride.

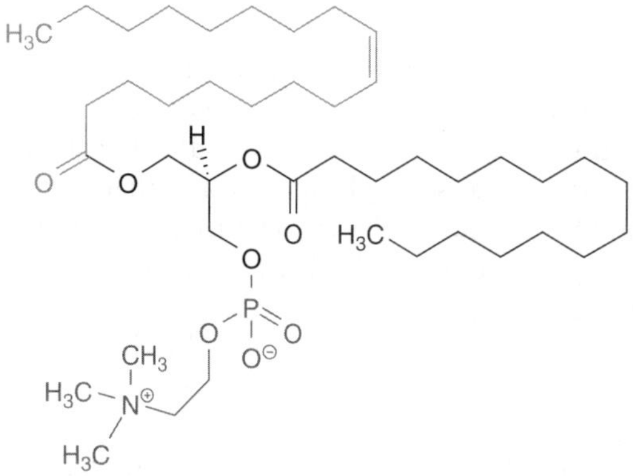

Figure 4.4 Phospholipid.

Phosphorus is common in nature and is vital element in the structure and make-up of many if not all biological systems, for instance, most of the structural elements of biological membranes are amphoteric amphiphiles, the so-called phospholipids, as, for instance, in the phospholipid found in lecithin (see Figure 4.4).

This complex structure is composed of choline and phosphate group (dark grey), glycerol (black), unsaturated fatty acid (light grey) and saturated fatty acid (grey). As can be seen many of these components are derived from natural products and are highly biodegradable.

4.1 Phosphates

In dispersed non-inhibited drilling fluid systems, chemical thinners are added to encapsulate sodium bentonite and reactive drilled solids. The systems do not contain inhibitive electrolytes; therefore, the cuttings are free to disperse as they are transported to the surface. Lignite–lignosulphonate muds are probably the most versatile exploratory drilling fluids in use. Their rheological properties are easily controlled with chemical thinners, and this reduces the risk of detrimental effects of contaminants, such as salt, anhydrite and cement, that may be encountered during drilling. Chemical thinners and filtration-control agents are used to control the high-temperature/high-pressure fluid loss. In this context phosphate salts are used in non-inhibited water-based muds in certain defined circumstances where temperatures are low, less than 55 °C, and salt concentrations are less than 500 ppm [2].

Calcium phosphate is used in cementing applications particularly to prevent CO_2 corrosion of cement. The calcium phosphate cement slurries are composed of calcium phosphate cement, retarder and fluid loss additive. These cement slurries have features such as good rheological properties, excellent anti-channelling and favourable overall performance [3]. These materials are being re-examined for their application in low-temperature Arctic environments [3] and conversely in higher-temperature environments [4]. Foamed high-temperature cements are already based on calcium phosphate.

Sodium and potassium phosphates, which include mono-, di- and tribasic forms, have been used as wetting agents in low permeability flooding applications, where it has been necessary to significantly improve the permeability in the natural flow channels [2].

Of environmental significance is the treatment of oil-based drill cuttings with potassium phosphate to decrease the amount of salt (sodium chloride) available for leaching into the environment. In water-based cuttings it has been found that aluminium phosphate is more efficient in this stabilisation process [5].

A variety of metal salts, including phosphates, and in particular zinc phosphate, are well known as corrosion inhibitors in cooling water systems [6] including applications in oilfield processing. Inorganic salts are discussed further in Chapter 5. Zinc in the form of other salts can provide excellent corrosion protection within these systems [7]; however zinc phosphate is used primarily in cooling water systems that will be operated at a pH of 8.4 or less. The reason for inclusion of phosphate is that as the pH of the circulating water drops, the solubility of zinc rises very rapidly, minimising its corrosion inhibitive properties. The inclusion of phosphate at these lower pH levels results in the formation of mixed zinc hydroxide/phosphate inhibitive films. The reason for this is the fact that zinc phosphate is even less soluble than zinc hydroxide.

This phosphate salt is generally added as orthophosphate, but occasionally some programmes utilising a polyphosphate are encountered. These materials have however come under increasing environmental scrutiny as will be discussed later in Section 4.8.

Sodium hexametaphosphate and sodium orthophosphate have been used as corrosion inhibitors in pipeline situations where they act in tandem or synergistically with drag-reducing polymers such as copolymers of acrylamide and acrylic acid. In this situation it is believed that the corrosion inhibitor contacts the metal surface more thoroughly, increasing its efficiency [8].

Where scaling is also an issue, polyphosphates are often used to provide both corrosion protection and scale control. In water treatment zinc polyphosphates offer a variety of benefits. In addition to some of the traditional benefits offered by zinc orthophosphates, the polyphosphate component prevents scale and stabilises soluble iron, manganese and calcium by sequestering these ions. Limited lead protection is also obtained through hydrolysis of polyphosphate to orthophosphate ion. Products in this category consist of a zinc salt combined with hexametaphosphate in a dry state.

The application of organic coatings to protect metals against corrosion is of great importance in the oil and gas industry. In the past the most common anticorrosive pigments contained lead or hexavalent chromium compounds, which are particularly hazardous and contribute to contamination of the environment. The toxicity of conventional pigments and the legal restrictions imposed on their application have driven manufacturers to undertake extensive research into development of non-toxic corrosion inhibitive compounds. World trends in eliminating toxic anticorrosive pigments mainly have focused on substitution with different phosphates [9]. The use of zinc phosphate, instead of zinc chromate, is now recommended for the formulation of environmentally compatible anticorrosive priming compositions. Ferrites and barrier pigments of lamellar structure also have been proposed. It is believed that the protective action of zinc phosphate results from phosphatisation of the metal substrate and the formation of complex substances with binder components [10]. These substances react with corrosion products to yield a layer that adheres strongly to the substrate. The metal substrate is phosphatised because humidity penetrating through the pores of the coating causes phosphate ions to solubilise.

4.2 Phosphonates and Phosphonic Acid Derivatives

Phosphonates are complexing agents that contain one or more $C-PO(OH)_2$ groups. They are used in numerous technical and industrial applications as chelating agents and scale inhibitors. They have properties that differentiate them from other chelating agents and that greatly affect their environmental behaviour. They also have a very strong interaction with surfaces. Due to this strong adsorption, there is little or no remobilisation of metals, and also little or no biodegradation of phosphonates during water treatment is observed. However, photodegradation of Fe(III) complexes is rapid. Aminopolyphosphonates are rapidly oxidised in the presence of Mn(II) and oxygen, and stable breakdown products are formed that have been detected in wastewater. The lack of quality information about phosphonates in the environment is linked to analytical problems of their determination at trace concentrations in natural waters. However, with the current knowledge on speciation, it can be concluded that phosphonates are mainly present as Ca and Mg complexes in natural waters and therefore do not affect metal speciation or transport. These are important process, which would suggested that phosphonates, when complexed, are not bioavailable.

This type of environmental profile does not suggest that phosphonates and related phosphorus compounds would be seen as green products for use as scale inhibitors in the North Sea. However, a number of phosphonates are sufficiently biodegradable, non-toxic and not bioaccumulating to meet the requirements for use and discharge in a number of regulated geographic areas and therefore are used in the UK sector and other regions of the North Sea and North East Atlantic.

Non-polymeric molecules with only carboxylate and/or sulphonate groups are known to be poor scale inhibitors; however this is not the case with molecules containing phosphonate groups. One drawback of phosphonates is that they tend to be less effective as scale inhibitors than polymers at higher temperatures (see Chapter 2).

There are a number of scale inhibitors with one phosphonate group and several carboxylic acid groups; the most common is 2-phosphonobutane-1,2,4-tricarboxylic acid (PBTCA) as represented in Figure 4.5, which is used mostly as a calcium carbonate inhibitor [11]. It is not inherently biodegradable and so is not normally used in regulated regions such as the North Sea.

Phosphonosuccinic acid (Figure 4.6) is a relatively poor scale inhibitor; however, mixtures with oligomers show good performance on carbonate and sulphate scales. They can also be transformed to phosphonocarboxylic acid esters, which are oil-soluble scale inhibitors.

Another common phosphonate scale inhibitor in the oilfield is 1-hydroxyethane-1,1-diphosphonic acid (HEDP) (Figure 4.7), which is inherently biodegradable.

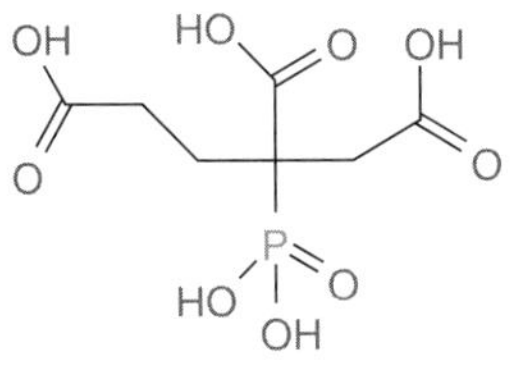

Figure 4.5 PBTCA.

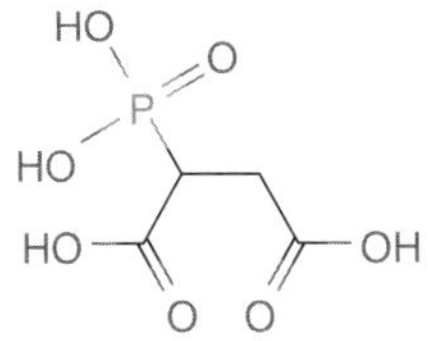

Figure 4.6 Phosphonosuccinic acid.

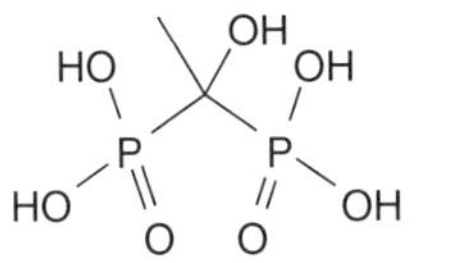

Figure 4.7 HEDP.

The introduction of an amine group into a phosphonate molecule to obtain the $-NH_2$-C-$PO(OH)_2$ group increases the metal binding capability of the molecule. There are a range of commercially available aminophosphonate scale inhibitors used for calcium carbonate and barium sulphate scale inhibition, and although not readily biodegradable, they do have biodegradation of over 20%, have low toxicity, and are not bioaccumulating [12, 13]. The most important chemistries are:

1) Ethanolamine-*N*,*N*-bis(methylenephosphonate) (EBMP) and its amine oxide are good oilfield scale inhibitors, and these have also been used to inhibit silica scales [14, 15].
2) Aminotris(methylenephosphonic acid) (ATMP) (Figure 4.8) is a cheap but not the most effective scale inhibitor [16].
3) 1,2-Diaminoethanetetrakis(methylenephosphonic acid) (EDTMP) (Figure 4.9) is a good all-round phosphonate scale inhibitor being effective on calcium carbonate and barium sulphate scales under a variety of conditions [17].

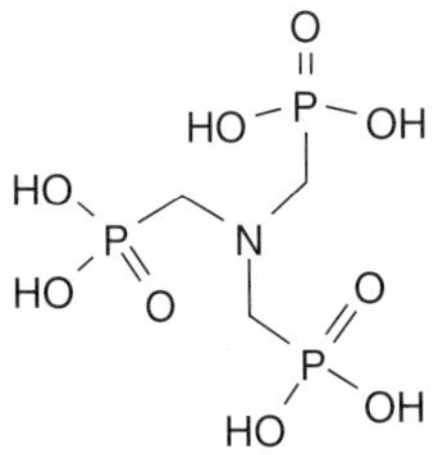

Figure 4.8 ATMP.

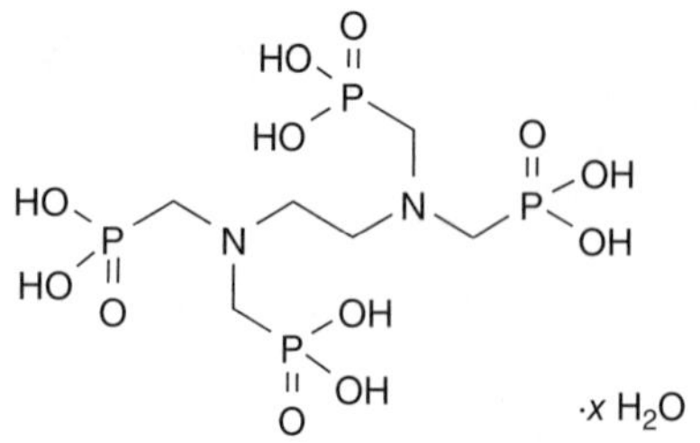

Figure 4.9 EDTMP.

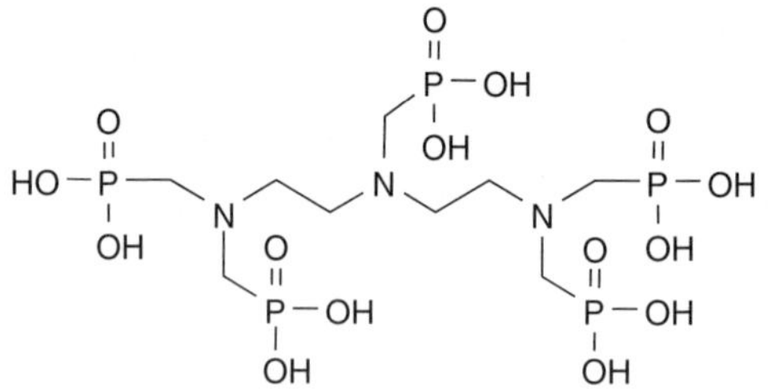

Figure 4.10 DTPMP.

4) Diethylenetriaminepentakis(methylenephosphonic acid) (DTPMP) (Figure 4.10) is analogous to EDTMP and is an excellent carbonate and sulphate scale inhibitor [18] and possibly the most used scale inhibitor in the oil and gas production sector.

5) Hexamethylenetriaminepentakis(methylenephosphonic acid) (Figure 4.11) is another highly effective aminophosphonate scale inhibitor. This molecule has excellent tolerance to high calcium ion concentrations and is good stability at high temperatures up to 140 °C [16].

In addition to the aforementioned, the phosphonate derivatives of *N,N'*-bis(3-aminopropyl) ethylenediamine are claimed to give superior barium sulphate inhibitor performance in high barium brines. Aminomethylene phosphonates, derived from small polyglycol diamines such as triethylene glycol diamine, are claimed to be good carbonate and sulphate scale inhibitors with good compatibility [19].

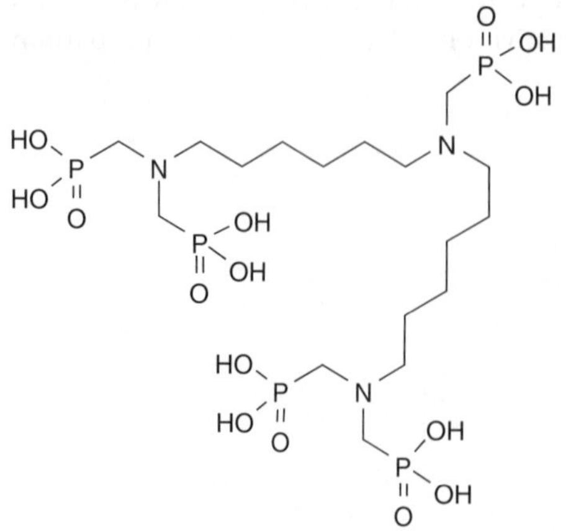

Figure 4.11 Hexamethylenetriaminepentakis(methylenephosphonic acid).

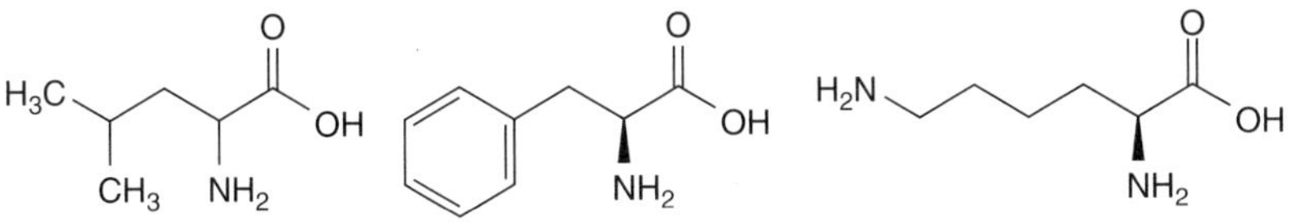

Figure 4.12 D,L-leucine, L-phenylalanine and L-lysine.

Environmentally acceptable phosphonates are obtained by the reaction of selected amino acids with formaldehyde and phosphorous acid. D,L-leucine, L-phenylalanine and L-lysine (see Figure 4.12) are the most common amino acids used.

Better environmental characteristics for these types of product can be obtained if the alkyl phosphonation is controlled to provide only partially alkyl phosphonated derivatives. Such derivatives are claimed to still exhibit equivalent scale inhibition properties as those of the fully substituted derivatives [20].

Phosphonates are applied in precipitation scale 'squeeze' treatments (see Section 4.3 for a general outline of scale squeeze treatments), and the chemistry selected is generally non-polymeric but of a similar chemical nature to that in adsorption squeezes. In particular the following types are preferred:

- Phosphonates
- Polyacrylates
- Phosphonates
- Other polycarboxylates

Critical factors in the selection of chemistry especially with regard to precipitation squeezes are the nature of the formation rock, types of carbonate; other divalent cations present such as magnesium or zinc, barite formations, etc.

Esterified phosphono or phosphino acids derived from long-chain alcohols are claimed to be effective as oil-soluble scale inhibitors and exhibit wax and asphaltene dispersant and inhibition properties [21].

The previously described phosphonic acid derivatives such as diethylene triamine tetra-methylene phosphonic acid (EDTMP) (Figure 4.9) or bis-hexamethylenetriaminepentakis-methylene phosphonic acid (Figure 4.11) are used as oil-soluble scale inhibitors when blended with amine compounds such as *tert*-alkyl primary amines [22].

Phosphonic acid is used in a number of well cement formulations, particularly ATMP, in combination with hydrazine hydrochloride to provide a corrosion-resistant formulation particularly in hydrogen sulphide environments and has a high adhesion to metal surfaces [2].

4.3   Polyphosphonates

There are two main classes of polyphosphonate, those with a polyamine backbone and those with a polyvinyl backbone. Those with a polyamine backbone are particularly useful in inhibiting barium sulphate and for downhole 'squeeze applications'.

'Scale squeeze' is the pre-emptive treatment of scale deposition problems at the near wellbore or indeed in the reservoir and has been deployed for more than four decades [23]. The basic premise in a scale inhibitor squeeze treatment is to protect the well downhole from scale deposition and subsequent formation damage, both of which will lead to production impairment.

In a scale inhibitor squeeze, the inhibitor is pumped into a water-producing zone. The inhibitor is attached to the formation matrix by chemical adsorption or by temperature-activated or pH-activated precipitation and returns with the produced fluid at sufficiently high concentrations to avoid scale precipitation. The inhibitor will of course continue to work above the wellhead, protecting the pipeline and subsequent systems from scale; however an additional top-up dose may be needed topside. In a squeeze treatment the inhibitor is injected into the well above the formation pressure whereby the inhibitor solution will be pushed into the near-well formation rock pores [1, 24]. The well is usually shut in for some hours to allow the scale inhibitor to be retained on the rock matrix by one of the mechanisms mentioned previously. When the well is put back on stream, the produced water will pass through the pores where the chemical has been retained dissolving some of it, and in this way the produced water should contain enough inhibitor to prevent scale deposition.

Traditionally adsorption scale inhibitor squeeze treatments use a water-based scale inhibitor solution. This is usually as a 5–20% active solution in potassium chloride brine or seawater. Often a pre-flush treatment is used composed of 0.1% selected inhibitor in potassium chloride brine or seawater, sometimes with a demulsifier or similar surfactant, to clean and prepare the near wellbore for the main treatment. There are three main types of polymer used in adsorption squeezes:

- Phosphonates
- Carboxylates including polyacrylates (see Chapter 2)
- Sulphonates – polyvinyl sulphonates (see Chapter 2)

Phosphonates generally adsorb better than carboxylates with sulphonates being the poorest absorbers; however the adsorption cannot be so strong that insufficient inhibitor is back-produced so as to be below the minimum or threshold inhibitor concentration. It is therefore a matter of design and balance in selecting the correct inhibitor chemistry for an adsorption squeeze. The ideal scenario is to have the scale inhibitor produced back at the minimum inhibitor concentration at a constant rate from day one.

A major challenge in regard to squeezing scale inhibitor remains the problem of ensuring a long and cost-effective lifetime for the squeeze. Polyphosphonates offer an extended life cycle of treatment, i.e. a greater time interval between treatment applications and therefore a great cost efficiency.

Polyphosphonates with a polyvinyl backbone are prepared from vinylphosphonic acid or vinyldiphosphonic acid monomers. These can be copolymerised with other monomers such as acrylic, maleic and vinylsulphonic acids and make good scale inhibitors, in particular for barium sulphate scales [25]. Vinylphosphonic acids such as in Figure 4.13 are expensive and are mostly used to make phosphonate end-capped phosphino polymers (see Section 4.4).

Other phosphonate polymers are produced from hypophosphorous acid by reaction with carbonyl or imine functionalities and then reacted with monomers such as acrylic acid to give polyphosphonates [26]. There are a number of variations of these copolymer derivatives, many of which are sufficiently biodegradable to be of potential interest for oilfield application in which ecotoxicological acceptance is a primary driver for use. Many also possess thermal

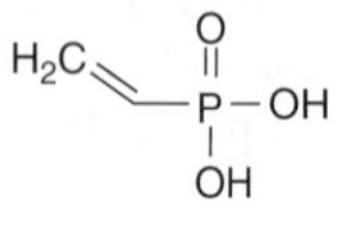

Figure 4.13 Vinylphosphonic acid.

stability for high-temperature and high-pressure application, including copolymers with acrylamido(methyl)propylsulphonic acid (AMPS) [27].

Polyphosphonates with a polyamine backbone such as the *N*-phosphonomethylated amino-2-hydroxypropylamine polymer is particularly useful against barium sulphate and widely used in squeeze applications [28]. They are also inherently biodegradable being mainly derived from amino acids [29].

4.4 Phosphino Polymers and Polyphosphinate Derivatives

This class contains the most common phosphino polymer used in the oil and gas industry, namely, polyphosphinocarboxylic acid (PPCA) (see structure in Figure 4.14). This is mainly mixtures containing oligomers with acrylic, maleic and succinic acids [30]. These mixtures, especially phosphinosuccinic acid oligomers, are useful calcium carbonate scale squeeze inhibitors [22]. The presence of the phosphorus atom makes PPCA easier to detect than the equivalent polycarboxylic acids, allows it to perform better, especially for barium sulphate scales, confers good calcium compatibility and excellent rock adsorption, giving enhanced 'squeeze' lifetimes [31].

Phosphino polymers with several phosphinate groups can be been derived from hypophosphorous acid and alkyne chemistry. If acetylene is used, a mixture of ethane-1,2-biphosphinic acid and diethylenetriphosphinic acid is obtained, and this mixture is used as a scale inhibitor [32]. These oligomers (or more correctly telomeres) can be reacted with vinyl monomers to give phosphino polymers with excellent scale inhibitor performance [33].

Phosphonate end-capped phosphino polymers are particularly useful barium sulphate inhibitors; they show good rock adsorption characteristics and good thermal stability. The end-capped vinyldiphosphonic acid polymers have been developed, and the key product structures are shown in Figure 4.15. Significantly they have ecotoxicological profiles, which show greater than 20% biodegradation [34].

Phosphinosuccinic acid adducts comprising mono-, bis- and oligomeric phosphinosuccinic acid adducts have been claimed to be useful corrosion inhibitors [35], particularly for aqueous systems. However as has been described in Chapter 3, phosphate esters (Section 3.2.9) are the most commonly used class of phosphorus-containing corrosion inhibitors used in the upstream oil and gas industry.

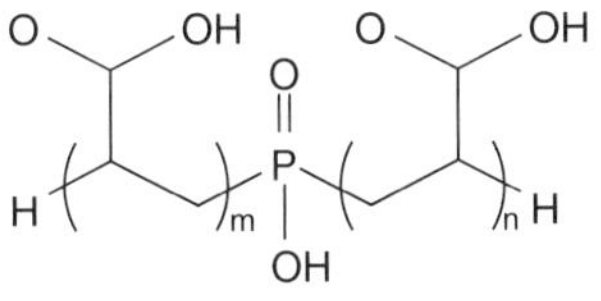

Figure 4.14 PPCA.

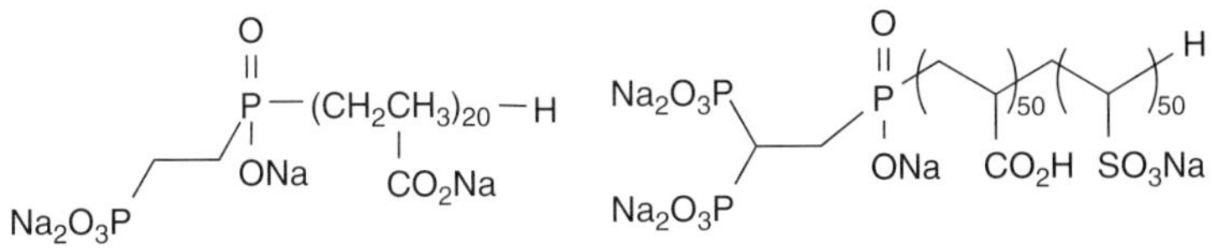

Figure 4.15 Phosphino polymers.

4.5 Phosphoric Esters (Phosphate Esters)

As has been described in Section 3.2.9, the —OH groups in phosphoric acid can condense with the hydroxyl groups of alcohols to form phosphate esters. Since orthophosphoric acid has three —OH groups, it can esterify with one, two or three alcohol molecules to form a mono-, di- or triester. The major use of phosphate esters, which are anionic surfactants, is as components of oilfield corrosion inhibitors, and phosphate esters are classified as film-forming corrosion inhibitors (FFCIs). In this section this chemistry will not be detailed as it has already been described previously; however some other applications will be exemplified.

During hydraulic fracturing operations, organic gels can be used for temporary plugging operations. These can be formed from a hydrocarbon liquid such as diesel or crude oil and aluminium phosphate diester. It is necessary in this process to have the reagents water-free and pH stable [36].

This process can also be utilised to prevent spillage of such hydrocarbons when storage or transportation containment is damaged [37]. In this application it is useful to use phosphoric esters, which are less volatile, such as in Figure 4.16, as volatile phosphorus-containing products can cause problems during subsequent distillation and refining.

From the environmental chemists' viewpoint, it is water solubility that is the critical issue for these phosphoric esters and related derivatives. If an ethoxylated alcohol is used, monoester can exhibit good water solubility even at acid pH at which the other acidic hydrogen atoms are not neutralised. These water-soluble esters can show good rates of biodegradation [38].

Phosphoric esters have also been used in the development of asphaltene dispersants, such as isooctyl acid phosphate (see Figure 4.17). Such phosphoric esters do not have aromatic groups that could aid interaction with asphaltenes via π—π bond interactions. Phosphoric acid esters, however, have highly acidic hydrogen ions, which can hydrogen bond to amines and hydroxyl groups in the asphaltenes or with metal ions and therefore destabilise the asphaltene agglomeration mechanism.

These phosphoric acid esters are normally blended with surfactants such as dodecylbenzene sulphonic acid (DDBSA) to give enhanced dispersant performance [39] (see Section 3.2.9), and synergistic blends with carboxylic acids and related derivatives have been claimed to be useful asphaltene dispersants and apparently perform better than the commonly used nonylphenol-formaldehyde polymers [40].

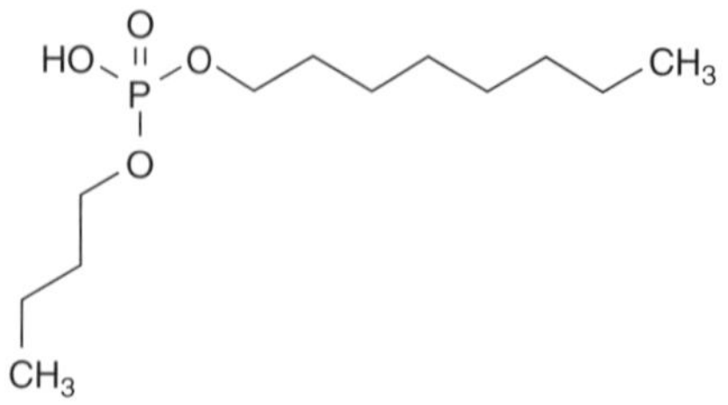

Figure 4.16 Phosphoric acid butyl octyl ester, a less volatile phosphate ester.

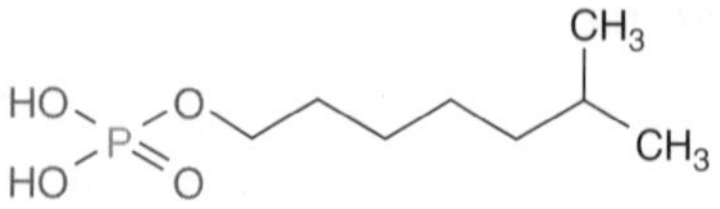

Figure 4.17 Isooctyl acid phosphate.

Phosphate esters have a variety of other oilfield uses, many of which have been described in Section 3.2.9, particularly as lubricants in drilling and completion operations [41] and commonly as inhibitors in corrosion control. This is particularly the case for general corrosion and stress-induced cracking corrosion where the reaction of a nitrogen base compound and a phosphate ester can provide suitable inhibitors [42].

The phosphate triesters of polyols and other esters such as orthoester are claimed to have useful activity as low dose naphthenate inhibitors [43]. Interestingly thiophosphorus compounds [44] and phosphorous acid with certain sulphur and phosphorus-free aromatic compounds [45] are claimed to be good naphthenic acid and other organic acid corrosion inhibitors; although this is rarely encountered in upstream oilfield operations, it can be a common downstream refinery problem, especially at high temperatures encountered during distillation and other operations.

Phosphate esters are well known and widely used as scale inhibitors across a range on industrial sectors including the upstream oil and gas sector. They are not the most effective class of scale inhibitors but generally exhibit good environmental profiles [46]. They are used primarily to control calcium carbonate and calcium sulphate scales but are also effective against barium sulphate if conditions are not too acidic. They are usually more tolerant of acidic conditions than polyphosphates, described previously in Section 4.3, and have good compatibility with high calcium concentration and are reasonably thermal stable [10]. Triethanolamine phosphate monoester has been used in scale control for over 20 years and is readily biodegradable; however it is less thermally stable than some other phosphate esters with a maximum temperature stability of 80 °C.

A phosphoric ester polymer has been evaluated as a potential wax inhibitor. The polymer was obtained by reacting a long-chain phosphoric ester with sodium aluminate, generating a molecule with relatively high molecular weight and amphiphilic character. Studies were carried out using a model system of petroleum paraffin (P140) dissolved in paraffin solvent. Rheological, calorimetric, chromatographic and optical and electron microscopy tests demonstrated that the additive acts as modifier of paraffin crystallisation, although the efficiency shown depended on the polymer molecular weight [47].

Aluminium phosphate ester salts are used to form liquid gels with diesel and crude oil, which can be used in temporary plugging during hydraulic fracturing operations. Such gels are substantially water-free and pH stable [48]. Further discussion on such aluminium salts is detailed in Section 5.2.

It is also worth restating that this class of phosphorus compound, the phosphate ester, is one of the most environmentally acceptable of all classes of surfactants, being highly susceptible to enzymatic hydrolysis and therefore having good rates of biodegradation. This is further discussed in Section 3.9. Comment is also made in Section 4.8.

4.6 Phosphonium Quaternary Salts and Related Compounds

Phosphonium quaternary compounds and in particular the salts of the tetrakis(hydroxymethyl) phosphonium [THP] ion are known to be excellent biocides. The sulphate salt (tetrakishydroxymethylphosphonium sulphate (THPS)) (Figure 4.18) is widely used across the upstream oil and gas industry, has good efficacy in a variety of applications, and is very effective in controlling all the major oilfield bacterial populations. It is also highly compatible with oilfield fluids, being non-oxidising and non-foaming [49, 50].

THPS is a water-soluble, short-chain quaternary phosphonium compound first used in the textile industry as an intermediate for the manufacture of flame retardants. The first

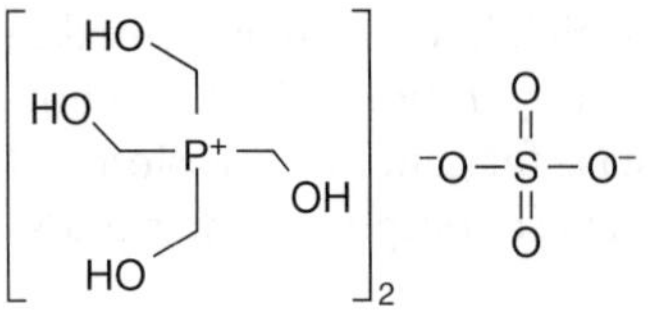

Figure 4.18 THPS.

documented use of THPS in an industrial biocide application was in 1983, and it was later examined for use in oilfield applications.

Since its first application in the UK sector of the North Sea in the late 1980s, THPS has developed a pre-eminent position for the control of problematic sulphate-reducing bacteria (SRB) and the operational problems associated with their metabolic activity. In the intervening decades, different 'THPS generations' have been developed [51] for a variety of biocidal and other applications such as iron sulphide removal [52] and control of schmoo [53].

It has been shown that THPS has multiple modes of action against a typical oilfield bacterium, which reduces the chances of the proliferation of resistant bacterial strains.

The biocidal action of THPS can be enhanced synergistically by the use of a number of additives including various surfactants, mutual solvents or biopenetrants such as poly[oxyethylene(dimethyliminio)ethylene(dimethyliminio)ethylene dichloride] [54, 55] (see structure in Figure 4.19).

In the last decade or so, another quaternary phosphonium biocide has been developed, which is based on tributyltetradecylphosphonium chloride (TTPC) (see Figure 4.20) [56]. TTPC has the advantage of being a surfactant and having biopenetrant and synergistic properties built in. Laboratory and field data has shown that TTPC is effective at low concentrations, is fast acting and is effective against both acid-producing and sulphate-reducing bacteria. It has outperformed both glutaraldehyde and THPS in comparative biocidal tests, giving complete kill of aerobic and anaerobic SRB in 1 h at 5 and 50 ppm active, respectively. TTPC is compatible with oxidising biocides, hydrogen sulphide and oxygen scavengers and has excellent thermal stability, which makes it broadly applicable to oilfield water systems. However it does not have iron sulphide removal characteristics, is low foaming, and cannot easily be deactivated. It

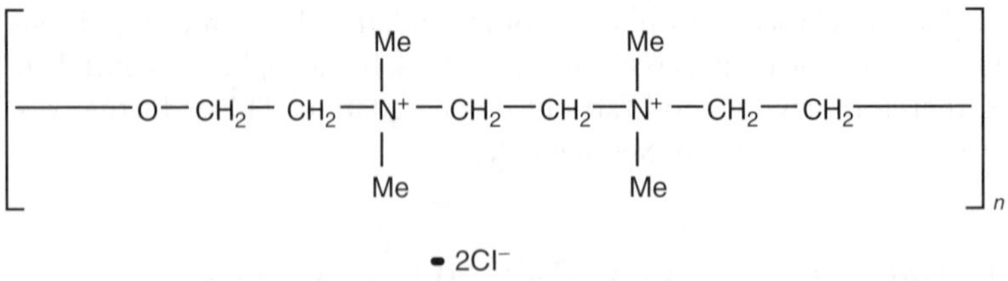

Figure 4.19 Poly[oxyethylene(dimethyl-iminio)ethylene (dimethyliminio)-ethylene dichloride].

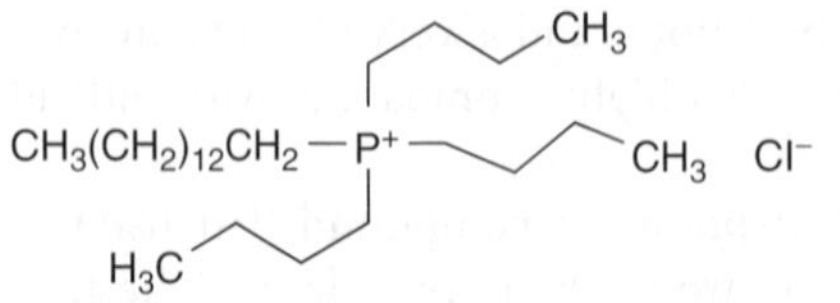

Figure 4.20 TTPC.

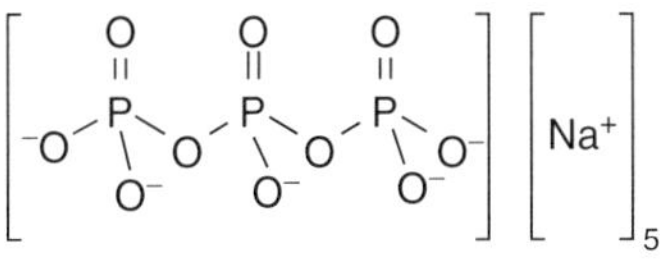

Figure 4.21 Sodium tripolyphosphate.

also has adsorption characteristics, which has been cleverly developed for reservoir treatments [57].

TTPC has also shown significant corrosion inhibition properties and has been used as such in oilfield applications [58, 59].

TTPC appears to have greater toxicity than other similar phosphonium salts. The mechanism of toxicity has been examined for alkyl-tributylphosphonium chlorides across a number of alkyl substituent groups. The elongation of one of the alkyl substituents resulted generally in higher toxicity, as defined by their inhibitory and lethal effects. The higher toxicity of phosphonium ionic liquids carrying long alkyl substituents is most likely due to their strong interaction with the cellular boundaries in the microbial species studied [60]. This would support the higher efficacies observed but may also result in less environmental acceptance against other toxicity criteria.

Phosphonium surfactant-type biocides have been developed for use in other industrial sectors and have been shown to have greater efficacy than traditional cationic surfactants such as the didecyldimethylammonium chlorides [61, 62]. Studies have discovered that for alkyltrimethylphosphonium salts, the biocidal activity increases with chain length, whereas for dialkyldimethylphosphonium salts, it was found to decrease with increasing chain length. Finally in this series the phosphate esters of tetraalkylphosphonium salts have also shown biocidal activity [63].

Biocides typically have an adverse environmental impact on produced water. THPS has a similar effect on produced water and treated seawater used for hydrotesting and bulk storage. The effect of temperature, pH, water depth, dissolved oxygen concentration and various ions in the system have been found to have important effects [64]. In general THP salts are only slowly biodegraded; however under alkaline conditions degradation can be faster. This has significant advantages in preservation applications [65].

However, overall and in comparison with other biocides, THP salts have reasonably favourable environmental characteristics but some problematic properties pertaining to mammalian toxicology, and due mainly to the latter, it has been prohibited from use in Norway and Norwegian waters in the last few years.

Phosphonium salts, which are used as chelating agents in cementing formulations, are classified as complexones. These materials are used in small concentrations within the cementing formulation to increase and improve adhesion to the rock surface. Such materials are exemplified by HEDP (Figure 4.7), ATMP (Figure 4.8) and sodium tripolyphosphate (Figure 4.21) [2].

4.7 Phospholipids

Phospholipids are part of the family of natural products, the lipids, which are a class of surfactants, and are more fully detailed in Section 3.7. The first phospholipid was identified in 1847 by the French chemist and pharmacist Theodore Gobley in the egg yolk of chickens. Phospholipids are a major component of all cell membranes and can form lipid bilayers because of their amphiphilic nature. The structure of the phospholipid molecule generally consists of

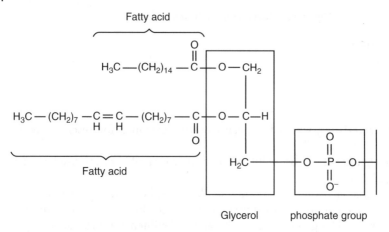

Figure 4.22 Typical phospholipid structure.

two hydrophobic fatty acid 'tails' and a hydrophilic phosphate 'head', joined together by a glycerol molecule (see structure in Figure 4.22).

These products are naturally present in living organisms, e.g. as in nucleic acids where the phosphate ester bond is prone to enzymatic hydrolysis that is catalysed by a variety of phosphatase enzymes, and these are also able to attack the ester linkages in these surfactants [66].

The phosphate groups can be modified with simple organic molecules, and in nature molecules such as choline (see Figure 4.4) are used to provide additional functionality and intermolecular binding.

Phospholipids, such as lecithin, are produced commercially by chemical extraction using non-polar solvents or mechanically from crop sources such as rapeseed, sunflower and soya bean. They have found applications in a number of industrial sectors such as pharmaceuticals, nanotechnology and materials science [67, 68]. Lecithin is sold as a food supplement and for medical uses. In cooking, it is sometimes used as an emulsifier and to prevent sticking, for example, in non-stick cooking sprays [69].

The application of phospholipids in the oil and gas industry has been very limited to date. They are known to be effective lubricants in aqueous drilling fluids, and because of their ionic nature, some phospholipids are water soluble [70]. Phospholipids are naturally occurring and environmentally benign materials, which seriously require further investigation as to their applicability in other oilfield applications, such as previously described in Chapter 3.

4.8 Biophosphorus Chemistry and Environmental Considerations

4.8.1 Phosphorus in Nature

Phosphorus is an essential element for all life forms and is a mineral nutrient. It is phosphate esters and diesters that perform essential biological functions. There are also some noteworthy phosphorous carbon bonds and phosphorous nitrogen bond natural products such as 2-aminoethylphosphonic acid (Figure 4.23).

Surprisingly, given its resemblance to other well-established phosphonates, this product and its potential derivatives have been little explored as to potential uses in oilfield chemistry. Phosphonates, as has been described, are well-known scale inhibitors, and 2-aminoethanol

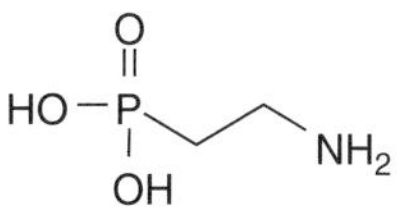

Figure 4.23 2-Aminoethylphosphonic acid.

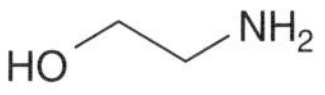

Figure 4.24 2-Aminoethanol.

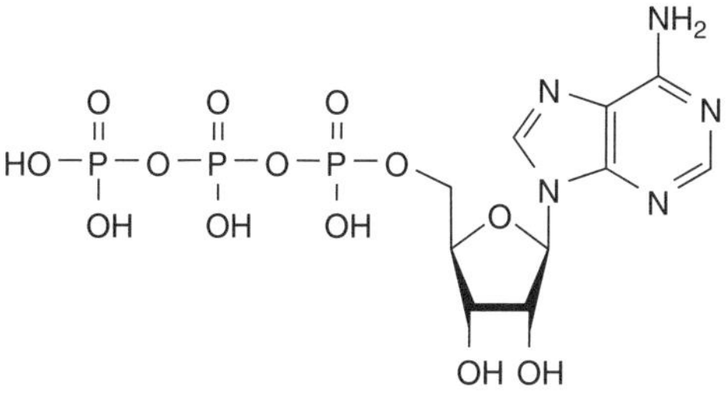

Figure 4.25 ATP.

(Figure 4.24) has corrosion inhibition properties that may be useful in cementing applications [71].

The single most important phosphorus-containing compound found in nature is adenosine triphosphate (ATP) that has the structure shown in Figure 4.25. ATP should be viewed in the context of nucleotide chemistry and is essential in the phosphorylation of a number of essential products including RNA and DNA and is an intermediate in providing catalysis and energy input to a vast number of metabolic pathways [72].

The use and application of nucleotides and polynucleotides, or rather their lack of use, has already be described in Section 2.2. In the natural world however, these biopolymers and the phosphate-containing monomers are widely distributed and have a wide variety of functions. As has been related before, oilfield chemists and environmental scientist could find a useful readily available chemistry set for exploration and application.

As a corollary to the critical function of phosphorus in the chemistry of life, it is important to note that some phosphorus-containing compounds are also highly toxic, which is often the basis of their biocidal properties, as has been described earlier in this chapter. The origin of this toxicity is the ability of certain phosphorus compounds to mimic naturally occurring carboxylic esters and hence inhibit essential enzymes. This property has been exploited in organophosphorus pesticide and insecticides and rather chillingly in agents of chemical warfare including neurotoxins [73]. This has led to highly regulated controls of specific phosphorus-containing agrochemicals, particularly pesticides with lethal mammalian toxic effects, after the true environmental impact of such agents was fully understood [74].

4.8.2 Environmental Fate and Impact

Phosphorus has long been recognised as the critical control factor in plant and algae growth for many aquatic environments, particularly lakes and streams. A minor increase in phosphorus can fuel substantial increases in both aquatic plant and algae growth, which can have severe impacts on the complete ecosystem. Extracellular enzymes can hydrolyse organic forms of phosphorus including many of the previously described products to phosphate. This can lead in certain aquatic environments to eutrophication, that is, the over-enrichment of receiving waters with mineral nutrients. The results are excessive production of autotrophs, especially algae and cyanobacteria. This high productivity leads to high bacterial populations and high respiration rates, leading to hypoxia or anoxia in poorly mixed bottom waters and at night in surface waters during calm, warm conditions. Low dissolved oxygen causes the loss of aquatic animals and release of many materials normally bound to bottom sediments including various forms of phosphorus. This further release of phosphorus reinforces the eutrophication process.

Phosphorus can originate from municipal and industrial facilities that discharge water, as well as runoff water from agricultural areas that makes its way into local water sources. Excessive concentration of phosphorus is the most common cause of eutrophication in freshwater lakes, reservoirs, streams and headwaters of estuarine systems.

It is also the case that it can stimulate marine algae growth particularly in shallow and enclosed or partly enclosed marine basins, such as areas of the North Sea or the Gulf of Mexico. However, in the ocean, nitrogen becomes the key mineral nutrient controlling primary production of autotrophs (i.e. an organism that is able to form nutritional organic substances from simple inorganic substance) such as algae. Estuaries and continental shelf waters are a transition zone where excessive phosphorus and nitrogen create problems [75].

Restrictive phosphorus discharge limits are being enacted or proposed, for example, in many areas of the United States where the receiving water streams are particularly sensitive to changes in phosphorus levels. Similar restrictions are being considered by a number of oil and gas regulatory bodies responsible for the control and discharge of production waters and associated chemicals [76].

As has been described, many phosphorus compounds are environmentally benign, being, in the main, readily biodegradable and relatively non-toxic. However their fate to compounds and salts of phosphorus that can lead to uptake by algae and related organisms is of serious concern. It is their bioavailability that is probably the overriding environmental impact factor, and environmental chemists and practitioners should consider this as the primary cause for environmental concern.

Added chemicals from oil and gas exploration development and production are only one factor affecting an imbalance of phosphorus in a specific environment; agricultural pesticides are a major source of pollution with regard to algae growth stimulation [77]. In general human activity continues to have a significant impact on this eutrophication phenomenon [78].

4.8.3 Degradation and Biodegradation

As has been described previously in this chapter and in Section 3.2.9, many phosphorus-containing compounds are water soluble and prone to enzymatic hydrolysis and therefore have good rates of biodegradation, and many possess low toxicity [79]. Many are susceptible to oxidative degradation, particularly phosphonates, which can be an undesirable property affecting the efficacy and efficiency of the chemical treatment [80]. The downside to the potential high rates of biodegradation is that smaller and less complex phosphorus-containing molecules are available to autotrophs for use and nutrition.

Most phosphates, phosphonates and related materials are however thermally stable with many polymer types, such as cellulose phosphates, being used as flame retardants [81].

4.8.4 Environmental Design

The properties of thermal stability and good biodegradation characteristics can be highly advantageous in the potential design of useful chemistries, which are both environmentally acceptable and efficient under high physically stressed environments, such as high pressure and high temperatures. As has been described, many phosphorus compounds afford these critical properties inherently. The main disadvantage is their bioavailability, where low molecular weight phosphates and related compounds are the results of degradation process; these products are available to certain microfauna, resulting in eutrophication and a disturbance in the balance of certain aquatic ecosystems. Hence, in designing products for effect when considering phosphorus chemistry, the environmental chemist should take due cognisance of bioavailability of the product and its degradation products.

Phosphorus-containing polymers such as polyphosphates can provide some answers in this respect.

The emerging fields of supramolecular chemistry, crystal engineering and materials chemistry have made significant advances in the last four decades. The chemistry of phosphonate ligands has played an important role in widening these areas of research [82]. Phosphonate ligands have attracted considerable attention in the context of fundamental research, but they have also been extensively used in several other technological and industrial significant applications, including the upstream oil and gas sector and in particular oilfield scale inhibition [83] and corrosion control [84]. There exists a vast number of metal phosphonate materials whose crystal structures exhibit attractive features that depend on several variables, such as nature of the metal oxidation state, ionic radius and coordination number in particular, the number of phosphonate groups on the ligand backbone, many of which may exhibit technically advance properties but many of which will all suffer from the same issues of biodegradation versus the products of such biodegradation being bioavailable.

In nature we have seen phosphorus used in polymeric chemistry and in other applications such as phospholipids. These chemistries have been little explored as to the suitability for application to oilfield problems.

References

1 Emsley, J. and Hall, D. (1976). *The Chemistry of Phosphorous*. London: Harper & Row.

2 Fink, J. (2013). *Petroleum Engineers Guide to Oilfield Chemicals and Fluids*. Elsevier.

3 Jianguo, Z., Fuquan, S., Aiping, L. and Yonghui, G. (2014). Study on calcium phosphate cement slurries. The 24th International Ocean and Polar Engineering Conference, Busan, Korea (15–20 June 2014), ISOPE-I-140-476.

4 Zeng, J., Xia, Y., Sun, F. et al. (2015). Calcium phosphate cement slurries for thermal production wells. The 25th International Ocean and Polar Engineering Conference, Kona, Hawaii (21–26 June 2015), ISOPE-I-15-156.

5 Filippov, L., Thomas, F., Filippova, I. et al. (2009). Stabilization of NaCl-containing cuttings wastes in cement concrete by in situ formed mineral phases. *Journal of Hazardous Materials* **171** (1–3): 731–738.

6 Hatch, G.B. (1977). Inhibition of cooling water. In: *Corrosion Inhibitors* (ed. C.C. Nathan). NACE.

7 Young, T.J. (1991). The use of zinc for corrosion control in open cooling systems. Association of Water Technologies, Inc. Spring Convention & Exposition, San Antonio, TX (3–5 April 1991).

8 Johnson, J.D., Fu, S.L., Bluth, M.J. and Marble, R.A. (1996). Enhanced corrosion protection by use of friction reducers in conjunction with corrosion inhibitors. GB Patent 2,299,331, assigned to Ondeo Nalco Energy Services L.P.

9 Romagnoli, R. and Vetere, V.F. (1995). Heterogeneous reaction between steel and zinc phosphate. *Corrosion* **51** (02): 116–123, NACE - 95 020 116.

10 del Amo, B., Romagnoli, R., Vetere, V.F., and Herna'ndez, L.S. (1998). Study of the anticorrosive properties of zinc phosphate in vinyl paints. *Progress in Organic Coatings* **33**: 28–35.

11 Holzner, C., Ohlendorf, W., Block, H.-D. et al. (1997). Production of 2-phosphonobutane-1,2, 4-tricarboxtlic acid and alkali metal salts thereof. US Patent 5,639,909.

12 Jordan, M.M., Feasey, N., Johnston, C. et al. (2007). Biodegradable scale inhibitors, laboratory and field evaluation of a 'Green' carbonate and sulphate scale inhibitor with deployment histories in the North Sea. Chemistry in the Oil Industry X (5–7 November 2007). Manchester, UK: Royal Society of Chemistry, p. 286.

13 Miles, A.F., Bodnar, S.H., Fisher, H.C. et al. (2009). Biodegradable phosphonate scale inhibitors. Chemistry in the Oil Industry XI (2–4 November 2009). Manchester, UK: Royal Society of Chemistry, p. 271.

14 Davis, K.P., Docherty, G.F. and Woodward, G. (2000). Water treatment. International Patent Application WO/2000/018695.

15 Davis, K.P., Otter, G.P. and Woodward, G. (2004). Scale inhibitor. International Patent Application WO/2004/078662.

16 Guo, J. and Severtson, S.J. (2004). Inhibition of calcium carbonate nucleation with aminophosphonates at high temperature, pH and ionic strength. *Industrial and Engineering Chemistry Research* **43**: 5411.

17 Amjad, Z. (1996). Scale inhibition in desalination applications: an overview. Corrosion 96, International Conference and Exposition, Denver, CO, NACE 96–230.

18 Fan, C., Kan, A.T., Zhang, P. et al. (2012). Scale prediction and inhibition for oil and gas production at high temperature/high pressure. *SPE Journal* **17** (02): 379–392.

19 Przybylinski, J.L., Rivers, G.T. and Lopez, T.H. (2006). Scale inhibitor, composition useful for preparing same, and method of inhibiting scale. US Patent Application 20060113505, assigned to Baker Hughes Inc.

20 Bodnar, S.H., Fisher, H.C., Miles, A.F. and Sitz, C.D. (2010). Preparation of environmentally acceptable scale inhibitors. International Patent Application WO/2010/002738, assigned to Champion Technologies Inc.

21 Jones, C.R., Woodward, G. and Phillips, K.P. (2004). Novel phosphonocarboxylic acid esters. Canadian Patent CA 2490931, Assigned to Authors.

22 Reizer, J.M., Rudel, M.G., Sitz, G.D. et al. (2002). Scale inhibitors. US Patent 6,379,612, assigned to Champion Technologies Inc.

23 Vetter, O.J. (1973). The chemical squeeze process some new information on some old misconceptions. *Journal of Petroleum Technology* **26** (3): 339–353, SPE 3544.

24 Sorbie, K.S. and Gdanski, R.D. (2005). A complete theory of scale-inhibitor transport and adsorption/desorption in squeeze treatments. SPE International Symposium on Oilfield Scale, Aberdeen, UK (11–12 May 2005), SPE 95088.

25 Herrera, T.L., Guzmann, M., Neubecker, K. and Gothlich, A. (2008). Process and polymer for preventing ba/sr scale with incorporated detectable phosphorus functionality. International Patent application WO/ 2008/ 095945, assigned to BASF Se.

26 Kerr, E.A. and Rideout, J. (1997). Telomers. US Patent 5,604,291, assigned to Fmc Corporation (UK) Ltd.

27 Greyson, E., Manna, J. and Mehta, S.C. (2011). Scale and corrosion inhibitors for high temperature and pressure conditions. US Patent Application 20110046023.

28 Singleton, M.A., Collins, J.A., Poynton, N. and Formston, H.J. (2000). Developments in phosphonomethylated polyamine (PMPA) scale inhibitor chemistry for severe BaSO scaling conditions. International Symposium on Oilfield Scale, Aberdeen, UK (26–27 January 2000), SPE 60216.

29 Tang, J. (2001). Biodegradable poly(amino acid)s, derivatized amino acid polymers and methods for making same. US Patent 6,184,336, assigned to Malco Chemical Company.

30 Benkbakti, A. and Bachir-Bey, T. (2010). Synthesis and characterization of maleic acid polymer for use as scale deposits inhibitors. *Journal of Applied Polymer Science* **116** (5): 3095.

31 Jordan, M.M., Sjuraether, K., Collins, I.R. et al. (2001) Life cycle management of scale control within subsea fields and its impact on flow assurance, Gulf of Mexico and the North Sea Basin. Chemistry in the Oil Industry VII (13–14 November 2001). Manchester, UK: Royal Society of Chemistry, p. 223.

32 Davis, K.P., Otter, G.P. and Woodward, G. (2005). Novel phosphorous compounds. International Patent Application WO/2001/057050, assigned to Rhodia Consumer Specialities Ltd.

33 Davis, K.P., Woodward, G., Hardy, J. et al. (2005). Novel polymers. International Patent Application WO/2005/023904, assigned to Rhodia UK Ltd.

34 Davis, K.P., Fidoe, S.D., Otter, G.P. et al. (2003). Novel scale inhibitor polymers with enhanced adsorption properties. International Symposium on Oilfield Scale, Aberdeen, UK (29–30 January 2003), SPE 80381.

35 Yang, B., Reed, P.E. and Morris, J.D. (2003). Corrosion inhibitors for aqueous systems. US Patent 6,572,789, assigned to Ondeo Nalco Company.

36 Jones, C.K., Williams, D.A. and Blair, C.C. (1999). Gelling agents comprising of Aluminium Phosphate compounds. GB Patent 2,326,882, assigned to Nalco/Exxon Energy Chemicals L.P.

37 Delgado, E. and Keown, B. (2009). Low volatile phosphorous gelling agent. US Patent 7,622,054, assigned to Ethox Chemicals Llc.

38 Saeger, V.W., Hicks, O., Kaley, R.G. et al. (1979). Environmental fate of selected phosphate esters. *Environmental Science & Technology* **13** (7): 840–844.

39 Miller, R.F. (1984). Method for minimizing fouling of heat exchangers. US Patent 4,425,223, assigned to Atlantic Richfield Company.

40 Miller, D., Volmer, A., Feustel, M. and Klug, P. (2001). Synergistic mixtures of phosphoric esters with carboxylic acids or carboxylic acid derivatives as asphaltene dispersants. US Patent 6,204,420, assigned to Clariant Gmbh.

41 Johnson, D.W. and Hils, J.E. (2013). Review: phosphate esters, thiophosphate esters and metal thiophosphates as lubricant additives. *Lubricants* **1**: 132–148.

42 Martin, R.L. (1993). The reaction product of nitrogen bases and phosphate esters as corrosion inhibitors. EP Patent 0567212, assigned to Petrolite Corporation.

43 Hellsten, M. and Uneback, I. (2008). A method for preventing the formation of calcium carboxylate deposits in the dewatering process for crude oil/water streams. International Patent Application WO/2008/155333, assigned to Akzo Nobel N.V.

44 Zetimeisi, M.J. (1999). Control of naphthenic acid corrosion with thiophosphorus compounds. US Patent 5,863,415, assigned to Baker Hughes Inc.

45 Sartori, G., Dalrymple, D.G., Blum, S.C. et al (2004). Method for inhibiting corrosion using phosphorous acid. US Patent 6,706,669, assigned to ExxonMobil Research and Engineering Company.

46 El Dahan, H.A. and Hegazy, H.S. (2000). Gypsum scale control by phosphate ester. *Desalination* **127** (2): 111–118.

47 Gentili, D.O., Khalil, C.N., Rocha, N.O. and Lucas, E.F. (2005). Evaluation of polymeric phosphoric ester-based additives as inhibitors of paraffin deposition. SPE Latin American and

Caribbean Petroleum Engineering Conference, Rio de Janeiro, Brazil (20–23 June 2005), SPE 94821.

48 Jones, C.K., Williams, D.A. and Blair, C.C. (1999). Gelling agents comprising aluminium phosphate agents. GB Patent 2,326,882, assigned to Nalco Exxon Energy Company.

49 Downward, B.L., Talbot, R.E. and Haack, T.K. (1997). Tetrakishydroxy- methylphosphonium sulfate (THPS) a new industrial biocide with low environmental toxicity. Corrosion97, New Orleans, LA (9–14 March 1997), NACE 97401.

50 Jones, C., Downward, B.L., Edmunds, S. et al. (2012). NACE 2012–1505. THPS: a review of the first 25 Years, lessons learned, value created and visions for the future. CORROSION 2012, Salt Lake City, Utah (11–15 March 2012).

51 Jones, C., Downward, B.L., Hernandez, K. et al. (2010) Extending performance boundaries with third generation THPS formulations. CORROSION 2010, San Antonio, TX (14–18 March 2010), NACE 105257.

52 Talbot, R.E., Gilbert, P.D., Veale, M.A. and Hernandez, K. (2002). Tetrakis hydroxymethyl phosphonium sulfate (THPS) for dissolving iron sulfides downhole and topsides - a study of the chemistry influencing dissolution, Publisher. CORROSION 2002, Denver, CO (7–11 April 2002), NACE 02030.

53 Blumer, D.J., Brown, W.M., Chan, A. and Ly, K.T. (1998). Novel chemical dispersant for removal of organic/inorganic schmoo scale in produced water injection systems. CORROSION 98, San Diego, CA (22–27 March 1998), NACE 98073.

54 Cooper, K.G., Talbot, R.E. and Turvey, M.J. (1998). Biocidal mixture of tetrakis (hydroxymethyl) phosphonium salt and a surfactant. US Patent 5,741,757, assigned to Albright and Wilson Ltd.

55 Jones, C.R. and Talbot, R.E. (2004). Biocidal compositions and treatment. US Patent 6,784,168, assigned to Rhodia Consumer Specialities Ltd.

56 Kramer, J.F., O'Brien, F. and Strba, S.F. (2008). A new high performance quaternary phosphonium biocide for microbiological control in oilfield water systems. CORROSION 2008, New Orleans, LA (16–20 March 2008), NACE – 08660.

57 Holtsclaw, J., Weaver, J.D., Gloe, L. and McCabe, M.A. (2012). Methods for reducing biological load in subterranean formations. US Patent 8,276,663, assigned to Halliburton Energy Services Inc.

58 Aiad, I.A., Tawfik, A.M., and Sayed, A. (2012). Corrosion inhibition of some cationic surfactants in oilfields. *Journal of Surfactants and Detergents* **15** (5): 577–585.

59 Aiad, I.A., Tawfik, A.M., Shaban, S.M. et al. (2014). Enhancing of corrosion inhibition and the biocidal effect of phosphonium surfactant compounds for oil field equipment. *Journal of Surfactants and Detergents* **17** (3): 391–401.

60 Petkovic, M., Hartmann, D.O., Adamova, G. et al. (2012). Unravelling the mechanism of toxicity of alkyltributylphosphonium chlorides in *Aspergillus nidulans* conidia. *New Journal of Chemistry* **36**: 56–63.

61 Kanazawa, A., Ikedo, T., and Endo, T. (1994). Synthesis and antimicrobial activity of dimethyl- and trimethyl-substituted phosphonium salts ith alkyl chains of various lengths. *Antimicrobial Agents and Chemotherapy* **38** (5): 945–952.

62 Jaeger, D.A. and Zelenin, A.K. (2001). Alkyltris(hydroxymethyl)phosphonium Halide surfactants. *Langmuir* **17** (8): 2545–2547.

63 Bradaric-Baus, C.J. and Zhou, Y. (2004). Phosphonium salts and methods of their preparation. International Patent Application WO/2004/094438, assigned to Cytec Canada Inc.

64 Annadori, K.M. and Darwin, A. (2010). Effect of THPS on discharge water quality: a lessons learned study. SPE International Conference on Health, Safety and Environment in Oil and Gas Exploration and Production, Rio de Janeiro, Brazil (12–14 April 2010), SPE 125785.

65 Willmon, J. (2010). THPS degradation in the long-term preservation of subsea flowlines and risers. NACE Corrosion 2010 Conference, San Antonio, TX (14–18 March 2010), NACE 10402.

66 Vincent, J.B., Crowder, M.W., and Averill, B.A. (1992). Hydrolysis of phosphate monoesters: a biological problem with multiple chemical solutions. *Trends in Biochemical Sciences* **17** (3): 105–110.

67 Gunstone, F. ed. (2008). *Phospholipid Technology and Applications*, 1ee. Elsevier.

68 van Hoogevest, P. and Wendel, A. (2014). The use of natural and synthetic phospholipids as pharmaceutical excipients. *European Journal of Lipid Science and Technology* **116** (9): 1088–1107.

69 Provost, J.J. and Colabroy, K.L. (2016). *The Science of Cooking*. Wiley-Blackwell.

70 Patel, A.D., Davis, E., Young, S. and Stamatkis, E. (2006). Phospholipid lubricating agents in aqueous based drilling fluids. US Patent 7,094,738, assigned to M-I LLC.

71 Gaidi, J.M. (2004). Chemistry of corrosion inhibitors. *Cement and Concrete Compositions* **26** (3): 181–189.

72 Knowles, J.R. (1980). Enzyme-catalyzed phosphoryl transfer reactions. *Annual Review of Biochemistry* **49**: 877–919.

73 https://www.epa.gov/sites/production/files/documents/rmpp_6thed_ch5_organophosphates.pdf

74 Carson, R. (2000). *The Silent Spring*. Penguin Modern Classics.

75 Correll, D.L. (1996). The role of phosphorus in the eutrophication of receiving waters: a review. *Journal of Environmental Quality Vol.* **27** (2): 261–266.

76 https://www.epa.gov/nutrientpollution/problem

77 Sharpley, A.N., Smith, S.J., and Waney, J.N. (1987). Environmental impact of agricultural nitrogen and phosphorus use. *Journal of Agricultural and Food Chemistry* **35** (5): 812–817.

78 Bennett, E.M., Carpenter, S.R., and Caraco, N.F. (2001). Human impact on erodable phosphorus and eutrophication: a global perspective. *BioScience* **51** (3): 227–234.

79 Saeger, V.W., Hicks, O., Kaley, R.G. et al. (1979). Environmental fate of selected phosphate esters. *Environmental Science & Technology* **13** (7): 840–844.

80 Demadis, K.D. and Ketsetzi, A. (2007). Degradation of phosphonate based scale inhibitor additives in the presence of oxidising biocides: "collateral damages" in industrial water systems. *Separation Science and Technology* **42**: 1634–1649.

81 Kaur, B., Gur, I.S., and Bahtnagar, H.L. (1987). Thermal degradation studies of cellulose phosphates and cellulose thiophosphates. *Macromolecular Materials and Engineering* **147** (1): 157–183.

82 Clearfield, A. (2011). *Metal Phosphonate Chemistry: From Synthesis to Applications*. Royal Society of Chemistry.

83 Dyer, S.J., Anderson, C.E., and Graham, G.M. (2004). Thermal stability of amine methyl phosphonate scale inhibitors. *Journal of Petroleum Science and Engineering* **43** (3–4): 259–270.

84 Fang, J.L., Li, Y., Ye, X.R. et al. (1993). Passive films and corrosion protection due to phosphonic acid inhibitors. *CORROSION* **49** (4): 266–271.

5

Metals, Inorganic Salts and Other Inorganics

This chapter is concerned with the chemistry of metals, their salts and other inorganic compounds as applied in the upstream oil and gas industry. Where a particular metal or other inorganic substance has a number of uses or is widely used across the sector, then a separate section will detail its properties and describe its action. Where materials or individual substances such as rare earth metals are used, then these are brought together in a separate section. The chapter is concerned with the simple salts and complexes of these metals and related products. It will not cover complex salts of organic compounds such as polymers (e.g. acrylates) or surfactants (e.g. sodium alkyl benzene sulphonate), as the primary substances are fully described in other chapters. Silicates are detailed in Section 7.2.

Metals particularly as metal ions are important to all living organisms and are present in the environment in various concentrations. They are usually bioavailable for uptake via the food chain, and many are essential for life at certain concentration ranges; however most are toxic at higher concentrations. This balance between essential presence and toxicity will be discussed throughout this chapter and in the closing Section 5.10, which will also consider the importance of bioavailability, toxicity and biodegradation.

5.1 Alkali Metal Salts and Related Materials

The alkali metals are a group (column) in the periodic table consisting of the chemical elements lithium (Li), sodium (Na), potassium (K), rubidium (Rb), cesium (Cs) and francium (Fr). The alkali metals provide the best example of group trends in properties in the periodic table, with elements exhibiting well-characterised homologous behaviour.

They are all shiny, soft and highly reactive, and as such they must be stored under oil to prevent reaction with air and are found naturally only in salts and never as the free elements. Cesium, the fifth alkali metal, is the most reactive of all the metals. In the modern IUPAC nomenclature, the alkali metals comprise the group 1 elements, excluding hydrogen (H), which is nominally a group 1 element but not normally considered to be an alkali metal as it rarely exhibits behaviour comparable to that of the alkali metals. All the alkali metals react with water, with the heavier alkali metals reacting more vigorously than the lighter ones.

All of the discovered alkali metals occur in nature, with sodium as the most abundant, followed by potassium, lithium, rubidium, cesium and finally francium, which is very rare due to its extremely high radioactivity.

Most alkali metals have many different applications. One of the best-known applications of the pure elements is the use of rubidium and cesium in atomic clocks, of which cesium atomic clocks are the most accurate representation of time. Table salt, or sodium chloride, has been

Oilfield Chemistry and Its Environmental Impact, First Edition. Henry A. Craddock.
© 2018 John Wiley & Sons Ltd. Published 2018 by John Wiley & Sons Ltd.

used since antiquity. Sodium and potassium are also essential elements, having major biological roles as electrolytes. Although the other alkali metals are not essential, they also have various effects on animal and human metabolism, both beneficial and harmful.

This section is mainly concerned with metal salts of sodium, potassium and lithium although some cesium salts are also considered. Primarily it will examine the halide salts, nitrates, sulphates and other similar compounds. The phosphates have already been described in Section 4.1. There are many sodium and potassium salts of polymers, surfactants, chelants and other complex organic molecules, many of which have been described in other chapters under their polymer, surfactant or other function. This section will only examine the relatively simple and mostly inorganic salts of alkali metals.

5.1.1 Sodium Salts

There are a large number of sodium salts used within the upstream oil and gas sector and the major ones alongside details on their application are now described.

5.1.1.1 Sodium Acetate

Sodium acetate is the salt of acetic acid and sodium, and it is widely used across a number of industrial sectors. Of interest in the oil and gas sector is its use to impede the vulcanisation of chloroprene in synthetic rubber production. It is similarly used as a vulcanisation retarder where latex is added to a water-based drilling fluid to reduce the rate at which drilling fluid permeates the borehole wall during the drilling operation [1].

Industrially, sodium acetate is prepared from glacial acetic acid and sodium hydroxide as shown in Equation (5.1):

$$CH_3COOH + NaOH \rightarrow CH_3COONa + H_2O. \tag{5.1}$$

Sodium acetate may be added to food as a seasoning and is often used to give potato chips a salt and vinegar flavour. It is also used in the construction industry to mitigate water damage to concrete by acting as a concrete sealant while also being environmentally benign and cheaper than the commonly used epoxy alternative. Although this property has been known for some time, it has recently been re-examined and developed [2]; no application appears to be recorded in oilfield cementing operations.

Sodium acetate is the conjugate base of acetic acid, a solution of sodium acetate and acetic acid, and can therefore act as a buffer to keep a relatively constant pH level. This is useful especially in formulation of products where the applications are pH dependent in a mildly acidic range (pH 4–6) (H.A. Craddock, unpublished results).

Sodium acetate is useful as a universal gel breaker for non-aqueous gels particularly gelled fracturing fluids (see also Section 5.1.1.3) [3].

5.1.1.2 Sodium Aluminate

Sodium aluminate is an important commercial inorganic chemical. It works as an effective source of aluminium hydroxide (see Section 5.2) for many industrial and technical applications. Pure sodium aluminate (anhydrous) is a white crystalline solid having a formula $NaAlO_2$, $NaAl(OH)_4$ (hydrated), $Na_2O \cdot Al_2O_3$ or $Na_2Al_2O_4$. Commercial sodium aluminate is available as a solution or a solid.

Sodium aluminate is manufactured by dissolving aluminium hydroxide in a caustic soda (NaOH) solution. Aluminium hydroxide can be dissolved in 20–25% aqueous NaOH solution at a temperature near the boiling point. The use of more concentrated NaOH solutions leads to a semi-solid product.

Sodium aluminate is used in water treatment as an adjunct to water softening systems, as a coagulant aid to improve flocculation, and for removing dissolved silica and phosphates.

In construction technology, sodium aluminate is employed to accelerate the solidification of concrete, mainly when working during frost. Sodium aluminate is also used in the paper industry and for firebrick production.

In the upstream oil and gas sector, sodium aluminate has some specialised uses. It has been used as a thickener in drilling fluid formulations [4]. It is formulated alongside other additives such as sodium silicate in drilling fluids to accelerate gel formation and reduce circulation losses [5]. Interestingly it is also part of a polymer complex with potential as a paraffin wax inhibitor [6]. The polymer was obtained by reacting a long-chain phosphoric ester with sodium aluminate, generating a molecule with relatively high molecular weight and amphiphilic character (see also Section 4.5).

5.1.1.3 Sodium Bicarbonate

Sodium bicarbonate, or more correctly sodium hydrogen carbonate, has the chemical formula $NaHCO_3$ and is an inorganic salt composed of sodium ions and bicarbonate ions. Sodium bicarbonate is a white solid that is crystalline but often appears as a fine powder. It is a food additive encoded by the European Union and identified as E 500. Since it has long been known and is widely used, the salt has many related names such as baking soda, bread soda, cooking soda and bicarbonate of soda. In colloquial usage, the names sodium bicarbonate and bicarbonate of soda are often truncated. Forms such as sodium bicarb, bicarb soda, bicarbonate and bicarb are common.

Sodium bicarbonate has a variety of uses both domestically and commercially. It is well known as a food additive especially in baking and is widely used as a cleaning agent. It has a number of uses in the oilfield, a number of which are now detailed.

Sodium bicarbonate (along with other salts such as sodium carbonate, potassium bicarbonate and potassium carbonate) when formulated can act as a delivery system for carbon dioxide [7].

Soda, which can be either sodium bicarbonate or sodium carbonate, is used as an antifreeze additive for wellbore cements [8].

Sodium bicarbonate, among other inorganic salts, such as sodium hydroxide, can be utilised as part of an aqueous buffering solution to activate natural surfactants in the crude oil in order to provide a stable emulsion of reduced viscosity for enhancing transportation via pipeline for further processing and refining. This method has been claimed to be very effective in certain heavy crudes such as the highly viscous crude oils found in the Orinoco Region of Venezuela [9].

A mixture containing sodium bicarbonate and calcium hydroxide (hydrated lime) is useful for breaking non-aqueous gels used in fracturing fluids or formed from the overall fluid composition [3].

In hydraulic fracturing at high temperatures, the efficiency of the fluid recovery from the formation can be increased and improved by the addition of a blowing agent. Such additives almost always contain sodium bicarbonate [10]. After fracturing the formation, the blowing agent decomposes and the filter cake becomes more porous, which provides a driving force for the removal of the fracturing fluid from the matrix [11].

5.1.1.4 Sodium Bisulphite and Metabisulphite

Sodium bisulphites, along with sodium metabisulphite, are commonly used oxygen scavengers in the upstream oil and gas sector [12] (see also Section 5.9.4.6). The bisulphite ion is oxidised to bisulphate as in the reaction in Equation (5.2):

$$2HSO_3^- + O_2 \rightarrow 2HSO_4^-. \tag{5.2}$$

This reaction is slow at temperatures below around 200 F (93 °C) and is often catalysed by the addition of small concentrations of transition metal ions [13]. Since seawater contains small amounts of such ions, it is not necessary to use catalysed materials [14]. This latter observation is particularly useful in that no addition of usually toxic transition metal ions is discharged to seawater. However, at low temperatures particularly cold seawater, oxygen scavengers such as sodium bisulphite have a reduced rate of reaction and therefore are usually catalysed under such conditions [15]. Bisulphite oxygen scavengers are also sensitive to pH and cannot be used in acidic conditions such as in acid stimulation and workover.

A major application for bisulphite scavengers is in the treatment of seawater for corrosion protection during hydrotesting applications. Removal of oxygen is essential as the static aqueous mixture may be left in place for several months, or even years. Microbial-induced corrosion can also occur, and therefore a biocide is added to the mixture. The compatibility of bisulphites (and sulphites) with many biocides is often problematic, and in many cases each reacts with the other to reduce or eliminate its activity. It is often advocated to add a sulphite oxygen scavenger and biocide in a stepwise manner so that the biocide is only present after all the bisulphite scavenger has been used up in oxygen removal; simultaneously the pH is adjusted to greater than 9.5 with a base such as sodium hydroxide [16]. This method has the major operational drawback of not being a single chemical package delivery system that is much preferred for logistic reasons during hydrotesting operations. The use of alternative oxygen scavengers and other combined action corrosion inhibitor/biocidal products has been investigated, and a number of proprietary formulations have been developed, where one in particular has good environmental characteristics (H.A. Craddock, unpublished results).

5.1.1.5 Sodium Borate

In the design of fracturing fluids for unconventional hydrocarbon recovery where hydratable polymers such as a water-soluble polysaccharide are used, e.g. galactomannan (see Eq. (5.3)), the sodium borate ($Na_3BO_3 \times nH_2O$), is used as a common cross-linking agent [17] (see also Section 5.4).

5.1.1.6 Sodium Bromide

In common with other halide salts of alkali metals and in particular potassium bromide and cesium bromide, various concentrations of aqueous solutions are utilised as brines for completion and preservation of wells.

Saline liquid (brine) is usually used in completion[1] operations and, increasingly, when penetrating a hydrocarbon-bearing ('pay') zone. Brines are preferred because they have higher densities than freshwater. It is also important that they lack solid particles that might damage producible formations. The common classes of brines include chloride brines (calcium and sodium), bromides and formates [18, 19].

Sodium bromide brine systems are used to form clear-brine workover and completion fluids with densities to a maximum of $1498\,kg\,m^{-3}$ where lower-crystallisation temperature or chloride-free solutions are required. Using sodium bromide brines eliminates the potential of formation damage due to the precipitation of carbonate, bicarbonate, or sulphate compounds associated with using calcium-based brines where formation waters contain high concentrations of bicarbonate and sulphate ions. The single-salt system is less likely to exhibit compatibility problems than the corresponding two-salt sodium bromide/sodium chloride fluid [20].

---

1 A generic term used to describe the assembly of downhole tubulars and equipment required to enable safe and efficient production from an oil or gas well.

Halide salt brines have serious corrosion problems at high temperature and to this end a number of other sodium salt systems based on carboxylic acids have been developed. These compounds are further detailed in Section 6.1.

5.1.1.7 Sodium Carbonate

Sodium carbonate is used where seawater is a primary make-up fluid for a water-based mud. Seawater has high hardness levels due to calcium and magnesium ions. The calcium ion concentration can be significantly reduced by the addition of sodium carbonate, which precipitates the calcium as calcium carbonate. Magnesium is similarly reduced in concentration by the addition of sodium hydroxide to precipitate magnesium hydroxide (see later). Similarly hydroxides and carbonates including sodium carbonate are used in drilling mud compositions to adjust the pH. This is particularly important when a water-based drilling fluid is designed to form a semipermeable membrane in order to increase wellbore stability [21].

In alkali surfactant flooding for enhanced oil recovery, sodium carbonate can be used as the alkali system. The sodium carbonate reacts with acidic components of the crude oil to form a secondary surfactant system. This secondary surfactant aids the primary surfactant to reduce the overall interfacial tension between any residual oil and the injected fluid, thereby improving and increasing overall oil recovery from the reservoir [22].

Sodium carbonate can be used as a controlling or delaying agent in certain filter cake degradation operations where ester or *ortho*-ester hydrolysis is required to produce the acid for the degradation to progress [23]. A similar procedure can be utilised in foamed fracturing fluids where *ortho*-esters are utilised as defoamers to convert the foaming surfactant to a non-foaming surfactant [24].

Finally sodium carbonate can be used in conjunction with some polymers such as low molecular weight polyacrylamides (PAMs). This forms a colloidal system that can be used as a blocking agent in water shut-off applications. The system has several advantages over more conventional polymer systems particularly as it has low viscosity and therefore is easily pumped and gives good placement selectivity within the treatment zones [25].

5.1.1.8 Sodium Chloride

Sodium chloride is an abundant halide salt present in its dissolved form in seawater. It has a number of applications as an additive in the upstream oil and gas sector mainly in drilling, cementing and completion operations. The salinity of associated water and concentration of sodium and chloride ions, in drilling, cementing, production and other operations, such as enhanced oil recovery, can have implications for the effectiveness of other additives; however in this section the main focus is on sodium chloride as an additive.

Seawater is used in drilling mud formulations but must have a concentration of greater than 10 000 ppm. The average concentration of seawater is 10 500 ppm, so in a large number of cases it is suitable for use. Although the vast majority of seawater has a salinity of between 3.1% and 3.8%, it is not uniformly saline throughout the world. Where mixing occurs with freshwater runoff from river mouths or near melting glaciers, seawater can be substantially less saline. The most saline open sea is the Red Sea, where high rates of evaporation, low rainfall and river inflow and confined circulation result in unusually salty water. The salinity in isolated bodies of water can be considerably greater still.

Sodium chloride brine is used as part of the main drilling fluid in a number of biodegradable drilling fluid formulations [26]. Similarly more traditional formulations use various amounts of sodium chloride, particularly where polysaccharides, such as starch or carboxymethyl cellulose (CMC), are also used in the formulation as fluid loss additives. However, it has been found that sodium chloride and other electrolytes can have both beneficial and damaging effects during drilling, depending on the rock geology [18].

It has been demonstrated that the addition of sodium chloride (and also potassium chloride; see later) to solutions of clay and shale inhibitors improves their ability to reduce the absorption of aqueous fluid into the shale rock [27].

As has been discussed under sodium bromide, solids-free brines provide well control during completion and workover operations that include such procedures as

- Perforating
- Gravel packing
- Fracturing
- Leaving a weighted fluid in the tubing/casing annulus as a packer fluid

Brines are selected on the basis of density, compatibility with the formation and corrosion control. When properly designed, these brines avoid the formation damage and reduced permeability characteristic of conventional weighted drilling fluids. Sodium chloride brines fit in the density range 8.4–10.0 ppg (pounds per gallon). In contrast sodium bromide has a wider range of action from 8.4 to 12.7 ppg.

Reservoir damage caused by drilling muds during completion operations is well documented [28, 29]. Because of this, solids-free brines are entering a fifth decade as the standard completion fluids in the oilfield. However, in some instances, the most commonly available brines can cause formation damage or corrosion in wellbore tubulars, and hence more exotic brines are considered, and these will be described as other metal salts are detailed.

Sodium chloride, along with calcium chloride (see Section 5.6) is the most common cement accelerator in well operations. It is used in formulations that are bentonite-free and to a maximum concentration of 5%. Above this concentration its effectiveness reduces, indeed saturated sodium chloride solutions have retarding properties [30].

It has been claimed that when 2% sodium chloride is incorporated with quaternary ammonium salts into the cement slurry, the cement rock increases in strength and in adhesion to the formation by over 50% [31].

As discussed with brines for completion, cementing can also be improved in wells containing clays or shales that are sensitive to freshwater cement filtrates by using sodium chloride (and potassium chloride) during cementing to give a less damaging filtrate [32].

The use of sodium chloride for shale control has certain advantages over potassium chloride. Sodium chloride solutions near saturation have elevated base viscosities and have lower water activities than those of concentrated potassium chloride solutions, giving rise to higher osmotic pressures. Therefore, they are better equipped to reduce filtrate invasion in shales. Although concentrated sodium chloride solutions do not make good shale drilling fluids by themselves, they are very effective when run in combination with systems that can enhance shale membrane efficiency such as silicates, polyols and methylglucoside [32].

Sodium chloride has inherent properties to depress freezing points in solution and therefore is applied as an antifreeze agent, for example, a 23% solution of sodium chloride will depress the freezing point of water by 21 °C. This property means that despite serious drawbacks, such sodium chloride solutions are still used as heat transfer agents in the offshore sector mainly as they are cheap and readily available from the surrounding seawater. Sodium chloride is also used as antifreeze agents for cement in well completions [33].

This antifreeze characteristic is also the basis of the use of sodium chloride as a gas hydrate inhibitor. The use of such kinetic hydrate inhibitors has been in place for over two decades; [34] however, sodium chloride is little used in this application as other products are more effective, treat at lower concentrations, and do not have inherent corrosive effects on metal surfaces.

As has been described earlier, sodium chloride has shale and clay inhibitor properties, and this characteristic is utilised in high permeability zones, which can be blocked with a reinforced swelling clay gel. Critical is the action of sodium chloride and in particular that the inhibitive cations bound to the clay are replaced by sodium ions, which attract water molecules and promote clay swelling [35]. Similar applications are utilised in terms of clay swelling in conjunction with clay stabilising additives (see Section 6.1).

The mineral form of sodium chloride is known as halite. The mineral is typically colourless or white and forms isometric crystals. It commonly occurs as an evaporite deposit mineral such as that seen in numerous large saltpan basins. It has been observed that some formation waters in reservoirs, particularly those with high pressure and high temperature, contain high concentrations of halite in solution. As the temperature of such formation water decreases in production operations, for example, then the saturated sodium chloride can precipitate. The kinetics of this process is fast, and therefore even in low water cut wells, salt deposits can rapidly form. Similarly release of water into the gas phase as pressure decreases can also concentrate the sodium chloride held in solution, leading to halite deposition [36].

The most common method of countering the deposition of halite scale is the injection of fresh or less saline water, via batch treatments to the affected zones or via continuous injection upstream of the deposition zones. This is a very valid approach that is both cost effective and efficient. However, this is not always the most practical solution as large volumes of freshwater may not be available, nor may it be possible to introduce the water in the required volume to the affected zones. In these instances, sub-stoichiometric chemical halite inhibitors are applied as a valid alternative. A number of these are discussed in Chapter 6 and later in Section 5.7, in particular, hexacyanoferrate salts.

The use of sodium chloride in a number of oilfield operations needs to be considered for its impact on corrosion. It is well known that chloride ion can accelerate a number of types of corrosion [37] particularly pitting corrosion.

The environmental concerns centre on increasing the salinity of the local environment as obviously the components are readily found naturally in seawater and other freshwater aquatic ecosystems and are unlikely to have any significant environmental impact. There is a general acceptance that freshwater ecosystems undergo little ecological stress when subjected to salinities of up to $1000\,mg\,l^{-1}$ [38]. However, much of our understanding of the effects of salinity on freshwater ecosystems comes from lowland rivers where exposure to significant salt concentrations already occurs; other systems may be more sensitive. For many aquatic species, sub-lethal effects may not be apparent at the community level for many generations. Other underlying factors such as habitat modification, loss of food resources or modification of predation pressure may also be causing changes within these systems. It is known that adult fish and macro-invertebrates appear to tolerate increasing salinity [39]; however freshwater algae, aquatic plants and micro-invertebrates appear to be less tolerant of increased salt concentrations. Clearly these systems are complex, and salinity is only one of many factors, which influence the overall ecological balance.

5.1.1.9 Sodium Chlorite, Hypochlorite and Related Substances

Sodium chlorite has been considered as a biocide in secondary oil recovery as it releases chlorous acid in situ when reacted with a low molecular weight organic acid, e.g. lactic acid [40]. A specific problem associated with 'secondary oil recovery' processes is the growth of micro-organisms in both the injection water system and the oil bearing formation. These micro-organisms are of many types and may be either anaerobic or aerobic in nature. A considerable advantage of the sodium chlorite system is that it allows for the introduction of an anti-bactericidal agent in the injection water system to rid both the injection water system and the oil-bearing formation of most known microorganisms. Furthermore, hydrogen sulphide is frequently produced in the water injection system and/or the oil-bearing formation, and both sodium chlorite and chlorous acid can provide a method of eliminating this problem by oxidising the hydrogen sulphide and/or killing the microbes that produce it. A considerable disadvantage to such a system is the high rates of corrosion experienced due to residual chlorite ion (H.A. Craddock, unpublished results).

Sodium hypochlorite, NaOCl, is a compound that is commonly used for water purification. It is used on a large scale for surface purification, bleaching, odour removal and water disinfection. Sodium hypochlorite was developed around 1785 by the French chemist Claude Berthollet, who manufactured liquid bleaching agents based on sodium hypochlorite. Sodium hypochlorite is a clear, slightly yellowish solution with a characteristic odour. As a bleaching agent for domestic use, it usually contains 5% sodium hypochlorite, with a pH of around 11. If it is more concentrated, it contains a concentration of 10–15% sodium hypochlorite, with a pH of around 13. Sodium hypochlorite is unstable. Chlorine evaporates from hypochlorite solutions at a rate of 0.75 g of active chlorine per day, if heated sodium hypochlorite disintegrates. This also happens when sodium hypochlorite comes in contact with acids, sunlight, certain metals and poisonous and corrosive gases, including chlorine gas. Sodium hypochlorite is a strong oxidising agent and reacts with flammable compounds and reducing agents. Sodium hypochlorite solution is a weak base that is inflammable.

Sodium hypochlorite can be produced in two ways:

1) By dissolving salt in softened water, which results in a concentrated brine solution. The solution is electrolysed and forms a sodium hypochlorite solution in water. This solution contains 150 g active chlorine (Cl_2) per litre. During this reaction the explosive hydrogen gas is also formed.
2) By adding chlorine gas (Cl_2) to caustic soda (NaOH). When this is done, sodium hypochlorite, water (H_2O) and salt (NaCl) are produced according to the reaction in the following equation:

$$Cl_2 + 2NaOH^+ \rightarrow NaOCl + NaCl + H_2O. \tag{5.3}$$

When sodium hypochlorite is added to water, hypochlorous acid (HOCl) is formed, as in Equation (5.4).

$$NaOCl + H_2O \rightarrow HOCl + NaOH. \tag{5.4}$$

Sodium hypochlorite is effective against bacteria, viruses and fungi and disinfects in the same manner as chlorine (see Section 5.6).

Sodium hypochlorite is a well-established water treatment biocide and is used widely in the upstream oil and gas industry. Indeed, hypochlorite is readily generated for seawater electrochemically [41], and many offshore installations use this process to treat water for injection or reinjection purposes. It is an oxidising biocide (as is chlorine). Such biocides cause irreversible oxidation and hydrolysis of protein groups within the microorganism and the polysaccharide biofilm that often attaches the microbial population to equipment surfaces. The result is cell death, and it is effective against all strains of bacteria and related microorganisms.

Although oxidising biocides such as sodium hypochlorite can be effective when applied properly, using them alone cannot guarantee a successful microbiological control. Many factors can adversely impact the performance of oxidising biocides [42].

A related substance is hypochlorous acid (HOCl), which is a weak acid that forms when chlorine (see Section 5.5) dissolves in water and is observed to be a more powerful biocide than the hypochlorite ions [43].

Finally sodium hypochlorate has been claimed for use as a gel breaker in fracturing fluid formulations [44].

These sodium salts are highly reactive and dissociate from their basic constituents very rapidly, and therefore the environmental impact, as with sodium chloride, results from additional concentration of the electrolytes in the aquatic environment.

5.1.1.10 Sodium Chromate and Dichromate

Sodium chromate has the chemical formula Na_2CrO_4. It exists as a yellow hygroscopic solid, which can form tetra-, hexa- and decahydrates. Sodium chromate, like other hexavalent chromium compounds, is toxic and has carcinogenic properties similar to those of chromium (see Section 5.10).

It is obtained on an industrial scale by roasting chromium ores in air in the presence of sodium carbonate (see Equation (5.5)):

$$Cr_2O_3 + 2Na_2CO_3 + 1.5O_2 \rightarrow 2Na_2CrO_4 + 2CO_2. \tag{5.5}$$

Subsequent to its formation, the chromate salt is converted with acid to sodium dichromate (see Eq. (5.6)), the precursor to most chromium compounds and materials:

$$2Na_2CrO_4 + 2H^+ \rightarrow^+ H_2O + Na_2Cr_2O_7. \tag{5.6}$$

In the upstream oil and gas sector, some sodium chromate has been used as a specialised cement biocide [45] and as a dispersant additive in certain cement slurries [33]. Sodium dichromate has been more commonly used particularly as a cross-linking agent in polymers, such as CMC, applied as gelling agents for water shut-off and enhanced oil recovery [46]. It is however fair to state that such materials are only sparingly used where no other technical or practical chemical alternative exist that is primarily driven by their extreme toxicity and potentially severe environment impact (see Section 5.11).

5.1.1.11 Sodium Hydroxide

Sodium hydroxide will undoubtedly be familiar to the reader by its common name, caustic soda. It is a white solid and highly caustic metallic base and alkali, which is available in a variety of solid forms as well as prepared solutions at different concentrations. Sodium hydroxide is soluble in water and alcohols such as ethanol and methanol. It is a deliquescent substance that readily absorbs moisture and carbon dioxide from air.

Sodium hydroxide is used in many industries, mostly as a strong chemical base in the manufacture of pulp and paper, soaps and detergents and in the textile industry. It is used mainly in drilling applications and enhanced oil recovery in the upstream oil and gas industry.

Sodium hydroxide was originally produced by treating sodium carbonate with calcium hydroxide in a metathesis reaction shown in Eq. (5.7). It is significant that sodium hydroxide is soluble, while calcium carbonate is not:

$$Ca(OH)_2(aq) + Na_2CO_3(s) \rightarrow CaCO_3 \downarrow + 2NaOH(aq). \tag{5.7}$$

This process was superseded in the late nineteenth century by the Solvay process [47], which in turn was supplanted by the chloralkali process used today. This is an industrial process for the electrolysis of sodium chloride. It is used to produce chlorine and sodium hydroxide (caustic soda). As the process gives equivalent amounts of chlorine and sodium hydroxide (2 mol of sodium hydroxide per mole of chlorine), it is necessary to find a use for these products in the same proportion. For every mole of chlorine produced, 1 mol of hydrogen is produced. Much of this hydrogen is used to produce hydrochloric acid or ammonia or is used in the hydrogenation of organic compounds.

Worldwide production is approximately 60 million tonnes, while consumption could be as much as 10 million tonnes lower. The chloralkali process is also an energy-intensive process, and their major environmental impact generated across all industries consuming and using sodium hydroxide lies in this energy use and its impact on carbon emissions.

In the oilfield sector, sodium hydroxide has a variety of uses. It can be used as a binder in cementing applications particularly in carbon fibre-based cements [48]. It is used also in some cement compositions that use humic acid to make it more water soluble [33].

It has a variety of uses in drilling mud formulations:

- Removal of hardness from seawater [49].
- Control of pH in silicate muds [33].
- pH adjustment and control for specific drilling fluids is shale applications [21].
- Additional fluid loss control [44].
- It is used as an inhibitor base in filter cake removal [23].

Sodium hydroxide has been applied as an aqueous buffer system to release natural surfactants in heavy crudes and allow effective transportation via pipeline of these viscous crudes without heating or dilution by solvents [9].

A major use of sodium hydroxide is in enhanced oil recovery particularly in alkaline surfactant and alkaline polymer surfactant flooding. In general it has been found that reduction in acid concentration lowers interfacial tension and improves recovery [50]. The interfacial tension between oil and water is one of the most important parameters for the efficiency of chemical enhanced oil recovery, and it has a strong time dependency especially under alkaline conditions. When sodium hydroxide is applied to floodwater, it improves the oil recovery because it activates the natural surfactants present in the crude oil [51].

The effectiveness of alkaline additives in enhanced oil recovery tends to increase with increasing pH; however in the reservoir such strong alkalis react with the rock minerals, giving serious deterioration in recovery performance, and therefore it is essential to optimise the operational pH to around 10 [52].

5.1.1.12 Minor use Sodium Compounds

A number of other simple sodium salts have intermittent and minor use in the upstream oil and gas sector.

Sodium cyanate can be used as a delaying agent in activated gels, particularly in systems where aluminium hydroxide (see later) is considered as a plugging agent [53].

It has been observed that sodium isocyanate along with other salts, such as sodium salicylate, can generate viscoelasticity in surfactant solutions, particularly those based on long-chain quaternary ammonium salts. This allows such solutions to act as fluid loss additives in drilling operations [54] and be used in fracturing fluids [33].

Sodium thiocyanate, along with other thiocyanates such as ammonium or calcium thiocyanate, is used as corrosion inhibitors in drilling, completion and workover fluids. This is particularly relevant in high scaling carbonate or sulphate wells where the fluid is essentially calcium-free [55].

Sodium molybdate has been used along with other alkali molybdates to aid graphite as a lubricant [56] (see Section 5.10).

After treatment with formed gels in water shut-off and fluid-based fracturing operations, it is often necessary to chemical 'break' or degrade these gel structures. Sodium persulphate is commonly used to break down polysaccharide-based gels [57]. It is also used as an activator in the 'SGN' heat-generating system for wax deposition treatment [58], which is further described later in Section 5.9.4.6.

Sodium sulphate can act as cement accelerator while also maintaining mobility pre-hardening for ease of application [59].

Sodium thiosulphate and sodium thioglycolate are used in formulated oilfield corrosion inhibitors to enhance the performance of other film-forming surfactant inhibitors particularly quaternary ammonium-based products [60]. It has been observed some time ago that thiosulphate salts and other sulphur-containing molecules have synergistic effects on corrosion inhibition [61, 62] and also that they show significant corrosion inhibition properties [63]. Their use has been limited to small additions to formulated products although they could be useful as environmentally acceptable inhibitors having low toxicity and dissociating to ionic species found in the aquatic environment.

Finally in this section concerned with sodium and its salts, the radioisotope [20]Na$^+$ has been considered for use as a tracer for water flow in carbonate reservoirs [64].

Other sodium salts such as phosphates (see Section 4.1), formates (see Section 6.1) and silicates (see Section 7.1) are discussed in the chapters relevant to the cationic species. Often alkali metal salts and in particular sodium (and potassium) are utilised to allow water solubility of complex organic molecules, and it is under the relevant chemical functionality that these have been detailed and described such as polymers (and monomers thereof), surfactants and related products.

5.1.2 Potassium Salts

Potassium salts can be found for most of the sodium salts described in Section 5.1.1; therefore in the main these shall not be re-examined. However some particular potassium salts, which are different or have different functionalities like potassium chloride, will be described.

5.1.2.1 Potassium Carbonate

Potassium has many similar functions to sodium carbonate such as being a drilling mud additive [21] and a provider of carbon dioxide for certain biocide applications [7]. Similar to potassium chloride, it can be used as a clay stabiliser [65] and is also useful in enhancing scale dissolver formulations [66].

5.1.2.2 Potassium Chloride

Potassium-based drilling muds are the most widely used water-based drilling fluid systems for drilling sensitive shales. This is because (due to atomic size) potassium ions are the most effective at attaching to clay surfaces to give stability to the shale that is exposed to the drilling fluid by the drill bit. Potassium ions are also the most efficient at aggregating the cuttings and minimising their dispersion as fine particles [67]. Potassium chloride is the most widely used source of potassium ions in such water-based mud formulations; however other salts like potassium acetate, potassium carbonate and potassium hydroxide are also used.

Swelling in shales is a common problem encountered in drilling operations, and this can be inhibited by adding large amounts of potassium chloride [68]. This and other inhibitors of clay swelling act by a chemical mechanism, changing the ionic strength and transport mechanism of fluids into the clays. Both the nature and amount of anions and cations are important to the efficiency of the inhibition process.

This problem is becoming more general as water-based drilling fluids are increasingly being used for oil and gas exploration and are generally considered to be more environmentally acceptable than oil-based or synthetic-based fluids. Their use, however, facilitates clay hydration and swelling. Clay swelling, which occurs in exposed sedimentary rock formations, can have an adverse impact on drilling operations and may lead to significantly increased oil well construction costs. To effectively reduce the extent of clay swelling, the mechanism by which clay minerals swell needs to be understood so that efficient swelling inhibitors may be developed. Acceptable clay swelling inhibitors must not only significantly reduce clay hydration but must also meet increasingly stringent environmental guidelines while remaining cost effective. The development of these inhibitors, which are generally based upon water-soluble polymers, therefore represents an ongoing challenge to oilfield geochemistry.

Similarly in wells with zones containing shales or clays that are freshwater sensitive, the cementing adhesion and integrity can be improved by the use of potassium chloride, which also leads to less filtrate damage in these zones.

Potassium chloride can be used to generate viscoelastic properties in surfactant solutions [54] (see also sodium isocyanate in previous section), and this can be used to effect in the design of fracturing fluids, particularly those containing long-chain quaternary ammonium salts [33].

Like sodium chloride, potassium chloride has inherent properties to depress freezing points in solution and therefore can also be applied as an antifreeze agent.

Potassium chloride can present some specific disposal problems in that certain regions prohibit or restrict the discharge of potassium salts to the environment, and similarly the use of potassium chloride in water-based drilling fluids can also present problems in onshore drilling due to possible contamination of groundwater [69].

5.1.2.3 Potassium Hexacyanoferrate

Hexacyanoferrate salts, particularly potassium hexacyanoferrate, have been used in the inhibition of halite (see earlier, Section 5.1.1.8). Unlike the more common carbonate and sulphate scales, halite contains only monovalent anions, making more traditional-scale inhibitors ineffective. Potassium hexacyanoferrate has been known for some time to modify the crystal morphology of halite, and at relatively high concentrations it increases the solution critical supersaturation, resulting in a significant inhibition effect [70]. Potassium hexacyanoferrate has been applied successfully in the field to control halite deposition in gas compression equipment albeit at a relatively high concentration [71].

There has been no further reported use of potassium hexacyanoferrate in the oil and gas industry, and other approaches and other chemistries to halite control have been explored [72, 73]. This is probably due to the high dose rates involved as potassium hexacyanoferrate has low toxicity [74].

5.1.2.4 Potassium Hydroxide

Potassium hydroxide is an inorganic salt and has the chemical formula KOH. It is commonly called caustic potash and is likely to be familiar to the reader. It has many industrial and niche applications, most of which exploit its corrosive nature and its reactivity towards acids. It is similar to sodium hydroxide as the archetypical strong base; however approximately 100 times more NaOH than KOH is produced annually [75].

As with sodium hydroxide potassium hydroxide was made by adding potassium carbonate (potash) to a strong solution of calcium hydroxide (slaked lime), leading to the precipitation of calcium hydroxide, leaving potassium hydroxide in solution:

$$Ca(OH)_2 + K_2CO_3 \rightarrow CaCO_3 + 2KOH. \tag{5.8}$$

It has been replaced since the nineteenth century by the current method of electrolysis of potassium chloride solutions (Eq. (5.9)). The method is analogous to the manufacture of sodium hydroxide (see earlier in this section):

$$2KCl + 2H_2O \rightarrow 2KOH + Cl_2 + H_2. \tag{5.9}$$

Hydrogen gas forms as a by-product on the cathode; concurrently, an anodic oxidation of the chloride ion takes place, forming chlorine gas as a by-product.

Potassium hydroxide and sodium hydroxide are often interchangeable for a number of applications including oilfield uses. Often sodium hydroxide is preferred due to its lower cost. Therefore potassium hydroxide can have uses in drilling fluids, enhanced oil recovery, fracturing fluids, cementing, lubricants and scale inhibitors, to name just a few. Generally when sodium hydroxide can be used, then it is favoured on a cost basis.

In drilling fluid formulation, potassium hydroxide can be used for pH adjustment and may be favoured where potassium is used to inhibit shale swelling. Potassium hydroxide is a preferred neutralising agent in glycerol-based drilling lubricants [76]. Similarly along with other agents including sodium hydroxide, it is used as a delaying agent in the generation of organic acids for filter cake removal [23].

In a somewhat analogous application in the inhibition of defoamers supplied as part of a fracturing fluid package, a small amount of a strong base such as potassium hydroxide can be preferred over larger volumes of a weak base [24]. This often provides better compatibility with other components in the fracturing fluid.

Potassium hydroxide is the preferred strong alkali in the formulation of scale dissolvers and also as a performance enhancer [66] (H.A. Craddock and R. Simcox, unpublished results).

5.1.2.5 Other Potassium Salts

There are a few other potassium salts that are used in the upstream oil and gas sector.

Potassium iodide has been used as a synergist in corrosion inhibitors based on quaternary ammonium salts and other products [77], in particular 1-(4-pyridyl)-pyridinium chloride hydrochloride, dodecylpyridiniumchloride, benzyldimethylstearylammonium chloride, and (dodecyltrimethylammonium bromide), alkynols (1-octyn-3-ol, propargyl alcohol) and *trans*-cinnamaldehyde. These products are used in acid corrosion inhibition, particularly for the protection of metal surfaces during acid workover and stimulation processes.

Potassium permanganate has been examined as an oxidising agent to promote dissolution of lead scales, but has difficulties with the subsequent precipitation of manganese oxide.

Like sodium molybdate, potassium molybdate has been used to aid graphite as a lubricant (see Section 5.10).

5.1.2.6 Potassium in the Environment

Potassium in the environment plays a central role in plant growth and often limits it. Potassium from dead plant and animal material is often bound to clay minerals in soils, before it dissolves in water. Consequently, it is readily taken up by plants again. Consequently, potassium fertilisers are often added to agricultural soils. Plants contain about 2% potassium (dry mass) on average, but values may vary from 0.1% to 6.8%. Potassium salts may kill plant cells because of high osmotic activity.

Potassium is weakly hazardous in water, but it does spread quite rapidly, because of its relatively high mobility and low transformation potential. Potassium salt toxicity is usually caused by other components in a compound, for example, cyanide in potassium cyanide.

5.1.3 Lithium and Its Compounds

Under normal conditions, lithium is the lightest metal and the least dense solid element. Like all alkali metals, lithium is highly reactive and flammable. For this reason, it is typically stored in mineral oil. Due to its high reactivity, lithium never occurs freely in nature and, instead, appears only in compounds, which are usually ionic salts. Lithium occurs in a number of minerals such as quartz, feldspar and mica; however due to its solubility as an ion, it is present in seawater and is commonly obtained from brines and shales. On a commercial scale, lithium is isolated electrolytically from a mixture of lithium chloride and potassium chloride.

Lithium and its compounds have several industrial applications, including heat-resistant glass and ceramics, lithium grease lubricants, flux additives for iron, steel and aluminium production, lithium and lithium ion batteries. Due to its relative isotopic instability, lithium is an important element in nuclear physics [78].

Trace amounts of lithium are present in all organisms. The element serves no apparent vital biological function, since animals and plants survive in good health without it, though non-vital functions have not been ruled out.

5.1.3.1 Lithium Greases

In the upstream oil and gas sector, lithium has been used as a complex in certain drilling lubricants and greases [79]. Such grease made with lithium soap ('lithium grease') adheres particularly well to metal surfaces, is non-corrosive, may be used under heavy loads and exhibits good temperature tolerance. For high-performance and higher-temperature applications, lithium greases have been superseded by other types of lubricants [80].

In cold climates and under subsea conditions, metal lubricants represent a significant challenge encountered by any moving components in cold climate. They need to demonstrate good viscosity properties over a wide temperature range and sustain such temperature extremes in storage. Furthermore, heavy duty greases for subsea connectors exposed to well stream heating have to maintain the required level of lubricity for a prolonged period under water to disconnect equipment any time [81].

From an environmental impact standpoint, most of lubricant compounds are not able to meet all requirements for use and discharge under most regulatory controls. Typically 'green' compounds to date present inferior performance characteristics and do not pass preliminary performance tests in cold climates. A compromise between lubricant properties and biodegradability/toxicity level needs to be found. It is highly possible that lithium-based lubricants can contribute in solving this dilemma.

5.1.3.2 Lithium Salts

Lithium bromide (LiBr) absorption chillers have been in use for more than half a century, mainly in the commercial air conditioning industry. However they are also operated with suitable modification in gas conditioning plants. It has been proven that such chiller technology can be operated economically and on a continuous basis in an oilfield environment with minimal operation and maintenance costs [82].

Lithium carbonate is used as a setting accelerator in cements with a high alumina content. Such cements, made from bauxite (aluminium ore of mixed composition) and limestone (calcium carbonate), are comparatively resistant to chemical attack and have rapid hardening properties [83].

Lithium salts of acrylic acid or methacrylic acid polymers and copolymers are used as drilling fluid dispersants [84].

5.1.3.3 The Environmental Impact of Lithium

The oil and gas exploration and development industry is not a major contributor to the introduction of lithium and lithium compounds into the environment and the effect of mining for lithium minerals. It is their use in lithium ion batteries found in most mobile phones, laptop computers, wearable electronics and almost anything else powered by rechargeable batteries.

Lithium is typically found in salt flats in areas where water is scarce. The mining process of lithium uses large amounts of water. Therefore, on top of potential water contamination as a result of its use, depletion or transportation costs are issues to be dealt with. Depletion results in less available water for local populations, flora and fauna.

The recovery rate of lithium ion batteries, even in first-world countries, is in the single-digit percent range. Most batteries end up in landfills.

The US Environmental Protection Agency (EPA) points out that nickel and cobalt, both also used in the production of lithium ion batteries, represent significant additional environmental risks [85].

A 2012 study commissioned by the European Union compares lithium ion batteries to other types of batteries available (lead–acid, nickel–cadmium, nickel–metal–hydride and sodium sulphur) [86]. It concludes that lithium ion batteries have the largest impact on metal depletion, suggesting that recycling is complicated. In their manufacture lithium ion batteries, together with nickel–metal–hydride batteries, are the most energy-consuming technologies using the equivalent of 1.6 kg of oil per kg of battery produced. They also ranked the worst in greenhouse gas emissions with up to 12.5 kg of CO_2 equivalent emitted per kg of battery.

Technology will of course improve, lithium supplies will be sufficient for the foreseeable future, and recycling rates will climb. Other issues like the migration of aging cars and electronic devices to countries with less developed infrastructures will, however, remain, as will the reality of lithium mining and processing. It is therefore conceivable that new battery technologies will gain more importance in the years to come, as well as hydrogen fuel cells.

5.2 Aluminium and Its Salts

Aluminium is a silver-white metal of group 13 (the boron group) of the periodic table. In the Earth's crust, it is the most abundant metallic element and the third most abundant element after oxygen and silicon, but is never found in nature in its elemental state. It occurs mainly as very stable oxides, hydroxides and silicates. The main naturally occurring form is the mineral bauxite [87].

All industrial production of aluminium today is by electrolytic reduction of alumina dissolved in molten cryolite, independently invented by C. M. Hall in the United States and P. Héroult in France. Aluminium has many desirable physical properties. It is malleable and lightweight. Pure aluminium is relatively soft and weak, but it forms many strong alloys. It has high thermal and electrical conductivity. Aluminium's adherent surface oxide film makes it corrosion resistant. It resists attack by most acids, but alkaline solutions dissolve the oxide film and cause rapid corrosion. The oxides and sulphate salts are the most widely used aluminium compounds. In the oil and gas sector, a number of aluminium products are used from cements, such as rapid setting high alumina cement [83], to lubricants and other additives, a number of which are further detailed below. Aluminium and its alloys are widely used in the construction of oil and gas facilities; however this is outside the scope of this study, and therefore their use and environmental effects and impacts are not considered.

Despite its abundance in the environment, no known form of life uses aluminium salts metabolically, but aluminium is thought to be well tolerated by plants and animals.

5.2.1 Aluminium Chloride and Polyaluminium Chloride (PAC)

Aluminium chloride, or more correctly, aluminium trichloride, is used as a gel-forming agent for reducing the permeability of water-conducting channels as it selectively forms such an inorganic gel based on aluminium hydroxide. Such a gel forms in response to change in temperature and pH. A gel forms as aluminium and hydroxide ions link to form a complex, amphorous, impenetrable and irregular network [88] (see also Section 5.2.2.1).

Aluminium chloride has been used in acidising systems that involve deployment of hydrofluoric acid (HF). This generates aluminium fluoride that has lower acidising rates than HF [89].

Aluminium chloride is essentially a waste product in the manufacture of other chemical products and is therefore relatively cheap.

Polyaluminium chloride is an extensively used coagulant for water treatment and is used in various industry sectors [90]. It can be manufactured in several ways [91].

Polyaluminium chloride is totally soluble in water and has the general formula $(Al_n(OH)_mCl_{(3n-m))x}$ and a polymeric structure. The length of the polymerised chain, molecular weight and number of ionic charges are determined by the degree of polymerisation. On hydrolysis, various mono- and polymeric species are formed, with $Al_{13}O_4(OH)^{7+}_{24}$ being a particularly important cation.

An important property of polyaluminium coagulants is their basicity. This is the ratio of hydroxyl to aluminium ions in the hydrated complex, and in general the higher the basicity, the lower the consumption of alkalinity in the treatment process and hence the impact on pH. The polyaluminium coagulants in general consume considerably less alkalinity than alum (potassium aluminium sulphate), and they are effective over a broader pH range compared with alum. They are also effective in cold-water applications.

Another important advantage of using polyaluminium coagulants in water treatment processes is the reduced concentration of sulphate added to the treated water. This directly affects sulphate levels in domestic waste water.

Surprisingly there are relatively few applications of this polymeric compound in waste water treatment in the oilfield sector [92]. It would seem that further exploration of the application of this material is long overdue especially as it has well-documented use in other water treatment areas and has low environmental impact. Indeed this and related aluminium salts have been used in the treatment of high phosphorous containing waste waters [93].

5.2.2 Aluminium Carboxylates

Aluminium carboxylates particularly aluminium citrate has a variety of actions in gel formation and application technology. Aluminium citrate is used as a cross-linker in low concentration polymer gels. This is particularly valuable for in-depth blockage of high permeability regions. However the performance of the gel depends strongly on the type of polymer used [94] (see also Chapter 2).

Aluminium carboxylates can also be used as activators for gelling agents and are able to raise the viscosity of hydrocarbon based fluids [95].

Aluminium and citric acid form a complex, which is a suitable aqueous drilling fluid dispersant particularly for bentonite dispersions [96].

Aluminium carboxylates have been described for application as non-polymeric drag-reducing agents (DRAs). These products are not subject to shear degradation (see Section 2.1.1) and do not effect or cause undesirable changes in the fluid quality. Examples are aluminium dioctoate and aluminium distearate and various mixtures thereof [97].

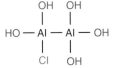

Figure 5.1 Aluminium hydroxychloride.

5.2.2.1 Aluminium Hydroxide

Sodium aluminate, $NaAl(OH)_4$, has been applied for water control in gas wells where on con-tact with water the pH drops and aluminium hydroxide is precipitated as a gel [98].

Aluminium hydroxide, $Al(OH)_3$, can be used as part of a self-destructive filter cake material [99]. These materials are used in drilling and completion operations where ideally they will disappear when no longer needed, without the need for mechanical intervention or injection of filter cake breaker chemicals.

A related material aluminium hydroxychloride (see Figure 5.1) can be used as a gel plugging agent [33]. This material is widely used in the formulation of cosmetics and personal care prod-ucts and is non-toxic.

5.2.3 Aluminium Isopropoxide

Aluminium isopropoxide is usually described with the formula $Al(O\text{-}i\text{-}Pr)_3$, where i-Pr is the isopropyl group $(CH(CH_3)_2)$. This colourless solid is a useful reagent in organic synthesis. The structure of this compound is complex, possibly time dependent, and may depend on the sol-vent. It is primarily used in the oil and gas industry in combination with phosphate esters to form an anionic association polymer in which the aluminium as the cross-linking agent. Such gelling agents are used as fluid loss control additives in drilling operations [100]. See also alu-minium phosphate esters later in this section.

5.2.4 Aluminium Oxide

Aluminium oxide, Al_2O_3, is an important component in the make-up of Portland cement and other cements used widely in wellbore completion [33], and in combination with aluminium sulphate, it can act as a setting accelerator.

Ceramic spheroids that have aluminium oxide as a critical ingredient have been described for use as a well proppant, [101]and bauxite, the ore of aluminium, is also used as a well proppant and in hydraulic fracturing [102].

5.2.5 Aluminium Phosphate Ester Salts

Organic liquid gels are used for temporary plugging and diversion during fracturing operations and the use of aluminium phosphate esters (see also Section 4.5) permits the in situ gelling of hydrocarbons, particularly those used in fracturing operations to enhance oil and gas produc-tion. A gel of crude oil or diesel can be produced using an aluminium isopropoxide (see earlier) and a phosphate diester [103].

Another similar polymer is formed by the reaction of triethyl phosphate and phospho-rous pentoxide to form a polyphosphate, which is then cross-linked with aluminium sul-phate [104].

5.2.6 Aluminium Sulphate

As mentioned in the previous section, aluminium sulphates are used as cross-linked in a polymeric aluminium phosphate ester. It is also, as previously described, a cement accelerator in combination with aluminium oxide.

Aluminium sulphate is also used as a coagulant in water treatment; however this is primarily focused on drinking water purification, waste water treatment and paper manufacture

5.2.7 Other Aluminium Compounds

A number of other aluminium compounds and materials are used in relatively small volumes by the upstream oil and gas industry.

Aluminium chlorohydrate in combination with a polyamine such as polydimethyldiallyl ammonium chloride (see Figure 5.2) gives a useful oil-in-water demulsifier particularly at elevated temperatures [105].

Aluminium gels can be blended with aqueous solutions of xanthan gum to form a product, which can be used to treat oil spills. Such a product exhibits shear thinning behaviour, a highly desirable effect for this application [106].

Alum (potassium aluminium sulphate) is used, alongside other components, in the formation of specific gel systems in the control of gas and oil-bearing zones [33].

Aluminium, along with other metals, is used as anti-seize agents. Such agents prevent and mitigate damage from high bearing stresses by providing a dissimilar metal surface between like substances. Such materials are under environmental and regulatory pressure, although aluminium with its environmental inertness may be more acceptable for use [107].

Aluminium/guanidine complexes in conjunction with modified starch can be utilised as a clay stabiliser as formulated in shale, stabilising drilling fluids [108].

Aluminium nitrate has been found to be a useful emulsion stabiliser in the transportation of heavy crudes [109]. Highly viscous crude oil can be transported as an emulsion that alleviates the need for heating or diluting the material in transit.

As can be seen, a variety of applications and compounds involve aluminium in the upstream oil and gas industry [110]. Given its apparent environmental inertness, it may be a useful avenue to examine further these and other aluminium products. This however may be required to be tempered by a more sustainable approach.

5.2.8 Environmental Impact

The process of transforming raw bauxite into aluminium is very energy intensive, requiring large amounts of power, water and resources to produce. Also the smelting process releases perfluorocarbons, which are much more harmful than carbon dioxide in terms of their effect on global warming.

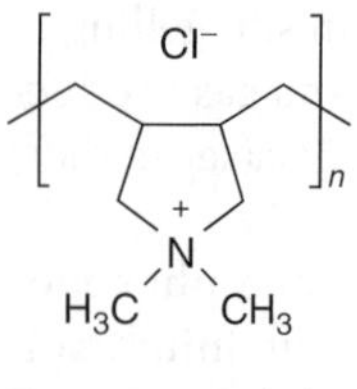

Figure 5.2 Polydimethyldiallyl ammonium chloride.

When bauxite is extracted, the strip-mining process removes all native vegetation in the mining region, resulting in a loss of habitat and food for local wildlife as well as significant soil erosion.

The caustic red sludge and toxic mine tailings that remain are commonly deposited into excavated mine pits where they could contaminate aquifers and local water sources.

Particulates are released during processing that are known to compromise air quality include combustion by-products, caustic aerosols, dust from bauxite, limestone, charred lime, alumina, etc.

Although there is general consensus that aluminium is environmentally inert, it has been established that when present in high concentrations, it can be toxic to aquatic freshwater organisms, i.e. downstream industrial point sources of aluminium-rich process water. It has also been observed that the environmental effects of aluminium are mainly a result of acidic precipitation where acidification of catchment areas leads to increased aluminium concentrations in soil solution and freshwaters. Large parts of both the aquatic and terrestrial ecosystems can be affected.

Aluminium has also been noted to accumulate in freshwater invertebrates and in plants and appears to affect the mycorrhiza and fine root systems. Aluminium-contaminated invertebrates and plants might thus be a link for aluminium to enter into terrestrial food chains [111].

5.3 Barium Salts

A restricted number of barium salts are utilised and encountered in the upstream oil and gas sector, and these are primarily barium sulphate (and its mineral form barite) and barium chloride. Barium, like aluminium, has been used in complexes as an anti-seize agent [107].

Many reservoir rocks consist of barite or related barium minerals. In developing and producing oil and gas from such geological formations, barium salts and in particular barium sulphate are often encountered as a problematic deposit. This is particularly the case where barium ions from production waters meet and are mixed with seawater. Seawater is rich in sulphate, and the resulting mixture due to cooling or change in temperature deposits barium sulphate. It is outside the scope of this study to detail and describe the mechanistic aspects of scale deposition in the oilfield; there are however numerous publications on the subject [112].

The economic impact of scale deposition can be severe on oilfield operations and therefore much effort is placed on understanding, predicting and mitigating scale deposition, in particular barium sulphate scale deposits, which are highly insoluble. Throughout this book, inhibitor chemistry has been described, particularly in Chapters 2 and 4. The environmental effect is usually minimal from the barium sulphate itself since it is a naturally occurring substance. Of course there is the environmental impact of disturbance in the environment. A more significant environmental impact can occur from naturally occurring radioactive material (NORM scale) often in oilfield operations where barium sulphate scale is an issue.

5.3.1 NORM Scale

Produced water, having been in contact with various rock strata at elevated pressure and temperature, contains many soluble components including barium and the radioactive intermediates of the uranium and thorium decay series. As water is produced, the temperature and pressure decrease, creating conditions under which the barium and the other radionuclides can co-precipitate inside separators, valves and pipework, forming an insoluble NORM scale. The scale is usually mechanically removed and macerated to a particle size of 1 mm or less and is then discharged by regulatory permission. Smaller items of the plant, such as valves and pipe-work, are normally transported to be cleaned at an authorised facility prior to being refurbished and reused.

Although certain parts of the produced water process system may be cleaned periodically, there is the potential for a quantity of NORM scale to remain within the system, which must be dealt with at the time of decommissioning of the installation.

The most significant radioactive element in NORM scale and produced water is radium and in particular the isotope [226]Ra, which is an alpha emitter with a half-life of 1620 years. The behaviour of radium in the marine environment and its effects on marine organism has been well studied [113]. It should be noted that the same radioactive salts found in NORM scale are found in all seawater. The sea has been naturally radioactive for millions of years and marine organisms have evolved with this background activity.

Discharge of radium would be of concern if it were to bioaccumulate in marine organisms and then pass up the food chain to humans. Generally there is an inverse relationship between taxonomic position and sensitivity to radiation – marine invertebrates are extremely tolerant to radiation toxicity, while humans are the most sensitive. Marine animals can potentially bio-accumulate radium from solution in the ambient seawater, from ingested seabed sediments, or from their food.

However, the NORM scale discharged from offshore installations is insoluble in seawater and when produced, water rich in barium and radium is discharged to sulphate-rich seawater; the radium precipitates rapidly as a complex of barium, radium and sulphate, which is also insoluble. As a result, radium has a very low concentration in solution in seawater and has a low bioavailability to marine organisms. It is also known that dissolved cations in seawater, particularly calcium and magnesium, inhibit the bioaccumulation of radium.

Similarly, any radium associated with precipitated barite in seabed sediments would not be bioavailable to benthic organisms. Because of its low concentration in seawater and sediments, radium is rarely accumulated to high concentrations in the tissues of marine plants and animals. Radium concentrations decrease with increasing trophic level, due to the poor assimilation efficiency, and studies in the North Sea and the Gulf of Mexico have shown that the risk to the human population through the consumption of fish and shellfish is not significant.

The radioactivity discharged from offshore oil and gas operations is not considered to have a significant environmental impact.

5.3.2 Barite

As well as being a naturally occurring mineral and a deposited scale in the oilfield, barite is used as a common additive in a number of applications, for example, a weighting agent in cements [114]. Such agents are used to increase the density of the cement to combat high bottom hole pressures. Barite as a cement additive has a number of issues; predominant is that of barite sag, which is a well-recognised but poorly understood phenomenon in the drilling and cementing sector of the upstream oil and gas industry. Within the last decade, however, field-proven technology has been developed to predict the potential for barite sag and to provide remedial measures through ultra-low shear rate viscosity modification [115].

Similarly Barite has been used as a drilling fluid weighting agent since the 1920s. It is generally preferred over other materials because of its high density, ease of handling and use and cheapness. Barite ore contains minerals other than barium sulphate such as iron carbonate, which can be detrimental to the drilling fluid, and so it is often characterised [116] and suitably modified [117].

5.3.3   Barium Sulphate

Barium sulphate ($BaSO_4$) scale is one of the most annoying and costly problems encountered in oilfield operations. It is common practice in the oilfield to determine the barium sulphate scaling tendency of an oilfield brine by analysis of water samples for their constituents, particularly for their barium and sulphate ion contents. Modern software programmes can use this data along with other data such as temperature, pressure, salt concentrations and others to determine the risk of scale deposition under normal operating conditions [118]. The control of this scale deposition by chemical treatments for dissolution and inhibition is extensively covered in other chapters, particularly Chapters 2 and 6.

Barium sulphate has been used as the weighting agent in a biodegradable drilling fluid [26] and as a sweep material in the removal of drilling cuttings. It is necessary to grind and sieve the barium sulphate so that is sufficiently small to be suspended in the drilling fluid. During circulation of the drilling fluid, this material can deliver small cuttings from the borehole to the surface for treatment and disposal [33].

5.3.4   Environmental Risk

Different barium compounds are known to exhibit vastly different toxicities; cleanup standards for barium in soil or sediment in the environment are typically based on total barium concentration without regard for the specific barium compounds present. However, the Texas Commission on Environmental Quality (TCEQ) among others has agreed that the insoluble barium compound barium sulphate (i.e. barite) is not a compound of concern with regard to human health, thereby eliminating the need for cleanup standards or site remediation for this compound [119]. Little or no other barium products are utilised in the upstream oil and gas industry.

5.4   Boron and Its Compounds

The main use of boron and related compounds, such as borates, is as cross-linking agents in fluid loss additives and fracturing fluids [120]. These are the most common fracturing fluid cross-linkers and can be formed by boric acid, borax, an alkaline earth metal borate or an alkali metal–alkaline earth metal borate. It is essential that the borate source has around 30% boric acid, which forms a complex with the hydroxyl units of the guar gum polysaccharide, cross-linking the polymer units. This process lowers the pH (hence the need for pH control) [121]. These fluids provide excellent rheological, fluid loss control and fracture conductivity properties in fluid temperatures up to 105 °C [122].

Clearly the environmental fate and toxicology of boron cross-linkers is more complex than food grade polysaccharides widely used in the fracturing fluid. Boric acid naturally occurs in air, water (surface and ground water), soil and plants, including food crops. It enters the environment through weathering of rocks, volatilisation from seawater and volcanic activity [123]. Most boron compounds convert to boric acid in the environment and the relatively high water solubility of boric acid results in the chemical reaching aquatic environments. Boric acid is therefore the boron compound of environmental significance [124].

It is assumed that boric acid is adsorbed in soil particles and aluminium and iron minerals and that this adsorption can be either reversible or irreversible, depending on soil characteristics. It is known that boric acid is mobile in soil [125]. The US EPA does not anticipate adverse effects to birds from the use patterns of boric acid products.

5.4.1 Borax

Borax, also known as sodium borate, sodium tetraborate or disodium tetraborate, is an important boron compound. Powdered borax is white, consisting of soft colourless crystals that dissolve easily in water. It has a wide variety of uses and is a component of many detergents and cosmetics. It is also used as a fire retardant and as an antifungal compound. Borax was first discovered in dry lake beds in Tibet and was imported via the Silk Road to Arabia. It also first came into common use in the late nineteenth century when it was popularised in a large variety of applications under the 20 Mule Team Borax trademark, named for the method by which borax was originally hauled out of the California and Nevada deserts in large enough quantities to make it cheap and commonly available. It is used in the upstream oil and gas industry as a cement retarder, particularly at high temperatures [33]. Its use has also been claimed as a transport aid when mixed with ethylene glycol.

5.5 Calcium and Its Salts

Calcium is one of the alkali earth metals, a group that also includes magnesium. In pure form, it is a silvery white metal that is highly reactive and fairly soft (softer than aluminium). Calcium is very reactive and is never found in its pure state in nature. It does occur in various compounds and in these forms it makes up 3–4% of all igneous rock, making it the fifth most abundant element in the Earth's crust. Among metals, calcium is the third most abundant and is found in every region of the Earth. It was first isolated as an element by Sir Humphry Davy in 1808.

The word 'calcium' is derived from the Latin *calx*, which means lime. Romans mixed slaked lime with sand and water to produce mortar, which they used to make buildings and roads. Mortar is still used today in bricklaying.

Calcium salts are predominantly found in the upstream oil and gas sector as problematic scale deposits, particularly calcium carbonate and calcium sulphate. In sandstones, which have been treated with the mineral acid and HF for stimulation, calcium fluoride deposition can occur. The mechanistic description of the occurrence of these scales and their deposition problems is beyond the scope of this book, and as with barium sulphate the reader is referred to other publications for more details [112].

The mitigation of these scales by chemical inhibitors and other agents for dissolution and removal is discussed in other chapters of this book, particularly Chapters 2 and 6.

The primary use of calcium salts are as additive in drilling and well completion operations and particularly in completion brines.

Calcium aluminate is found as a component of cements with high strength, low permeability and carbon dioxide resistance [126].

Mixed metal hydroxides and in particular calcium aluminium hydroxide, $Ca_3Al_2[OH]_{12}$, are able to impart shear thinning properties on ordinary bentonite muds. At rest these muds are highly viscous but are thinned to water-like consistency when pumping shear forces are applied [127].

5.5.1 Calcium Bromide, Calcium Chloride and Calcium Carbonate

These salts of calcium are widely used individually or mixed as brine systems to provide a range of relative densities of completion fluids [128].

As has been briefly described earlier in this chapter under sodium chloride, these brines have a number of functions, some of which are listed below:

- Mud displacement prior to cementing.
- Debris removal.
- Controlling formation pressures during completion and intervention operations.
- Enabling repair operations as a circulating or kill fluid medium.
- As packer fluids.
- In some stimulations and workovers as base fluids.
- Enable cleanup of the zone prior to running screens.
- Reduce friction while running screens and equipment.
- Avoid damaging the well after completion, stimulation or repair
- Allow other well operations to be conducted.

The variation in composition and types of calcium salt allows a wide range of brine densities to be achieved, which in turn allows a range of formation pressures to be controlled. Of these the control of formation pressures is perhaps the most critical and calcium salts and combinations thereof, afford a wide range of control parameters and operating pressures [126].

Calcium chloride and calcium carbonate also have a number of other additive functions. Calcium carbonate is one of the main additives of cement for well completion, and along with sodium chloride (see earlier) calcium chloride is used as a common cement accelerator [30]. Calcium carbonate has been used as a successful alternative to barite in drilling fluids as a weighting agent and also as an additive to aid in filter cake formation to prevent migration of fine particles into the reservoir [129]. Calcium carbonate is also acid reactive and can be selectively degraded, creating voids in the filter cake to allow production fluids to move more freely [99].

Graded and selected calcium carbonate has been used in combination with modified starch as a fluid loss additive. It is important that the calcium carbonate particles have a large size distribution to prevent filtration or loss of fluid to the formation [130]. Calcium carbonate has also been used as a weighting agent in surfactant-based fluid loss formulations [33]. The use of calcium carbonate in fluid loss additive formulation also extends to a comparable function of hydraulic fracturing fluids [130].

In the formation of CMC gels, calcium chloride offers an alternative control agent for gelation times as opposed to the more toxic chromium salts [33]. This gelation control is applicable to water shut-off applications, and it is worthy to note that calcium chloride is a critical component in plugging solutions for cementing applications.

Calcium chloride, along with other salts as has been described earlier, is an effective antifreeze agent [34]. However its use in the oilfield is restricted as it is highly corrosive. It is however widely used in road surface treatment, and its environmental impact has been well researched; see later in this section.

5.5.2 Calcium Fluoride

As described earlier, a number of metal salts are used an anti-seize agents and calcium fluoride afford a useful alternative [107]. It is either used in combination with molybdenum disulphide, where it is utilised as a thickening agent or as a replacement for it [131].

5.5.3 Calcium Hydroxide

Calcium hydroxide, traditionally known as slaked lime, has the chemical formula $Ca(OH)_2$. It is a colourless crystal or white powder and is obtained when calcium oxide (known as lime or quicklime) is mixed with water. Calcium hydroxide is used in many applications, including food preparation. Calcium hydroxide like other calcium salts is used in a range of drilling additive applications.

A lime mud is a type of water-based drilling fluid that is saturated with lime, $Ca(OH)_2$, and has excess, undissolved lime solids maintained in reserve. Lime muds are classified according to excess lime content and all lime muds have a pH of 12. The filtrate from these fluids is saturated with calcium hydroxide. The ability to carry very high mud alkalinity (as excess lime) to neutralise acid gases is one reason lime muds are used. H_2S zones can be drilled with more safety, and copious amounts of CO_2 can be neutralised by a large excess of lime. Such lime muds that have high contents of potassium chloride have been successfully used for drilling hydratable shales.

In the compositional make-up of some speciality mud compositions that involve the formation of a semipermeable membrane in situ in specific shale formations, calcium hydroxide is a preferred agent for pH adjustment [21].

Like calcium carbonate, calcium hydroxide is also used as a degradable bringing agent in cementing [99].

The addition of calcium hydroxide to a viscoelastic surfactant (VES)-based fluid loss additive improves its performance [132]. This behaviour has also been applied to similar VESs in fracturing fluid application for fluid loss control. Initially in these applications when calcium hydroxide saponifies the fatty acids in the VES, there is an increase in viscosity, thought to be due to the formation of viscosity enhancing co-surfactants [133].

5.5.4 Calcium Oxide

Calcium oxide (CaO), commonly known as quicklime, is a widely used in a variety of applications and most importantly is the major component of Portland cement. It is a white, alkaline, crystalline solid at room temperature. An important attribute of calcium oxide is that it survives processing without reacting in building products such as cement. It is relatively inexpensive and is therefore an important commodity.

In the oil and gas industry, the amount of calcium oxide present in cementing applications can allow the calculation of the setting time of the cement. Also the amount of water for the correct level of hydration of the cement can be calculated. This behaviour and its quantification are important so that correct placement of the cement in the wellbore can be achieved [33]. Also the expansion of the cement depends on the time at which the expansion additive, mainly calcium oxide, is added. The expansion characteristics of the calcium oxide depend on its thermal history during its processing, and this can be optimised to give specific characteristics such as high strength [134].

Similarly calcium oxide has an additive presence in water shut-off applications, for example, where formaldehyde resins are used as the base polymer component, calcium oxide is the expansion additive [33]. Cements that contain fibre reinforcement, particularly mineral fibres that will degrade under controlled conditions, contain a large proportion of calcium oxide [135].

Calcium peroxide, CaO_2, is a yellow-coloured solid, which is insoluble in water but does dissolve in organic acids such as acetic acid to form hydrogen peroxide. This therefore provides a method of delivery of hydrogen peroxide for filter cake degradation. Such products can be designed to give a delayed release of hydrogen peroxide [136].

5.5.5 Other Calcium Salts

The calcium salts of lignosulphonates provide a more environmentally acceptable alternative to chromium lignosulphonates [137]. Calcium lignosulphonate is also used as a cement

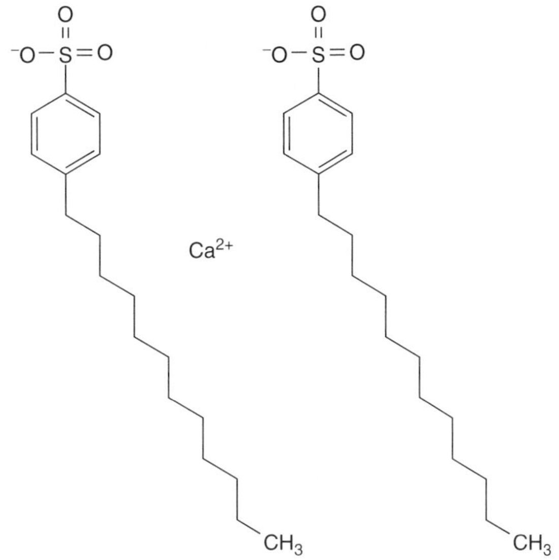

Figure 5.3 Calcium dodecylbenzene sulphonate.

viscosity control additive to prevent settling of solids during pumping operations [33]. The lignosulphonates are discussed in more detail in Section 2.2.2.

Calcium sulphonates (see Figure 5.3, calcium dodecylbenzene sulphonate, as an example) form the basis of a class of drilling operations greases. During drilling operations, threaded connections in the drill pipe and other assemblies are exposed to the drilling fluids and cuttings from the drilling. These fluids and other debris dissolve and erode the grease on the connections. Calcium sulphonate-based grease formulations offer a higher performance in terms of withstanding dissolution and erosion during drilling and related operations [138].

Calcium thiocyanate, as mentioned previously with sodium salts, can be used as corrosion inhibitors in drilling, completion and workover fluids [55].

The calcium salts of organic acids such as calcium citrate are discussed in Chapter 6. Similarly the silicates, such as dicalcium silicate, are discussed in Chapter 7.

5.5.6 Environmental Impact of Calcium and Its Salts

Calcium is a major constituent of bone and teeth, the shells of marine organisms and coral reefs; it gives these structures their mechanical strength. When the organisms making up a coral reef die, their calcium-rich bodies are deposited as sediment, which eventually becomes limestone. Limestone deposits on land originate from these marine deposits and provide an important part of the marine fossil record.

Spectacular limestone caves are formed when groundwater that is slightly acidic (due to the presence of carbon dioxide) seeps through cracks in the ground, gradually dissolving the limestone and hollowing out a cavern. The dripping process often produces stalagmites and stalactites, long icicle-like projections from the ground and ceiling, respectively.

Limestone, like the coral and mollusc shells it is derived from, is primarily calcium carbonate ($CaCO_3$). Calcium carbonate is also used as an antacid; because it is a base, it neutralises stomach acid. Marble is the same substance, with certain impurities present. Concrete is made from a mixture of sand, gravel, water and Portland cement (named after the natural limestone found on Britain's Isle of Portland). Gypsum, another calcium compound, is the material used to make Sheetrock or drywall. Its chemical name is calcium sulphate dihydrate. Other important compounds are lime (calcium oxide, CaO) and 'slaked' lime (calcium hydroxide, CaOH).

Calcium plays an important role in human biology. Ninety-nine percent of the calcium in the human body is in the skeleton. Calcium in the skeleton is largely in crystalline form – a compound of calcium, phosphate and hydroxyl atoms called hyroxylapatite. This calcium is continuously being replaced as bones are destroyed and renewed; the minerals in the bones undergo resorption, balanced by the formation and mineralisation of a new bone. This process is controlled by the body's hormones to keep bone density and volume constant.

The biological importance of calcium extends beyond its role in the skeleton, as calcium is also present in soft tissues. Calcium ions in extra-cellular fluids help maintain the integrity and permeability of cell membranes. They also play a role in blood coagulation, ion transport, maintenance of heart rhythm and excitation of nerves and muscles.

Calcium is an important part of the human diet, and most populations get about half their dietary calcium from milk and other dairy products. Soils in humid regions generally have less calcium than those in dry regions, and calcium is often added to the soil in humid regions to reduce acidity. Calcium levels in soil do not seem to affect human nutrition, because the amount of calcium taken up by a plant depends much more on the nature of the plant species than it does on the calcium content of the soil. The amount of calcium in one's diet depends on what types of plants one eats (not where they were grown) and, just as importantly, what kinds of plants are eaten by the cows whose milk one drinks.

Calcium is thought to be environmentally benign, and its salts, which are discharged by industries including the oil and gas industry, are thought to have little or no risk to the environment. Nonetheless a lot of research has been conducted on the potential environmental effects of calcium salts and particularly calcium chloride, which is widely used as a roadway deicer [139]. It has generally been found that increased application of chemical deicers for winter maintenance has resulted in increased concentrations of deicer constituents in the environment. The runoffs from the deicing operation can have a deteriorating effect on soil and water quality. But the degree of impact is localised, and it depends on various climatic factors and can also be attributed to the type of salts used and their storage conditions. In the main, calcium salts and in particular calcium chloride have a low environmental impact.

Calcium cements have been examined as binders for synthetic drilling fluid cuttings wastes and primarily to treat synthetic drill cuttings as a pre-treatment to landfilling or for potential reuse as construction products [140]. Two synthetic mixes have been examined based on average concentrations of specific contaminates present in typical drill cuttings from the North Sea and the Red Sea areas. A number of conventional binders were tested including Portland cement, lime, and blast furnace slag, in addition to novel binders such as micro-silica and magnesium oxide cement. Despite the significant difference in the hydrocarbon content in the two synthetic cuttings, the measured uniaxial compressive strength (UCS) values of the mixes with the same binder type and content were similar. Importantly the leachability results showed the reduction of the synthetic drill cuttings to a stable non-reactive hazardous waste, compliant with the UK acceptance criteria for non-hazardous landfills.

5.6 Halogens

This section will examine the use of a variety of chemical compounds in the oil and gas industry based on halogens. It will not include the halide salts of metals as these are considered in other sections of this chapter under the appropriate metal or its salt. Along with salts and other chemical compounds, the free elements are also considered as they can also be used in chemical applications for the upstream oil and gas sector.

The halogens or halogen elements are a group in the periodic table consisting of five chemically related elements, namely, fluorine (F), chlorine (Cl), bromine (Br), iodine (I) and astatine (At). The last element astatine is radioactive, and is not used at all in the oil and gas sector, and is therefore not discussed. The name 'halogen' means 'salt producing'. When halogens react with metals, they produce a wide range of salts.

All of the halogens form acids when bonded to hydrogen and these products will be considered in this section. Most halogens are typically produced from minerals or salts. The middle halogens, that is, chlorine, bromine and iodine, are often used as disinfectants. Organo-bromides are an important class of flame retardants. Elemental halogens are classified as dangerous and can be lethally toxic.

5.6.1 Bromine and Bromine Compounds

5.6.1.1 Biocides, Bromine and Organo-Bromides

Biocides are often applied in oilfield functional fluids that are used for well construction and well stimulation. These help reduce bacterial populations and thereby increase the stability of a well treatment. Moreover, biocides can minimise inadvertent well inoculation and can reduce or eliminate the harmful effects of bacteria, for example, the production of H_2S (souring) due to the activity of anaerobic organisms such as sulphate-reducing bacteria (SRBs).

Biocides can also control biofilms, the thin layers of bacteria and slime attached to pumps, pipework, heat exchangers and filters associated with oilfield systems. Biofilms are also found downhole in formations. Biofilms can lead to system fouling, contribute to under-deposit corrosion, and reduce heat transfer efficiency. Moreover, biofilms can block formation and filter permeability.

Oxidising chlorine-based biocides, discussed later in this section, have a long history in the oilfield for effecting microbiological control. Bromine-based technologies have been developed for a number of industrial and recreational applications, and now they are being applied in the oilfield as well. Bromine is more expensive than chlorine, and therefore its application as a microbiocide must necessarily be based upon microbiocidal performance, cost-effectiveness and other attributes.

Elemental bromine is not used in the upstream oil and gas industry; however a stabilised version of bromine chloride is utilised as a biocide. It has the same disadvantageous related to stability as hypochlorite salts, discussed later, but has been made easier to handle by stabilising it as a sulphamate salt. This technology effects microbiological control, and the product penetrates and removes biofilm. It is from a new class of biocides offering the efficacy of oxidisers and the stability of non-oxidisers [141].

There are also a number of organic bromide biocides, and the most common are 2-bromo-2-nitro-propanediol (BNPD) (see Figure 5.4), 2,2-dibromo-3-nitrolopropionamide (DBNPA) (see Figure 5.5), 1-bromo-1-(bromomethyl)-1,3-propanedicarbonitrile (see Figure 5.6) and 2,2-dibromo-2-nitro-ethanol (see Figure 5.7). Strictly these products are neither oxidising nor non-oxidising biocides. BNPD or Bronopol has been applied in the oilfield for several decades [142] and was first used in the cosmetic and personal care sector [143].

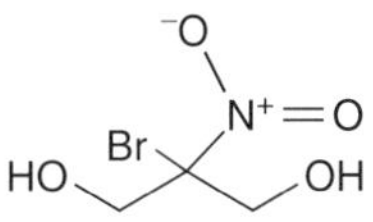

Figure 5.4 Bronopol.

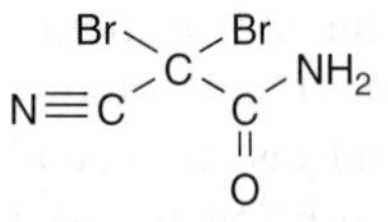

Figure 5.5 DBNPA.

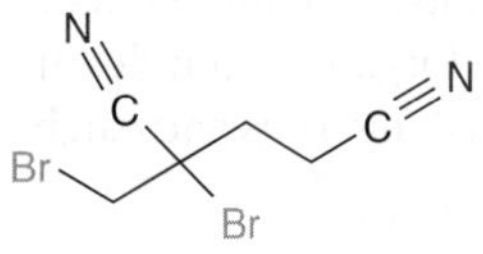

Figure 5.6 1-Bromo-1-(bromomethyl)-1,3-propanedicarbonitrile.

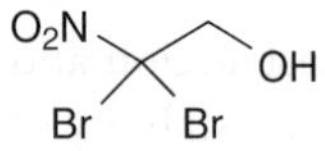

Figure 5.7 2,2-Dibromo-2-nitroethanol.

Of the above DNPA is by far the most commonly used. It is preferred due to its instability in water where it quickly hydrolyses and degrades to ammonia and bromide ions. It is pH sensitive, hydrolysing under both acidic and alkaline conditions. DNPA is also believed to act synergistically with non-oxidising and oxidising biocides [144].

5.6.1.2 Hypobromite and Hypobromous Acid

Stable hypobromite and hypobromous acid solutions have been examined for oilfield use and claimed to be superior to hypochlorite and hypochlorous acids [145]. These solutions are powerful oxidising agents.

5.6.2 Chlorine and Chlorine Compounds

Primarily two compounds are used in this class, namely, chlorine and chlorine dioxide. Elemental chlorine gas has been applied in the upstream oil and gas industry as a biocide to water injection systems particularly in seawater lift systems. It is an oxidising biocide and is toxic, corrosive and difficult to handle; however it can be generated on site in an offshore facility by electrolytic oxidation of the chloride ions in seawater using direct current [146]. In this process the produced chlorine reacts with water to produce hydrochloric and hypochlorous acids.

5.6.2.1 Hypochlorous Acid

Hypochlorous acid (HOCl) is a weak acid, which partially dissociates in water, into hypochlorite and hydronium ions. It is these ions that are the primary agents for disinfection when chlorine is used to disinfect water. Hypochlorous acid cannot be isolated in pure form due to rapid equilibration with its precursor.

Hypochlorous acid is a powerful oxidising agent and is usually supplied, either as sodium hypochlorite or calcium hypochlorite, to be used as a bleach or disinfectant.

5.6.2.2 Chlorine Dioxide

Chlorine dioxide has been evaluated as a replacement for chlorine gas [147], and it is an extremely powerful oxidising agent and biocide. Chlorine dioxide is known to selectively oxidise sulphide-based material and has successfully removed iron sulphide plugging material from the near-wellbore area of injection and producing wells, increasing injection rates and reducing pressure losses. The reduction of hydrogen sulphide-induced corrosion is also an added benefit to any production improvement [148]. While the exact chemistry between the sludge removed from the water distribution system and the chlorine dioxide has been little investigated, the destabilisation of the emulsion holding the sludge components together has been observed in the field on a number of occasions. A major disadvantage in the use of chlorine dioxide is that it is highly corrosive as held in solution. A number of corrosion inhibitors have been examined, wherein sodium dichromate, which is toxic and environmentally persistent, is highly effective. An alternative and an effective but more environmentally acceptable corrosion inhibitor package, consisting of an alcohol, an acid, a fatty acid and an ethoxylated fatty diamine, has been claimed [149].

Chlorine dioxide has also been used in combination with hydrochloric acid (see following paragraphs) for stimulation of wells, which have been impaired by iron sulphide or bacterial-based material [150]. The use of chlorine dioxide with organic acids to provide a chlorous acid treatment has also been described [151].

5.6.2.3 Hydrochloric Acid

Hydrochloric acid is primarily used in oilfield application for the stimulation of calcium carbonate reservoirs [152]. It is not within the scope of this book to give detailed synopsis of the use and application of hydrochloric acid; however some key points are now described. The concentrations of hydrochloric acid used in such stimulation workovers vary between 15 and 28 wt% in normal practice. Hydrochloric acid is manufactured and supplied at 37 wt%. Calcium carbonate rocks, predominantly chalk and limestone, dissolve in acid to release carbon dioxide and form calcium chloride in solution; see reaction in Eq. (5.10):

$$CaCO_3 + 2HCl \rightarrow CO_2 + CaCl_2 + H_2O. \tag{5.10}$$

A variety of additives can be added to the acid to retard the reaction, allowing slower and more diverse dissolution of the rock [152]. Also calcium chloride and calcium bromide have been deliberately added to the hydrochloric acid to aid in stimulation of high temperature and high pressure wells [153]. Hydrochloric acid can simply be applied under controlled conditions to remove calcium carbonate scale deposits by dissolution [66].

Hydrochloric acid can be generated in situ from halogenated organic compounds such as trichloroacetic acid. This generated hydrochloric acid can be used to activate gel formation for water shut-off applications [98]. In other water shut-off applications, hydrochloric acid can play an important role in gel formation, for example, in the use of hydrolysed polyacrylamide (HPAM) and in the formation of polymer silicate gels [46, 88].

In sandstone formations hydrochloric acid is normally used in combination with HF (see next section, 'Fluorine and Fluorine Compounds') [152]. Hydrochloric acid is also used to dissolve iron sulphide scale; however there should be a recognition that hydrogen sulphide (a highly toxic gas) is likely to be released [154].

Hydrochloric acid is often used as a component of a breaker system for well clean-up and/or filter cake degradation and removal. This is particularly the case for horizontal well completions [155, 156].

Until recently there was thought to be little environmental impact from hydrochloric acid (this is further discussed under environmental effects later in this section on halogens); however in order to mitigate its aggressive corrosive effects, a number of corrosion inhibitors that can have environmental toxicological effects are used. These corrosion inhibitors are mainly organic in nature, some are surfactant-type molecules that are discussed in Chapter 3, and

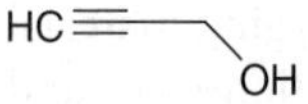

Figure 5.8 Propargyl alcohol.

others are discrete low molecular weight organic molecules such as propargyl alcohol (see Figure 5.8) and will be further discussed in Chapter 6.

5.6.3 Fluorine Compounds and Fluoropolymers

Fluorine is the lightest halogen and has atomic number 9. It exists as a highly toxic, pale yellow gas under standard conditions of temperature and pressure. It is extremely reactive: almost all other elements, including some noble (inert) gases, form compounds with fluorine. Fluorite, the primary mineral source of fluorine, was first described in 1529, as it was added to metal ores to lower their melting points for smelting. Only in 1886 did French chemist Henri Moissan isolate elemental fluorine using low-temperature electrolysis, a process still employed for modern production.

Owing to the expense of refining pure fluorine, most commercial applications use fluorine compounds, with about half of mined fluorite used in steelmaking. The rest of the fluorite is converted into corrosive hydrogen fluoride en route to various organic fluorides, or into cryolite, which plays a key role in aluminium refining. Organic fluorides have very high chemical and thermal stability; their major uses are as refrigerants, electrical insulation and cookware, as well as polytetrafluoroethylene (PTFE) (Teflon). The fluoride ion inhibits dental cavities and so finds use in toothpaste and water fluoridation.

In the oil and gas industry, the main fluorine compounds used are HF in workover and stimulation, a variety of fluoro-surfactants and some fluoropolymers, in different applications. The latter was not fully described in Chapter 2, 'Polymers', and is covered here; however fluoro-surfactants are covered in Chapter 3, particularly in Sections 3.2.4 and 3.6.

5.6.3.1 Hydrofluoric Acid

HF is primarily used as a stimulation agent, either on its own or in combination with hydrochloric acid to enhance oil recovery from sandstone reservoirs [152]. Most sandstone reservoirs are composed of over 70% quartz, i.e. silica that is bound together by other minerals and salts such as carbonates and silicates. The injection of HF and formulations containing HF for remediation of permeability impairment in sandstone reservoirs is a common treatment method. HF exhibits a high reactivity towards siliceous materials especially above 65 °C and is therefore able to dissolve such material; however the presence of other metal species such as potassium, calcium, etc., can lead to precipitation of their fluoride salts. As described earlier the addition of phosphonates (see Chapter 4) and boron compounds (see earlier section in this chapter) can mitigate against these undesirable effects [157].

Often combinations of hydrochloric and HF are used to maximise dissolution of 'fines' and scales causing reduction in permeability. In order to mitigate against the formation of salts of incompatible anions such as potassium, magnesium and others, as well as the previously described additives, ammonia and/or ammonium chloride is often added to the injected acid mix [158].

HF is highly corrosive and difficult to transport and handle and therefore attention in the last few decades has focused on its in situ generation and a number of useful formulations and injection methods have been developed [159].

5.6.3.2 Fluoropolymers

Fluoropolymers are polymers that contain carbon and fluorine. They are in the main high-performance plastic materials used in harsh chemical and high temperature environments, primarily where a critical performance specification must be met. They are used by defence-related industries and in automotive, aerospace, electronics and telecommunications. They are also used in many consumer products most notably PTFE that is used in non-stick coatings used on cookware and small appliances.

Fluoropolymers were discovered in 1938 by Dr Roy J. Plunkett while working on freon for the DuPont Corporation. He accidentally polymerised tetrafluoroethylene and the result was PTFE (commonly known by its DuPont brand name 'Teflon'). PTFE turned out to have the lowest coefficient of friction of any known solid material. It also has a very low 'surface energy' (which is what gives cookware coatings the non-stick feature), and possesses the quality of being inert to virtually all chemicals [160].

Sealing materials for the oil and gas sector have to resist chemically aggressive environments including seawater, high salinity brines, sour gas and steam as well as high pressures and high temperatures. The industry has established specific testing and qualification standards to ensure that materials used in the harsh oil and gas drilling and production environments meet the critical demands of these applications. PTFE is widely used as a basis of elastomers for seals.

In coating applications the combination of extreme low and extreme high temperature stability favours the use of PTFE over a wider temperature range than any other organic coating material. PTFE coatings for the oil and gas sector provide reduced torque and galling levels even after long-term exposure to corrosion. Fasteners, valves and subsea connectors all benefit from low friction characteristics, good load bearing and anti-corrosion properties [161].

5.6.4 Other Halogen Compounds

The other main halogen is iodine, which is the heaviest of the stable halogens and exists as a lustrous, purple-black metallic solid under standard conditions of temperature and pressure. It however sublimes readily to form a violet gas. The elemental form was discovered by the French chemist Bernard Courtois in 1811.

It does not have a wide range of uses; however it has found favour as a non-toxic radio-contrast material and because of the specificity of its uptake by the human body, radioactive isotopes of iodine can also be used to treat thyroid cancer. Iodine is also used as a catalyst in the industrial production of acetic acid and certain polymer products.

In the oil and gas sector, there are also relatively few uses of iodine and its salts. Iodides, primarily, sodium, potassium or copper, have been used as corrosion inhibitor intensifiers in acidising formulations [162]. As described earlier in the section on potassium, potassium iodide is used as a synergist in certain corrosion inhibitor formulations [77]. Somewhat esoterically *S*-alkylthiouronium iodides (see formula (5.11)) have been examined as corrosion inhibitors particularly under aggressive acid conditions [163].

S-alkylthiouronium iodide:

$$\left\{ (NH_2)_2 \text{-} C \text{-} S \text{-} R \right\}^+ I^-. \tag{5.11}$$

In sour wells there is a tendency for iron sulphide to precipitate. Acidisation can stimulate these wells by dissolving the iron sulphide; however as the acid treatment becomes spent, the iron can precipitate as ferric hydroxide or reprecipitate as ferric sulphide. In order to prevent this, iron control agents are added to the formulated acid. Iodide salts are one of the agents that can be used and act as a reducing agent on the iron, changing the oxidation state from iron(III) (ferric) to iron(II) (ferrous) that in turn form more soluble salts [164].

Radioactive iodine has been used as a water tracer for the accurate evaluation of reservoir performance characteristics in the secondary recovery of petroleum by water flooding. The water tracer is injected into water input wells and detected at oil production wells to supplement data obtained from core analyses, wellhead tests and subsurface measurements. Radioactive iodine has been used successfully as a water tracer in field tests to determine (i) relative rates and patterns of flow of injected water between water input and oil production wells and (ii) zones of excessive water entry into oil production wells. The radioactive tracer method, using radioactive iodine, allows measurement of either the relative rates and patterns of flow or zones of excessive water entry, into wells under conditions of comparatively rapid transit time between wells [165].

5.6.5 The Environmental Effects of Halogen Containing Compounds and Polymers

Halogen elements and particularly their halide salts are generally considered as environmentally benign, as many are naturally occurring compounds of low toxicity. Indeed often they are elementally essential to the environment and to human health. It is the author's view that this is misleading, and although on a macro-ecological scale, this may hold true that there are some definite and serious effects on local ecologies and ecosystems. Most regulators including the US EPA have set various concentration levels of halides, as it has been shown that above such levels environmental toxicity can occur [166].

Fluorides are strongly attracted to soil particles and can be dissolved into running water sources from nearby soil samples. Aquatic animals can be exposed directly to fluoride in natural water sources, while terrestrial animals are also at risk due to consumption of vegetation containing high fluoride concentrations, which are harmful to all organisms, resulting in stunted growth, bone degradation, birth defects in animals and low crop yields. The US EPA has set 2 ppm as the maximum acceptable concentration of fluoride in drinking water [167].

Chlorides are found in natural waterways from inorganic chloride containing salts, such as sodium chloride and chlorinated tap water. They may also be introduced through drainage, precipitation, runoff, soil and clay sources, pollution and waste disposal. Once such ions come in contact with natural water sources, it is possible to form chloroform and other carcinogens by reacting with other naturally occurring compounds. Due to its high reactivity, it also contributes to the corrosion of many metals, including man-made structures. The EPA standard for chloride in drinking water is 250 ppm, considerably higher than fluoride reflecting its lower risk [167].

As has been stated earlier in the section under Calcium, a lot of research has been carried out in the potential environmental effects of calcium chloride (and other chlorides) that is widely used as a roadway deicer [139]. It has generally been found that increased application of chemical deicers for winter maintenance has resulted in increased concentrations of deicer constituents in the environment. The runoffs from the deicing operation can have a deteriorating effect on soil and water quality. But the degree of impact is localised, and it depends on various climatic factors and can also be attributed to the type of salts used and their storage conditions.

Bromides exist naturally as inorganic salts, both in the Earth's crust and in natural water sources. Industrial dumping and run-off can contribute to increased levels. It is believed that high bromide concentrations can have lethal effects on aquatic organisms, and bromides can also be absorbed by plants, with adverse effects such as stunted growth and poor germination.

Most freshwater sources tend to reflect the halogen concentration due to the local precipitation, but likely have elevated levels if human activity occurs nearby. Freshwater lakes and ponds typically contain $100-300\,\mathrm{mg\,l^{-1}}$ of chloride, while streams contain less than $100\,\mathrm{mg\,l^{-1}}$. The concentration of chloride in residential and urban areas fluctuates greatly, with concentrations ranging from 200 to $700\,\mathrm{mg\,l^{-1}}$ as common. The importance of bromide is in combination with chloride as a Cl^-/Br^- ratio, which can be used to identify the history of freshwater groundwater systems based on the pollution present. This is possible because the ratio is conserved in potable groundwater until a source of water with another ratio is added. This allows researchers to follow the source of pollution to an origin point in groundwater systems [168].

The effects of concentrations of halide ions can be complex and, for instance, can interfere with the oxidation process in seawater due to hydroxyl radicals. It could be expected that additional halide ions, i.e. above normal concentrations, could interfere with such processes. However it has been shown that the hydroxyl radical converts the halide to a reactive halide species, which can aid in degradation of contaminants and pollutants [169].

Hydrochloric acid and its acidifying effect have come under re-evaluation. Research on the ecosystem impacts of acidifying pollutants, and measures to control them, have focused almost exclusively on sulphur (S) and nitrogen (N) compounds. Hydrochloric acid (HCl) has been somewhat overlooked as a driver of ecosystem change because most of it was considered to redeposit close to emission sources rather than in remote natural ecosystems [170].

It is well documented that fluorocarbon gases are greenhouse gases with considerably higher global warming potential than carbon dioxide [171]. Despite the international agreement in 1987, the Montreal Protocol [172], to phase out these compounds, there is an increase in use and consumption due to dramatic rises globally in the use of air conditioning.

Organo-fluorine compounds are known to be persistent in the environment due to the strength of the carbon–fluorine bond. Natural organo-fluorine compounds are rare, with only a few known to be produced by some species of plant (mainly as poisons to deter herbivores) and microorganisms. Consequently, the mechanism of enzymatic carbon–fluorine bond formation is poorly understood. However studies have shown that there is a common intermediate in the biosynthesis of the fluoro-metabolites, which has been identified as fluoro-acetaldehyde [173]. Fluorine has no known metabolic role in mammals.

Related to the strength of the carbon–halogen bond is the observation that organic molecules containing fluorine and chlorine have lower rates of biodegradation than the non-halogenic analogues. This is especially the case where more than three chlorine or fluorine atoms are present in the molecule [174].

5.7 Iron and Its Salts

Iron is by mass the most common element, forming much of the Earth's core. It is the fourth most common element in the Earth's crust and is a metal in the first transition series of the periodic table. Iron exists in a wide range of oxidation states from -2 to $+6$, although $+2$ (ferrous) and $+3$ (ferric) are the most common. Elemental iron is reactive to oxygen and water and is found mainly as iron oxide. Fresh iron surfaces appear as lustrous silvery grey, but oxidise in normal air to give hydrated iron oxides, commonly known as rust. Unlike metals, such as aluminium described earlier that form passivating oxide layers, iron oxides occupy more volume than the metal and thus flake off, exposing fresh surfaces for oxidation.

Pure iron is relatively soft, but cannot be obtained by smelting because it is significantly hardened and strengthened by impurities, in particular carbon, from the smelting process. A certain proportion of carbon (between 0.002% and 2.1%) produces steel, which may be up to 1000 times harder than pure iron. Steels and iron alloys formed with other metals (alloy steels) are by far the most common industrial metals because they have a great range of desirable properties and iron-bearing rock is abundant. It is a pre-eminent material of choice for construction in the oil and gas industry. In itself iron or steel is not used as an additive in the upstream oil and gas industry; however hematite, the oxide mineral ore of iron, can be used as a weighting agent in cement [33].

Iron chemical compounds have many uses from construction to pharmaceuticals and are also used in a number of functions in the upstream oil and gas industry, particularly in fracturing and enhanced oil recovery where a number of iron salts and complexes are used as crosslinkers in polymer gel formation and as chelating agents. Importantly a number of iron salts are the sources of severe and problematic issues in the upstream oil and gas sector, and these are also discussed in this section.

Iron is involved in important biological processes, forming complexes with molecular oxygen in haemoglobin and myoglobin; these proteins are common vehicles for oxygen transport in vertebrate animals. Iron is also the metal at the active site of many important redox enzymes dealing with cellular respiration and oxidation and reduction in plants and animals.

5.7.1   Iron Carbonate

Iron carbonate can be a problematic scale in oilfield production but is remediated by acid dissolution or prevented by application of a suitable scale inhibitor such as those described in Chapters 2 and 4. Iron carbonate can form a stable protective scale on carbon steels that can be an effective means of preventing and retarding corrosion, particularly carbon dioxide promoted corrosion [175].

In the upstream oil and gas industry, this has been used with particular effectiveness in gas production systems and transportation where monoethylene glycol (MEG) is used as a thermodynamic hydrate inhibitor [176]. In such systems pH-stabilised MEG can promote the deposition of a stable film of iron carbonate scale that acts as a physical barrier to corrosion from acidic carbon dioxide-based species.

5.7.2   Ferric Chloride

Ferric or iron(III) chloride can be used as a hardening agent in the composition of certain resin-based plugging agents [33] and in certain acid-resistant cement formulations particularly at high temperatures, i.e. temperatures up to 160 °C [177]. As will be illustrated in the next section of ferric hydroxide, inorganic gels can be somewhat controlled in their formation and application by the careful addition of blocking agents such as ferric chloride [178].

5.7.3   Ferric Hydroxide

Iron(III) ions can be a source of problems particularly during and after acidisation of wells for stimulation of production. Particularly common is the precipitation of ferric hydroxide, which is a colloidal precipitate, which can cause formation damage. For this reason iron control agents are usually used in acidisation workovers [179]. The chemistry of these products tends to be composed of small organic molecules and is discussed more fully in Chapter 6.

Ferric hydroxide gel has been deliberately used for water shut-off applications in mature oilfields. This gel treatment technique is based on transformation of water-soluble Fe(III) compounds into gel-like precipitates by in situ hydrolysis and flocculation. The blocking material has excellent stability under field conditions, and yet simple remediation is possible in case of placement failures. Further, the method is characterised by self-controlling chemical mechanism, and using the technique injectivity problems do not arise even in low permeability and tough formations. The application has a 40–50% success rate in providing an economic benefit [24].

5.7.4 Iron Sulphides

Both ferric and ferrous sulphides are found as problematic species in oil and gas production, particularly as iron sulphide scales in sour production where hydrogen sulphide is prevalent. These deposits are a major source of economic loss in the upstream oil and gas industry. The simplest way to remove these deposits is to dissolve them, which unfortunately produces large volumes of toxic hydrogen sulphide gas. Another method would be to treat any deposits with strong oxidising agents that, despite avoiding the toxicity issues, produces oxidation products, including elemental sulphur, which are generally highly corrosive, and is therefore not usually applied.

A number of molecular species have been used as iron sulphide dissolution or chelation agents including low molecular weight organics that are further discussed in Chapter 6 and tetrakishydroxymethyl phosphonium sulphate (THPS) that has already been detailed in Chapter 4.

5.7.5 Other Iron Salts and Compounds

There are a small number of other iron salts and related compounds, which have applications in the upstream oil and gas industry. Iron(II) oxalate is an excellent scavenger of hydrogen sulphide and has been applied to remove hydrogen sulphide contamination in drilling fluids [180]. Ferrous sulphate and lignosulphonates have been used as dispersants in cement compositions and cementing applications [181]. Also iron oxide is part of most standard cement compositions including Portland cement [33].

Ferric acetylacetonate (see Figure 5.9) is used as a complexing agent with water-soluble polymers, particularly polyacrylamides to form a temporary gel that is useful for temporary plugging of specific zones, particularly for water shut-off within the formation. These gels have a lifetime of approximately 6 months [182].

5.7.6 Potential Environment Effects

Iron salts and other iron additives are used in very small quantities compared with the iron in the construction of facilities and the iron available in the natural environment. Nonetheless iron ores and salts can have an effect in the environment especially as they are often readily bioavailable and uptake into flora and fauna is facilitated due to the use of iron in biological processes.

The effects of iron compounds on the macro-invertebrate fauna have been investigated in a lowland river system [183]. As a result of industrial farming activities, water levels have been lowered and iron has leached from this system. Concentrations of annual average dissolved iron from 28 localities varied from about 0 to 32 mg Fe^{2+}/litre, while pH varied from 6.7 to 8.8. A correlation between numbers of taxa and concentration of dissolved iron has

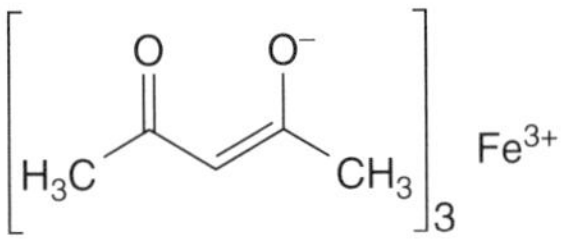

Figure 5.9 Iron(III) acetylacetonate.

been suggested. Sixty-seven taxa of macro-invertebrates are from sites where concentrations are below 0.2 mg Fe^{2+}/litre; however, between 0.2 and 0.3 mg Fe^{2+}/litre, an abrupt drop to 53 taxa was recorded. Elimination of taxa continued up to 10 mg Fe^{2+}/litre, where about 10 taxa were left. Numbers of individuals declined at concentrations >1 mg Fe^{2+}/litre. Overall the conclusion has been that high Fe^{2+} concentrations are detrimental to the overall diversity of species in the freshwater environment.

Iron plays an important role in the uptake of foodstuffs by aquatic animals, indeed crucial is the interchange between the two oxidation states of Fe(II) and Fe(III). Fe(II) enters the food chain mainly as dissolved salts and its uptake is directly via water. Fe(III) is mainly taken up as a direct dietary input from Fe(III) precipitates. Such precipitates may induce the formation of iron hydroxides on the gut walls of aquatic animals and subsequently reduce their ability to take in important iron salts [184].

More important may be the effects that iron concentrations have on the toxicity, bioavailability, and fate of other trace metals and organic pollutants. Iron is continuously flushing through aquatic ecosystems both physically by sedimentation diffusion and chemically by reduction and oxidation processes. Therefore direct effects can be difficult to measure and assess [184].

In the overall impact assessment of effects from the oil and gas industry, however, given the concentration effects of iron in the macro-ecology would not appear to be very significant.

5.8   Zinc and Its Salts

Zinc is the first element in group 12 of the periodic table. In some respects zinc is chemically similar to magnesium (see earlier in this chapter) in that both elements exhibit only one normal oxidation state (+2), and the Zn^{2+} and Mg^{2+} ions are of similar size. The most common zinc ore is sphalerite, a zinc sulphide mineral. The largest workable lodes are in Australia, Asia and the United States.

Brass, an alloy of copper and zinc in various proportions, was used as early as the third millennium BC; however zinc metal was not produced on a large scale until the twelfth century in India and was unknown to Europe until the end of the sixteenth century.

The German chemist Andreas Marggraf is credited with discovering pure metallic zinc in 1746, and later work uncovered the electrochemical properties of zinc in 1800. Corrosion-resistant zinc plating of iron (hot dip galvanisation) is the major application for zinc. Other applications are in electrical batteries, small non-structural castings, and alloys such as brass. A variety of zinc compounds are commonly used, such as zinc carbonate and zinc gluconate (as dietary supplements), zinc chloride (in deodorants), zinc pyrithione (anti-dandruff shampoos) and zinc sulphide. Some these products and other are used in small amounts in the upstream oil and gas industry.

Zinc is an essential mineral for good health and well-being especially regarding prenatal and postnatal development. Enzymes with a zinc atom in the reactive centre are widespread in biological systems.

5.8.1   Zinc Products in Upstream Oil and Gas Use

In general zinc compounds has fairly specialised uses within the oilfield sector and are used in relatively small amounts. The one exception would be the use of certain zinc salts such as zinc carbonate [185] and zinc carboxylates [186] as hydrogen sulphide scavengers in drilling fluid formulations and in drilling operations. Zinc products have a high reactivity with respect to hydrogen sulphide and therefore are suitable for the quantitative removal of even small amounts of hydrogen sulphide. As is observed with organic compounds (Chapter 6), this quantitative removal of hydrogen sulphide especially from gas streams can be very challenging.

Basic zinc carbonate is a complex compound containing both zinc carbonate and zinc hydroxide. Treatments of $1\,lb\,bbl^{-1}$ can remove about $500\,mg\,l^{-1}$ of sulphide. Although basic zinc carbonate has a high zinc content, it also contains $20\,wt\%$ carbonate, which can flocculate lightly treated, high solids muds, and therefore lime must be added to precipitate the carbonate ions.

In cementing operations, zinc carbonate and zinc oxide can be used as bridging agents [136].

Zinc oxide is also used as a weighting agent in drilling mud formulations. It is particularly suitable as it has a high density, 5.6 versus $4.5\,g\,ml^{-1}$ for barite, the most common weighting agent. It can be sorted for particle size and is soluble in acid. The high density means less weighting agent is required per unit volume of mud. The acid solubility is useful as dissolved zinc oxide can pass through a production screen without plugging it. The particle size of $10\,\mu m$ is desired as it means that zinc oxide particles do not invade the formation, but they are also not large enough to settle out of suspension [33].

Zinc oxide has also been considered as a coating material for proppants used in hydraulic fracturing. Such materials are used to reduce friction between the proppant particles [187].

Zinc metal, along with other metals (see Aluminium), can be used as an anti-seize agent; [107] however, due to its poorer environmental properties compared with other alternatives, it is little used.

Zinc peroxide has been used in acidic aqueous solution for filter cake removal especially filter cakes containing polymeric materials [188]. It has also been used in a similar fashion to aid in the breaking of guar-based polymers used in fracturing fluids [189].

Zinc halides, primarily chloride and bromide, have been used in completion brines. They tend to be used as a mixture with calcium brines and give a clear brine solution. Zinc halides have a high density, but the pH of the resulting brine is low and can lead to acid-based corrosion issues and handling problems. [190]

The use of zinc phosphate as a corrosion inhibitor has been discussed in Chapter 4. Other phosphate type zinc salts, namely, zinc dialkyl dithiophosphate and dioctyl phenyl dithiophosphate are used in lubricating grease formulations and have antioxidant properties [191].

Where zinc bromide has been used as a completion or workover brine and the reservoir contains hydrogen sulphide, zinc sulphide scales can be formed [192]. Zinc sulphide scales have also been found in several fields along the Gulf Coast of the United States and in fields within the North Sea Basin. Scale deposition has caused significant pressure and rate reductions in high temperature and high-rate gas, condensate, and black oil wells [193].

Acid washes are usually applied to remove zinc sulphide scale (and other acid-soluble solids); however, new scale deposits are often formed and retreatments are required. Therefore scale inhibitor applications are often considered [192, 193].

5.8.2   Environmental Impact of Zinc and Its Compounds

From natural weathering processes, soluble compounds of zinc are formed and lead to traces of zinc in the aquatic environment and surface waters. It is of note that as pH rises, the concentration of zinc decreases; however highly alkaline conditions are unlikely in natural waters and do not occur in normal seawater.

The refinement of zinc ores produces large volumes of sulphur dioxide and cadmium vapour. Smelter slag and other residues contain significant quantities of metals including zinc. The dumps of the past mining operations can leach zinc and cadmium. Zinc emissions from mining and refinement peaked at 3.4 million tonnes per year in the 1980s and declined to 2.7 million tonnes in the 1990s, although a 2005 study of the Arctic troposphere found that the concentrations there did not reflect the decline. Anthropogenic and natural emissions occur at a ratio of 20 to 1 [194].

Soils contaminated with zinc from mining, refining or fertilising with zinc-bearing sludge can contain several grams of zinc per kilogram of dry soil. Levels of zinc in excess of 500 ppm in soil interfere with the ability of plants to absorb other essential metals such as iron and manganese.

Zinc is easily detected by direct reading spectrometry, and the mean concentration of zinc in UK groundwater is reported at 110 $\mu g \, l^{-1}$. This however belies a widespread of results from 1.27 $\mu g \, l^{-1}$ to 155.07 $mg \, l^{-1}$. The zinc content in seawater is widely accepted to be 0.005 $mg \, kg^{-1}$ [195].

The European Union and World Health Organisation have set a permissible or recommended limit of 5.0 $mg \, l^{-1}$ for zinc in drinking water [196]. Zinc however has no known adverse physiological effects on humans except at very high concentrations where with over 700 $mg \, l^{-1}$, it can have an emetic effect. Indeed zinc is essential to human nutrition and a normal intake should be 10–15 mg per day. This is easily achieved in a normal balanced diet.

An adverse factor, which zinc has in common with other heavy metals, is that of taint. The astringent taste can be detected at concentrations as low as 2 $mg \, l^{-1}$. Traces of zinc are common in most foods; however pertinent to the impacts from the oilfield is its toxicity towards aquatic organisms and in particular fish where it displays its greatest effects [197].

It is well known that the water quality particularly, hardness, pH, temperature and dissolved oxygen have an important effect on heavy metal toxins with respect to fish and other marine organisms [198]. With respect to zinc both calcium and magnesium have complicating competitive effects, and indeed it is recognised that calcium has an antagonistic effect.

The result of these and other factors is that fish toxicity is not from internal poisoning but from interaction of the zinc (or other heavy metal) ion and the mucus secreted by the gills. This results in a film of coagulated mucus formed on gill membranes, impairing the fish so that it is asphyxiated.

Finally with respect to heavy metal poisoning of marine species, the effect of zinc on the level of dissolved oxygen has a critical impact. The body of work in this area concludes that slight variations in the concentrations of zinc and other heavy metals can have a serious impact on the level of dissolved oxygen in a given body of water, and even in the marine environment where this is relatively constant and there are still effects [199].

The uptake and accumulation in zinc in plants and animals has been well understood for some time with the definitive work for marine species conducted in the late 1950s. It was ascertained that the amounts of zinc found in marine organisms particularly shellfish such as oysters, clams, and scallops was thousands of times more than the seawater per unit of weight. This means that biomagnification of increased rates of bioaccumulation occur for zinc in marine organisms such as shellfish and fish more than other organisms [200].

It can be concluded that the environmental impacts of zinc (and other heavy metals) is more complex than a straightforward dose correlation to concentration and that bioconcentration and biomagnification can be very important in a particular ecosystem especially in respect to the aquatic environment. This and other factors are further considered in Chapter 9.

5.9 Other Metals, Other Inorganics and Related Compounds

A number of other metals, metal salts and complexes, inorganics and other related compounds are or have been used in the upstream oil and gas industry. However, their use is relatively small and so their environmental impact, in the main, proportionally less than the previously described classes of compound, for example, arsenic was used historically as a corrosion inhibitor in acid stimulation packages; however this has been completely phased out and replaced with less toxic alternatives.

When crude oil is produced, there are some heavy metals and other contaminates particularly mercury and arsenic that can become concentrated and cause both production processing and environmental problems. These are discussed later in this section.

5.9.1 Other Metals and Related Compounds

5.9.1.1 Antimony

Antimony tri-bromide, and other antimony halides, have good corrosion inhibitor properties particularly at high temperatures; [201] however it is highly toxic and has been replaced in oil and gas use by other less hazardous compounds.

Antimony trioxide has been considered as a friction-reducing material for coating proppant beads used in hydraulic fracturing operations [187], however, less toxic alternatives exist and are used in preference.

5.9.1.2 Bismuth

Bismuth salts have also been proposed as acid corrosion inhibitor intensifiers [202], and also with antimony as coating for friction reduction on proppant beads [187].

Bismuth citrate has been claimed for use as a dissolvable bridging agent [203].

Extreme pressure agents are additives that when added to lubricants help to minimise wear on gears and other parts that are exposed to high pressures. Many of the products used until the last decade have been made of toxic materials such as lead dithiocarbamate. Bismuth 2-ethyl hexanoate (see Figure 5.10) provides a non-toxic alternative, which can be added in ratios of up to 20%.

The scientific literature concurs that bismuth and most of its compounds are less toxic compared with other heavy metals (lead, antimony, etc.) and that it is not bioaccumulative. Bismuth salts have low solubilities in the blood, are easily removed with urine, and show no carcinogenic, mutagenic or teratogenic effects in long-term tests on animals. Bismuth metal and its compounds are not considered a threat to the environment and have minimal environmental impact. These considerations should lead to further investigation of bismuth compounds for

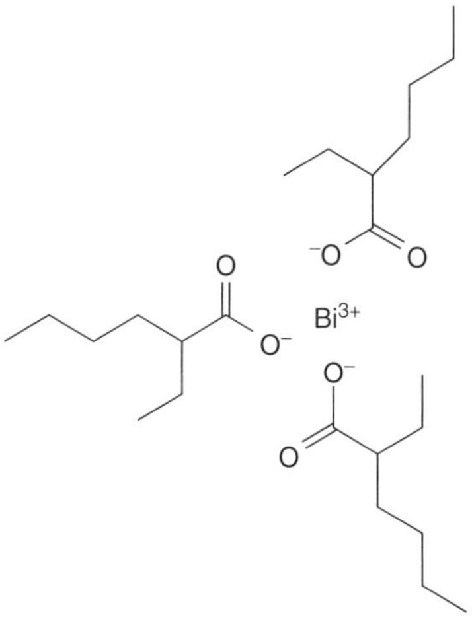

Figure 5.10 Bismuth 2-ethyl hexanoate.

other applications in the oilfield to give potentially more environmentally acceptable replacements for other less suitable chemistries.

5.9.1.3 Chromium

Chromium is the first element in group 6 of the periodic table. It has the chemical symbol Cr and atomic number 24. It is a steely grey, lustrous, hard and brittle metal, which takes a high polish, resists tarnishing and has a high melting point. Chromium metal is of high value for its high corrosion resistance and hardness. A major development was the discovery that steel could be made highly resistant to corrosion and discoloration by adding metallic chromium to form stainless steel. Stainless steel and chrome plating (electroplating with chromium) together comprise the majority of the commercial use of chromium.

The most common oxidation states of chromium are +6, +3 and +2; however, a few stable compounds of the +5, +4 and +1 states are known. In the +6 oxidation state (chromic), the most important ion species formed by chromium are the chromate, CrO_4^{2-}, and dichromate, $Cr_2O_7^{2-}$. Chromate and dichromate salts have been discussed earlier in the section for sodium and potassium.

The trivalent chromium (Cr(III)) ion is essential in trace amounts for humans, insulin, sugar and lipid metabolism. While chromium metal and Cr(III) ions are not considered toxic, hexavalent (Cr(VI)) is toxic and carcinogenic. Unfortunately Cr(VI) compounds tend to be the ones most in use in the upstream oil and gas industry.

Chromic acetate carboxylate salts have been used as cross-linking agents for the formation of polymer gels for conformance control, sweep efficiency improvement and water and gas shut-off treatments to enhance oil recovery applications [204]. Chromic triacetate was until the late 1990s, usually the preferred cross-linking agent in polymer gel technology; this is because it forms robust gels that are insensitive to the environment of the petroleum reservoir and a wide range of interferences including temperature and pH [204].

Chromium(III) (Cr(III)) gel technology has been reported. The gels can be used in conjunction with a number of oilfield treatments. The single-fluid acrylamide polymer/Cr(III) carboxylate aqueous gels are formed by cross-linking acrylamide polymer with a Cr(III)-carboxylate complex cross-linking agent [205]. Chromium(III) acetate has been much used in the field to cross-link partially hydrolysed polyacrylamides (PHPAMs) [98]. Chromium(III) salts have also been derived in situ from chromium(IV) by incorporation of a reducing agent. This avoids a delay in the gelling agent working as it is there at hand where it is needed [206]. A chromium(III) propionate system has been used with polymers to create gels in hard water brines [207], and other chromium salts such as chromium sulphate have been used as cross-linking agents in gel formation.

Chromium lignosulphonates have been found to be highly effective dispersants in drilling fluids and are useful in controlling viscosity of the drilling fluid package [33].

All of these applications suffer from perceived or actual toxicity of chromium and the attention placed on possible persistence and environmental toxicity. Chromium toxicity is however governed by two major factors, namely, oxidation state and solubility. Chromium(VI) compounds, which are powerful oxidising agents and thus tend to be irritating and corrosive, appear to be much more toxic systemically than chromium(III) compounds, given similar amounts and solubilities. Although mechanisms of biological interaction are uncertain, variation in toxicity may be related to the ease with which chromium(VI) can pass through cell membranes and its subsequent intracellular reduction to reactive intermediates [208–210].

Despite the body of evidence showing that Cr(III) is poorly absorbed by any route and that the toxicity of chromium is mainly attributable to the Cr(VI) form. The use of all chromium products is under regulatory pressure and is often prescribed from use in the upstream oil and gas industry. This has led to seeking out of alternatives such as zirconium salts (see later in this chapter) for cross-linking agents in polymer gel systems.

5.9.1.4 Copper

A small number of copper compounds are utilised in the upstream oil and gas sector, and copper metal has been used as a proppant coating to minimise friction between the proppant beads [187] (see also Antimony and Zinc).

Cuprous iodide has been used as a corrosion inhibitor intensifier and is effective up to 160 °C but has limited solubility in acid solutions and therefore limited use in acid stimulation packages [211].

Along with other previously described materials, such as aluminium and barium salts, copper has been considered as an anti-seize agent [107].

Copper is an essential nutrient at low concentrations, but is toxic to aquatic organisms at higher concentrations. In addition to acute effects such as mortality, chronic exposure to copper can lead to adverse effects on survival, growth, reproduction and alterations of brain function, enzyme activity, blood chemistry and metabolism. For these and other toxicological reasons, copper compounds are little used in the oil and gas industry.

5.9.1.5 Lead

In common with other metals, lead has also been considered as an anti-seize agent; [107] however, its toxicity usually precludes its use. Lead dithiocarbamate becomes active at high temperature and has been used as an extreme pressure lubricant; however this has been replaced by other agents such as bismuth 2-ethylhexanoate that is non-toxic (see Section 5.9.1.2). Also lead oxide and lead sulphide have been considered as a proppant coating material to reduce friction between the beads [187].

Lead sulphide scale can occur as a problematic issue in oil and gas production, and it is common that mixed lead and zinc sulphide scales appear together. Lead sulphide deposition is much more difficult to inhibit chemically than the commonly found calcium carbonate or barium sulphate scales. A number of common chemical inhibitors have been examined [212], and most are ineffective or at best partially effective. Lead (and zinc) sulphide can be removed by strong acids; [213] however as has been described in Section 5.7.4, this also presents a number of challenges not least of which is the potential release of hydrogen sulphide.

Lead of course from an environmental impact viewpoint is dominated by its toxicity, and this is the major reason for its minimum use as an additive. Lead moves into and throughout ecosystems. Atmospheric lead is deposited in vegetation, ground and water surfaces. The chemical and physical properties of lead and the biogeochemical processes within ecosystems will influence the movement of lead through ecosystems. The metal can affect all components of the environment and can move through the ecosystem until it reaches equilibrium. Lead accumulates in the environment, but in certain chemical environments, it will be transformed in such a way as to increase its solubility (e.g. the formations of lead sulphate in soils), its bioavailability or its toxicity.

5.9.1.6 Magnesium and Magnesium Salts

Magnesium is essential to both plant and animal life. Chlorophyll, the molecule that allows plants to photosynthesise light, has a magnesium ion at the centre of its chemical structure. Magnesium is also critical for the proper function of the human nervous system, the skeletal structure and the brain.

Magnesium is one-third lighter than aluminium. This quality makes it useful as structural materials for aerospace, aircraft, automobiles and machinery. Magnesium is alloyed with aluminium to improve its mechanical, fabrication and welding characteristics. Magnesium oxide is frequently used as a refractory material in furnace linings for producing iron and steel, nonferrous metals, glass and cement. Other applications include agricultural, chemical and construction industries.

Some magnesium salts and related products are used as functional additives in the upstream oil and gas sector. Magnesium oxide in particular has various uses in drilling, cementing and stimulation activities. It is a minor component of a variety of cements, including Portland cement [33]. It is also found in cement slags that are recycled for use.

It is used in the cement as an expansion additive, which aids in the flow of cement to fill all the existing cavities. The behaviour of magnesium oxide in this respect depends on its thermal history that can be modified to fit specific requirements [33].

Magnesium oxide and magnesium carbonate can be used in combination with hydroxyethyl cellulose or xanthan as a suspension aid in solid particle bridging of the cement to the rock surface [203]. Magnesium peroxide is the most suitable oxidising agent for degradation in this scenario [136].

Magnesium peroxide is very stable in alkaline environments, and therefore in polymer-based drilling fluids, it remains inactive when present. This is also the case in completion and workover fluids. As it is a powdered solid, it becomes an integral part of the filter cake and can be activated by a mild acid wash and therefore promotes filter cake degradation by release of hydrogen peroxide [214]. Magnesium peroxide has become the breaker system of choice in alkaline water-based systems especially in wells with bottom hole temperatures of 150 °F or less. Magnesium peroxide is also used as a delayed release oxidising agent and is encapsulated in such applications [136]. Similarly it is used as an inorganic breaking agent for fracturing fluid gels [215].

It has been found that use of magnesium oxide in borate cross-linked guar-based fracturing fluids allow the fluids to be used at higher temperatures [216].

Mixed metal hydroxides such as magnesium aluminium hydroxide are added to bentonite-based muds to provide an extremely shear thinning fluid. At rest such fluids exhibit a high viscosity but are water-like in rheology when a shear force is applied [217].

Magnesium mixed phosphate salts are used in ceramic particulate bridging agents [218], which are used in fluid loss control in drilling operations.

Magnesium cement [33] or magnesium oxychloride cement is a completely acid-soluble cement which has allowed the placement of a cement slurry in a given reservoir, establish a seal, and then remove the blockage by dissolving the cement in acid. Magnesium oxide and magnesium chloride are the major components of Magnesia cement.

Magnesium chloride, along with other chloride salts previously described, has been considered as an antifreeze agent in cements [34].

Magnesium nitrate, in common with a number of other polyvalent metal salts, can be used as a cross-linking agent for polymers in water shut-off applications. It has a distinct advantage of being less toxic and environmentally hazardous as a number of other alternatives [33].

Magnesium sulphate has been claimed as a dehydration additive in oil-based muds [219]. Magnesium sulphate and other magnesium salts are also found in the oilfield as problematic scales; however there inhibition and removal is fairly straightforward and not as difficult as the more common calcium carbonate and barium sulphate scales [98].

The major effect of magnesium on the environment results from the emission of hazardous air pollutants from magnesium industrial plants. Processing magnesium for industrial uses produces greenhouse gases such as sulphur hexafluoride (SF_6), hydrochloric acid (HCl), carbon monoxide, particulate matter and dioxin/furan. The major regulatory agencies have set limits on the release of these substances, and some have banned some of these substance especially sulphur hexafluoride.

Recycling can significantly reduce magnesium's impact on the environment and used beverage can recycling represents most of the recycled magnesium scrap used today.

5.9.1.7 Manganese Salts

Manganese is not found as an elemental metal in nature; it is often found in minerals in combination with iron. Manganese is a metal with important industrial metal alloy uses, particularly in stainless steels.

Manganese phosphating is used for rust and corrosion prevention on steel. Ionised manganese is used industrially as pigments of various colours, which depend on the oxidation state of the ions. The permanganates of alkali and alkaline earth metals are powerful oxidisers (see Section 5.1.2).

In biology, manganese(II) ions function as cofactors for a large variety of enzymes with many functions [220]. Manganese enzymes are particularly essential in detoxification of superoxide free radicals in organisms that must deal with elemental oxygen. Manganese also functions in the oxygen-evolving complex of photosynthetic plants. The element is a required trace mineral for all known living organisms but is also a neurotoxin.

Manganese compounds are rarely used in the upstream oil and gas sector with only a few specialised cases of note. Manganese hydroxide has been used in the composition of degradable cement bridging agents [136]. Manganese tetroxide, Mn_3O_4, has been considered as a weighting agent as it has a high specific gravity, making it suitable for muds used in drilling deep gas wells; however the filter cake formed by this mud also contains manganese tetroxide that has a very fine particle size and allows the filtrate to invade the formation [221].

Although it has been shown that waterborne manganese has a greater bioavailability than that of dietary manganese, the amount of manganese entering the environment from oilfield use is very small if measurable, and therefore its environmental impact is negligible.

5.9.1.8 Molybdenum and Molybdates

Molybdenum does not occur naturally as a free metal on Earth; it is found only in various oxidation states in minerals. The free element, a silvery metal with a grey cast, and readily forms hard, stable carbides in alloys, and for this reason most of world production of the element (about 80%) is used in steel alloy production.

Most molybdenum compounds have low solubility in water, but when molybdenum-bearing minerals contact oxygen and water, the resulting molybdate ion MoO_4^{2-} is quite soluble.

Molybdenum-bearing enzymes are by far the most common bacterial catalysts for breaking the chemical bond in atmospheric molecular nitrogen in the process of biological nitrogen fixation. At least 50 molybdenum enzymes are now known in bacteria and animals, although only bacterial and cyanobacterial enzymes are involved in nitrogen fixation. These nitrogenases contain molybdenum in a form different from other molybdenum enzymes, which all contain fully oxidised molybdenum in a molybdenum cofactor. These various molybdenum cofactor enzymes are vital to the organisms, and molybdenum is an essential element for life in higher organisms, though not in all bacteria.

This characteristic of molybdenum as an enzymatic inhibitor has been used in the oil and gas industry in the enhancement of nitrate and nitrite treatments of SRBs to inhibit them from producing hydrogen sulphide gas [222].

Molybdenum and in particular molybdenum disulphide are used as lubricants and anti-wear agents in a variety of applications. Molybdenum disulphide has been traditionally used in greases for drill bit application [223]. It has also been applied as an extreme pressure lubricant and as an anti-seize agent [107]. Again in common with other similar compounds, it has been examined as a coating for proppants to reduce friction [187]. However, due to environmental concerns, molybdenum products and particularly molybdenum disulphide have been replaced by less hazardous alternatives. A number of these alternatives employ the much less bioavailable and less hazardous molybdate salts, and as stated earlier in Section 5.1.1, sodium molybdate has been used along with other alkali metal molybdates to aid graphite as a lubricant [55].

The performance of some corrosion inhibitors, particularly those based on quaternary ammonium surfactants (see Chapter 3), can be enhanced by adding a source of molybdate ions to the formulated product [224]. In general molybdate ions act as good corrosion inhibitors in freshwater or condensate water systems. They are passivating or anodic corrosion inhibitors and work by forming a non-reactive thin surface film on the metal that inhibits the access of corrosive species to the metal surface. Some passivating corrosion inhibitors, such as phosphates (see Chapter 4) require oxygen to work effectively; however, molybdate ions can work anaerobically and work even better in the presence of oxygen [225]. Until recently molybdate was not widely used in oil and gas applications; however in the use of more environmentally acceptable polymers, molybdate ion has been considered to enhance the technical performance of water-soluble polymers as corrosion inhibitors [226].

5.9.1.9 Titanium and Salts

Titanium is a transition metal with a lustrous silver colour, low density and high strength. It is highly resistant to corrosion in seawater and other media. It occurs within a number of mineral deposits, which are widely distributed in the Earth's crust and lithosphere, and it is found in almost all living things, rocks, water bodies and soils.

Titanium can be alloyed with iron, aluminium and other metals, to produce strong, light-weight alloys for aerospace, military use and a wide range of other industrial applications. The two most useful properties of the metal are corrosion resistance and the highest strength-to-density ratio of any metallic element [227].

The most common compound of titanium is titanium dioxide, which is a photocatalyst and is used in the manufacture of white pigments. It is also found in Portland cement, which is a common cement in wellbore completion operations.

Titanates usually refer to titanium(IV) compounds, and these are used in the oil and gas sector as polymer cross-linking agents and have similar applications as the borate complexes (see Section 5.4). They often have specific uses in cross linking, fracturing fluid polymers [228].

The oxy salt of titanium (as well as other exotic metals such as hafnium, niobium tantalum, vanadium and zirconium) is useful in corrosion inhibition when combined with mercaptopyrimidine species, [229] to provide a sulphur synergistic environment.

5.9.1.10 Tungsten Compounds and Ions

Tungsten, also known as wolfram, is a hard, rare metal under standard conditions; however, tungsten is found naturally on Earth, almost exclusively in chemical compounds. Tungsten's many alloys have numerous applications, including incandescent light bulb filaments, electrodes in welding, superalloys and radiation shielding. Its hardness and high density are useful in military applications, and tungsten compounds are also often used as industrial catalysts.

Tungsten is the heaviest element known to be essential to any living organism; however, it interferes with molybdenum and copper metabolism and is somewhat toxic to animal life [230]. This toxicity plus economic considerations makes the application of tungsten compound in the oil and gas sector small and specialised.

Tungsten disulphide has been used as a lubricant and anti-wear agent in drilling operations [223].

Tungstate compounds are useful passivating or anodic corrosion inhibitors and are similar to the molybdate ions described earlier. However, like phosphates (see Chapter 4), they require oxygen to work effectively [98].

Tungstate ions in common with other transition metal oxyanions (permanganate, vanadate, etc.) can act as SRBs growth inhibitors [231].

5.9.1.11 Zirconium and Salts

Zirconium takes its name from the mineral zircon, its most important source. It is a lustrous, grey-white, strong transition metal and small amounts are used as an alloying agent for its strong resistance to corrosion [227].

Zirconium forms a variety of inorganic and organometallic compounds, a number of which have specialised uses in the upstream oil and gas sector, with utilisation of a number of organo-metallic complexes as polymer and gel cross-linking agents as its main application.

As has been described earlier, in Section 5.4, borates are common cross-linkers of, in particular, hydroxypropyl guar gum; however zirconates are also employed to do this function [232], particularly for prevention of dynamic fluid loss at high temperatures.

Various zirconium systems are used to ensure delayed cross-linking. These complexes are formed initially with low molecular weight compounds, which are then intramolecularly exchanged with the polysaccharides; this process causes the delayed cross-linking. The zirconium complexes that are used are usually initiated from compounds with diamine-based molecules such as hydroxyethyl-tris-(hydroxypropyl) ethylenediamine [233], or hydroxyl acids such as glycolic acid, lactic acid, citric acid, etc. or polyhydroxy compounds such as arabitol, glycerol, sorbitol, etc. These materials can then form suitable gels with polysaccharides. Zirconium halide chelates [234] and boron zirconium chelates [235] are also suitable delayed cross-linking agents.

Complexes of tetravalent zirconium with ligands of organic acid such as citric acid are suitable aqueous drilling fluid dispersants particularly bentonite-based suspensions [236].

Zirconium acetylacetonate has been described as suitable cross-linking agent where chromium salts have previously been used [237].

Zirconium 2-ethylhexanoate has been described as an extreme pressure agent; however as detailed earlier in this chapter, bismuth 2-ethylhexanoate is a suitable non-toxic alternative.

Similar to zinc oxide (see Section 5.8.1), zirconium oxide has been considered as a specialised weighting agent.

Zirconium compounds have no known biological role; however, zirconium and its salts generally have low systemic toxicity. Zirconium is unlikely to present a hazard to the environment. While aquatic plants have a rapid uptake of soluble zirconium, land plants have little tendency to absorb it, and indeed 70% of plants that have been tested showed no zirconium at all. Zirconium is generally considered to have low mobility in soils. Zirconium emissions to the atmosphere are increasing due to anthropogenic activities such as its use in different industries. Zirconium forms various complexes with soil components, which reduces its soil mobility and phytoavailabilty. The mobility and phytoavailabilty of Zr in soil depends on its speciation and the physicochemical properties of soil that include soil pH, texture and organic contents [238].

Despite having low soil mobility and phytoavailability, zirconium may have impact on the biosphere and deserves to be further evaluated.

5.9.2 Miscellaneous Small Use Metals and Metal Products

This penultimate section on metals and metal salts describes some miscellaneous uses of certain metals and metal salts, in the upstream oil and gas sector, not previously described.

Generally these are used intermittently and in low volumes so their environmental impact, if any, is minimal.

Cerium and tin lignosulphonates offer a less toxic alternative to chromium and mixed metal lignosulphonates [137] Cerium bromide and other bromides are used as high density brines in completion applications [239]. However, in this regard cesium brines and particularly cesium formate brines are more commonly used due to their comparative cost. Formate brines are more fully described in Section 6.1.1.1.

As has been previously described in this chapter, a number of metals and related substances have been considered as friction-reducing materials in fracturing fluid proppant coatings; in addition to those already considered indium, silver, tin and niobium diselenide are the most prominent [187].

Nickel has been considered as a lubricant component for drill bit greases [223], anti-seize agents and extreme pressure agents [107].

In addition to zirconium complexes, hafnium compounds have been considered as cross-linkers for guar-based fracturing fluids [10].

Finally in this section, and with regard to metal elements utilised in oil and gas upstream operations, vanadate ions are useful as passivating corrosion inhibitors and like molybdate do not require oxygen to be present [225]. They also show the same biocidal activity with molybdate and tungstate ions as growth inhibitors of SRBs [231].

As stated earlier the environmental impact generated from these extremely low used specialised products is very minimal if at all measurable. Of far greater concern are the toxic metals and related elements, which are present naturally in the crude oil and the reservoir rock and which are disturbed due to intervention by oil and gas production.

5.9.3 Toxic Metals and Related Elements in Crude Oil

The two elements of greatest concern are mercury and arsenic. Mercury is found in the volcanic rocks that underlie most oil and gas formations and consequently is associated with the production of hydrocarbons. It is soluble in the hydrocarbons as organometallic species such as dialkylated mercury, inorganic mercury slats and elemental mercury. Mercury associated with asphaltenes is found in a number of crude oils as a significant fraction of the total amount of mercury present [240]. Mercury, as it is well known, is a toxic element; however the organometallic derivatives are even more toxic. Mercury and its related derivatives are usually volatile and follow the liquid and in particular the gas hydrocarbons at significant concentrations. They cause a number of processing problems and in most regions they are regulated when discharged to the environment [241].

There are a number of technologies designed for the removal of mercury from oil and gas production streams and most involve the use of adsorbants [242], and some of which are discussed in other chapters.

Inorganic mercury salts have been found in some produced waters and as such can pose a serious environmental threat if discharged. This is because mercury readily bioaccumulates ending up in higher trophic species such as fish. Fish and shellfish in particular concentrate mercury in their bodies, often in the form of methyl mercury. Species of fish that are long lived and above the food chain such as tuna, shark and swordfish contain higher concentrations of mercury than others do [243].

A number of remediation treatments have been adopted including chemical precipitation, adsorption methods and reinjection of the produced water into disposal or depleted production wells [244].

Arsenic is found in crude oils and condensates at concentrations of less than 10 parts per million. This is in the form of arsenic compounds such as highly toxic hydrides, e.g. arsine (AsH_3). Arsenic compounds like mercury compounds can be found in produced water. Again there are a number of methods to remove arsenic and arsenic compounds including adsorption and pyrolysis, and many of which are used also for mercury [245].

Toxic metals, such as mercury, and metalloids like arsenic, have no biological role; however, they sometimes imitate the action of an essential element in the body, interfering with the metabolic process to cause illness.

There are also a number of NORMs associated with oil and gas production. Uranium and thorium isotopes are present but not mobile within the reservoir, whereas radium isotopes can be transported by chloride-rich formation waters. Radium isotopes can also co-precipitate with sparingly soluble alkaline cations such as sulphate or carbonate, and therefore formation waters can become radioactive due to the transportation of radium isotopes. This leads to the deposition of radium-contaminated scales during processing and production, and the so-called NORM scale has been discussed in Section 5.3.1.

5.9.4 Other Inorganic Compounds

A variety of other inorganic materials and compounds are used in upstream oil and gas operations and processing, mostly in relatively low amounts. These materials are also mainly inert in the environment in that they are not bioavailable or are rapidly biodegraded so as also not to be bioavailable. A relatively few are toxic and of environmental concern.

A number of inert and reactive gases are deployed in the upstream oil and gas industry, and the major ones of interest are discussed in the following sections.

5.9.4.1 Nitrogen

Nitrogen is utilised as its diatomic species and has a number of applications. It is the most common gas used in foamed cement where the foaming is generated by the direct application of gas to the cement slurry [246]. Nitrogen is also a commonly used gas in the generation of foamed fracturing fluids where the gas is injected in the presence of a surfactant to create a stable foam. However the stability and quality of the foam depends on how the gas is injected, and unstable foams tend to have large slugs of gas within the foam structure [247].

Clays in water-sensitive subterranean formations can be stabilised by injecting a sufficient amount of an unreactive gas, such as nitrogen, into the formation to mobilise the clays, including any fine particles, thus enabling their removal from the formation. This then allows subsequent treatments with salts such as potassium chloride to stabilise the clays and reduce swelling. The resulting process protects against formation damage by the presence in the formation of 'fresher' water than the connate water. Both injection and production wells have been treated in this manner [248].

Nitrogen (as well as carbon dioxide) is used in gas floods for enhanced oil recovery [249].

5.9.4.2 Carbon Dioxide

Carbon dioxide, as has previously been discussed, is a problematic corrosive gas in the oil and gas industry; however it has a major use in gas flooding in enhanced oil recovery primarily through carbon dioxide injection, which accounts for over 50% of the enhanced oil recovery in North America, with carbon dioxide flooding as one of the most proven methods. Almost pure CO_2 (>95% of the overall composition) can be mixed with oil to swell it, make it lighter, detach it from the rock surfaces, and cause the oil to flow more freely within the reservoir so that it can be 'swept up' in the flow from injector to producer well [250].

Much has been written about the need to capture CO_2 from industrial plants, especially coal power plants, and other processes [251]. There has been a growing attention to this due primarily to concerns of accelerating accumulation of greenhouse gases in the atmosphere, and this will be discussed in further detail in Chapter 10. However, the oil and gas industry gives an almost unique opportunity to utilise this gas for a considerable energy advantage. Currently a number of factors restrict the application of CO_2 to enhanced oil recovery, namely:

- Dependence on natural sources of CO_2
- Transportation of CO_2
- Safety and environmental issues
- Breakthrough of CO_2 to production wells
- Corrosion of well and field equipment

In carbonate reservoirs at high temperatures, chemical reactions can occur, resulting in the formation of carbon dioxide. This process can be stimulated by the injection of superheated steam [252].

Like nitrogen, carbon dioxide is a commonly used gas in the generation of foamed fracturing fluids [247].

Effervescent biocides are often used as a delivery method for specialised biocide applications to give a greater dispersant of the biocide at a low dose rate and as has been described in Section 5.1.1, sodium bicarbonate as well as other sodium salts can act as a delivery system for the carbon dioxide [7].

It has been found that the efficiency of drilling operations can be improved by the use of a supercritical fluid or a dense gas, which occurs at temperature and pressure conditions in the drill site. Carbon dioxide is such a gas in that it can be readily compressed under borehole conditions to form a liquid. The very low viscosity of this supercritical fluid gives efficient cooling at the drill head and excellent cuttings removal [253].

The storage of carbon dioxide has been extensively studied so that alongside its capture it can be effectively stored for use later and removed from the emission cycle. This is again further discussed in Chapter 10 [254].

5.9.4.3 Hydrogen

Hydrogen gas is used in the formation of foamed cement. In such cases the hydrogen gas is generated in situ by the reaction of a high pH cement slurry and finely powdered aluminium [33].

The thermal conversion of organic waste material can yield a hydrogen gas product that will crack long hydrocarbon chains of plastics and also help in improved oil recovery [255].

5.9.4.4 Hydrogen Sulphide

Like carbon dioxide, hydrogen sulphide is a problematic, toxic and highly corrosive gas. Unlike carbon dioxide it is not deliberately applied in the upstream oil and gas sector, and all emphasis is on its prevention and removal [256].

The primary focus on controlling hydrogen sulphide is to prevent or minimise its generation from SRBs and sulphate-reducing archaea (SRA), resident in many oil-bearing formations. Much study has been conducted in this area, and although much is understood about these microbes, there is still a lot being discovered and recently learned [257].

There are three primary methodologies for attacking SRB and SRA populations, these are the following:

- Control through biocide application

- Removal of food sources
- Manipulation of the SRB (and SRA) populations

The main application of biocides is to process systems where they can be very effective in control and indeed eliminating microbial populations [258]. They have several drawbacks:

i) They need to be continuously dosed or at least batch dosed to maintain a level of biocide in the system. Continuous dosing is usually not an economic option and batch dosing is common practice, leading to serious failures when either biocide is not maintained or is deliberately cut back as no problems were observed.
ii) They are selectively more effective against planktonic (waterborne, free-floating organisms) bacteria.
iii) They are high regulated due to their toxicity.
iv) Many are persistent and are not permitted for use.
v) They are not very effective in the reservoir, as they cannot be guaranteed to reach all areas. Under attack microorganisms migrate to more preferential regions.

There are two main applications involving food source removal or manipulation. These are sulphate removal from injection water (seawater) by reverse osmosis and deliberate nitrate feeding of the resident SRB population.

Sulphate Removal. Where a reservoir is supported by seawater, produced water and/or aquifer-formation water, it is possible to remove or substantially reduce the concentration of ions, which may be problematic. This is particularly focused on sulphate ion removal, which has a high concentration in seawater. This is a primary food source for SRB organisms, and the waste product is H_2S gas. The introduction of sulphate removal by reverse osmosis is however costly, and this capital requires to be designed. As no sulphate will be encountered, until there is injection water breakthrough, a number of operating companies do not spend this capital preferring to react to the situation at a later date. If however, in addition to potential reservoir souring, barium sulphate deposition is identified at injection water breakthrough, due to high concentration of barium ion in the formation water, then there are supportive arguments for capital investment in sulphate removal or reduction [259].

Nitrate Feeding. In the last two decades, there have been a number of applications, in particular in the Norwegian sector of the North Sea, in the use of nitrate addition via injection water to feed resident SRB (and perhaps SRA) populations in the reservoir. This has the effect of changing these SRBs to nitrate-reducing bacteria (NRBs) whose end reduction pathway and waste product is nitrite ion; therefore no H_2S is produced. This requires continuous addition of calcium nitrate to the injection water for the field lifetime, as studies have shown aggressive reversion to SRB life cycle and even higher production rates of H_2S being observed. Also recently several fields, where nitrate feeding has been ongoing for several years and now seawater breakthrough is being experienced, are observing increases in H_2S levels [260].

Nitrate feeding does also not address resident populations producing H_2S prior to injection water breakthrough or fields where such an application is not possible or economic (which are many).

The manipulation of SRBs is a little used technique, and the main application is the use of biostatic agents, in particular anthraquinone (see Chapter 6). These agents inhibit sulphate respiration in SRBs by uncoupling electron transfer from ATP syntheses. This effectively stops the SRB population growing and also reduces and/or eliminates H_2S production. In the North Sea region, anthraquinone and other electron transfer agents do not meet the environmental criteria for use and discharge of chemicals and therefore are not used [261].

All of this means that considerable amounts of hydrogen sulphide gas is generated and has to be controlled by chemical scavenging or other means. In many onshore environments the H_2S

is captured and converted to elemental sulphur by further reduction and used as a chemical feedstock. However in offshore environments this practice is neither permitted nor economic. The main method of H_2S reduction is therefore by chemical scavenging.

In oil and gas production, the three phases of hydrogen sulphide are generally present: gas, oil and condensate. In many parts of the production process, these three phases coexist, and therefore it is important to understand the distribution of the H_2S over these three phases. Figure 5.11 represents this dissociation behaviour where P_{H_2S} is the partial pressure of H_2S in the gas phase and C_{H_2S} is the concentration in either the oil/condensate or water phase.

The dissociation of H_2S in the water phase, at any given temperature and pressure, is dictated by the pH that is dictated by the concentration of acidic gases present, primarily CO_2 and H_2S [262]. Therefore H_2S can be removed from the system using a suitable base such as monoethanolamine; however this requires the pH of the water to be raised high enough to shift the dissociation of dissolved H_2S to the right and at a pH greater than 8. Such neutralisers are used in onshore environments where absorber equipment of sufficient size can be constructed. In systems where both CO_2 and H_2S are present, raising the pH with a neutraliser means that the protons generated by the dissolved CO_2 must be reacted before the H_2S can be neutralised, as CO_2 generates a stronger acid than H_2S. Therefore large amounts of amine neutraliser are needed, and these are applied via percolating tower systems. As a result in the offshore environment, therefore only chemical scavengers are considered. Inorganic scavengers such as sodium chlorite, iron and zinc carboxylates have already been described earlier in this chapter. Other scavengers are usually low molecular weight organic compounds and are considered in various sections in Chapter 6.

5.9.4.5 Hydrogen Peroxide

Hydrogen peroxide has the chemical formula H_2O_2 and in its pure form, it is a colourless liquid, which is slightly more viscous than water. Hydrogen peroxide is the simplest peroxide, i.e. a compound with an oxygen–oxygen single bond. It finds uses as a weak oxidiser, bleaching agent and disinfectant. Concentrated hydrogen peroxide is a reactive oxygen species and has been used as a propellant.

Pure hydrogen peroxide will explode if heated to boiling, will cause serious contact burns to the skin, and can set materials alight on contact. For these reasons it is usually handled as a dilute solution (household grades are typically 3–6% w/w). About 60% of the world's production of hydrogen peroxide is used for pulp and paper bleaching [263]. It is applied as a solution in the upstream oil and gas sector and finds uses mainly as an oxidising agent.

The decomposition of hydrogen peroxide can generate heat, and this property has been used to create a situation in situ for the removal of wax deposits. This methodology has the added

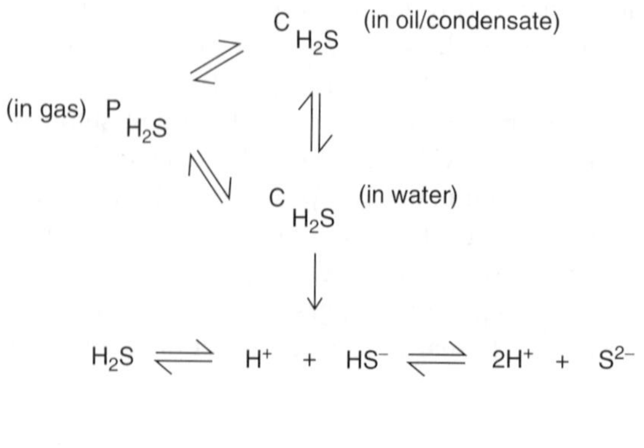

Figure 5.11 The dissociation behaviour of hydrogen sulphide.

advantage that in addition to the heat 'melting' the deposited wax, the hydoxyl free radical released also oxidises and decomposes the long-chain hydrocarbons [264].

It has been found that difficult-to-remove lead scales can be successfully treated with a blend of acetic acid and hydrogen peroxide [98].

Hydrogen peroxide along with other peroxides and inorganic oxidising agents have been considered and used as H_2S scavengers in oilfield and geothermal operations [265].

The oxidising nature means that hydrogen peroxide has been used as a gel breaker and in combination with sodium carbonate has been reported as a reagent for the removal of clay deposits [33]. However one of its main uses lies in cementing and stimulation applications where it is used to degrade and destroy the filter cake often in controlled conditions [99].

Hydrogen peroxide offers considerable advantages in enhanced oil recovery by chemical injection being able to offer the more favourable aspects of steam and carbon dioxide injection; its wider use would appear to be economically driven [266].

5.9.4.6 Ammonia and Ammonical Salts

Ammonia is a chemical compound composed of nitrogen and hydrogen with the formula NH_3. Ammonia is a colourless gas with a characteristic pungent smell. It contributes significantly to the nutritional needs of most terrestrial organisms by serving as a precursor to food and fertilisers. Although common in nature and in wide use, ammonia is both caustic and hazardous in its concentrated form.

NH_3 boils at $-33.34\,°C$ ($-28.012\,°F$) at a pressure of one atmosphere, so the liquid form must be stored under pressure or at low temperature. Household ammonia or ammonium hydroxide is a solution of NH_3 in water.

A combination of ammonia and a low molecular weight alcohol has been proposed as a cost-effective chemical enhanced oil recovery treatment [267]. Aqueous ammonia is also used as a buffer to adjust and maintain the pH of aqueous-based fracturing fluids [268].

In addition to ammonia solutions, salts of ammonia are also used in the upstream oil and gas sector. Primary among these is ammonium bisulphite which is the most commonly used an oxygen scavenger for water treatment within the sector (see also sodium bisulphite in Section 5.1.1) [12].

Ammonium bifluoride is used as a precursor or releasing agent for HF in acid stimulation of sandstone reservoirs. This improves safety and handling concerns [269].

Ammonium carbonate decomposes in acid medium to release carbon dioxide, which is useful in enhanced oil recovery [270].

Ammonium chloride, similar to potassium chloride, can be used in surfactant solutions to generate viscoelasticity and be useful as fluid loss additives [54]. Also similar to potassium chloride, ammonium chloride can be used as a temporary clay stabiliser during drilling, completion, and well servicing operations [33]. A number of ammonium salts, including ammonium chloride, are used in cleanup of degraded bridging agents in filter cake removal [271].

Ammonium hydroxide can be used, in conjunction with a silica cement, to alter the permeability profile of the producing formation where reservoir temperatures are over 90 °C. This process substantially closes off the higher permeability zone to fluid flow, allowing stimulation techniques to enhance recovery from the lower permeability zone where hydrocarbons are 'trapped' [272].

The complex salt, ammonium ferric oxalate (Figure 5.12), potentially offers more acceptable environmental alternative to sodium dichromate as a polymer and gel cross-link agent [46], and also for ferric acetylacetonate used in cross-linking water-soluble polacryamides [182].

Ammonium persulphate and ammonium peroxydisulphate have been examined for use in encapsulated gel breaking system that allows delayed action in breaking the gel system [215].

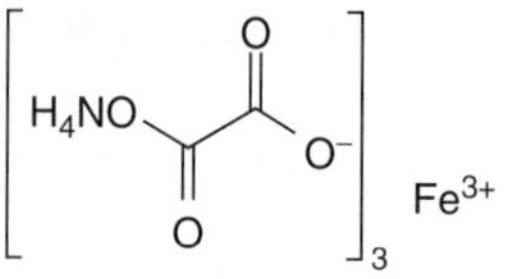

Figure 5.12 Ammonium ferric oxalate.

As described earlier in Section 5.1.1, thiocyanates such as ammonium or sodium thiocyanate, are used as corrosion inhibitors in drilling, completion and workover fluids [55].

The thioglycolates also form interesting ammonium salts; however these are discussed in Section 6.2.2.

The nitrates and nitrites have a specialised application in decomposition to give heat, and this is discussed further in Section 5.9.4.7.

The use of ammonium salts as applied to the upstream oil and gas sector would seem to be underexplored and may offer a number of viable and more environmentally acceptable alternatives to heavy metal and other salts with more persistent and toxic properties. This has to be balanced against the following. When in gaseous form, ammonia has a short atmospheric lifetime of about 24 h and usually deposits near its source. In particulate form, ammonia can travel much further, impacting a larger area. Both gaseous and particulate ammonia contribute to eutrophication of surface waters, soil acidification, fertilisation of vegetation and formation of smog in cities.

Since ammonia is one of the only basic species in the atmosphere, it readily reacts with strong acidic species in the atmosphere such as nitric and sulphuric acids, which are by-products of combustion process including vehicle and industrial sources to form ammonium salts.

5.9.4.7 Nitric Acid and Nitrates

In general nitric acid is too strong an acid to be used in the upstream oil and gas sector; however it has been used in a controlled fashion when mixed with hydrochloric acid to dissolve and remove metal and metal tools from wellbores [33]. In contrast to this, nitrates and nitrites are applied and have some particular uses.

Nitrates and nitrites, mainly sodium and calcium salts, are used as competitor feedstocks to modify SRBs (see earlier Section 5.9.4.4). These materials are cheap and environmentally benign. However such treatments have shown to have had mixed results in inhibiting SRBs from producing hydrogen sulphide and have led to increased rates of corrosion [273].

Ammonium nitrite can be formed in situ from an ammonium salt such as ammonium chloride and sodium nitrite. The addition of a small amount of acid causes the thermochemical decomposition of the ammonium nitrite to nitrogen, water and sodium chloride, plus a large amount of heat [274]. Such a system is the basis of a number of packages used for the downhole and subsea removal of wax deposits [275] and also to remove subsea hydrate plugs [276].

Plants get nitrogen from water and from the soil. They get nitrogen by absorbing it in the form of nitrates and ammonium. Nitrates are the major source of nitrogen for aquatic plants. Nitrates are not utilised by aquatic organisms such as fish and aquatic insects.

Increases in nitrate, like an increase in any salt will increase the osmotic concentration of the soil solution. The roots of the plant then have to take up minerals from a more concentrated solution. If the solution outside gets too concentrated, there will come a point where the plant is not able to take up any water against the concentration gradient, and the plant will start to wilt. Even before this point is reached, the plant will grow slower.

Application of too much fertiliser means that plants cannot absorb them fast enough, or retained by the soil, so there is a danger that it will be washed through into the drainage water and onto river systems. If this happens, then algal growth (and the growth of other water plants) in rivers can be stimulated and the result is eutrophication. This is particularly the case for ammonium nitrate [277].

5.9.4.8 Hydrazine and Related Products

Hydrazine is an inorganic compound having the chemical formula N_2H_4 (see Figure 5.13). It is a colourless flammable liquid with an ammonia-like odour, highly toxic and dangerously unstable unless handled in solution. Hydrazine is usually handled as hydrazine hydrate, which is about a 64% solution of hydrazine in water by weight. Hydrazine is mainly used as a foaming agent in preparing polymer foams, but significant applications also include its uses in polymer manufacture and pharmaceuticals. It has also been used as a rocket fuel. Hydrazine is used within both nuclear and conventional electrical power plants as an oxygen scavenger to control concentrations of dissolved oxygen in an effort to reduce corrosion, and historically it has been used not only in the oilfield as such in boiler water feeds but also for corrosion control in drilling, workover and cementing operations [278].

Hydrazine and its direct derivatives, mainly hydrazine hydrochloride, have been mostly removed from oilfield use due to concerns over toxicity. Traditionally they have been used as corrosion inhibitors where oxygen removal has been required and also as iron control agents in conjunction with chelants such as ethylenediamine tetraacetic acid (EDTA) [279], chelants are described in more detail in Chapter 6. Iron control agents are required under acidic conditions during stimulation or acid fracturing. Hydrazine hydrochloride has been claimed as a key ingredient in a corrosion-resistant cement, and when combined with certain phosphonic acid derivatives affords increased plugging inhibition [33].

In general, most of these applications utilising hydrazine and its derivatives can be successfully replaced by alternative chemistries, including close analogues in hydroxylamine (Figure 5.14) and hydroxylamine hydrochloride [279]. It may have been that in certain of the applications, and particularly in oxygen scavenging, larger amounts of related oximes are required [280].

Oxime derivatives, guanidine salts, and the related semi-carbazides are further discussed in Chapter 6.

Carbonic dihydrazide is related to hydrazine, and urea (see Chapter 6) although fully inorganic in nature (see Figure 5.15) has been proposed as a non-corrosive scale inhibitor for geothermal wells [281].

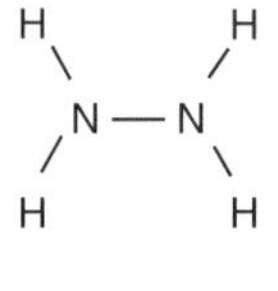

Figure 5.13 Hydrazine.

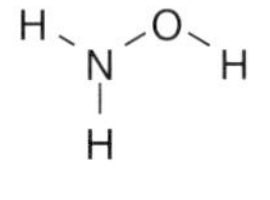

Figure 5.14 Hydroxylamine.

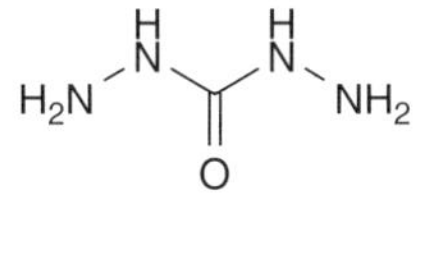

Figure 5.15 Carbonic dihydrazide.

The inorganic salt sodium azide (NaN_3) acts as a biostatic agent and has been suggested could be used to prevent biofouling in oil and gas wells [282]. It is known to inhibit the enzyme cytochrome oxidase in gram negative bacteria [283].

5.9.4.9 Sulphuric Acid, Sulphates and Related Salts

A number of organic sulphur-containing products are used in the oilfield sector, primarily as synergists in corrosion inhibitor formulations, and these are discussed in Chapter 6. Occasionally, however, elemental sulphur can be deposited in pipes and conduits in which sulphur-containing gas has been transported, and this can be mitigated using a suitable dispersant [284].

Sulphuric acid has been used in enhanced oil recovery particularly in thermal treatments. In contrast with hydrochloric acid, the sulphuric acid reacts with the crude oil, causing a viscosity reduction of the crude, and it is thought that the sulphuric acid thermal treatment alters the fundamental aggregation properties of the components of the crude responsible for its high viscosity [285].

As has already been described under various metal salts, sulphates, sulphites, bisulphites and thiosulphates are used in a variety of applications in the upstream oil and gas sector.

The dithionite salts such as sodium dithionite (see Figure 5.16) are oxygen scavengers and have been proposed for use in drilling and completion operations [286]. However, they are particularly underutilised (H.A. Craddock, unpublished results).

The major environmental impact of these small inorganic sulphur compounds is in the main due to their oxidative degradation to sulphur dioxide (and sulphur trioxide). When sulphur dioxide combines with water and air, it forms sulphuric acid, which is the main component of acid rain. Acid rain can

- Cause deforestation
- Acidify waterways to the detriment of aquatic life
- Corrode building materials and paints

The amount of sulphur dioxide produced from these additive is very small in comparison with other sources, primarily the combustion of fossil fuels, including crude oil. This paradox of balance between the environmental impact from chemical inputs to exploiting oil and gas and its use as a fossil fuel is considered more fully in Chapter 10.

5.9.4.10 Inorganic Carbon and Its Compounds

Carbon black is a material produced by the incomplete combustion of heavy petroleum products, mainly various tars, and a small amount from vegetable oil. It is dissimilar to soot in its much higher surface-area-to-volume ratio and significantly lower (negligible and non-bioavailable) polyaromatic hydrocarbon content. Carbon black is mainly used as a reinforcing filler in tyres and other rubber products. In plastics, paints and inks, carbon black is used as a colour pigment [287].

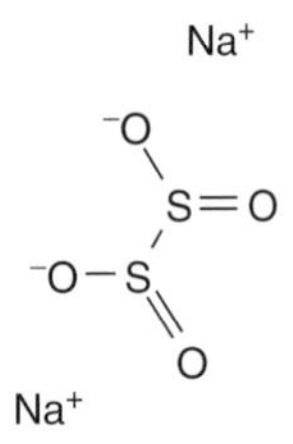

Figure 5.16 Sodium dithionite.

The incorporation of carbon black into a non-aqueous drilling fluid at levels between 0.2% and 10% by volume allows the mud to exhibit electrical conductivity, and therefore information from electrical logging tools can be obtained while drilling [288].

Carbon black can also be a constituent of high alumina cement composites [289]. It can also serve as a low cost additive for the control of gas migration into cementing completions and act as an alternative to latex and silica products [290]. Recently oxidised carbon black has been deployed as a nano-additive (concentrations between 0.05 and 0.5 wt%) in aqueous solutions of PAMs to improve viscosity and stability in aqueous and saline solutions and therefore lead to application in enhanced oil recovery [291].

Carbon black is considered to be carcinogenic to humans, its major disadvantage for extensive use [292]. It appears to be environmentally inert. It has been considered as a possible alternative to molybdenum disulphide as lubricant for drilling and other similar applications [293]. However graphite and in particular polarised graphite are more effective. Polarised graphite can be used as a lubricant for rock bits since it exhibits excellent load carrying and anti-wear performance.

Ordinary graphite consists of carbon in a layered structure and has a laminar hexagonal crystal structure. The closed rings of carbon atoms do not have any electrical polarisation, and this lack of polarity prevents graphite powder from forming a lubricant film and adhering to metal surfaces. Polarisation of the graphite allows it to adhere to metals and form good lubricating films, which can carry head loads without failing [56]. Graphite has also been examined as a proppant coating material to reduce friction in the fracturing fluid mixture [187].

Reinforced cements containing various fibres were initially developed as a high-strength material for bore hole lining, and carbon fibres offer the greatest toughness and increased compressive strength and have potential use at high temperatures [48, 294]. Carbon fibres have also been used in hydraulic fracturing to build a porous pack to filter out unwanted fines and other particles. Also, pumping fibres together with the proppant reduces the frictional forces, which limit the pumping of the fluids containing proppant materials [295]. Carbon fibres are however a relatively expensive material and this has tended to limit their use.

Related to the use of carbon fibres is that of carbon nanotube (CNT) particles. Nanotechnology is one of the most active research areas that encompass a number of disciplines including civil engineering and construction materials. Currently, the most active research areas dealing with cement and concrete are understanding of the hydration of cement particles and the use of nano-size ingredients such as CNT particles. CNTs are expected to have several distinct advantages as a reinforcing material for cements as compared with more traditional fibres. First, they have significantly greater strengths than other fibres, which should improve overall mechanical behaviour. Second, CNTs have much higher aspect ratios, requiring significantly higher energies for crack propagation around a tube as compared with across it than would be the case for a lower aspect ratio fibre. Third, the smaller diameters of CNTs means both that they can be more widely distributed in the cement matrix with reduced fibre spacing and that their interaction with the matrix may be different from that of the larger fibres. Using the CNT reinforced cement can reduce probability of casing collapse in oil and gas wells. Gas migration is a serious problem of cementing in gas wells, in which using CNT in cement can reduce amount of gas migration [296].

These CNTs are part of the class of carbon compounds known as fullerenes in which a molecule of carbon is in the form of a hollow sphere, ellipsoid, tube and many other shapes. Fullerenes are similar in structure to graphite. They have to date been little explored for oilfield application. Some substituted fullerenes have been proposed as additives to affect the cold flow properties of base oil [297]. Although to date very few real applications of these materials have been developed, their potential is tantalising and may lead to new highly effective materials, which could reduce the overall environmental burden of chemical applications [298].

These inorganic carbon materials are, in the main, environmentally inert and offer promise in some critical areas of oilfield application to reduce and replace more environmentally hazardous materials.

5.9.4.11 Others Non-classified Inorganics

In the main the remaining inorganics not yet classified such as glass, quartz and minerals such as kaolin, bentonite and zeolites are either silica based or inorganic silicates, e.g. kaolin is aluminium silicate. These products are covered under Chapter 7.

A critical material not covered elsewhere are radioactive inorganic tracer materials based on radioactive isotopes of ions. The most relevant of these is S [13]CN$^-$ as the tracer is not an isotope of a non-radioactive species, which could be naturally occurring in connate and produced water [299]. Such materials are used in very small quantities and relatively infrequently and therefore have no measurable environmental impact.

5.10   Environmental Issues Relating to Metals and Inorganics

Throughout this chapter, a number of biological factors and environmental impacts relating to metals, their salts and other inorganic compounds such as halogens have been described. The issues surrounding biodegradation as described in previous chapters and encountered by the majority of oilfield additives do not arise, as in general these molecules do not biodegrade, other than through oxidation to their oxides, which are usually relatively stable. Also factors affecting their stability such as chemical compatibility, mechanical degradation, and others are relatively minor in terms of changes, leading to enhanced environmental effects.

The critical factors surrounding the environmental impact of these products are driven by their bioavailability and their inherent toxicity. This has focused the 'greening' of chemistry in this area to be dominated by substitution strategies. As has been described previously, such a strategy inevitably compromises the reduced environmental impact with reduced efficacy. This in turn can lead to greater quantities of product being used as higher dose rates are required and lead to higher environmental loadings.

5.10.1   Inherent Toxicity

Many metals, particularly heavy metals, are toxic, but some heavy metals are essential, and some, such as bismuth, have a low toxicity. Most often the definition of 'heavy metals' includes cadmium, lead, mercury and the radioactive metals. Metalloids, such as arsenic, may also be included in the definition. Radioactive metals have both radiological toxicity and chemical toxicity. Metals in an oxidation state abnormal to the body may also become toxic, e.g. chromium(III) is an essential trace element, but chromium(VI) is a carcinogen. These inherent toxicities particularly as related to water-soluble metal salts can put a serious burden on the environment and can lead to accumulation of lethal toxic amounts of heavy metals and related toxins in the food chain. This is particularly the case for aquatic animals, particularly shellfish and fish [300].

Throughout this chapter those metals of particular concern from a toxic viewpoint have been highlighted; however it is worth noting that a number of metals and metal salts in common usage are not toxic. It is also noteworthy in particular that bismuth, which has low toxicity, is relatively little used in oilfield applications.

5.10.2 Other Toxic Effects

A major environmental effect is often the action of the metal salts on other factors in the ecosystem such as salinity, pH, available oxygen, osmotic effects due to changes (usually increases) in salt concentrations, etc. These effects are usually gradual in their impact until a tipping point is reach, and the balance in a specific ecosystem or the effect on a specific species or genus can be catastrophically affected [38].

It is well known that water quality, particularly the amount of dissolved oxygen has an important effect on heavy metal toxins with respect to fish and other marine organisms. Many metal ions can have competing and conflicting effects, and this can result in an overall depletion in the amount of dissolved oxygen in the aquatic environment, leading to unsuitable conditions for many forms of aquatic life particularly fish [198].

Some of inorganics, e.g. carbon dioxide, previously described are either precursors to, or are themselves agents such as greenhouses gases that are affecting the overall balance of the biospheres climate [301]. However the corollary to their use is that they can be controlled and not emitted to impact on the environment. The recycling of carbon dioxide in enhanced oil recovery would be such an example [251].

In general regulatory authorities classify metals and inorganic compounds as being toxic in aquatic environments if they have an LC_{50} or EC_{50} of less than $1\,mg\,l^{-1}$. These are measurements of concentration where 50% of the population is killed or detrimentally harmed and have been established by laboratory testing and quantitative extrapolation.

5.10.3 Bioavailability and Bioaccumulation

Bioavailability in the context of environmental science is generally accepted as a molecule being bioavailable when it is available to cross an organism's cellular membrane from the environment where the organism has access to the chemical [302]. It has been generally accepted by the scientific community and regulatory authorities that most organic molecules having a molecular weight greater than 700 will not cross cellular membranes and are therefore not bioavailable. This criterion is further discussed in Chapters 9 and 10..

Most metals, metals salts and other inorganic complexes discussed in this chapter have molecular weights below 700 and the mechanisms for being bioavailable vary. Receptors in soil, such as plants and microbiota, are directly exposed to metals such as cadmium, copper, mercury, lead and zinc via the soil solution, and this should ideally be reflected in environmental quality objectives that, presently, are commonly based on 'total' rather than bioavailable metal contents of soils. An understanding of the factors influencing the partition of metals between the solid and solution phases of soils is an essential prerequisite of any attempt to apply a critical load approach in a manner that relates to the effects of metals on soil organisms [303].

Other metals can have a rapid uptake in soluble forms generally considered to have low mobility in soils as they can form various complexes with soil components, which reduces their soil mobility and bioavailability. The mobility and bioavailability of such metals in soil depends on their speciation and the physicochemical properties of soil that include soil pH, texture and organic contents [238].

These solubility limiting factors in soil-based environments are a key limiting factor in plant growth. Adsorption of plant nutrients and other materials by soil colloids and in the removal of toxic substances from the food chain by microorganisms are also critical (due to sorption to, or partitioning of, otherwise degradable substances into inaccessible phases in the environment) [304].

The complexity of these pathways and partitioning of metals and other inorganics can lead to both their removal from biological pathways and their accumulation. The bioaccumulation of heavy metals and other toxic inorganic compounds in soils and marine sediments can lead to their entrance to the food chain through processing of such sediments by microorganisms and algae [305].

Clearly it is important to consider both the acute toxic effects and also the potential long-term bioaccumulative effects of metals, metal salts and other inorganic compounds, particularly when considering a substitution program to utilise and develop more environmentally acceptable materials. To date the oil and gas industry and regulatory authorities have taken a fairly simplistic approach, involving substitution to perceived less environmentally hazardous materials based on toxicity and mainly toxic properties inherent to the substance.

Going forward it would seem reasonable to consider the development of models to include bioaccumulation potential; however there are complications in this, and again a simplistic approach can be misleading as to potential ecological impacts.

Bioaccumulation factor data has been compiled from field studies that accounts for combined waterborne and dietary metal exposures. Trophic transfer factor data for metals have also been compiled from laboratory studies in which aquatic food chains were simulated. Results indicate that field bioaccumulation factors tend to be significantly inversely related to exposure concentration. This can be attributed to both lower exposure levels in the field. Trophic transfer factors have also been observed to be inversely related to exposure concentration, particularly at lower exposure concentrations. These inverse relationships have important implications for environmental regulations (e.g. hazard classification and tissue residue-based water quality criteria) and for the use of metal bioaccumulation data in site-specific environmental evaluations, such as ecological and human health risk assessments [306].

The science to date would indicate that for metals and metalloids, unlike many organic substances, no one data point can be used to express bioaccumulation and/or trophic transfer without consideration of the exposure concentration.

In conclusion, however, the compounds and elements described in this chapter are highly useful in the development and production of oil and gas resources and often with little environmental impact.

References

1 Reddy, B.R. and Palmer, A.V. (2009). Sealant compositions comprising colloidally stabilized latex and methods of using the same. US Patent 7,607,483, assigned to Haliburton Energy Services.

2 Song, Y.-K., Jo, Y.-H., Lim, Y.-J. et al. (2013). Sunlight-induced self-healing of a microcapsule-type protective coating. *ACS Applied Materials & Interfaces* **5** (4): 1378–1384.

3 Syrinek, A.R. and Lyon, L.B. (1989). Low temperature breakers for gelled fracturing fluids. US Patent 4,795,574, assigned to Nalco Chemical Company.

4 Patel, B.B. (1994). Fluid composition comprising a metal aluminate or a viscosity promoter and a magnesium compound and process using the composition. EP Patent 617106, assigned to Phillips Petroleum Company (US).

5 Xian, T. (2007). Drilling fluid systems for reducing circulation losses. US Patent 7,226,895, assigned to Baker Hughes Inc.

6 Gentili, D.O., Khalil, C.N., Rocha, N.O., and Lucas, E.F. (2005). Evaluation of polymeric phosphoric ester-based additives as inhibitors of paraffin deposition, SPE 94821. SPE Latin American and Caribbean Petroleum Engineering Conference, Rio de Janeiro, Brazil (20–23 June 2005).

7 Smith, K., Persinski, L.J., and Wanner, M. (2008). Effervescent biocide compositions for oilfield application. US Patent Application 2008/0004189, assigned to Weatherford/Lamb Inc.

8 Kunzi, R.A., Vinson, E.F., Totten, P.L., and Brake, B.G. (1993). Low temperature well cementing compositions and methods. CA Patent 2088897, assigned to Haliburton Company (US).

9 Padron, A. (1999). Stable emulsion of viscous crude hydrocarbon in aqueous buffer solution and method for forming and transporting same. CA Patent 2113597, assigned to Maravan S.A.

10 Brannon, H.D., Hodge, R.M., and England, K.W. (1989). High temperature guar-based fracturing fluid. US Patent 4,801,389, assigned to Dowell Schlumberger Inc.

11 Jennings Jr., A.R. (1995). Method of enhancing stimulation load fluid recovery. US Patent 5,411,093, assigned to Mobil Oil Corp.

12 Matsuka, N., Nakagawa, Y., Kurihara, M., and Tonomura, T. (1984). Reaction kinetics of sodium bisulfite and dissolved oxygen in seawater and their applications to seawater reverse osmosis. *Desalination* **51** (2): 163–171.

13 Ulrich, R.K., Rochelle, G.T., and Prada, R.E. (1986). Enhanced oxygen absorption into bisulphite solutions containing transition metal ion catalysts. *Chemical Engineering Science* **41** (8): 2831–2191.

14 McMahon, A.J., Chalmers, A., and MacDonald, H. (2001). Optimising oilfield oxygen scavengers. Chemistry in the Oil Industry VII (13–14 November 2001), p. 263. Manchester, UK: Royal Society of Chemistry.

15 Mitchell, R.W. (1978). The forties field sea-water injection system, SPE 6677. *Journal of Petroleum Technology* **30** (6): 877–884.

16 Prasad, R. (2004). Chemical treatment for hydrostatic test. US Patent 6815208, assigned to Champion Technologies Inc.

17 McCabe, M.A., Harris, P.C., Slabaugh, B. et al. (2000). Methods of treating subterranean formation using borate cross-linking compositions. US Patent 6,024,170, assigned to Haliburton Energy services Inc.

18 Paul, J.R. and Plonka, J.H. (1973). Solids free completion fluids maintain formation permeability, SPE 4655. Fall Meeting of the Society of Petroleum Engineers of AIME, Las Vegas, Nevada (30 September to 3 October 1973).

19 Bridges, K.L. (2000). *Completion and Workover Fluids*. SPE Monograph Series.

20 Caenn, R., Hartley, H.C.H., and Gray, G.R. (2011). *Composition and Properties of Drilling and Completion Fluids*, 6the. Gulf Professional Publishing.

21 Schlemmer, R.F. (2007). Membrane forming in-situ polymerization for water based drilling fluids. US Patent 7,279,445, assigned to MI L.L.C.

22 Schramm, L.L. ed. (2000). *Surfactants: Fundamentals and Applications in the Petroleum Industry*. Cambridge University Press.

23 K. Schriener and Munoz Jr., T. (2006). Methods of degrading filter cakes in a subterranean formation. US Patent 7,497,278, assigned to Haliburton Energy Services Inc.

24 Welton, T.D., Todd, B.L., and McMechan, D. (2010). Methods for effecting controlled break in pH dependent foamed fracturing fluid. US Patent 7,662,756, assigned to Haliburton Energy Services Inc.

25 Kosztin, B., Palasthy, G., Udvari, F. et al. (2002). Field evaluation of iron hydroxide gel treatments, SPE 78351. European Petroleum Conference, Aberdeen, UK (29–31 October 2002).

26 Goncalves, J.T., De Oliveira, M.F., and Aragao, A.F.L. (2007). Compositions of oil-based biodegradable drilling fluids and process for drilling oil and gas wells. US Patent 7,285,515, assigned to Petroleo Brasileiro S.A. – Petrobras.

27 Valenziano, R., Harris, K.L., and Dixon, M.D. (2009). Servicing a wellbore with an aqueous based fluid comprising a clay inhibitor. US Patent 7,549,474, assigned to Haliburton Energy Services Inc.

28 Baijal, S.K., Houchin, L.R., and Bridges, K.L. (1991). A practical approach to prevent formation damage by high density brines during the completion process. SPE Production Operations Symposium, Oklahoma City, OK (7–9 April 1991).

29 Houchin, L.R., Baijal, S.K., and Foxenberg, W.E. (1991). An analysis of formation damage by completion fluids at high temperatures, SPE 23143. Offshore Europe, Aberdeen, UK (3–6 September 1991).

30 Ludwig, N.C. (1951). Effect of sodium chloride on setting properties of oil-well cements, API 51020. American Petroleum Institute, Drilling and Production Practice, New York, NY (1 January 1951).

31 Tsytsymuskin, P.F., Tarnaviskiy, A.P., Mikhailov, B.V. et al. (1991). Grouting mortars for fixing wells of salt deposits. SU Patent 1,700,201, assigned to Volga Urals Hydrocarbon.

32 van Oort, E. (2003). On the physical and chemical stability of shales. *Journal of Petroleum Science and Engineering* **38**: 213–223.

33 Fink, J. (2013). *Petroleum Engineers Guide to Oilfield Chemicals and Fluids*. Elsevier.

34 Kelland, M.A., Svartaas, T.M., and Dybvik, L. (1995). Studies on new gas hydrate inhibitors, SPE 30420. Offshore Europe, Aberdeen, UK (5–8 September 1995).

35 Zhou, Z. (2000). Process for reducing permeability in a subterranean formation. US Patent 6,143,699, assigned to Alberta Oil Sands Technology and Research Authority.

36 Wylde, J.J. and Slayer, J.L. (2013). Halite scale formation mechanisms, removal and control: a global overview of mechanical, process, and chemical strategies, SPE 164081. SPE International Symposium on Oilfield Chemistry, The Woodlands, TX (8–10 April 2013).

37 Foley, R.T. (1970). Role of the chloride ion in iron corrosion. *Corrosion* **26** (2): 58–70.

38 Nielsen, D.L., Brock, M.A., Rees, G.N., and Baldwin, D.S. (2003). Effects of increasing salinity on freshwater ecosystems in Australia. *Australian Journal of Botany* **51**: 655–665.

39 Williams, M.D. and Williams, W.D. (1991). Salinity tolerance of four species of fish from the Murray-Darling basin river system. *Hydrobiologia* **210** (1): 145–150.

40 Mason, J.A. (1988). Use of chlorous acid in oil recovery. US Patent 4,892,148, assigned to the Inventor.

41 Kuhn, A.T. and Lartey, R.B. (1975, 1975). Electrolytic generation "in-situ" of sodium hypochlorite. *Chemie Ingenieur Technik* **47** (4): 129–135.

42 Nalepa, C.J., Howarth, J., and Azomia, F.D. (2002). Factors to consider when applying oxidizing biocides in the field, NACE-02223. CORROSION 2002 (7–11 April 2002) Denver, Colorado.

43 Clementz, D.M., Patterson, D.E., Aseltine, R.J., and Young, R.E. (1982). Stimulation of water injection wells in the Los Angeles basin by using sodium hypochlorite and mineral acids, SPE 10624. *Journal of Petroleum Technology* **34** (9): 2,087–2,096.

44 Todd, B.L., Slabaugh, B.F., Munoz Jr., T., and Parker, M.A. (2008). Fluid loss control additives for use in fracturing subterranean formations. US Patent 7,096,947, assigned to Haliburton Energy Services Inc.

45 Lyons, W., Plisga, G.J., and Lorenz, M. ed. (2015). *Standard Handbook of Petroleum and Natural Gas Engineering*, 3rde. Gulf Publishing.

46 Han, M., Alshehri, A.J., Krinis, D., and Lyngra, S. (2014). State-of-the-art of in-depth fluid diversion technology: enhancing reservoir oil recovery by gel treatments, SPE 172186. SPE Saudi Arabia Section Technical Symposium and Exhibition, Al-Khobar, Saudi Arabia (21–24 April 2014).

47 http://www.webcitation.org/5WGM2f75m (accessed 12 December 2017)

48 Williams, R., Therond, E., Dammel, T., and Gentry, M. (2010). Conductive cement formulation and application for use in well. US Patent 7772166, assigned to Schlumberger Technology Corporation.

49 Guichard, B., Wood, B., and Vongphouthone, P. (2008). Fluid loss reducer for high temperature high pressure water based-mud application. US Patent 7,449,430, assigned to Eliokem S.A. S.

50 Thibideau, L., Sakanoko, M., and Neale, G.H. (2003). Alkaline flooding processes in porous media in the presence of connate water. *Powder Technology* **132** (2-3): 101–111.

51 Horvath-Szabo, G., Czarnecki, J., and Masliyah, J.H. (2002). Sandwich structures at oil–water interfaces under alkaline conditions. *Journal of Colloid and Interface Science* **253** (2): 427–434.

52 Lorenz, P.B. (1991). The effect of alkaline agents on the retention of EOR chemicals. *Technical Report NIPER-535*, National Institue for Petroleum and Energy Research, Bartlesville, OK.

53 Parker, A. (1988). Process for controlling and delaying the formation of gels or precipitates derived from aluminium hydroxide and corresponding compositions together with its applications particularly those concerning the operation of oil wells. EP Patent 266808, assigned to Pumptech N.V.

54 Sullivan, P., Christanti, Y., Couillet, I. et al. (2006). Methods for controlling the fluid loss properties of viscoelastic surfactant based fluids. US Patent 7,081,439, assigned to Schlumberger Technologies Corporation.

55 Dadgar, A. (1988). Corrosion inhibitors for clear, calcium-free high density fluids. US Patent 4,784,779, assigned to Great Lakes Chemical Corp.

56 Denton, R.M. and Lockstedt, A.W. (2006). Rock bit with grease composition utilizing polarized graphite. US Patent 7,121,365, assigned to Smith International Inc.

57 Mondshine, T. (1993). Process for decomposing polysaccharides in alkaline aqueous systems. US Patent 5,253,711, assigned to Texas United Chemical Corp.

58 De Souza, C.R. and Khallil, C.N. (1999). Method for the thermo-chemical dewaxing of large dimension lines. US Patent 6,003,528, assigned to Petroleo Brasiliero S.A. – Petrobras.

59 Kumar, M., Singh, N.P., Singh, S.K., and Singh, N.B. (2010). Combined effect of sodium sulphate and superplasticizer on the hydration of fly ash blended Portland cement. *Materials Research* **13** (2).

60 Jenkins, A., Grainger, N., Blezard, M., and Pepin, M. (2012). Quaternary ammonium corrosion inhibitor. EP Patent 2429984, assigned to M-I Drilling Fluids UK Limited and Stepan UK Limited.

61 Phillips, N.J., Renwick, J.P., Palmer, J.W., and Swift, A.J. (1996). The synergistic effect of sodium thiosulphate on corrosion inhibition. Proceedings of 7th Oilfield Chemistry Symposium, Geilo, Norway.

62 Graham, G.M., Bowering, D., MacKinnon, K. et al. (2014). Corrosion inhibitors squeeze treatments-misconceptions, concepts and potential benefits, SPE 169604. SPE International Oilfield Corrosion Conference and Exhibition, Aberdeen, Scotland (12–13 May 2014).

63 Srinivasan, S., Veawab, A., and Aroonwilas, A. (2013). Low toxic corrosion inhibitors for amine-based CO_2 capture process. *Energy Procedia* **37**: 890–895.

64 Bjornstad, T., Haugen, O.B., and Hundere, I.A. (1994). Dynamic behavior of radio-labelled water tracer candidates for chalk reservoirs. *Journal of Petroleum Science and Engineering* **10** (3): 223–228.

65 Alford, S.E. (1991). North sea field application of an environmentally responsible water-base shale stabilizing system, SPE 21936. SPE/IADC Drilling Conference, Amsterdam, Netherlands (11–14 March 1991).

66 Jones, T.G.J., Tustin, G.J., Fletcher, P., and Lee, J.C.-W. (2008). Scale dissolver fluid. US Patent 7,343,978, assigned to Schlumberger Technology Corporation.

67 Steiger, R.P. (1982). Fundamentals and use of potassium/polymer drilling fluids to minimize drilling and completion problems associated with hydratable clays, SPE 101100. *Journal of Petroleum Technology* **34** (8): 1661–2136.

68 Anderson, R.L., Ratcliffe, I., Greenwell, H.C. et al. (2010). Clay swelling – a challenge in the oilfield. *Earth Science Reviews* **98** (304): 201–216.

69 Reid, P.I., Craster, B., Crawshaw, J.P., and Balson, T.G. (2003). Drilling fluid. US Patent 6,544,933, assigned to Schlumberger Technology Corporation.

70 Rodriguez-Navarro, C., Linares-Fernandez, L., Doehne, E., and Sebastian, E. (2002). Effects of ferrocyanide ions on NaCl crystallization in porous stone. *Journal of Crystal Growth* **243** (3-4): 503–516.

71 Frigo, D.M., Jackson, L.A., Doran, S.M., and Trompert, R.A. (2000). Chemical inhibition of halite scaling in topsides equipment, SPE 60191. International Symposium on Oilfield Scale, Aberdeen, UK (26–27 January 2000).

72 Herbert, J., Leasure, J., Saldungaray, P., and Ceramics, C. (2016). Prevention of halite formation and deposition in horizontal wellbores: a multi basin developmental study, SPE 181735. SPE Annual Technical Conference and Exhibition, Dubai, UAE (26–28 September 2016).

73 Wyldeand, J.J. and Slayer, J.L. (2013). Halite scale formation mechanisms, removal and control: a global overview of mechanical, process, and chemical strategies, SPE 164081. SPE International Symposium on Oilfield Chemistry, The Woodlands, TX (8–10 April 2013).

74 European Commission (2001). Opinion of the scientific committee for animal nutrition on the safety of potassium and sodium ferrocyanide used as anticaking agents. Health and Consumer Protection Directorate General, Directorate C – Scientific Opinions (adopted 3 December 2001).

75 Schultz, H., Bauer, G., Schachl, E. et al. (2002). Potassium compounds. In: *Ullman's Encyclopedia of Industrial Chemistry*. Wiley-VCH.

76 Breedon, D.L. and Meyer, R.L. (2005). Ester-containing downhole drilling lubricating composition and processes therefor and therewith. US Patent 6,884,762, assigned to Newpark Drilling Fluids L.L.C.

77 Trabanelli, G., Zucchi, F., and Brunoro, G. (1988). Inhibition of corrosion resistant alloys in hot hydrochloric acid solutions. *Materials and Corrosion* **39** (12): 589–594.

78 Tilley, R.J.D. (2013). *Understanding Solids: The Science of Materials*, 2nde. Wiley Blackwell.

79 Eberan-Eberhorst, G.A., Haitzman, P.F., McConnell, G., and Stribley, F.T. (1983). Recent developments in automotive, industrial and marine lubricants, Paper RP9 WPC 20338. 11th World Petroleum Congress, London, UK (28 August to 2 September 1983).

80 Orazzini, S., Kasirin, R.S., Ferrari, G. et al. (2012). New HT/HP technology for geothermal application significantly increases on-bottom drilling hours, SPE 150030. IADC/SPE Drilling Conference and Exhibition, 6-8 March, San Diego, CA.

81 Akulichev, A. and Thrkildsen, B. (2014). Impact of lubricating materials on arctic subsea production systems, OTC 24538. OTC Arctic Technology Conference, 10-12 February, 2014, Houston, TX.

82 Huey, M.A. and Leppin, D. (1995). Lithium bromide chiller technology in gas processing, SPE 29486. SPE Production Operations Symposium, Oklahoma City, OK (2–4 April 1995).

83 Scian, A.N., Porto Lopez, J.M., and Pereira, E. (1991). Mechanochemical activation of high alumina cements – hydration behaviour. I. *Cement and Concrete Research* **21** (1): 51–60.

84 Garvey, C.M., Savoly, A., and Weatherford, T.M. (1987). Drilling fluid dispersant. US Patent 4711731, assigned to Diamond Shamrock Chemicals Company.

85 US Environmental Protection Agency (2012). Lithium-ion batteries and nanotechnology for electric vehicles: a life cycle assessment DRAFT.

86 European Commission (2012). Environmental impacts of batteries for low carbon technologies compared. Science for Environmental Policy, Issue 303.

87 Frank, W.B., Haupin, W.E., Vogt, H. et al. (2009). Aluminium. In: *Ullmann's Encyclopedia of Industrial Chemistry*. Wiley VCH.

88 Kabir, A.H. (2001). Chemical water and gas shutoff technology – an overview, SPE 72119. SPE Asia Pacific Improved Oil Recovery Conference, Kuala Lumpur, Malaysia (6–9 October 2001).

89 Frenier, W. and Chang, F.F. (2004). Composition and method for treating a subterranean formation. US Patent 6,806,236, assigned to Schlumberger Technology Corporation.

90 Gebbie, P. (2001). Using polyaluminium coagulants in water treatment. 64th Annual Water Industry Engineers and Operators Conference, Bendigo, Australia (5–6 September 2001).

91 Soderlund, M. and Gunnarsson, S. (2010). Process for the production of polyaluminium salts. EP Patent 2158160, assigned to Kemira Kemi AB.

92 Xu, Y., Zhijun, W., and Lizhi, Z. (1997). Study of high-sulfur natural gas field water treatment, PETSOC-97-122. Annual Technical Meeting (8–11 June 1997). Calgary, Alberta: Petroleum Society of Canada.

93 Smith, D.R., Moore, P.A., Griffis, C.L. et al. (2000). Effects of alum and aluminum chloride on phosphorus runoff from swine manure. *Journal of Environmental Quality* **30** (3): 992–998.

94 Smith, J.E. (1995). Performance of 18 polymers in aluminum citrate colloidal dispersion gels, SPE 28989. SPE International Symposium on Oilfield Chemistry, San Antonio, TX (14–17 February 1995).

95 Samuel, M. (2009). Gelled oil with surfactant. US Patent 7,521,400, assigned to Schlumberger Technology Corporation.

96 Burrafato, G. and Carminati, S. (1996). Aqueous drilling muds fluidified by means of zirconium and aluminum complexes. US Patent 5,532,211, assigned to Eniricerche S.P.A., Agip S.P.A.

97 Jovancicevic, V., Campbell, S., Ramachandran, S. et al. (2007). Aluminum carboxylate drag reducers for hydrocarbon emulsions. US Patent 7,288,506, assigned to Baker Hughes Incorporated.

98 Kelland, M.A. (2014). *Production Chemicals for the Oil and Gas Industry*, 2nde. CRC Press.

99 Willberg, D. and Dismuke, K. (2009). Self-destructing filter cake. US Patent 7,482,311, assigned to Schlumberger Technologies Corporation.

100 Reid, A.L. and Grichuk, H.A. (1991). Polymer composition comprising phosphorous-containing gelling agent and process thereof. US Patent 5,034,139, assigned to Nalco Chemical Company.

101 Laird, J.A. and Beck, W.R. (1987). Ceramic spheroids having low density and high crush resistance. EP Patent 207688.

102 Andrews, W.H. (1987). Bauxite proppant. US 4,713,203, assigned to Comalco Aluminium Limited.

103 Jones, C.K., Williams, D.A., and Blair, C.C. (1999). Gelling agents comprising of aluminium phosphate compounds. GB Patent 2,326,882, assigned to Nalco/Exxon Energy Chemicals L.P.

104 Huddleston, D.A. (1989). Hydrocarbon geller and method for making the same. US Patent 4,877,894, assigned to Nalco Chemical Company.

105 Hart, P.R. (2005). Method of breaking reverse emulsions in a crude oil desalting system. CA Patent 2,126,889, assigned to Betz Dearborn Inc.

106 Gioia, F., Urciuolo, M. et al. (2004). *Journal of Hazardous Materials* **116** (1–2): 83–93.

107 Oldiges, D.A. and Joeseph, A.W. (2003). Methods for using environmentally friendly anti-seize/lubricating systems. US Patent 6,620,460, assigned to Jet-Lube Inc.

108 Branch III, H. (1988). Shale-stabilizing drilling fluids and method for producing same. US Patent 4,719,021, assigned to Sun Drilling Products Corporation.

109 Rodriguez, D., Quintero, L., Terrer, M.T. et al. (1990). Hydrocarbon dispersion in water. GB Patent 2231284, assigned to Intevep S.A.

110 Benaissa, S., Clapper, D.K., Parigot, P., and Degouy, D. (1997). "Oilfield applications of aluminum chemistry and experience with aluminum-based drilling-fluid additive, SPE 37268. SPE International Symposium on Oilfield Chemistry, Houston, TX (18–21 February 1997).

111 Rosseland, B.O., Eldhuset, T.D., and Staurnes, M. (1990). Environmental effects of aluminium. *Environmental Geochemistry and Health* **12** (1–2): 17–27.

112 Fernier, W.W. and Zauddin, M. (2008). *Formation, Removal and Inhibition of Inorganic Scale in the Oilfield Environment*. SPE Publications.

113 Neff, J.M. (2002). *Bioaccumulation in Marine Organisms: Effect of Contaminants from Oil Well Produced Water*. Elsevier Science.

114 Johnson, R.M. and Garvin, T.R. (1972). Cementing practices – 1972, SPE 3809. Joint AIME-MMIJ Meeting, Tokyo, Japan (25–27 May 1972).

115 Dye, W.M., Mullen, G.A., and Gusler, W.J. (2006). Field-proven technology to manage dynamic barite sag, SPE 98167. IADC/SPE Drilling Conference, Miami, FL (21–23 February 2006).

116 API RP 13K (R2016). *Recommended Practice for Chemical Analysis of Barite*, 3rde. Standard by American Petroleum Institute.

117 Shen, W., Pan, H.F., and Qin, Y.Q. (1999). Advances in chemical surface modification of barite. *Oilfield Chemistry* **16** (1): 86–90.

118 OLI ScaleChem (2010). Version 4.0 (revision 4.0.3). OLI Systems, Inc.

119 McGinty, J., McHugh, T.E., and Higgins, E.A. (2007). Barium sulfate: a protocol for determining higher site-specific barium cleanup levels, SPE 106802. E&P Environmental and Safety Conference, Galveston, TX (5–7 March 2007).

120 Dobson, J.W., Hayden, S.L., and Hinojosa, B.E. (2005). Borate crosslinker suspensions with more consistent crosslink times. US Patent 6,936,575, assigned to Texas United Chemical Company Llc.

121 Ainley, B.R. and McConnell, S.B. (1993). Delayed borate cross-linked fracturing fluid. EP Patent 528461, assigned to Pumptech N.V.

122 Brannon, H.D. and Ault, M.G. (1991). New delayed borate-crosslinked fluid provides improved fracture conductivity in high-temperature applications, SPE 22838. SPE Annual Technical Conference and Exhibition, Dallas, TX (6–9 October 1991).

123 World Health Organization (1998). Boron. Environmental Health Criteria, 204, Geneva, Switerland.

124 Eisler, R. (1990). Boron hazards to fish, wildlife, and invertebrates: a synoptic review. *U.S. Fish and Wildlife Service: Biological Report* **82**: 1–32.

125 U.S. Environmental Protection Agency, Office of Pesticide Programs (1993). Reregistration eligibility decision document: boric acid and its sodium salts, EPA 738-R-93-017, September 1993. Washington, DC: U.S. Government Printing Office.

126 Karcher, J.D., Brenneis, C., and Brothers, L.E. (2013). Calcium phosphate cement compositions comprising pumice and/or perlite and associated methods. US Patent Application 20130126166, assigned to Haliburton Energy Services Inc.

127 Mueller, H., Breuer, W., Herold, C.-P. et al. (1997). Mineral additives for setting and/or controlling the rheological properties and gel structure of aqueous liquid phases and the use of such additives. US Patent 5,663,122, assigned to Henkel Ag.

128 GEO Drilling Fluids Inc. (1997). Brine fluids. http://www.geodf.com/store/files/24.pdf (accessed 12 December 2017).

129 Lyons, W., Pilsga, G.J., and Lorenz, M. ed. (2015). *Standard Handbook of Petroleum and Natural Gas Engineering*, 3rde. Gulf Professional Publishing.

130 Johnson, M. (1996). Fluid systems for controlling fluid losses during hydrocarbon recovery operations. EP Patent 0691454, assigned to Baker Hughes Inc.

131 Landry, D.K. and Kollermann, T.J. (1991). Bearings grease for rock bit bearings. US Patent 5,015,401, assigned to Hughes Tool Company.

132 Huang, T., Crews, J.B., and Tredway Jr., J.H. (2009). Fluid loss control agents for viscoelastic surfactant fluids. US Patent 7,550,413, assigned to Baker Hughes Incorporated.

133 Crews, J.B. (2010). Saponified fatty acids as breakers for viscoelastic surfactant-gelled fluids. US Patent 7,728,044, assigned to Baker Hughes Incorporated

134 Ghofrani, R. and Werner, C. (1993). Effect of calcination temperature and the durations of calcination on the optimisation of expanding efficiency of the additives, CaO and MgO, in swelling (expanding) cements. *Erdoel, Erdgas, Kohle* **109** (1): 7–9.

135 Montgomery, F., Montgomery, S., and Stephens, P. (1994). Method of controlling porosity of well fluid blocking layers and corresponding acid soluble mineral fiber well facing product. US Patent 5,354,456, assigned to Inventors.

136 Todd, B.L. (2009). Filter cake degradation compositions and methods of use in subterranean operations. US Patent 7,598,208, assigned to Haliburton Energy Services Inc.

137 Patel, B.B. (1994). Tin/cerium compounds for lignosulphonate processing. EP Patent 600343, assigned to Phillips Petroleum Company.

138 Olson, W.D., Muir, R.J., Eliades, T.I., and Steib, T. (1994). Sulfonate greases. US Patent 5,308,514, assigned to Witco Corporation.

139 Ramakrishna, D.M. and Viraraghavan, T. (2005). Environmental impact of chemical deicers – a review. *Water, Air, and Soil Pollution* **166** (1): 49–63.

140 Al-Ansary, M.S. and Al-Tabba, A. (2007). Stabilisation/solidification of synthetic petroleum drill cuttings. *Journal of Hazardous Materials* **141** (2): 410–421.

141 Carpenter, J.F. and Nalepa, C.J. (2005). Bromine-based biocides for effective microbiological control in the oil field, SPE 92702. SPE International Symposium on Oilfield Chemistry, The Woodlands, TX (2–4 February 2005).

142 Kleina, L.G., Czechowski, M.H., Clavin, J.S. et al. (1997). Performance and monitoring of a new nonoxidizing biocide: the study of BNPD/ISO and ATP, Nace 97403. Corrosion 97, New Orleans, Louisiana (9–14 March 1997).

143 Bryce, D.M., Crowshaw, B., Hall, J.E. et al. (1978). The activity and safety of the antimicrobial agent bronopol (2-bromo-nitropan-1,3-diol). *Journal of the Society of Cosmetic Chemists* **29** (1): 3–24.

144 Cronan Jr., J.M. and Mayer, M.J. (2006). Synergistic biocidal mixtures. US Patent 7,008,545, assigned to Hercules Incorporated.

145 Yang, S. (2004). Stabilized bromine and chlorine mixture, method of manufacture and uses thereof for biofouling control. World Patent Application, WO/2004/026770, assigned to Nalco Company.

146 Flatval, K.B., Sathyamoorthy, S., Kuijvenhoven, C., and Ligthelm, D. (2004). Building the case for raw seawater injection scheme in Barton, SPE 88568. SPE Asia Pacific Oil and Gas Conference and Exhibition, Perth, Australia (18–20 October 2004).

147 Romaine, J., Strawser, T.G., and Knippers, M.L. (1996). Application of chlorine dioxide as an oilfield facilities treatment fluid, SPE 29017. *SPE Production & Facilities* **11** (1): 18–21.

148 McCafferty, J.F., Tate, E.W., and Williams, D.A. (1993). Field performance in the practical application of chlorine dioxide as a stimulation enhancement fluid, SPE 20626. *SPE Production & Facilities* **8** (1): 9–14.

149 Ohlsen, J.R., Brown, J.M., Brock, G.F., and Mandlay, V.K. (1995). Corrosion inhibitor composition and method of use. US Patent 5,459,125, assigned to BJ Services Company.

150 Cavallaro, A., Curci, E., Galliano, G. et al. (2001). Design of an acid stimulation system with chlorine dioxide for the treatment of water-injection wells, SPE 69533. SPE Latin American and Caribbean Petroleum Engineering Conference, Buenos Aires, Argentina (25–28 March 2001).

151 Mason, J.A. (1993). Use of chlorous acid in oil recovery. GB Patent 2239867.

152 Kalfayan, L. (2008). *Production Enhancement with Acid Stimulation*, 2nde. Oklahoma: PennWell.

153 Cheng, X., Li, Y., Ding, Y. et al. (2011). Study and application of high density acid in HPHT deep well, SPE 142033. SPE European Formation Damage Conference, Noordwijk, The Netherlands (7–10 June 2011).

154 Nasr-El-Din, H.A. and Al-Humaidan, A.Y. (2001). Iron sulfide scale: formation, removal and prevention, SPE 68315. International Symposium on Oilfield Scale, Aberdeen, UK (30–31 January 2001).

155 Ali, S.A., Sanclemente, L.W., Sketcher, B.C., and Lafontaine-McLarty, J.M. (1993). Acid breakers enhance open-hole horizontal completions. *Petroleum Engineer International* **65** (11): 20–23.

156 Chan, A.F. (2009). Method and composition for removing filter cake from a horizontal wellbore using a stable acid foam. US Patent 7,514,391, assigned to Conocophillips Company.

157 Ke, M. and Qu, Q. (2010). Method for controlling inorganic fluoride scales. US Patent 7,781,381, assigned to BJ services Company Llc.

158 H.K. Kotlar, F. Haavind, M. Springer et al. (2006). Encouraging results with a new environmentally acceptable, oil-soluble chemical for sand consolidation: from laboratory experiments to field, SPE 98333. SPE International Symposium and Exhibition on Formation Damage Control, Lafayette, Louisiana (15–17 February 2006).

159 Qu, Q. and Wang, X. (2010). Method of acid fracturing a sandstone formation. US Patent 7,704,927, assigned to BJ Services Company.

160 Teng, H. Overview of the development of the fluoropolymer industry. *Applied Sciences* **2**: 492–512.

161 B. Jones, "*Fluoropolymers for Coating Applications*", JCT Coatings Tech Magazine, 2008.

162 Brezinski, M.M. and Desai, B. (1997). Method and composition for acidizing subterranean formations utilizing corrosion inhibitor intensifiers. US Patent 5,697,443, assigned to Haliburton Energy Services.

163 Arab, S.T. and Noor, E.A. (1993). Inhibition of acid corrosion of steel by some S-alkylisothiouronium iodides, NACE-93020122. *Corrosion* **49** (2).

164 Brezinski, M.M. (2002). Electron transfer system for well acidizing compositions and methods. US Patent 6,653,260, assigned to Haliburton Energy Services.

165 Watkins, J.W. and Mardock, E.S. (1954). Use of radioactive iodine as a tracer in water-flooding operations, SPE 349-G. *Journal of Petroleum Technology* **6** (09): 117–124.

166 Liang, L. and Sanger, P.C. (2003). Factors influencing the formation and relative distribution of haloacetic acids and trifluoromethanes in drinking water. *Environmental Science & Technology* **37** (13): 2920–2928.

167 https://www.epa.gov/dwstandardsregulations (accessed 12 December 2017).

168 Davis, S.N., Whittemore, D.O., and Fabryka-Martin, J. (1998). Uses of chloride/bromide ratios in studies of potable water. *Ground Water* **36** (2): 338–350.

169 Grebel, J.E., Pignatello, J.J., and Mitch, W.A. (2010). Effect of halide ions and carbonates on organic contaminant degradation by hydroxyl radical-based advanced oxidation processes in saline waters. *Environmental Science & Technology* **44** (17): 6822–6828.

170 Evans, C.D., Monteith, D.T., Fowler, D. et al. (2011). Hydrochloric acid: an overlooked driver of environmental change. *Environmental Science & Technology* **45** (5): 1887–1894.

171 IIF-IIR (1997). Fluorocarbons and global warming. 12th Informatory Note on Fluorocarbons and Refrigeration. International Institute of Refrigeration, Intergovernmental organization for the development of refrigeration.

172 http://ozone.unep.org/en/treaties-and-decisions/montreal-protocol-substances-deplete-ozone-layer (accessed 12 December 1997).

173 Murphy, C.D., Schaffrath, C., and O'Hagan, D. (2003). Fluorinated natural products: biosynthesis of fluoroacetate and 4-fluorothreonine in *Streptomyces cattleya*. *Chemosphere* **52** (2): 455–461.

174 Boethling, R.S., Sommer, E., and DiFiore, D. (2007). Designing small molecules for biodegradability. *Chemical Reviews* **107**: 2207–2227.

175 Sun, W. and Nesic, S. (2006), Basics revisited: kinetics of iron carbonate scale precipitation in CO_2 corrosion, NACE 06365. CORROSION 2006, San Diego, CA (12–16 March 2006).

176 Halvorsen, A.K., Andersen, T.R., Halvorsen, E.N. et al. (2007). The relationship between internal corrosion control method, scale control and meg handling of a multiphase carbon steel pipeline carrying wet gas with CO_2 and acetic acid, NACE-07313. CORROSION 2007, Nashville, TN (11–15 March 2007).

177 Aitken, P.A., Kasatkina, N.N., Makeev, N.M., and Vantsev, V.Y. (1992). Oilwell composition. SU Patent 1776761.

178 Lakatos, I., Lakatos-Szabo, J., Kosztin, B. et al. (2000), Application of iron-hydroxide-based well treatment techniques at the Hungarian oil fields, SPE 59321. SPE/DOE Improved Oil Recovery Symposium, Tulsa, OK (3–5 April 2000).

179 Feraud, J.P., Perthuis, H., and Dejeux, P. (2001). Compositions for iron control in acid treatments for oil wells. US Patent 6,306,799, assigned to Schlumberger Technology Corporation.

180 Sunde, E. and Olsen, H. (2000). Removal of H_2S in drilling mud. WO Patent Application 20000023538, applicant Den Norske Stats Oijeselskap AS.

181 Lindstrom, K.O. and Riley, W.D. (1994). Soil cement compositions and their use. EP Patent 605075, assigned to Haliburton Company.

182 Moradi-Araghi, A. (1995). Gelling compositions useful for oil field applications. US Patent 5,432,153, assigned to Phillips Petroleum Company.

183 Rasmussen, K. and Lindegaard, C. (1988). Effects of iron compounds on macroinvertebrate communities in a Danish lowland river system. *Water Research* **2** (9): 1101–1108.

184 Vuori, K.-M. (1995). Direct and indirect effects of iron on river systems. *Annales Zoologici Fennici* **32**: 317–329.

185 Browning, W.C. and Young, H.F. (1975). Process for scavenging hydrogen sulfide in aqueous drilling fluids and method of preventing metallic corrosion of subterranean well drilling apparatuses. US Patent 3,928,211, assigned to Milchem Inc.

186 Ramachandran, S., Lehrer, S.E., and Jovancicevic, V. (2014). Metal carboxylate salts as H_2S scavengers in mixed production or dry gas or wet gas systems. US Patent Application 20140305845, applicant Baker Hughes Incorporated.

187 de Grood, R.J.C. and Baycroft, P.D. (2010). Use of coated proppant to minimize abrasive erosion in high rate fracturing operations. US Patent 7,730,948, assigned to Baker Hughes Incorporated.

188 Mondshine, T.C. and Benta, G.R. (1993). Process and composition to enhance removal of polymer-containing filter cakes from wellbores. US Patent 5,238,065, assigned to Texas United Chemical Corporation.

189 Daniel, S. and Dessinges, M.N. (2010). Method for breaking fracturing fluids. US Patent 7,857,048, assigned to Schlumberger Technology Corporation.

190 Leth-Olsen, H. (2005). CO_2 corrosion in bromide and formate well completion brines, SPE 95072. SPE International Symposium on Oilfield Corrosion, Aberdeen, UK (13 May 2005).

191 Wiley, T.F., Wiley, R.J., and Wiley, S.T. (2007). Rock bit grease composition. US Patent 7,312,185, assigned to Tomlin Scientific Inc.

192 Wang, X., Qu, Q., and Ke, M. (2008). Method for inhibiting or controlling inorganic scale formations with copolymers of acrylamide and quaternary ammonium salts. US Patent 7,398,824, assigned to BJ Services Company.

193 Collins, I.R. and Jordan, M.M. (2003). Occurrence, prediction, and prevention of zinc sulfide scale within Gulf Coast and North Sea high-temperature and high-salinity fields, SPE 84963. *SPE Production & Facilities* **18** (3): 200–209.

194 Broadly, M.R., White, P.J., Hammond, J.P. et al. (2007). Zin in plants. *New Phytologist* **173** (4): 677–702.

195 Bass, J.A.B., Blust, R., Clarke, R.T. et al. (2008). Environmental Quality Standards for trace metals in the aquatic environment. Science Report SC030194, Environment Agency, April 2008.

196 http://dwi.defra.gov.uk/consumers/advice-leaflets/standards.pdf (accessed 12 December 2017).

197 Matthiessen, P. and Brafield, A.E. (1977). Uptake and loss of dissolved zinc by the stickleback *Gasterosteus aculeatus* L. *Journal of Fish Biology* **10** (4): 399–410.

198 Skidmore, J.F. (1964). Toxicity of zinc compounds to aquatic animals, with special reference to fish. *The Quarterly Review of Biology* **39** (3): 227–248, The University of Chicago Press.

199 Alabaster, J.S. ed. (2013). *Water Quality Criteria for Freshwater Fish*, 2nde. Butterworth-Heinemann.

200 U.S. Environmental Protection Agency (1972). Hazardous of zinc in the environment with particular reference to the aquatic environment.

201 Frenier, W.W. (1992). Process and composition for inhibiting high-temperature iron and steel corrosion. US Patent 5,096,618, assigned to Dowell Schlumberger Incorporated.

202 Brezunski, M.M. and Desai, B. (1997). Method and composition for acidizing subterranean formations utilizing corrosion inhibitor intensifiers. US Patent 5,697,443, assigned to Haliburton Energy Services Inc.

203 Todd, B.L. (2009). Methods and fluid compositions for depositing and removing filter cake in a well bore. US Patent 7,632,786, assigned to Haliburton Energy Services.

204 Sydansk, R.D. and Southwell, G.P. (2000). More than 12 years of experience with a successful conformance-control polymer gel technology, SPE 62561. SPE/AAPG Western Regional Meeting, Long Beach, CA (19–22 June 2000).

205 Sydansk, R.D. (1990). A newly developed chromium(III) gel technology, SPE 19308. *SPE Reservoir Engineering* **5** (3): 346–352.

206 Clampitt, R.L. and Hessert, J.E. (1979). Method for acidizing subterranean formations. US Patent 4,068,719, assigned to Phillips Petroleum Company.

207 Mumallah, N.A. (1988). Chromium(III) propionate: a crosslinking agent for water-soluble polymers in hard oilfield brines , SPE 15906. *SPE Reservoir Engineering* **3** (1): 243–250.

208 Nriagu, J.O. and Niebor, E. ed. (1988). *Chromium in the Natural and Human Environments*, Advances in Environmental Science and Technology. Wiley.

209 Tchounwou, P.B., Yedjou, C.G., Patolla, A.K., and Sutton, D.J. (2012). Heavy metal toxicity and the environment. *Molecular, Clinical and Environmental Toxicology* **101**: 133–164.

210 Rai, D., Eary, L.E., and Zachara, J.M. (1989). Environmental chemistry of chromium. *Science of the Total Environment* **86** (1-2): 15–23.

211 Cassidy, J.M., Kiser, C.E., and Wilson, J.M. (2009). Corrosion inhibitor intensifier compositions and associated methods. US Patent Application 20090156432, assigned to Haliburton Energy Services Inc.

212 Chen, T., Chen, P., Montgomerie, H., Hagen, T.H., and Jefferies, C. (2010). Development of test method and inhibitors for lead sulfide, SPE 130926. SPE International Conference on Oilfield Scale, Aberdeen, UK (26–27 May 2010).

213 Jordan, M.M., Sjursaether, K., Edgerton, M.C., and Bruce, R. (2000). Inhibition of lead and zinc sulphide scale deposits formed during production from high temperature oil and condensate reservoirs, SPE 64427. SPE Asia Pacific Oil and Gas Conference and Exhibition, Brisbane, Australia (16–18 October 2000).

214 Mahapatra, S.K. and Kosztin, B. (2011). Magnesium peroxide breaker for filter cake removal, SPE 142382. SPE EUROPEC/EAGE Annual Conference and Exhibition, Vienna, Austria (23–26 May 2011).

215 Gulbis, J., King, M.T., Hawkins, G.W., and Brannon, H.D. (1992). Encapsulated breaker for aqueous polymeric fluids, SPE 19433. *SPE Production Engineering* **7** (1): 9–14.

216 Nimerick, K.H., Crown, C.W., McConnell, S.B., and Ainley, B. (1993). Method of using borate crosslinked fracturing fluid having increased temperature range. US Patent 5,259,455, assigned to Inventors.

217 Fraser, L.J. (1992). Unique characteristics of mixed metal hydroxide fluids provide gauge hole in diverse types of formation. International Meeting on Petroleum Engineering, Beijing, China (24–27 March 1992).

218 Munoz Jr., T. and Todd, B.L. (2008). Treatment fluids comprising starch and ceramic particulate bridging agents and methods of using these fluids to provide fluid loss control. US Patent 7,462,581, assigned to Haliburton Energy Services Inc.

219 Smith, R.J. and Jeanson, D.R. (2001). Dehydration of drilling mud. US Patent 6,216,361, assigned to Newpark Canada Inc.

220 Roth, J., Ponzoni, S., and Aschner, M. (2013). Manganese homeostasis and transport. In: *Metal Ions in Life Sciences*, vol. **12**, 169–201. Springer.

221 Al-Yami, A.S. and Nasr-El-Din, H.A. (2007). An innovative manganese tetroxide/KCl water-based drill-in fluid for HT/HP wells, SPE 110638. SPE Annual Technical Conference and Exhibition, Anaheim, CA (11–14 November 2007).

222 Dennis, D.M. and Hitzman, D.O. (2007). Advanced nitrate-based technology for sulfide control and improved oil recovery, SPE 106154. International Symposium on Oilfield Chemistry, Houston, TX (28 February to 2 March 2007).

223 Denton, R.M. and Fang, Z. (1996). Rock bit grease composition. US Patent 5,589,443, assigned to Smith International Inc.

224 Walker, M.L. (1995). Hydrochloric acid acidizing composition and method. US Patent 5,441,929, assigned to Haliburton Company.

225 El Din, A.M.S. and Wang, L. (1996). Mechanism of corrosion inhibition by sodium molybdate. *Desalination* **107** (1): 29–43.

226 Fan, L.G. and Fan, J.C. (2001). Inhibition of metal corrosion. US Patent 6,277,302, assigned to Donlar Corporation.

227 Heidersbach, R. (2011). *Metallurgy and Corrosion Control in Oil and Gas Production*. Wiley.

228 Putzig, D.E. and Smeltz, K.C. (1986). Organic titanium compositions as useful cross-linkers. EP Patent 195531, assigned to DuPont De Nemours and Company.

229 Ramanarayanan, T.A. and Vedage, H.L. (1994). Inorganic/organic inhibitor for corrosion of iron containing materials in sulfur environment. US Patent 5,279,651, assigned to Exxon Research and Engineering Company.

230 Hille, R. (2002). Review: molybdenum and tungsten in biology. *Trends in Biochemical Sciences* **27** (7): 360–367.

231 Ollivier, B. and Magot, M. ed. (2005). *Petroleum Microbiology*. Washington, DC: ASM Press.

232 Miller II, W.K., Roberts, G.A., and Carnell, S.J. (1996). Fracturing fluid loss and treatment design under high shear conditions in a partially depleted, moderate permeability gas reservoir, SPE 37012. SPE Asia Pacific Oil and Gas Conference, Adelaide, Australia (28–31 October 1996).

233 Putzig, D.E. (1988). Zirconium chelates and their use for cross-linking. EP Patent 278684, assigned to DuPont De Nemours and Company.

234 Ridland, J. and Brown, D.A. (1990). Organo-metallic compound. CA Patent 2002792, assigned to Tioxide Group.

235 Sharif, S. (1993). Process for preparation and composition of stable aqueous solutions of boron zirconium chelates for high temperature frac fluids. US Patent 5,217,632, assigned to Zirconium Technology Corporation.

236 Burrafato, G., Guameri, A., Lockhart, T.P., and Nicora, L. (1997). Zirconium additive improves field performance and cost of biopolymer muds, SPE 32788. International Symposium on Oilfield Chemistry, Houston, TX (18–21 February 1997).

237 Fox, K.B., Moradi-Araghi, A., Brunning, D.D., and Zomes, D.R. (1999). Compositions and processes for oil field applications. WO Patent 1999047624, assigned to Phillips Petroleum Company.

238 Shahid, M., Ferrand, E., Schreck, E., and Dumat, C. (2013). Behavior and impact of zirconium in the soil-plant system: plant uptake and phytotoxicity. *Reviews of Environmental Contamination and Toxicology* **221**: 107–127.

239 Perez, G.P. and Deville, J.P. (2013). Novel high density brines for completion applications. WO Patent 2013059103, assigned to Haliburton Energy Services Inc.

240 Wilhelm, S.M., Liang, L., and Kirchgessner, D. (2006). Identification and properties of mercury species in crude oil. *Energy and Fuels* **20** (1): 180–186.

241 Naerheim, J. (2013). Mercury guideline for the Norwegian oil and gas industry, SPE 164950. European HSE Conference and Exhibition, London, UK (16–18 April 2013).

242 Sainal, M.R., Uzaini Tg Mat, T.M., Shafawi, A.B., and Mohamed, A.J. (2007). Mercury removal project: issues and challenges in managing and executing a technology project, SPE 110118. SPE Annual Technical Conference and Exhibition, Anaheim, CA (11–14 November 2007).

243 U.S. Food and Drug Administration (2012). Mercury levels in commercial fish and shellfish (1990–2010).

244 Rongponsumrit, M., Athichanagorn, S., and Chaianansutcharit, T. (2006). Optimal strategy of disposing mercury contaminated waste, SPE 103843. International Oil and Gas Conference and Exhibition in China, Beijing, China (5–7 December 2006).

245 Gallup, D.L. and Strong, J.B. (2006). Removal of mercury and arsenic from produced water. Proceedings of 13th Intern. Petrol. Environ. Conf. Paper 91, p. 24, http://ipec.utulsa.edu/Conf2006/Papers/Gallup_91.pdf (accessed 12 December 2017).

246 Davie, D.R., Hartog, J.J., and Cobbett, J.S. (1981). Foamed cement – a cement with many applications, SPE 9598. Middle East Technical Conference and Exhibition, Bahrain (9–12 March 1981).

247 Harris, P.C. and Heath, S.J. (1996). High-quality foam fracturing fluids, SPE 35600. SPE Gas Technology Symposium, Calgary, Alberta, Canada (28 April to 1 May 1996).

248 Sloat, B.F. (1989). Nitrogen stimulation of a potassium hydroxide wellbore treatment. US Patent 4,844,169, assigned to Marathon Oil Company.

249 Naylor, P. and Frorup, M. (1989). Gravity-stable nitrogen displacement of oil, SPE 19641. SPE Annual Technical Conference and Exhibition, San Antonio, TX (8–11 October 1989).

250 Martin, D.F. and Taber, J.J. (1992). Carbon dioxide flooding, SPE 23564. *Journal of Petroleum Technology* **44** (4): 396–400, Society of Petroleum Engineers.

251 Leung, D.Y.C., Caramanna, G., and Maroto-Valer, M.M. (2014). An overview of current status of carbon dioxide capture and storage technologies. *Renewable and Sustainable Energy Reviews* **39**: 426–443.

252 Metwally, M. (1990). Effect of gaseous additives on steam processes for Lindbergh Field, Alberta, PETSOC 90-06-01. *Journal of Canadian Petroleum Technology* **29** (6).

253 Gupta, A.P., Gupta, A., and Langlinais, J. (2005). Feasibility of supercritical carbon dioxide as a drilling fluid for deep underbalanced drilling operation, SPE 96992. SPE Annual Technical Conference and Exhibition, Dallas, TX (9–12 October 2005).

254 Kang, S.M., Fathi, E., Ambrose, R.J. et al. (2011). Carbon dioxide storage capacity of organic-rich shales, SPE 134583. *SPE Journal* **16** (4): 842–855.

255 Fink, M. and Fink, J. (1998). Usage of pyrolysis products from organic materials to improve recovery of crude oil. 60th EAGE Conference and Exhibition, Leipzig, Germany (8–12 June 1998).

256 Sunder, E. and Torsvik, T. (2005). Microbial control of hydrogen sulphide production in oil reservoirs. In: *Petroleum Microbiology*, Chapter 10 (ed. B. Ollivier and M. Magot). ASM Press.

257 Corduwisch, R., Kleintz, W., and Widdel, F. (1987). Sulphate reducing bacteria and their activities in oil production, SPE 13554. *Journal of Petroleum Technology* **39**: 97.

258 Keasler, V., Bennet, B., Diaz, R. et al. (2009). Identification and analysis of biocides effective against sessile organisms, SPE 121082. SPE International Symposium on Oilfield Chemistry, The Woodlands, TX (20–22 April 2009).

259 Hitzman, D.O. and Dennis, D.M. (1997). New technology for prevention of sour oil and gas, SPE 37908. SPE/EPA Exploration and Production Environmental Conference, Dallas, TX (3–5 March 1997).

260 Stott, J.F.D. (2005). Modern concepts of chemical treatment for the control of microbially induced corrosion in oilfield water systems. Chemistry in the Oil Industry IX (31 October to 2 November 2005). Manchester, UK: Royal Society of Chemistry, p. 107.

261 Burger, E.D., Crewe, A.B., and Ikerd III, H.W. (2001). Inhibition of sulphate reducing bacteria by anthraquinone in a laboratory biofilm colum under dynamic conditions, Paper 01274. NACE Corrosion Conference, Houston, TX (11–16 March 2001).

262 van Dijk, J. and Bos, A. (1998). An experimental study of the reactivity and selectivity of novel polymeric 'triazine type' H_2S scavengers. Chemistry in the Oil Industry, Recent developments, p. 170. Royal Society of Chemistry Cambridge, UK.

263 Hage, R. and Lienke, A. (2005). Review: Applications of transition-metal catalysts to textile and wood-pulp bleaching. *Angewandte Chemie, International Edition* **45** (2): 206–222.

264 Clarke, D.G. (2002). Apparatus and method for removing and preventing deposits. US Patent 6,348,102, assigned to BP Exploration and Oil Inc.

265 Castrantas, H.M. (1981). Use of hydrogen peroxide to abate hydrogen sulfide in geothermal operations, SPE 7882. *Journal of Petroleum Technology* **33** (5): 914–920, Society of Petroleum Engineers.

266 Bayless, J.H. (1998). Hydrogen peroxide: a new thermal stimulation technique. *World Oil* **219** (5): 75–78.

267 Cobb, H.G. (2010). Composition and process for enhanced oil recovery. US Patent 7,691,790, assigned to Coriba Technologies LLC.

268 Nimerick, K. (1996). Improved borate crosslinked fracturing fluid and method. GB Patent 2291907, assigned to Sofitech N.V.

269 Malate, R.C.M., Austria, J.J.C., Sarimieto, Z.F. et al. (1998). Matrix stimulation treatment of geothermal wells using sandstone acid. Proceedings: Twenty Third Workshop on Geothermal Reservoir Engineering Stanford University, Stanford, CA (26–28 January 1998).

270 Stepanova, G.S., Rosenberg, M.D., Bocsa, O.A. et al. (1994). Invention relates to use of ammonium carbonate in enhanced oil recovery. RU Patent 2021495, assigned to Union Oil and Gas Research Institute.

271 Todd, B.L., Reddy, B.R., Fisk Jr., J.V. and Kercheville, J.D. (2002). Well drilling and servicing fluids and methods of removing filter cake deposited thereby. US Patent 6,422,314, assigned to Haliburton Energy Services Inc.

272 Shu, P., Phelps, C.H., and Ng, R.C. (1993). In-situ silica cementation for profile control during steam injection. US Patent 5,211,232, assigned to Mobil Oil Corporation.

273 Martin, R.L. (2008). Corrosion consequences of nitrate/nitrite additions to oilfield brines, SPE 114923. SPE Annual Technical Conference and Exhibition, Denver, CO (21–24 September 2008).

274 Richardson, E.A. and Scheuerman, R.F. (1979). Method of starting gas production by injecting nitrogen-generating liquid. US Patent 4,178,993, assigned to Shell Oil Company.

275 Ashton, J.P., Kirspel, L.J., Nguyen, H.T., and Creduer, D.J. (1989). In-situ heat system stimulates paraffinic-crude producers in Gulf of Mexico, SPE 15660. *SPE Production Engineering* **4** (02): 157–160.

276 Marques, L.C.C., Pedroso, C.A., and Neumann, L.F. (2002). A new technique to troubleshoot gas hydrates buildup problems in subsea Christmas-trees, SPE 77572. SPE Annual Technical Conference and Exhibition, San Antonio, TX (29 September to 2 October 2002).

277 Ahlgren, S., Backy, A., Bernesson, S. et al. (2008). Ammonium nitrate fertiliser production based on biomass – environmental effects from a life cycle perspective. *Bioresource Technology* **99** (17): 8034–8041.

278 Ahmad, Z. (2006). *Principles of Corrosion Engineering and Corrosion Control*, 2nde. Butterworth and Heinemann.

279 Walker, M.L., Ford, W.G.F., Dill, W.R., and Gdanski, R.D. (1987). Composition and method of stimulating subterranean formations. US Patent 4,683,954, assigned to Haliburton Company.

280 Shimura, Y. and Takahashi, J. (2006). Oxygen scavenger and the method for oxygen reduction treatment. US Patent 7,112,284, assigned to Kurita Water Industries Ltd.

281 Mouche, R.J. and Smyk, E.B. (1995). Noncorrosive scale inhibitor additive in geothermal wells. US Patent 5,403,493, assigned to Nalco Chemical Company.

282 Grimshaw, D. (2006). Method and apparatus for treating biofouled wells with azide compounds and/or preventing biofouling in a well. US Patent Application 20060185851.

283 Lichstein, H.C. and Soule, M.H. (1944). Studies on the effect of sodium azide on microbic growth and respiration. *Journal of Bacteriology* 231–239.

284 Emmons, D.H. (1993). Sulfur deposition reduction. US Patent 5,223,160, assigned to Nalco Chemical Company.

285 Varadaraj, R. (2008). Mineral acid enhanced thermal treatment for viscosity reduction of oils (ECB-0002). US Patent 7,419,939, assigned to Exxonmobil Upstream Research Company.

286 Watson, J.L. and Carney, L.L. (1977). Oxygen scavenging methods and additives. US Patent 4,059,533, assigned to Haliburton Company.

287 http://www.ceresana.com/en/market-studies/chemicals/carbon-black/ (accessed 12 December 2017).

288 Sawdon, C., Tehrani, M., and Craddock, P. (2000). Electrically conductive invert emulsion wellbore fluid. GB Patent 2345706, assigned to Sofitech N.V.

289 Villar, J., Baret, J.-F., Michaux, M., and Dargaud, B. (2000). Cementing compositions and applications of such compositions to cementing oil (or similar) wells. US Patent 6,060,535, assigned to Schlumberger Technology Corporation.

290 Calloni, G., Moroni, N., and Miano, F. (1995). Carbon black: a low cost colloidal additive for controlling gas-migration in cement slurries, SPE 28959. SPE International Symposium on Oilfield Chemistry, San Antonio, TX (14–17 February 1995).

291 Silva, G.G., De Oliveira, A.L., Caliman, V. et al. (2013). Improvement of viscosity and stability of polyacrylamide aqueous solution using carbon black as a nano-additive, OTC 24443, OTC Brasil, Rio de Janeiro, Brazil (29–31 October 2013), Society of Petroleum Engineers.

292 US Department of Health and Human Services (1988). Occupational safety and health guideline for carbon black, potential human carcinogen.

293 Runov, V.A., Subbotina, T.V., Mojsa, Y.N. et al. (1992). The lubricant additive to clayey mud. SU Patent 1726491, assigned to Volga Don Br. Sintez Pav.

294 J.P.M. van Vliet, R.P.A.R. van Kleef, T.R. Smith et al. (1995). Development and Field Use of Fibre-containing Cement, OTC 7889. Offshore Technology Conference, Houston, TX (1–4 May 1995).

295 Card, R.J., Howard, P.R., Feraud, J.-P., and Constien, V.G. (2001). Control of particulate flowback in subterranean wells. US Patent 6,172,011, assigned to Schlumberger Technology Corporation.

296 Rahimirad, M. and Baghbadorani, J.D. (2012). Properties of oil well cement reinforced by carbon nanotubes, SPE 156985. SPE International Oilfield Nanotechnology Conference and Exhibition, Noordwijk, The Netherlands (12–14 June 2012).

297 Schriver, G.W., Patil, A.O., Martella, D.J., and Lewtas, K. (1995). Substituted fullerenes as flow improvers. US Patent 5,454,961, assigned to Exxon Research & Engineering Co.

298 Kapusta, S., Balzano, L., and Te Riele, P.M. (2011). Nanotechnology Applications in Oil and Gas Exploration and Production, IPTC 15152. International Petroleum Technology Conference, Bangkok, Thailand (15–17 November 2011).

299 Bjornstad, T., Haugen, O.B., and Hundere, I.A. (1994). Dynamic behaviour of radio-labelled water tracer candidates for chalk reservoirs. *Journal of Petroleum Science and Engineering* **10** (3): 2323–2238.

300 Forstner, U. and Witman, G.T.W. (1981). *Metal Pollution in the Aquatic Environment Hardcover*. Springer.

301 Hegerl, G.C. and Cubasch, U. (1996). Greenhouse gas induced climate change. *Environmental Science and Pollution Research* **3** (2): 99–102.

302 Semple, K.T., Doick, K.J., Jones, K.C. et al. (2004). Peer reviewed: defining bioavailability and bioaccessibility of contaminated soil and sediment is complicated. *Environmental Science & Technology* **38** (12): 228A–231A.

303 Rieuwerts, J.S., Thornton, I., Farago, M.E., and Ashmore, M.R. (1998). Factors influencing metal bioavailability in soils: preliminary investigations for the development of a critical loads approach for metals. *Chemical Speciation and Bioavailability* **10** (2): 61–75.

304 McKinney, J. and Rogers, R. (1992). ES&T metal bioavailability. *Environmental Science & Technology* **26** (7): 1298–1299.

305 Haritonidis, S. and Malea, P. (1999). Bioaccumulation of metals by the green alga *Ulva rigida* from Thermaikos Gulf, Greece. *Environmental Pollution* **104** (3): 365–372.

306 DeForest, D.K., Brix, K.V., and Adams, W.J. (2007). Assessing metal bioaccumulation in aquatic environments: the inverse relationship between bioaccumulation factors, trophic transfer factors and exposure concentration. *Aquatic Toxicology* **84** (2): 236–246.

6

Low Molecular Weight Organic Chemicals and Related Additives

This chapter is concerned with the use of low molecular weight chemicals used in the upstream oil and gas industry and their environmental risks and impacts. It excludes low molecular weight chemical compounds whose primary use is as solvents such as alcohols, glycols and aromatic hydrocarbons, e.g. toluene. These materials are examined in Chapter 8. The chapter focuses on materials that are chemically organic and simple in structure, i.e. are non-polymeric in nature and usually, but not exclusively, have molecular weights of less, usually much less than 700 Da. Polymers are described in Chapter 2. Other materials that may overlap with this chapter will be referenced, and these include surfactants, more fully detailed in Chapter 3, and phosphorus products, which are described in Chapter 4.

This chapter is not an attempt to provide a comprehensive catalogue of all organic chemicals used in the upstream oil and gas industry. Rather it will focus on a number of classes of organic chemicals and describe the more important ones from an oil and gas industry perspective. It will also review the potential hazards and risks from the use of such chemicals with particular regard to their environmental impact. Of critical importance here is the fact that most of these materials having a low molecular weight are bioavailable. Environmental bioavailability refers to the fraction of total chemical in the animal or plants' exposure to the environment that is available for absorption. It has been generally accepted that materials, mainly polymers having molecular weights greater than 700, pose a reduced environmental risk as such materials cannot be transported across cell membranes by absorption [1, 2]. In general, however, for lower molecular weight materials, bioavailability is governed by the properties of the chemical (including molecular weight and structure), the physiology of the exposed species and the environmental conditions.

Due to the presumed bioavailability of such compounds, the inherent toxicity and the susceptibility of small organic molecules to biodegrade become important and will be referenced throughout the chapter. There has been a study of the factors affecting the design of small molecules for biodegradability [3], and this will be illustrated along with other key references throughout the description of particular molecules in this chapter. In general, the critical factors that increase resistance to aerobic biodegradation are listed as follows:

- Halogens, which are bonded to a carbon atom, especially fluorine and chlorine. Also, if more than three halogen atoms are in the molecule, it is more resistant to aerobic biodegradation.
- Carbon chain branching if extensive quaternary carbon atoms are particularly stable to biodegradation processes.
- Tertiary amines and quaternary ammonium centres.
- Nitro, nitroso, azo and arylamine groups.
- Polycyclic molecules, e.g. polyaromatic hydrocarbons.

Oilfield Chemistry and Its Environmental Impact, First Edition. Henry A. Craddock.
© 2018 John Wiley & Sons Ltd. Published 2018 by John Wiley & Sons Ltd.

- Heterocycles, e.g. imidazoles.
- Ether bonds but not ethoxylate groups.

Conversely certain features generally increase biodegradation rates for small molecules, namely:

- Groups that are prone to enzymatic hydrolysis, mainly esters and amides.
- Groups containing oxygen atoms, such as hydroxyl, aldehyde, ketone and carboxylic acid, but not ethers as stated previously.
- Unsubstituted linear alkyl chains, especially those with more than four carbon atoms, and phenyl rings. These are the next best structural feature after oxygen moieties to promote improved biodegradation as they are attacked by oxygenase (i.e. oxygen-inserting) enzymes.

It should be recognised, however, that these are only 'rules of thumb' and many anomalies and exceptions exist, some of which will be exemplified in this chapter. There are probably well over 1000 individual low molecular weight organic molecules and similar substance in use in the oilfield. It is not the intention of this chapter to provide a comprehensive listing rather to act as a source of reference for the individual classes of product as delineated and to explore relevant and important examples.

6.1 Organic Acids, Aldehydes and Related Derivatives

6.1.1 Organic Acids and Their Salts

A number of organic acids are utilised in the oilfield, of which formic acid (Figure 6.1) is possibly the most important. Many weak organic acids such as formic acid, acetic acid, succinic acid and fumaric acid are common aqueous buffers. Such materials are used in control and maintenance of pH during hydraulic fracturing operations, particularly during relative permeability modifications [4].

6.1.1.1 Formate Salts

The salts of formic acid, such as formates, particularly potassium and caesium formate, are used for completion brines. In the upstream oil and gas industry, formate brines are the main ingredients of drilling and completion fluids used for reservoir sections in high-pressure high-temperature (HPHT) drilling. The drilling and completion fluids used in conventional drilling such as potassium chloride, caesium bromide, etc. (see Chapter 5) are often unsuitable for HPHT drilling. In the last two decades, high-density caesium formate brine has provided unique solids-free, non-damaging clear brines for multiple applications. It has been proven in numerous fields across the world, and its properties have translated into real benefits by cutting well construction costs; reducing safety, health and environmental risk; and helping optimise revenues. Such brines have been used successfully in HPHT fields at temperatures as high as 235 °C/455 °F and pressures up to 1126 bar/16 331 psi with no fluid-related well control incidents [5].

Formic acid is the starting material for the production of potassium formate and caesium formate. When dissolved in water, these formate salts give a solution (formate brines) of high density. They exhibit a number of useful properties:

- *Low Corrosivity*. Caesium formate brine is highly compatible with downhole metals. With its monovalent and alkaline properties, caesium formate brine buffered with carbonate/bicar-

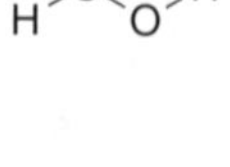

Figure 6.1 Formic acid.

bonate provides corrosion protection in harsh CO_2 and H_2S environments, even maintaining favourable pH after large influxes of acid gas. The formate ion is an antioxidant that limits the need for oxygen scavengers and antioxidants. A high concentration of formate ions in the brine – up to $14 \, mol \, l^{-1}$ – prevents problems that often occur when these additives become depleted. Furthermore, concentrated formate brines are naturally biocidal, which reduces additive costs further. Caesium formate brine's stable nature means that the risk of catastrophic, fast-acting localised corrosion associated with acidic halide brines is negligible [6].

- *Non-critical Health, Safety and Environmental (HSE) Profile.* High-density caesium formate brine is much safer to handle than alternative acidic halide brines. Rig crews can work more effectively as no specialist personal protective equipment is needed [7].
- *Environmentally Friendly.* Caesium formate brine is environmentally non-damaging as the formate ion biodegrades entirely if discharged to the sea. It meets the demanding environmental standards set by Centre for Environment, Fisheries and Aquaculture Science (CEFAS) in the United Kingdom, the Norwegian Environment Agency and other such authorities across the world. It can be recycled with up to 90% reclamation, further enhancing its environmental profile [7].
- *No Solids Problems.* Caesium formate brine is naturally heavy caesium formate salt dissolved in water. With a natural high density, ceasium formate brine offers a clean and solids-free environment for problem-free operation of tools, valves and packers, ensuring operations go smoothly from start to finish [5].
- *Thermally Stable and Highly Durable.* Caesium formate brine has been safely deployed in over 150 HPHT wells at temperatures up to 235 °C/455 °F. It is proved thermally stable over time with ceasium formate brine, successfully withstanding HPHT conditions for as long as 2 years [5].

High-density caesium formate brine is ideal for creating hydrostatic pressure and therefore ideally suited as a packer fluid. It lowers differential pressure across sealing elements and on the wellbore and casing to prevent collapse and gives metal corrosion protection [8].

In polymer-based aqueous drilling fluids, formate salts such as potassium and sodium formate have been added to aid viscosity reduction and improve thermal stability [9].

Potassium formate can be used as a treatment additive in conjunction with a cationic formation control chemical, such as dimethyldiallylammonium chloride, to promote clay stabilisation during drilling operations [10].

Formic and acetic acids are used in scale dissolver applications, particularly against deposits of calcium carbonate [11]. They are less aggressive, slower acting and less corrosive than the mineral acids that are usually confined to simulation processes where scale deposition has occurred in the formation reservoir and/or in the rock of the near wellbore area (see Chapter 5). They still however often require to have a corrosion inhibitor applied in the formulated product as they can be corrosive to the surrounding metal tubing or other equipment, especially at high temperature. The use of a mixture of acetic acid and formic acid as buffered blends with formate ions can be useful in reducing corrosion effects [12].

Interestingly lactic acid (Figure 6.2) is relatively stable at temperatures under 60 °C and therefore shows reduced corrosivity. It has been applied in the oilfield in scale deposit dissolution without corrosion inhibitor in applications where scale is deposited in the lower wellbore,

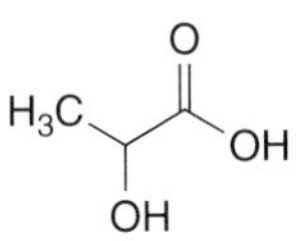

Figure 6.2 Lactic acid.

downhole temperatures are greater than 60 °C and temperatures in the upper wellbore are less (H.A. Craddock, unpublished results).

Acetic acid and formic acid along with other organic acids have been considered and applied in carbonate reservoir stimulation for many years as alternatives to hydrochloric acid, due to the retarded reaction rate, low corrosivity and reduced tendency to form acid–oil sludge in asphaltene-rich crudes [13]. Acetic acid has been used as a stimulation agent in certain unstable and weakly consolidated carbonate formations where contact with mineral acids such as hydrochloric acid would produce a large quantity of fine particles, leading to serious formation damage and loss of productivity. In such situations, a large-scale application of acetic acid has been developed and used to remove the drilling mud filter cake from vertical wells [14]. In high-pressure and high-temperature wells, particularly gas wells, blends of acetic and formic acid have been found to be particularly useful [15].

Esters of both acetic acid and formic acid have been examined to allow targeted acidisation in the subterranean formation or in a specific well, particularly horizontal and deviated completions, with an enzyme being subsequently added to hydrolyse the ester and release the acetic or formic acid [16].

A number of carboxylic acids including acetic acid, citric acid (see Figure 6.5) and erythorbic acid (Figure 6.3) have been used as iron control agents during acid stimulation operations to prevent the deposition of iron-containing compounds [17]. These materials act as chelation and binding agents trapping the iron(III) ions and not making them available for salt formations, particularly iron(III) hydroxide.

Traditionally naphthenate deposition is a problem in the oilfield that has been managed by lowering the pH of the production fluids by adding an acid of a lower pK_a value than the naphthenic acids present. These naphthenic acids are weak acids and exist in equilibrium with naphthenate ions. The higher the pH, the more dissociated the acid and the more likely it will form a soap at the oil–water interface. Field experience has shown that the optimal pH is around 6.0; further lowering the pH has no additional control benefits and increases corrosion [18].

Acetic acid in combination with hydrogen peroxide has been shown to remove lead sulphide scale, which is a particularly insoluble material [19]. It is likely that the active species is peracetic acid (Figure 6.4).

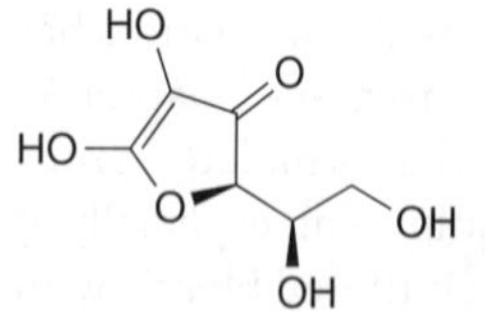

Figure 6.3 Erythorbic acid.

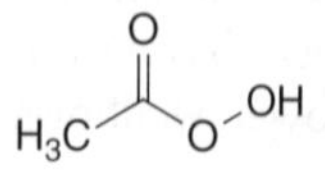

Figure 6.4 Peracetic acid.

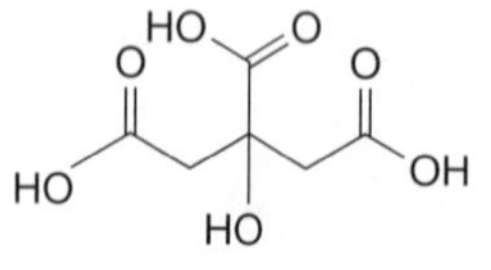

Figure 6.5 Citric acid.

Other sulphide scales can be removed by organic acids such as formic acid, maleic acid and glyoxylic acid if temperatures are sufficiently high [20, 21].

Polyhydroxyacetic acid is a low molecular weight condensation product of hydroxyacetic acid or glycolic acid. Its action as a fluid loss additive has already been described in Section 2.2.4. However, under formation conditions the condensation product can hydrolyse to produce hydroxyacetic acid that can be utilised as aqueous gel breaker in hydraulic fracturing operations [22]. As this is also been used as a fluid loss additive in these situations, the polymer provides the gel breaking capability without the need to add a separate gel breaker.

In aqueous drilling muds, a complex of aluminium or zirconium with citric acid (Figure 6.5) can be used as a low molecular weight dispersant particularly in bentonite-based muds [23].

Citric acid in conjunction with alkaline metal salts such as potassium chloride has been reported as a useful compositional product for the dissolution of filter cake deposits left by the drilling mud in the wellbore after drilling operations are complete [24]. It is also claimed that such products are useful in clearing 'stuck pipe' in wellbores and as a stimulation fluid package for dissolving filter cake that is blocking pores in the production formation. It should be noted that lactic acid can also perform some similar functions to these described for citric acid.

Salicylic acid and its sodium salt (Figure 6.6) generate viscoelasticity in surfactant solutions. In such materials, a viscoelastic surfactant (VES) is used as fluid diverting agent to aid in stimulation applications in both fracturing and matrix acidising [25]. Similar surfactant systems containing sodium salicylate has also been applied as drag-reducing agents to reduce turbulent flow in transporting fluids in pipelines [26].

Mercaptocarboxylic acids and in particular thioglycolic acid (TGA) (see Figure 6.7) are known to be useful synergists in film-forming corrosion inhibitor (FFCI) formulations [27]. Such materials are of low toxicity but exhibit an unpleasant odour, and therefore alternatives are often applied. Such alternatives are other sulphur-containing molecules that are described later in this chapter.

The glutamic acid derivative glutamic acid diacetic acid (GLDA) has found usefulness as an environmentally acceptable chelating agent, which is discussed in Section 6.2.4.

Gallic acid (see Figure 6.8), which is the main constituent of tannins and forms condensed polymeric structures with glucose, have been used as flocculant in produced water cleanup (H.A. Craddock, unpublished results). Gallic acid is also a useful oxygen scavenger [28].

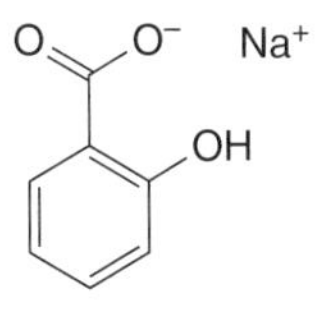

Figure 6.6 Sodium salicylate.

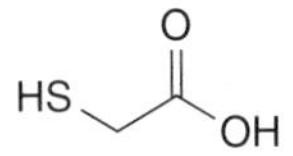

Figure 6.7 Thioglycolic acid.

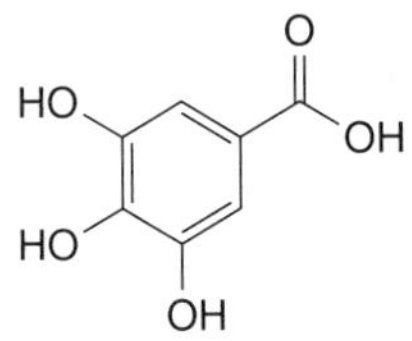

Figure 6.8 Gallic acid.

A number of other organic carboxylic acids such as maleic acid, glyoxylic acid and others have found a number of minor uses, mainly in blends or as alternative substrates in polymer manufacture or related uses.

An important class of organic acid materials not discussed in this section are the tetraacetic acid derivatives such ethylenediaminetetraacetic acid (EDTA), which are discussed in Section 6.2.

Acrylic acid and its derivatives are widely used in the oil and gas industry in polymeric forms, which have been extensively discussed in Chapter 2. Some amide derivatives, acrylamides, of a non-polymeric nature are discussed later in this chapter (see Section 6.1.6). Some non-polymeric derivatives are discussed in other sections as they have been reacted to form other molecules, e.g. imidazolines.

6.1.2 Environmental Impacts of Organic (Carboxylic) Acids and Their Salts

As has been described a number of low molecular weight organic acids, mainly simple carboxylic acids and their salts, are used in the upstream oil and gas industry. These are generally of low toxicity particularly in the dilute concentrations that are generally applied. Many are found in foodstuffs (acetic acid in vinegar), and many are essential to plant and animal life (ascorbic acid is vitamin C). Some carboxylic acids, such as formic acid, have biocidal properties and others conversely are required by bacteria for growth [29].

Carboxylic acids undergo various decomposition reactions particularly in aqueous environments [30]. In general the consensus is that two decomposition pathways predominate, namely, (i) decarboxylation, which leads to carbon dioxide and hydrogen, and (ii) dehydration, which leads to carbon monoxide and water. In the gaseous phase evidence suggests that pathway 2, dehydration, is the predominant mechanism of decomposition and in the aqueous phase pathway 1, decarboxylation, predominates (see Eq. (6.1) for formic acid).

Decomposition pathways of formic acid:

$$
\begin{aligned}
HCOOH &\rightarrow CO_2 + H_2 \quad \text{pathway } 1: \text{decarboxylationo} \\
HCOOH &\rightarrow CO + H_2O \quad \text{pathway } 2: \text{dehydration}
\end{aligned}
\tag{6.1}
$$

Such decomposition would infer that enzymatic biodegradation is likely to follow similar pathways and give rise to reasonable rates of biodegradation. Indeed, formic acid has been shown to be readily biodegradable. Compared with other organic acids, formic acid has the lowest chemical oxygen demand (COD). COD is a measure of all oxidisable substances found in the aqueous environment. Consequently, it indicates the mass of oxygen (in $mg\,l^{-1}$) that would be needed to oxidise them if oxygen were the oxidant. This means that the degradation of formic acid requires less oxygen from the given environment. The result is a lower cost of treating wastewater in sewage treatment plants or reduced consumption of the oxygen that is in limited supply in natural surface waters [31].

Moreover, unlike phosphoric acid (see Chapter 4), formic acid does not load wastewater or groundwater with phosphate, which is one of the chemical species responsible for eutrophication.

Similar characteristics exist for other low molecular weight organic acids and their salts, and although rates of biodegradation are slower than for formic acid and COD is higher, they are in the main still readily biodegradable. This is also the case for many simple aldehydes and ketones, although their toxicity is considerably different.

The use of organic carboxylic acids is certainly of low environmental impact, and their ultimate biodegradation into the ecological food chain would suggest that any long-term effects are negligible. There is possibly a small addition to carbon dioxide in the ecosystem but is insignificant compared with other man-made sources.

6.1.3 Aldehydes and Ketones

A number of aldehydes and ketones are used in the upstream oil and gas industry; however the majority of volume is dominated by two products, namely, formaldehyde and glutaraldehyde (Figure 6.9a,b).

These products are primarily used for their biocidal action.

Formaldehyde is particularly effective against sulphate-reducing bacteria (SRB), and it has been shown that it can be used for *in situ* control of biogenic reservoir souring [32]. This can be accomplished by periodic treatment, provided that the formaldehyde biocide can be transported through the reservoir. It is also claimed to be a useful agent in hydrogen sulphide removal [33].

It is thought that formaldehyde reacts with H_2S in water to give 1,3,5-trithiane (see Figure 6.10) [34]. The reaction is slow and the trithiane is poorly soluble in water and may lead to operational problems if applied in the production system; however if used in a reservoir situation where environmental temperatures are higher, reaction times can be faster and formaldehyde may be a suitable H_2S removal agent.

Formaldehyde has also been claimed as an effective agent for killing sessile bacteria in established biofilms [35].

The use of formaldehyde has declined as it has been classified as a 'probable human carcinogen'. The National Cancer Institute researchers have concluded that based on data from studies in people and from lab research, exposure to formaldehyde may cause leukaemia, particularly myeloid leukaemia, in humans [36]. To combat this potential exposure problem, a number of formaldehyde releaser molecules have been deployed both as biocides and as hydrogen sulphide scavengers. Some of these molecules are discussed in a later section of this chapter.

The main alternative aldehyde to formaldehyde as a microbial control agent in the oilfield is glutaraldehyde. It is thermally unstable and can be unpleasant to handle and indeed is classified as a respiratory sensitising agent [37]. It is probably along with tetrakis(hydroxymethyl)phosphonium sulphate (THPS) (see Chapter 4), the most commonly applied biocide in the upstream oil and gas sector [38]. Glutaraldehyde products and THPS possess similar modes of action, achieving microbial control by disabling cell wall amino acids. However, THPS requires an activation step to release formaldehyde, which then generates the active species. The resulting action of these biocides in killing and controlling microbial species, especially SRB, has the effect of reducing biofilm formation to ensure production throughput and minimising corrosion especially from microbially induced corrosion (MIC) and the effects of biogenically produced hydrogen sulphide.

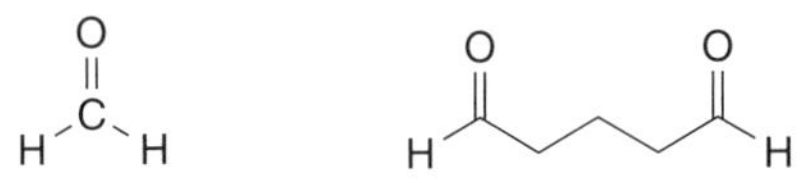

(a) Formaldehyde   (b) Glutaraldehyde

Figure 6.9 Aldehydes used in upstream oil and gas industry.

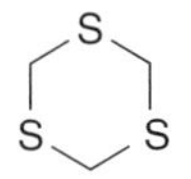

Figure 6.10 1,3,5-Trithiane.

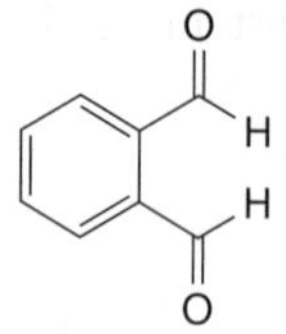

Figure 6.11 *o*-Phthalaldehyde.

The more exotic aldehyde *o*-phthalaldehyde (Figure 6.11) has been claimed as a biocide for oilfield use [39]. The product cost and potentially less favourable environmental profile have meant that its use has been limited to date although it does have a very similar performance profile to glutaraldehyde. Also, similar to glutaraldehyde, there is evidence that this material is also a respiratory and skin sensitiser, having potential in the treatment of bronchial asthma and dermatitis [40].

Cinnamaldehyde (see Figure 6.12) is a useful additive to a number of corrosion inhibitor packages for acidising treatments [41]. They are generally more environmentally acceptable than quaternary ammonium surfactants and less toxic than acetylenic alcohols [42]. It has also been shown that small amounts of *trans*-cinnamaldehyde can enhance corrosion inhibitor effects of quaternary ammonium salts particularly in acid environments [43].

The di-aldehyde glyoxal (Figure 6.13) can be applied as a hydrogen sulphide scavenger (H.A. Craddock, unpublished results) [44]. Glyoxal is supplied as a 40% aqueous solution and usually applied as a 20–30% aqueous solution.

It is much less toxic than formaldehyde and has no known carcinogenic properties. It has a good environmental profile and is permitted for use in highly regulated regions such as the North Sea. The reaction products are water-soluble sulphide polymers [45]. Like formaldehyde the reaction kinetics are slow; however at above 40 °C the reaction proceeds more rapidly, and it will also proceed in a large pH range, although the optimal range is pH 5–8. This would indicate that the aldehyde reacts with the HS⁻ ion.

In oil and gas production three phases are generally present: gas, oil and condensate. In many parts of the production process, these three phases coexist, and therefore it is important to understand the distribution of the H_2S over these three phases. Figure 6.14 represents this

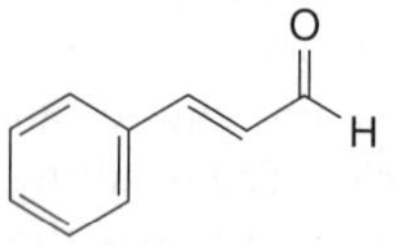

Figure 6.12 *trans*-Cinnamaldehyde.

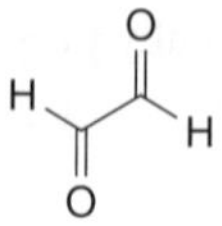

Figure 6.13 Glyoxal.

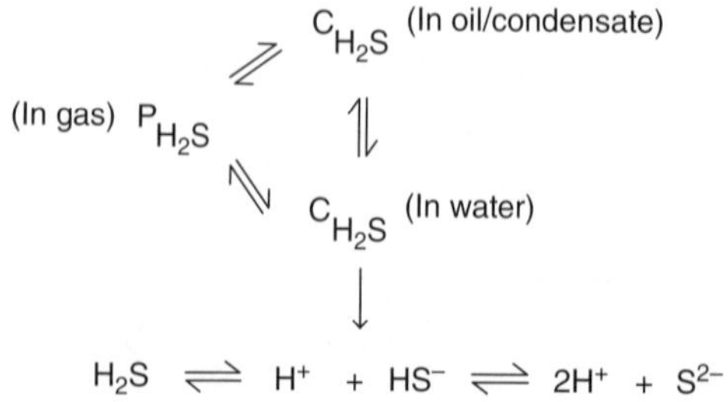

C_{H_2S} (In oil/condensate)

(In gas) P_{H_2S}

C_{H_2S} (In water)

Figure 6.14 Dissociation behaviour of H_2S.

$$H_2S \rightleftharpoons H^+ + HS^- \rightleftharpoons 2H^+ + S^{2-}$$

dissociation behaviour where P_{H_2S} is the partial pressure of H_2S in the gas phase and C_{H_2S} is the concentration in either the oil/condensate or water phase.

The dissociation of H_2S in the water phase, at any given temperature and pressure, is dictated by pH, which is dictated by the concentration of acidic gases present, primarily CO_2 and H_2S. This is the main mechanism to explain the action of other H_2S scavengers, which are detailed later in this chapter.

Glyoxal also has biocidal properties but is not as efficient as the other aldehydes previously described. It is also not registered in Europe as a biocidal product, an important requirement for manufacture and use of biocides (see Chapter 9 for further details).

Glyoxal and cinnamaldehyde blends have been claimed as useful corrosion inhibitors for mineral acids in stimulation applications. [46]

A final aldehyde to consider is acrolein (Figure 6.15), which is growing in use in the oilfield. It provides a one chemical solution to the numerous problems attributed to bacterial activity, associated hydrogen sulphide (H_2S), biogenic iron sulphides, and mercaptan production, in that it is a potent biocide that scavenges H_2S and mercaptans and dissolves iron sulphides [47].

The mechanism of action of acrolein as hydrogen sulphide scavenger is thought to be different from that of formaldehyde and glyoxal, previously described. It is believed that acrolein reacts with hydrogen sulphide to form a thioaldehyde, which in turn reacts with a further molecule of acrolein to form a water-soluble thiopyran [48]. Like other aldehydes, when deployed as an H_2S scavenger, acrolein suffers from the reaction being slow, especially at low temperatures. Hexahydrotriazines, which are discussed in Section 6.3, have much faster reaction kinetics. However, aldehydes have the advantage of not raising the pH of the system under treatment that can enhance possible carbonate scaling.

Ketones are seldom used as functional additives other than solvents such as acetone (see Chapter 8). Like cinnamaldehyde, acetophenone (Figure 6.16) and similar ketones in terms of compositions have been utilised as additives for acid corrosion inhibitor packages [49].

6.1.4 Environmental Impacts of Simple Aldehydes

As has been described a primary use of many simple aldehydes in the oilfield is as a biocidal agent. Therefore, such molecules are inherently toxic. Both formaldehyde and acrolein are highly toxic and suspected carcinogens. Glutaraldehyde is a skin and respiratory sensitiser and a corrosive agent. However, they are all also readily biodegradable. This is a major factor in considering their use in highly regulated regions such as the North Sea [50, 51].

The biodegradation of formaldehyde has been extensively studied both aerobically and anaerobically [52, 53]. The usual pathways are degradation by oxidation to the carboxylic acid and formic acid and then as described previously further degradation or dehydration to the gases carbon dioxide and carbon monoxide and water. This is also believed to be the main biodegradation pathways of all simple aldehydes.

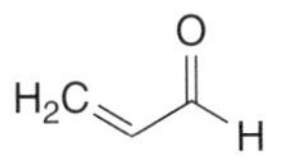

Figure 6.15 Acrolein.

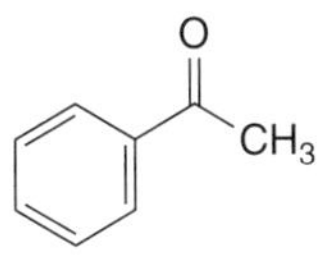

Figure 6.16 Acetophenone.

The environmental impacts of these simple molecules appear to be restrained to their high toxicity, which is counterbalanced by their ready biodegradation, and overall appear to have a minimal impact of ecosystems and the flora and fauna therein [54].

6.1.5   Acetals and Thioacetals

The acetal functional group has the following connectivity $R_2C(OR')_2$, where both R′ groups are organic fragments. The central carbon atom has four bonds to it and is therefore saturated and has tetrahedral geometry. The two R′O (R^2 and R^3 groups in Figure 6.17) may be equivalent to each other or not. The two R groups can be equivalent to each other (a 'symmetric acetal') or not (a 'mixed acetal'), and one (as in Figure 6.17) or both can even be hydrogen atoms rather than organic fragments. Acetals are formed from and convertible to aldehydes and ketones.

Acetals are stable compounds, but their formation is a reversible equilibrium. They are, however, stable with respect to hydrolysis by bases and with respect to many oxidising and reducing agents.

Various specific carbonyl compounds have special names for their acetal forms. For example, an acetal formed from formaldehyde is called formal. The acetal formed from acetone is sometimes called an acetonide.

Unfortunately, despite some useful properties and good environmental characteristics, these products have been little exploited in the upstream oil and gas industry. They have been examined for application in invert emulsion and emulsion-based drilling muds to replace diesel, base oil and other less biodegradable base materials [55].

Thioacetals are derived from the reaction of aldehydes with thiols. Such products, and in particular, that are derived from the reaction of cinnamaldehyde and thioethanol (Figure 6.18) are useful corrosion inhibitors particularly in heavy brine solutions where many commonly used corrosion inhibitors such as quaternary ammonium products, phosphate esters and filming amines are ineffective outside of particular temperature and pH ranges [56].

As stated earlier this class of derivatives is somewhat under-exploited, with a few examples in drilling mud applications, and applications as diverse as corrosion inhibition, water clarification and scale control through potential chelation effects. They are in the authors' opinion worthy of further investigation.

6.1.6   Esters, Amides and Related Derivatives

The reaction of alcohols and amines with carboxylic acids forms esters and amides, respectively, through acid- or base-catalysed dehydration reaction. The classic synthesis of an ester is the Fischer esterification, which involves treating a carboxylic acid with an alcohol in the

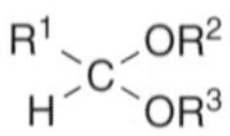

Figure 6.17 General structure for acetals.

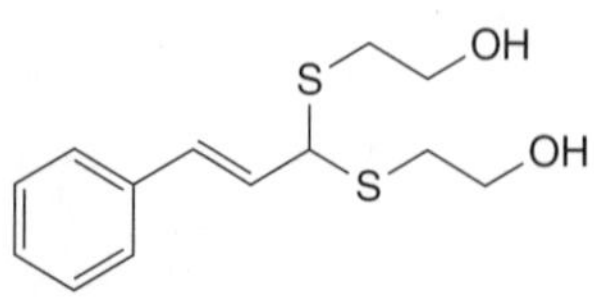

Figure 6.18 Thioacetal derivative of cinnamaldehyde.

presence of a dehydrating agent: sulphuric acid is a typical catalyst for this reaction. Since esterification is highly reversible, the yield of the ester can be improved using the following:

- Using the alcohol in large excess (i.e. as a solvent).
- Using a dehydrating agent: Sulphuric acid not only catalyses the reaction but also sequesters water (a reaction product). Other drying agents such as molecular sieves are also effective.
- Removal of water by physical means such as distillation as a low-boiling azeotrope with toluene.

Many methods exist to synthesise amides. It would appear that an analogous method to esterification, namely, the reaction of a carboxylic acid with an amine, would be the simplest method. However, this reaction has a high activation energy, largely due to the amine first deprotonating the carboxylic acid, which reduces its reactivity, and therefore such a direct reaction often requires high temperatures. In general, therefore amides are usually formed from 'activated' carboxylic acid derivatives such as esters, acid chlorides and anhydrides [57]. Recently there has also been the development of boron-based reagents for amide bond formation [58].

Both esters and amides have however similarly biodegradation pathways via oxidative hydrolysis to form the carboxylic acid and ethanol or amine, respectively. This is further detailed in Section 6.1.7.

6.1.6.1 Esters

As has been detailed in Chapter 3, glycerides, which are fatty acid esters of glycerol, are important esters in biology, being one of the main classes of lipids (see Section 6.6) and making up the bulk of animal fats and vegetable oils. Esters with low molecular weight are commonly used as fragrances and found in essential oils. Their biological profile makes them of great interest for use in green chemical applications due to their potentially low environmental impact.

Several carboxylic esters are used in the composition of ester-based oils that are suitable as drilling lubricants. Such materials have good toxicity profiles and are readily biodegradable [59]. Many of these materials are based on fatty acid esters and have been discussed in Chapter 3 as they are surfactants; however, some are also based on long-chain carboxylic acid esters derived from saturated and unsaturated acids, e.g. isononanoic acid (Figure 6.19), and long-chain alcohols, e.g. *n*-octanol (Figure 6.20). These materials are less sensitive to hydrolysis but are still biodegradable [60].

As has been described in Chapter 3, many esters of natural product materials such as the triglycerides and fatty acids from rapeseed oil (mainly oleic and linoleic acids) are surfactants. Such esters are very biodegradable; however they are also prone to hydrolysis that often makes their application in oilfield areas problematic, particularly in drilling fluid compositions.

In order to improve thermal stability, dibasic acid esters and esters derived from polyols, such as neopentyl glycol (Figure 6.21), have been considered. These materials have been primarily focused on replacement and enhancement of synthetic-based fluids in drilling lubricants. They provide better biodegradation characteristics and have better resistance to oxidation and hydrolysis than many other ester-type materials [61].

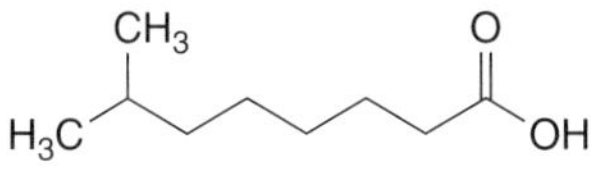

Figure 6.19 Isononanoic acid.

Figure 6.20 *n*-Octanol.

Figure 6.21 Neopentyl glycol.

Phosphate esters have a number of oilfield functions and have been detailed and discussed in Chapter 4.

A series of succinic esters of alkylaminoalkyl derivatives (discussed further in Section 6.2) have been found to be useful kinetic hydrate inhibitors (KHIs), which exhibit good biodegradation characteristic and activity as corrosion inhibitors [62].

Esters are often an optional component in fracturing fluid compositions, usually esters of polymeric carboxylic acids discussed in Chapter 2; a preferred ester is the citrate as it can be further derived by acylation on its hydroxyl groups. Such derivation makes the overall polymer less susceptible to hydrolysis and allows greater control as degradation is slowed [63].

An aqueous solution of cyclohexylammonium benzoate (Figure 6.22) and a benzoate salt, which is preferably prepared by reacting cyclohexylamine with excess sodium benzoate in an aqueous alcohol solvent, results in a useful liquid anti-corrosion composition adapted for application to ferrous tubular goods [64].

Orthoesters, particularly tripropyl orthoformate (Figure 6.23), are used as agents for controlled degradation of filter cakes, in that such orthoester compositions are capable of attacking the acid-soluble portion of the filter cake [65].

6.1.6.2 Amides

A variety of amides are applied in the upstream oil and gas sector, and some critical examples are now detailed. Acrylamides particularly in polymeric form (see Chapter 2) and as sulphonic acid derivatives (see Section 6.1.8) are widely applied. Polymeric acrylamides have a variety of functions, which are detailed in Chapter 2. The sulphonic acid derivatives are described in the next section of this chapter and are mainly used in scale control.

Amide derivatives constructed from long-chain amines have been examined as environmentally acceptable FFCI [66]. These although exhibiting good environmental characteristics have been difficult to formulate and apply as they upset the water/oil separation process. Other polymeric amides and diamides such as polymethylenepolyamine dipropionamides have been successfully applied [67]. These diamides are low molecular weight polymers and display FFCI characteristics. They are particularly effective when blended with mercapto acids such as TGA (see Figure 6.7). These materials, as previously stated, are also covered in Chapter 2.

Amides derived from alkylphenol amines and fatty acids or long-chain diacids have been used effectively in oilfield corrosion control [68]. These fatty acids can be replaced by tall oil, which is a mixture of resinic acids (abietic acid (see Figure 6.24) and other similar acids) and conjugated C18 fatty acids, linoleic acid, oleic acid and other saturated fatty acids. Such reaction products are mixtures of not only amides but also imidazolines through further reaction

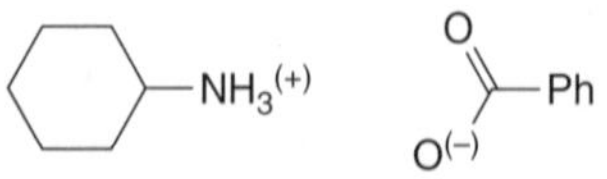

Figure 6.22 Cyclohexylammonium benzoate.

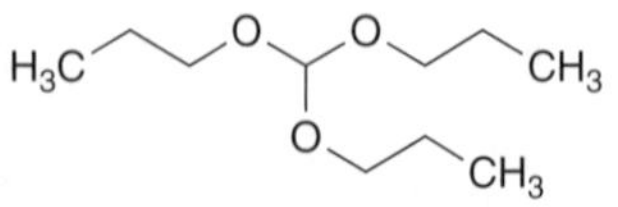

Figure 6.23 Tripropyl orthoformate.

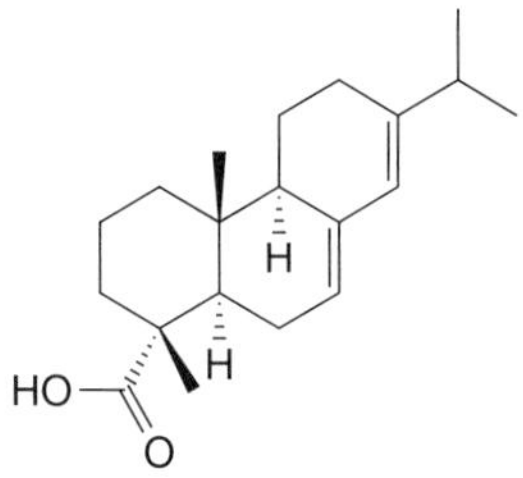

Figure 6.24 Abietic acid.

and dehydration. These imidazolines, which are very effective corrosion inhibitors, are described in more detail in Section 6.3.

Vinyl amides such as *N*-vinylformamide (Figure 6.25) have been used as the key monomer building blocks in synthetic drilling muds [69].

Polyamides are also used in demulsifier products across a range of oilfield applications. The less polymeric dialkyl amide adduct of fatty acids can be used when suitably formulated for sludge and emulsion prevention during drilling operations [70].

Low molecular weight amidines, with a general structure as in Figure 6.26, have been shown to be reasonable hydrogen sulphide scavengers [71].

Acetamides, which have a related structure, have been found to be useful as halite inhibitors. Halite deposition can be a problem where supersaturation of the produced water occurs. Nitrilotriacetamide (see Figure 6.27) and nitrilotripropionamide are commercially available products that are claimed to inhibit halide precipitation.

These acetamides are also related to ureas (see later in this section).

Amidoamines are a class of chemical compounds that are formed from fatty acids and diamines. They are used as intermediates in the synthesis of surfactants (see Chapter 3). In the upstream oil and gas industry, a water-soluble fatty acid/amidoamine corrosion inhibitor has been reported [72]. These products are reported to have a high-temperature stability as high as 230 °C, of which there are a limited number of corrosion inhibitors capable of working at these temperatures and those that are, are in the main, oil soluble. A water-soluble corrosion inhibitor has many advantages such as greater partitioning in the water phase, higher flash point and potentially a better environmental profile than the oil-soluble counterparts, due to their

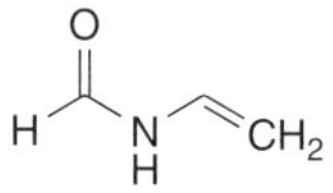

Figure 6.25 *N*-Vinylformamide.

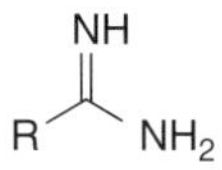

Figure 6.26 Simple amidine.

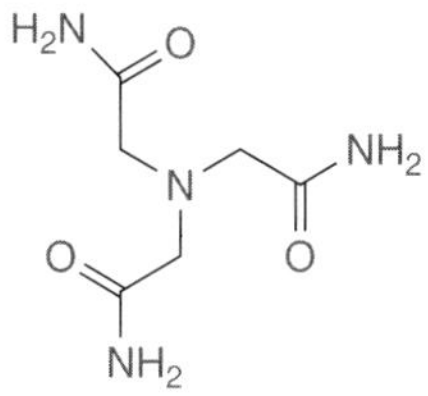

Figure 6.27 Nitrilotriacetamide.

preferential partitioning to the water phase. Such inhibitors are primarily formulated in water and will transport more easily to low spots in the transport lines, effecting a greater overall corrosion protection.

Similarly, the amidoamine salts have been found to very effective corrosion inhibitors with good environmental characteristics. They are generally used as formate solutions, provided that the high-pH environment does not promote hydrolysis of the amido group. Again, these are the reaction products of a fatty acid and a diamine with the product being solubilised as a salt with formic or acetic acid.

A commercially available inhibitor [73] is derived from maleic anhydride and a fatty acid with subsequent reaction with an amine.

Adol-amine adducts are useful both as corrosion inhibitors and in the prevention of metal sulphide precipitation by aqueous acids, such as in acidisation for stimulation. Aldols are aldehyde alcohols and have the general structure shown in Figure 6.28 and are derived from an aldehyde and a ketone by the aldol addition [74]. Aldols in themselves or as aqueous solutions are not easy to handle or have good storage characteristics, often separating into two layers that are not redispersible.

Aldols can be reacted with a primary amine such as alkanolamine monoethanolamine (MEA) (see Section 6.2) to form an aldol-amine adduct that has much improved stability. These aldol-amine adducts preferentially react with sulphide ions in solution in acidising situations, preventing the dissolved sulphide ions from reacting with dissolved metal ions and forming an insoluble sulphide salt [75].

6.1.6.3 Urea and Thiourea and Other Derivatives

Similar to the previously described acetamides, ureas and thiourea have also been used as halite inhibitors in the upstream oil and gas industry [76]. Such products have the general structure as shown in Figure 6.29.

Dialkylthioureas such as 1,3-dibutylthiourea (see Figure 6.30) are a common class of sulphur-containing corrosion inhibitor [77]. These are particularly useful for mitigating acid corrosion in combination with other inhibitors such as acetylenic alcohols (see Section 6.2) and cinnamaldehyde. They are particularly useful in reducing pitting corrosion [78] (see Section 6.1.3).

The related dithiocarbamates such as potassium dimethyl dithiocarbamate (Figure 6.31 shows the dimethyldithiocarbamate anion) are also good corrosion inhibitors. These products

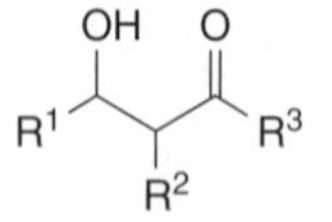

Figure 6.28 An aldol.

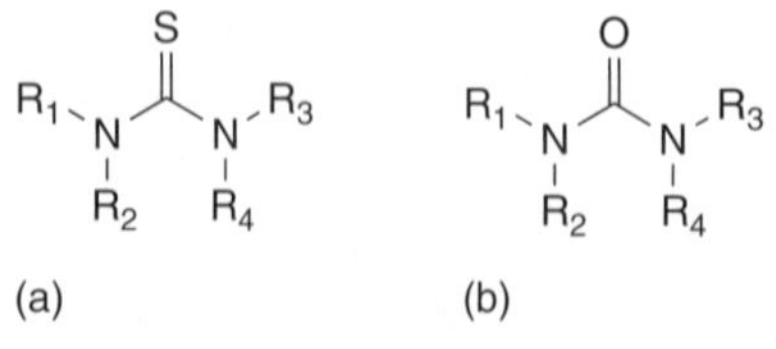

Figure 6.29 General structure of (a) thioureas and (b) ureas.

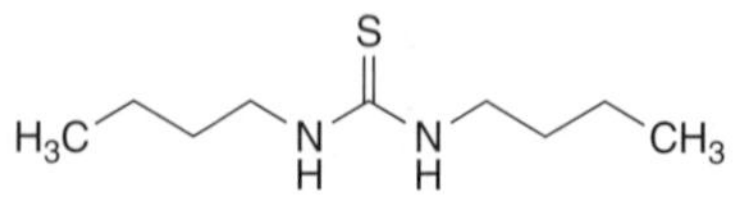

(a)   (b)

Figure 6.30 1,3-Dibutylthiourea.

are known as anti-biofoulant corrosion inhibitors as they provide a persistent and smooth protective film that resists the adhesion of sulphide salts, sessile bacteria and other solids.

Lead dithiocarbamate has also been utilised as an extreme pressure agent. These additives when added to lubricants help to decrease wear on gears and other parts that are exposed to high pressures. Lead dithiocarbamate is a highly toxic material and has largely been replaced by less toxic additives such as bismuth 2-ethylhexanoate (see Section 5.9.1.2).

Oil-soluble organic dithiocarbamates have been proposed as mercury removal agents from crude oil [79], and sodium dithiocarbamate has been found to be an effective precipitation agent for mercury (Hg)II [80]. In a similar vein dithiocarbamate additives to sequester iron have been claimed. These products are proposed to reduce asphaltene precipitation in asphaltenic subterranean reservoirs treated with aqueous solutions of strong acid, which in turn is achieved by utilising these novel dithiocarbamate compositions to sequester iron [81].

Despite the high toxicity of dithiocarbamates, they have found favourable application in some regions due to their better environmental properties as compared with other additives. This is particularly true when compared with high molecular weight cationic polymers (see Section 2.1.2.3) used as flocculants and deoilers in produced water treatment. Dithiocarbamates are claimed to have a lower toxicity than polymers and are reasonably biodegradable, and being lipophilic will not end up in the water phase and therefore into the receiving environment [82]. In effect these molecules are low molecular weight anionic polymers that complex with iron(II) ions in situ to make them lipophilic. There are usually adequate ferrous ions available in the produced water to form the complexes [83]. A number of amines have been used in the synthesis of dithiocarbamates [84], and the salts made from these products have found uses as scale and corrosion inhibitors as well as biocides. In particular sodium dimethyldithiocarbamate (see Figure 6.31) and disodium ethylene-1,2,bisdithiocarbamate are particularly effective biocides particularly against SRBs [85] and also in the presence of bisulphite-based oxygen scavengers (H.A. Craddock, unpublished results).

Recently regulatory opinion has shift to not allow the use of these materials primarily due to their toxicity, and evidence for this has come from their use as agricultural pesticides [86]. A number of studies have also illuminated that their toxicity to fish is rather high [87]. As a consequence the dithiocarbamates are no longer applied in the European oilfield sector.

6.1.7 Environmental Impacts of Esters, Amides and Related Compounds

The environmental impact of simple esters and amides is driven by their biodegradation, which in general is through enzymatic hydrolysis to the acid.

In the case of simple methyl carboxylates, bacteria hydrolyse these esters to the corresponding carboxylic acid and methanol under either growing or resting cell conditions. Methanol can further oxidised to formate under growing but not resting conditions. No hydrogen requirement appears to be necessary for ester biotransformation. The hydrolysis of methyl carboxylates is thermodynamically favourable under standard conditions, and the mixotrophic metabolism of ester/CO_2 allows for bacterial growth [88].

This appears to be a general pathway for most simple ester derivatives. The carboxylic acid biodegradation has been described in Section 6.1.2. The released alcohols undergo an oxidative process that is described in Section 6.2.5. Generally, most if not all of the degraded metabolites

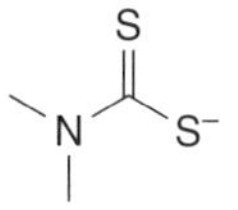

Figure 6.31 Dimethyldithiocarbamate anion.

have a negligible impact on the environment, have low toxicity and are ultimately biodegraded to carbon dioxide, carbon monoxide and water.

Similarly, many simple amides undergo enzymatic hydrolysis by amidases to the carboxylic acid derivative and an amine or ammonia. Amidases are widespread, being found in both prokaryotes and eukaryotes. Amidases catalyse the hydrolysis of the amide bond (CO—NH2).

The difference in environmental impact lies in the potential for amines to undergo complex oxidative degradation via amine oxides. Much study has been directed at the oxidative pathways of amines as their use in carbon capture could prove to be highly significant [89]. Amines are subject to further degradation processes, and possibly most importantly these degradation processes due to their volatile nature will occur in the atmosphere. Many amines are considered to pose a relatively low environmental risk. However, a number of the suggested degradation products belong to classes of chemicals of risk, with established human health and environmental impacts [90]. This is especially so for nitrosamines (see general structure Figure 6.32) and nitramines, which are molecules containing an —NO_2 group bound to a nitrogen atom. Nitramines may be primary, $RNHNO_2$, or secondary, $RR'NNO_2$ (where R and R' are alkyl or aryl groups).

Therefore, these pathways can lead to toxic species being formed that have an impact on the receiving environment. There are still a number of identifiable knowledge gaps in this area of research. The data so far would indicate a wide range of possible degradation products, including ammonia, nitrosamines, nitramines, alkylamines, aldehydes and ketones [91]. The products of most concern, namely, the nitrosamines and nitramines, however, are likely to be formed in very small amounts. Nitrosamines have been detected in measurement campaigns, but currently not nitramines. It is the general consensus that ammonia would be the main degradation product for most amines.

Potential long-term effects associated with mutagenicity, genotoxicity/carcinogenicity, and reproduction are documented for nitrosamines, volatile aldehydes and alkylamines.

There is still ambiguity surrounding degradation product composition, and concentration means for most terrestrial and aquatic environment studies have focused on amines only.

However, most degradation compounds examined so far appear to be biodegradable (30–100%), while ecotoxicity can vary significantly.

The combination of expected low emission, water solubility (low bioaccumulation potentials), biodegradability and low to moderate acute ecotoxicity for most of the post-oxidative degradation products indicate that the environmental risks should be moderate or low. However, some degradation products may persist in the environment due to low biodegradability, posing a possible risk if accumulated in aquatic systems [91].

What is the case for amines and amides is even more so for the more complex derivatives, such as ureas and thioureas (described in Section 6.1.6), and for their derivatives, such as dithiocarbamates.

The biodegradation of urea in the aquatic environment has been evaluated, and it has been found that urea will degrade to ammonia at a rate depending on the bacterial state of the environment and on the water temperature [92]. For substituted ureas of the type described in the previous section, this means that the most likely biodegradation pathway will be to amines and their subsequent biodegradation has been briefly discussed and is further detailed in Section 6.2.5.

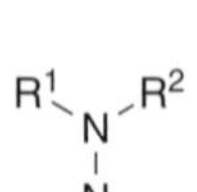

Figure 6.32 Nitrosamine.

Thioureas are and have been widely used as agricultural pesticides, and so there is a body of evidence on their toxicology and ecotoxicity, as well as their biodegradation and bioaccumulation potential. Usually these materials are highly water soluble, and unlike amines, they are not volatile. From water solubility and vapour pressure data, thiourea and related derivatives are not expected to volatilise from aqueous solutions. Based on the physicochemical properties of thiourea and its use pattern, the hydrosphere is expected to be the main target compartment for this compound. Soil and ground adsorption is also a likely area of bioaccumulation of these derivatives [93].

It has been shown that a main route of degradation of dithiocarbamates is to thioureas [94]. And has been previously stated, regulatory opinion is now not to allow the use of these materials, as evidence indicates that their toxicity and persistence is high [86], particularly with respect to fish [87].

6.1.8 Sulphonic Acids

Sulphonic acids are analogous to carboxylic acids in that the carbon atom has been replaced by a sulphur. A sulphonic acid is a member of the class of organosulphur compounds with the general structure shown in Figure 6.33, where R is an organic alkyl or aryl group and the $S(=O)_2$—OH group a sulphonyl hydroxide. A sulphonic acid can be also thought of as sulphuric acid with one hydroxyl group replaced by an organic component.

Although not all sulphonic acids, or more correctly their salts, are surfactants, many are, particularly those derived from the alkyl aryl derivatives. These are powerful and versatile detergents and surfactants that are widely used in the oil and gas industry and have been previously described in Section 3.2. A number of other sulphonic acids also have uses in the oilfield sector, which for the sake of completeness are detailed here.

One of the simplest derivatives is methane sulphonic acid (Figure 6.34), and this and other alkane sulphonic acids have been proposed as acidisation agents for stimulation of carbonate reservoirs, having excellent dissolution properties and low corrosivity [95]. It has also been shown that methane sulphonic acid is useful in carbonate scale deposit removal, particularly in tandem with other sulphonate surfactants where it is thought to act synergistically [96].

The alkane sulphonic acids, particularly alkyl aryl sulphonic acids such as dodecylbenzene sulphonic acid (DDBSA) (Figure 6.35), are a common class of asphaltene dispersant in the upstream oil and gas industry [97]. This and other alkyl aryl sulphonates have been extensively discussed in Chapter 3, as to their widespread application in surfactant-type uses.

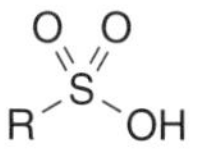

Figure 6.33 A sulphonic acid.

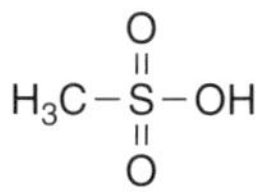

Figure 6.34 Methane sulphonic acid.

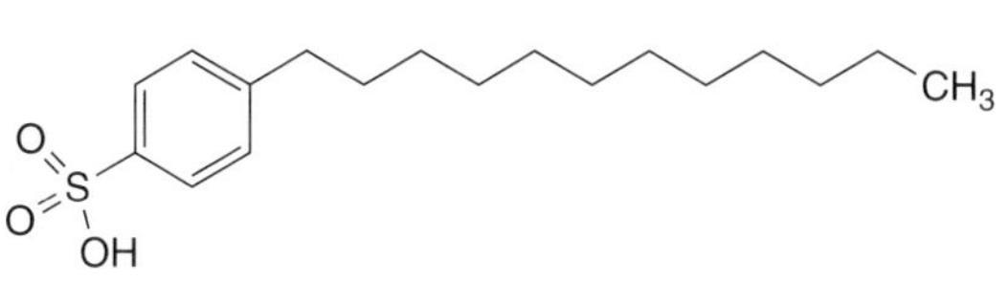

Figure 6.35 Dodecylbenzene sulphonic acid.

DDBSA has been improved as an asphaltene dispersant by the following methods:

- Removal of the aryl group leaving the alkyl chain directly bonded to the sulphonic acid group. For optimal efficiency, it is claimed that these molecules are secondary alkane sulphonic acids having chain lengths of preferably 8–11 carbon atoms [98].
- The other improvement involves using a branched alkyl polyaromatic sulphonic acid such as a sulphonated alkyl naphthalene with at least one branching alkyl chain [99].

These synthetic dispersants can greatly increase the solubility of asphaltenes in crude oils at low concentrations. They have one or more head groups that complex with the polynuclear aromatic structures in asphaltenes and long paraffinic tails that promote solubility in the rest of the oil. As a result, it appears that synthetic dispersants can be much more effective than the natural dispersants in the oil, the resins. It has been found that a straight-chain paraffinic tail is not effective above 16 carbons [98]. This is due to the decreased solubility in the oil caused by crystallisation with other tails and with waxes in the oil. In addition, *n*-alkyl-aromatic sulphonic acids can lose their ability to disperse asphaltenes with time. By using two branched tails of varying length proportions between the two tails, these problems can be alleviated, and as a result, the effectiveness of the dispersant increases with total tail length, well above 30 carbons, and it remains effective with time [99]. These latter products really fall out of the remit of this chapter and are covered in Chapters 2 and 3.

DDBSA has also been deployed in the control of naphthenate soap deposition [100].

Benzene sulphonic acid and its derivatives have a number of other uses in the upstream oil and gas sector, for example dihydroxybenzene sulphonate is used as a cement accelerator [101]. In addition, a number of surfactant applications particularly in water flooding for enhanced oil recovery have been developed and are further illustrated in Chapter 3.

Sulphonated naphthols have been used as dispersants in aqueous cement slurries [102].

Quaternised carboxylic sulphonic acid salts have been found to be useful in promoting desalting properties when formulated in demulsifier products [103].

Benzylsulphinyl and benzylsulphonyl acetic acids (Figure 6.36) have been proposed as corrosion inhibitors [104], particularly for inhibition of acidic medium such as mineral acid deployed during acidisation workover and stimulation.

Derivatives, particularly acyl derivatives, of 2-aminoethanesulphonic acid or taurine (see Figure 6.37), have found to be useful in filter cake removal [105].

2-Aminoethanesulphonic acid is a natural product occurring in a variety of foodstuffs. It is used in a number of industrial sectors. Some have good surfactant properties, which have been described in Chapter 3; however the simple salts and derivatives have been rarely explored in relation to the oil and gas sector despite their potentially good environmental profile.

The molecule 2-acrylamido-2-methyl-1-propanesulphonic acid (AMPS) (Figure 6.38) is a highly useful monomer for the synthesis of a number of polymers and related copolymers used

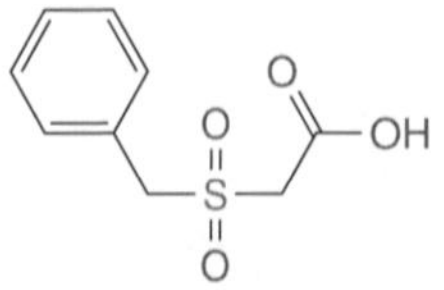

Figure 6.36 Benzylsulphonyl acetic acid.

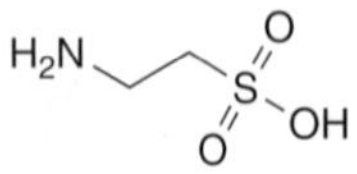

Figure 6.37 2-Aminoethanesulphonic acid.

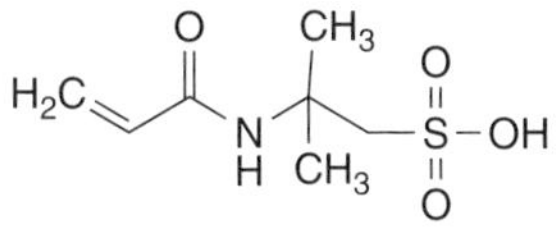

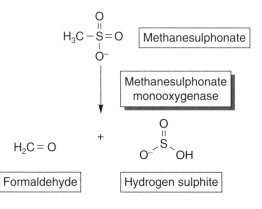

Figure 6.38 2-Acrylamido-2-methyl-1-propane sulphonic acid (AMPS).

in scale inhibitors, drag-reducing agents, naphthenate inhibitors and other polymeric oilfield additives. The acrylamide polymers and related materials, as well as polysulphonates, have already been illustrated in Chapter 2.

In general, sulphonic acids and their derivatives display good environmental properties, particularly branched alkyl sulphonic acids and their salts [106]. They are usually reasonably non-toxic and have good rates of biodegradation. As an example, a copolymer prepared from epoxysuccinic acid (ESA) and AMPS has been reported as a useful scale inhibitor and has good biodegradation performance with a biodegradation rate at 65% after 28 days [107].

Simple sulphonic acids, such as methanesulphonic acid, unlike other sulphur-containing species such as dimethyl sulphide, do not undergo photochemical oxidation. They do however biodegrade by enzymatic oxidation to formaldehyde and bisulphite ion [108]. The diagram shown in Figure 6.39 illustrates the metabolic pathway.

In comparison with many other simple organic classes, the sulphonic acids offer a green alternative with a number of interesting functionalities, a number of which remain relatively unexploited.

6.2 Alcohols, Thiols, Ethers and Amines

A number of simple alcohols such as methanol, monoethylene glycol and triethylene glycol along with some ethers such as glycol ethers are used as solvents in the formulation of additive packages. This is discussed in more detail in Chapter 8. In this chapter their use as functional additives is considered.

6.2.1 Alcohols and Related Derivatives

6.2.1.1 Alcohols and Glycols
Alcohols of low molecular weight have a multitude of functions other than acting as solvents. Methanol and glycols, especially monoethylene glycol (ethylene glycol) (see Figure 6.40), are

Figure 6.39 Biodegradation of methanesulphonic acid.

$$H_3C - \overset{\overset{\displaystyle O}{\|}}{\underset{\underset{\displaystyle O^-}{|}}{S}} = O$$

Methanesulphonate

Methanesulphonate monooxygenase

$$H_2C = O \quad + \quad O^- \overset{\overset{\displaystyle O}{\|}}{S} OH$$

Formaldehyde   Hydrogen sulphite

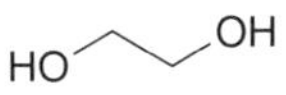

Figure 6.40 Ethylene glycol.

$$HO\diagup\diagdown OH$$

used as thermodynamic hydrate inhibitors and for the prevention of hydrate plugging in gas wells.

Hammerschmidt first explained gas hydrates in oil and gas production in 1934. These are ice-like structures that form from water and small hydrocarbons at elevated pressures and low temperatures [109]. Later it became known that in nature, antifreeze proteins (AFPs) or ice structuring proteins (ISPs), refering to a class of polypeptides produced by certain vertebrates, plants, fungi and bacteria that permit their survival in sub-zero environments [110]. AFPs bind to small ice crystals to inhibit growth and recrystallisation of ice that would otherwise be fatal. Unlike the widely used automotive antifreeze ethylene glycol, AFPs do not lower freezing point in proportion to concentration but work in a non-colligative manner instead. This allows them to act as antifreeze at concentrations 1/300–1/500th of those of other dissolved solutes. This minimises their effect on osmotic pressure. The unusual capabilities of AFPs are attributed to their binding ability at specific ice crystal surfaces, and this is the type of mechanistic approach used in the study of gas hydrate clathrate structures [111]. Polymeric KHIs and anti-agglomerate chemistries have already been illustrated in Chapter 2.

Triethylene glycol (see Figure 6.41) is also used to control water in gas streams but mainly as a dehydrating agent in gas processing [112].

In drilling operations, aqueous-based fluids in many areas are preferred over oil-based drilling muds due to their acceptable environmental profile. However, such aqueous-based fluids tend to suffer from lubricity problems and can cause more damage to the subterranean formation than oil-based materials. Clay swelling is a particular issue for most aqueous drilling fluids due to the absorption of water by clays. Silicate-based fluids (see Chapter 7) inhibit the formation damage caused by water but do not provide the lubricity characteristics of an oil-based mud. Alcohols, in combination with alkyl glucosides, can provide this lubricity when formulated in a silicate-based aqueous drilling fluid. Suitable alcohols are 2-octyldecanol, oleyl alcohol and stearyl alcohol [113].

These fatty acid alcohols are highly biodegradable and, as has been previously described in Chapter 3, are the basic feedstocks of a number of fatty acid and fatty ester-derived surfactants having a number of functions as oilfield chemicals.

It has also been found that the lower molecular weight amino alcohols in conjunction with natural oils such as linseed oil can also be used in silicate-based muds to provide lubricity [114]. These are particularly useful at high pH where lubricating additives based on esters would hydrolyse. These products are also claimed to be useful in enhancing the performance of a wide range of biocides [115]. It is also thought that aminoethanol or MEA (Figure 6.42) provides some corrosion inhibitor properties although this is disputed. MEA and other alkanolamines are further described in Section 6.2.4.

The use of alcohols in acid fluids for stimulation can increase the corrosivity of the fluids and reduce the effectiveness of any corrosion inhibitor in the package, in particular fatty amine-based corrosion inhibitors [116].

Acetylenic alcohols, particularly propargyl alcohol (Figure 6.43), have been known for some time to have good corrosion inhibitor properties [117] and have become one of the most

HO~O~~O~OH **Figure 6.41** Triethylene glycol.

H_2N~OH **Figure 6.42** Aminoethanol

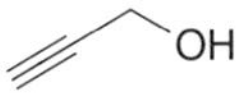

Figure 6.43 Propargyl alcohol.

common corrosion inhibitors for acid stimulation packages. This has been driven by the fact that traditionally, acid corrosion inhibitors have been some of the most toxic chemicals routinely pumped downhole [118]. It is believed that these unsaturated oxygen molecules polymerise the metal surface due to reaction in acid to form oligomers that have film-forming properties. Often quaternary ammonium or amine compounds are also added to hold the polymers on the metal surface as well as adding some film-forming characteristics of their own [119].

tert-Amyl alcohol (Figure 6.44) can be used to enhance foam degradation where defoamers are used in the process of removal of hydrogen sulphide and carbon dioxide from natural gas by processing the gas stream through an amine solution.

As previously illustrated in Section 6.1.5, thioethanol (Figure 6.45) can be used in combination with aldehydes to give a useful corrosion inhibitor product [56].

Aside from being a solvent, diluent or carrier solvent in formulated products, alcohols such as butanol are used to construct esters such as polysilicate esters (see Chapter 7) or phosphoric acid esters (see Chapter 4). Octanol and decanol, and mixtures of both, are particularly useful in this latter case [120]. These phosphate ester blends are used in enhanced oil recovery operations and due to their low volatility and good stability and are easily separated and recovered in the refining process.

Alcohols, particularly *n*-butanol (Figure 6.46) and isopropanol (Figure 6.47), are used in alcohol-water flooding operations. They have been found suitable for hot water flooding in heavy oil reservoirs at depths greater than 1500 m [121].

Isopropanol has been used in conjunction with ammonia in an aqueous carrier for water flooding operations; such a system has a reduced environmental impact, as it can be recycled and reused, due to their being no reaction with the recovered oil [122].

Alcohols, particularly butanol, can be used in the formation of microemulsions (see Chapter 8). Such systems are useful in the delivery of oil-soluble inhibitors for corrosion, asphaltene and scale.

Some alcohols or glycols, in particular the butyl glycols, 2-butoxyethanol (Figure 6.48) and 2-isobutoxy ethanol, that act as the carrier solvent have been observed to act synergistically and enhance the performance of polymeric KHIs [123].

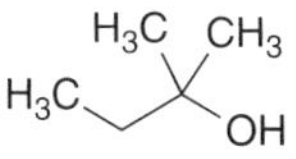

Figure 6.44 *tert*-Amyl alcohol.

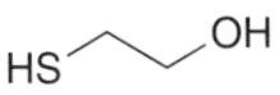

Figure 6.45 Thioethanol.

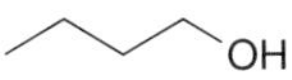

Figure 6.46 *n*-Butanol.

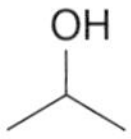

Figure 6.47 Isopropanol.

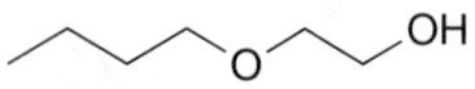

Figure 6.48 2-Butoxyethanol.

6.2.1.2 Phenols and Phenolic Compounds

A number of phenol derivatives and related phenolic products are used in the upstream oil and gas industry, and their polymeric derivatives such as the alkylphenol ethoxylates have been extensively discussed in Section 3.3.2. Although these phenolic compounds are found in nature, it is usually in structural resins such as lignin where their bioavailability is low. In using them as chemical additives and feedstocks, they have become more bioavailable, and their toxicity can be pronounced on the receiving ecosystems. This balance between bioavailability and environmental toxicity will be discussed in more detail in Chapter 10.

Non-polymeric amino alkylphenol derivatives when added in small amounts can enhance the compressive strength of well cements without affecting other properties [124].

The phenolic derivatives nonylphenol and dodecyl resorcinol (Figure 6.49) have been utilised as asphaltene dispersants; indeed the latter product was found to have a better performance than the previously described DDBSA (see Section 6.1.8).

Alkylphenols, in particular nonylphenol and its ethoxylate, have been found to be good asphaltene inhibitors although various studies show somewhat conflicting results [125].

In an attempt to reduce the toxicity of alkylphenols and circumvent the oestrogenic activity promoted by the alkylphenol, products such as resorcinol and the natural product cardanol (see Figure 6.50) have been utilised as substitutes [126]. Cardanol is found in cashew nut shell liquid.

Resorcinol has also been examined as a more environmentally friendly precursor to the phenol-formaldehyde polymers, particularly in gel systems for water shut-off [127].

Phenolic materials also exhibit powerful biocidal activity, and although they are low cost materials, they have in the main been withdrawn from the oil and gas market due to human health and environmental concerns. A common example is 4-hydroxybenzoic acid (Figure 6.51), a member of the parabens class of biocides. They are commonly used in cosmetics at low levels as a preservative.

Parabens easily penetrate the skin [128]. The European Commission on endocrine disruption has listed parabens as category 1 priority substances, based on evidence that they interfere with hormone function. [129] Parabens can mimic oestrogen, the primary female sex hormone and may also interfere with male reproductive functions [130].

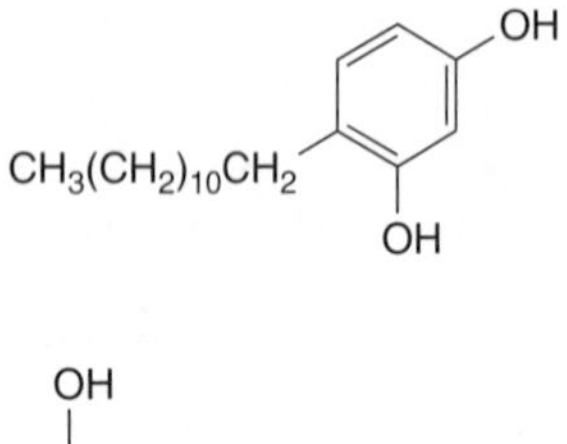

Figure 6.49 Dodecyl resorcinol.

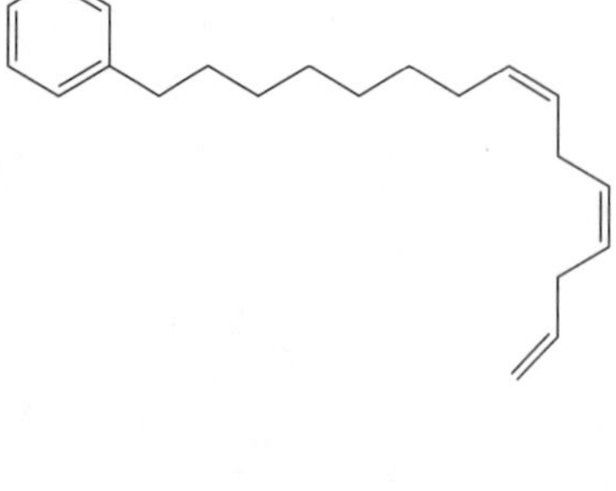

Figure 6.50 Cardanol.

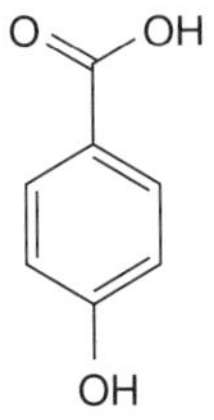

O⤸OH   **Figure 6.51** 4-Hydroxybenzoic acid.

OH

Parabens occur naturally at low levels in certain foods, such as barley, strawberries, currents, vanilla, carrots and onions, although a synthetic preparation derived from petrochemicals is used in cosmetics.

The environmental impact of phenolic derivatives and other alcohols is further discussed in Section 6.2.5.

6.2.2 Thiols and Mercaptans

A number of low molecular weight thiols, mercaptans and related sulphur products are used in the oilfield sector. Aside from having intrinsic functional properties, such as biocidal activity, in their own right, they are also added to formulated products in relatively small amounts to enhance the overall performance of the additive formulation, e.g. many formulated corrosion inhibitors contain sulphur-containing additives including thiols to enhance the overall corrosion protection performance (see also thioethanol earlier in this chapter where it forms an aldol adduct).

6.2.2.1 Thiols and Mercaptans

A variety of mercaptans including thioethanol (Figure 6.45) are used as iron control agents during acid stimulation operations [131]. The mineral acids used in such treatments cause the iron oxide present on the well-casing top dissolve and also any iron-containing minerals present in the formation. This leads to the presence of iron(III) ions in the acid that can cause the precipitation of asphaltenes in the crude oil, leading to formation damage. The presence of iron(III) ions can also lead to the precipitation of colloidal ferric hydroxide that also lead to formation damage. Therefore, it is common practice for iron control agents including thiols and mercaptans to be added in most acid stimulation treatments. In effect this mercaptans act as reducing agents converting the iron(III) to iron(II).

Most of these products such as thioglycerol (Figure 6.52) and the thiol-containing amino acids cysteine and cysteamine (see Section 6.6) are environmentally benign.

A variety of simple mercaptans have been found to perform as corrosion inhibitors, and it has been shown that under acidic oxidative conditions (high CO_2 content) of the production fluids, the thiol (—SH) group forms a disulphide (—S—S—) grouping, which complexes with iron ions at the metal surface [132]. As has been described earlier in this chapter, TGA (see Figure 6.7) is a well-documented corrosion inhibitor synergist, and as will be shown later, sulphide and thione derivatives exhibit similar properties.

A number of sulphur-containing compounds including volatile thiols are used as odorisation agents for natural gas, with *tert*-butyl mercaptan being the most commonly used. However, this area is outside the scope of this current work.

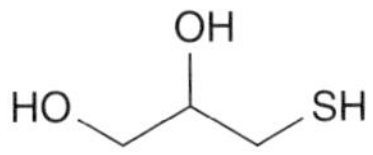

OH   **Figure 6.52** Thioglycerol.

HO⤸⤸SH

6.2.2.2 Sulphides and Disulphides

As mentioned in the previous section, sulphides also have a corrosion inhibitor effect, in particular in high shear environments [133]. It is difficult to achieve corrosion inhibition under high shear conditions without impacting water quality. However, the development of a new high shear corrosion inhibitor described showed an improved effect on water quality and superior high temperature stability relative to many traditional high shear corrosion inhibitors.

The interesting product 2-mercaptoethyl sulphide (Figure 6.53) is claimed to be a sulphur-only FFCI [2]; however it is the experience of the author that although it showed good synergistic effects with other corrosion inhibitor molecules such as imidazolines, this sulphide did not show good corrosion inhibitor properties when solely formulated (H. Craddock, unpublished results).

However, longer alkyl chain mercaptans and in particular *tert*-dodecyl mercaptan (Figure 6.54) did show some film-forming characteristics (H. Craddock, unpublished results).

The acetylenic sulphides have been shown to exhibit corrosion inhibition properties against the corrosive effects of acids similar to the acetylic alcohols discussed previously in Section 6.2.1 [2].

6.2.2.3 Thiones

The thiones and in particular the tri-thiones (see general structure in Figure 6.55) have been claimed to exhibit good corrosion inhibitor properties in high CO_2 environments [134]. Tri-thiones, in particular 4-neopentyl-5-*t*-butyl-1,2-dithiole-3-thione, have been shown to particularly useful as a corrosion inhibitor where stress cracking-induced corrosion is a concern [135].

Other thiones are used in the oilfield as biocides and are derived from a variety of heterocyclic groups such as isothiazolinones, pyrithiones and thiadiazinethiones, such as 3,5-dimethyl-1, 35-thiadiazine-2-thione (dazomet) (Figure 6.56). These chemistries are further explored in the following sections of this chapter on the relevant heterocyclic grouping.

The recent development of hydraulic fracturing to produce gas and oil from tight shale-based formations has seen a great deal of attention placed on biocide selection. The stimulation fluids are based on polysaccharide and highly biodegradable polymers and biopolymers. These polymers are also susceptible to oxidative degradation, which means that both oxygen scavengers and biocides are utilised in the chemical treatment of these fluids in order to improve their

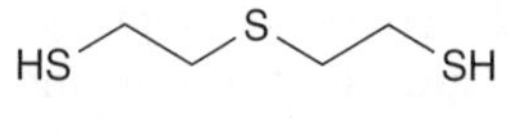

Figure 6.53 2-Mercaptoethyl sulphide.

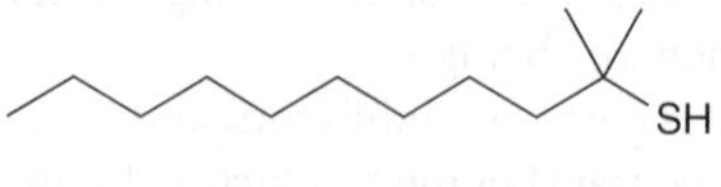

Figure 6.54 *tert*-Dodecyl mercaptan.

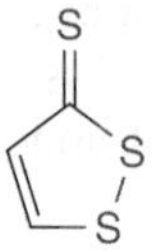

Figure 6.55 Tri-thiones.

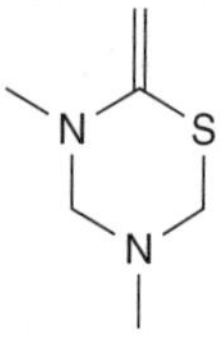

Figure 6.56 Dazomet.

stability. Bisulphite-based oxygen scavengers are normally the main choice for oxygen removal, but these are not compatible with glutaraldehyde or THPS, the two most widely used oilfield biocides. Therefore, other more compatible biocides have been examined, including thiones that are also potentially less environmentally hazardous [136].

6.2.3 Ethers

A number of ether-based derivatives are used in the development and production of oil and gas, mainly as solvents, co-solvents or mutual solvents. These functionalities are explored and discussed in Chapter 8. This section will address those molecules that are used as additives (sometimes additionally to their primary solvent function) to aid oil and gas drilling, development, production and related upstream operations.

The use of the small cyclic ether ethylene oxide and its homologue propylene oxide has been extensively described in Chapter 2, where they are used in the formation of a number of polymers and their alkoxylation to modify their surfactant and other properties such as solubility. Other small ethers, such as allyl ether and carboxylic acid ethers, are also used as monomers in polymers and copolymers for a variety of functional polymeric additives, such as pour point depressants and asphaltene dispersants. Again, these are described in Chapter 2.

A non-polluting ether additive or spotting fluid that lubricates, releases and/or prevents differentially stuck drill strings and casings in the wellbore has been developed. The reaction product between 2-ethylhexanol and the epoxide of 1-hexadecene is claimed to be the preferred compound. This enhances the lubricity of a drilling fluid to prevent drill string sticking and when utilised as a spotting agent reduces the time required to release a stuck pipe. This product also helps reduce or prevent foaming. It is non-toxic to marine life, biodegradable and environmentally acceptable, with the capability of being disposed of at the drill site without costly disposal procedures [137].

Ether derivatives of propargyl alcohol (Figure 6.43) have been proposed as acid corrosion inhibitors [138].

Water-wetting agents such as non-ionic surfactants are needed in acid stimulation operations to remove any oily film from the formation rock or deposited scale so that the aqueous acid has good contact for dissolution. These agents also clean up the well and leave the formation water wet, allowing improved oil or gas flow. As an alternative to surfactants, mutual solvents such as monobutyl glycol ether and dipropylene glycol methyl ether can be considered. These mutual solvents can also remove trapped water caused by all aqueous treatments [139]. These and other mutual solvents are further described in Chapter 8.

The introduction of aliphatic ether bonds can enhance and improve the overall biodegradation of a given chemical moiety [3]. However, the overall degradation of ethers can be complex and lead to toxic pollutants. The ethers used in the upstream oil and gas sector tend to be either alkyl ethers or alkyl aryl ethers. These have distinct degradation pathways. The non-aromatic cyclic ethers such as 1,4-dioxane (Figure 6.57) are rarely used, and the aromatic ethers such as furan are considered in Section 6.4.

The alkyl ethers undergo a well-understood and established degradation pathway as illustrated in Figure 6.58. Enzymatic oxidation to the hemiacetal followed by spontaneous oxidative cleavage to the alcohol and aldehyde is the dominant degradation pathway [140, 141].

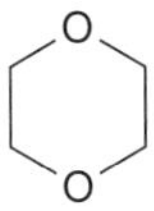

Figure 6.57 1,4-Dioxane.

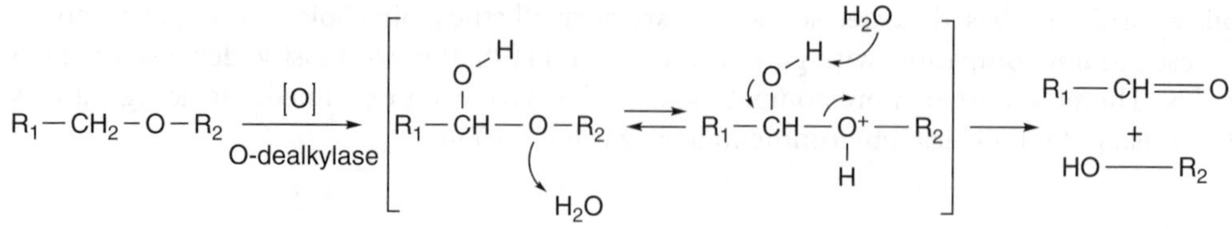

Figure 6.58 The biodegradation of alkyl ethers.

This ether bond scission is of great importance with regard to many ethoxylated and alkoxylated surfactants and polymers, both with regard to their stability and their biodegradation and supposed ecotoxicological acceptability. As has been discussed (see Chapter 3), many surfactants are now being preferred for used due to their biodegradation characteristics [142].

By contrast the degradation, through ether cleavage, of aryl ethers can be more complex. Much work has been done in the study of the chemical and biological degradation of the naturally occurring polymer lignin that contains many O-aryl bonds (see also Figure 2.60) [143]. The oxidative cleavage of these molecules tends to lead to a variety of phenolic and other aromatic compounds of various complexities.

The O-aryl bond is far stronger than the O-alkyl bond, so in mixed ethers, O-alkyl cleavage is likely to dominate and/or be preferential [140].

The evidence would suggest that ethers are relatively environmentally benign, having low ecotoxicity and good biodegradation, and also such molecules are not, based on their physical properties expected to persist or bioaccumulate. Studies on *n*-butyl glycidyl ether (Figure 6.59), a substance that many regulatory authorities have targeted as it is a known human carcinogen, have also concluded that such products are not entering ecosystem in quantities that would have an immediate or long-term effect [144].

6.2.4 Amines and Related Derivatives

A wide variety of low molecular weight amines are used across the upstream oil and gas sector.

The reaction products of amines with aldehydes to form hexahydrotriazines are considered in Section 6.3.

6.2.4.1 Amines and Diamines
A number of amines and diamines have been examined as gas hydrate inhibitors [145]. These cover not only simple amines but also substituted amines, alkylene diamines and polyamines. Simple amines are also used as copolymer units with maleic anhydride to form complex amino polyols, which are also claimed as hydrate inhibitors [146]. These and a number of other polymeric derivatives for gas hydrate inhibition are described further in Chapter 2.

Small volatile amines such as trimethylamine (see Figure 6.60) have been deployed as vapour-phase corrosion inhibitors. These are essentially organic chemicals that have sufficient vapour

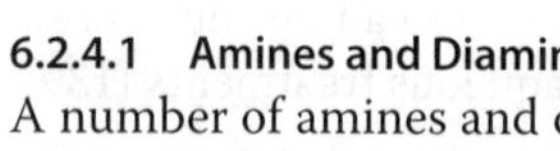

Figure 6.59 *n*-Butyl glycidyl ether.

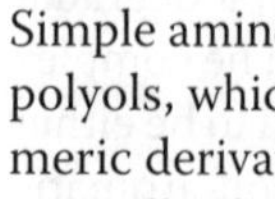

Figure 6.60 Trimethylamine.

pressure under ambient temperature and pressure conditions to travel to a metal surface by gas diffusion and adsorb onto that surface [147].

The fatty amines and their derivatives have been covered to a large extent in Sections 3.3.4 and 3.5.3, and are widely used to manufacture imidazoline- and amidoamine-based surfactants. The fatty amines themselves have also been deployed as filming corrosion inhibitors [148]. It has also been reported that oleylamine (Figure 6.61) in a reaction product with two moles of acrylic acid can form a tertiary amine, which is a useful asphaltene dispersant [2].

Alkylamines and their derivatives have a variety of functional uses in the oilfield sector. Again, many of these have been described in Chapter 3, as they form surfactant molecules, e.g. alkylamines when reacted with fatty carboxylic acids form amine salts that are well-established FFCI, a number of which are claimed to be environmentally friendly [149].

Ethoxylation of an amine molecule makes it more water soluble, and fatty amines have been ethoxylated to make them more biodegradable while retaining good corrosion inhibitor properties [150]. Alternatively hydroxyl groups have been introduced into the overall molecule such as in the general structure (Figure 6.62) and is claimed to be a useful FFCI [151].

Amino-substituted heterocycles such as aminopyridines are claimed to be useful corrosion inhibitors, and these are further described in Section 6.3.

Alkylamines are blended with phosphonic acid scale inhibitors for 'oil-soluble' scale inhibitor application packages. Such blends have been developed to avoid the practice of injecting aqueous-based scale inhibitor products or calcium-sensitive scale inhibitors. These products exist as an ion pair that is oil soluble. Such products have now been widely applied in water-sensitive wells [152], with the preferred amine being 2-ethylhexylamine (Figure 6.63) as it is environmentally accepted in regulated areas.

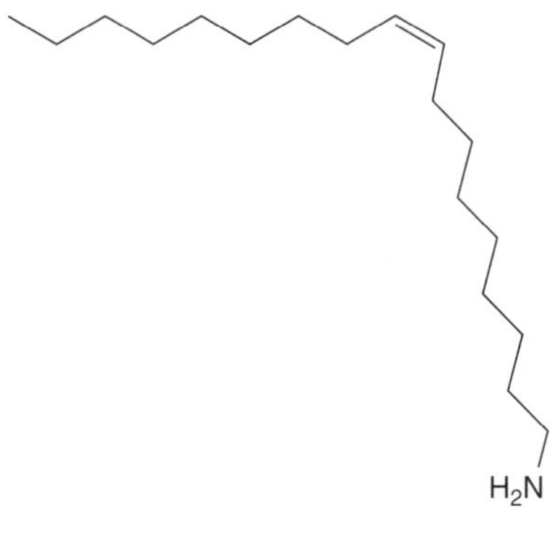

Figure 6.61 Oleylamine.

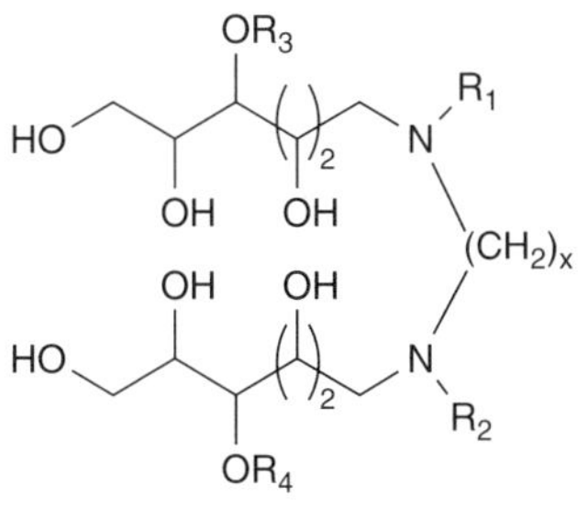

(I)

Figure 6.62 *N,N′*-Dialkyl derivatives of polyhydroxyalkyl alkylenediamines.

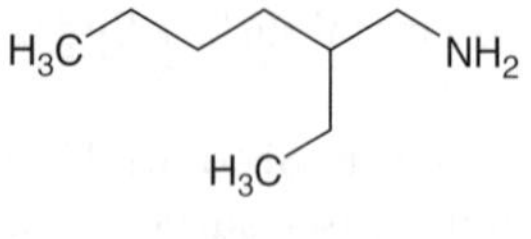

Figure 6.63 2-Ethylhexylamine.

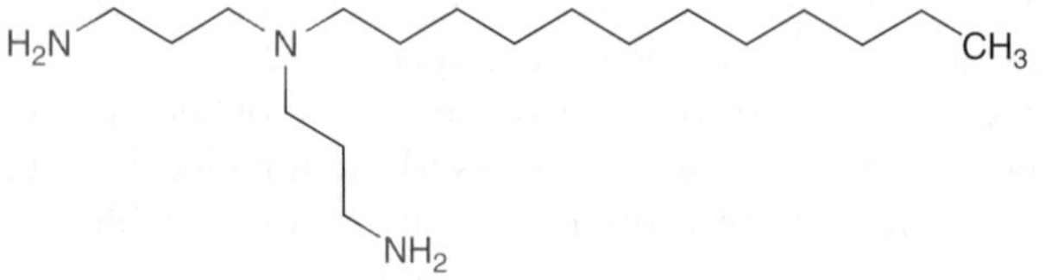

Figure 6.64 *N,N*-bis(3-Aminopropyl) dodecylamine.

The surfactant cocodiamine is an established oilfield biocide [153], and triamines such as *N,N*-bis(3-aminopropyl)dodecylamine (Figure 6.64) also have been shown to have useful antimicrobial properties [154, 155].

A number of alkylamines and related diamines, e.g. *n*-butylamine, cyclohexylamine, laurylpropyldiamine, etc., have been used as feedstock monomers for gas hydrate inhibitors [156], and the esters of these product are also claimed to have improved biodegradability [62]. These and related polymeric molecules are discussed in more detail in Chapter 2.

Alkylamines (and alkanolamines; see later in this section) are also used to neutralise various polymers or copolymers of acrylic or methacrylic acid. The resulting products are used as dispersing agents in drilling fluids [157]. Alkylamines provide important building blocks in a number of corrosion inhibitors. In general, such molecules are surfactants, and many, e.g. the imidazolines, are covered extensively in Chapter 3. They are also important in the formation of quaternary ammonium salts, e.g. diethylamino groups are known to provide good corrosion inhibition characteristics when quaternised [158]. Again these quaternised molecules have been extensively covered in Chapter 3.

Exotic amines, e.g. isophorone diamine (Figure 6.65), in combination with fatty acid amides have been shown to be good high-temperature and high-pressure corrosion inhibitors [159].

Amidoamines, particularly acyclic amidoamines, are used in combination with fatty acids as corrosion inhibitors [72] as reported earlier in Section 6.1.6.

An aldol (see Figure 6.28) such as that derived from acetaldehyde can be reacted with a primary amine such as MEA or aminoethanol (see Figure 6.66) to give an aldol-amine adduct, which is a useful sulphide ion scavenger [75] as described again in Section 6.1.6.

Polyalkylenepolyamine, derived from di(C-alkyl)-diethylenetriamine having alkyl groups between 8 and 22 carbon atoms, is claimed as a corrosion inhibitor [160].

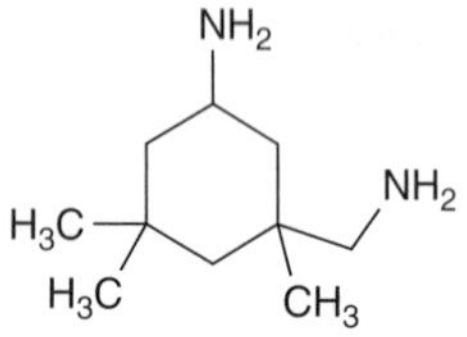

Figure 6.65 Isophorone diamine.

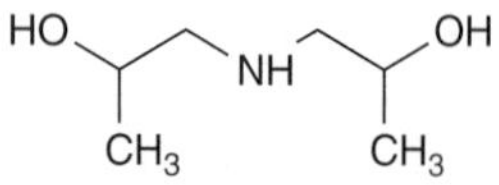

Figure 6.66 Diisopropanolamine.

6.2.4.2 Alkanolamines

MEA is one of the simplest alkanolamines that have a number of applications in the oilfield sector. Alkanolamines on their own have surfactant activity; however they are usually derivatised to tailor or enhance particular functionality or activity. These surfactants have been extensively described in Chapter 3. They are a common surfactant additive to drilling muds [161], and in combination with tartaric acid, ethanolamine gives an adduct that acts a cement retarder [162]. Like the amines described earlier in this section, they are also used in the construction of gas hydrate inhibitors [145], particularly diisopropanolamine (Figure 6.66).

Diisopropanolamine has a number of other functionalities. It is used in combination with succinic anhydride and various fatty esters to solubilise asphaltenes [2]. It has also been used as a hydrogen sulphide removal agent in gas sweetening plants; however this action is common to most alkanolamines including diethanolamine (DEA) and triethanolamine and also a number of diamines such as diglycolamine (DGA) (Figure 6.67). However, the preferred amine of choice for this action is methyl diethanolamine (MDEA) (Figure 6.68).

MDEA is commonly used in the 20–50% weight solution range. MDEA, being a tertiary amine, has significant advantages over primary and secondary amines such as MEA and DEA, as it is less basic and can be used in significantly higher concentrations. This results in MDEA having a higher capacity to react with acid gases. Also, MDEA is highly selective for hydrogen sulphide, reacting with all of the H_2S present, whereas MEA and DEA will also react with any carbon dioxide (CO_2) present; this is because being a tertiary amine, carbamate formation does not take place, whereas MEA and DEA will form carbamates with CO_2. MDEA also has a higher resistance to degradation and exhibits much better corrosivity than MEA and DEA, allowing longer refill times. These advantages lead to the following:

- Increased capacity for existing units
- Decreased capital costs for new installations
- Lower operating energy costs.

DGA is generally used as a 50–70 wt% solution in water. As with MEA, the corrosion problems with DGA prevent high solution loadings. DGA has a tendency to preferentially react with CO_2 over H_2S. It has a higher pH than MEA and thus can easily achieve low concentrations of H_2S, except in some cases where large amounts of CO_2 are present relative to H_2S. DGA has some definite advantages over other amines, in that higher DGA concentrations in the solution result in lower circulation rates and also in lower freezing points. In addition, DGA is not as likely to react irreversibly with other acid gases, which is a significant disadvantage of MDEA [163].

Various alkanolamine polymers have been shown to be good pour point depressants, in particular hexa-triethanolamine oleate esters [164]. Triethanolamine (Figure 6.69) and its derivatives have a number of functionalities in the upstream oil and gas sector. Its phosphate esters are useful as scale inhibitors in scale 'squeeze' applications [165], and the monophosphate ester has good rates of biodegradation, as it is reasonably prone to hydrolysis [166]. However, this characteristic means it is also thermally unstable and therefore only useful at temperatures below 80 °C.

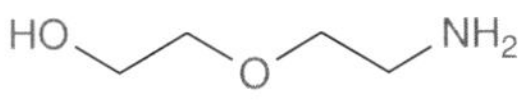

Figure 6.67 Diglycolamine.

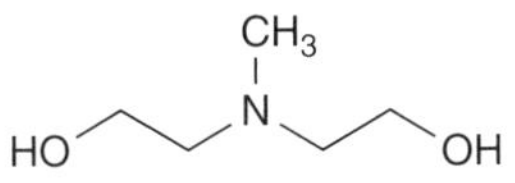

Figure 6.68 Methyl diethanolamine.

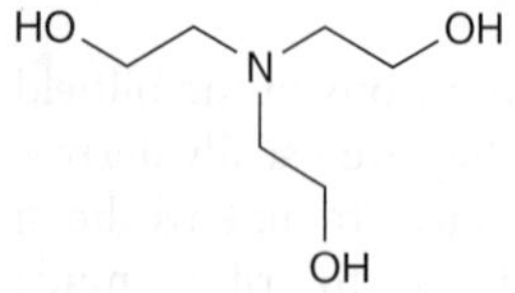

Figure 6.69 Triethanolamine.

These phosphate esters are also useful corrosion inhibitors [167], and more detail is given in Chapter 4.

Similar to the esters mentioned earlier, the propoxylated derivatives of triethanolamine have been shown to have reasonable activity as a kinetic gas hydrate inhibitor, and although lower than other products, they can be used as useful synergists [168].

Alkanolamines are readily condensed to form polyalkanolamines that are useful oil and water demulsifiers [169]. Triethanolamine, in particular, can be polymerised to give molecules with many peripheral hydroxy (OH) groups, which can be further derivatised to confer specific properties and cross-linking [170].

The related ethanolamides, particularly as carboxylic acid derivatives, such as lauric acid diethanolamide (Figure 6.70), have been employed as hydrate inhibitors in drilling fluid compositions [69].

Similar products have also been reported in blends with phosphate esters of alkylphenol ethoxylates as useful asphaltene dispersants [171].

6.2.4.3 Complexing and Chelating Amine-Based Products

As has been described in Chapter 4, a number of aminophosphonates and aminophosphonic acid derivatives, such as aminotris(methylenephosphonic acid) (ATMP), are commonly employed as scale inhibitors in the upstream oil and gas industry. Most of these molecules are poorly biodegradable but do have low toxicity and do not bioaccumulate. Nonetheless there is a general move, particularly in highly regulated regions, to replace these with other greener alternatives.

A number of products derived from ethylenediamine and in particular EDTA (see Figure 6.71) are widely used as chelating agents.

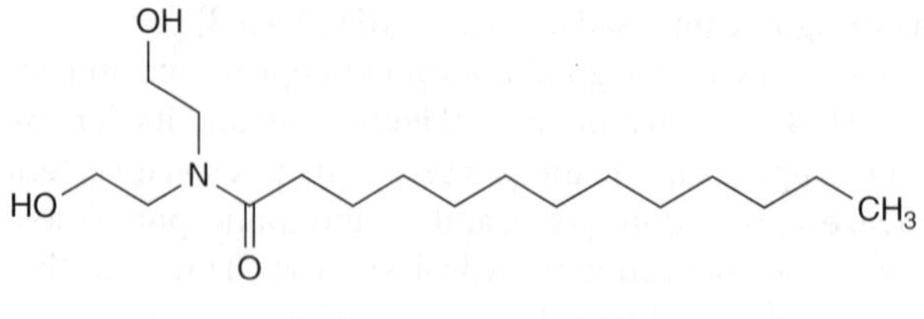

Figure 6.70 Lauric acid diethanolamide.

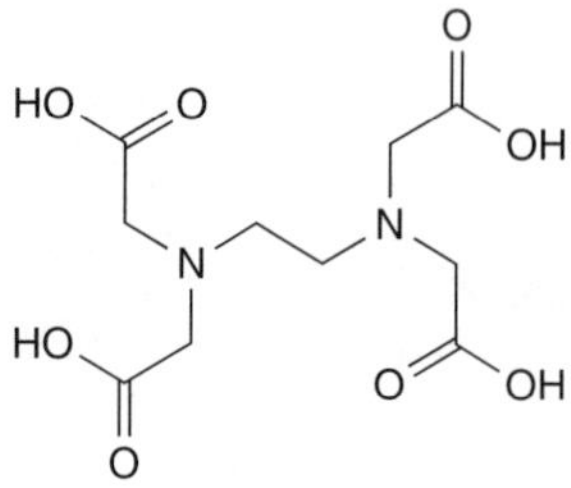

Figure 6.71 Ethylenediaminetetraacetic acid (EDTA).

Such agents are widely used as scale dissolvers across a number of industrial and household sectors including the oilfield sector. This is due to its ability to 'sequester' metal ions such as calcium, barium and iron. After being bound by EDTA into a metal complex, metal ions remain in solution but exhibit diminished reactivity. Its usefulness arises because of its role as a hexadentate ('six-toothed') ligand and chelating agent.

EDTA is produced as several salts, notably disodium EDTA and calcium disodium EDTA. The tetrapotassium salt is preferred as an oilfield scale dissolver for deposited barite and its capacity and efficiency can be enhanced by the use of carbonate and formate salts [172]. Carbonate scales are usually removed by mineral or organic acids (see Chapter 5).

It has been found that EDTA is a synergist for a number of biocides [173], particularly with respect to the control of biofouling due to biofilm production and consequential microbial induced corrosion.

EDTA is poorly biodegradable and therefore its use is restricted in environmental terms. Ethylenediamine disuccinate (the trisodium salt is illustrated in Figure 6.72) is a biodegradable chelating agent and has been found to enhance the efficacy of glutaraldehyde. It has a similar chelation capacity as EDTA but produces no persistent metabolic residual products during biodegradation [174].

Other amine-type chelating agents based on amino diacetic and triacetic acids are available as well as some based on iminoacetic acids. However, in oilfield applications these are rarely used, being less efficient than EDTA. Although EDTA is used as a barite scale dissolver, the most commonly used chelating agent for this task in oilfield use is diethylenetriamine pentaacetic acid (DTPA) (see Figure 6.73) [175].

DTPA is the only chelant that dissolves barite scale at an appreciable rate at pH higher than 12. Again, however, DTPA also suffers from low rates of biodegradation. An alternative diester dissolver has been described, which is reported to be readily biodegradable [176], as well as another material with reportedly good biodegradation [177]. The structures of these materials are however not fully described.

A more environmentally acceptable alternative to these materials is based on GLDA (see Figure 6.74).

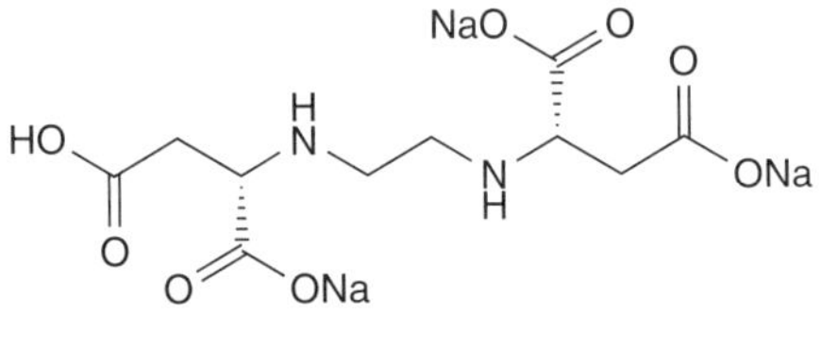

Figure 6.72 Ethylenediamine-*N,N'*-disuccinic acid trisodium salt.

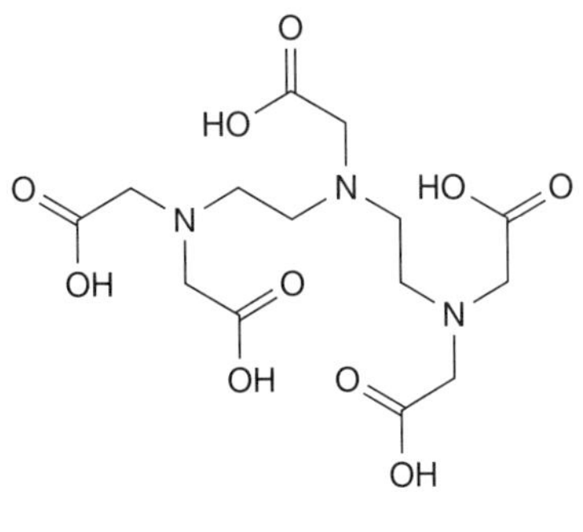

Figure 6.73 Diethylenetriamine pentaacetic acid.

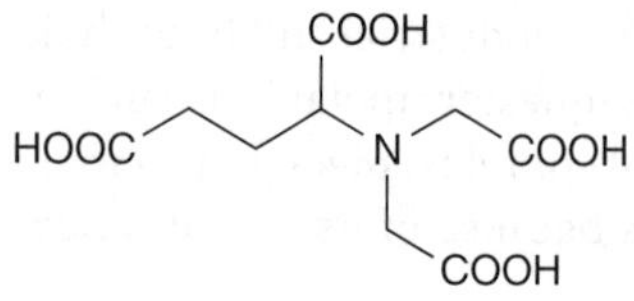

Figure 6.74 Glutamic acid diacetic acid (GLDA).

Ammonium salts of GLDA and the related methylglycine-*N*,*N*-diacetic acid have been claimed as dissolvers for a variety of carbonate sulphate and sulphide scales [178]. This technology has also been used to stimulate wells such as high-temperature gas wells where matrix acidising is a difficult task, especially if these wells are sour or if they are completed with high chrome content tubulars. These harsh conditions require high loadings of corrosion inhibitors and intensifiers in addition to hydrogen sulphide scavengers and iron control agents. Selection of these chemicals to meet the strict environmental regulations adds to the difficulty in dealing with such wells. GLDA, being non-damaging in carbonate and sandstone formations, has shown significant permeability improvements over a wide range of conditions [179].

6.2.4.4 Other Amine Derivatives

The main categories of amines and related molecules have been covered in the preceding sections. In this final short section, a few difficult-to-categorise amine derivatives are described.

Amine oxides have been extensively detailed in Chapter 3, as they are an important class of surfactants with uses in the oilfield sector in the following applications:

- As VES in water and gas control and acid stimulation
- As gas hydrate dispersants and inhibitors.

Also, amidoamine oxides have been used as VES in acid stimulation. Both this class of products and amine oxides are used to form a viscous fluid diverting agent to assist in the stimulation operations. Such VES molecules are also used in hydraulic fracturing [180].

Hydroxylamines and the related oximes have a speciality use as oxygen scavengers in particular applications. Diethylhydroxylamine (Figure 6.75) has a use as an oxygen scavenger for water in a closed-loop system. This molecule is also a useful corrosion inhibitor and minimises pitting corrosion. The reaction with oxygen is slow, so its general use even when catalysed is not recommended as it requires to be added in large amounts [181].

Oximes, such as methyl ethyl ketoxime or 2-butanone oxime (Figure 6.76), have been proposed to be used in gas stripping operations to reduce the amount of amine or glycol degradation products.

The guanidine salts such as guanidine acetate are also possible oxygen scavengers, and these along with semicarbazides and related carbohydrazides have been proposed for boiler water treatments [182]. These products are also closely related to hydrazine and its derivatives, which are detailed in Section 5.9.4. These products are also well-known oxygen scavengers.

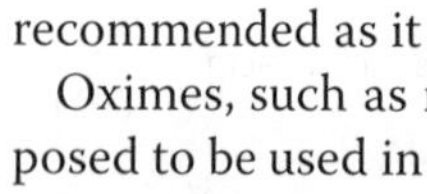

Figure 6.75 Diethylhydroxylamine.

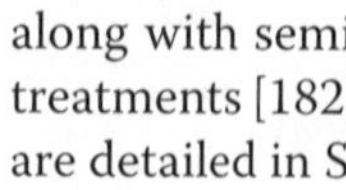

Figure 6.76 Methyl ethyl ketoxime.

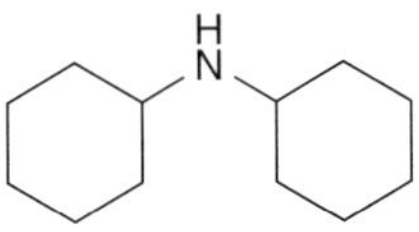

· HNO₂ **Figure 6.77** Dicyclohexylamine nitrite.

As mentioned earlier some volatile amines are used as vapour-phase corrosion inhibitors; in addition to the simple amines, the dicyclohexylamine nitrite (Figure 6.77) and carbonate salts are also used [147].

A class of short polymers, which show biocidal activity and are amine related, are the biguanides [183]. Polyaminopropyl biguanide illustrated in Figure 6.78 is a typical example.

Salts of amido-methionines, such as *n*-octanoylmethionine (Figure 6.79), are claimed to be useful FFCI. This is not surprising given that the amino acid methionine exhibits corrosion inhibitor properties [184]. The use of amino acids in the oil and gas sector and their potential is discussed in Section 6.6.

Such materials are described as having improved rates of biodegradation and low levels of toxicity [2].

As has been detailed in Chapter 2, a large number and variety of amines, diamines and related derivatives, such as imides, are used as monomers in polymer construction. Amines, diamines and other functional amides are used in the construction of surfactants, particularly the imidazolines that are further described in Section 6.3 and Chapter 3.

6.2.5 Environmental Impacts of Alcohols, Mercaptans, Amines and Their Derivatives

The environmental impact of simple alcohols, thiols and amines is primarily as a consequence of their inherent toxicity, ready bioavailability and low molecular weight. For higher molecular weight and more complex molecules, their environmental impact is usually as a consequence of their biodegradation.

There is much ambiguity and lack of coherence with regard to the definition of bioavailability, and this is further discussed in Chapter 10. Suffice to say, in terms of environmental risk, the bioavailability of introduced contaminants can have a significant impact on a wide variety of ecosystem receptors, i.e. the plants and animals living in and contributing to a specific ecosystem.

Many low molecular weight organic molecules have been regarded by regulatory authorities as having little or no risk to the environment, and these have been listed and designated [185]. It is stated that these substances do not need to be strongly regulated as from an assessment of

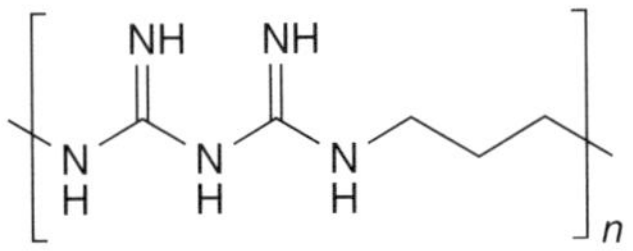

Figure 6.78 Polyaminopropyl biguanide.

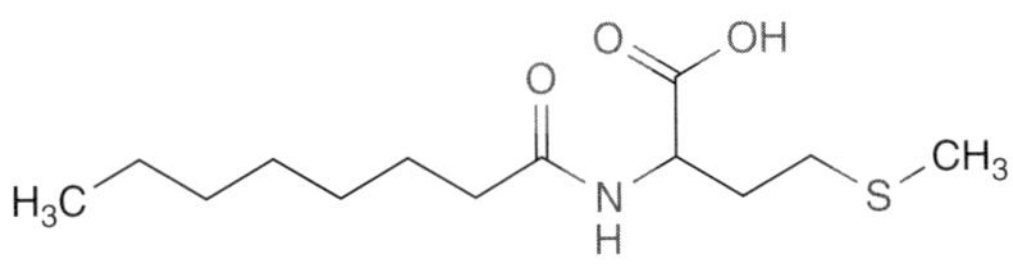

Figure 6.79 *N*-Octanoylmethionine.

their intrinsic properties; primarily due to their toxicity, they pose little or no risk to the environment.

Of course, many of these molecules are readily biodegradable as well, and as has been described earlier, the primary first step in biodegradation is by enzymatic oxidation to esters and sulphonates for alcohols and mercaptans, respectively [88]. The subsequent biodegradation of these species has been previously detailed in Sections 6.1.7 and 6.18.

It has been shown that most linear alcohols, and their ethoxylate esters, biodegrade in a stepwise fashion with the terminal hydroxyl or ethoxylate group being microbially attacked to release acetaldehyde. This can be achieved through both aerobic and anaerobic conditions [186]. The conditions and pathways will vary to accommodative both oxidative cleavage and hydrolytic degradation. The end result is a process of degradation to simple aldehydes and other metabolites that have a low environmental impact and eventually are further degraded to carbon monoxide or dioxide and water. This also seems to hold for most branched alcohols, whose rates of degradation can be faster, and also for glycols.

Phenols, as has been described in Section 6.1.2, can pose a higher environmental risk. This is due to their inherent toxicity and the risk of toxic degradation products or metabolite formation. Inherent toxicity is primarily due to phenoxy radical formation [187] that in turn can lead to metabolites such as quinones being formed. The quinones are more environmentally toxic than the originating phenols [188]. An important consideration of phenol toxicity is their potential to be endocrine disrupters, and this is particularly true of the alkylated phenols such as nonylphenol and their ethoxylates.

The alkylphenol ethoxylates have been described in detail in Section 3.3.2. It is also noted that certain alkyl phenols, particularly octyl and nonylphenols, adsorb readily to suspended solids and their fate in the environment is not as well established [189]. Nonylphenol has been demonstrated to be toxic to both marine and freshwater species, and to induce oestrogenic responses. It has been shown to be poorly biodegradable and to bioaccumulate in aquatic species [190]. This has led to OSPARCOM Treaty signatures banning the use of alkylphenols and their ethoxylates in the offshore oil and gas sector [191].

The biological oxidation of amines, as has been described for amides in Section 6.1.7, can lead to degradation products that pose a greater environmental risk than the amines themselves, primarily from nitramines and nitrosamines [90].

In conclusion, the majority of simple alcohols and mercaptans have little impact on the environment and pose little environmental risk in the concentrations applied by the industry. However, care and caution needs to be exercised in the use of amines and their derivatives and in the application of phenols and related phenolics, as their inherent toxicities and the toxicity of their metabolites coupled with their poor biodegradation and propensity to bioaccumulate could lead to severe and chronic environmental impacts.

6.3 Nitrogen Heterocycles

A number of low molecular weight nitrogen-based heterocyclic molecules are used in the upstream oil and gas sector; however, probably only two classes are used in any volume, namely, imidazolines and hexahydrotriazines. The imidazolines, especially those of a higher molecular weight such as the alkyl fatty acid and fatty amine derivatives, have been substantially described in Sections 3.5 and 3.5.3 and therefore will only briefly be described in this section. A number of other heterocyclic molecules are used in much lower quantities, and some important examples will be described and their environmental impacts discussed.

6.3.1 Imidazolines

As stated above the high molecular weight fatty nitrogenous heterocyclic compounds, the imidazolines, are able to provide corrosion inhibiting properties at low dilutions, which may be widely utilised in a number of industrial sectors including the upstream oil and gas industry. They are the basis of probably the most commonly used type of general FFCI, and their mode of action has been extensively studied [192] and is detailed in Chapter 3. In general, however, without modification they exhibit a high degree of environmental toxicity.

Alkylimidazolines are described as a component of an ion pair type of asphaltene dispersant. The alkylimidazoline is in combination with an organic acid, usually an alkyl aryl sulphonic acid. The acidic proton is conjugated with the imidazoline, and the anion–cation pair is formed [2]. Other ion pair asphaltene dispersants are claimed between alkylimidazoline derivatives of the structure illustrated in Figure 6.80 and organic acids such as ascorbic or oxalic acid [193].

These amidoimidazolines are well-known FFCI [194]. The main route to making these molecules more environmentally acceptable has been to enhance their water solubility and improve their biodegradation and reduce their toxicity through being more susceptible to hydrolysis.

Ethoxylation of N-nitrogen in the imidazoline ring or the side-chain amine or alcohol can provide more water-soluble products with lower bioaccumulation potential and toxicity. Another method to make the imidazoline more water soluble and less toxic is to react a pendant alkylamine group of an imidazoline intermediate with stoichiometric amounts of acrylic acid, giving the structure shown in Figure 6.81.

This type of product has been shown to act synergistically with oligophosphate esters, i.e. the phosphate esters of ethoxylated polyols, especially at low dose rates [195].

In deriving more water-soluble imidazolines, there is conflict as to the mechanism of hydrolysis with regard to protonation of the imidazoline ring structure; [196] however there is agreement that the rate of hydrolysis is dependent upon the pH with hydrolytic degradation of the imidazoline ring occurring at pH greater than 8. Such hydrolysis gives rise to amide structured metabolites [192]. Such conditions although not commonly occurring in the natural environment do occur during oilfield operations and so may promote the biodegradation of imidazoline products prior to their discharge to the receiving environment.

6.3.2 Azoles–Imidazoles, Triazoles and Other Related Molecules

The related imidazoles are much less commonly used than the imidazolines, as they are also much more chemically stable materials and less prone to biodegradation. Many of them are also more toxic; indeed the molecule metronidazole (Figure 6.82) is used as a biocide in the oil

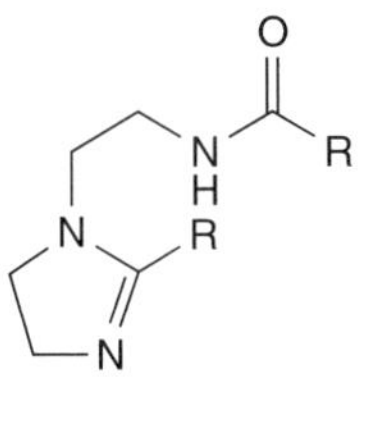

Figure 6.80 Alkylimidazoline derivative.

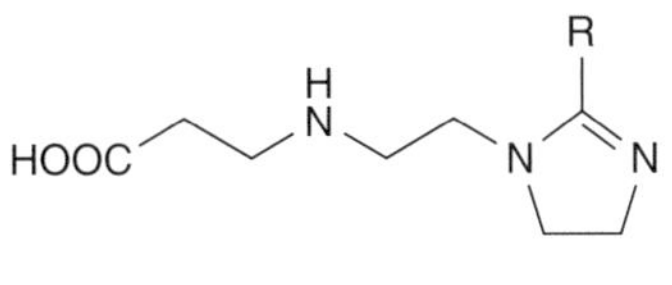

Figure 6.81 Aminoethyl imidazoline reacted with acrylic acid.

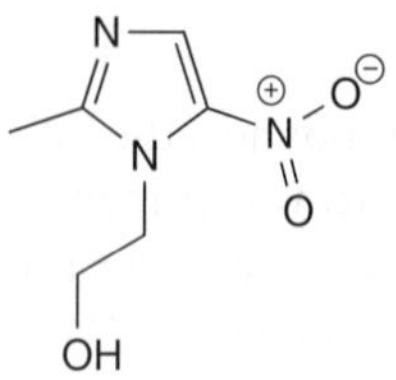

Figure 6.82 Metronidazole.

and gas sector and has shown to be effective in controlling bacterially produced hydrogen sulphide [197].

The related azoles, in particular benzimidazole and benzotriazole derivatives, are widely used in metal work as corrosion inhibitors, particularly against the corrosion of copper [198]. Such products have also been examined for use as corrosion inhibitors for mild steel in up to 15% hydrochloric acid solutions and hence have potential as corrosion inhibitors in acid stimulation and workover operations. 2-Phenylbenzimidazole (Figure 6.83) was found to be among the best-performing products [199].

Benzotriazoles and tolyltriazole, the latter being a mixture of two geometric isomers (Figure 6.84), are well-known corrosion inhibitors for yellow metals [200]. These corrosion inhibitors, however, are seldom used in the oil and gas sector. This is primarily due to the major use of carbon steel and other iron ore-based alloys where such corrosion inhibitors have been shown to be more expensive and less effective than other film-forming materials, particularly surfactant-based corrosion inhibitors. The triazoles are used in the protection of carbon steels under acidic conditions where they can be very effective [201]. They have also been combined with the fatty acids to form some interesting derivatives; again these have been primarily directed at corrosion mitigation under acidic conditions [202].

Copper alloys are a common component of cooling systems, especially at the heat exchangers that are used in oil and gas processing. The corrosion of these surfaces and resulting galvanic deposition of copper onto existing ferrous metal surfaces can have detrimental effects on the structural integrity and operation of the cooling system. As a result, copper corrosion inhibitors have always been a staple ingredient in most water treatment formulations, in particular tolyltriazoles. The film formed by tolyltriazoles has been found to be more resistant to breakdown in aqueous environments. The methyl group on the molecule is believed to sterically hinder the film thickness, as well as offer more hydrophobicity. Both of these properties are contributory to its greater resistance. However, studies have shown that triazoles' thin film is not as forgiving if breakdown occurs and accelerated corrosion can occur. This lack of per-

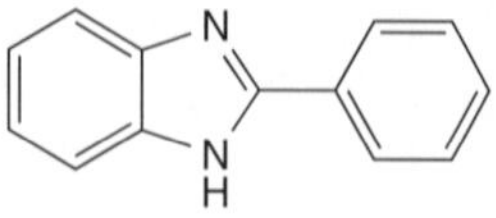

Figure 6.83 2-Phenylbenzimidazole.

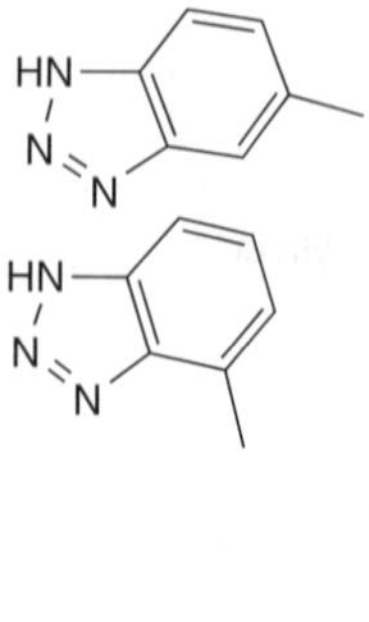

Figure 6.84 Mixture of isomers of tolyltriazole.

sistency is undoubtedly a reason for its lack of general use in the upstream oil and gas industry [203].

It has also been shown that chlorine or bromine present from halogenated oxidising biocides can penetrate the thin triazole film, causing accelerated corrosion rates. At elevated levels, both chlorine and bromine have been found to attack and break down the formed film, causing corrosion inhibition failure and rapid corrosion at the sites of film breakdown [204].

Outside the oil and gas industry, benzotriazole and tolyltriazole are widely applied as anti-corrosive agents, as well as in a number of other applications such as for so-called silver protection in dishwasher detergents, and therefore their environmental impact and fate have been fairly extensively studied.

They have been shown to have low biodegradability and limited sorption tendency and are therefore only partly removed in wastewater treatment. Residual concentrations of benzotriazoles and tolyltriazole have been determined in the ambient surface waters of major European river systems including the Rhine. Indeed, it has been observed that 277 kg benzotriazole per week flows through the Rhine river. In most cases, tolyltriazole was about a factor 5–10 less abundant. The observed environmental occurrences indicate that these triazoles are ubiquitous contaminants in the aquatic environment and that they belong to the most abundant individual water pollutants. This evidence further supports their poor biodegradation profile and environmental persistency particularly in the aquatic environment [205], and as stated in the introduction to this chapter, the azo group has been identified as having resistance to enzymatic biodegradation and hydrolysis [3].

6.3.3 Pyrroles and Pyrrolidines and Related Heterocycles

Pyrrole, a heterocyclic aromatic compound (Figure 6.85), and its derivatives are rarely used in the upstream oil and gas industry. These are however important biological molecules, being the components of more complex macrocyclic molecules such as the porphyrins and chlorins, the latter group includes chlorophyll.

There has been a claim in the patent literature as to the use of tetrapyrrolic materials as asphaltene stabilising and dispersing agents, which is based on their ability to conjugate with metal ions, particularly iron [206].

Some pyrrolidones, particularly *N*-vinylpyrrolidinone (Figure 6.86), have uses in polymer applications where they act as a vinyl monomer. These polymers and copolymers include water control agents for enhanced oil recovery [207], as copolymers in acrylamido and other polymer systems for fluid loss additives in water-based drilling fluids at high temperatures and pressures [208] and low molecular weight polymers for downhole corrosion control [209].

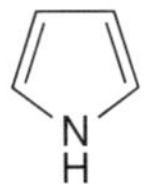

Figure 6.85 Pyrrole.

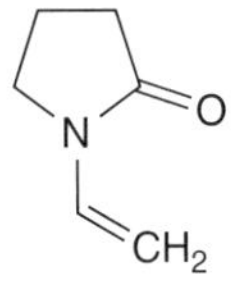

Figure 6.86 *N*-Vinylpyrrolidinone.

These pyrrolidinones belong also to a class of heterocyclic compound known as lactams, which are in effect cyclic amides. In the oilfield sector their vinyl derivatives are used in a variety of polymer applications.

Polyvinylpyrrolidinones have gas hydrate inhibition activity; however it is in general of low performance, and the vinylcaprolactam (Figure 6.87) derivatives, the so-called VCap polymers, are those that are in field use [210]. It has been shown that performance is improved with increasing lactam ring size [211]. It should be noted that even these VCap polymers have considerable technical limitations on their use in terms of their application to control hydrate formation across a wide range of temperature and pressure conditions [212].

Furthermore, these vinyl-lactam polymers exhibit poor rates of biodegradation; however a graft polyethylene derivative with vinylcaprolactam shows reasonable biodegradation rates [213].

Polymeric amide derivatives such as the 1-vinylpyrolidinone/α-olefin copolymers have been shown to be useful for asphaltene inhibition. The pyrrolidone group has a strong hydrogen bonding characteristic and has structural similarities to the pyrrole groups found in asphaltenes [2].

The related maleimide/α-olefin polymers exhibit wax inhibition characteristics [214]. These polymers are derived by using maleimides (Figure 6.88) as the key monomer in the grafting process instead of maleic anhydride.

All of the previously described examples of pyrrolidinones and related heterocycles are applied in polymer products, and such products are more fully detailed in Section 2.1.1.

1-Aminopyrrolidine (Figure 6.89) is related to hydrazine (see Section 5.9.4), being a cyclic dialkyl hydrazine. Like hydrazine it has good oxygen scavenging properties and therefore has been used in boiler water applications [215].

6.3.4 Pyridines and Related Heterocycles

Pyridine and the related quinoline (see Figure 6.90) derivatives have historically been used in a few oilfield applications, particularly as corrosion inhibitors; however their inherent toxicity

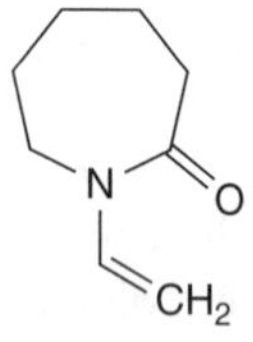

Figure 6.87 *N*-Vinylcaprolactam.

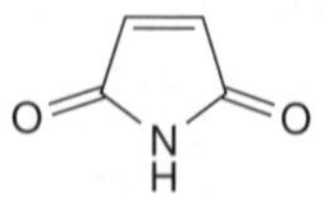

Figure 6.88 Maleimide.

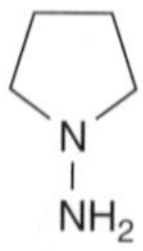

Figure 6.89 1-Aminopyrrolidine.

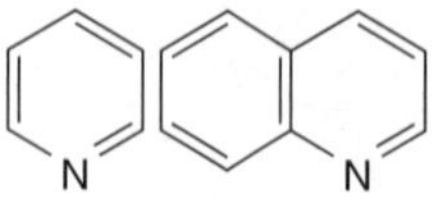

Figure 6.90 Pyridine and quinoline.

has made them less attractive, and their use has declined in the last two decades. In general, these compounds have been applied as quaternary ammonium salts and are particularly useful in corrosion mitigation in acidic media, such as 15% hydrochloric acid used in well stimulation, at temperatures up to 400 °F (>200 °C) [216].

Alkyl pyridinium and quinolinium salts in the presence of a sulphur-containing compound have been claimed to be more general corrosion inhibitors [217].

Mono and diesters of pyridinium and quinolinium compounds are claimed to be environmentally friendly FFCI [218].

The quarternised and polymerised pyridines and quinolones that are cationic polymers have been claimed to act as demulsifiers, flocculants, corrosion inhibitors and biocides [219].

The cationic surfactant cetylpyridinium salicylate has a dual function in that it can act as a drag-reducing agent as well as a corrosion inhibitor [220]. This is also a general property of many cationic surfactants that are FFCI, and such surfactants have large anions that form rod-like micelles and tend to exhibit drag-reducing capability above the critical micelle concentration [221].

Pyridines are good solvents (see Chapter 8) and are useful asphaltene dissolvers; [222] however their application can be problematic as they are not compatible with many of the commonly used polymers and elastomers found in the upstream oil and gas sector.

Pyridine is also a toxic agent in man although little is known on its environmental toxicology. Pyridine and quinolone are soluble in water, and pyridine will not volatise into the air quickly although this is pH dependent [223]. Due to its water solubility and low partitioning coefficients, pyridine and related derivatives are unlikely to bioaccumulate.

2,4,6-Trimethylpyridine or collidine (see Figure 6.91) when reacted with an aldehyde such as 1-dodecanal forms a useful oil or gas well inhibitor [224]. A similar product is also formed from the related 2,3,5,6-tetramethylpyrazine (see Section 6.3.5).

Pyrithione is the common name of the compound that exists as a pair of tautomers, as illustrated in Figure 6.92. The major form is the thione 1-hydroxy-2(1*H*)-pyridinethione and the minor form is the thiol 2-mercaptopyridine N-oxide. It is commercially available as either the neutral compound or its sodium salt. The zinc complex will be familiar to the reader being used to treat dandruff in a component of some shampoos. It is used as a biocidal agent in some specialist biocide formulations in the upstream oil and gas industry [225].

6.3.5 Azines: Pyrazines, Piperazines, Triazines and Related Heterocycles

As mentioned in the previous section, 2,3,5,6-tetramethylpyrazine (Figure 6.93) when reacted with aldehydes or ketones forms a useful oil or gas well inhibitor [224].

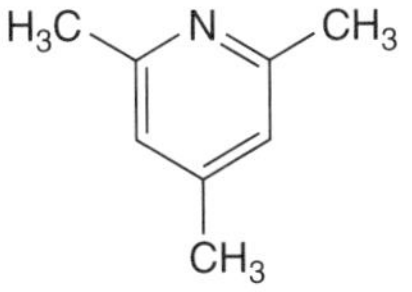

Figure 6.91 Collidine.

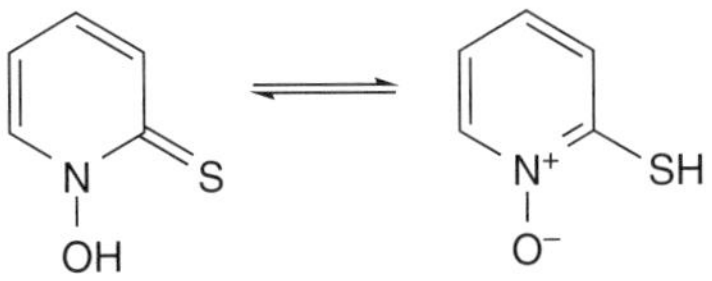

Figure 6.92 Pyrithione.

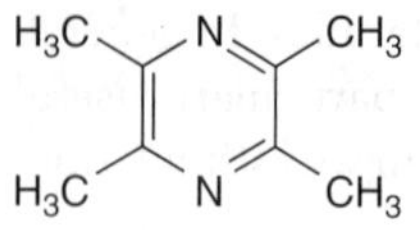

Figure 6.93 Tetramethylpyrazine.

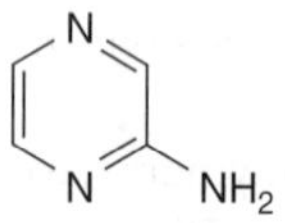

Figure 6.94 Aminopyrazine.

Amino-substituted pyrazines, such as aminopyrazine (Figure 6.94) in reaction with epoxides, such as the glycidyl ether of a mixture of alkanols having from 12 to 14 carbon atoms, can form useful corrosion inhibitors [226].

Pyrazine compounds appear to be little used in the upstream oil and gas industry with the examples above being some of the known few. They are far less toxic than pyridine compounds often being allowed as food additives [227]. Unlike the pyridines they are also readily found in nature in plants such as coffee and even as free agents in soil [228].

A few compounds based on the saturated derivative of pyrazine and piperazine are used in the upstream oil and gas industry, and it would appear exclusively in the production operations sector. Aminoethylpiperazines can be used to make dithiocarbamates that have been described earlier as highly effective flocculants (see Section 6.1.6). A triazine-type derivative with formaldehyde and the highly hydrophilic aminoethylpiperazine has been reported as a hydrogen sulphide scavenger that avoids solid deposits arising from elemental sulphur [229]. As will be described in the following paragraphs, triazine-type derivatives and more explicitly hexahydrotriazine derivatives are widely used as hydrogen sulphide scavengers in wet gas. Piperazinone (Figure 6.95) or an alkyl derivative such as 1,4-dimethylpiperazinone has been claimed as good hydrogen sulphide scavengers [230].

By far the most commonly used hydrogen sulphide scavengers are the triazine-based derivatives, particularly the hexahydrotriazines. Although there are a large number of potential molecules, only three are used in the oilfield sector. The most popular is hexahydro-1,3,5-tri(hydroxyethyl)-*s*-triazine (see Figure 6.96). The hexahydro-1,3,5-tri(methyl)-*s*-triazine (Figure 6.97) has also been used, and also more recently the hexahydro-1,3,5-tri (2-hydroxypropyl)-*s*-triazine has been promoted.

All are applied as an aerosol in the produced gas stream, where it is assumed that they react in the water droplets with the HS⁻ ion. It should be noted that all have a tendency to return spent reaction products to the water phase, increasing the pH and promoting scale deposition espe-

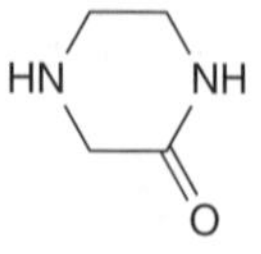

Figure 6.95 Piperazinone.

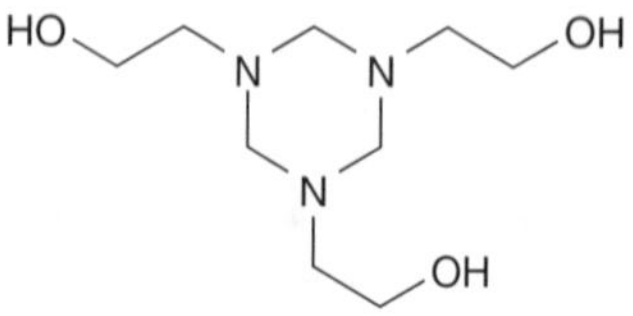

Figure 6.96 Hexahydro-1,3,5-tri(hydroxyethyl)-*s*-triazine.

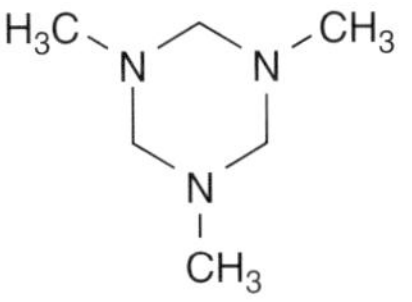

Figure 6.97 Hexahydro-1,3,5-tri(methyl)-*s*-triazine.

cially with respect to calcium carbonate. These 'triazines' also have problems with self-polymerisation and these deposits can lead to fouling issues. Despite these drawbacks the hexahydro-1,3,5-tri(hydroxyethyl)-*s*-triazine can often be the largest single selling chemical product used by a number of geographical regions including the North Sea basin.

1,3,5-Tris(hydroxyethyl)-*s*-triazine is the condensation product of MEA and formaldehyde or paraformaldehyde. It is supplied to the industry for formulation as a 75–8% w/w aqueous solution that is normally further diluted to around 40% w/w for application. The reaction is not stoichiometric and is dependent on a number of factors to ensure a reasonable reactivity and efficiency, particularly pH and temperature. The industry accepted standard is 10 kg of scavenger for every 1 kg of H_2S that can often lead to large volumes of the 'triazine' being deployed in order to achieve a specification of below 3 ppm hydrogen sulphide in the export gas stream.

Most if not all of these hexahydrotriazines exhibit antimicrobial properties mainly due to the nature of their aqueous degradation in releasing small amounts of formaldehyde at a steady rate over time [231]. However, some such as 2-(*tert*-butylamino)-4-chloro-6-(ethylamino)-*s*-triazine, commonly known as terbuthylazine (Figure 6.98), exhibit good biocidal activity in their own right and are marketed in a number of industrial sectors. Terbuthylazine is mainly marketed in the agricultural sector as a selective herbicide.

Melamine (Figure 6.99) derivatives, particularly sodium poly-melamine sulphate, may be used as dispersing agents for oilfield cement slurries. It is of note that this molecule is a fully aromatic triazine.

The majority of the triazines described and commonly applied are highly biodegradable by hydrolysis to their constituent amine and aldehyde building blocks; indeed under test conditions the commonly applied hexahydro-1,3,5-tri(hydroxyethyl)-*s*-triazine has been found under approved test conditions to be readily biodegradable and not expected to be bioaccumulating [232]. On the other hand, the highly functionalised *s*-triazines such as terbuthylazine have been found not to be particularly biodegradable and persistent in the soil environment [233].

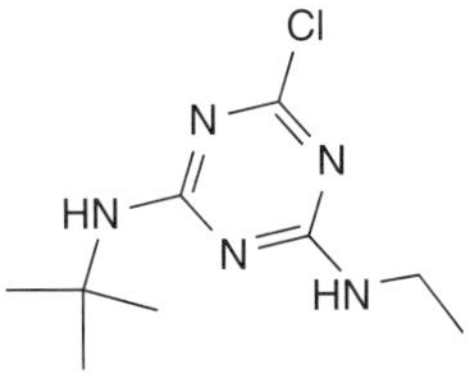

Figure 6.98 Terbuthylazine.

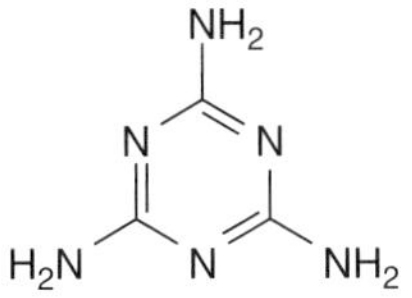

Figure 6.99 Melamine.

6.3.6 Hydantoins

The hydantoins are somewhat structurally related to the imidazoles being a fully saturated non-aromatic form (see Figure 6.100 for the structure of hydantoin).

As has been described in Section 5.1.1, the halogen-based ions hypochlorite and hypobromite and their associated hypochlorous and hypobromous acids are excellent oxidising biocides. Such biocides cause irreversible oxidation and hydrolysis of the proteins in the microbial cell and of the polysaccharides in the cell wall and those that are excreted to attach the microbe to surfaces, the so-called biofilm [234]. This action is almost universal across a broad spectrum of bacteria and other microorganisms and has the advantage of the microorganism not being able to form any sort of resistance strategy to the biocide as the action is very severe, which can be a distinct advantage of non-oxidising biocides.

Some bacteria are able to become resistant to some non-oxidising biocides. That is to say, some bacteria will have a higher tolerance of certain biocides than others and will survive a biocide dosage at certain concentrations and contact time. With other less tolerant bacteria succumbing (dying) under the same conditions, the surviving bacteria become a greater percentage of the bacterial population and will continue multiply under consistent biocide conditions (e.g. type of biocide, concentration and contact time). They may be encouraged to grow because of less competition for food and nutrients with the other bacteria dying off.

Of course, there are some distinct disadvantages in using oxidising biocides, and these have been described in the cited paper [235]. Many are highly corrosive, e.g. sodium chlorite. Many are difficult to handle and are unstable, e.g. bromine chloride. Others can be explosive under certain conditions, e.g. chlorine dioxide. The hydantoin derivative 1-bromo-3-chloro-5, 5-dimethylhydantoin (BCDMH) (see Figure 6.101) is a stable solid that is an excellent source of both chlorine and bromine. It reacts in a slow-release fashion with water to provide a controlled concentration of hypochlorous and hypobromous acid. It has been found to be more active than sodium hypochlorite at the same dosage, is more effective on biofilm removal, and is less corrosive [236]. Dichloro and dibromo derivatives are also commercially available and have been deployed as biocides.

The 5,5-dialkylhydantoins in combination with other species such as aldehydes and/or amines can form stable formulations that are claimed to be rapid, non-corrosive, environmentally friendly hydrogen sulphide scavengers [237]. However, to date the author knows of no commercial or field application of these formulated products.

Thiohydantoin (Figure 6.102) and its derivatives have been developed as both corrosion inhibitors and more applicably corrosion inhibitor synergists and have been shown to be syn-

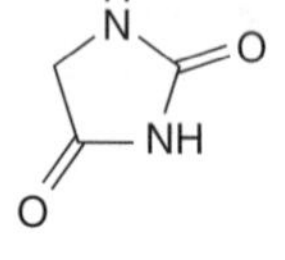

Figure 6.100 Hydantoin.

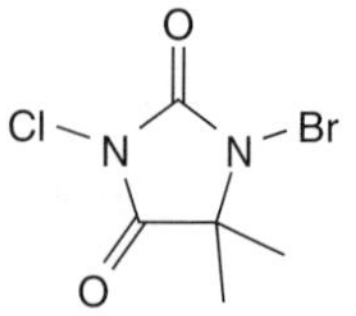

Figure 6.101 BCDMH.

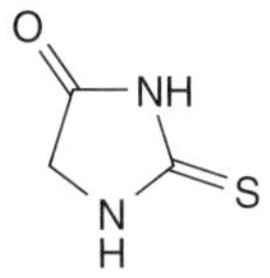

Figure 6.102 Thiohydantoin.

ergistic with a number of commonly applied corrosion inhibitors such as the alkyl quaternary ammonium salts, imidazolines and phosphate esters [238].

These materials have distinct advantages over other corrosion inhibitor synergists in that they are very effective at low dosages, do not cause or enhance pitting or stress cracking corrosion (as thiosulphate salts can be prone to do) [239], and are readily biodegradable.

The hydantoins and related molecules have very low environmental impacts and are low in toxicity and highly biodegradable [240].

6.3.7   The Environmental Impacts of *N*-Heterocycles

Throughout this section there has been an attempt to illustrate the variety of environmental toxicology and potential environmental impacts of the nitrogen-containing heterocyclic molecules. Some are inherently toxic such as pyridine, and others like the pyrazines are used as food flavourings. In general, compared with other low molecular weight molecules other than the imidazolines, there is relatively little use and application of products or formulations using these materials in the upstream oil and gas sector. Indeed, examples of the use of pyrimidines (see Figure 6.103) are scant and really only briefly mentioned in other work for comparison [241].

Pyrimidines are the building units of some critical biologically important molecules such as nucleic acids and polynucleotides (see Section 2.2). These have been ignored as a source of chemical function with the upstream oil and gas industry and may hold some promise in providing interesting chemicals directly for application and for chemical feedstocks for further synthesis and/or formulation. Furthermore, due to their natural product heritage, they are likely to be low on toxicity and environmental impact.

A number of *N*-heterocycles such as pyridine do have high levels of toxicity, and again as has been described with other low molecular weight compounds, it is a matter of balance between inherent toxicity, biodegradation and bioaccumulation that is important in their overall environmental impact. This makes those products and compounds derived from natural product sources even more interesting, some of which are further described in Section 6.6.

6.4   Sulphur and Oxygen Heterocycles

This section is concerned with heterocyclic molecules containing only sulphur, e.g. thiophenes, or only oxygen, e.g. furan. The use of such molecules in the oilfield sector is relatively rare; however a few examples are known and some are now described.

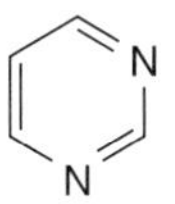

Figure 6.103 Pyrimidine.

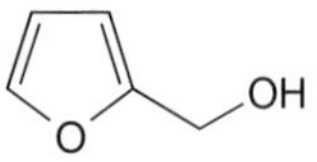

Figure 6.104 Furfuryl alcohol.

As described earlier in Section 6.2.2, the heterocyclic tri-thiones (Figure 6.55) have been claimed to be good corrosion inhibitors in high CO_2 environments [134] and where stress cracking-induced corrosion is a concern [135]. Other heterocycles containing only sulphur are rarely used, if at all, as additives to aid in oilfield operations. Some sulphur heterocycles such as thiophenes are found in certain crude oils and are usually derived from the formation of biogenic hydrogen sulphide [242].

The oxygen heterocycles are a little more common. Furfuryl alcohol (Figure 6.104), a derivative of furan, is used as an agent to cross-link polyacrylamides in situ in the oil and gas reservoir for water shut-off and control applications [127]. The use of furfuryl alcohol is as a replacement for more toxic phenol or formaldehyde cross-linkers. Such gels are also used in sand consolidation treatments [243].

Furan resins based on furfuryl alcohol have also been used in sand consolidation, and it is claimed that they have a greater interval of retreatment than conventional resin treatments [244]. These furan resins are also used as sealing compositions for cements that can be prone to corrosive attack from hydrogen sulphide and carbon dioxide gases at temperatures above 95 °C [245]. It is also claimed that hydrofurfuramide (Figure 6.105), generated in situ from furfuryl alcohol and ammonia, when added to plugging cement increases the corrosion resistance of the cement as well as reducing the water permeability [246].

Another plugging material based on a polymer from furfurylidene acetone (Figure 6.106) and silicone oligomers has also been described [246].

Coumarone or 2,3-benzofuran (Figure 6.107) is used as a constituent of dispersants in cementing technology; the related polycyclic molecule indene is also used.

As can be seen, there are relatively few examples of these types of heterocyclic molecule applied in the oilfield sector. The combinations that contains both sulphur and nitrogen,

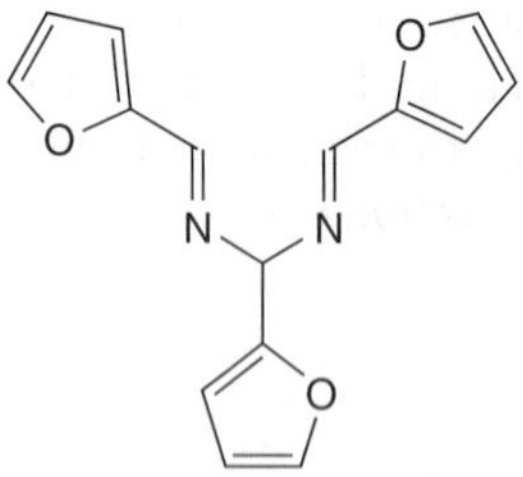

Figure 6.105 Hydrofurfuramide.

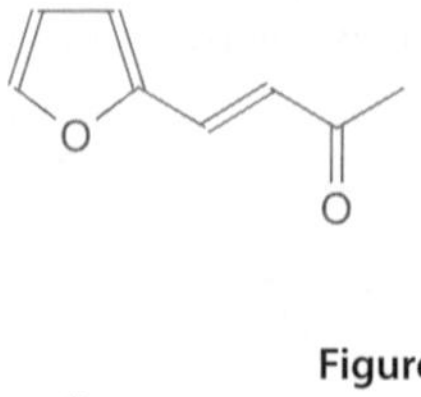

Figure 6.106 Furfurylidene acetone.

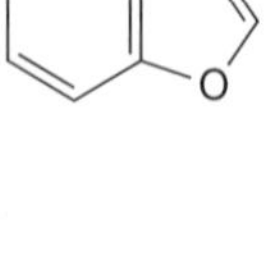

Figure 6.107 Coumarone.

nitrogen and oxygen and combinations thereof are more commonly applied (see following Section 6.5).

All of these types of molecules biodegrade to some extent although depending on conditions this can be relatively slow [247]. Thiophenes and its derivatives are in general not persistent in the environment and do not bioaccumulate, but they have some toxicity towards both aquatic and terrestrial plants and animals. This toxicity varies in line with persistency according to structure [248]. Although these molecules do not offer the best environmental profiles, it seems that their potential applications in the upstream oil and gas sector are somewhat unexplored.

6.5 Other Heterocycles

In this section, some examples of heterocyclic molecules containing combinations of nitrogen, sulphur and oxygen are explored. By far the most commonly used are the biocides based on thiazolines and isothiazolines.

6.5.1 Thiazoles and Related Products

Sulphur-containing biocides, in particular the benzothiazoles and the thiazolines (see next section), are applied to hydraulic fracturing fluids based on guar gum or other natural products [249].

When selecting such biocides, it is recommended in industry practice to consider the following physicochemical and toxicological aspects:

- Uncharged species will dominate in the aqueous phase and be subject to degradation and transport, whereas charged species will sorb to soils and be less bioavailable.
- Many biocides are short lived or degradable through abiotic and biotic processes, but some may transform into more toxic or persistent compounds.
- The understanding of the fate of these biocides under downhole conditions (high-pressure, temperature, and salt and organic matter concentrations) is limited.
- Several biocidal alternatives exist, but high cost, high energy demands and/or formation of disinfection by-products limits their use.

A popular additive is 2-mercaptobenzothiazole (Figure 6.108), which acts as a biocide to stabilise and prevent degradation of the fracturing fluid, i.e. reduction in its rheological properties at high temperatures and permits smoothly carrying out the fracturing operation [250].

The dilemma here is that 2-mercaptoimidazole is relatively stable to hydrolysis and therefore poorly biodegradable while being toxic and bioaccumulating [251]. However few other biocides, none of which are readily biodegradable, meet the temperature criteria required to perform in the fracturing fluid, without disturbing the rheology of the fluid.

The benzothiazoles are also useful corrosion inhibitors and due to their stability are applied in acidic conditions. Aminobenzothiazoles (see Figure 6.109) are preferred as the positive charge on the amine at low pH is believed to contribute to the overall corrosion inhibitor performance [252].

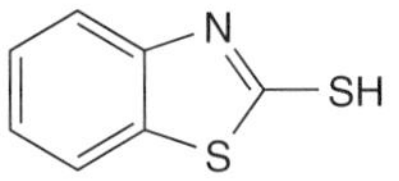

Figure 6.108 2-Mercaptobenzothiazole.

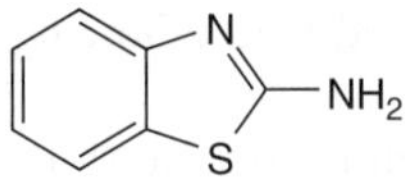

Figure 6.109 2-Aminobenzothiazole.

The benzothiazoles are most commonly applied as corrosion inhibitors for yellow metals such as copper; however, alkoxylated benzothiazoles derived from mercaptobenzothiazoles are claimed to be useful corrosion inhibitors for ferrous metals [2].

Thiadiazole compounds, in particular 2,5-bis(4-dimethylaminophenyl)-1,3,4-thiadiazole (DAPT), have also been found to be good cathodic corrosion inhibitors for mild steel in acidic conditions [253], as is also the case for the equivalent oxadiazole (DAPO) (see Figure 6.110).

A heterocyclic thione in this type of product class is the non-aromatic 3,5-methyl-1,3,5-thia-diazine-2-thione (dazomet) (see Figure 6.56), which is shown in Section 6.2.2, is well-known biocidal agent and is used in a number of industrial sectors, e.g. wood preservation and water treatment, as well as in oilfield applications [254].

6.5.2 Thiazolines, Isothiazolinones and Related Products

Thiazolines (or dihydrothiazoles) are a group of isomeric heterocyclic compounds containing both sulphur and nitrogen in the ring. Although unsubstituted thiazolines are rarely encountered themselves, their derivatives are more common, and some have biocidal properties. This is most common in the group known as the isothiazolinones (see Figure 6.111), which includes methyli-sothiazolinone (MIT), chloromethyisothiazolinone (CMIT) and benzisothiazolinone (BIT).

These isothiazolinones, particularly CMIT and MIT, are widely used for microbial control in a variety of water treatment applications. They are broad-spectrum biocides with good biodeg-radation rates and are effective against sessile bacteria [255]. Despite what seem to be good characteristics for oilfield application, they are similar to aldehydes, such as glutaraldehyde and THPS.

A chloromethyl-methylisothiazolone (CMIT/MIT) combination biocide along with slug doses of glutaraldehyde has been shown to control microbially influenced corrosion and bio-fouling by SRB [256]. However other evidence suggests that their biocidal performance is reduced by hydrogen sulphide produced by SRBs [257]. They are, however, very effective bio-stats, which prevent hydrogen sulphide from forming by ensuring SRB populations are con-

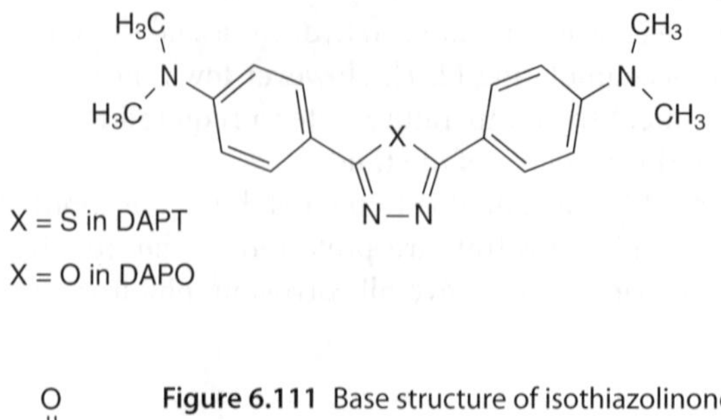

X = S in DAPT

X = O in DAPO

Figure 6.110 Thiadiazole and oxadiazole structure.

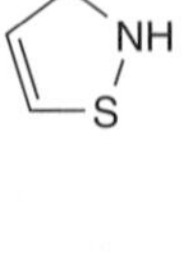

Figure 6.111 Base structure of isothiazolinones.

trolled at a low level. Therefore, they are often used in combination with other biocides [256, 258].

Other dichloro and dibromo derivatives of alkyl isothiazolinones have been claimed in the patent literature and have been proposed for oilfield use [254].

The BITs are less water soluble, and benzothiazolinone that is isomeric with 2-hydroxybenzothiazole (Figure 6.112) has been described as a biocide for drilling fluid preservation [259].

The highly alkylated 2,5-dihydrothiazoles are claimed to exhibit good corrosion control when applied as volatile inhibitors in gas wells; [260] however, they are not very water soluble and therefore of limited application. They can, however, be converted to the water-soluble thiazolines by reaction with formic acid and an aldehyde. These thiazolidines (see Figure 6.113) are claimed as FFCI [261].

Somewhat similar products, the related dithiazines, are also proposed as corrosion inhibitors for oil and gas wells [262]. These are usually alkylated or esterified with a straight chain of branched alkyl group or alkyl aryl grouping. They can also be quaternised by forming a salt at the available nitrogen of the six-membered ring system. The parent dithiazine in these products is isolated from the reaction of hydrogen sulphide with hexahydro-1,3,5-tri(hydroxyethyl)-*s*-triazine (Figure 6.96) (see Section 6.3.5) [263]. This reaction is the main reaction in the scavenging process described earlier in Section 6.3.5 (see reaction scheme in Figure 6.114).

6.5.3 Oxazoles, Oxazolidines and Related Heterocycles

Oxazole (Figure 6.115) is the parent compound for a large class of heterocyclic aromatic organic compounds. These are aromatic compounds but less so than thiazoles.

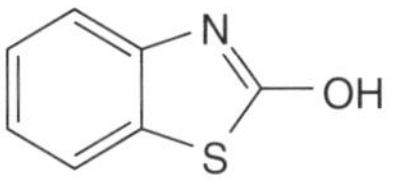

Figure 6.112 2-Hydroxybenzothiazole.

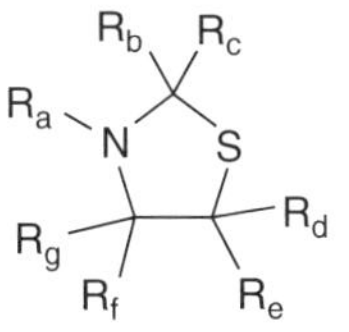

Figure 6.113 Alkylated thiazolidines.

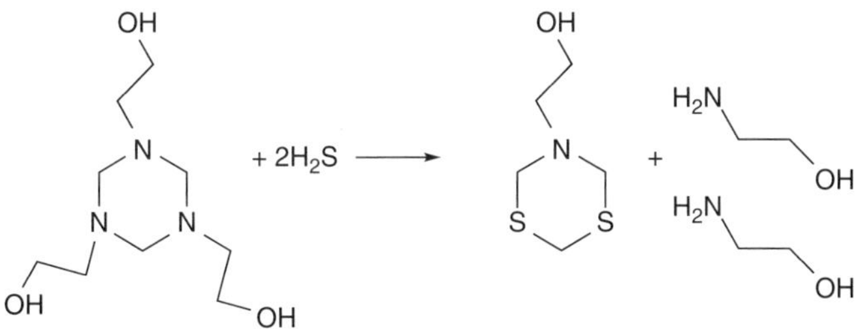

Figure 6.114 Reaction of H_2S with hexahydrotriazine to form dithiazine and monoethanolamine.

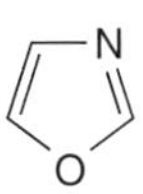

Figure 6.115 Oxazole.

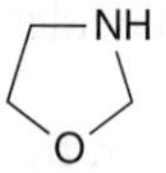

Figure 6.116 Oxazolidine.

An oxazolidine is a five-membered ring compound as illustrated in Figure 6.116. In oxazolidine derivatives, there is always a carbon between the oxygen and the nitrogen, and if the nitrogen and oxygen are adjacent, then it is isoxazolidine. All of the carbons in oxazolidines are reduced compared with oxazole.

There is no recoded use of oxazole derivatives in the upstream oil and gas industry; however some oxazolidines such as 4,4′-dimethyloxazolidine (DMO) (Figure 6.117) are used as biocides for treatment of production fluids [264] and in fracturing fluids in shale gas production [265]. Generally DMO is used in conjunction with or blended with glutaraldehyde for application.

This oxazolidine as well as other related derivatives such as methylene bis-oxazolidine has a broad spectrum of biocidal activity [266]. Bis-oxazolidines are chemical compounds that contain two oxazolidine rings bridged by usually a methylene group as in *N,N′*-methylene-bis-oxazolidine derivative shown in Figure 6.118.

Some derivatives of these bis-oxazolidines have been explored for use as asphaltene stabilising agents [267]. The compound illustrated in Figure 6.118 also shows activity as a hydrogen sulphide scavenger [268] and as a corrosion inhibitor (H.A. Craddock, unpublished results).

The oxadiazole illustrated in Figure 6.110, as stated earlier, also has corrosion inhibitor properties. Finally, the related molecule 1,3-oxazin-6-one (Figure 6.119) when derivatised with *N*-alkyl, *N*-alkenyl or *N*-aryl groups have been claimed as FFCI [269].

6.5.4 Environmental Effects and Impacts

The heterocyclic compounds examined in this section exhibit a wide variety of physical properties and environmental impacts, as they show a variety of stabilities and toxicities, often due to whether or not displaying aromaticity.

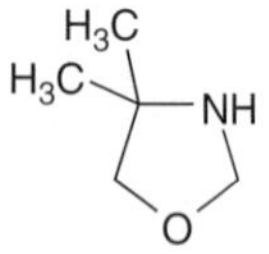

Figure 6.117 4,4′-Dimethyloxazolidine.

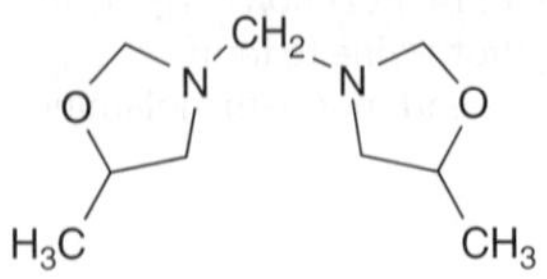

Figure 6.118 *N,N′*-Methylenebis-oxazolidine.

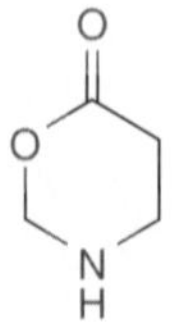

Figure 6.119 1,3-Oxazin-6-one.

The thiazoles and benzothiazoles display high stabilities that allow them to be used under acidic conditions at high temperatures and pressures as corrosion inhibitors. The biodegradation of benzothiazoles by pure and mixed microbial cultures derived from activated sludge has been studied [270]. The degradation of 2-aminobenzothiazole resulted in high yields of ammonia and sulphate (87% and 100 %, respectively, of the theoretical yield). Evidence suggests that the benzothiazole moiety degrades via the meta-cleavage pathway. However, 2-mercaptobenzothiazole was shown to be unable to act as a growth substrate for any of the cultures studied, but some could cause some biotransformation, suggesting that the mercaptobenzothiazole is more biologically stable. A number of thiazoles and benzothiazole-based herbicides have been studied for persistency in soils and found to be persistent [271].

The environmental toxicity of thiazoles and benzothiazoles is sparsely studied, and their overall environmental impact is somewhat unknown.

Isothiazolines, which are probably the most widely type of product used in the oil and gas industry from this group of heterocycles, have a high aquatic toxicity [272]. This is unsurprising given their biocidal and fungicidal properties; however they are also shown to be persistent in many species of fish. However, isothiazolinones do exhibit good rates of biodegradation, as stated earlier [255].

Bis-oxazolidine rings hydrolyse in the presence of moisture to give amine and hydroxyl groups, which can undergo further hydrolysis and biodegradation. There are, however, no definite studies on biodegradation rates.

DMO is very soluble in water and, when introduced, will have a tendency to remain in water. It has minimal tendency to bind to soil or sediment and is unlikely to persist in the environment. DMO is readily biodegradable and susceptible to rapid hydrolysis. The compound will biodegrade in water and soil and will not persist in the environment. Formaldehyde, a hydrolysis product of DMO, as has been earlier described is also susceptible to rapid biodegradation (see Section 6.1.4). DMO is not likely to accumulate in the food chain (its bioconcentration potential is low). It is, however, toxic to aquatic organisms, particularly algae, on an acute basis [273].

Overall this group of mixed heterocyclic compounds has a variety of environmental impacts very much dependent on their stability, water solubility and rate of hydrolysis.

6.6 Natural Products and Biological Molecules

In this section, low molecular weight molecules of natural origin or synthesised to mimic natural products or have been derived from natural products are described as to their use and potential environmental impacts. Many natural products or related derivatives are polymeric in nature, which have been detailed in Section 2.2. As described before with natural and biopolymers, the use of natural products offers the oilfield chemist the possibility of improved environmental acceptability without compromising efficacy.

6.6.1 Saccharides and Related Materials

A number of sugar and sugar-like derivatives are used in the upstream oil and gas industry. Many are polymers and are covered in Chapter 2; however a smaller proportion are applied as monomers or as identifiable entities in their own right.

Alkyl glycosides with C_8–C_{18} alkyl groups on a glucose or fructose ring have been considered for hydrate control in oil and gas transportation pipelines [274]. The derivation of methyl glucoside (Figure 6.120) with ethylene or propylene oxide results in a clay stabiliser that is added

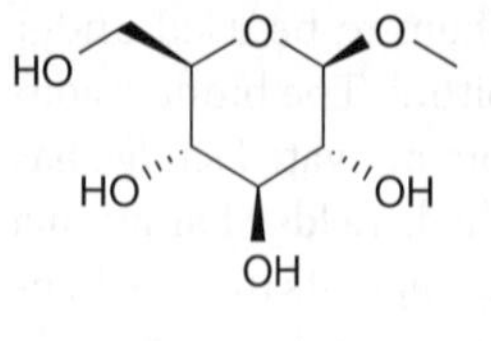

Figure 6.120 Methyl glucoside.

to drilling fluids. Such products are water soluble at ambient temperature but become insoluble in water at elevated temperatures. This means that these products concentrate at important surfaces such as the drill bit cutting surface, the borehole surface and the drill cuttings themselves.

The aldose group of antioxidants can be used as corrosion inhibitors in combination with a thiocyanate salt. These are applied to calcium-free drilling, completion and workover fluids in carbonate- or sulphate-containing wells [275]. This group of aldoses includes arabinose, gluconic acid, ascorbic and isoascorbic acids (Figure 6.121). These molecules can also be combined with thiol-group bearing molecules such as ammonium thioglycolate with which they act synergistically to give another type of corrosion inhibitor [276].

A number of polyalcohols such as gluconic acid (Figure 6.121) and others such as arabitol and sorbitol (Figure 6.122) are substrates for cross-linked gels, usually complexed with zirconium compounds that are used as rheological modifiers and suspension agents in fracturing fluids [277].

The polyols are fully described in Section 2.1.2. The zirconium complexes are further detailed in Section 5.9.1.11.

The polyol dextrin (Figure 6.123) can be used as part of an additive formulation to water-based drilling fluids that can provide a semipermeable membrane in specific shale formations.

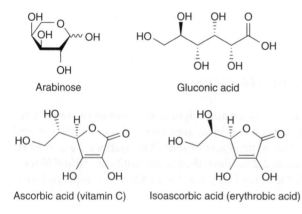

Arabinose

Gluconic acid

Figure 6.121 The aldose antioxidants.

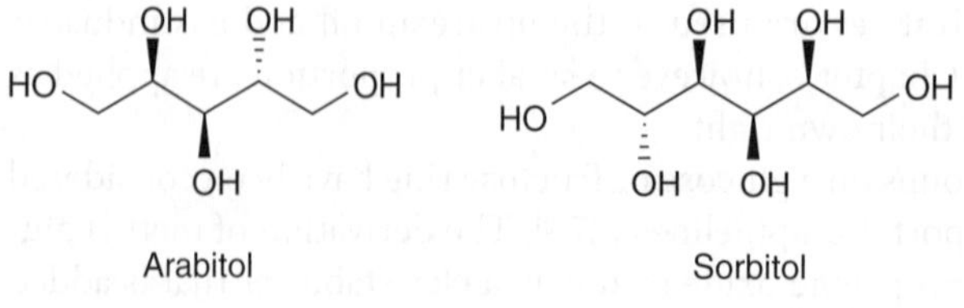

Ascorbic acid (vitamin C)   Isoascorbic acid (erythrobic acid)

Arabitol

Sorbitol

Figure 6.122 Polyols for gel complex formation.

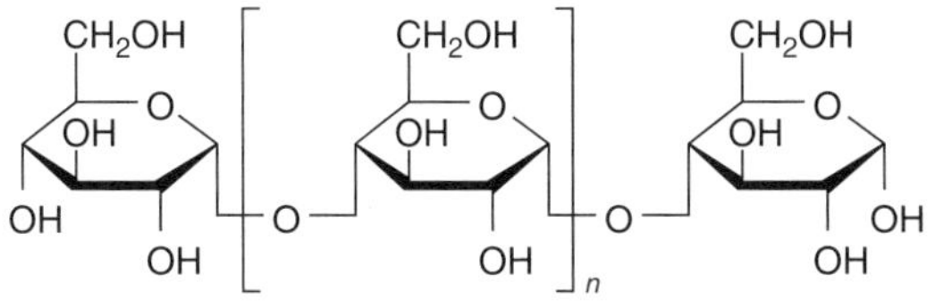

Figure 6.123 Dextrin.

This membrane allows the free movement of water through the shale but restricts ion flow across the membrane, thus ensuring stability of the shale and the wellbore [278].

Although polymeric in nature, dextrins are low molecular weight carbohydrates produced by the hydrolysis of starch or glycogen. Traditionally starches and modified starches such as dextrin have been used as an additive in aqueous drilling fluids for improved dispersion and high temperature stability [279]. However their applicability in oilfield use has been sparsely explored.

Diglycidal compounds, such as diglycidal ether (Figure 6.124), are used in the construction of polymers for demulsifiers [280].

Carboxyl-containing fructans, in particular carboxymethyl inulin (Figure 6.125), have been shown to be capable scale inhibitors of calcium, barium and strontium sulphate and carbonate salts [281].These products have been preliminarily investigated as to their potential for corrosion inhibition and show some capability of protection from corrosion under mild conditions (H.A. Craddock, unpublished results). They possess excellent ecotoxicological properties being derived from sugars extracted from chicory roots and used mainly in food applications [282]. They are worthy of further investigation as corrosion inhibitors, especially in formulated products and with known synergists.

Other commonly used saccharide derivatives are in the main polymeric materials such as starches that are derived from amylose and amylopectin and cellulose derivatives. These have been extensively discussed in Section 2.2.

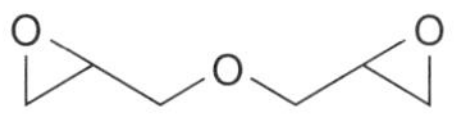

Figure 6.124 Diglycidal ether.

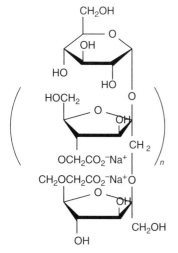

Figure 6.125 Carboxymethyl innulin.

6.6.2 Amino Acids and Proteins

Amino acids have been relatively little explored as oilfield additives, although in the last two decades there has been more activity in this area.

It had been noted some time ago that aspartic acid (see Figure 6.126) has some reasonable corrosion inhibitor properties [283], and it was later found that polypeptide derivatives gave good CO_2 corrosion protection. The polymeric derivatives, polyaspartates, are fully detailed as to their oilfield use in Section 2.2.1.

A derivative of aspartic acid, *S*-aspartic acid-*N*-monoacid, has been claimed as a chelating agent and scale dissolver for barium and calcium sulphate [284], and also the derivative *N*-(2-aminoethyl) aspartic acid has similar properties [2]. Glutamic acid (see Figure 6.74), a related compound, has already been described earlier in Section 6.2.4 and is suitably derivatised to give glutamic acid *N,N*-diacetic acid (GDLA), a highly effective and biodegradable chelating agent. The amino acid sarcosine or methylglycine (Figure 6.127), again when derivatised to methylglycine *N,N*-diacetic acid, is also claimed as an effective chelating agent for barium and calcium sulphate [178].

A more complex derivative of glutamic acid pteroyl-L-glutamic acid is a type of water-soluble vitamin, folic acid (Figure 6.128), that widely exists in nature and has an excellent environmental profile, is non-toxic, and has a high rate of biodegradation. Static and dynamic studies have shown that folic acid could be a potential scale inhibitor for produced water with high scaling tendency and has potential for continuous injection into the wellbore below bubble point region for controlling carbonate scale [285].

The sulphur-containing amino acids cysteine and cystine (formed from two molecules of cysteine) and their decarboxylated analogues cysteamine and cystamine (see Figure 6.129) are good corrosion inhibitor synergists for polyamino acid-based FFCI [286], such as the polyaspartates as described in Section 2.2.1.

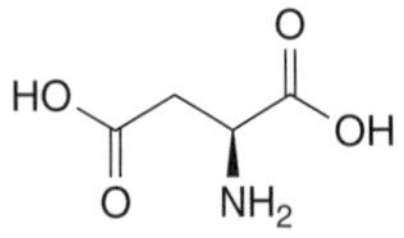

Figure 6.126 Aspartic acid.

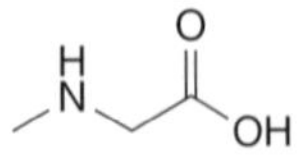

Figure 6.127 Sarcosine or methylglycine.

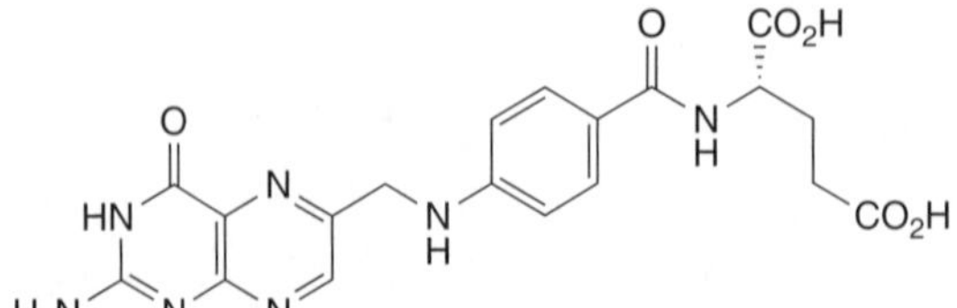

Figure 6.128 Folic acid.

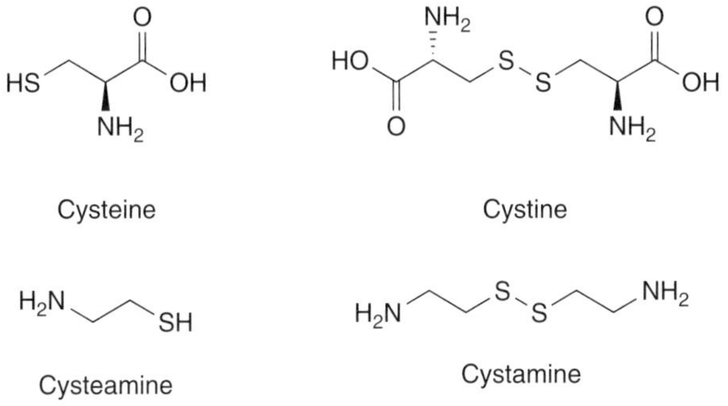

Figure 6.129 Sulphur amino acids and derivatives.

Cysteine

Cystine

Cysteamine

Cystamine

These amino acids and decarboxylated derivatives have also been used as iron control additives in acid stimulation treatments [131]. Here they act as reducing agents to ensure that the presence of Fe^{3+} ions is minimised, as during such treatments the presence of these ions leads to the formation of colloidal ferric hydroxide that can cause formation damage.

As described earlier in Section 6.2.4, some derivatives of methionine (Figure 6.130) exhibit corrosion inhibition properties [184].

The amino acid tyrosine (Figure 6.131) and its methyl ester have been considered as additives for hydrate control in transportation pipelines [287].

As discussed in Section 2.2.1, a few protein and protein-like polymers are used in the upstream oil and gas industry, such as extracted glycoproteins from soya and other plant material; however the use of such compounds is rare.

The milk protein casein (Figure 6.132), which has a relatively low molecular weight at 784, has been claimed as a biodegradable dispersing and fluid loss agent for well cements [288].

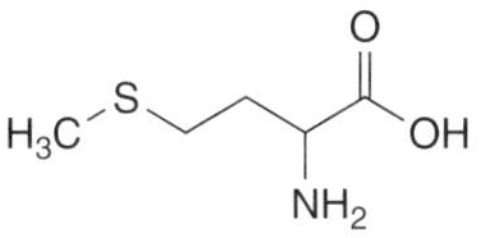

Figure 6.130 Methionine.

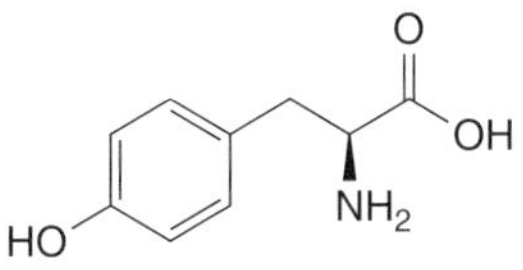

Figure 6.131 Tyrosine.

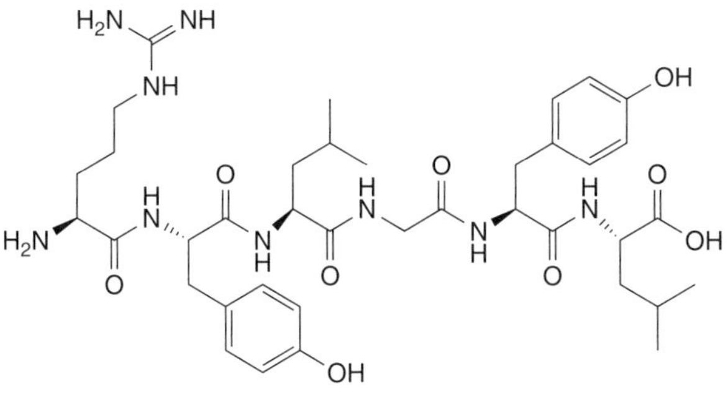

Figure 6.132 Casein.

6.6.3 Fats, Oils, Lipids and Other Natural Products

The section explores the use of a variety of relatively low molecular weight non-polymeric products that are either used directly or derived from natural sources.

The lipids are a group of naturally occurring molecules that include fats, waxes, sterols, fat-soluble vitamins (such as vitamins A, D, E and K), monoglycerides, diglycerides, triglycerides, phospholipids and others. A number of these compounds and related products have been described and discussed in Chapter 3.

Tall oil, a mixture of resin acids and fatty acids, in particular the conjugated C18 fatty acids, is widely used as a feedstock in a number of corrosion inhibitor products. It has the advantage of being a waste product from paper and pulp processing, and therefore its use has considerable environmental advantage in its reuse as a useful product.

Such fatty acids when converted to amides by condensation with ethoxylated or propoxylated alkylphenol amines or alkoxylated amines provide useful corrosion inhibitors against both sweet and sour types of corrosion [68, 289].

As has been described in Chapter 3, the fatty acids and their related esters, alcohols and amines, are the basis of a wide range of oleochemicals that are inherently biodegradable and non-toxic. However, their toxicity profile can change depending on their derivation, and their surfactant nature makes assessing their potential to bioaccumulate problematic.

The fatty alcohols can be condensed with glucose to give the alkyl glucoside (see Figure 3.54). These alkyl glucosides have been little explored for oil and gas application in their own right although the polymeric forms find application in wellbore cleaners [290] and corrosion inhibition [291]. Decyl glucoside (Figure 6.133) has corrosion inhibitor properties and is applied in other industries [292]. Decyl glucoside is a popular ingredient in detergents and cleansers as it is a non-ionic surfactant that is derived from natural sources such as corn and coconuts. It is non-toxic and is highly biodegradable.

As described in detail in Chapter 3, fatty alcohols, fatty esters and fatty amines all act as surfactants and are used in a variety of oilfield applications.

The ester compounds are of interest as biodegradable alternatives in drilling fluids composition, in particular as lubricant components. A variety of materials derived from vegetable oils and related feedstocks have been considered, including a glycerol waste stream [293] that lends a sustainable theme to the use of such products.

The complex molecule rotenone occurs naturally in the roots, seeds and stems of several plants and was the first described member of the family of chemical compounds known as rotenoids. It is an isoflavone having the structure shown in Figure 6.134 and used as a broad-spectrum insecticide, piscicide (fish poisoning agent) and pesticide.

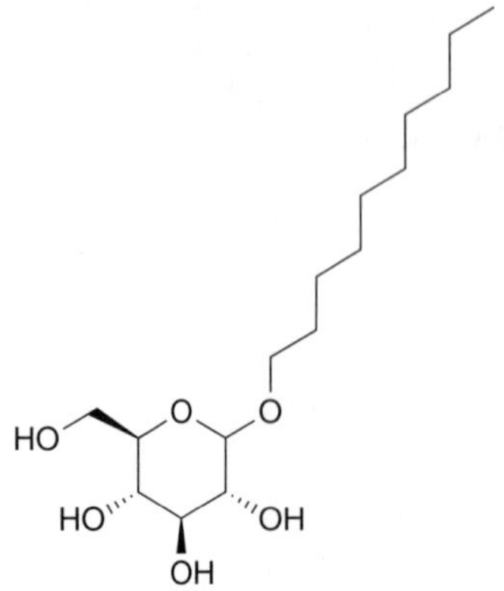

Figure 6.133 Decyl glucoside.

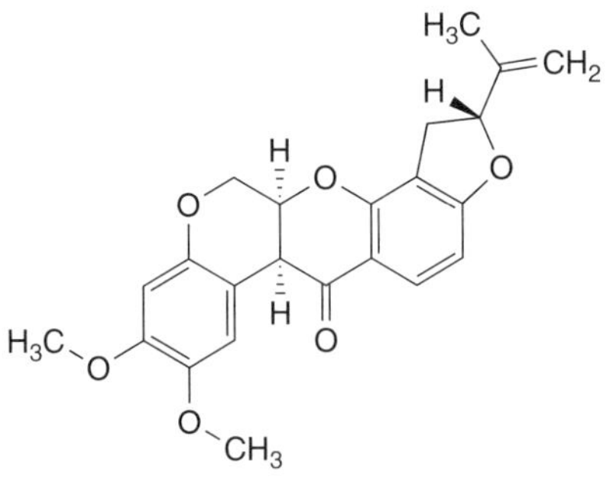

Figure 6.134 Rotenone.

It is claimed to have an effect on SRB by interfering with the electron transport chain in their mitochondria. The overall effect is a reduction in the production of hydrogen sulphide and related waste products and consequential effects on the reduction of corrosion [294]. The flavonoids are related to the tannins described in the following paragraphs.

Tannins are polyphenolic biomolecules that bind to and precipitate proteins and various other organic compounds including amino acids and alkaloids. The tannin compounds are widely distributed in many species of plants, where they play a role in protection from predation, and perhaps also as pesticides, and in plant growth regulation. The astringency from the tannins is what causes the dry and puckered feeling in the mouth following the consumption of unripe fruit or red wine or tea. Tannins have molecular weights ranging from 500 to over 3000 and are complex in structure (see Figure 6.135 [295]).

Although unpleasant to taste, they are non-toxic and are biodegradable; indeed they may have therapeutic effects. They have been little used in the oil and gas industry although they can be applied as a deoiler in water clarification (H.A. Craddock, unpublished results). They have also been examined as a deflocculating agent in water-based drilling fluids that have been contaminated with calcium sulphate [296]. Yet again a class of biological molecule that appears to be under-exploited in potential application in the upstream oil and gas sector.

Finally, in the section with the classification of natural products, the terpenes and their derivatives, the terpenoids, are considered.

Terpenes are a large and diverse class of organic compounds produced by a variety of plants, particularly the conifers. They often have a strong odour and may protect the plants that produce them by deterring herbivores and by attracting predators and parasites of herbivores. They are the major components of resin and of turpentine produced from resin, and the name 'terpene' is derived from the word 'turpentine'. Terpenes and terpenoids are also the primary constituents of the essential oils of many types of plants and flowers.

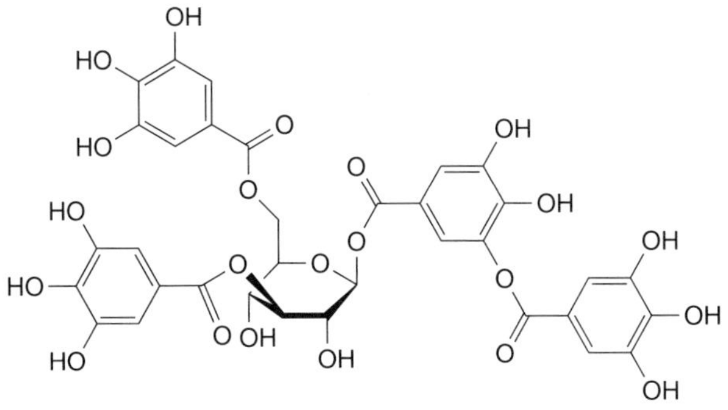

Figure 6.135 Tannic acid.

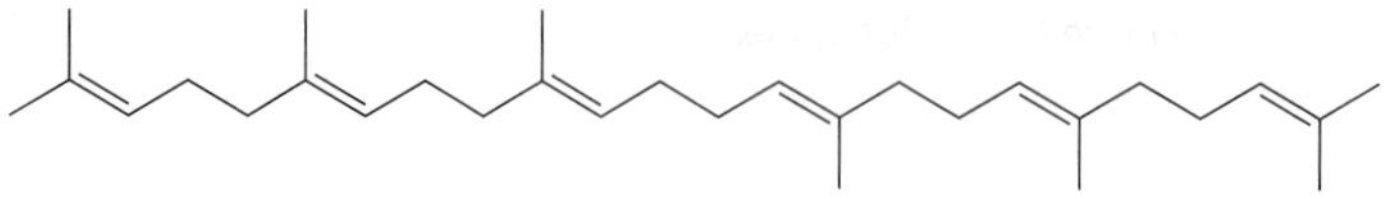

Figure 6.136 Squalene.

In addition to their roles as end products in many organisms, terpenes are major biosynthetic building blocks within nearly every living creature. Steroids for example are derivatives of the triterpene, squalene (Figure 6.136).

As can be seen this type of molecule is clearly constituted from units of isoprene (see Section 2.2.4.1).

When terpenes are modified chemically, such as by oxidation or rearrangement of the carbon skeleton, the resulting compounds are generally referred to as terpenoids and are also known as isoprenoids as derived from isoprene. Vitamin A (Figure 6.137) is a terpenoid, and the similarities in structural units between it and squalene are obvious.

In the upstream oil and gas sector, the simpler terpenes such as D-limonene (Figure 6.138) are utilised primarily for the dissolution of wax. They can be considered also as solvents; however, in the authors' view they appear to behave in a more complex way than just providing solvency particularly with respect to petroleum wax dissolution [297, 298].

This is supported by the fact that it has been considered as a pour point depressant in a biodegradable drilling fluid [299].

Terpene compounds have been claimed to act as corrosion inhibitor intensifiers, particularly with products such as quaternary ammonium salts [300]. An aqueous organic acid composition containing a terpene as corrosion inhibitor intensifier is especially claimed to be suitable for use in acidising subterranean formations and wellbores. The composition substantially claims to reduce the corrosive effects of the acidic solution on metals in contact with the acidic solution. Recent work has shown that terpene alkoxylates in combination with alkyl polyglucosides can afford a surfactant-based corrosion inhibitor [301].

Cyclic monoterpenes have been used as anti-sludging agents or dispersants in the recovery of sludging crudes during acid stimulation [302].

In general, other applications describe the terpenes as solvents, and this is further discussed in Chapter 8, where due to their high rates of biodegradation and natural product heritage the terpenes are used as an environmentally acceptable solvent. Overall this class of molecules has been relatively unexplored as to its potential applications in the upstream oil and gas industry. It offers many variations in structure, all based on the basic isoprene unit. These are natural

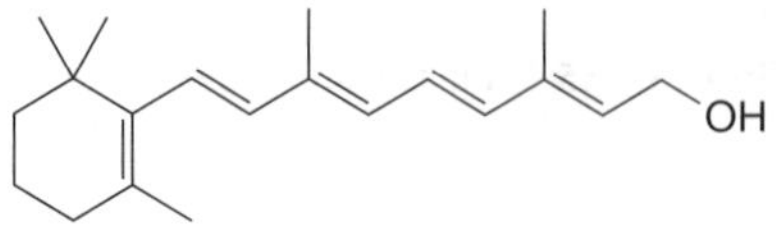

Figure 6.137 Vitamin A (retinol).

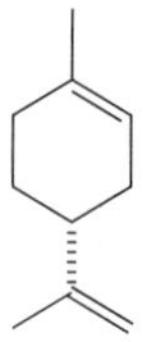

Figure 6.138 D-Limonene.

products that could easily be further derived and chemically manipulated to give improved effectiveness and offer good environmental properties.

6.7 Other Non-classified Products

There are a number of low molecular weight molecules used as oilfield additives that do not fit into any of the categories so far described, and therefore the more important examples are now illustrated.

Many of these compounds are used as biocides or biostatic agents. Bromo-2-nitropropane and associated derivatives, such as 2-bromo-3-nitro-1,3-propanediol, generically known as Bronopol, and the commonly used 2,2-dibromo-3-nitrilopropionamide (DBNPA) are all used in the oilfield sector (Figure 6.139).

Bronopol and DNPA are used in fracturing fluids used for unconventional shale gas development. In this application, such biocides are a preservative against microbial degradation of the guar and other natural viscosifiers and to control and kill bacterial population that can use such viscosifiers as growth media. These bacteria may cause the production of biofilms resulting in inhibiting gas extraction, produce toxic hydrogen sulphide and induce corrosion, leading to downhole equipment failure. [230] They have also been used in the composition of effervescent biocides that have been found to be a useful delivery method in a number of oilfield applications [225]. In general oilfield operations, although Bronopol has been used, DBNPA is preferred as it is pH sensitive and hydrolyses quickly in both acidic and alkaline conditions to ammonia and bromide ions that allows quick kill rates. It has a drawback in that it is poorly water soluble; however methods have been proposed to circumvent this by formulating it as a suspension [303]. It has also been claimed to act synergistically with oxidising and non-oxidising biocides [304]. A related molecule, 1-bromo-1-(bromomethyl)-1,3-propanedicarbonitrile, commonly known as methyldibromo glutaronitrile (MDBGN) (see Figure 6.140), has primarily been used as a preservative in skin care products such as lotions, shampoos and liquid soaps, and it has been discovered to have skin sensitisation problems. It has been claimed for oilfield use; [305] however there is no recorded incidence of its application.

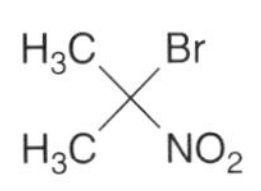

Figure 6.139 Propane-based biocides.

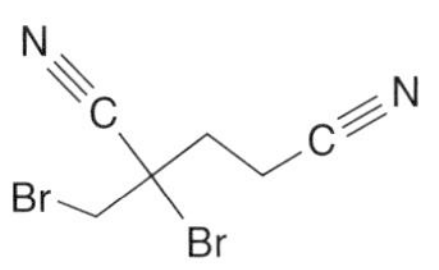

Bromo-2-nitropropane

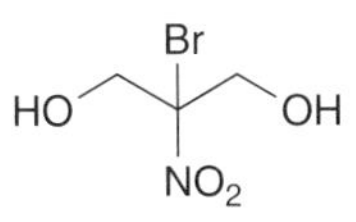

2-Bromo-3-nitro-1,3-propanediol (bronopol)

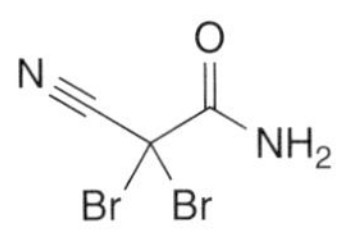

2,2-Dibromo-3-nitrilopropionamide (DBNPA)

Figure 6.140 1-Bromo-1-(bromomethyl)-1,3-propanedicarbonitrile.

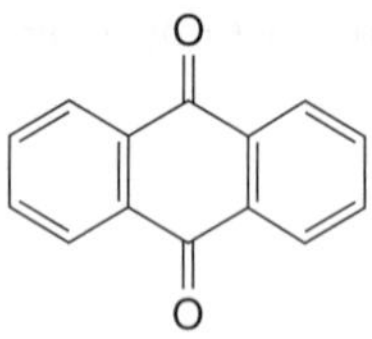

Figure 6.141 Anthraquinone.

The molecule anthraquinone (Figure 6.141) has been applied as a biostat in oilfield operations for over 20 years. It acts by inhibiting the sulphate respiration of the SRB, effectively shutting down the production of hydrogen sulphide [306].

Anthraquinone is not water soluble, but the disodium salt of the anthracenediol is and behaves as if it were anthraquinone [307]. A simpler benzoquinone, *para*-benzoquinone (Figure 6.142), is reported as being good hydrogen sulphide scavenger, particularly in alkaline pH [308]. Other benzoquinone derivatives are claimed to be useful as multifunctional scavengers of sulphides, mercaptans and other species from hydrocarbon fluids [309].

The simple aromatic molecules of benzene, toluene and xylene are mainly used as solvents for the formulation of oil-soluble products (see Chapter 8); however on occasion they are applied for a specific effect. Toluene as a solvent for asphaltenes also has some dispersant properties [310]. Xylene, a mixture of dimethyl benzene isomers, is probably the most common asphaltene solvent/dissolver; [311] toluene is also used but is less effective [312]. These aromatic solvents interact with the π–π orbitals of the asphaltene aromatic components, interfering with and replacing the asphaltene–asphaltene π–π interactions, thereby solubilising the agglomerated complex. Some studies have shown that the bicyclic aromatics such as tetralin and 1-methylnaphthalene (see Figure 6.143) perform better than the monocyclic aromatics [313].

Both toluene and xylene are commonly used solvents in the dissolution of wax deposits, particularly xylene and often as hot xylene [314, 315]. The critical issue with these aromatic solvents is that they are classified as marine pollutants [316], and therefore their use is controlled and often prohibited in certain oil- and gas-producing regions.

There are a number of other more exotic molecules that have been consider and are used as specialist dissolvers. In the previous section, terpenes have already been detailed as wax dissolvers. Pyridines are good asphaltene dissolvers; however they are rarely used as they are toxic

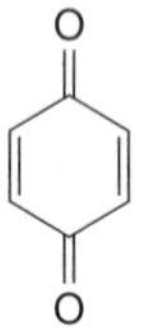

Figure 6.142 *para*-Benzoquinone.

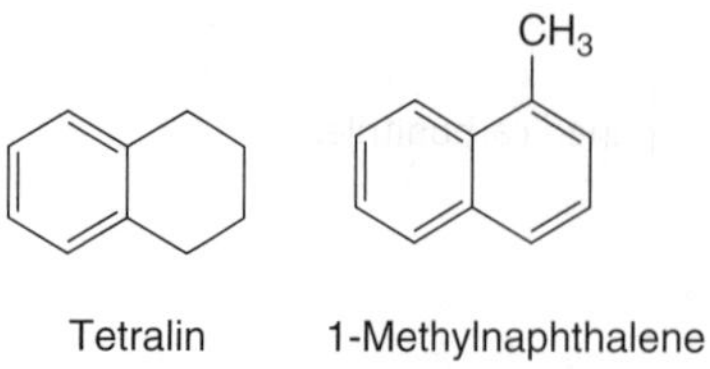

Tetralin   1-Methylnaphthalene

Figure 6.143 Bicyclic aromatics.

and are in general incompatible with most commonly used oilfield elastomers [317]. However certain aromatic esters in particular isopropyl benzoate (Figure 6.144) are good asphaltene dissolvers and also are relatively of low toxicity and environmentally acceptable [318].

t-Butylhydroxytoluene (Figure 6.145), a sterically hindered phenol, has been used as an alternative in grease formulations to zinc dialkyl dithiophosphate as an antioxidant, where such compounds are incompatible with the rock bit metal [319]. This molecule has the added advantage of being non-toxic and of low environmental impact.

Bisphenol A and epichlorohydrin are the basis of a number of resin polymers used in a variety of oilfield functions including sand consolidation, demulsification, water clarification and asphaltene control [2].

The molecule 2-(decylthio)ethanamine and its hydrochloride salt (Figure 6.146) have been shown to have biocidal properties. It is normally used in the control of biofouling at high pH [320].

Dyes are deployed in the subsea hydrotesting of pipelines and other equipment. The standard dye applied is fluorescein, which has the structure shown in Figure 6.147. It is only required to be available for detection at concentrations of around 10 ppm where it is optically visible. It is however a toxic molecule that has high persistency and is poorly biodegradable. It is of note that the free acid is poorly water soluble, but the disodium and dipotassium salts are readily soluble in water.

Rhodamine B (Figure 6.148), a chemically related species, is less toxic and more biodegradable and offers a similar fluorescent capability [321].

There are a number of other unusual molecules utilised in single instances in the upstream oil and gas industry; however these are somewhat exotic and offer no significant environmental advantages and are therefore not covered.

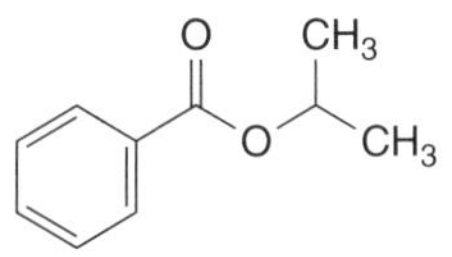

Figure 6.144 Isopropyl benzoate.

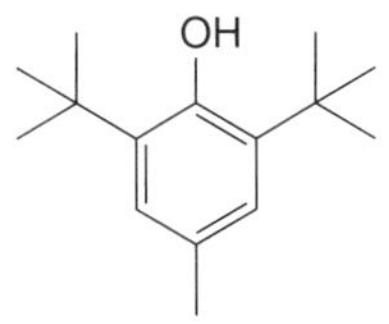

Figure 6.145 *t*-Butylhydroxytoluene.

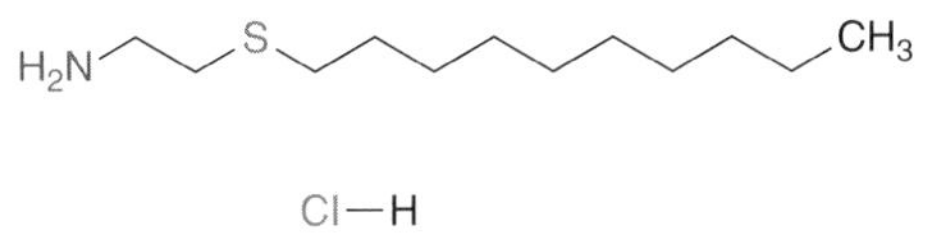

Figure 6.146 2-(Decylthio)ethanamine hydrochloride.

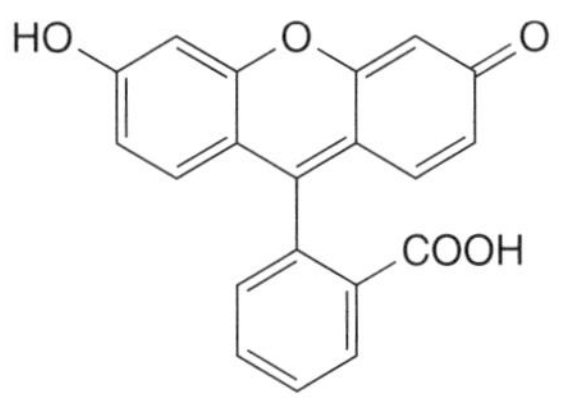

Figure 6.147 Fluorescein.

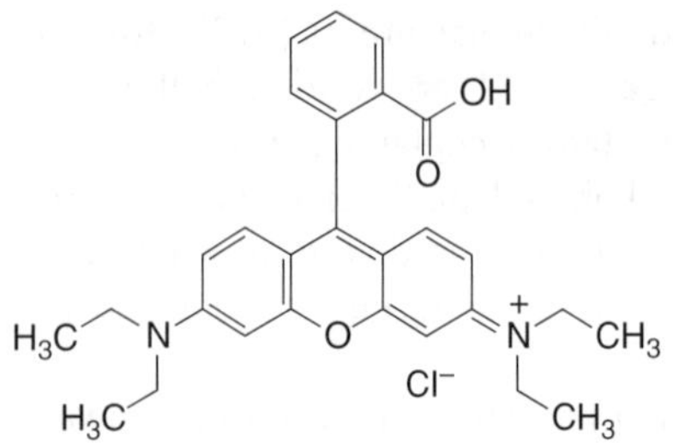

Figure 6.148 Rhodamine B.

6.8 Environmental Impact of Low Molecular Weight Organic Molecules

The environmental hazard profile, properties, impact and fate of low molecular weight organic compounds offer a myriad of outcomes across a broad spectrum of environmental effects. Low molecular weight molecules offer a versatility of function that is unrivalled against other classes of compounds described in this book, and although in volume and application terms they are much less frequently used than surfactants and polymers, they are used against a wide range of oilfield challenges from drill bit lubrication to mitigation of hydrogen sulphide gas. They are also often formulated with surfactants and polymers to give boosted effects through synergistic enhancement of the desired functional properties, and, as is further described in Section 8.4.5.4, can aid in the stability and homogeneity of an overall formulation. All this means that these types of molecule are widely used and their elimination from oilfield practices is difficult if not impossible.

Low molecular weight organics pose a significant environmental challenge as often they have an inherent toxicity; many as has been described previously in this chapter are biocides, e.g. aldehydes, phenols, various amines, etc., and are bioavailable mainly as they have low molecular weights. This being stated, many also readily biodegrade to ecotoxicological benign compounds although some of these such as carbon dioxide also offer an environmental burden. In the introduction of this chapter, the factors affecting the biodegradation of small organic molecules were described [3], and although these are generally important, they are 'rules of thumb', and a number of exceptions have been illustrated, particularly with regard to biologically active molecules, which have a number of features that may suggest a retarded rate of biodegradation that is not observed. The complex molecule folic acid, which however is widely used in foodstuffs and is cheap and readily available, despite contain the azo-heterocyclic pethidine unit, is prone to biodegradation [322]and conversely is chemical and physically a stable unit [323]. Such chemistry is observed to be underutilised in the upstream oil and gas industry and has, to the authors' knowledge, been relatively unexplored and underdeveloped.

Throughout the chapter the environmental impacts and properties of several small organic molecules have been discussed, and as has been stated there are no obvious general patterns in structure and effects. However, where materials do not readily biodegrade to benign materials or biodegrade to potentially toxic materials, the bioaccumulation potential of such contaminants should be considered as a potential environmental concern. Such materials entering the ecosystem food chain can therefore accumulate in tissues of many organisms including those at higher levels such as fish in the aquatic environment and animals, e.g. birds and mammals, and in land environments. It could therefore be argued that emphasis for ecotoxicological assessment regarding such molecules should be placed on their toxicity and on their bioaccumulation potential. Most assessment systems currently place a great emphasis on biodegradation rates over short time frames. Such assessments where used for regulatory compliance can lead to the use of products that although biodegrading can have a great impact on the environment and conversely disallow materials that do not readily biodegrade but are not bioavailable. This dysfunctional dilemma is discussed in more detail in Chapters 9 and 10; however with regard to low molecular weight materials, most are bioavailable. Despite their ready biodegradation they are strongly regulated against or prohibited from use. A case in point is the control of formaldehyde that although seriously detrimental to human health has low environmental impact. This is due to its ready and rapid biodegradation to environmentally benign substances, although it could be argued some additional carbon dioxide/monoxide burden is being generated. Despite this it is prohibited from use in a number of areas of oil and gas exploration and production. In contrast the acute nerve toxin acrolein that has similar efficacy, and performs a similar biocide role, is allowed.

Bioaccumulation of low molecular weight organic materials can be highly complex as many can be used as substrates for organic growth [324]. Also some organisms can bioremediate substances, and this can be used advantageously in sewage treatment [325]. In general, low molecular weight substances do not lend themselves to an easy categorisation by environmental hazard, which is what most regulatory systems seek to do. They can have simple biodegradation pathways to benign substances, can have complex and slow biodegradation mechanisms, can be stable but also bioavailable, can have biodegradation products that are environmentally toxic, and can have many combinations and variations. This can make the overall determination of the environmental fate of such molecules more complex compared with more complex but regular polymers. The use, as in other areas of oilfield chemistry, of natural substances, biological molecules and derivatives thereof is relatively unexplored and under-exploited, presumably for economic reasons, as many of these chemistries offer both good efficacy and good environmental profiles.

In an even more complex twist to the environmental fate of these molecules, work in the last two decades or so has shown that in soils and sediments (the sea bed, for instance), there is preferential and extensive sorption of organic compounds to ketogenic carbon sources such as coal and hydrocarbon reserves [326]. This in retrospect is not surprising as it has been known for almost a century that small molecules and in particular volatile molecules and gases can be adsorbed by activated carbon [327]. Indeed, the adsorption of organic molecules onto carbon materials is a well-studied process that is now widely used in the treatment of gaseous and aqueous waste streams [328]. It can therefore be argued that in order to fully understand the environmental impact of small organic molecules, a more comprehensive understanding of their receiving environment is required.

These themes and discussion points are more fully developed in Chapters 9 and 10; suffice to say in concluding this chapter that with regard to the state of the art in assessing the environmental impact of low molecular weight organic compounds, there is still much to learn.

References

1 Semple, K.T., Doick, K.J., Jones, K.C. et al. (2004). Peer reviewed: defining bioavailability and bioaccessibility of contaminated soil and sediment is complicated. *Environmental Science & Technology* **38** (12): 228A–231A.

2 Kelland, M.A. (2014). *Production Chemicals for the Oil and Gas Industry*. 2e: CRC Press.

3 Boethling, R.S., Sommer, E., and DiFiore, D. (2007). Designing small molecules for biodegradability. *Chemical Reviews* **107**: 2207–2227.

4 Mahajan, M., Rauf, N., Gilmore, T.G. and Maylana, A. (2006). Water control and fracturing: a reality. SPE Asia Pacific Oil & Gas Conference and Exhibition, Adelaide, Australia (11–13 September 2006), SPE 101019.

5 Downs, J.D. (2006). Drilling and completing difficult HP/HT wells with the aid of cesium formate brines-a performance review. IADC/SPE Drilling Conference, Miami, FL (21–23 February 2006), SPE 99068.

6 Howard, S.K. and Downs, J. (2008). The hydrothermal chemistry of formate brines and its impact on corrosion in HPHT wells. SPE International Oilfield Corrosion Conference, Aberdeen, UK (27 May 2008), SPE114111.

7 Gilbert, Y.M., Nordone, A., Downs, J.D. et al. (2007). REACH and the HSE case for formate brines. International Petroleum Technology Conference, Dubai, UAE (4–6 December, 2007), IPTC-11222-MS, IPTC-11222.

8 Downs, J.D. (2011). Life without barite: ten years of drilling deep HPHT gas wells with cesium formate brine. SPE/IADC Middle East Drilling Technology Conference and Exhibition, Muscat, Oman (24–26 October 2011), SPE 145562.

9 Maresh, J.L. (2009). Wellbore treatment fluids having improved thermal stability. US Patent 7,541,316, assigned to Halliburton Energy Services Inc.

10 Smith, K.W. (2009). Well drilling fluids. US Patent 7,576,038, assigned to Clearwater International LLC.

11 Clemmit, A.F., Balance, D.C. and Hunton, A.G. (1985). The dissolution of scales in oilfield systems. Offshore Europe, Aberdeen, UK (10–13 September 1985), SPE 14010.

12 Proctor, S.M. (2000). Scale dissolver development and testing for HP/HT systems. International Symposium on Oilfield Scale, Aberdeen, UK (26–27 January 2000), SPE 60221.

13 Buijse, M., de Boer, P., Breukel, B. et al. (2003). Organic acids in carbonate acidizing. SPE European Formation Damage Conference, The Hague, Netherlands (13–14 May 2003), SPE 82211.

14 Nasr-El-Din, H.A., Lynn, J.D. and Taylor, K.C. (2001). Lab testing and field application of a large-scale acetic acid-based treatment in a newly developed carbonate reservoir. SPE International Symposium on Oilfield Chemistry, Houston, TX (13–16 February 2001), SPE 635036.

15 da Motta, E.P., Quiroga, M.H.V., Arago, A.F.L. and Pereira, A. (1998). Acidizing gas wells in the Merluza field using an acetic/formic acid mixture and foam pigs. SPE Formation Damage Control Conference, Lafayette, LA (18–19 February 1998), SPE 39424.

16 Harris, R.E., McKay, I.D., Mbala, J.M. and Schaaf, R.P. (2001). Stimulation of a producing horizontal well using enzymes that generate acid in-situ - case history. SPE European Formation Damage Conference, The Hague, Netherlands (21–22 May 2001), SPE 68911.

17 Taylor, K.C. and Nasr-El-Din, H.A. (1999). A systematic study of iron control chemicals - Part 2. SPE International Symposium on Oilfield Chemistry, Houston, TX (16–19 February 1999), SPE 50772.

18 Turner, M.S. and Smith, P.C. (2005). Controls on soap scale formation, including naphthenate soaps - drivers and mitigation. SPE International Symposium on Oilfield Scale, Aberdeen, UK (11–12 May 2005), SPE 94339.

19 Eylander, J.G.R., Frigo, D.M., Hartog, F.A. and Jonkers, G. (1998). A novel methodology for in-situ removal of NORM from E & P production facilities. SPE International Conference on Health, Safety, and Environment in Oil and Gas Exploration and Production, Caracas, Venezuela (7–10 June 1998), SPE 46791.

20 Buske, G.R. (1981). Method and composition for removing sulphide containing scale from metal surfaces. US Patent 4,289,639, Assigned to The Dow Chemical Company.

21 Lawson, M.B. (1982). Method for removing iron sulfide scale from metal surfaces. US Patent 4,351,673, assigned to Halliburton Company.

22 Casad, B.M., Clark, C.R., Cantu, L.A. et al. (1991). Process for the preparation of fluid loss additive and gel breaker. US Patent 4,986,355, Assigned to Conoco Inc.

23 Burrafalo, G. and Carminati, S. (1996). Aqueous drilling muds fluidified by means of zirconium and aluminum complexes. US Patent 5,532,211, Assigned to Eniricerche SPA and Agip SPA.

24 Kirstiansen, L.K. (1994). Composition for use in well drilling and maintenance. WO Patent 9,409,253, Assigned to Gait Products Limited.

25 Ye, Z., Han, L., Chen, H. et al. (2010). Effect of sodium salicylate on the properties of gemini surfactant solutions. *Journal of Surfactants and Detergents* **13** (3): 287–292.

26 Bewersdorff, H.-W. and Ohlendorf, D. (1988). The behaviour of drag-reducing cationic surfactant solutions. *Colloid and Polymer Science* **266** (10): 941–953.

27 Watson, J.D. and Garcia, J.G. Jr. (1998). Low toxic corrosion inhibitor. GB Patent 2,319,530, Assigned to Ondeo Nalco Energy Services LP.

28 Soderquist, C.A., Kelly, J.A. and Mandel, F.S. (1990). Gallic acid as an oxygen scavenger. US Patent 4,5689,783, Assigned to Nalco Chemical Company.

29 Vazquez, J.A., Duran, A., Rodriquez-Amado, I. et al. (2011). Evaluation of toxic effects of several carboxylic acids on bacterial growth by toxicodynamic modelling. *Microbial Cell Factories* **10** (1): 100.

30 Akiya, N. and Savage, P.E. (1998). Role of water in formic acid decomposition. *AIChE Journal* **44** (2): 405–415.

31 Eckenfelder, W.W. ed. (1993). *Chemical Oxidation: Technologies for the Nineties*, vol. **2**. CRC Press.

32 Cheung, C.W.S., Beech, I.D., Campbell, S.A. et al. (1994). The effect of industrial biocides on sulphate-reducing bacteria under high pressure. *International Biodeterioration & Biodegradation* **33** (4): 299–310.

33 Galloway, A.J. (1995). Method for removing sulphide(s) from sour gas. US Patent 5,405,591, assigned to Galtech Canada Ltd.

34 van Dijk, J. and Bos, A. (1998). An experimental study of the reactivity and selectivity of novel polymeric triazine type H2S scavengers. In: *Chemical in the Oil Industry: Recent Developments* (ed. L. Cookson and P.H. Ogden), 17–181. Cambridge: RSC Publishing.

35 Bird, A.F., Rosser, H.R., Worrall, M.E. et al. (2002). Sulfate reducing bacteria biofilms in a large seawater injection system. SPE International Conference on Health, Safety and Environment in Oil and Gas Exploration and Production, Kuala Lumpur, Malaysia (20–22 March 2002), SPE 73959.

36 https://www.cancer.gov/about-cancer/causes-prevention/risk/substances/formaldehyde/formaldehyde-fact-sheet

37 Cochrane, S.A., Arts, J.H.E., Ehnes, C. et al. (2015). Thresholds in chemical respiratory sensitization. *Toxicology* **33** (3): 179–194.

38 Craddock, H.A. (2011–2016). *Various Marketing Studies Commission by Oil Operating Clients.* HC Oilfield and Chemical Consulting.

39 Theis, A.B. and Leder, J. (1991). Method for the control of biofouling. US Patent 5,128,051, Assigned to Union Carbide Chemicals & Plastics Technology Corporation.

40 Fujita, H., Sawada, Y., Ogawa, M., and Endo, Y. (2007). Health hazards from exposure to othro-phthalaldehyde, a disinfectant for endoscopes, and prevent measures for health care workers. *Sango Eiseigaku Zasshi* **49** (1): 1–8.

41 Frenier, W.W. and Growcock, F.B. (1988). Mixtures of α,β-unsaturated aldehydes and surface active agents used as corrosion inhibitors in aqueous fluids. US Patent 4,734,259, assigned to Dowell Schlumberger Incorporated.

42 Frenier, W.W. and Hill, D.G. (2002). Well treatment fluids comprising mixed aldehydes. US Patent 6,399,547, assigned to Schlumberger Technology Corporation.

43 Singh, A. and Quraishi, M.A. (2015). Acidizing corrosion inhibitors: a review. *Journal of Materials and Environmental Science* **6** (1): 224–235.

44 Karas, L.J. and Goliaszewski, A.E. (2010). Process for removing hydrogen sulfide in crude oil. International Patent Application WO 2010 027353, applied for by General Electric Company.

45 Bedford, C.T., Fallah, A., Mentzer, E., and Willianson, F.A. (1992). The first characterisation of a glyoxal–hydrogen sulphide adduct. *Journal of the Chemical Society, Chemical Communications* 1035–1036.

46 Cassidy, J.M., Kiser, C.E. and Lane, J.L. (2008). Corrosion inhibitor intensifier compositions and associated methods. US Patent Application 20080139414, applied for by Halliburton Energy Services Inc.

47 Horaska, D.D., San Juan, C.M., Dickinson, A.L. et al. (2011). Acrolein provides benefits and solutions to offshore oilfield production problems. Annual Technical Conference and Exhibition, Denver, CO, USA (30 October – 2 November 2011), SPE 146080SPE.

48 Reed, C., Foshee, J., Penkala, J.E. and Roberson, M. (2005). Acrolein application to mitigate biogenic sulfides and remediate injection well damage in a gas plant water disposal system. SPE International Symposium on Oilfield Chemistry, The Woodlands, TX (2–4 February 2005), SPE 93602.

49 Frenier, W.W. (1992). Process and composition for inhibiting high-temperature iron and steel corrosion. US Patent 5,096,618, assigned to Dowell Schlumberger Incorporated.

50 Brandon, D.M., Fillo, J.P., Morris, A.E. and Evans, J.M. (1995). Biocide and corrosion inhibition use in the oil and gas industry: effectiveness and potential environmental impacts. SPE/EPS Exploration and Production Environmental Conference, Houston, TX (27–29 March 1995), SPE 29735.

51 McGinley, H.R., Enzein, M., Hancock, G. et al. (2009). Glutaraldehyde: an understanding of its ecotoxicity profile and environmental chemistry. Paper 090405, NACE Corrosion Conference, Atlanta (22–29 March 2009).

52 Omil, F., Mendez, D., Vidal, G. et al. (1999). Biodegradation of formaldehyde under aerobic conditions. *Enzyme and Microbial Technology* 24 (5–6): 255–262.

53 Qu, M. and Bhattacharya, S.K. (1997). Toxicity and biodegradation of formaldehyde in anaerobic methanogenic culture. *Biotechnology and Bioengineering* 55 (5): 727–736.

54 McLelland, G.W. (1997). Results of using formaldehyde in a large north slope water treatment system. *SPE Computer Applications* 2 (9): 55–60, SPE 35675.

55 Hille, M., Wittkus, H. and Weinett, F. (1998). Use of acetal-containing mixtures. US Patent 5,830,830, assigned to Clariant Gmbh.

56 Welton, T.D. and Cassidy, J.M. (2007). Thiol/aldehyde corrosion inhibitors. US Patent 7,216,710, assigned to Halliburton Energy Services Inc.

57 https://chem.libretexts.org/Core/Organic_Chemistry/Amides/Synthesis_of_Amides

58 Lanigan, R.M., Starkov, P., and Sheppard, T.D. (2013). Direct synthesis of amides from carboxylic acids and amines using $B(OCH_2CF_3)_3$. *Journal of Organic Chemistry* 78 (9): 4512–4523.

59 Genuyt, B., Janssen, M., Reguerre, R. et al. (2001). Composition lubrifiante biodegradable et ses utilisations, notamment dans un fluide de forage. WO Patent 0183640, assigned to Totalfinaelf France.

60 Muller, H., Herold, C.P., von Tapavicza, S. et al. (1990). Esters of medium chain size carboxylic acids as components of the oil phase of invert emulsion drilling fluids. EP Patent 0386636, assigned to Henkel KGAA.

61 Rudnick, L.R. ed. (2013). *Synthetics, Mineral Oils and Bio-based Lubricants: Chemistry and Technology*, 2ee. CRC Press.

62 Dahlmann, U. and Feustal, M. (2008). Corrosion and gas hydrate inhibitors having improved water solubility and increased biodegradability. US Patent 7,435,845, assigned to Clariant Produkte (Deutschland) Gmbh.

63 Dawson, J.C. and Le, H.V. (2004). Fracturing using gel with ester delayed breaking. US Patent 6,793,018, assigned to BJ services Company.

64 Johnson, D.M. and Ippolito, J.S. (1995). Corrosion inhibitor and sealable thread protector end cap for tubular goods. US Patent 5,452,749, assigned to Centrax International Corp.

65 Schriener, K. and Munoz, T. Jr. (2009). Methods of degrading filter cakes in a subterranean formation. US Patent 7,497,278, assigned to Halliburton Energy Services Inc.

66 Darling, D. and Rakshpal, R. (1998). Green chemistry applied to corrosion and scale inhibitors. Corrosion `98, San Diego, CA (22–27 March 1998). (see also *Materials Performance* **37** (12), 42-47.)

67 Pou, T.E. and Fouquay, S. (2002). Polymethylenepolyamine dipropionamides as environmentally safe inhibitors of the carbon corrosion of iron. US Patent 6,365,100, assigned to Ceca SA.

68 Valone, F.W. (1989). Corrosion inhibiting system containing alkoxylated alkylphenol amines. US Patent 4,867,888, assigned to Texaco Inc.

69 Fink, J. (2013). *Petroleum Engineers Guide to Oilfield Chemicals and Fluids*. Elsevier.

70 Romocki, J. (1996). Application of N,N-dialkylamides to control the formation of emulsions or sludge during drilling or workover of producing oil wells. US Patent 5,567,675, assigned to Buck Laboratories of Canada Ltd.

71 Weers, J.J. and Thomasson, C.E. (1993). Hydrogen sulfide scavengers in fuels, hydrocarbons and water using amidines and polyamidines. US Patent 5,223,127, assigned to Petrolite Corporation.

72 Ramachandran, S., Jovancicevic, V. and Long, J. (2009). Development of a new water soluble high temperature corrosion inhibitor. CORROSION 2009, Atlanta, GA (22–26 March 2009), NACE 09237.

73 Miksic, B.A., Furman, A., Kharshan, M. (2004). Corrosion resistant system for performance drilling fluids utilizing formate brine. US Patent 6,695,897, assigned to Cortec Corporation.

74 Mahrwald, R. (2004). *Modern Aldol Reactions*, vol. **1 and 2**, 1218–1223. Weinheim: Wiley-VCH.

75 Brezinski, M.M. (2001). Methods and acidizing compositions for reducing metal surface corrosion and sulfide precipitation. US Patent 6,315,045, Halliburton Energy Services Inc.

76 Keatch, R. and Guan, H. (2011). Method of inhibiting salt precipitation from aqueous streams. US Patent 20110024366.

77 Anderson, J.D., Hayman, E.S. Jr. and Radzewich, E.A. (1976). Acid inhibitor composition and process in hydrofluoric acid chemical cleaning. US Patent 3,992,313, assigned to Amchem Products Inc.

78 Jenkins, A. (2011). Organic corrosion inhibitor package for organic acids. US Patent 20110028360, assigned to MI Drilling Fluids (UK) Ltd.

79 Fankiewicz, T.C. and Gerlach, J. (2004). Process for removing mercury from liquid hydrocarbons using a sulfur-containing organic compound. US Patent 6,685,824, assigned to Union Oil Company of California.

80 Tang, T., Xu, J., Lu, R. et al. (2010). Enhanced Hg^{2+} removal and Hg^0 re-emission control from wet fuel gas desulfurization liquors with additives. *Fuel* **89** (12): 3613–3617.

81 Jacobs, I.C. and Thompson, N.E.S. (1992). Certain dithiocarbamates and method of use for reducing asphaltene precipitation in asphaltenic reservoirs. US Patent 5,112,505, assigned to Petrolite Corporation.

82 Hart, P.R. (2001). The development and application of dithiocarbamate (DTC) chemistries for use as flocculants by North Sea operators. Chemistry in the Oil Industry VII, Manchester, UK (13–14 November 2001), pp. 149–162.

83 Bellos, T.J. (2000). Polyvalent metal cations in combination with dithiocarbamic acid compositions as broad spectrum demulsifiers. US Patent 6,019,912, assigned to Baker Hughes Incorporated.

84 Durham, D.K., Conkie, U.C. and Downs, H.H. (1991). Additive for clarifying aqueous systems without production of uncontrollable floc. US Patent 5,006,274, assigned to Baker Hughes Incorporated.

85 Musa, O.M. (2011). International Patent Application WO 2011/163317, International Specialty Product Inc.

86 http://www.inchem.org/documents/ehc/ehc/ehc78.htm

87 Haendel, M.A., Tilton, F., Bailey, G.S., and Tanguay, R.L. (2004). Developmental toxicity of the dithiocarbamate pesticide sodium metam in zebrafish. *Toxicological Sciences* **81** (2): 390–400.

88 Liu, S. and Suflita, J.M. (1994). Anaerobic biodegradation of methyl esters by *Acetobacterium woodii* and *Eubacterium limosum*. *Journal of Industrial Microbiology* **13** (5): 321–327.

89 Neilsen, C.J., D'Anna, B. and Karl, M. et al. (2011). Atmospheric degradation of Amines (ADA) Summary Report: Photo-Oxidation of Methylamine, Dimethylamine and Trimethylamine. CLIMIT Project No. *201604*. Norwegian Institute for Air Research.

90 Booth, A., da Silva, E. and Brakstad, O.G. (2011). Environmental impacts of amines and their degradation products: current status and knowledge gaps. 1st Post Combustion Capture Conference, Abu Dhabi (17–19 May 2011).

91 Brakstad, O.G., da Silva, E.F. and Syversen, T. (2010). TCM Amine Project: Support on Input to Environmental Discharges. Evaluation of Degradation Components. SINTEF 2010.

92 Evans, W.H., David, E.J., and Patterson, S.J. (1973). Biodegradation of urea in river waters under controlled laboratory conditions. *Water Research* **7** (7): 975–985.

93 Saltmiras, D.A. and Lemley, A.T. (2000). Degradation of ethylene thiourea (ETU) with three fenton treatment processes. *Journal of Agricultural and Food Chemistry* **48** (12): 6149–6157.

94 Watts, R.R., Storherr, R.W., and Onley, J.H. (1974). Effects of cooking on ethylene bisdithiocarbamate degradation to ethylene thiourea. *Bulletin of Environmental Contamination and Toxicology* **12** (2): 224–226.

95 Bertkua, W. and Steidl, N. (2012). Alkanesulfonic acid microcapsules and use thereof in deep wells. US Patent Application 20120222863, assigned to BASF De.

96 Shank, R.A. and McCartney, T.R. (2013). Synergistic and divergent effects of surfactants on the kinetics of acid dissolution of calcium carbonate scale. CORROSION 2013, Orlando, FL (17–21 March 2013), NACE 2013-2762.

97 De Boer, R.B., Leerlooyer, K., Eigner, M.R.P., and van Bergen, A.R.D. (1995). Screening of crude oils for asphalt precipitation: theory, practice, and the selection of inhibitors. *SPE Production & Facilities* **10** (01): 55–61, SPE 24987.

98 Miller, D., Volimer, A. and Feustal, M. (1999). Use of alkanesulfonic acids as asphaltene-dispersing agents. US Patent 5925233, Clariant Gmbh.

99 Weiche, I. and Jermansen, T.G. (2003). Design of synthetic dispersants for asphaltene constituents. *Petroleum Science and Engineering* **21** (3–4): 527–536.

100 Goldszal, A., Hurtevent, C. and Rousseau, G. (2002). Scale and naphthenate inhibition in deep-offshore fields. International Symposium on Oilfield Scale, Aberdeen, UK (30–31 January 2002), SPE 74661.

101 Fry, S.E., Totten, P.L., Childs, J.D. and Lindsey, D.W. (1990). Chloride-free set accelerated cement compositions and methods. US Patent 5,127,955, assigned to Halliburton Company.

102 Patel, B. and Stephens, M. (1991). Well cement slurries and dispersants therefor. US Patent 5,041,630, assigned to Phillips Petroleum Company.

103 Varadaraj, R., Savage, D.W. and Brons, C.H. (2001). Chemical demulsifier for desalting heavy crude, US Patent 6,168,702, assigned to Exxon Research and Engineering Company.

104 Lindstrom, M.R. and Mark, H.W. (1987). Inhibiting corrosion: benzylsulfinylacetic acid or benzylsulfonylacetic acid. US Patent 4,637,833, assigned to Phillips Petroleum Co.

105 Walele, I.I. and Syed, S.A. (1995). Process for making N-acyl taurides. US Patent 5,434,276, assigned to Finetex Inc.

106 The Soap and Detergent Association. (1996). Monograph on Linear Alkylbenzene Sulphonate.

107 Liu, Z., Wang, X., and Liu, Z.F. (2012). Synthesis and properties of the ESA/AMPS copolymer. *Applied Mechanics and Materials* **164**: 194–198.

108 Higgins, T.P., Davey, M., Trickett, J. et al. (1996). Metabolism of methanesulfonic acid involves a multicomponent monooxygenase enzyme. *Microbiology* **142** (Pt. 2): 251–260.

109 Hammerschmidt, E.G. (1934). Formation of gas hydrates in natural gas transmission lines. *Industrial and Engineering Chemistry* **26** (8): 851–855.

110 DeVries, A.L. and Wohlschlaq, D.E. (1969). Freezing resistance in some Antarctic fishes. *Science* **163** (3871): 1073–1075.

111 Sloan, E.D. Jr. (1998). *Clathrate Hydrates of Natural Gasses*, 2ee. CRC Press.

112 Anyadiegwu, C.I.C., Kerunwa, A., and Oviawete, P. (2014). Natural gas dehydration using triethylene glycol. *Petroleum & Coal* **56** (4): 407–417.

113 Fisk, J.V., Kerchevile, J.D. and Prober, K.W. (2006). Silicic acid mud lubricants. US Patent 6,989,352, Halliburton Energy Services Inc.

114 Argiller, J.-F., Demoulin, A., Aidibert-Hayet, A. and Janssen, M. (2004). Borehole fluid containing a lubricating composition—method for verifying the lubrification of a borehole fluid—application with respect to fluids with a high pH. US Patent 6,750,180, assigned to Institute Francais Du Petrole.

115 Coburn, C.E., Pohlman, J.L., Pyzowski, B.A. et al. (2015). Aminoalcohol and biocide compositions for aqueous based systems. US Patent 9,034,929, assigned to Angus Chemical Company.

116 Mainier, F., Saliba, C.A. and Gonzalez, G. (1990). Effectiveness of acid corrosion inhibitors in the presence of alcohols. Unsolicited paper published in e-Library, Society of Petroleum Engineers, SPE 20404.

117 Sullivan, D.S. III, Strubelt, C.E. and Becker, K.W. (1977). Diacetylenic alcohol corrosion inhibitors. US Patent 4,039,336, assigned to Exxon Research and Engineering Company.

118 Vorderbruggen, M.A. and Kaarigstad, H. (2006). Meeting the environmental challenge: a new acid corrosion inhibitor for the Norwegian sector of the North Sea. SPE Annual Technical Conference and Exhibition, San Antonio, TX (24–27 September 2006), SPE 102908.

119 Frenier, W.W., Growcock, F.B., and Lopp, V.R. (1988). Mechanisms of corrosion inhibitors used in acidizing wells. *SPE Production Engineering* **3** (4): 584–590, SPE 14092.

120 Delgado, E. and Keown, B. (2009). Low volatile phosphorous gelling agent. US Patent 7,622,054, assigned to Ethox Chemicals Llc.

121 Richardson, W.C. and Kibodeaux, K.R. (2001). Chemically assisted thermal flood process. US Patent 6,305,472, assigned to Texaco Inc.

122 Cobb, H.G. (2010). Composition and process for enhanced oil recovery. US Patent 7,691,790, assigned to Coriba Technologies L.L.C.

123 Fu, B. (2001). The development of advanced kinetic hydrate inhibitors. Chemistry in the Oil Industry VII, Manchester, UK (13–14 November 2001), pp. 264–276.

124 Gartner, E.M. and Kreh, R.P. (1993). Cement additives and hydraulic cement mixes containing them. CA Patent 2071080, assigned to Inventors and W.R. Grace and Co.

125 Barker, K.M. and Newberry, M.E. (2007). Inhibition and removal of low-pH fluid-induced asphaltic sludge fouling of formations in oil and gas wells. International Symposium on Oilfield Chemistry, Houston, TX (28 February–2 March, 2007), SPE 102738.

126 Holtrup, F., Wasmund, E., Baumgartner, W. and Feustal, M. (2002). Aromatic aldehyde resins and their use as emulsion breakers. US Patent 6,465,528, assigned to Clariant Gmbh.

127 Al-Anazi, M., Al-Mutairi, S.H., Alkhaldi, M. et al. (2011). Laboratory evaluation of organic water shut-off gelling system for carbonate formations. SPE European Formation Damage Conference, Noordwijk, The Netherlands (7–10 June 2011), SPE 144082.

128 Pozzo, A.D. and Pastori, N. (1996). Percutaneous absorption of parabens from cosmetic formulations. *International Journal of Cosmetic Science* **18** (2): 57–66.

129 http://ec.europa.eu/environment/chemicals/endocrine/

130 Darbre, P.D. and Harvey, P.W. (2008). Paraben esters: review of recent studies of endocrine toxicity, absorption, esterase and human exposure, and discussion of potential human health risks. *Journal of Applied Toxicology* **28** (5): 561–578.

131 Feraud, J.P., Perthius, H. and Dejeux, P. (2001). Compositions for iron control in acid treatments for oil wells. US Patent 6,306,799, assigned to Schlumberger Technology Corporation.

132 Jovanciecevic, V., Ahn, Y.S., Dougherty, J.A. and Alink, B.A. (2000). CO_2 corrosion inhibition by sulfur-containing organic compounds. CORROSION 2000, Orlando, FL (26–31 March 2000), NACE Paper 7.

133 Ramachandram, S., Jovanciecevic, V., Williams, G. et al. (2010). Development of a new high shear corrosion inhibitor with beneficial water quality attributes. CORROSION 2010, San Antonio, TX (14–18 March 2010), NACE 10375.

134 Hausler, R.H., Alink, B.A., Johns, M.E. and Stegmann, D.W. (1988). Carbon dioxide corrosion inhibiting composition and method of use thereof. European Patent Application EP027651, assigned to Petrolite Corporation.

135 Redmore, D. and Alink, B.A. (1972). Use of thionium derivatives as corrosion inhibitors. US Patent 3,697,221, assigned to Petrolite Corporation.

136 Starkey, R.J., Monteith, G.A. and Wilhelm, C.A. (2008). Biocide for well stimulation treatment fluids. US Patent Application 20080004189, assigned to Kemira Chemicals Inc.

137 Alonso-Debolt, M.A., Bland, R.G., Chai, B.J. et al. (1995). Glycol and glycol ether lubricants and spotting fluids. WO Patent 1995028455, assigned to Baker Hughes Incorporated.

138 Karaev, S.F.O., Gusejnov, S.O.O., Garaeva, S.V.K. and Talybov, G.M.O. (1996). Producing propargyl ether for use as a metal corrosion inhibitor. RU Patent 2056401.

139 Collins, I.R., Goodwin, S.P., Morgan, J.C. and Stewart, N.J. (2001). Use of oil and gas field chemicals. US Patent 6,225,263, assigned to BP Chemicals Ltd.

140 White, G.F., Russel, N.J., and Tideswell, E.C. (1996). Bacterial scission of ether bonds. *Microbiological Reviews* **60** (1): 216–232.

141 Kim, Y.-H. and Engesser, K.-H. (2004). Degradation of alkyl ethers, aralkyl ethers, and dibenzyl ether by rhodococcus sp. strain DEE5151, isolated from diethyl ether-containing enrichment cultures. *Applied and Environmental Microbiology* **70** (7): 4398–4401.

142 Tideswell, E.C., Russel, N.J., and White, G.F. (1996). Ether bond scission in the biodegradation of alcohol ethoxylate non-ionic surfactants by pseudomonas sp. strain SC25A. *Microbiology* **142**: 1123–1131.

143 Reid, I.D. (1995). Biodegradation of lignin. *Canadian Journal of Botany* **73** (S1): 1011–1018.

144 The Ministers of The Environment and Health. (2010). Screening assessment for n-Butyl glycidyl ether (CAS Registry Number 2426-08-6) Environment Canada/Health Canada, March 2010.

145 Meier, I.K., Goddard, R.J. and Ford, M.E. (2008). Amine-based gas hydrate inhibitors. US Patent 7,452,848, assigned to Air Products and Chemicals Inc.

146 Klug, P. and Kelland, M.A. (2003). Additives for inhibiting gas hydrate formation. US Patent 6,544,932, assigned to Clariant Gmbh and Rf-Rogaland Research.

147 Boyle, B. (2004). A look at developments in vapour phase corrosion inhibitors. *Metal Finishing* **102** (5): 37–41.

148 Papir, Y.S., Schroder, A.H. and Stone, P.J. (1989). New downhole filming amine corrosion inhibitor for sweet and sour production. SPE International Symposium on Oilfield Chemistry, Houston, TX (8–10 February 1989), SPE 18489.

149 Miksic, B.A., Furman, A. and Kharshan, M. (2004). Corrosion inhibitor barrier for ferrous and non-ferrous metals. US Patent 6,800,594, assigned to Cortec Corporation.

150 Hellberg, P.-E. (2009). Structure-property relationships for novel low-alkoxylated corrosion inhibitors. Chemistry in the Oil Industry XI, Manchester, UK (2–4 November 2009).

151 Ford, M.E., Kretz, C.P., Lassila, K.R. et al. (2006). N,N'-dialkyl derivatives of polyhydroxyalkyl alkylenediamines. European Patent EP 1,637,038, assigned to Air Products and Chemicals Inc.

152 Jenvey, N.J., MacLean, A.F., Miles, A.F. and Montgomerie, H.T.R. (2000). The application of oil soluble scale inhibitors into the Texaco galley reservoir. A comparison with traditional squeeze techniques to avoid problems associated with wettability modification in low water-cut wells. International Symposium on Oilfield Scale, Aberdeen, UK (26–27 January 2000), SPE 60197.

153 Jorda, R.M. (1962). Aqualin biocide in injection waters. SPE Production Research Symposium, Tulsa, OK (12–13 April 1962), SPE 280.

154 Ludensky, M., Hill, C. and Lichtenberg, F. (2003). Composition including a triamine and a biocide and a method for inhibiting the growth of microorganisms with the same. US Patent Application 20030228373, assigned to Lonza Inc. and Lonza Ag.

155 Greene, E.A., Brunelle, V., Jenneman, G.E., and Voordouw, G. (2006). Synergistic inhibition of microbial sulfide production by combinations of the metabolic inhibitor nitrite and biocides. *Applied and Environmental Microbiology* **72** (12): 7897–7901.

156 Dahlmann, U. and Feustal, M. (2007). Additives for inhibiting the formation of gas hydrates. US Patent 7,183,240, assigned to Clariant Product Gmbh.

157 Garvey, C.M., Savoly, A. and Weaterford, T.M. (1987). Drilling fluid dispersant. US Patent 4,711,731, assigned to Diamond Shamrock Chemicals Company.

158 Meyer, G.R. (1999). Corrosion inhibitor compositions including quaternized compounds having a substituted diethylamino moiety. US Patent 6,488,868, Assigned to Ondeo Nalco Energy Services L.P.

159 Kissel, C.L. (1999). Process and composition for inhibiting corrosion. EP Patent 9,069,69, assigned to Degussa AG.

160 Young, L.A. (1993). Low melting polyalkylenepolyamine corrosion inhibitors. WO Patent 9,319,226, assigned to Chevron Research and Technology Company.

161 Hatchman, K. (1999). Drilling fluid concentrates. EP Patent 9,033,90, assigned to Albright and Wilson UK Ltd.

162 Chatterji, J., Morgan, R.L. and Davis, G.W. (1997). Set retarded cementing compositions and methods. US Patent 5,672,203, assigned to Halliburton Company.

163 Polasek, J. and Bullin, J.A. (1994). Process considerations in selecting amines for gas sweetening. Proceedings Gas Processors Association Regional Meeting, Tulsa, OK (September 1994).

164 Hafiz, A.A. and Khidr, T.T. (2007). Hexa-triethanolamine oleate esters as pour point depressant for waxy crude oils. *Journal of Petroleum Science and Engineering.* **56** (4): 296–302.

165 Przybylinski, J.L. (1989). Adsorption and desorption characteristics of mineral scale inhibitors as related to the design of squeeze treatments. SPE International Symposium on Oilfield Chemistry, Houston, TX (8–10 February 1989), SPE 18486.

166 Darling, D. and Rakshpal, R. (1998). Green chemistry applied to corrosion and scale inhibitors. CORROSION 98, San Diego, CA (22–27 March 1998), NACE 98207.

167 Hollingshad, W.R. (1976). Corrosion inhibition with triethanolamine phosphate ester compositions. US Patent 3,932,303, assigned to Calgon Corporation.

168 Burgazli, C.R., Navarette, R.C., and Mead, S.L. (2005). New dual purpose chemistry for gas hydrate and corrosion inhibition. *Journal of Canadian Petroleum Technology* **44** (11): doi: 10.2118/05-11-04.

169 Bellos, T.J. (1984). Block polymers of alkanolamines as demulsifiers for O/W emulsions. US Patent 4,459,220, assigned to Petrolite Corporation.

170 Fikentscher, R., Oppenlaender, K., Dix, J.P. et al. (1993). Methods of demulsifying employing condensates as demulsifiers for oil-in water emulsions. US Patent 5,234,626, assigned to BASF Ag.

171 von Tapavicza, S., Zoeliner, W., Herold, C.-P. et al. (2002). Use of selected inhibitors against the formation of solid organo-based incrustations from fluid hydrocarbon mixtures. US Patent 6,344,431, assigned to Inventors.

172 Keatch, R.W. (1998). Removal of sulphate scale from surfaces. GB Patent 2,314,865, assigned to Inventor.

173 Raad, I. and Sheretz, R. (2001). Chelators in combination with biocides: treatment of microbially induced biofilm and corrosion. US Patent 6,267,979, assigned to Wake Forrest University.

174 Schowanek, D., Feijtel, T.C.J., Perkins, C.M. et al. (1997). Biodegradation of [S,S], [R,R] and mixed stereoisomers of ethylene diamine disuccinic acid (EDDS), a transition metal chelator. *Chemosphere* **34** (11): 2375–2391.

175 Putnis, A., Putnis, C.V. and Paul, J.M. (1995). The efficiency of a DTPA-based solvent in the dissolution of barium sulfate scale deposits. SPE International Symposium on Oilfield Chemistry, San Antonio, TX (14–17 February 1995), SPE 29094.

176 Rebeschini, J., Jones, C., Collins, G. et al. The development and performance testing of a novel biodegradable barium sulphate scale dissolver. Paper presented at 19th Oilfield Chemistry Symposium Geilo, Norway (9–12 March 2008).

177 Boreng, R., Chen, P., Hagen, T. et al. (2004). Creating value with green barium sulphate scale dissolvers - development and field deployment on statfjord unit. SPE International Symposium on Oilfield Scale, Aberdeen, UK (26–27 May 2004), SPE 87438.

178 de Wolf, C.A., Lepage, J.N., Nasr-El-Din, H. (2013). Ammonium salts of chelating agents and their use in oil and gas field applications. US Patent Application 20130281329, Applicant Akzo Nobel Chemicals International B.V.

179 Braun, W., de Wolf, C.A. and Nasr-El-Din, H.A. (2012). Improved health, safety and environmental profile of a new field proven stimulation fluid. SPE Russian Oil and Gas Exploration and Production Technical Conference and Exhibition, Moscow, Russia (16–18 October 2012), SPE 157467.

180 McElfresh, P.M. and Williams, C.F. (2007). Hydraulic fracturing using non-ionic surfactant gelling agent. US Patent 7,216,709, assigned to Akzo Nobel.

181 Shimura, Y. and Takahashi, J. (2007). Oxygen scavenger and the method for oxygen reduction treatment. US Patent 7,112,284, assigned to Kurita Water Industries Ltd.

182 Slovinsky, M. (1981). Boiler additives for oxygen scavenging. US Patent 4,269,717, assigned to Nalco Chemical Company.

183 Colclough, V.L. (2001). Fast acting disinfectant and cleaner containing a polymeric biguanide. US Patent 6,303,557, assigned to S.C. Johnson commercial Markets Inc.

184 Oguzie, E.E., Li, Y., and Wang, F.H. (2007). Corrosion inhibition and adsorption behavior of methionine on mild steel in sulfuric acid and synergistic effect of iodide ion. *Journal of Colloid and Interface Science* **10** (1): 90–98.

185 http://www.cefas.co.uk/media/1384/13-06e_plonor.pdf

186 Huber, M., Meyer, U., and Rys, P. (2000). Biodegradation mechanisms of linear alcohol ethoxylates under anaerobic conditions. *Environmental Science and Technology* **34** (9): 1737–1741.

187 Shadina, H. and Wright, J.S. (2008). Understanding the toxicity of phenols: using quantitative structure-activity relationship and enthalpy changes to discriminate between possible mechanisms. *Chemical Research in Toxicology* **21** (6): 1197–1204.

188 Bolton, J.L., Trush, M.A., Penning, T.M. et al. (2000). Role of quinones in toxicology. *Chemical Research in Toxicology* **13** (3): 135–160.

189 Scott, M.J. and Jones, M.N. (2000). The biodegradation of surfactants in the environment. *Biochimica et Biophysica Acta* **1508** (1–2): 235–251.

190 Pachura-Bouchet, S., Blaise, C., and Vasseur, P. (2006). Toxicity of nonylphenol on the cnidarian Hydra attenuata and environmental risk assessment. *Environmental Toxicology* **21** (4): 388–394.

191 Jaques, P., Martin, I., Newbigging, C., and Wardell, T. (2002). Alkylphenol based demulsifier resins and their continued use in the offshore oil and gas industry. In: *Chemistry in the Oil Industry VII* (ed. T. Balson, H. Craddock, J. Dunlop, et al.), 56–66. The Royal Society of Chemistry.

192 Tyagi, R., Tyali, V.K., and Pandey, S.K. (2007). Imidazoline and its derivatives: an overview. *Journal of Oleo Science* **56** (5): 211–222.

193 Chheda, B.D. (2007). Asphaltene dispersants for petroleum products. US Patent Application 20070124990.

194 Feustel, M. and Klug, P. (2002). Compound for inhibiting corrosion. US Patent 6,372,918, assigned to Clariant Gmbh.

195 Durnie, W.H. and Gough, M.A. (2003). Characterisation, isolation and performance characteristics of imidazolines: Part II development of structure-activity relationships. Paper 03336, NACE Corrosion 2003, San Diego, CA (March 2003).

196 Watts, M.M. (1990, 1990). Imidazoline hydrolysis in alkaline and acidic media - a review. *Journal of the American Oil Chemists Society* **67** (12): 993–995.

197 Littman, E.S. and McLean, T.L. (1987). Chemical control of biogenic H2S in producing formations. SPE Production Operations Symposium, Oklahoma City, OK (8–10 March 1987), SPE 16218.

198 Thompkins, H.G. and Sharma, S.P. (1982). The interaction of imidazole, benzimidazole and related azoles with a copper surface. *Surface and Interface Analysis* **4** (6): 261–266.

199 Samant, A.K., Koshel, K.C. and Virmani, S.S. (1988). Azoles as corrosion inhibitors for mild steel in a hydrochloric acid medium. Unsolicited paper published in e-Library, Society of Petroleum Engineers, SPE 19022.

200 Ward, E.C., Foster, A., Glaser, D.E. and Weidner, I. (2004). Looking for an alternative yellow metal corrosion inhibitor. CORROSION 2004, New Orleans, LA (28 March-1 April 2004), NACE 04079.

201 Finsgar, M. and Jackson, J. (2014). Application of corrosion inhibitors for steels in acidic media for the oil and gas industry: a review. *Corrosion Science* **86**: 17–41.

202 Quarishi, M. and Jamal, D. (2000). Fatty acid triazoles: novel corrosion inhibitors for oil well steel (N-80) and mild steel. *Journal of American Oilfield Chemists* **77** (10): 1107–1111.

203 Antonijevic, M.M. and Petrovic, M.B. (2008). Copper corrosion inhibitors: a review. *International Journal of Electrochemical Science* **3**: 1–28.

204 Ward, E.C. and Glaser, D.E. (2007). A new look at azoles. NACE International, CORROSION 2007, Nashville, TN (11–15 March 2007), NACE 07065.

205 Giger, W., Schaffner, C., and Kohler, H.-P.E. (2006). Benzotriazole and tolyltriazole as aquatic contaminants. 1. Input and occurrence in rivers and lakes. *Environmental Science and Technology* **40** (23): 7186–7192.

206 Rouet, J., Groffe, D. and Saluan, M. (2011). Asphaltene-stabilising molecules having a tetrapyrrolic pattern. European patent EP 2,097,162, assigned to Scomi Anticor.

207 Eoff, L.S., Raghava, B. and Dalrymple, E.D. (2002). Methods of reducing subterranean formation water permeability. US Patent 6,471,69, assigned to Halliburton Energy Services Inc.

208 Jarret, M. and Clapper, D. (2010). High temperature filtration control using water based drilling fluid systems comprising water soluble polymers. US Patent 7,651,980, assigned to Baker Hughes Inc.

209 Wu, Y. and Gray, R.A. (1992). Compositions and methods for inhibiting corrosion. US Patent 5,118,536, assigned to Phillips Petroleum Company.

210 Fu, S.B., Cenegy, L.M. and Neff, C.S. (2001). A summary of successful field applications of A kinetic hydrate inhibitor. SPE International Symposium on Oilfield Chemistry, Houston, TX (13–16 February 2001), SPE 65022.

211 O'Reilly, R., Ieong, N.S., Chua, P.C., and Kelland, M.A. (2011). Missing poly(*N*-vinyl lactam) kinetic hydrate inhibitor: high-pressure kinetic hydrate inhibition of structure II gas hydrates with poly(*N*-vinyl piperidone) and other poly(*N*-vinyl lactam) homopolymers. *Energy Fuels* **25** (10): 4595–4599.

212 Sloan, D., Koh, C., and Sum, A.K. ed. (2011). *Natural Gas hydrates in Flow Assurance*. Burlington, MA: Gulf Professional Publishing.

213 Angel, M., Nuebecker, K. and Saner, A. (2005). Grafted polymers as gas hydrate inhibitors. US Patent 6,867,662, assigned to BASF Ag.

214 Balzer, J., Feustal, M., Matthias, M. and Reimann, W. (1995). Graft polymers, their preparation and use as pour point depressants and flow improvers for crude oils, residual oils and middle distillates. US Patent 5,439,981, assigned to Hoechst Ag.

215 Shimura, Y., Uchida, K., Sato, T. and Taya, S. (2000). The performance of new volatile oxygen scavenger and its field application in boiler systems. CORROSION 2000, Orlando, FL (26–31 March 2000), NACE 00327.

216 Frenier, W.W. (1989). Acidizing fluids used to stimulate high temperature wells can be inhibited using organic chemicals. SPE International Symposium on Oilfield Chemistry, Houston, TX (8–10 February 1989), SPE 18468.

217 Kennedy, W.C. Jr. (1987). Corrosion inhibitors for cleaning solutions. US Patent 4,637,899, assigned to Dowell Schlumberger Incorporated.

218 Tiwari, L., Meyer, G.R. and Horsup, D. (2009). Environmentally friendly bis-quaternary compounds for inhibiting corrosion and removing hydrocarbonaceous deposits in oil and gas applications. International Patent Application WO.2009/076258, assigned to Nalco Company.

219 Quinlan, P.M. (1982). Use of quaternized derivatives of polymerized pyridines and quinolones as demulsifiers. US Patent 4,339,347, assigned to Petrolite Corporation.

220 Campbell, S.E. and Jovancicevic, V. (2001). Performance improvements from chemical drag reducers. SPE International Symposium on Oilfield Chemistry, Houston, TX (13–16 February 2001), SPE 65021.

221 Schmitt, G. (2001). Drag reduction by corrosion inhibitors – a neglected option for mitigation of flow induced localized corrosion. *Materials and Corrosion* **52** (5): 329–343.

222 Kokal, S.L. and Sayegh, S.G. (1995). Asphaltenes: the cholesterol of petroleum. Middle East Oil Show, Bahrain (11–14 March 1995), SPE 29787.

223 Roper, W.L. (1992). *Toxicological Profile for Pyridine*. Agency for Toxic Substances and Disease Registry U.S. Public Health Service.

224 Treybig, D.S. (1987). Novel compositions prepared from methyl substituted nitrogen-containing aromatic heterocyclic compounds and an aldehyde or ketone. US Patent 4,676,834, assigned to The Dow Chemical Company.

225 Smith, K., Persinski, L.J. and Wanner, M. (2008). Effervescent biocide compositions for oilfield applications. US Patent Application 20080004189, assigned to Weatherford/Lamb Inc.

226 Fischer, G.C. (1990). Corrosion inhibitor compositions containing inhibitor prepared from amino substituted pyrazines and epoxy compounds. US Patent 4,895,702, The Dow Chemical Company.

227 Sundh, U.B., Binderup, M.-L., Brimer, L. et al. (2011). Scientific opinion on flavouring group evaluation 17, revision 3 (FGE.17Rev3): pyrazine derivatives from chemical group 24, EFSA panel on food contact materials, enzymes, flavourings and processing aids (CEF). *EFSA Journal* **9** (11).

228 Schulten, H.-R. and Schnitzer, M. (1997). The chemistry of soil organic nitrogen: a review. *Biology and Fertility of Soils.* **26** (1): 1–15.

229 Schieman, S.R. (1999). Solids-free H2S scavenger improves performance and operational flexibility. SPE International Symposium on Oilfield Chemistry, Houston, TX (16–19 February 1999), SPE 50788.

230 Bozelli, J.W., Shier, G.D., Pearce, R.L. and Martin, C.W. (1978). Absorption of sulfur compounds from gas streams. US Patent 4,112,049, The Dow Chemical Company.

231 Craddock, H.A. (2014). The use of formaldehyde releasing biocides and chemicals in the oil and gas industry. Presented to the Formaldehyde Biocide Interest Group (FABI), Vienna (9 December 2014).

232 https://echa.europa.eu/registration-dossier/-/registered-dossier/15014/5/3/2

233 Pinto, A.P., Serrano, C., Pires, T. et al. (2012). Degradation of terbuthylazine, difenoconazole and pendimethalin pesticides by selected fungi cultures. *Science in the Total Environment* **435–436**: 402–410.

234 Finnegan, M., Linley, E., Denyer, S.P. et al. (2010). Mode of action of hydrogen peroxide and other oxidizing agents: differences between liquid and gas forms. *Journal of Antimicrobial Chemotherapy* **65**: 2108–2115.

235 Nepla, C.J., Howarth, J. and Azomia, F.D. (2002). Factors to consider when applying oxidizing biocides in the field. CORROSION 2002, Denver, CO (7–11 April 2002), NACE 02223.

236 Sook, B., Harrison, A.D. and Ling, T.F. (2003). A new thixotropic form of bromochlorodimethylhydantoin: a case study. CORROSION 2003, San Diego, CA (16–20 March 2003), NACE 03715.

237 Janek, K.E. (2012). Beyond triazines: development of a novel chemistry for hydrogen sulfide scavenging. CORROSION 2012, Salt Lake City, UT (11–15 March, 2012), NACE 2012-1520.

238 Craddock, H.A. (2004). Method and compounds for inhibiting corrosion. European Patent EP 1,457,585, assigned to TR Oil Services Ltd.

239 Haruna, T., Toyota, R., and Shibata, T. (1997). The effect of potential on initiation and propagation of stress corrosion cracks for type 304l stainless steel in a chloride solution containing thiosulfate. *Corrosion Science* **39** (10–11): 1873–1882.

240 Himpler, F.J., Sweeny, P.G. and Ludensky, M.L. (2001). The benefits of a hydantoin - based slimicide in papermaking applications. 55th Appita Annual Conference, Hobart, Australia (30 April-2 May 2001), Proceedings. Carlton, Victoria: Appita Inc., pp. 99–103.

241 Lukovits, I., Kalmn, E., and Zucchi, F. (2001). Corrosion inhibitors correlation between electronic structure and efficiency. *Corrosion* **57** (01): 3–8, NACE 01010003.

242 McLeary, R.R., Ruidisch, L.E., Clarke, J.T. et al. (1951). Thiophene from hydrocarbons and hydrogen sulfide. 3rd World Petroleum Congress, The Hague, The Netherlands (28 May-6 June 1951), WPC-4406.

243 James, S.G., Nelson, E.B. and Guinot, F.J. (2002). Sand consolidation with flexible gel system. US Patent 6,450,260, assigned to Schlumberger Technology Corporation.

244 Ayoub, J.A., Crawshaw, J.P. and Way, P.W. (2003). Self-diverting resin systems for sand consolidation. US Patent 6,632,778, assigned to Schlumberger Technology Corporation.

245 Reddy, B.R. and Nguyen, P.D. (2010). Sealant compositions and methods of using the same to isolate a subterranean zone from a disposal well. US Patent 7,662,755, assigned to Halliburton Energy Services Inc.

246 Leonov, Y.R., Lamosov, M.E., Rayabokon, S.A. et al. (1993). Plugging material for oil and gas wells. SU Patent 1,818,463, assigned to Borehole Consolidation Mu.

247 Dyreborg, S., Arvin, E., Broholm, K., and Christensen, J. (1996). Biodegradation of thiophene, benzothiophene, and benzofuran with eight different primary substrates. *Environmental Toxicology and Chemistry* **15** (12): 2290–2292.

248 Koleva, Y. and Tasheva, Y. (2012). The persistence, bioaccumulation and toxic estimation of some sulphur compounds in the environment. *Petroleum and Coal* **54** (3): 220–224.

249 Kahrilas, G.A., Blotevogel, J., Stewart, P.S., and Borch, T. (2015). Biocides in hydraulic fracturing fluids: a critical review of their usage, mobility, degradation, and toxicity. *Environmental Science and Technology* **49** (1): 16–32.

250 Kanda, S., Yanagita, M. and Sekimoto, Y. (1987). Stabilized fracturing fluid and method of stabilizing fracturing fluid. US Patent 4,681,690, assigned to Nito Chemical Industry C. Ltd.

251 Hansen, H.W. and Henderson, N.D. (1991). *A Review of the Environmental Impact and Toxic Effects of 2-MBT*. British Columbia: Environmental protection Agency.

252 Jafari, H., Akbarzade, K., and Danaee, I. (2014). Corrosion inhibition of carbon steel immersed in a 1 M HCl solution using benzothiazole derivatives. *Arabian Journal of Chemistry*, Available online 13 November 2014, http://www.sciencedirect.com/science/article/pii/S1878535214002664.

253 Bentiss, F., Traisnel, M., Vezin, H. et al. (2004). 2,5-Bis(4-dimethylaminophenyl)-1,3,4-oxadiazole and 2,5-bis(4-dimethylaminophenyl)-1,3,4-thiadiazole as corrosion inhibitors for mild steel in acidic media. *Corrosion Science* **46** (11): 2781–2792.

254 Starkey, R.J., Monteith, G.A. and Aften, C.A. (2008). Biocide for well stimulation and treatment fluids. US Patent Application 20080032903, assigned to applicants.

255 Kessler, V., Bennet, B., Diaz, R. et al. (2009). Identification and analysis of biocides effective against sessile organisms. SPE International Symposium on Oilfield Chemistry, The Woodlands, TX (20–22 April 2009), SPE 121082.

256 Williams, T.M., Hegarty, B. and Levy, R. (2001). Control of SRB biofouling and MIC by chloromethyl-methylisothiazolone. CORROSION 2001, Houston, TX (11–16 March 2001), NACE 01273.

257 Williams, T.M. (2009). Efficacy of isothiazolone biocide versus sulfate reducing bacteria (SRB). CORROSION 2009, Atlanta, GA (22–26 March 2009), NACE 09059.

258 Clifford, R.P. and Birchall, G.A. (1985). Biocide. US Patent 4,539,071, assigned to Dearborn Chemicals Ltd.

259 Morpeth, F.F. and Greenhalgh, M. (1990). Composition and use. EP Patent 3,903,94, assigned to Imperial Chemicals Industries PLC.

260 Alink, B.A.M.O., Martin, R.L., Dogherty, J.A. and Outlaw, B.T. (1993). Volatile corrosion inhibitors for gas lift. US Patent 5,197,545, assigned to Petrolite Corporation.

261 Alink, B.A.M.O. and Outlaw, B.T. (2002). Thiazolidines and use thereof for corrosion inhibition. US Patent 6,419,857, Baker Hughes Incorporated.

262 Taylor, G.N. (2013). Method of using dithiazines and derivatives thereof in the treatment of wells. US Patent 8,354,361, assigned to Baker Hughes Incorporated.

263 Taylor, G.N. (2013). The isolation and formulation of highly effective corrosion inhibitors from the waste product of hexahydrotriazine based hydrogen sulphide scavengers. Chemistry in the Oil Industry XIII (4–6 November 2013). Manchester, UK: The Royal Society of Chemistry, pp. 61–78.

264 Corrin, E., Rodriguez, C. and Williams, T.M. (2015). A case study evaluating a co-injection biocide treatment of hydraulic fracturing fluids utilized in oil and gas production. CORROSION 2015, Dallas, TX (15–19 March 2015), NACE 2015-5998.

265 Enzien, M.V., Yin, B., Love, D. et al.. (2011).Improved microbial control programs for hydraulic fracturing fluids used during unconventional shale-gas exploration and production. SPE International Symposium on Oilfield Chemistry, The Woodlands, TX (11–13 April), SPE 141409.

266 Eggensperger, H. and Diehl, K.-H. (1979). Preserving and disinfecting method employing certain bis-oxazolidines. US Patent 4,148,905, assigned to Sterling Drug Inc.

267 Mena-Cervantes, V.Y., Hernandez-Altamirano, R., Buenrostro-Gonzalez, E. et al. (2013). Development of oxazolidines derived from polyisobutylene succinimides as multifunctional stabilizers of asphaltenes in oil industry. *Fuel* **110**: 293–301.

268 Dillon, E.T. (1990). Composition and method for sweetening hydrocarbons. US Patent 4,978,512, assigned to Quaker Chemical Company.

269 V.Y. Mena-Cervantes, R. Hernandez-Altamirano, E. Buenrostro-Gonzalez et al. (2011). Multifunctional composition base 1,3-oxazinan-6-ones with corrosion inhibition and heavy organic compounds inhibition and dispersants and obtaining process. US Patent Application20110269650, assigned to Instituto Mexicano Del Petroleo.

270 Gaja, M.A. and Knapp, J.S. (1997). The microbial degradation of Benzothiazoles. *Journal of Applied Microbiology* **83** (3): 327–334.

271 Wang, B.Y. ed. (2008). *Environmental Biodegradation Research Focus*. Nova Science Publishers Inc.

272 Hu, K., Li, H.-R., and Ou, R.-J. (et al., 2014). Tissue accumulation and toxicity of isothiazolinone in *Ctenopharyngodon idellus* (grass carp): association with P-glycoprotein expression and location within tissues. *Environmental Toxicology and Pharmacology* **37** (2): 529–535.

273 https://www.pharosproject.net/uploads/files/cml/1400084689.pdf

274 Reynhout, M.J., Kind, C.E. and Klomp, U.C. (1993). A method for preventing or retarding the formation of hydrates. EP Patent 5,269,29, assigned to Shell Int. Research, B.V.

275 Dadgar, A. (1988). Corrosion inhibitors for clear, calcium-free high density fluids. US Patent 4,784,779, assigned to Great Lakes Chemical Corp.

276 Shin, C.C. (1988). Corrosion inhibiting composition for zinc halide-based clear, high density fluids. WO Patent 1988002432, assigned to Great Lakes Chemical Corp.

277 Almond, S.W. (1984). Method and compositions for fracturing subterranean formations. US Patent 4,477,360, assigned to Halliburton Company.

278 Schlemmer, R.F. (2007). Membrane forming in-situ polymerization for water based drilling fluids. US Patent 7,279,445, assigned to M.I. L.L.C.

279 Walker, C.O. (1967). Drilling fluid. US Patent 3,314,883, assigned to Texaco Inc.

280 Buriks, R.S. and Dolan, J.G. (1986). Demulsifier composition and method of use thereof. US Patent 4,626,379, assigned to Petrolite Corporation.

281 Kuzee, H.C. and Raajmakers, H.W.C. (1999). Method for preventing deposits in oil extraction. WO Patent 1999064716, assigned to Cooperatie Cosun U.A.

282 Johannsen, F.R. (2003). Toxicological profile of carboxymethyl inulin. *Food and Chemical Toxicology* **41**: 49–59.

283 Kalota, D.J. and Silverman, D.C. (1994). Behavior of aspartic acid as a corrosion inhibitor for steel. *Corrosion* **50** (2): 138–145.

284 Yamamoto, H., Takayanagi, Y., Takahashi, K. and Nakahama, T. (2001). Chelating agent and detergent comprising the same. US Patent 6,221,834, assigned to Mitsubishi Rayon Co. Ltd.

285 Kumar, T., Vishwanatham, S., and Kundu, S.S. (2010). A laboratory study on pteroyl-l-glutamic acid as a scale prevention inhibitor of calcium carbonate in aqueous solution of synthetic produced water. *Journal of Petroleum Science and Engineering* **71** (1–2): 1–7.

286 Fan, J.C. and Fan, L.-D.G. (2001). Inhibition of metal corrosion. US Patent 6,277,302, assigned to Donlar Corporation.

287 Duncum, S.N., Edwards, A.R. and Osborne, C.G. (1993). Method for inhibiting hydrate formation. EP Patent 5,369,50, assigned to The British Petroleum Company plc.

288 Vijn, J.P. (2001). Dispersant and fluid loss control additives for well cements, well cement compositions and methods. US Patent 6,182,758, assigned to Halliburton Energy Services Ltd.

289 Valone, F.W. (1987). Corrosion inhibiting system containing alkoxylated amines. US Patent 4,636,256, assigned to Texaco Inc.

290 Knox, D. and McCosh, K. (2005). Displacement chemicals and environmental compliance-past present and future. Chemistry in the Oil Industry IX 31, Manchester, UK (October - 2 November 2005).

291 Craddock, H.A., Berry, P. and Wilkinson, H. (2007). A new class of. green. Corrosion inhibitors, further development and application. Transactions of the 18th International Oilfield Chemical Symposium, Geillo, Norway (25–28 March 2007).

292 Deyab, M.A. (2016). Decyl glucoside as a corrosion inhibitor for magnesium-air battery. *Journal of Power Sources* **325**: 98–103.

293 Breedon, D.L. and Meyer, R.L. (2005). Ester-containing downhole drilling lubricating composition and processes therefor and therewith. US Patent 6,884,762, assigned to Nupark Drilling Fluids LLC.

294 Wallace, J. (2011). Composition and method for inhibiting the deleterious effects of sulphate reducing bacteria. US Patent Application 2011/0056693, assigned to Weatherford US.

295 Ashok, K. and Upadhyaya, K. (2012). Tannins are astringent. *Journal of Pharmacognosy and Phytochemistry* **1** (3): 45–51.

296 Perez, M.A. and Collins, R.A. (2015). Rheological behavior of water-based drilling fluids contaminated with gypsum ($CaSO_4$) using unmodified dividivi tannins (*Caesalpinia coriaria*) as deflocculant agent. SPE Latin American and Caribbean Petroleum Engineering Conference, Quito, Ecuador (18–20 November 2015), SPE 177032.

297 Craddock, H.A., Mutch, K., Sowerby, K. et al. (2007). A case study in the removal of deposited wax from a major subsea flowline system in the gannet field. International Symposium on Oilfield Chemistry, Houston, TX (28 February-2 March 2007), SPE 105048.

298 Craddock, H.A., Campbell, E., Sowerby, K. et al. (2007). The application of wax dissolver in the enhancement of export line cleaning. International Symposium on Oilfield Chemistry, Houston, TX (28 February-2 March, 2007), SPE 105049.

299 Goncalves, J.T., De Oleveira, M.F. and Aragao, A.F.L. (2007). Compositions of oil-based biodegradable drilling fluids and process for drilling oil and gas wells. US Patent 7,285,515, assigned to Petroleo Brasileiro S.A.

300 Penna, A., Arias, G. and Rae, P. (2006). Corrosion inhibitor intensifier and method of using the same. US Patent Application 20060264335, assigned to BJ Services company.

301 Hatchman, K., Fellows, A., Jones, C.R. and Collins, G. (2013). Corrosion inhibitors. International Patent Application WO 2013113740, assigned to Rhodia Operations.

302 Ford, W.G.F. and Hollenbeck, K.H. (1987). Composition and method for reducing sludging during the acidizing of formations containing sludging crude oils. US Patent 4,663,059, assigned to Halliburton Company.

303 Gartner, C.D. (1997). Suspension formulations of 2,2-dibromo-3-nitrilopropionamide. US Patent 5,627,135, assigned to Dow Chemical Company.

304 Cronan, J.M. Jr. and Mayer, M.J. (2008). Synergistic biocidal mixtures. US Patent 7,008,545, assigned to Hercules Incorporated.

305 Jakubowski, J.A. (1986). Admixtures of 2-bromo-2-bromomethylglutaronitrile and 2,2-dibromo-3-nitrilopropionamide. US Patent 4,604,405, assigned to Calgon Corporation.

306 Burger, E.D. and Odom, J.M. (1999). Mechanisms of anthraquinone inhibition of sulphate reducing bacteria. International Symposium on Oilfield Chemistry, Houston, TX (16–19 February 1999), SPE 50764.

307 Burger, E.D. and Crews, A.B. (2001). Inhibition of sulfate reducing bacteria by anthraquinone in a laboratory biofilm column under dynamic conditions. CORROSION 2001, Houston, TX (11–16 March 2001), NACE -1274

308 Cattanach, K.C., Jovancicevic, V., Philippe Prince, S. et al. (2012). Water-based formulation of H2S/mercaptan scavenger for fluids in oilfield and refinery applications. International Patent Application WO 2012003267, assigned to Baker Hughes Incorporated.

309 Yang, J., Salma, T., Schield, J.A. et al. (2009). Multifunctional scavenger for hydrocarbon fluids. International Patent Application WO 2009052127, assigned to Baker Hughes Incorporated.

310 Al-Sahhaf, T.A., Fahim, M.A., and Elkilani, A.S. (2002). Retardation of asphaltene precipitation by addition of toluene, resins, deasphalted oil and surfactants. *Fluid Phase Equilibria* **194–197**: 1045–1057.

311 Trbovich, M.G. and King, G.E. (1991). Asphaltene deposit removal: long-lasting treatment with a co-solvent. SPE International Symposium on Oilfield Chemistry, Anaheim, CA (20–22 February 1991), SPE 21038.

312 Galoppini, M. (1994). Asphaltene deposition monitoring and removal treatments: an experience in ultra deep wells. European Production Operations Conference and Exhibition, Aberdeen, UK (15–17 March 1994), SPE 27622.

313 Canonico, L.B., del Bianco, A., Piro, G. et al. (1994). A comprehensive approach for the evaluation of chemicals for asphaltene deposit removal. In: *Recent Advances in Oilfield Chemistry*, 220–233. Royal Society of Chemistry.

314 Straub, T.J., Autry, S.W. and King, G.E. (1989). An investigation into practical removal of downhole paraffin by thermal methods and chemical solvents. SPE Production Operations Symposium, Oklahoma City, OK (13–14 March 1989), SPE 18889.

315 Bailey, J.C. and Allenson, S.J. (2008). Paraffin cleanout in a single subsea flowline environment: glycol to blame? Offshore Technology Conference, Houston, TX (5–8 May 2008), OTC 19566.

316 Kentish, M.J. (1996). *Practical Handbook of Estuarine and Marine Pollution*. CRC Marine Science.

317 http://www.saltech.co.il/_uploads/dbsattachedfiles/chemical.pdf

318 Scovell, E.G., Grainger, N. and Cox, T. (2001). Maintenance of oil production and refining equipment. International Patent Application WO 2001074966, assigned to Imperial Chemical Industries PLC.

319 Willey, T.F., Willey, R.J. and Willey, S.T. (2007). Rock bit grease composition. US Patent 7,312,185, assigned to Tomlin Scientific Inc.

320 Walter, R.W. Jr., Relenyi, A.G. and Johnson, R.L. (1989). Control of biofouling at alkaline pH and/or high water hardness with certain alkylthioalkylamines. US Patent 4,816,061, assigned to The Dow Chemical Company.

321 Baldev, E., Ali Mubarak, D., Ilavarasi, A. et al. (2013). Degradation of synthetic dye, Rhodamine B to environmentally non-toxic products using microalgae. *Colloids and Surfaces B: Biointerfaces* **105**: 207–214.

322 Rappold, H. and Bacher, A. (1974). Bacterial degradation of folic acid. *Journal of General Microbiology* **85**: 283–290.

323 Indrawati, C., Arroqui, C., Messagie, I. et al. (2004). Comparative study of pressure and temperature stability of 5-methyltetrahydofolic acid in model systems and in food products. *Journal of Agricultural and Food Chemistry* **52**: 485–492.

324 Strobel, B.W. (2001). Influence of vegetation on low-molecular-weight carboxylic acids in soil solution: a review. *Geoderma* **99** (3–4): 169–198.

325 Yoshida, K., Ishii, H., Ishihara, Y. et al. (2009). Bioremediation potential of formaldehyde by the marine microalga nannochloropsis oculata ST-3 strain. *Applied Biochemistry and Biotechnology* **157** (2): 321–328.

326 Cornelissen, G., Gustaffsson, O., Buchelli, T.D. et al. (2005). Extensive sorption of organic compounds to black carbon, coal, and kerogen in sediments and soils: mechanisms and consequences for distribution, bioaccumulation, and biodegradation. *Environmental Science and Technology* **39** (18): 6881–6895.

327 Brunauer, S., Emmett, P.H., and Teller, E. (1938). Adsorption of gases in multimolecular layers. *Journal of the American Chemical Society* **60**: 309–319.

328 Moreno-Castilla, C. (2005). Adsorption of organic molecules from aqueous solutions on carbon materials. *Carbon* **42** (1): 83–94.

7

Silicon Chemistry

The use and application of silicon chemistry in the upstream oil and gas industry can be designated into two main types of chemistries, namely, silica (silicon dioxide) and the silicates, including the zeolites mainly used in drilling operations and the silicones, which includes silanes, siliconates, silicones with and without functionalisation, silicon resins and polymers and silicon surfactants. This latter group has found uses across all the main operations in the oilfield sector.

There are also examples of products straddling across both functionalities such as the poly-silicate esters, and these shall be included where most suited in application; in this case these are used as gelling agents that can selectively reduce the permeability of highly permeable regions of reservoir formations [1].

The environmental acceptance and impact of these products possess some interesting dilemmas and challenges both to the oilfield chemistry practitioner and to the environmental scientist, and these will be discussed throughout the chapter.

Silicon has the atomic number 14 and is a hard and brittle crystalline solid with a blue-grey metallic lustre. It is a tetravalent metalloid. It is a member of group 14 in the periodic table, which is interestingly the same group as carbon. However, unlike carbon it is not very reactive, but does have has great chemical affinity for oxygen. It was first purified and characterised in 1823 by Jöns Jacob Berzelius. However, it very rarely occurs as the pure element in the Earth's crust. It is most widely distributed as various forms of silicon dioxide (silica) or silicates. Indeed over 90% of the Earth's crust is composed of silicate minerals, making silicon the second most abundant element in the Earth's crust after oxygen [2].

Most silicon is used commercially without being separated and often with little processing of the natural minerals. Such use includes industrial construction with clays, stone and sand. Silicates, primarily sodium silicate, is used in Portland cement for mortar and mixed with silica sand and gravel to make concrete for all sorts of construction. Silicates are used in ceramics, traditional soda-lime glass and many other specialty glasses. Silicon is also the basis of the widely used synthetic polymers called silicones.

Silicon is an essential element in biology, although only tiny traces are required by animals. However, various microorganisms, such as diatoms, and also sea sponges secrete skeletal structures made of silica.

The use of sand, quartz, glass and related materials has not been covered in this chapter, and as these materials are chemically inert, they are not within the scope of this book. These materials are used in drilling and completion work and also in cementing and hydraulic fracturing in a variety of applications [3].

Oilfield Chemistry and Its Environmental Impact, First Edition. Henry A. Craddock.
© 2018 John Wiley & Sons Ltd. Published 2018 by John Wiley & Sons Ltd.

7.1   Silica

In addition to being a regular component of many cements including oil well cements, silica powder has been examined as a stabiliser for such cements at high temperatures [3]. Both increased temperature stability and improved compressive strength of the cement have come from the use of silica flour and silica fume [4].

Silica flour is simply made by grinding pure silica sand to a very fine powder. Fumed silica is made in a very unusual way, by reacting silicon tetrachloride in an oxyhydrogen flame that has an excess of oxygen. A very light, fluffy, pourable solid results.

Varying the amounts of silica flour and fumed silica can have a positive effect on the amount of compressive strength regression and high temperatures [5], and also in wells containing high levels of carbon dioxide, the use of fumed silica was found not to significantly affect the permeability of the cement [6].

Cement slurry formulations containing colloidal silica have been used in a number of offshore operations, including the North Sea, such as in primary cementing operations to help control annular gas migration, enhance cement slurry characteristics without solids settling and aid cement bonding to formation and casing. Colloidal silica slurries have been successfully placed as lightweight lead cements in intermediate casing operations and have effectively sealed annuli in gas storage wells and in horizontal completions.

Colloidal silica is more efficiently handled in offshore operations than usual silica cement additives, because less bulk material is required. The reduced volumes also reduce overall slurry cost. Because colloidal silica particles are so fine, there is virtually no settling of the base material in containment offshore [7].

It is worthy of note that silica is one of the most common components of sand, particularly sands from inland continental and non-tropical coastal sources. As such it is widely used in hydraulic fracturing operations as it is cheap and the simplest type of proppant materials.

Silica particles are also known to be utilised in mixture with silicones such as polydimethylsiloxane (PDMS) (see Section 7.3.1) to improve anti-foam behaviour and efficiency and have a strong deteriorating effect on foam stability, even at low concentrations [8].

7.1.1   Silica Gel

Silica gel is a granular, vitreous, porous form of silicon dioxide. It is manufactured from sodium silicate and is composed of a porous silica microstructure, suspended inside a liquid. It is also a naturally occurring mineral that is purified and processed into either granular or beaded form. Its main function is as a desiccant, where it has an average pore size of 2.4 nm and has a strong affinity for water molecules. It is probably most commonly encountered by the reader as beads in a small (typically 2 cm × 3 cm) paper packet, where it is used as a desiccant to control local humidity and avoid spoilage or degradation of certain goods.

A colloidal suspension of silica gel is used in a number of oilfield applications. It has been used for water production control and well casing repair and in well workovers for water injection profile modification with mixed success [9].

7.1.2   Nano-silica

Silicon dioxide nanoparticles, also known as silica nanoparticles or nano-silica, are divided into two types according to their structure: P type (porous particles) and S type (spherical particles). P-type nano-silica surface contains a number of nanopores with the pore rate of $0.611 \, \mathrm{ml \, g^{-1}}$;

therefore, P type has much larger specific surface area (SSA) compared with S type (see US3440). US3436 is an S type. Nano-silica particles are the basis for a great deal of biomedical research due to their stability, low toxicity and ability to be functionalised with a range of molecules and polymers [10].

Recently the application of nano-silica particles has been examined for oilfield use.

As has been described in other chapters of this book, scale deposition in producing wellbores is a serious problem in the upstream oil and gas industry. The problem gets worse when the scale is caused by barium and strontium salts. These salts are difficult to clean as they are not easily soluble in any kinds of solution or chelating agents. Formation of scale changes the surface roughness of the production tubing, thereby increasing the frictional pressure drop, leading to a decreased production rate. Further deposition clogs the production tubing, creating hindrances for lowering tools into lower sections of the production string. In worst cases tubing replacement needs to be carried out, which is a capital-intensive activity.

Creating a super-hydrophobic surface with multi-scale nanostructures on the inside of the production tubing can greatly reduce the chances of scale deposition. This surface is created on epoxy paint surfaces using a feasible dip coating process. Microstructures are created on this surface using sandblasting; then nanostructures are introduced onto the micro surface by anchoring 50–100 μm silicon dioxide particles and finally completed by dip coating with nano-silicon dioxide/epoxy adhesive solution.

The produced water experiences an increased contact angle that significantly reduces the chances of deposition of the ions that contribute to scale formation and growth [11].

A recent interesting application of nano-silica is its use as a cross reference in a microfluidic approach to enhanced oil recovery in heavy oil. The authors claim that this approach saves time and considerable expense in conducting and examining core flood studies before committing to the field application of surfactant- and polymer-based enhanced oil recovery. Additionally, it was shown that the incorporation of silica nanoparticles in the injected solution (i.e. nanofluid) can dramatically enhance oil production [12].

7.1.3   Silica Scale

As can be observed, the breadth of application, so far explored in the use of silica, is quite specialised, and this may be due to the fact that colloidal silica is one of the most unwanted deposits in produced water as its control can be difficult and problematic. This is particularly true where silica is becoming the critical limiting factor in the reuse of produced water, which of course would be both economically and environmentally useful. Silica deposits are a result of polymerisation, co-precipitation with other minerals, precipitation with other multivalent ions and biological activity in the water. Several of these processes may take place concurrently, making it difficult to predict equilibrium solubility. The solubility of amorphous silica is also dependent on many other factors such as pH, temperature, particle size, particle hydration and the presence of other ions, such as iron, aluminium, etc. A hard silica scale is formed when calcium carbonate or other mineral precipitate provides a crystalline matrix in which silica can be entrapped [13].

All of these factors make silica scale a difficult scale to control and remove once deposited. In the last decade or so, there have been a number of chemical inhibitors examined to assess their effectiveness in the prevention of silica deposits [14]. Work has shown that a number of polymers, many of which have been described in Chapter 2, and also blends of polymers containing carboxymethyl inulin, described in Section 6.6.1 (see Figure 6.126), exhibit excellent performance in preventing silica scale formation.

7.1.4 Health and Safety Issues around Silica Particles

Undoubtedly this is a major factor if not the most predominant cause in restricting the use and application of silica more widely in the upstream oil and gas sector. Crystalline silica has been one of the most widely studied substances in the history of occupational disease and industrial hygiene. Crystalline silica is a basic component of soil, sand, granite and many other minerals. Quartz is the most common form of crystalline silica. Cristobalite and tridymite are two other forms of crystalline silica. All three forms may become respirable size particles when workers chip, cut, drill or grind objects that contain crystalline silica.

Silica exposure remains a serious threat to nearly 2 million US workers, including more than 100 000 workers in high-risk jobs such as abrasive blasting, foundry work, stonecutting, rock drilling, quarry work and tunnelling. Crystalline silica has been classified as a human lung carcinogen. Additionally, breathing crystalline silica dust can cause silicosis, which in severe cases can be disabling or even fatal. The respirable silica dust enters the lungs and causes the formation of scar tissue, thus reducing the lungs' ability to take in oxygen. There is no cure for silicosis. Since silicosis affects lung function, it makes one more susceptible to lung infections like tuberculosis [15].

It is believed that the recent development and increased activity in hydraulic fracturing could lead to a rise in occupational health concerns related to crystalline silica exposure and that this could reverse the decreasing fatality trend associated with silica exposure in the last decade, particularly in the United States [16].

7.1.5 Environmental Impacts and Effects of Silica

In opposition to the serious safety concerns, silica is one of the most benign and low environmental impact substance used as an additive across a large number of industrial sectors. Its main environmental impact as is described in the previous section is due to its airborne emission as dust and crystalline particles, resulting in occupational exposure, causing serious debilitating and fatal lung disease.

Extensive studies have proven that silicon is an essential element required in humans and other animals for normal growth, development and integrity of bones, joints and connective tissue, as well as for hair, skin, nails, arteries and cartilage. Literally every part of the body that requires strength and elasticity requires silicon. Specifically, silicon is associated with collagen formation, the fibrous protein matrix that provides support for body structures such as cartilage and bones. This is why optimal bone health depends upon silicon as well as calcium. Animal bodies process silicon inefficiently, and the only bioavailable form of silicon is orthosilicic acid, $Si(OH)_4$, not silica SiO_2. Silica is not bioavailable. Silicon is found in food as silicates, which are likewise not bioavailable. All dietary silicon must be dissolved in the stomach into orthosilicic acid, the monomeric (single unit) correct form of silicic acid, which is absorbed and used by the body. Silicic acid, however, is unstable. In concentrations over 1 ppm (the amount typically found in mineral water), orthosilicic acid readily polymerises into long chains, converting in the process back into a non-bioavailable silicate [17].

Therefore, as can be seen from an environmental impact viewpoint, silica and its associated materials are not bioaccumulating, being not bioavailable, and also do not biodegrade, being highly unreactive, except under extreme acid conditions.

In terms of silica as used in nanotechnology, the evidence is still somewhat circumspect, and caution should be taken as to the possible effects of all chemically based nanotechnology [18].

There is a potential for positive environment effects to be utilised from using silica due to its unreactive nature and capability of encapsulating other materials. Oil and hydrocarbon drill cuttings have been proposed to be microencapsulated to allow their disposal in an environmentally safe manner [19].

7.2 Silicates

Silicates both as synthesised materials such as sodium silicate and as natural minerals, for example, bentonite, have a number of uses across the drilling and cementing areas or the upstream oil and gas industry. Their use as additives in production operations is much less although there are some that will also be described.

7.2.1 Sodium (and Potassium) Silicate

Sodium silicate is the common name for compounds with the formula $(Na_2SiO_2)_nO$. It is formed by dissolving silica or silicate minerals in NaOH solutions. The silicate anion, SiO_4^{-4}, is found in solutions of sodium silicate. Silicate anions form polysilicates or colloidal silica gel. Sodium silicate is one of a number of related compounds that include sodium orthosilicate (Na_4SiO_4), sodium pyrosilicate ($Na_6Si_2O_7$) and sodium metasilicate (Na_2SiO_3), also known as water glass. All are glassy, colourless solids that are soluble in water.

Anhydrous sodium silicate contains a chain polymeric anion composed of corner-shared $\{SiO_4\}$ tetrahedral, not a discrete SiO_3^{2-} ion; an idealised structure is shown in Figure 7.1. In addition to the anhydrous form, there are hydrates with the formula $Na_2SiO_3 \cdot nH_2O$ (where $n = 5, 6, 8, 9$), which contain the discrete, approximately tetrahedral anion $SiO_2(OH)_2^{2-}$ with water of hydration. For example, the commercially available sodium silicate pentahydrate $Na_2SiO_3 \cdot 5H_2O$ is formulated as $Na_2SiO_2(OH)_2 \cdot 4H_2O$, and the nonahydrate $Na_2SiO_3 \cdot 9H_2O$ is formulated as $Na_2SiO_2(OH)_2 \cdot 8H_2O$.

Sodium silicate is stable in neutral and alkaline solutions. In acidic solutions, the silicate ion reacts with hydrogen ions to form silicic acid, which when heated and roasted forms silica gel, a hard, glassy substance.

One of the main uses of silicates is in the composition of drilling fluids, but this is normally restricted to areas that are environmentally sensitive and highly regulated regions such as the North Sea basin. In comparison with oil-based muds and/or synthetic-based muds, silicate-based muds are usually deselected on the basis of their high coefficients of friction against rock. Such muds are based on potassium and sodium silicate and have the added advantage of being somewhat cheaper than synthetic-based fluids. These materials are preferred in high-temperature, high-pressure drilling and highly deviated drilling. Although at present in the North Sea this type of completion is uncommon, this is likely to change as developments in the Arctic and the North East Atlantic, west of Shetland, are brought into production. In other areas where drilling in deep water is more common, such fluids are more often used, in particular the Gulf of Mexico, offshore Brazil and deepwater West Africa [20].

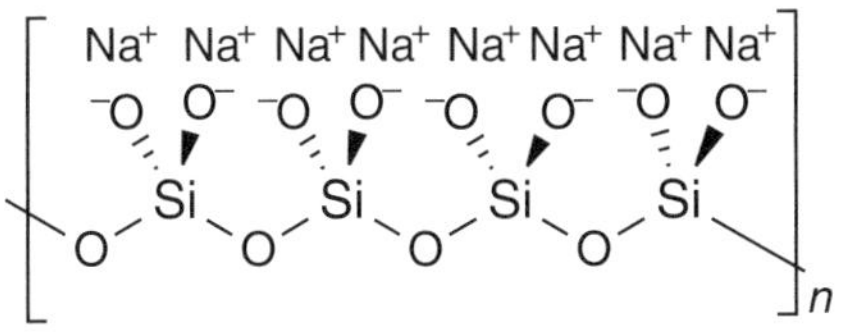

Figure 7.1 Sodium silicate.

A silicate mud is a type of shale-inhibitive water-based drilling fluid that contains sodium silicate (or potassium) silicate polymeric ions. Sodium (or potassium) silicates provide levels of shale inhibition comparable with oil-based muds. These inhibitive properties are due to the gelation of silicates that occurs with divalent ions in the formulation, providing an effective water barrier that prevents hydration and dispersion of the shales. This physicochemical barrier helps improve wellbore stability and provides in-gauge holes through troublesome shale sections that otherwise might require a non-aqueous drilling fluid, hence their preference in environmentally sensitive areas [21].

In formulating silicate-based drilling fluids, it is often necessary to add a lubricant enhancer, and in recent years these have been based around complex thiophosphates. The lubricants that are commonly used in water-based muds do not provide good lubricity in silicate-based drilling fluids [22]. The addition of silicic acid polymers and related products has been shown to afford drilling fluids with good lubricity; however these, especially high-pH fluids, are not receptive to traditional lubricants [23].

Sodium silicate has also been used in combination with sodium hydroxide to activate natural surfactants in heavy crude oils to aid in their transportation [24].

Sodium silicate has been used in sealing composition to successfully control loss of circulation and stop undesirable fluid production. This occurs at a change of pH where the sodium silicate solution is polymerised or cross-linked to form a gel [25]. Similarly sodium (and potassium) orthosilicate has been utilised repeatedly in conjunction with polymers for a variety of water shut-off treatments [26]. The use of alkaline metal silicates in conjunction with aminoplastic resins has also been described for plugging and diversion materials [27].

Sodium orthosilicate is also used as an additive of choice in maintaining or improving steam flooding performance, a method of enhanced oil recovery. Steam flooding can suffer reduced performance due to channelling and gravity segregation, and the addition of the orthosilicate helps to maintain and/or improve performance [28].

Silicates, mainly sodium silicate, have been used in corrosion inhibition for over half a century primarily in cooling water systems as they are ineffective at high ionic strength [29].

There is little direct application of silicates as corrosion inhibitors in the oilfield; however the use of silicate-based drilling fluids has been shown to have the added advantage of protection of metals against corrosion. Aside from providing an alkaline environment, silicate has been shown to deposit a protective film on various metal surfaces. Drill string corrosion coupons were used to gauge the amount of wear and corrosion while running potassium silicate-based drilling fluids, and gravimetric analysis confirmed that the rate of metal corrosion was minimal under a variety of drilling conditions [30].

7.2.2 Aluminium Silicate, Aluminosilicates and Related Minerals (e.g. Bentonite)

Aluminium silicate has been used to improve the compressive strength of well cements [31]. The use of aluminosilicates, particularly minerals such as quartz, are claimed to provide increased chemical resistance [32], and the use of aluminosilicate microspheres has been utilised in calcium phosphate cements to produce lightweight materials that can be used under hydrothermal conditions at high temperatures between 200 and 1000 °C and therefore particularly useful in geothermal applications [33]. This latter type of silicate belongs to a class known as zeolites.

7.2.2.1 Zeolites

Zeolites are a large group of minerals consisting of hydrated aluminosilicates of sodium, potassium, calcium and barium. They can be readily dehydrated and rehydrated and are used as cation exchangers and molecular sieves in a wide variety of industrial applications. They are widely used in catalysis of petrochemicals in the downstream oil and gas sector; however their use in upstream applications has been confined to cement additives particularly in compressive strength development, rheology and density [34]. The zeolites chabazite and clinoptilonite are claimed to be particularly useful [35].

7.2.2.2 Bentonite and Kaolinite

Bentonite is an absorbent aluminium phyllosilicate clay consisting mostly of montmorillonite that has the chemical formula $(Na, Ca)_{0.33}(Al, Mg)_2Si_4O_{10}(OH)_2 \cdot nH_2O$. It was named by Wilbur C. Knight in 1898 after the Cretaceous Benton Shale found in Wyoming and adjacent states. The different types of bentonite are each named after the respective dominant element, such as potassium (K), sodium (Na), calcium (Ca) and aluminium (Al). Bentonite is usually formed by weathering of volcanic ash, most often in the presence of water. For industrial purposes, two main classes of bentonite exist: sodium and calcium bentonite. In addition to montmorillonite another common clay species that is sometimes dominant and used in the oilfield is kaolinite. Kaolinite-dominated clays are commonly referred to as tonsteins and are typically associated with coal.

Bentonite has a variety of uses in the upstream oil and gas sector, the majority of which are associated with drilling, cementing and well completion. Water-based muds, primarily fresh-water-based muds, usually contain bentonite that is present as a viscosity enhancer and fluid loss control additive. A number of bentonite-based muds are used primarily when drilling the upper-hole sections and are usually referred to as dispersed non-inhibited systems [3]. The modification of bentonite with alkyl silanes can improve their dispersing properties [36].

These bentonite clays have been widely used in drilling fluids for many decades due to their dispersant, rheological and fluid control properties; they have a capacity for ion exchange and their properties can thus be modified. For example, their surface can be modified by acid treatment, and their organophilic nature can be increased by treatment with quaternary ammonium compounds [3].

Bentonite and kaolinite clays have been found to be suitable for preparing a solids-stabilised oil-in-water emulsion. This low viscosity oil-in-water emulsion can be used to enhance production of oil from subterranean reservoirs and can also be used to enhance the transportation of oil through a pipeline [37]. However, in enhanced oil recovery by alkali water flooding, clays such as kaolinite can be consumed by dissolution, making them ineffective as a useful additive. The dissolution rates of the silica are dependent on a variety of factors including temperature, salinity and pH [38].

Bentonite clay in combination with polyacrylamide (PAM) can form a gel-like material when added to the circulating drilling fluid that expands in water up to 30–40 times its initial volume and enters cracks in the natural rock strata. This material changes to a strongly adhering insulating film and plugging material within 40 min of addition [3].

Bentonite is used in well cementing operations as it can reduce the over cement slurry weight. The quality and quantity of bentonite used is critical in ensuring effective performance [39].

In hydraulic fracturing lightweight and high strength proppants derived from kaolinite have been described [40].

Bentonite compositions have been described and used in water shut-off applications. Bentonite particles covered in a natural resin coating are claimed as an environmentally friendly well plugging material [41].

High permeability zones in a reservoir can be blocked by the use of a bentonite clay that can be selectively placed in an organic solvent so that in contact with water, or when the ionic strength changes, it swells [42].

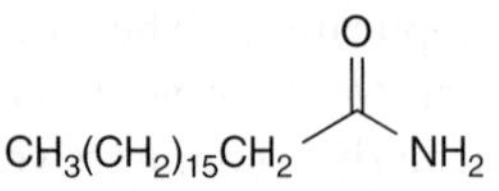

Figure 7.2 Stearamide.

As stated earlier bentonite-type clays can have their organophilic nature be increased by treatment with quaternary ammonium compounds. If such surfactants have an amide linkage such as those based on stearamide (Figure 7.2), then this type of surfactant is usually biodegradable, and drilling fluid compositions with clays treated with such surfactants are classified as biodegradable.

7.2.2.3 Diatomaceous Earth

This has lower SG than bentonite and is a useful cement additive as it will not increase the slurry viscosity and concentrations of up to 40% w/w have been used [3].

The typical chemical composition of oven-dried diatomaceous earth is 80–90% silica, with 2–4% alumina (attributed mostly to clay minerals) and 0.5–2% iron oxide. Diatomaceous earth consists of fossilised remains of diatoms, a type of hard-shelled algae.

Other related mineral such as mica and feldspar are occasionally used in drilling and cementing; however their use is infrequent.

7.2.3 Calcium Silicate

Dicalcium silicate and tricalcium silicate react with water to form calcium silicate hydrate. The initial reaction is exothermic, and then the hydration rests over several hours that is useful behaviour in the placement of completion cements. After placement, the hydration restarts and cement develops strength and other properties [3, 43].

7.2.4 Fluorosilicates

There are a number of fluorosilicate salts that generally have the ion SiF_6^{2-}, and many are used in a variety of industrial functions. In the upstream oil and gas sector, they are primarily encountered when acidising sandstone or other siliceous formations with hydrofluoric acid where they are a problematic deposit as an undesirable scale formed by the acidisation process. As with silica scales described in Section 7.1.3, additives such as chelants and other molecules, a number of which are described in Chapter 6, are added in the acidisation process to prevent and/or inhibit their formation [44].

7.2.5 Polysilicates

Polysilicate esters can be used to control the permeability of certain oil-bearing formations. Such polysilicates are built up from simple alcohols, such as methanol or ethanol, or from diols such as ethylene glycol or polyols, such as glycerol. After injection, the polysilicate ester forms a gel that selectively decreases the permeability of the high permeable regions of the formation [1]. An example of the methyl silicate polymer is illustrated in Figure 7.3.

7.2.6 Silicate Gels

It has been known for some time that relatively inexpensive sodium silicate gels are useful for large-scale well conformance treatments. For a successful treatment, gelation must be delayed

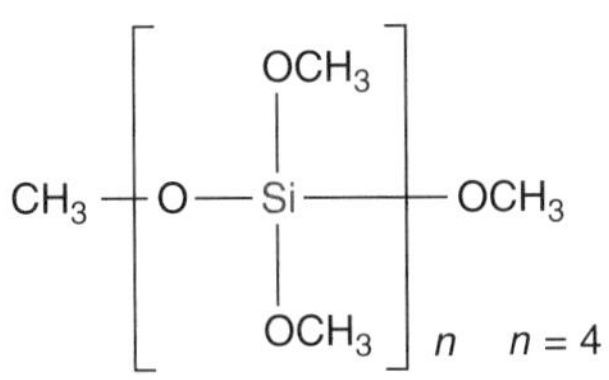

Figure 7.3 Methyl silicate polymer.

to permit proper placement of the silicate solution within the reservoir, to minimise rock/fluid interactions during placement and to ensure long-term gel stability. The addition of an acid or alcohol will cause these high-pH sodium silicate solutions to gel. To regulate the gelling time, an organic material that reacts with water to produce an alcohol and/or acid is often mixed with the silicate solution. An organic rather than inorganic compound is selected because the slow organic reaction rates yield a controlled gelation.

Long-term stability studies showed that, upon standing for several weeks, silicate gels tend to contract, expelling water. This process, called syneresis, clearly will affect the long-term effectiveness of a silicate treatment. Gels formed from the high silicate concentrations exhibited the greatest degree of syneresis. Increasing the temperature increased the syneresis. Some degree of syneresis may be desirable because all treated zones will retain residual permeability [45].

Silicate gels can also enhance the sweep efficiency in water flooding or similar operations by reducing the permeability of the high permeability zones. The gel generation rate can be controlled by the addition of a weak acid such as ammonium sulphate [46].

Recently this established technology has been revisited, and new silicate gel polymers have been examined [47]. Silicate gels and related silicate polymer systems have also been used for water shut-off and casing repairs [48].

7.2.7 Environmental Impacts of Silicates

Due to the physico-chemical properties of soluble silicates, a release into the atmosphere during their application and use is generally not to be expected, and direct emissions from soluble silicates are considered to be negligible. Consequently, no environmental risk assessment related to the use of soluble silicates to soil and air has been performed. In the oil and gas sector, most soluble silicates are mainly discharged to the aquatic environment, either directly to the freshwater or marine environment or via wastewater treatment plants or autonomous wastewater systems.

As soluble silicates are inorganic substances, biodegradation studies are not applicable. However, the removal of silica in several sewage treatment plants has been measured, and an average removal of 10% was determined. In addition, it has been found that silica is continuously removed from water by biochemical processes: diatoms, radiolarians, silicoflagellates and certain sponges serve as a sink for silica by incorporating it into their shells and skeletons as amorphous biogenic silica, referred to as opal ($SiO_2 \cdot nH_2O$) [49, 50].

The primary environmental hazard of commercially used soluble silicates is their moderate-to-strong alkalinity. Soluble silicates with a low molar ratio, such as sodium metasilicate and its hydrates, exhibit a higher alkalinity than the soluble silicates of higher molar ratio. However, most of natural aquatic ecosystems are slightly acidic or alkaline, and usually their pH values fall within the range of 6–9, and due to the high buffer capacity of these ecosystems, pH effects of released soluble silicates to aquatic organisms are very unlikely [51]. Consequently, the predicted no effect concentration (PNEC) derived from artificial laboratory test systems tend to overestimate the effects of soluble silicates to aquatic organisms in realistic natural ecosystems. Therefore, the PNEC derived from the ubiquitous SiO_2 background concentration in the environment with a mean of 7.5 mg SiO_2 per litre in European rivers has been described. This conservative PNEC of 7.5 mg SiO_2 per litre has been used for risk characterisation. Based on freshwater experiments, a PEC/PNEC (predicted effect concentration/predicted no effect concentration) ratio of well below 1 has been observed, indicating that there is no risk to aquatic organisms, at least in riverine systems, after an input of silicates [52]. This would seem even more the case in the larger dilution of the marine environment.

In addition, the amount of soluble silicates introduced into the environment must be seen in the context of the background level due to geochemical weathering processes of silicate minerals. The overall anthropogenic contribution to this total flux is only about 4%, indicating that the natural background concentration/fluctuation is of much higher significance for the silica content of aquatic ecosystems than the use of silicate in commercial or domestic use. For this reason, it can be concluded that the SiO_2, which originates from the use of soluble silicates, has a negligible effect on aquatic ecosystems [53].

There is some evidence of an impact of silicates on eutrophication of surface waters due to nutrient enrichment as a result of the use of silica. A high input of dissolved silicate from Rhine water was rapidly depleted in receiving water systems through vigorous phytoplankton (diatom) growth. In reservoirs ca. 50% of the silicate input was retained over several years. Regeneration of silicate immobilised by diatoms was accelerated by very dense blooms of phytoplankton in reservoirs and associated lakes that increased the pH value over 9 [54]. However the overall effect is controlled by the ratio of silicon to nitrogen and phosphorus [55, 56]. Also the growth of diatoms and their seasonal fluctuation (blooms) are not influenced significantly by the additional anthropogenic silica input, taking into account that the input of silica from the use of commercial silicates is negligible as compared to geochemical weathering processes. Such effects are dependent on many factors varying spatially and temporally (temperature, light, concentrations of phosphates and of other nutrients, activity of grazer population, etc.). Based on the available data, the use of soluble silicates in oil and gas upstream use is not expected to have adverse effects on the aquatic ecosystem. Indeed the use of these materials is often preferred due to their environmental profile and relatively economic use, provided that they can achieve the technical requirements of the particular application.

7.3   Silicones and Silicone Polymers

Silicone products offer low surface tension, low surface viscosity, low cohesive strength and high gas permeability. Combined with the additional properties of thermal and chemical stability and low sensitivity to temperature variations, silicones are excellent candidates to consider for use in the upstream oil and gas sector and the petroleum processing industry.

The term 'silicone' is very generic and covers the family of organosiloxane polymers. The polyorganosiloxanes are polymeric materials containing Si—O—Si bonds of the same nature as in silicates, but with organic radicals fixed on the silicon. They occupy, therefore, an intermediate position between organic and inorganic compounds, or more precisely between silicates and organic polymers, and, through this dual nature, have a number of interesting properties [57].

Historically, the primary use of silicone products in the oil and gas industry has been as foam control agents in applications from exploration and drilling (drilling muds, cements) through production (gas separation, gas treatment, well water injection) to refining (gas treatment, distillation, delayed coking – these treatments are outside the scope of this book). Silicone effectiveness in these applications is so high that it usually only requires extremely low concentrations of active material to achieve the desired results [58].

Silicone-based materials are also used in the upstream oil and gas sector industry for other purposes, such as in demulsification, where the unique properties of silicone polyethers have allowed cost-effective formulations to be developed for use in some demanding applications [59].

7.3.1 Polydimethylsiloxane (PDMS)

Of all the silicone materials in use in the upstream oil and gas industry, the most common is the silicone polymer PDMS (see Figure 7.4); however, the methyl radicals along the chain can be substituted by many organic groups, and in particular, hydrogen, alkyl, allyl, trifluoropropyl, glycol ether, hydroxyl, epoxy, alkoxy, carboxy and amino are the most useful and give variations in properties and effects.

PDMS is a mainstay of anti-foam application in gas processing plants throughout the world. It is a highly effective low dosage treatment in preventing and remediating foaming problems associated with gas processing. It is also used in oil and oil/water/gas separation systems as an anti-foam agent, again at very low dose rates [60]. A polymer with a peak molecular weight distribution of 15 000–130 000 Da has been described to give the best performance in crude oil systems [61].

PDMS is a highly effective defoamer in non-aqueous systems but shows almost no effect in wholly aqueous systems, but when it is blended with hydrophobically modified silica, a highly effective defoamer for aqueous systems is created. This is due to the dual nature of silicone defoamer materials. Soluble silicones can concentrate at the oil and air interface to stabilise gas bubbles, while dispersed drops of silicone can accelerate the coalescence process by rapidly spreading the gas–liquid interface of a bubble that causes the film to thin across the surface and disperse [62].

PDMS is also used in silicone rubber compositions that in the oil and gas industry are used in elastomers as they are chemically compatible with many common solvents, in particular ethylene glycols [58]. Recent studies have also shown them to be compatible with biofuels [63].

7.3.2 Environmental Effects of Polydimethylsiloxane (PDMS)

It had been assumed that as all silicones contain a covalent bond between the silicon atom and an organic group, such as PDMS having a siloxane (Si—O—Si) repeat unit and two methyl groups on each silicon atom. Since the organosilicon linkage is not found in nature, it has been assumed that these polymers do not degrade naturally in the environment. It has been found however that minor degradation takes place by hydrolysis of PDMS to dimethylsilandiol, followed by oxidation of the methyl group to aldehyde and ultimately to CO_2 by a number of bacteria; the major degradation processes are abiotic. High molecular weight PDMS are initially depolymerised by soil hydrolysis of the siloxane bonds to yield organosilanol-terminated oligomers. These organosilanols and low molecular weight linear PDMS and cyclics are evaporated into the atmosphere and are oxidised there by hydroxyl radicals to benign silica, water and CO_2 [64–66].

It is believed that a similar process will be happening in the marine environment although further research is required. In many regulated offshore regions such as the North Sea basin, despite the poor biodegradation of many silicone materials, they are becoming increasingly

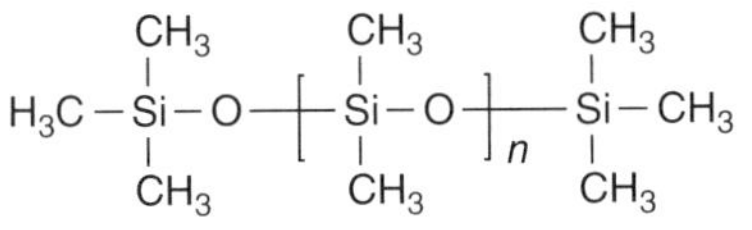

Figure 7.4 Polydimethylsiloxane (PDMS), $n = 200$–1500.

recognised as a potential environmentally acceptable alternative to many toxic and bioaccumulating materials. In general silicones are not bioavailable and although often persistent in the environment are not bioaccumulating and are often non-toxic [67].

7.3.3 Other Organosilicones

Outside of the application of PDMS, there are a number of other organosilicones used in the upstream oil and gas sector that cover a variety of functions.

In cementing and particularly with regard to well abandonment, silicones have been shown to give enhanced properties [3]. Silicone rubber/Portland cement plugging materials have been developed for this application significantly in conjunction with the relevant legislative body [68].

Gas production is often accompanied by unwanted water production, and silicone compounds particularly as microemulsions have been shown to be effective in controlling this unwanted water production. Various silanes and siloxanes are mostly formulated, as microemulsions, in isopropanol or isooctane and shown to affect permeability and reduce water mobility mainly due to a shift in wettability to oil-wet character in the reservoir rock [69].

Gas channelling can occur during the setting of the cement slurry. The formation of these channels is dependent on the setting characteristics, and additives can influence this. During the setting period of a cement, two time cycles of expansion and contraction are observed. This is due to the individual contribution of each component in the cement mixture. To obtain optimal contraction and tightness of the mixture, the final contraction of the cement is critical for blocking gas migration. To aid in this a large number of additives have been tested and used. Critically one of the best and probably most commonly used is a mixture of lignosulphonates containing organic silicon compounds such as ethyl silicone. It is also claimed that such additives retard the setting rates up to 200 °C and increase corrosion resistance [58].

Sodium ethyl siliconate-based materials have been described as drilling lubricants [3].

As described earlier in Section 7.2, most fines in sandstone are not clay but quartz, feldspar or other minerals. The usual clay stabilisers will not control these. 3-Aminopropyltriethoxysilane has been injected during acidisation workovers to control these fines. It forms polysiloxanes in situ with water, which in turn bind the siliceous fines [70, 71].

7.3.4 Silicon Surfactants

The introduction of an organosilicone moiety in a surfactant molecule tends to increase its hydrophobicity. This is because the silicon is heavier than the carbon atom, and hence less silicon atoms are required to give a similar hydrophobic effect. Essentially all surfactants can be made by replacing a carbon-based hydrophobic tail with one silicon atom, usually a dimethylsiloxane group.

Silicone surfactants in the preparation of emulsions have been established in the cosmetics industry for some time [72]. These and similar materials have also been examined for application in the oil and gas industry, particularly in the drilling sector as emulsifying agents.

In the oil and gas exploration and drilling sector, extreme conditions are encountered in offshore deepwater, and this has required adaptations of the drilling fluid's composition particularly with temperatures close to 0 °C and pressures not uncommonly at 400 bar. Under such conditions gas hydrates can readily form and can have serious implications to drilling operations with operational interruptions and even catastrophic effects as potentially in the BP Macondo well disaster. The oil-based muds required in the drilling of these deepwater wells are usually formulated as invert emulsions that can be difficult to stabilise over large ranges of temperature and pressure, and hence there is the need to add additional emulsifiers, which can include silicone-based surfactants, which are also environmentally acceptable additives [73].

A growing use of silicon surfactants in the oilfield sector is as demulsifiers, particularly in combination with other surfactants [59], or as part of the polymeric structure [74, 75]. Polyoxyalkylene-polysiloxane block copolymers are claimed as primary demulsifier bases. Indeed some are known to be used as the primary base in low-temperature applications and for stable emulsions in heavy crude oils. A number of polysilicones such as dimethylmethyl (polyethylene oxide) siloxane and ethoxylate dimethylsiloxane that has been terminated with a 3-hydroxypropyl group are used as demulsifier performance boosters. These are particularly synergistic with a number of more traditional demulsifier copolymers. There is a growing use of these silicone polymers in demulsifier blends particularly in highly regulated areas, including in the North Sea, as they are sufficiently biodegradable to pass required test criteria, are non-toxic and are not bioaccumulating [76].

The synergistic use of silicon surfactants has been claimed and investigated to enhance the use of polymer-based wax control agents, particularly ethyl vinyl acetate-based pour point depressants [77].

7.3.5 Silicone Oils

Drilling fluids consisting of silicone oil as the base liquid have been developed and deployed. The silicone will form the continuous phase of an invert emulsion, preferably with water or brine forming the internal phase. In practice dimethylsiloxane polymers, having a low viscosity of about 2.0 cSt at 25 °C, are generally used. These fluids are regarded as minimally toxic, or substantially non-toxic, essentially non-polluting, and are functionally capable of carrying out additional wellbore functions such as those performed by a spotting fluid, packer fluid, completion fluid, workover fluid and coring fluid [78].

7.3.6 Silicone Resins

Silicone resins are a type of silicone material that is formed by branched, cage-like oligosiloxanes with the general formula of $R_nSiX_mO_y$, where R is a non-reactive substituent, usually Me or Ph and X is a functional group H, OH, Cl or OR. These groups are further condensed in many applications to give highly cross-linked, insoluble polysiloxane networks as in Figure 7.5.

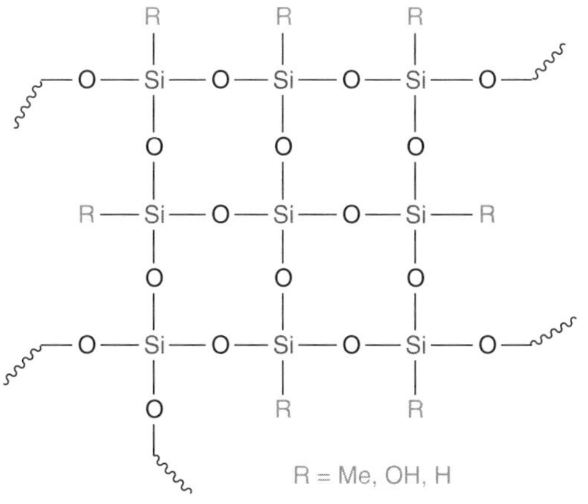

Figure 7.5 Silicone resin.

When R is methyl, the four possible functional siloxane monomeric units are described as follows:

- 'M' stands for Me_3SiO
- 'D' for Me_2SiO_2
- 'T' for $MeSiO_3$
- 'Q' for SiO_4.

Note that a network of only Q groups becomes fused quartz.

The most abundant silicone resins are built of D and T units (DT resins) or from M and Q units (MQ resins); however many other combinations (MDT, MTQ, QDT) are also used in a variety of industries. The MQ resins have been proposed in the oil and gas sector for use in water-based muds as fluid loss control agents and have been shown to be non-damaging [79].

Furan-silicone resins based on a 2-furaldehyde–acetone monomer and silicone oligomers have been described as a plugging material for water shut-off in cementing operations [80].

The stimulation of production through hydraulic fracturing has long been applied to oil and gas wells, and its application to the production of shale gas is now a matter of record, particularly in North America. In hydraulic fracturing, fluid containing solid particles (proppant) is injected into the oil- or gas-bearing (or both) formation to fracture the rock in the hydrocarbon-bearing zone. By such means, 'clear' passages are created that allow the hydrocarbon and associated water, if present, to flow more rapidly to the wellbore. This technique is becoming more widely used in shale formations where vertical movement through shale layers is very low [81].

The proppants used in the fracturing fluid to ensure that the fractures remain open to flow are normally coarse sand or ceramic proppants. In scenarios of high pressure, these materials can be crushed, thus reducing the fracture width as well as generating fines that can plug open spaces between remaining proppant particles. This reduces the conductivity of fractures; the proppant pack pore throats are plugged by displaced formation fines, and the pack permeability is therefore reduced. Resin-coated proppant materials including silicon polymers, added in the manufacturing of the proppants, can lead to improvements in crush resistance and a reduction in fines development. Such materials also have higher temperature stability [82].

7.3.7 Fluorosilicones

Fluorosilicones are primarily used in the oilfield sector as defoamers or anti-foams, but due to their expense only where other silicones such as PDMS are ineffective. Although PDMS has some solubility in hydrocarbons, this can be greatly reduced by substituting some of the methyl groups, as in Figure 7.6, with fluorinated alkyl groups, making it more effective as an anti-foam in crude oil systems [83].

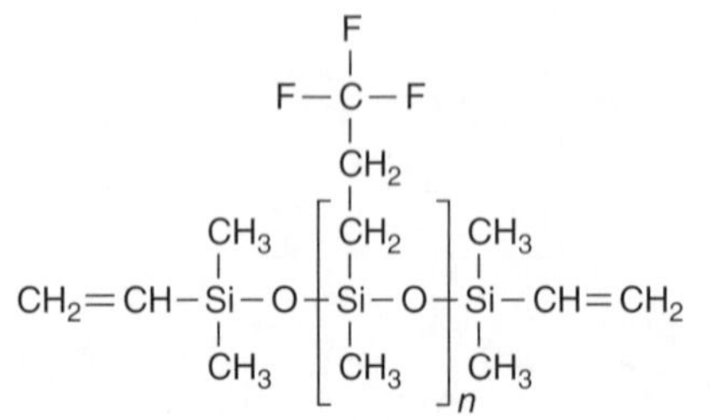

Figure 7.6 Fluorosilicone polymer.

The use of a fluorosilicone in a blend with a non-fluorinated silicone is claimed to produce a synergistic effect that performs better at a lower dose rate than either silicone by itself. The blended product is claimed to reduce liquid carry-over into the gas stream and gas carry-under into a liquid stream in a separation process [84].

A water-continuous emulsion containing a fluorosilicone and a surfactant has also been claimed to act as an anti-foam in the separation of crude oil and associated gas [85]. Also, in a similar vein, degassed crude oil can be treated with fluorinated norbornylsiloxanes in the suppression of foam [86]. These products are highly effective at low dose rates but are seldom applied probably due to economic reasons.

7.3.8 The Environmental Effects and Fate of Silicones

As has been described in Section 7.3.2, all silicones contain a covalent bond between the silicon atom and an organic group. The organosilicon linkage is not found in nature, and it has been assumed that these polymers do not degrade naturally in the environment as they are very stable and resistant to all but extreme conditions of temperature and pressure and chemical attack [57].

It has been shown, however, that silicones will degrade in soils with silanols being the main type of degradation product, and these silanols can undergo further biodegradation [64]. It has found, however, that this is minor degradation by hydrolysis of PDMS to dimethylsilanediol followed by oxidation of the methyl groups to aldehyde and ultimately to CO_2 by *Arthrobacter* and *Fusarium oxysporum Schlechtendahl*. The major degradation processes are abiotic. High molecular weight PDMS are initially depolymerised by soil hydrolysis of the siloxane bonds to yield organosilanol-terminated oligomers. These organosilanols and low molecular weight linear PDMS and cyclics are evaporated into the atmosphere and are oxidised there by hydroxyl radicals to benign silica, water and CO_2 [87].

Studies have mainly centred on PDMS, and it has been shown that PDMS behaves as an inert material during wastewater treatment, with no significant effect on wastewater treatment processes (other than the expected benefit of foam control) [88].

In earlier studies, in order to assess reliably and safety, and the potential threats posed by the environmental presence of silicones, a number of key parameters were examined. Based on extensive toxicological and environmental fate studies, it was concluded that commercially important organosilicon materials do not appear to present any ecologically significant threat [67].

Studies have also been conducted on the bioavailability of water-soluble methylsilanes on aquatic biota, particularly in freshwater fish species. Early work concluded that there was an uptake of PDMS is silver carp and that this increased with increasing molecular weight [89], which is contradictory to established thinking that high molecular weight compounds and molecules are not bioavailable and therefore do not bioaccumulate [90]. However this early work has been repudiated. It has been conclusively shown that there is no significant uptake of organosilicones in fish species and that an inverse relationship is present between molecular weight and bioconcentration [91].

As stated earlier in this section, it has been shown that organosilicones such as PDMS can be present in soils, and in this respect lake sediments have also been examined [92]. *Lumbriculus variegatus*, also known as the blackworm or California blackworm, is a species of worm found in the shallow part of pond and lakes in North America and Europe. It feeds on microorganisms and organic material found in the sediment. In a study *L. variegatus* were exposed to Lake Michigan sediment spiked in the laboratory with PDMS, and the accumulation of PDMS was monitored. Only very low concentrations of PDMS were found associated with the worms, which suggests some surface sorption or association with material in the gut.

The PDMS was excreted within 10 h both in sediment and water-only depuration exposures, indicating that most of the measured body burden was due to the sediment-associated material inside the organisms' gut. The study suggests little or no ecological effect from PDMS in the sediments.

Overall it can be concluded that silicon products are among the most environmental acceptable chemical additives used in a number of industrial sectors including the oil and gas industry. Although they are not biodegradable, they are generally non-toxic and non-polluting and do not bioaccumulate. Indeed, where bioavailable they are utilised by a number of microorganisms in building skeletal structures. Their use in this respect seems to be very low and restricted to a few specialist areas such as drilling muds and foam control. It is hoped that further studies on their potential applications will be examined as they could play a significant role in many areas of problem control within the oilfield.

References

1 Hoskin, D.H. and Rollmann, L.D. (1988). Polysilicate esters for oil reservoir permeability and control. EP Patent 2,836,02, assigned to Mobil Oil Corporation.
2 http://hyperphysics.phy-astr.gsu.edu/hbase/Tables/elabund.html
3 Fink, J. (2013). *Petroleum Engineers Guide to Oilfield Chemicals and Fluids*. Elsevier.
4 De Larrard, F. (1989). Ultrafine particles for making high strength concretes. *Cement and Concrete Research* **19** (2): 161–172.
5 Milestone, N.B., Bigley, C.H., Durant, A.T., and Sharp, M.D.W. (2012). Effects of CO_2 on geothermal cements. *GRC Transactions* **36**: 301–306.
6 Banthia, N. and Mindess, S. (1989). Water permeability of cement paste. *Cement and Concrete Research* **19** (5): 727–736.
7 Bjordal, A., Harris, K.L. and Olaussen, S.R. (1993). Colloidal silica cement: description and use in North Sea operations. Offshore Europe, Aberdeen, UK (7–10 September 1993), SPE 26725.
8 Marinova, K.G., Denkov, N.D., Branlard, P. et al. (2002). Optimal hydrophobicity of silica in mixed oil–silica antifoams. *Langmuir* **18** (9): 3399–3403.
9 Jurinak, J.J. and Summers, L.E. (1991). Oilfield applications of colloidal silica gel. *SPE Production Engineering* **6** (4): 406–412, SPE 18505.
10 Wang, H.-C., Wu, C.-Y., Chung, C.-C. et al. (2006). Analysis of parameters and interaction between parameters in preparation of uniform silicon dioxide nanoparticles using response surface methodology. *Industrial & Engineering Chemistry Research* **45**: 8043–8048.
11 Kumar, D., Chishti, S.S., Rai, A. and Patwardhan, S.D. (2012). Scale inhibition using nano-silica particles. SPE Middle East Health, Safety, Security, and Environment Conference and Exhibition, Abu Dhabi, UAE (2–4 April 2012), SPE 149321.
12 Bazazi, P., Gates, I.D., Nezhad, A.S. and Hajazi, S.H. (2017). Silica-based nanofluid heavy oil recovery a microfluidic approach. SPE Canada Heavy Oil Technical Conference, Calgary, Alberta, Canada (15–16 February 2017), SPE 185008.
13 Gill, J.S. (1998). Silica scale control. CORROSION 98, San Diego, CA (22–27 March 1998), NACE 98226.
14 Demadis, K.D., Stathoulopou, A. and Ketsetzi, A. (2007). Inhibition and control of colloidal silica: can chemical additives untie the "Gordian Knot" of scale formation? CORROSION 2007, Nashville, TN (11–15 March 2007), NACE 07085.
15 https://www.osha.gov/Publications/osha3176.html
16 Cyrs, W.D., Le, M.H., Hollins, D.M., and Henshaw, J.L. (2014, 2014). Settling the dust: silica past, present & future. *Professional Safety* **59** (04): 38–43, ASSE-14-04-38.

17 Cuomo, J. and Rabovsky, A. (2000). *Bioavailability of Silicon from Three Sources*. USANA Health Sciences Clinical Research Bulletin.

18 Colvin, V.L. (2003). The potential environmental impact of engineered nanomaterials. *Nature Biotechnology* **21**: 1166–1170.

19 Quintero, L., Lima, J.M. and Stocks-Fisher, S. (2000). Silica micro-encapsulation technology for treatment of oil and/or hydrocarbon-contaminated drill cuttings. IADC/SPE Drilling Conference, New Orleans, LA (23–25 February 2000), SPE 59117.

20 Van Ourt, E., Ripley, D., Ward, I. et al. (1996). Silicate-based drilling fluids: competent, cost-effective and benign solutions to wellbore stability problems. SPE/IADC Drilling Conference, New Orleans, LA (12–15 March 1996), SPE 35059.

21 Alford, S.E. (1991). North sea field application of an environmentally responsible water-base shale stabilizing system. SPE/IADC Drilling Conference, Amsterdam, Netherlands (11–14 March 1991), SPE 21936.

22 Cheng, Z.-Y., Breedon, D.L. and McDonald, M.J. (2011). The use of zinc dialkyl dithiophosphate as a lubricant enhancer for drilling fluids particularly silicate-based drilling fluids. SPE International Symposium on Oilfield Chemistry, The Woodlands, TX (11–13 April 2011), SPE 141327.

23 Fisk, J.V. Jr., Krecheville, J.D. and Pober, K.W. (2006). Silicic acid mud lubricants. US Patent 6,989,352, assigned to Halliburton Energy Services Inc.

24 Padron, A. (1995). Stable emulsion of viscous crude hydrocarbon in aqueous buffer solution and method for forming and transporting same. EP Patent 6,728,60, assigned to Maraven SA.

25 Nasr-El-Din, H.A. and Taylor, K.C. (2005). Evaluation of sodium silicate/urea gels used for water shut-off treatments. *Journal of Petroleum Science and Engineering* **48** (3–4): 141–160.

26 Laktos, I., Laktos-Szabo, J., Munkacai, I., and Tromboczki, S. (1993). Potential of repeated polymer well treatments. *SPE Production & Facilities* **8** (04): 269–275, SPE 20996.

27 Soreau, M. and Siegel, D. (1986). Injection composition for sealing soils. DE Patent 3,506,095, assigned to Hoechst France.

28 Mohanty, S. and Khantaniar, S. (1995). Sodium orthosilicate: an effective additive for alkaline steamflood. *Journal of Petroleum Science and Engineering* **14** (1–2): 45–49.

29 Sastri, V.S. (2011). *Green Corrosion Inhibitors*. New Jersey: John Wiley & Sons, Inc.

30 McDonald, M.J. (2007). The use of silicate-based drilling fluids to mitigate metal corrosion. International Symposium on Oilfield Chemistry, Houston, TX (28 February-2 March 2007), SPE 100599

31 Mueller, D.T., Boncan, V.G. and Dickersen, J.P. (2001). Stress resistant cement compositions and methods for using same. US Patent 6,230,804, assigned to BJ Services Company.

32 Baret, J.-F., Villar, J., Darguad, B. and Michaux, M. (1997). Cementing compositions and application of such compositions to cementing oil (or similar) wells. CA Patent 2,207,885, assigned to Schlumberger Ca. Ltd.

33 Sugama, T. and Wetzel, E. (1994). Microsphere-filled lightweight calcium phosphate cements. *Journal of Material Science* **29** (19): 5156–5176.

34 Flyten, G.C., Luke, K. and Rispler, K.A. (2006). Cementitious compositions containing interground cement clinker and zeolite. US Patent 7,303,015, assigned to Halliburton Energy Services Inc.

35 Luke, K., Fitzgerald, R.M. and Zamora, F. (2008). Drilling and cementing with fluids containing zeolite. US Patent 7,448,450, assigned to Halliburton Energy Services Inc.

36 Kondo, M. and Sawada, T. (1996). Readily dispersible bentonite. US Patent 5,491,248, assigned to Hojun Kogyo Co.Ltd.

37 Bragg, J.R. and Varadaraj, R. (2006). Solids-stabilized oil-in-water emulsion and a method for preparing same. US Patent 7,121,339, assigned to Exxonmobil Upstream Research Company.

38 Drillet, V. and Difives, D. (1991). Clay dissolution kinetics in relation to alkaline flooding. SPE International Symposium on Oilfield Chemistry, Anaheim, CA (20–22 February 1991), SPE 21030.

39 Grant, W.H. Jr., Rutledge, J.M., and Gardner, C.A. (1990). Quality of bentonite and its effect on cement-slurry performance. *SPE Production Engineering* **5** (04): 411–414, SPE 19940.

40 Bienvenu, R.L. Jr. (1996). Lightweight proppants and their use in hydraulic fracturing. US Patent 5,531,274, assigned to Inventor.

41 Ryan, R.G. (1995). Environmentally safe well plugging composition. US Patent 5,476,543, assigned to Inventor.

42 Zhou, Z. (2000). Process for reducing permeability in a subterranean formation. US Patent 6,143,699, assigned to Alberta Oil Sands Technology and Research Authority.

43 Nakashima, S., Bessho, H., Tomizawa, R. et al. (2014). Calcium silicate hydrate formation rates during alkaline alteration of rocks as revealed by infrared spectroscopy. ISRM International Symposium – 8th Asian Rock Mechanics Symposium, Sapporo, Japan (14–16 October 2014), ISRM-ARMS8-2014-238

44 De Wolf, C.A., Bang, E., Bouwman, A. et al. (2014). Evaluation of environmentally friendly chelating agents for applications in the oil and gas industry. SPE International Symposium and Exhibition on Formation Damage Control, Lafayette, LA (26–28 February 2014). SPE 168145.

45 Vinot, B., Schechter, R.S., and Lake, L.W. (1989, 1989). Formation of water-soluble silicate gels by the hydrolysis of a diester of dicarboxylic acid solubilized as microemulsions. *SPE Reservoir Engineering* **4** (03): 391–397, SPE 14236.

46 Chou, S. and Bae, J. (1994). Method for silica gel emplacement for enhanced oil recovery. US Patent 5,351,757, assigned to Chevron Research and Technology Company.

47 Oglesby, K.D., D'Souza, D., Roller, C. et al. (2016). Field test results of a new silicate gel system that is effective in carbon dioxide enhanced recovery and waterfloods. SPE Improved Oil Recovery Conference, Tulsa, OK (11–13 April 2016), SPE 179615.

48 Burns, L.D., McCool, C.S., Willhite, G.P. et al. (2008). New generation silicate gel system for casing repairs and water shutoff. SPE Symposium on Improved Oil Recovery, Tulsa, OK (20–23 April 2008), SPE 113490.

49 DeMaster, D.J. (1981). The supply and accumulation of silica in the marine environment. *Geochimica et Cosmochimica Acta* **45** (10): 1715–1732.

50 DeMaster, D.J. (2002). The accumulation and cycling of biogenic silica in the Southern Ocean: revisiting the marine silica budget. *Deep Sea Research Part II: Topical Studies in Oceanography* **49** (16): 3155–3167.

51 CEFIC (2014). *Soluble Silicates: Chemical, Toxicological, Ecological and Legal Aspects of Production, Transport, Handling and Application*. Centre European delude des Silicates.

52 CEFIC (2005). *Soluble Silicates: Human& Environmental Risk Assessment on Ingredients of European Household Cleaning Products*. Centre European delude des Silicates.

53 Laurelle, G.G., Roubeix, V., Sferratore, A. et al. (2009). Anthropogenic perturbations of the silicon cycle at the global scale: key role of the land-ocean transition. *Global Biogeochemical Cycles* **23** (4): 18–24.

54 Admiraal, W. and van der Vlugt, J.C. (1990). Impact of eutrophication on the silicate cycle of man-made basins in the Rhine delta. *Hydrobiological Bulletin* **24** (1): 23–26.

55 Correll, D.L. (1996). The role of phosphorus in the eutrophication of receiving waters: a review. *Journal of Environmental Quality* **27** (2): 261–266.

56 Ryther, J.H. and Dunstan, W.M. (1971). Nitrogen, phosphorus, and eutrophication in the coastal marine environment. *Science* **171** (3975): 1008–1013.

57 Noll, W. (1968). *Chemistry and Technology of Silicones*. New York: Academic Press.

58 Pape, P.G. (1983). Silicones: unique chemicals for petroleum processing. *Journal of Petroleum Technology* **35** (06): 1197–1204, SPE 10089.

59 Koczo, K. and Azouani, S. (2007). Organomodified silicones as crude oil demulsifiers. Chemistry in the Oil Industry X, Royal Society of Chemistry, Manchester, UK (5–7 November 2007), p. 323.

60 Callaghan, I.C. (1993). Antifoams for non-aqueous systems in the oil industry. In: *Defoaming: Theory and Industrial Applications* (ed. P.R. Garrett), 119–150. New York: Marcel Dekker.

61 Callaghan, I.C., Fink, H.-F., Gould, C.M. et al. (1985). Oil gas separation. US Patent 4,357,737, assigned to British Petroleum Company PLC.

62 Mannheimer, R.J. (1992). Factors that influence the coalescence of bubbles in oils that contain silicone antifoams. *Chemical Engineering Communications* **113** (1): 183–196.

63 Weltschev, M., Heming, F. and Werner, J. (2014). Compatibility of elastomers with biofuels. CORROSION 2014, San Antonio, TX (9–13 March 2014), NACE 2014-3745.

64 Lehmann, R.G., Millar, J.R., and Kozerski, G.E. (2000). Degradation of a silicone polymer in a field soil under natural conditions. *Chemosphere* **41**: 743–749.

65 Smith, D.M., Lehmann, R.G., Narayan, R. et al. (1998). Fate and effects of silicone polymer during the composting process. *Compost Science and Utilization* **6** (2): 6–12.

66 Stevens, C. (1998). Environmental degradation pathways for the breakdown of polydimethylsiloxanes. *Journal of Inorganic Biochemistry* **69** (3): 203–207.

67 Frye, C.L. (1988). The environmental fate and ecological impact of organosilicon materials: a review. *The Science of the Total Environment* **73** (1–2): 17–22.

68 Bosma, M.G.R., Cornelissen, E.K., Reijrink, P.M.T. et al. (1998). Development of a novel silicone rubber/cement plugging agent for cost effective Thru' Tubing well abandonment. IADC/SPE Drilling Conference, Dallas, TX (3–6 March 1998), SPE 39346.

69 Lakatos, I., Toth, J., Baur, K. et al. (2003). Comparative study of different silicone compounds as candidates for restriction of water production in gas wells. International Symposium on Oilfield Chemistry, Houston, TX (5–7 February 2003), SPE 80204.

70 Watkins, D.R., Kaifayan, L.J. and Hewgill, G.S. (1991). Acidizing composition comprising organosilicon compound. US Patent 5,039,434, assigned to Union Oil Company of California.

71 Stanley, F.O., Ali, S.A. and Boles, J.L. (1995). Laboratory and field evaluation of organosilane as a formation fines stabiliser. SPE Production and Operations Symposium, Oklahoma City, OK (2–4 April 1995), SPE .29530.

72 Thiminuer, R.J. and Traver, F.J. (1988). Volatile silicone-water emulsions and methods of preparation and use. US Patent 4,784,844, assigned to General Electric Company.

73 Zakharov, A.P. and Konovalov, E.A. (1992). Silicon-based additives improve mud rheology. *Oil and Gas Journal* **90** (32): http://www.ogj.com/articles/print/volume-90/issue-32/in-this-issue/drilling/silicon-based-additives-improve-mud-rheology.html.

74 Koerner, G. and Schaefer, D. (1991). Polyoxyalkylene-polysiloxane block-copolymers as demulsifier for water containing oils. US Patent 5,004,559, assigned to Th. Goldschmidt Ag.

75 Koczo, K., Falk, B., Palambo, A. and Phikan, M. (2011). New silicone copolymers as demulsifier boosters chemistry in the oil industry XII. Royal Society of Chemistry, Manchester, UK (7–9 November 2011), pp. 115–131.

76 Dalmazzone, C. and Noik, C. (2001). Development of new green demulsifiers for oil production. SPE International Symposium on Oilfield Chemistry, Houston, TX (13–16 February 2001), SPE 65041.

77 Craddock, H.A. (2010). Silicon materials as additives in wax inhibitors. Patent Application 1020439.4, Filing date 2nd December 2010.

78 Patel, A.D. (1998). Silicone oil-based drilling fluids. US Patent 5,707,939, assigned to MI Drilling Fluids.

79 Berry, V.L., Cook, J.L., Gelderbloom, S.J. et al. (2008). Silicone resin for drilling fluid loss control. US Patent 7,452,849, assigned to Dow Corning Corporation.

80 Leonov, Y.R., Lamosov, M.E., Ryabokon, S.A. et al. (1993). Plugging material for oil and gas wells. SU Patent 1,821,550, assigned to Borehole Consolidation Mu.

81 Craddock, H.A. (2012). Shale Gas: The Facts about Chemical Additives. www.knovel.com (accessed May 2012).

82 Fourneir, F. (2014). Oil and gas well proppants of silicone-resin-modified phenolic resins. WO Patent 2014067807, assigned to Wacker Chemie Ag.

83 Kobayashi, H. and Masatomi, T. (1995). Fluorosilicone antifoam. US Patent 5,454,979, assigned to Dow Corning Toray Silicon Co. Ltd.

84 Gallagher, C.T., Breen, P.J., Price, B. and Clement, A.F. (1998). Method and composition for suppressing oil-based foams. US Patent 5,853,617, assigned to Baker Hughes Corporation.

85 Taylor, A.S. (1991). Fluorosilicone anti-foam additive. GB Patent 2,244,279, assigned to British petroleum Company PLC.

86 Berger, R., Fink, H.-F., Koerner, G. et al. (1986). Use of fluorinated norbornylsiloxanes for defoaming freshly extracted degassing crude oil. US Patent 4,626,378, assigned to Th. Goldschmidt Ag.

87 Graiver, D., Farminer, K.W., and Nraayan, R. (2003). A review of the fate and effects of silicones in the environment. *Journal of Polymers in the Environment* **11** (4): 129–136.

88 Watts, R.J., Kong, S., Haling, C.S. et al. (1995). Fate and effects of polydimethylsiloxanes on pilot and bench-top activated sludge reactors and anaerobic/aerobic digesters. *Water Research* **29** (10): 2405–2411.

89 Wanatabe, N., Naskamura, T., and Wanatabe, E. (1984). Bioconcentration potential of polydimethylsiloxane (PDMS) fluids by fish. *Science in the Total Environment* **38**: 167–172.

90 Connell, D.W. (1988). Bioaccumulation behavior of persistent organic chemicals with aquatic organisms. *Reviews of Environmental Contamination and Toxicology* **102**: 117–154.

91 Annelin, R.B. and Frye, C.L. (1989). The piscine bioconcentration characteristics of cyclic and linear oligomeric permethylsiloxanes. *The Science of the Total Environment* **83**: 1–11.

92 Kukkonen, J. and Landrum, P.F. (1995). Effects of sediment-bound polydimethylsiloxane on the bioavailability and distribution of benzo[a]pyrene in lake sediment to *Lumbriculus variegatus*. *Environmental Toxicology and Chemistry* **14** (3): 523–531.

8

Solvents, Green Solvents and Formulation Practices

This chapter is concerned with the use of solvents in the upstream oil and gas industry, the growing use of the so-called 'green' solvents and the practices adopted in formulating products for use in the oilfield. Most, if not all, products used in the upstream oil and gas industry are formulated in a carrier solvent, and often the carrier is water. Even straightforward products are normally diluted for application as they are often required to be transported from the injection or application point to a further point of use, and therefore the viscosity of the product applied is of critical importance, especially if a distance of several kilometres is to be travelled by pumping pressure to the point of application. The formulated products must also be stable under a variety of environmental conditions, in particular in relation to storage and application, which can be at high temperature and pressure. These properties may also be at odds with regulatory requirements that require products to have a good rate of biodegradation, which can affect the stability of the product, particularly if storage is required under hot conditions for several months. The challenge for the formulator is to develop formulated products, often to meet very specific requirements.

The global solvents market is driven by the demand from end-use applications such as paints and coatings, agricultural chemicals, printing inks, adhesives, rubber and polymer, personal care, metal cleaning, pharmaceuticals and others, including the oil and gas sector. Solvents are chemicals used to dissolve a substance (a solute), thereby forming a solution. In oilfield practice, they are also used as a carrier to deliver active chemical components to the point of application and to produce a stable and useful mixture of a variety of component compounds, which may have synergistic properties. In the oil and gas sector, the main types of organic solvent used are alcohols, glycol ethers and aromatics, many of which have been described in Chapter 6 for their non-solvent use. Their solvent use is explored in this chapter alongside their use in formulation and their environmental impacts. The main solvents, as well as examples of other less used materials, are separately described and their use and function detailed.

8.1   How Solvents Work

A solvent can be defined as a liquid that serves as the medium for a reaction or, in the case of oilfield products, a medium to carry the active chemical ingredients [1]. It can serve two major purposes:

1) *Non-participatory*: To dissolve the reactants, with polar and non-polar solvents being best for dissolving polar reactants (such as ions) and non-polar reactants (such as hydrocarbons), respectively.

Oilfield Chemistry and Its Environmental Impact, First Edition. Henry A. Craddock.
© 2018 John Wiley & Sons Ltd. Published 2018 by John Wiley & Sons Ltd.

2) *Participatory*: As a source of acid (proton), base (removing protons), or nucleophile (donating a lone pair of electrons).

In the oilfield only non-participatory solvents are generally of concern.

Polar solvents have large dipole moments (i.e. 'partial charges'); they contain bonds between atoms with very different electronegativities, such as oxygen and hydrogen (e.g. water and methanol).

Non-polar solvents contain bonds between atoms with similar electronegativities, such as carbon and hydrogen (e.g. hydrocarbons). Bonds between atoms with similar electronegativities will lack partial charges; it is this absence of charge that makes these molecules 'non-polar'.

There are two common ways of measuring this polarity. One is through measuring a constant called 'dielectric constant' or permittivity. The greater the dielectric constant, the greater the polarity (water has a high dielectric constant, whereas a hydrocarbon such an *n*-octane has a low dielectric constant). The second measurement comes from directly measuring the dipole moment.

There is a final distinction to be made that often causes confusion. Some solvents are called 'protic', whereas some are called 'aprotic'.

Protic solvents have O—H or N—H bonds. Importantly protic solvents can participate in hydrogen bonding, which is a powerful intermolecular force. Additionally, these O—H or N—H bonds can serve as a source of protons (H^+).

Aprotic solvents may have hydrogen atoms as a constituent but lack O—H or N—H bonds and therefore cannot form hydrogen bonds with or between themselves.

Overall there is a general principle that states *like dissolves like*, i.e. non-polar solutes do not dissolve in polar solvents like water because they are unable to compete with the strong attraction that the polar solvent molecules have for each other. For example, water is a polar solvent, and it will dissolve salts and other polar molecules, but not non-polar molecules such as hydrocarbons.

8.1.1 Polymer Dissolution

It is well known that the dissolution of polymers depends not only on their physical properties but also on their chemical structure, such as polarity, molecular weight, branching, cross-linking degree and crystallinity [2]. The *like dissolves like* general principle is also appropriate in the case of polymers. Thus, polar macromolecules like poly(acrylic acid), poly(acrylamide) and polyvinyl alcohol, among others, are soluble in water. Conversely, non-polar polymers or polymers showing a low polarity such as polystyrene, poly(methyl methacrylate), poly(vinyl chloride) and poly(isobutylene) are soluble in non-polar solvents.

On the other hand, the molecular weight of polymers plays an important role in their solubility. In a given solvent at a particular temperature, as molecular weight increases, the solubility of a polymer decreases. This same behaviour is also noticed as the cross-linking degree increases, since strongly cross-linked polymers will inhibit the interaction between polymer chains and solvent molecules, preventing those polymer chains from being transported into solution.

A similar situation occurs with crystalline macromolecules, although in such a case the dissolution can be forced if an appropriate solvent is available or the polymer is warmed up to temperatures slightly below its crystalline melting point (T_m). For example, highly crystalline linear polyethylene ($T_m = 135\,°C$) can be dissolved in several solvents above 100 °C. Nylon 6.6 ($T_m = 265\,°C$), a crystalline polymer that is more polar than polyethylene, can be dissolved at room temperature in the presence of solvents with enough ability to interact with its chains through, for example, hydrogen bonding. Branched polymer chains generally increase solubility, although the rate at which this solubility occurs depends on the particular type of branching. Chains containing long branches cause dense entanglements, making the penetration of solvent molecules difficult. Therefore, the rate of dissolution in these cases becomes slower than if it was short branching, where the interaction between chains is practically nonexistent.

These general rules of dissolution are important in formulating oilfield products, since, as has been described in previous chapters, a wide variety of chemical compounds and polymers are used in their formulation.

8.2 Oilfield Solvents

In general, the use of oilfield solvents is governed by the following criteria:

- Local availability
- Price
- Environmental impact
- Technical performance

These criteria vary in importance depending on the region in which the formulated product and/or solvent is being used and the regulatory pressures on their use. For example, methanol is widely used around the world; however, in Brazil ethanol is preferred as it is cheap and readily available from sugar cane distillation. Similar economic preferences prevail in other countries and regions; however, these can lead to formulation and compatibility challenges, and this will be explored later in the chapter, Section 8.4.

8.2.1 Water

Where water is the solvent or carrier medium of choice, then its availability is usually the most pressing requirement. This is very much the case in onshore hydraulic fracturing for shale gas and oil, where in each horizontal well, 3–5 million gallons of water can be used [3].

The availability of water in large quantities can be critical to the deployment of hydraulic fracturing methods, particularly in desert regions such as those found in the continental United States, where shale gas predominates. In such areas, a suitable local aquifer is often required to supply water needs, which will usually require treatment to ensure particulate removal and elimination of microbial organisms and other aquatic life forms. This water is used to create the hydraulic fracturing fluid with the suspension of proppant and the addition of a variety of chemicals. [4] In other countries (e.g. China), the water source may be abundant but can be under pressure from other uses, particularly domestic use. China possesses some of the most abundant technically recoverable shale gas resources in the world, and water availability could be a limiting factor for hydraulic fracturing operations. In addition to geological, infrastructural and technological barriers, the baseline water availability for the next 15 years in Sichuan Basin, one of the most promising shale gas basins in China, shows that continued water demand for the domestic sector in this region could result in high to extremely high water stress in certain areas where intensification of hydraulic fracturing and water use might compete with other water utilisation [5].

A similar situation of concern is that in the Marcellus and Utica shales, which stretch across a wide area of the Eastern United States, including the Appalachian watershed and associated drainage basin. Of the water used, only around 10% is recovered and reused, with the remaining 90% being lost to the hydrological cycle. A number of sustainable water practices have been implemented as part of the permit-to-operate process. There remain, however, a number of critical questions surrounding the sustainability of water management associated with unconventional gas extraction and the possible environmental impacts [6].

Water is of course the major constituent of oilfield brines. Saline liquid is usually used in completion operations and, increasingly, when penetrating a pay zone. Brines are preferred because they have higher densities than fresh water but lack solid particles that might damage producible formations. Classes of brines include chloride brines (calcium and sodium), bromides and formates and have been extensively discussed in Chapter 5 under the halides of a number of different metal cations [7].

Outside of the large volumes of water used in hydraulic fracturing associated with unconventional gas and oil production, water is used as a carrier solvent in two main areas in the upstream oil and gas sector: (i) as a base solvent for water-based muds and (ii) as a carrier solvent for production chemicals that are required to function in the water phase. The latter class of products usually includes both organic and inorganic salts and comprises a wide range of chemicals whose function is directed at a similarly wide range of applications such as scale inhibitors (phosphonates; see Chapter 4), corrosion inhibitors (quaternary ammonium surfactants; see Chapter 3) and oxygen scavengers (ammonium bisulphite; see Chapter 5). Most of these applications use water both as a solvent for creating a stable solution of components and as a diluent. In Section 8.3 the formulation practice around water-based drilling fluids and its use in formulating production and other oilfield chemicals will be described.

Water is also used to deliberately create emulsions (and microemulsions) in order to have a transportable medium to carry water-insoluble chemicals via the water phase to the oil phase.

As has been stated, water in itself has no discernable environmental impact; it is however a carrier of potential pollutants, and it is likely that bioavailable and potentially toxic materials that are formulated in water are more likely to cause harm to organisms, particularly aquatic organisms, than if they are in another medium where they may not be so readily accessible.

Water of course can have a beneficial effect on pollution, particularly in the marine environment. When pollutants are introduced into the marine environment, they are subject to a number of physical processes, which result in their dilution in the receiving water. Dilution is one of the main processes for reducing the concentration of substances away from the discharge point. It is more important for reducing the concentration of conservative substances (those that do not undergo rapid biodegradation, e.g. metals) than of non-conservative substances (those that do undergo rapid biodegradation, e.g. some organic substances). This is a subject that will be discussed further in Chapter 10.

Also, the dilution of seawater by large quantities of fresh water may have serious consequences for the marine ecosystem, and there is already evidence of changes associated with increased carbon dioxide emission. This has led to an increase in ocean acidification and sea ice melting, which in turn has decreased the saturation state of calcium carbonate in the Canada Basin of the Arctic Ocean. This undersaturation has been found to be a direct consequence of the extensive melting of sea ice in the Canada Basin. In addition, the retreat of the ice edge well past the shelf break has produced conditions favourable to enhanced upwelling of subsurface, aragonite-undersaturated water onto the Arctic continental shelf. Undersaturation will affect both planktonic and benthic calcifying biota and therefore the composition of the Arctic ecosystem [8].

Again, these environmental impacts will be returned to in more detail in Chapters 9 and 10.

8.2.2 Alcohols

In Chapter 6 the non-solvent use of alcohols, particularly methanol, was described. Methanol is used as a thermodynamic hydrate inhibitor in both drilling operations and pipeline transport of gas and liquids [9]. Low molecular weight alcohols can also be beneficial in gas well stimulation by lowering the surface tension to release the spent acid from the formation [10].

8.2.2.1 Methanol

As a solvent methanol is a polar solvent with protic characteristics. It is miscible with water and is particularly useful in oilfield applications for solvating a number of polymers and other materials, either on its own or in conjunction with other solvents, particularly water.

Methanol (and also monoethylene glycol (MEG)) is used as winterisation additive in depressing the overall freezing point of the carrier solvent system. This is particularly the case with many corrosion inhibitor formulations that are formulated primarily in water.

Methanol is used in dewatering pipelines, particularly gas transmission pipelines, again as a form of hydrate inhibitor. Pipelines are flooded for hydrotest purposes, and once the line has been successfully hydrotested, a dewatering pig train is run. The pipeline and pre-commissioning specification usually states that the line must contain no more than 4% water at the end of the dewatering operation. This enables the export of sales-quality gas. To achieve this water content specification, methanol (and glycols; see Section 8.2.3) and methanol mixtures are used. The methanol content is calculated to avoid any hydrate formation [11].

Methanol is used both as a solvent in formulating and more often as a co-solvent primarily with water. Methanol has to be used with care in formulating as it can worsen scale deposition problems. This is particularly the case where methanol is being used to control hydrate risk, which in turn presents new challenges in terms of increased scale tendency (dehydration of brine) and interactions of the hydrate control chemicals with scale inhibitor performance [12]. Also methanol in large concentrations can cause downstream issues such as polluting catalysts used at the refinery or lowering the overall value of the hydrocarbons in the crude oil. This and other issues associated with compatibility will be further detailed in Section 8.4.

Methanol can have beneficial effects in formulations however, and performance improvements of anionic viscoelastic surfactants as diverting agents have been noted in systems mixed with a small amount of methanol. In such cases the surfactant concentration is insufficient to divert fluid flow in the formation and only diverts fluid as it flows through the formation due to loss of the methanol from the solution [13].

In most, if not all, of these examples of methanol use as a solvent, ethanol can be substituted, however normally methanol is considerably cheaper, and it is not denatured or subject to licensing to prevent its illegal consumption.

Methanol occurs naturally during the decomposition of different plants and animals, and we come into contact with it every day in fruits, juices and even wine. Though larger quantities of methanol can be toxic if ingested [14], it has a very low impact when released into the environment because of how quickly it biodegrades.

When methanol is released into the environment, it rapidly breaks down into other compounds, is completely miscible in water and serves as a food substrate for a number of different bacteria [15].

8.2.2.2 Isopropanol (Isopropyl Alcohol (IPA))

Isopropanol (Figure 8.1) is a slightly larger molecule than methanol or ethanol and is often cheaper to manufacture. It is much less poisonous than methanol and just as biodegradable.

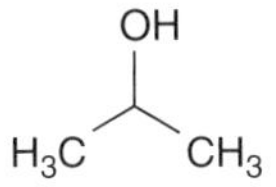

Figure 8.1 Isopropanol.

H_3C CH_3 OH

Like methanol, isopropanol is miscible with water; however, unlike methanol, it is not miscible with salt solutions and can be separated from aqueous solutions by adding a salt such as sodium chloride [16]. It will however solubilise a number of polymers and resins and is more soluble in hydrocarbons than in methanol [2].

These properties can be both advantageous and problematic when used in formulating oilfield products. Hydrocarbon solubility can be advantageous in ensuring oil-active chemicals are delivered to the oil phase, especially in saltwater environments. However, this solubility could be highly incompatible with polymers and plastics used in construction materials, such as umbilical pipelines and seals for pumps and other equipment.

Isopropanol, along with other alcoholic solvents, is used as the solvent of choice for formulating gas hydrate inhibitors and corrosion inhibitors together (see also Section 8.4) [17].

Alcohols can be used in enhanced oil recovery (EOR) operations, and a combination of isopropanol and ammonia in an aqueous carrier has been considered to be a cost-effective EOR methodology. It is claimed that the mixture does not react with the oil in the formation and that there is no significant amount trapped in the formation rock, therefore the additive mixture can be separated from the oil, recycled and reused [18].

Isopropanol is the solvent of choice for silicone microemulsions that are utilised for the restriction of water production. This system makes the rock surfaces water-wet, and any negative effect of silicone on wettability is absent. A shift to oil-wet character is observed with siloxane solutions in isopropanol [19].

8.2.2.3   2-Ethyl Hexanol (Butoxyethanol) and Hexanol

2-Ethylhexanol is a branched, eight-carbon chiral alcohol and is also a fatty alcohol; see structure in Figure 8.2. This isooctanol is a colourless liquid that is poorly soluble in water but soluble in most organic solvents. It has a low volatility, making it a useful solvent where low emission exposure is required. It is produced on a large scale for use in numerous applications such as flavours and fragrances and especially as a precursor for the production of other chemicals such as emollients and plasticisers. It is encountered in the natural world as plant fragrances, and the odour has been reported as 'heavy, earthy and slightly floral' for the R enantiomer and 'a light, sweet floral fragrance' for the S enantiomer [20].

2-Ethylhexanol is a useful solvent for copolymers derived from *N*-vinyl caprolactam, which are used as kinetic hydrate inhibitors; [21] see Section 2.1.3. Also, in highly regulated areas such as the North Sea basin, they are used as the solvent of choice in a number of product formulations that require to be oil soluble or are surfactant based, such as demulsifiers (H.A. Craddock, unpublished results).

2-Ethylhexanol can be a critical co-solvent in wellbore cleaner formulations where it is very effective at dispersing oil-based mud (OBM) residues (H.A. Craddock and R. Simcox, unpublished Results). Generally the solvent system in well cleaners is based on xylene, toluene, ketones and low molecular weight esters. These have proved to be very effective; however they can both be environmentally hazardous and pose additional health and safety problems to personnel. These have largely been replaced in highly regulated areas such as the North Sea by a range of biodegradable esters, middle distillates and terpene-based solvents.

Hexanol is used as a solvent in some gel systems for temporary plugging during fracturing operations as it can result in a high viscosity at higher temperatures while maintaining a suitable

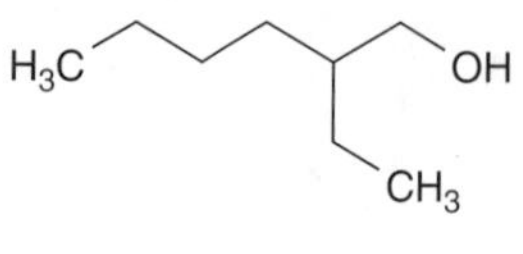

Figure 8.2 2-Ethylhexanol.

viscosity at ambient temperatures for pumping [22]. It is also used as a co-solvent in a number of surfactant formulations; see Section 8.4.

8.2.2.4 Butanol and Isobutanol

Butanol (also called butyl alcohol) is a four-carbon alcohol with a chemical formula of C_4H_9OH. It occurs in five isomeric structures, from a straight-chain primary alcohol *n*-butanol to a branched-chain tertiary alcohol. All are a butyl or isobutyl group linked to a hydroxyl group: *n*-butanol, 2 stereoisomers of 2-butanol, *tert*-butanol and isobutanol. Butanol is primarily used as a solvent across a number of industrial sectors, whereas in the upstream oil and gas industry, *n*-butanol and isobutanol are primarily used.

n-Butanol (and the other isomers) is suitable for alcohol–water flooding in medium to heavy reservoirs [23]. This EOR process uses a chemical mixture containing water and high boiling point alcohol, with the alcohol present in concentrations up to 20%. The mixture is injected at a temperature of between about 100 °F and about 500 °F to ensure a thermal element to the recovery process.

Butanol is also a solvent carrier of choice in the use of microemulsions carrying a corrosion inhibitor; however such systems can also be used to deliver a wide variety of oil-soluble chemical additives [24].

Isobutanol has reportedly been used to increase the strength of liquid alloy cements [25].

Like many alcohols, butanol is considered toxic. It has, however, been shown to have a low order of toxicity in single-dose experiments to laboratory animals and is considered safe enough for use in cosmetics.

8.2.2.5 Other Alcohols

A number of other alcohol and alcohol-type compounds are used, mainly, as solvents, in the upstream oil and gas industry. This primarily centres around octanol and decanol and related derivatives; indeed, a blend of octanol and decanol, known as Epal 810 (mainly 1-decanol), is used as carrier fluid for gelling agents based on phosphate esters (see Section 4.5) [26].

2-Octyldecanol and also other long-chain fatty alcohols such as oleyl alcohol and stearyl alcohol are used in lubricant compositions for silicic acid-based drilling fluids [27]. It has also been claimed that fatty alcohols C_{10}–C_{20} aid in the efficiency of fatty acid detergents in the removal and dispersal of oil spillages from surface waters. It is usually applied in the form of an aqueous emulsion that is sprayed onto the contaminated surface [25].

Phenoxyethanol (Figure 8.3) has been considered for oilfield use as a biocide particularly when combined with effervescent materials to deliver the biocide to the production fluids. Generally, the effervescent material is a solid that includes an acid and a base that react in aqueous medium to produce a gas, such as carbon dioxide.The effervescent material can be tablets, powder, flakes and the like. The methods and compositions are claimed to be particularly suited for treating fracturing fluids [28].

As an alternative carrier solvent, a blend of octanol and decanol, known as Epal 810, can be used for gelling agents based on phosphate esters. Phenoxyethanol, also known as ethylene glycol phenyl ether, can also be used [26].

8.2.3 Glycols and Glycol Ethers

The glycols and glycol ethers are widely used across a number of industrial sectors as chemical intermediates, refrigerants or antifreeze agents and as solvents. MEG or ethane-1,2-diol

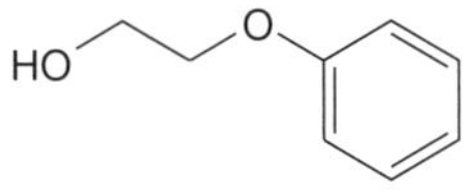

Figure 8.3 Phenoxyethanol.

HO\~\~/OH   **Figure 8.4** Monoethylene glycol.

(see Figure 8.4), the simplest in this class of molecule, has two main uses, as a raw material in the manufacture of polyester fibres and for antifreeze formulations. It is an odourless, colourless, sweet-tasting syrup and is moderately toxic [29]. The glycols with regard to their use as hydrate inhibitors have been detailed in Section 6.2.1.

8.2.3.1 Monoethylene Glycol (MEG), Diethylene Glycol (DEG) and Triethylene Glycol (TEG)

These are the most commonly used by volume of all the glycols used in the upstream oil and gas industry. In particular, large quantities of MEG are used in hydrate inhibition and large volumes of triethylene glycol (TEG) in gas dehydration. Diethylene glycol (DEG) is the least used by volume, being the most expensive. MEG is not as active a freezing point depressant as methanol but has a low vapour pressure, which means in coolant systems fluid loss is mainly due to the evaporation of the water, and also unlike methanol, MEG is not flammable, which is a significant advantage in oilfield use. Often MEG is formulated with DEG and/or TEG for antifreeze use.

In low-temperature formations where the cement in the well completion can be subject to a freeze–thaw cycle, MEG is added as a freezing point depressant [30].

As solvents, these materials are widely used as co-solvents, particularly in water-based formulations. They are also used to impart winterisation characteristics, particularly MEG. In both these uses, care must be given when formulating as glycols can react with some active ingredients of the formulations. This is further detailed in Section 8.4.

In certain clay stabilisation additives, MEG is the solvent of choice, but it is also in certain cases where maleic acid polymers are involved in *in situ* reactant forming maleic imides, which have been found to be particularly suitable for water-based drilling fluids [31].

Similar to hexanol mentioned earlier in this chapter, MEG when added to aqueous-based guar gum gels can increase the viscosity of the gel and stabilise the associated brines. These materials are commonly used in hydraulic fracturing, and it means that such fluids exhibit greater stability at higher temperatures. This in turn minimises formation damage associated with hydraulic fracturing as less guar polymer can be used to achieve the same viscosity through the addition of MEG [32].

As stated earlier these glycols, and additionally propylene glycol, have good heat loss control characteristics, having low thermal conductivities. To this effect, they are used as solvents in products used to reduce undesired heat loss from production tubing and uncontrolled heat transfer to the annulus. This heat loss can result in productivity impairment due to paraffin or asphaltene deposition and potentially gas hydrate formation. Where pipelines are in the arctic regions, the heat loss can result in the destabilisation of the permafrost. [33]

Various materials and formulations have been proposed in order to free stuck drill pipes, the most common of which is diesel oil added directly to the drilling mud. However, this can be of limited success, and oil and water emulsions of dodecylbenzenesulphonate surfactants in glycols have been claimed to be more effective. In these products, MEG and DEG are mainly used and described as co-surfactants [34].

MEG is one of the solvents used both in water-based lubricants and oil-based materials, which are used in water-based muds. Their function is to reduce friction and torque on metal-to-metal surfaces during drilling operations. Hexylene glycol or 2-methylpentane-2,4-diol (Figure 8.5) is also used for this function [35].

MEG has been used, along with others such as methanol, as a carrier solvent for anti-sludging agents. Such agents are added to acid packages for well stimulation [36].

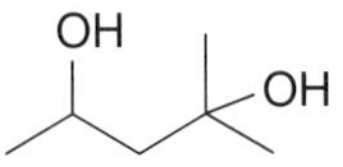

OH

OH

Figure 8.5 Hexylene glycol.

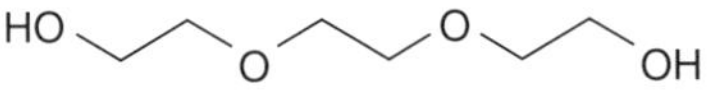

HO O O OH

Figure 8.6 Triethylene glycol.

TEG (Figure 8.6) is mainly used in the oil and gas sector for gas drying, i.e. for absorbing water from gas in flowlines or in processing plants. Its use as a solvent, as that of DEG, is rare and ill defined because presumably it has no differential benefit compared with the use of MEG, and both are more expensive.

8.2.3.2 Glycol Ethers and Mutual Solvents

A number of glycol ethers are utilised in the upstream sector of the oil and gas industry, and some of the more prominent are exemplified in this section. Many, such as monobutyl glycol ether or 2-butoxyethanol (Figure 8.7), commonly known as butyl glycol, are mutual solvents. A mutual solvent is miscible with both water and oil and is used in conjunction with surfactants to aid their water-wetting characteristics.

Monobutyl glycol ether is a butyl ether of ethylene glycol, or ethylene glycol monobutyl ether (EGMBE), and is a relatively nonvolatile, inexpensive solvent of low toxicity. It is used in many domestic and industrial products because of its surfactant properties.

It has been found that polyvinylcaprolactam (PVCap) polymers synthesised in butyl glycol as opposed to isopropanol gave a better kinetic hydrate inhibitor performance [37].

Mutual solvents are very effective in sandstone acidising, in which it is important to keep all solids water-wet. Here the mutual solvents are either butyl glycol or other modified glycol ethers. They improve the solubility of corrosion inhibitors in the spent acid in the formation and compatibility of inhibitors with emulsion preventers and other additives. The most important property is to reduce the adsorption of corrosion inhibitors on residual clay particles in the formation and to help maintain water-wetting for maximum oil/gas flow after acidising. A mutual solvent also reduces residual water saturation (spent acid) following a treatment. Gas wells clean up better by keeping surfactants in solution rather than adsorbing on sand and clay that is too near the wellbore [38].

Butyl diglycol ether (BDE) or DEG monobutyl ether (Figure 8.8) is used as water-based mud lubricant.

Ethylene glycol dimethyl ether (EGDE) or dimethoxyethane (Figure 8.9) is used as a solvent system with toluene in establishing the relative solubility number (RSN) of demulsifier polymers and surfactants [39].

Dipropylene glycol methyl ether (Figure 8.10) is another mutual solvent and can be used to remove water blocks, or trapped water, caused by the use of all aqueous treatments [40].

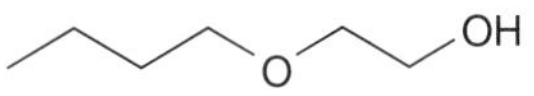

OH   **Figure 8.7** Monobutyl glycol ether.

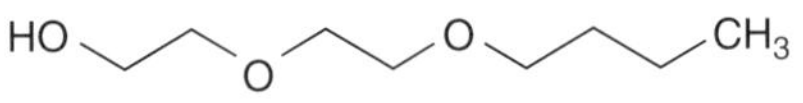

O

HO O O CH$_3$ **Figure 8.8** Butyl diglycol ether.

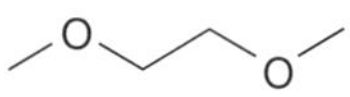

O O   **Figure 8.9** Ethylene glycol dimethyl ether.

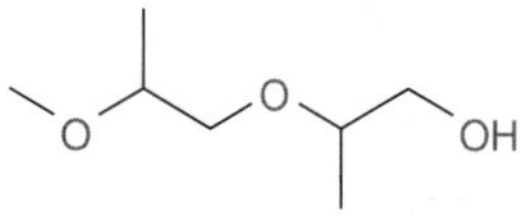

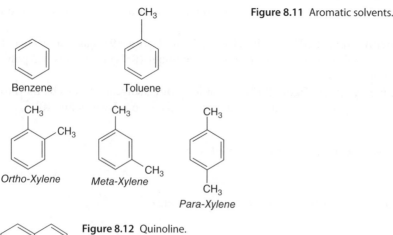

Figure 8.10 Dipropylene glycol methyl ether.

Polymeric glycols have been considered in detail in Chapter 2, and it is fair to state here that a main use of polyethylene glycol (PEG) and various ether derivatives is as lubricants in aqueous drilling fluids. The PEG is designed to an optimal molecular weight and its compatibility can be affected by other components in the fluid, in particular divalent cations such as calcium [41].

8.2.4 Aromatic Solvents

Aromatic solvents are organic chemicals that have a benzene ring structure such as benzene, toluene, ethylbenzene, mixed xylenes (BTEX) and high flash aromatic naphthas. They are used in a wide range of industrial sectors; however, it is estimated that more than 50% of the global aromatic solvents demand comes from the paints and coatings industry.

They have higher rate of solvency than aliphatic solvents, which makes them a better choice in many industries. With benzene becoming less common due to its toxicity, toluene and xylene are primarily and widely used in the upstream oil and gas sector, particularly in the formulation of oil-soluble materials such as wax inhibitors and asphaltene dispersants. Increasingly heavy aromatic naphtha (HAN) is also utilised although it also is under some threat due to its environmental toxicity. Meanwhile low aromatic naphtha (LAN) is used where possible.

8.2.4.1 Toluene, Xylene and Benzene

Generally, it has been found that heavy asphaltenic crudes are soluble in aromatic solvents such as benzene, toluene and xylene, of which there are three isomers (see Figure 8.11), and these are often used for asphaltene deposit dissolution and removal [42]. Xylene is usually used as a mixture of all three isomers.

It has also been observed that mixtures of aromatic solvents, e.g. xylene and toluene, and quinoline (Figure 8.12) or alkylquinoline were better dissolvers of asphaltene deposits than the aromatic solvents alone [43].

Figure 8.11 Aromatic solvents.

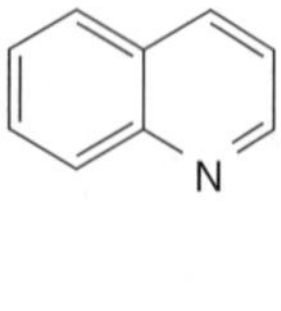

Benzene

Toluene

Ortho-Xylene

Meta-Xylene

Para-Xylene

Figure 8.12 Quinoline.

Most asphaltene dissolvers are combinations based on aromatic solvents. Sometimes the so-called enhancers are added for solvency [43]. It has been assessed that effective dissolvers can dissolve up to their own weight in asphaltene at downhole temperatures in a few hours; however after acid stimulation and workover, any asphaltenic sludge can be difficult to remove with solvents or dissolvers as the asphaltene is chemically bonded through interaction with water molecules to charged minerals on the rock surface [44, 45].

Xylene (as a mixture of all three isomers), despite its low flash point at 28 °C, is probably the most commonly used aromatic solvent. It is used both in asphaltene dissolution [46] and wax dissolution [47] and in formulation of oil-soluble materials.

It has been also shown that bicyclic molecules such as tetralin (1,2,3,4-tetrahydronaphthalene) and 1-methylnaphthalene (see Figure 8.13) performed better than the monocyclic aromatics such as xylene, toluene and benzene in regard to both the amount of asphaltene dissolved and rate of dissolution [48]. These aromatic solvents interact with the asphaltene–asphaltene π–π interactions via π–π orbital overlap, thereby replacing them and solubilising the deposited asphaltene agglomerate.

Most commercial asphaltene dissolvers are formulated using monocyclic aromatics, such as xylene, and small amounts of bicyclic aromatics to ensure low cost.

High flash point emulsion-based asphaltene dissolvers have been developed in which the base solvent is an aromatic solvent, but these do not contain any of the previously stated monocyclic aromatics, the so-called BETX group (benzene, ethylbenzene, toluene and xylene) [49]. However, acids emulsified with aromatic solvents such as xylene are used to enhance well productivity in situations where asphaltene dissolution is required downhole [50, 51]. In general, it would appear that higher asphaltene solubility is observed in emulsion systems, where the aqueous phase is water or 15% hydrochloric acid, than with xylene alone [52]. Many of these emulsion systems are also an attempt to make the application of asphaltene dissolvers more environmentally acceptable while retaining the use of an aromatic solvent component. Other hydrocarbon materials, however, have better environmental profiles, and these are discussed in Section 8.2.5.

All three of the main aromatic solvents, benzene, toluene and xylene, have been used to remove wax deposits; however current practice mainly uses toluene and xylene. Other distillates containing a high aromatic content have also been used. Often a surfactant is incorporated into the blend with the aromatic solvent [53]. It has also been found that miscible xylene/acid formulations, used in stimulation and workover, can be used to dissolve wax deposits [54]. These miscible micellar acidising solvents can solve paraffin deposit problems in the well and near-wellbore scenarios more safely and effectively than xylene alone. They also have other advantages over solvent-only systems in that they are less flammable, reduce surface tension to remove water blocks and water-wet formation matrix and are fully miscible in acid systems.

In formulating microemulsions, a small amount of aromatic solvent, usually toluene, is required. A microemulsion is a thermodynamically stable fluid that differs from other kinetic emulsions in that it does not readily separate, into its components oil and water, over time. This stability is due to the small emulsion particle sizes of 10–300 nm, which have an ultralow interfacial tension between the water phase and the oil phase. They are used in the oil and gas

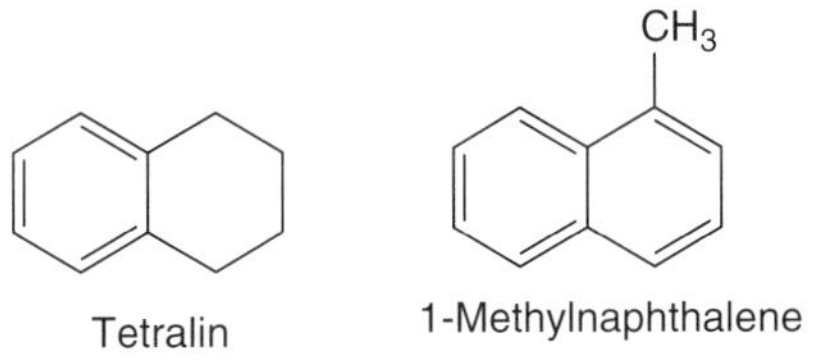

Figure 8.13 Bicyclic aromatics.

Tetralin   1-Methylnaphthalene

industry to deliver a variety of chemical actives to their point of application [24] (see also earlier on butanol in Section 8.2.2).

Other aromatics, such as phenols and aromatic heterocyclic compounds (e.g. pyridine) have been used in specialist applications, but this is however becoming rarer due to their usually high toxicity.

Phenols (see Figure 8.14) are used as a cross-linker with polyacrylamides (see Section 2.1.1) where, in conjunction with formaldehyde, they react with the amide groups of the polymer. Although they are toxic, they are the most suitable cross-linking agents for normal hydrophilic polymers such as the polyacrylamide/polyacrylate copolymer [55].

Phenol ethers can form specialist hybrid ionic/nonionic surfactants (see also Chapter 3). These materials can be useful in high salinity formations where nonionic and anionic surfactants are inefficient due to salting out. They can be used in conjunction with alkyl sulphonates to give particularly low interfacial tensions [56].

The main simple aromatic heterocyclic molecule considered for oilfield application is pyridine (Figure 8.15). It is a very powerful solvent and has been used in both wax and asphaltene dissolution and removal. However, due to its toxicity and environmental persistence, it is now rarely used. Pyridine and substituted pyridines have been used in small concentrations as non-radioactive tracers for monitoring the flow of subterranean fluids [57].

8.2.5 Other Organic Solvents

There are a number of other organic materials that are used as solvents in the upstream oil and gas industry; however these are in much lower quantities than the previously described categories despite many of them having significant environmentally better characteristics. Their use is, in many circumstances, dependent on economics and the availability of supply.

8.2.5.1 Hydrocarbons

A number of hydrocarbons could be used as solvents, and α-olefins (see dec-1-ene as an example, Figure 8.16) in particular are useful in the solvation of asphaltenes and as paraffin wax dissolvers, often in conjunction with an aromatic additive [58]. α-Olefins are a family of organic compounds, which are alkenes (also known as olefins) with a chemical formula C_xH_{2x}, distinguished by having a double bond at the primary or alpha (α) position. This location of a double bond enhances the reactivity of the compound and makes it useful for a number of applications.

Linear α-olefins are also used in the removal of pipe dope, which is a chemical sealing compound that is used to make a drill pipe thread joint leak proof and pressure tight. The pipe dope also acts as a lubricant and helps prevent seizing of the mating parts, which can later cause difficulty during disassembly [25]. Related products, such as hydrotreated middle distillate paraffins, are used in drilling fluids [25] and often as the base solvent in wellbore cleaners [59].

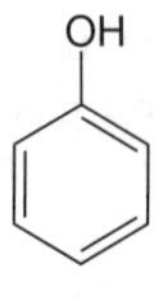

OH   **Figure 8.14** Phenol.

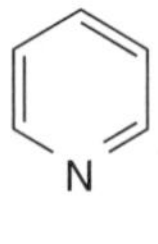

Figure 8.15 Pyridine.

N

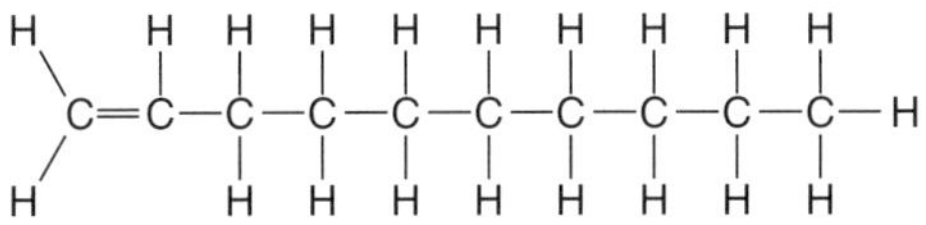

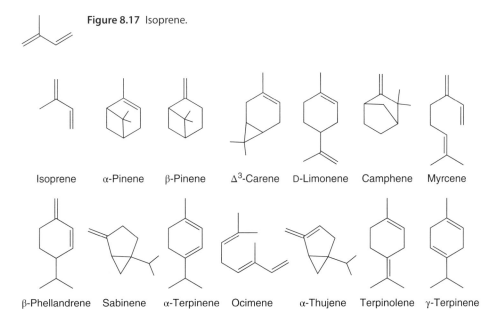

Figure 8.16 Dec-1-ene, an α-olefin.

One of the most effective drilling fluids is OBM, which historically was formulated with diesel oil. As the potential environmental implications of the discharge of OBMs became apparent, this led to the introduction of synthetic or non-mineral base oils with improved biodegradability. α-olefins have a major use in the formulation of synthetic-based drilling fluids, and linear α-olefins and polyalphaolefins are used in the formulation of synthetic-based muds (SBMs) [60, 61].

8.2.5.2 Natural Products

The natural products used as solvents and dissolvers centre mainly around molecules based on isoprene units. Isoprene, or 2-methyl-1,3-butadiene (Figure 8.17), is a common organic compound. In its pure form, it is a colourless volatile liquid and is produced by many plants. Isoprene polymers are the main component of natural rubber [62] and were isolated from the thermal decomposition of natural rubber in 1860 by C.G. Williams. Isoprene and its polymers are further discussed in Chapter 2.

The main isoprene-derived products with solvent use are the cyclic monoterpenes. These are a large group of volatile unsaturated hydrocarbons found in the essential oils of plants, especially conifers and citrus trees. They are based on a cyclic molecule with the formula $C_{10}H_{16}$. A number of examples are shown in Figure 8.18.

The most commercial and useful of these in the upstream oil and gas sector are the pinenes and D-limonene. However, their use is often limited due to price pressure and supply constrictions. These essential oils are usually a by-product of other processes and are in demand by other industrial sectors, especially the personal care and perfumery sectors.

Figure 8.17 Isoprene.

Isoprene   α-Pinene   β-Pinene   Δ³-Carene   D-Limonene   Camphene   Myrcene

β-Phellandrene   Sabinene   α-Terpinene   Ocimene   α-Thujene   Terpinolene   γ-Terpinene

Figure 8.18 The terpenes.

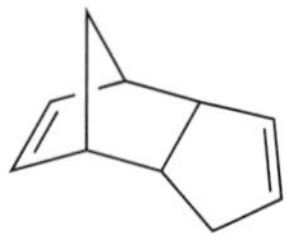

Figure 8.19 Dicyclopentadiene.

During acid stimulation work in carbonate reservoirs, some crude oils especially those that have high paraffinic contents or are asphaltenic in nature can produce difficult-to-handle sludges. A mixture of dicyclopentadiene (Figure 8.19) and cyclic monoterpenes has been found to be a useful anti-sludging agent [63]. The mixture acts as a dispersant and enhances the stimulation effect.

Dicyclopentadiene is not a product derived from nature but rather is a co-product from the cracking of naphtha and gas oils to produce ethylene.

The main terpene employed as a solvent or dissolver is D-limonene, either in a pure state but more commonly as a mixture of citrus fruit terpenes where D-limonene constitutes over 50% of the mixture. It has been used as a constituent of a biodegradable oil-based drilling fluid where it acts as a pour point depressant (PPD) [64]. This property is very useful when used as a wax dissolver and is a useful solvent or part of a solvent system when formulating PPDs (H.A. Craddock, unpublished results).

Some water-in-oil emulsions are very efficient in breaking residual emulsions inside a filter cake during its removal by decreasing the cohesion of the filter cake and its adhesion to the rock formation face [65]. They act like a demulsifier in breaking the oil-based drilling mud or synthetic-based drilling mud water-in-oil emulsions and therefore change the adherence properties of the filter cake to the wellbore and formation. The internal phase of the water-in-oil emulsion is water, and the external phase is a hydrophobic organic solvent and D-limonene is the preferred substance. A number of surfactants have been used. Such systems have also been claimed as asphaltene dissolvers [66]. Limonene and pinene have been used as corrosion inhibitor intensifiers particularly with regard to enhancing the performance at higher temperatures of quaternary salts [67].

The main use of D-limonene and related terpenes is in the dissolution of paraffin wax deposits. These materials have come to prominence in certain regions as aromatic solvents are mostly classified as marine pollutants. A dissolver based on orange terpenes that has a high D-limonene content has been successfully used in field application [68]. A blend of a limonene and an alkyl glycol ether or other polar solvents has been claimed as a useful wax dissolver [69]. A considerable technical advantage for the use of D-limonene and related terpenes is its capability as a wax dissolver at low temperatures, in particular seabed temperatures at 4 °C [68], although there is a conflicting evidence on this [39]. It has also been shown that low doses of D-limonene-based terpene solvent mixtures can soften wax deposits to allow more efficient mechanical cleaning in pipelines [70]. This work also gave supporting evidence that D-limonene can act as a PPD.

8.2.5.3 Other Solvents

There are a number of other solvents that have been deployed in the upstream oil and gas sector. Most are no longer used due to toxicity and persistency concerns. Historically carbon disulphide and dimethyl sulphoxide (DMSO) (see Figure 8.20) have been used in wax removal.

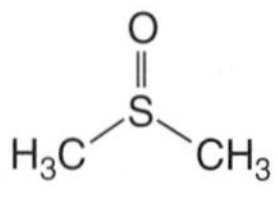

Figure 8.20 Dimethyl sulphoxide.

Figure 8.21 Sulpholane.

DMSO is a colourless liquid and is a polar aprotic solvent that dissolves both polar and non-polar compounds. It is miscible in a wide range of organic solvents as well as water.

Sulpholane, 2,3,4,5-tetrahydrothiophene-1,1-dioxide (Figure 8.21), is an organosulphur compound and is similar to DMSO. It is formally a cyclic sulphone and is a colourless liquid commonly used in the chemical industry as a solvent for extractive distillation and chemical reactions. Like DMSO, sulpholane is a polar aprotic solvent, and it is readily soluble in water. It has been used in the upstream oil and gas sector as a commercial hydrogen sulphide solvent, particularly in combination with amines. It is widely used in upstream applications to purify natural gas as part of the sulphinol process [71].

Sulpholane is highly stable and can therefore be reused many times; however, it does eventually degrade into acidic by-products. A number of measures have been developed to remove these by-products, allowing the sulpholane to be reused, and increase the lifetime of a given supply. Some methods that have been developed to regenerate spent sulpholane include vacuum and steam distillation, back extraction, adsorption and anion–cation exchange resin columns.

Both sulpholane and DMSO are relatively non-toxic; the issue is their ability to carry toxic materials through skin adsorption. Also, their high miscibility with water makes them particularly adept in carrying environmental toxic agents through the aquatic environment as a co-solvent for environmental hazardous agents. Sulpholane in particular has been widely released to the environment over the last several decades. It does not volatilise from water or soil, nor does it readily adsorb to organic matter or soils. The primary attenuation mechanism appears to be biodegradation in an aerobic environment. There is little information on viable approaches to remediation. Wide distribution and low cleanup levels add a level of complexity to practicable remediation. Treatability studies have demonstrated the most effective treatment options for groundwater sulpholane contamination, which are aeration and UV irradiation combined with chemical oxidation using hydrogen peroxide or sodium persulphate. Biotreatment and *in situ* chemical oxidation are commonly used methods for soil contamination remediation [72].

Some ketones are used in the upstream oil and gas sector as solvents, particularly methylisobutylketone (MIBK); however, compared with other industrial sectors, their volume use is small.

The unusual chemical isopropyl benzoate (see Figure 8.22) has been found to be particularly useful as a solvent for asphaltene dissolution [73].

Interestingly these benzoate esters have been found to have good rates of biodegradation and a reduced environmental impact compared with other aromatic solvents. This has allowed them to be used in domestic fabric softeners [74].

Compared with other industrial sectors, crown ethers have been little explored and developed for use in the upstream oil and gas industry. They have been examined as scale dissolvers [75] and have been shown to dissolve a wide variety of oilfield scales, including sulphates of

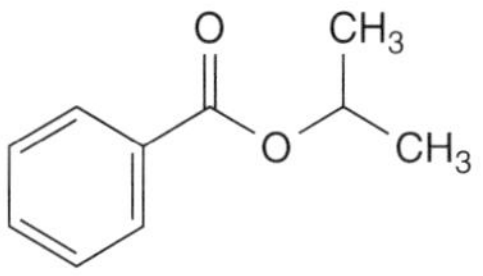

Figure 8.22 Isopropyl benzoate.

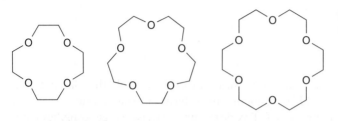

Figure 8.23 Examples of crown ethers.

barium, strontium and calcium, as well as calcium carbonate. Also, a naturally occurring radio-active material (NORM), which is usually co-precipitated with barium and strontium sulphates, is removed by this solvent.

Crown ethers are cyclic chemical compounds that consist of a ring containing several ether groups. The most common crown ethers are cyclic oligomers of ethylene oxide (EO), the repeating unit being ethyleneoxy (i.e. $-CH_2CH_2O-$). Important members of this series are the tetramer ($n = 4$), the pentamer ($n = 5$) and the hexamer ($n = 6$) (see Figure 8.23). The term 'crown' refers to the resemblance between the structure of a crown ether bound to a cation and a crown sitting on a person's head.

It is unclear why these products have not been examined and exploited more, perhaps for economic reasons, as they have interesting solvation properties. They do however form stable complexes with cations, which can make their degradation slow.

8.2.6 Ionic Liquids

An ionic liquid is a salt in the liquid state. While ordinary liquids such as water and hydrocarbons are predominantly made of electrically neutral molecules, ionic liquids are largely made of ions and short-lived ion pairs. They have become known as 'solvents of the future' as well as 'designer solvents'.

Ionic liquids have a number of interesting properties. They are powerful solvents and electrically conducting fluids (electrolytes). Salts that are liquid at near-ambient temperature are important for electric battery applications and have been considered as sealants due to their very low vapour pressure.

The ionic bond is usually stronger than the van der Waals forces between the molecules of ordinary liquids. For that reason, common salts tend to melt at higher temperatures than other solid molecules. Some salts are liquid at or below room temperature.

Low-temperature ionic liquids can be compared with ionic solutions, liquids that contain both ions and neutral molecules, and in particular with the so-called deep eutectic solvents, mixtures of ionic and nonionic solid substances that have much lower melting points than the pure compounds. Certain mixtures of nitrate salts can have melting points below 100 °C [76].

A wide range of physical and chemical properties of ionic liquids have been developed, and not just as solvents, in a wide range of industries, including catalysis of chemical reactions, gas handling and processing, nuclear waste treatment and pharmaceutical applications.

In the upstream oil and gas sector, the use of novel hydrophobic ionic liquid solvents has been considered for the removal of organics from produced water [77]. Although this work was successful at the laboratory scale, many of the ionic liquid candidates were in themselves environmentally hazardous. However, in certain regions of the world, mercury has become a significant contaminant of natural gas and is much more hazardous to the aquatic environment and associated wildlife; therefore its removal is highly desirable. Most available technologies for mercury removal are located downstream of the acid gas removal and dehydration units and thus leading to contamination of these units as well as mercury release to the environment.

Besides its potential to be located upstream of these units to better protect the plant operations, ionic liquid technology is capable of treating the full range of mercury species by a single adsorbent or treatment step to meet the current targeted mercury level at the outlet of mercury removal unit of less than 0.01 μgm^{-3}. In addition, it can be applied to the existing mercury removal units without requiring any plant modifications or retrofitting.

Efficient scrubbing of mercury vapour from natural gas streams has been demonstrated both in the laboratory and on an industrial scale using chlorocuprate(II) ionic liquids impregnated on high surface area porous solid supports. This material has been tested for use within the petroleum gas production industry and has currently been running continuously at a number of gas processing plants in Malaysia for several years [78].

Other areas of application have been explored in recent times in the oil and gas sector, namely, the upgrading of heavy crudes [79] and the application in EOR [80].

These latter applications are of particular note, as the upstream petroleum industry faces operational and technical challenges due to the production of crude oil containing waxes, asphaltenes and aromatic compounds and also due to the formation of gas hydrates resulting in their deposition in the surface and production equipment, and in offshore pipelines affecting the safety of operations, in turn resulting in huge production losses, and threatening the environment.

It is also estimated that all over the globe there are more than double the number of reserves of heavy and extra-heavy crude oil than the lighter ones. In spite of this, the production of heavy oil is still low. As the world's demand for light crude oil continues to increase tremendously, the supplies of these easily extractable crude oils continue to decrease although efforts have already been made to extract the heavy and extra-heavy oil that was previously considered uneconomical to produce and process. The enhancement in the solubility of heavy crude oil in solvent and ionic liquids mixture has been investigated [80]. Ionic liquids also aid in the reduction of surface forces between oil–water systems, thereby helping for the recovery of entrapped oil from exhausted reservoirs, which have failed to produce the residual oil.

Ionic liquids have also been investigated as a potential clay and shale stabiliser in drilling applications. As has been described earlier in Chapter 5, a number of inorganic salts particularly potassium and sodium chloride have been effectively used for many years in workover fluids for temporary clay stabilisation. However, due to potential environmental issues and the logistics of using large quantities of salts for this application, many oil and gas operators have begun to search for alternative clay stabiliser products. Ionic liquids offer an alternative with the advantage of much reduced volumes required in application [81]. However, to date no further work or known field application is in progress.

Many imidazolium salts are ionic liquids (e.g. 1-ethyl-3-methylimidazolium chloride (Figure 8.24)) and are obvious candidates for investigation as corrosion inhibitors given their structural similarity to the imidazoline film-forming surfactant corrosion inhibitors (see Chapter 3). They do exhibit some inhibitory properties, but further investigation would be required before field application could be considered [82].

The ionic liquids also offer useful properties for carbon capture applications, and this is explored further in Chapter 10, 'Sustainability and Green Chemistry'.

Figure 8.24 1-Ethyl-3-methylimidazolium chloride.

8.3 The Environmental Impact of Solvents and Green Solvents

8.3.1 Water and Produced Water

Water is known as the universal solvent and is environmentally neutral. However, it can have serious environmental impacts when there is too much of it or too little. Flooding and drought are natural disasters with consequences to the environment and plant and animal life. Flooding is often part of the natural cycle; however its frequency and severity in certain regions of the planet alongside extended periods of drought in other regions are, as many scientists believe, a consequence of man-made accelerated climate change. It is well known that the water contained on planet Earth is at a constant volume at around about 332 500 000 cubic miles (mi^3) or 1 386 000 000 cubic kilometres (km^3) [83]. About 71% of the Earth's surface is water covered, and the oceans hold about 96.5% of all the Earth's water. Water also exists in the air as water vapour, in rivers and lakes, in icecaps and glaciers in the ground as soil moisture and in aquifers. It is also contained in most plants and animals, and indeed a human consists of more than 55% water.

The Earth's water supply is constantly moving from one place to another and from one form to another due to the water cycle.

The vast majority of water on the Earth's surface (over 70% of the planet is covered with it), over 96%, is saline water in the oceans. The freshwater resources account for less than 4% of the total volume of water and come from water falling from the skies as rainwater and moving into streams, rivers, lakes and groundwater. It is this water that provides people with the water they need every day to live. Throughout this book, concern has been described with the contamination, pollution and conservation of water resources from chemical activities originating from the oil and gas industry. In the later chapters, the importance on how we use and conserve this resource, particularly with regard to fresh water, will also be explored.

As an example, the use of saline water has been explored and developed in irrigation for some years, and the shortage of water resources of good quality is becoming an important issue in the arid and semi-arid zones. For this reason, the availability of water resources of marginal quality such as drainage water, saline groundwater and treated wastewater has become an important consideration. Consequently, the sustainable use of saline water in irrigated agriculture has been extensively examined. This requires the control of soil salinity at the field level, a decrease in the amount of drainage water and the disposal of the irrigation return flows in such a way that minimises the side effects on the quality of downstream water resources [84].

In general, in the oil and gas industry, the volumes of water used are relatively low; however in drilling and hydraulic fracturing operations, these can be significant. The later operations can use millions of gallons of water under pressure to achieve fracture promotion, and the quality, use and reuse of this water is of concern to both industry practitioners, environmentalists and regulators and the general public, who have concerns over possible groundwater contamination [4].

The upgrading and treatment of this 'frac' water is of major interest to the industry as it requires a reusable supply of 'clean' water for shale gas and oil exploitation, which often occur in arid and desert regions [85]. Well stimulation flowback water generally contains the chemicals and/or by-products of a hydraulic fracturing process used on a specific well. This water has been considered a waste by-product of oil and gas production and typically presents logistical difficulties for the operators. Some of the challenges include transportation of wastewater over long distances as well as local government and environmental regulations related to the safe disposal of the water from oilfields.

The ability to recycle flowback water provides great opportunities for service providers and producers to help reduce the total amount of fresh water that is used in their operations.

By reducing the volumes of fresh water that is used in hydraulic fracturing and, at the same time, reducing the amount of flowback water that has to be transported and disposed, operators are able to show their commitment to the community and the environment and can potentially minimise logistics.

Produced water is a different category, which is a naturally occurring formation brine that is produced along with the hydrocarbons from the well, and can contain large quantities of dissolved salts, dispersed hydrocarbons and other materials, particularly organic and inorganic chemical additives. Produced water is the largest waste stream generated in oil and gas industries. Due to the increasing volume of waste all over the world, the outcome and effect of discharging produced water on the environment has become a significant issue of environmental concern. Produced water is conventionally treated through different physical, chemical and biological methods. In offshore platforms because of space constraints, compact physical and chemical systems are used. However, current technologies cannot remove small-suspended oil particles and dissolved elements. Besides, many chemical treatments, whose initial and/or running cost is high, produce hazardous sludge. In onshore facilities, biological pretreatment of oily wastewater can be a cost-effective and environmentally friendly method. As high salt concentration and variations of effluent characteristics have direct influence on the turbidity of the effluent, it is appropriate to incorporate a physical treatment (e.g. membrane to refine the final effluent) [86].

Produced water management is now a major specialist discipline within the upstream oil and gas sector and is of growing concern in a number of regions where mature fields often produce considerably more water than hydrocarbons. This associated production water from usual production streams is often recycled through reinjection into the reservoir or into a disposal well. The reuse of this resource can have serious environmental and economic benefits.

The management, recycling and reuse of water resource will be further developed in Chapter 10.

8.3.2   Environmental Impacts of Solvents

Solvents are used in several industrial processes, including, as has been illustrated, the upstream oil and gas sector, and in a large section of economic activities worldwide, and the volume of these materials released into the environment is of growing concern.

Their generally high vapour pressure allows for their easy dispersion in air, promoting contact with living organisms and the pollution of the environment. The high volume, the particular properties of each solvent and the volatility factor in high-risk solvents can have a serious impact on the environment and threaten animal and plant life.

A number of international treatises on the subject have generated guidelines to ban the usage of highly volatile chlorinated solvents such as carbon tetrachloride and the more toxic aromatic solvents such as benzene. The oil and gas sector does not use chlorinated solvents and has substantially replaced solvents such as benzene with safer alternatives.

However, the use of many solvents despite evidence of harm both to humans, wildlife and the environment continues, and although contradictory, this is at some level reasonable as their properties and applications are economically important that banning them may cause the collapse of many industries. Production of essentials like shoes, bathtubs, pharmaceuticals, polymer fabrication, manufacturing of light bulbs, extraction of cooking oils, glues, etc., are just a few examples that put us in this paradoxical situation of having to survive within the demands of modern life while impacting on the fragile equilibrium of the ecosystems we inhabit.

As has been illustrated in this chapter, some solvents have specific environmental impacts, while others are more severe. Many volatile solvents have detrimental effects on atmospheric quality and can disturb the climate, and many others are water miscible and can pollute the aquatic environment, both surface waters and the marine environment. Many of these substances are difficult to remove from waste streams due to the very characteristics that are utilised for their use, namely, their solvency. These characteristics often also make many solvents poorly biodegradable and therefore persistent in the environment [87].

In order to ensure continued growth in economic activity and minimise environmental impacts, it is necessary to consider the following alternatives to increase solvent use and the upstream oil and gas sector requires to play its part:

- Minimisation of solvent use
- Reuse and recycling of solvents
- Replacement with green alternatives including water
- Elimination of solvent

Currently the research efforts in the green chemistry area are focused on the design and search of the materials to replace environmentally harmful toxic solvents for more acceptable and less toxic materials, some of which are discussed in the next section.

8.3.3 Green Solvents

Green solvents are generally defined as environmentally friendly solvents or bio-solvents that are derived from the processing of agricultural crops. The use of petrochemical solvents is key to the majority of chemical processes but not without severe implications on the environment.

The Montreal Protocol [88] identified the need to re-evaluate chemical processes with regard to their use of volatile organic compounds (VOCs), which in the main are solvents, and the impact these compounds have on the environment. Green solvents were developed as a more environmentally friendly alternative to petrochemical solvents.

In considering the environmental impact of solvents it has been shown that simple alcohols (methanol, ethanol) or alkanes (heptane, hexane) are preferred, whereas the use of solvents such as dioxan, acetonitrile and tetrahydrofuran are not recommended for application from an environmental perspective. Results also indicate that methanol–water or ethanol–water mixtures have better environmental profiles compared with pure alcohol or propanol–water mixtures. A framework that can be used to select green solvents or environmentally sound solvent mixtures for processes in the chemical industry and this same framework can be used for a comprehensive assessment of new solvent technologies, providing sufficient data is available [89]. However, to date this tool has not be applied in the upstream oil and gas sector.

The oil and gas sector has begun to introduce a number of green solvents into applications, and some have been described in earlier sections of this chapter. The terpenoids are a good example of a sustainable natural product-based class of solvents; indeed D-limonene is in some ways a secondary product from citrus fruit groups, which are primarily grown for fruit juice manufacture.

Other similar solvents are utilised in the upstream oil and gas sector and some examples are described below.

8.3.3.1 Hydrolysable Esters

A number of bioderived materials from soya, castor and palm oils have been considered in a number of applications in the upstream oil and gas sector, primarily in synthetic drilling fluids [90, 91]. The esters of these fatty acids reduce their surfactant behaviour and increase solvency. The oleates, particularly methyl oleate (Figure 8.25), have been considered as co-solvents in formulating a number of production chemical additives and in particular corrosion inhibitors where they can also impart some corrosion inhibitory properties.

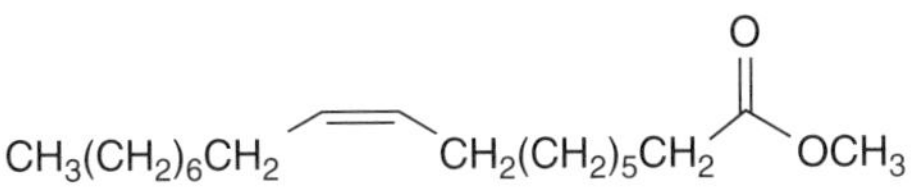

Figure 8.25 Methyl oleate.

These materials are prone to enzymatic hydrolysis, and as has been described for other chemical molecules, such properties make them biodegradable.

A much simpler molecule, ethyl lactate, the ester of lactic acid, whose structure is shown in Figure 8.26, is a green solvent derived from processing corn.

Lactate ester solvents are commonly used solvents in the paint and coating industry and have numerous attractive advantages including being inherently biodegradable, easy to recycle, non-corrosive, non-carcinogenic and non-ozone depleting.

Ethyl lactate is a particularly attractive solvent as a result of its high solvency power, high boiling point, low vapour pressure and low surface tension. In the paint and coating sector, ethyl lactate has replaced more hazardous solvents, including toluene, acetone and xylene [92].

Other applications of ethyl lactate include being an excellent cleaner for the polyurethane industry. Ethyl lactate has a high solvency power, which means it has the ability to dissolve a wide range of polyurethane resins. The excellent cleaning power of ethyl lactate also means it can be used to clean a variety of metal surfaces, efficiently removing greases, oils, adhesives and solid fuels. The use of ethyl lactate is highly variable, and it has eliminated the use of chlorinated solvents in these industries. Ethyl lactate however is not, to date, used in the upstream oil and gas industry.

8.3.3.2 Orthoformates

Orthoformates, in particular tripropyl orthoformate (see Figure 8.27), have been applied in filter cake removal. The delayed acid-releasing system, which is based on *ortho*esters in combination with alkaline inhibitors, can remain seemingly inactive for days before becoming strongly activated. The acid that is generated can then be used to break polymers, remove filter cakes or activate other chemical processes [93, 94].

Orthoformates are highly biodegradable materials; indeed this hydrolytic action is used as described above to release the formic acid, which is used as the solvent *in situ*.

8.3.3.3 Other Bioderived Materials

The obvious bioderived solvents are alcohols, and through brewing, ethanol has been a useful bioderived material for centuries. In the last few decades, a lot of interest has focused on bio-alcohols as fuels, but they of course can also be used as solvents. In the right circumstances, they can be a sustainable material; however replacing food substrates or using these starch substances as substrates for biofuels in particular can lead to both environmental and economic

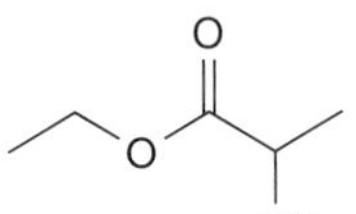

Figure 8.26 Ethyl lactate.

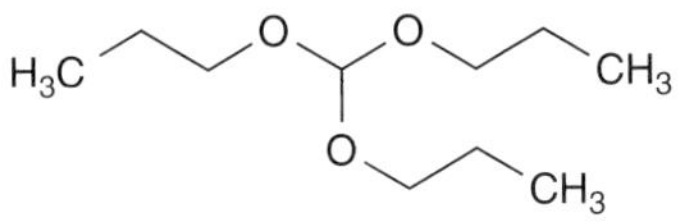

Figure 8.27 Tripropyl orthoformate.

issues. Ethanol however is a widely used solvent and thermodynamic hydrate inhibitor in Brazil [95] where sugar cane is an abundant crop.

To date very few truly green solvents, i.e. those that are derived from natural substrates through extraction or fermentation, have been deployed in the upstream oil and gas industry. The 'green' solvents that are used have inherent green properties such as methanol, but are not necessarily derived from a sustainable source. This supply chain dilemma is further discussed in Chapter 10.

8.4 Formulation Practices

In the supply of chemicals to the upstream oil and gas sector, the use of formulations is of paramount importance. Nearly all chemical products deployed are mixtures of a variety of chemical substances, even if it is only dilution in a solvent such as water. Very few chemicals are applied as a neat 100% material, a notable exception being certain solvents such as alcohols and glycols used in hydrate inhibition and pipeline dewatering. Sometimes neat aromatic solvents can be applied in dissolution of wax and asphaltenes. In this section, some of the main practices will be described in particular for their use of solvents and similar carrier materials. It is however outside the scope of this work to give a list of all the types of formulation; nonetheless some examples will be provided.

8.4.1 Drilling Fluids

In engineering terms a drilling fluid is used to aid in the drilling of boreholes into the Earth, particularly in drilling oil and natural gas wells. Liquid drilling fluid is often called drilling mud. There are two main categories of drilling fluids, namely, water-based muds (which can be dispersed and non-dispersed) and non-aqueous muds, usually called OBM. There is also another variant, a gaseous drilling fluid, in which a wide range of gases can be used to create a foam.

The main functions of drilling fluids include providing hydrostatic pressure to prevent fluids from the formation from entering into the wellbore, keeping the drill a bit cool and clean during drilling, carrying out drill cuttings, and suspending the drill cuttings while drilling is paused and when the drilling assembly is brought in and out of the hole. The drilling fluid used for a particular job is selected to avoid formation damage and to limit corrosion.

These drilling fluids are formulated to achieve the best performance suited to the well conditions, and three key factors affect this [96]:

- A change of drilling fluid viscosity
- A change of drilling fluid density
- A change of mud pH

Drilling fluids are classified based on their fluid phase, alkalinity, dispersion and the type of chemicals used.

8.4.1.1 Water-Based Muds

As the name implies, these drilling fluids have water as their continuous phase, and they are most commonly composed of bentonite clay with additives such as barium sulphate (barite) or calcium carbonate. Various thickeners are used to alter the viscosity of the fluid (e.g. xanthan gum, guar gum, ethylene glycol carboxymethyl cellulose (CMC), polyanionic cellulose (PAC) or starch). In turn, deflocculants are used to reduce viscosity of clay-based muds; anionic polyelectrolytes, e.g. acrylates, polyphosphates lignosulphonates or tannic acid derivatives, are among those frequently used. Other components are added to provide various specific functional characteristics including lubricants, shale inhibitors and fluid loss additives (to control loss of drilling fluids into permeable formations). A weighting agent such as barite is added to increase the overall density of the drilling fluid so that sufficient bottom-hole pressure can be maintained, thereby preventing an unwanted (and often dangerous) influx of formation fluids.

Water-based muds generally fall into two main types:

1) Dispersed systems:
   Freshwater mud – Low-pH mud (7.0–9.5) that includes spud, bentonite, natural and phosphate-treated muds, organic mud and organic colloid-treated mud. High-pH mud, for example alkaline ternate-treated muds, are above 9.5 in pH.
   Water-based drilling mud that represses hydration and dispersion of clay – There are four types: high-pH lime muds, low-pH gypsum, seawater and saturated saltwater muds.
2) Non-dispersed systems:
   Low solids mud – These muds contain less than 3–6% solids by volume and weigh less than 9.5 lb gal^{-1}. Most muds of this type are water based, with varying quantities of bentonite and a polymer.

8.4.1.2 Oil-Based Muds

These drilling fluids have oil as their continuous phase, usually diesel, a mineral oil or a low toxicity base oil. These fluids typically contain less than 5% water; therefore in effect these materials are required to be emulsions and must contain emulsifying agents. This leads to the classification of two types of emulsion:

- Oil in water (oil emulsion muds).
- Water in oil (invert oil emulsion muds) – here additional water is deliberately added usually for economic reasons.

OBMs offer unrivalled performance characteristics in terms of rate of penetration, shale inhibition, wellbore stability, thermal stability, lubricity and salt tolerance. However, their use and discharge is subject to strict environmental regulations, which now apply almost globally. In effect the contaminated cuttings must be removed from the well site or offshore location and shipped for processing and reuse (see also Chapter 9).

A number of SBMs, where the 'oil' has been replaced, have been used as environmental alternatives such as palm oil [97] or hydrated castor oil [98]. Synthetic oil base drilling muds were developed to maintain the performance of OBMs while reducing environmental impact. The organic fluids used are esters, polyolefins, acetals, ethers and various others. Also, biodegradable oil-based drilling fluids have been developed [64], where the main oil phase is a mixture of esters of biodegradable fatty acids, and a typical formulation is shown in the table in Figure 8.28.

The use of such fluids, which are more expensive, has in most areas been discontinued as the regulations have become just as great for these types of fluids as for OBMs, and many regulatory regimens still enforce a requirement for shipment away from the well site for processing. In some instances, reinjection of cuttings is permitted.

8.4.1.2.1 *Invert Emulsions* 10–20% of all wells drilled involve invert emulsions, which historically like emulsion-based OBMs used diesel and mineral oils to formulate the drilling fluid. Due to the toxicity and persistency in the environment of these materials, alternatives have been developed, primarily based on esters, particularly C_6–C_{11} monocarboxylic acids [99], acetals [100] and branched chained poly α-olefins (PAOs). The latter however are less biodegradable and more expensive but can be formulated to higher densities [101].

Compound	Amount w/w (%)	Functionality
Methyl ester of soya oil	55–70	Main oil phase
D-Limonene	1–5	PPD
Hydrogenated castor oil	<1	Oil component
Fatty acid salts	3–6	Soap/emulsifiers
Magnesium oxide	1–3	Saponifier
Sodium chloride brine	26–30	Aqueous phase
Organophilic clay	1	Viscosifier
Sodium polyacrylate	<1	Fluid loss agent
Citric acid	1	pH buffer
Barium Sulphate	Up to 25	Weighting agent

Figure 8.28 A biodegradable drilling fluid.

8.4.2 Well Cleaners for Completion

The process of displacing the drilling fluid from a well and its replacement with completion fluid is one of the most critical phases in the well construction process. In order to ensure that the well is clean, a series of small volume 'pills' are pumped around the well on removal of the drilling fluid in order to clean the tubulars and well casing prior to the addition of the completion brine [102]. Not all applications require the materials to be pumped as 'pills', and some have been formulated as a single-application multi-effect material (H.A. Craddock, unpublished results).

These well cleaners must function across a broad range of temperatures from 4 °C at the seabed to up to 200 °C at the total well depth. In addition, these cleaning agents may have to be weighted to maintain a safe hydrostatic pressure in the well, in which case the cleaning formulation will have to function in high salinity carrier fluids such as calcium bromide or potassium formate.

Once these wash pills are spent and returned to the annulus, they contain significant quantities of well contaminants such as drilling mud, barite, hydrocarbon drill cuttings, sand, pipe dope, etc. Historically, it was permitted to discharge these chemicals and the dirty water, which followed. Now, with increased regulatory standards, there has to be a clean-up of the water to permit its discharge, and the chemicals used have to be of an environmentally acceptable standard.

The categories of chemicals used in wellbore cleaning are itemised below, and these can be used as individual pills or formulated as a complete one-pill package [59]. These are

- Surfactants
- Solvents
- Flocculants

Some example formulations of well cleaners are given in Figure 8.29.

8.4.2.1 Surfactants

These constitute about 80–90% of the materials used in this area, and of the four basic groups of surfactants, anionic and nonionic are mainly used as wash fluids. Amphoteric surfactants do not possess the detergency required under the demands of oilfield conditions, and cationics tend to have poor environmental properties. These materials are formulated at an average concentration of 7% but can vary between 2% and 30%. The concentration depends on the type of surfactant being used and the material to be cleaned, which is usually OBM and related cuttings.

Material	V/V (%)
2-Ethyl Hexanol	70–75
Surfactant	20–24
Butyl diglycol ether	1–2

Material	V/V (%)
Middle distillate	56–59
Surfactant	10–12
Butyl diglycol ether	20–22
2-Ethyl hexanol	7–9

Material	V/V (%)
2-Ethyl hexanol	70–80
Monoethylene glycol (MEG)	10–20
Sodium dioctyl sulphosuccinate	7–10

Material	V/V (%)
2-Ethyl hexanol	70–80
Ethoxylated Isotridecanol	20–30

Figure 8.29 Some typical well cleaner formulations.

The carrier fluid is usually aquifer or drill water; however, seawater is almost always used in offshore wells. Brines may also be used to increase carrier fluid density in unstable well completions.

The following surfactant chemistries are used in well cleaning operations:

Alcohol ethoxylates. These are the most common surfactants used in well cleaners, and the ones having a hydrophilic–lipophilic balance (HLB) value of 10–15 are preferred. Indeed, in research that the author has conducted, HLB values between 11.5 and 13.8 are preferred, depending on the mud type to be cleaned (H.A. Craddock and J. McPherson, unpublished results). Many are highly biodegradable and are therefore widely used. These materials have replaced alkylphenol ethoxylates and in particular nonylphenol ethoxylates, which were widely used until the mid-1990s. These phenol ethoxylate materials have been phased out due to poor environmental properties and claimed oestrogenic mimic behaviour [103].

Alkanolamides. These products have been used in drilling fluid removal and have suitable environmental properties. They, however, are not widely used, and this may be due to cost.

Phosphates. The use of these materials, in particular the cheap *ortho*phosphates, is common where it is necessary to reduce the cloud point of the surfactant in solution. Many surfactants are most effective at a temperature where they are just about to come out of solution (cloud point). This temperature may be adjusted by addition of a 'builder' such as these phosphates. The downside is that the use of these materials in formulations reduces the temperature operating range of the surfactant, which is an issue when the material is required to be effective over a broad temperature range.

Alkyl polyglucosides (APGs). These materials are used in well cleaning operations in particular the octyl-substituted APG, which is used at 5–10% v/v concentrations. They are however not suitable for heavy brine applications and in particular those with a high level of divalent cations.

There are a number of properties that are important in constructing an efficient and acceptable well cleaner, and the following are the most important:

- *HLB* – An HLB number in the range 10–15 (and as stated earlier probably 11.5–14.0) is the best at 'dirt' removal. The HLB is dependent on structural properties in the surfactant, such as degree of ethoxylation or substitution, lengths of side chains, etc.
- *Cloud point* – Surfactants precipitate out of solution at a specific temperature, their cloud point. This is usually measured in MEG. At just below this point, it is widely held that surfactants operate at their optimum. However, using solely the cloud point to design well

cleaners, in the author's experience, is exceedingly difficult due to the wide variations in temperature that the cleaning formulation is likely to encounter. Nonetheless, it is a good indicator within the range above as to those products likely to be effective.

- *Foaming characteristics* – It is desirable that excessive foaming is not present as this can cause considerable operational problems at the rig site.
- *Biodegradability* – In this respect, the main factors under consideration, which cause surfactant molecules to biodegrade, are their molecular size and the degree of side chain branching. Small linear molecules tend to biodegrade more efficiently than those that are highly branched.

The properties above require to be balanced as the HLB efficiency and biodegradability act against each other, and therefore commercial products are a blend of chemicals that take into account the interdependency of these properties.

8.4.2.2 Solvents

The solvents in these cleaning packages can be run as separate pills or included in the formulation, and these are utilised to remove hydrocarbon deposits and pipe dopes. They are also widely used where invert emulsions form the drilling fluid; indeed in these cases they are exclusively used. The application of these materials is a balance of their solvency versus their environmental profile and most importantly their toxicology and flammability [102].

The main solvents used in formulating well cleaners, especially those that can effectively remove pipe dopes, have been xylene, toluene, ketones and low molecular weight esters. These have proved to be very effective; however they can be both environmentally hazardous and pose additional health and safety problems to the personnel. These have largely been replaced in highly regulated areas such as the North Sea by a range of biodegradable esters, middle distillates and terpene-based solvents.

Glycols and glycol ethers are also used although these are not effective against pipe dope, but they will reduce and disperse drilling mud residues and 2-butoxyethanol; (2-ethylhexanol) is widely used for this purpose.

8.4.2.3 Flocculants

These are the smallest group of materials used in well cleaners. They are usually based around fatty alcohols and fatty acids that are lipophilic in nature. They are added at 3–10% v/v active concentration. During the cleaning process, they disperse through the aqueous phase as lipophilic droplets and readily adsorb on oil and oil-wet materials. In the main these materials are used either as filter aids or as post-displacement 'polishing' fluids to coalesce any fine material following treatment with main wash pill(s) [59].

Although a number of typical formulations are illustrated in Figure 8.29, unlike many other product lines, it is difficult to target specific product details as most are tailored to the well conditions of a specific site and supplied as part of the drilling fluids package [104].

8.4.3 Acidising Formulations

Acid stimulation is the most common method used to increase the permeability of both production and injector wells. It is applied to both carbonate and sandstone reservoirs and in effect dissolves acid-soluble solids naturally present or as formation damage [105]. There are three basic methods by which acids are used to stimulate oil and gas production; these are

- *Acid washing* – This is usually a soak over a small amount of time with a suitable mineral or organic acid to 'wash' or dissolve some rock face and/or formation damage in the wellbore.

- *Fracture acidising* – This is placing the acid treatment under pressure to induce or extend a fracture in the rock and supplement this with acid dissolution to extend the fracture as deep into the near wellbore as possible.
- *Matrix acidising* – This is the most common form of acidisation treatment. Here the acid is pumped into the production well at below the formation-fracturing pressure. The objective is to deliberately create wormholes in the near wellbore region, reaching as far into the formation as possible without fracturing. This technique can double the production rate in undamaged formations, and in damaged formations even higher rates are obtainable.

In general, there are three key chemical additives that should always exist within acid packages, and these are as follows:

- An acid stable corrosion inhibitor, which may also include an intensifier for well temperatures above 120 °C.
- An iron control agent, of which there are three types (see below) that can be used in a variety of combinations [106]:
  - Reducing agents, which reduce iron(III) to iron(II)
  - Complexing agents or chelates to bind iron from further reaction and make it water soluble
  - Anti-sludging surfactants, which are only required where possibility of asphaltenic sludges may be formed
- A surfactant (water-wetting surface agent); this retards the acid reaction with the rock face, and to some extent the path of the reaction at the near wellbore. This is often premixed with the acid to form a semi-stable emulsion, which breaks once the acid is spent and allows easier fluid flowback.

A wide variety of chemical agents are used in these formulations, and critically they need to be stable to the acidic conditions of the base material [107].

8.4.3.1 Acids Used in Stimulation

In carbonate reservoirs, hydrochloric acid is the main acid deployed in these treatments. The concentration is usually 15 wt%; however up to 28 wt% can be used. Organic acids, in particular acetic acid and formic acid, are sometimes used for high-temperature applications. These 'retarded' acids exhibit failings due to the formation of acetate and formate salts. They also exhibit high rates of corrosion at high temperatures. It is of note that all of these acids are identified as causing no harm for the marine environment [108].

High-pH chelating agents have also been looked at for matrix acidising, particularly ethylenediaminetetraacetic acid (EDTA) [109] and hydroxyaminocarboxylic acids [110]. This technique also has the advantage of significantly reduced corrosion rates. As EDTA has somewhat disputed biodegradation rates, other products such as glutamic acid diacetic acid (GLDA) have been promoted, particularly in the North Sea, Gulf of Mexico and Canada [111].

Unless the formation is damaged, treating sandstone formations with matrix acid (HCl or similar) is ineffective in stimulating production. This is because the sandstone reservoir is composed mainly of quartz and aluminosilicates, the fines of which can migrate into the pores at the near-wellbore area, causing production impairment. Hydrofluoric (HF) acid is therefore used to dissolve this type of formation damage.

Long-chain carboxylic acids have also been examined as they offer low corrosion rates and good dissolving power at high temperatures, are highly biodegradable and are safe and easy to handle [112].

Although very corrosive HF acid is classed as a weak acid, it is also very toxic and has extreme safety risks due to its potential to react rapidly with calcium. For these reasons, HF acid is usually deployed as an HF acid-releasing chemical such as ammonium bifluoride, and this is usually combined with HCl. This HF/HCl blend is known as 'mud acid' as it will also dissolve clays from drilling operations. Generally, the maximum concentration of HF acid is 3 wt% due to the possibility of deconsolidating the sandstone near wellbore.

In using HF formulations, one has to be careful of reprecipitation of reaction products leading to new formation damage. This is due to the HF reacting with aluminosilicates to form fluorosilicates, which can then precipitate as their potassium or sodium salts.

8.4.3.2 Corrosion Inhibitors

Acid corrosion inhibitors are part of a chemical package that is added to the acidisation and stimulation fluid to minimise and retard the corrosion rate of the acid and spent acid as it is pumped into the well and near wellbore. It has to be seen in the context of the well and formation conditions, geology and other chemicals used to maximise the stimulation effects of the acidisation [113].

Temperature has an adverse effect on corrosion rates particularly in acidic medium by increasing them dramatically. Very few chemical corrosion inhibitors are effective at temperatures over 120 °C and almost none at over 180 °C.

The major base chemistry primarily used in the composition and formulation of acid corrosion inhibitor packages is propargyl alcohol (Figure 8.30) or derivatives thereof; see also Section 6.2.1. In general, the main acid corrosion inhibitor packages have been mixtures of quaternary ammonium halides and in particular the iodide and propargyl alcohol. The iodide quats act as intensifiers as well as imparting corrosion inhibition properties.

It has been found that the presence of a small amount of hexamethylenetetramine (Figure 8.31) dramatically improves the performance of the acetylenic alcohols, which allows reduced concentrations to be used and improves performance at higher temperatures.

A number of the chemistries used in the formulation, including propargyl alcohol, of acid corrosion inhibitors are highly toxic and are classified as marine pollutants. Over the last decade or so, there has been considerable work investigating potential alternatives that are environmentally acceptable. It has been shown that 1,3-unsaturated aldehydes with surfactants are effective corrosion inhibitors for steel in HCl [113]. Based on this technology, the aromatic compound cinnamaldehyde, a natural product (Figure 8.32), has been proposed as an ingredient in low toxicity inhibitor formulations [114].

A product line formulated from propargyl alcohol and a pyridinium or quinolium quaternary salts is another variant on the acid corrosion inhibitor package.

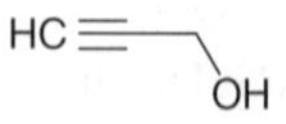

Figure 8.30 Propargyl alcohol.

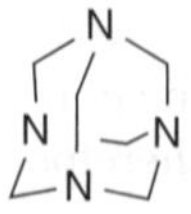

Figure 8.31 Hexamethylenetetramine.

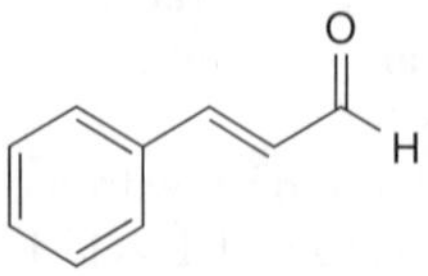

Figure 8.32 Cinnamaldehyde.

8.4.3.3 Controlling Iron

Iron control agents are always added and the chemistries here are discussed in a following section. Iron control agents are added at thousands of parts per million to percentage levels. Some specific concentrations, of particular for chelants and others, are tabulated in Figure 8.33.

Iron, which is probably present in all oilfield operations, is now recognised as a significant problem in acidisation. Tanks, lines, equipment and well tubulars all have iron in contact to a lesser or greater degree, and this will to some extent, at least, dissolve in the applied acid. Even minimal amounts can amount to several thousand milligram per litres. Both Fe^{2+} and Fe^{3+} species can be found, the latter coming from the conversion of Fe^{2+} by dissolved oxygen.

Where there is a tendency for iron sulphide scale to form in the well or near wellbore region, which is particularly the case in sour production, then this can cause further significant problems [115]. This is because when acid is used to stimulate the well, the iron sulphide will dissolve and release hydrogen sulphide, which is very toxic and can stimulate corrosion. In addition, the dissolved iron tends to precipitate as ferric hydroxide or ferric sulphide as the acid treatment is spent. These products can cause further formation damage and production impairment, compromising the stimulation of the well. It has also been observed that the release of these Fe^{3+} ions can lead and/or contribute to asphaltenic flocculation and subsequent sludge deposition [116].

Several iron reaction species can precipitate from the acid as it is spent and the pH rises. The most common is ferric hydroxide. As mentioned above, iron sulphide is a common deposit in sour wells, and even elemental sulphur is known to deposit when the pH is around 1.9 [117].

8.4.3.3.1 Reducing Agents
There are a number of reducing agents [106] that have been deployed, in particular:

- Iodine or iodide salts; these also help to prevent corrosion.
- Metal ions such as tin(II) and copper(I) and transition metal ions; however these are not used in the North Sea.
- Formic acid and hypophosphorous acid; a hypophosphorous precursor such as a metal phosphinate has also been used.
- Erythorbic acid and ascorbic acid.
- Reducing thioacids such as thioglycolic acid with a catalyst such as iodide ions or ketones that react with iron sulphide.

Some reducing agents such as iodide and some inorganic acids can help to prevent corrosion.

Chelant	Dose	Comments
EDTA	0.8 wt%	Not very effective and has a low iron carrying capacity
NTA	5.0–5.5 wt%	High concentrations required to keep Fe(III) in solution
Citric acid	5.0–6.0 wt%	
Acetic acid	1.5–2.0 wt%	Precipitates iron at high concentrations
Erythorbic acid	1.6%	This is added as a reducing agent. Degrades at above 65 °C

Figure 8.33 Chelant and reducing agent concentrations.

8.4.3.3.2 Chelating Agents More commonly used are chelating/complexing agents such as citric acid [106], EDTA and nitrilotriacetic acid (NTA). The latter is most commonly deployed in the North Sea due to its reasonable sequestering power and its superior biodegradation compared with EDTA.

In sour wells [118] it has been reported that organotin products and other carbonyl products have been successfully used. Also, phosphonic acids and similar products such as tetrakis(hydroxymethyl)phosphonium sulphate (THPS) have been added to afford additional protection against iron precipitation, as they form water-soluble complexes and operate at sub-stoichiometric levels, effectively inhibiting potential precipitates.

Recently, a new environmentally friendly chelating agent, GLDA, has been developed and extensively tested for carbonate and sandstone formations. Significant permeability improvements have been shown over a wide range of conditions. This chelating agent has subsequently been applied in the field to acidise a sour, high-temperature, tight-gas wells completed with high-chrome-content tubulars [119].

Two other systems have been proposed to sequester iron, namely, the use of dithiocarbamates [120] and oximes such as acetaldoxime [121].

8.4.3.3.3 Anti-Sludging Agents These are used to prevent or mitigate against sludge formation, and a number of surfactants have been deployed in acid treatment packages. A sludge can be formed by the reaction of iron ions with polar groups of asphaltenes present in the crude oil. Ferric ions in particular induce sludging as they act as phase-transfer catalysts for HCl [122].

Anionic surfactants, such as the alkylbenzene sulphonic acids, and in particular dodecylbenzenesulphonic acid (DDBSA), are the most common anti-sludge agents. The ideal anti-sludge agent would contain a head group with high polarity, such as the acidic functional group on DDBSA, which can bond onto acid/base groups on the asphaltene molecule in addition to potential π–π bonding of the aromatic groups. In addition to the high polarity, a longer tail is ideal, even though a longer tail group would marginally reduce the overall polarity of the molecule. If the molecule had a tail smaller than a six-carbon chain, the surfactant would become a part of the asphaltene flocculant instead of inhibiting its formation. Longer tails on the anti-sludge agents will provide steric hindrances to asphaltene aggregation and therefore are often described as forming a steric stabilisation layer [123].

Additionally, anionic anti-sludge surfactants have been shown to minimise the transfer of acid into the oil phase, therefore minimising the formation of acid induced sludge. Although in the presence of ferric iron, these surfactants are ineffective at preventing the phase transfer of acid, and therefore anionic surfactant and HCl will compete for acid/base sites on the asphaltene. To prevent sludging, iron-reducing agents are needed in combination with anionic anti-sludge agent [124].

Anionic anti-sludge agents often have limited dispersability in hydrochloric acid and often need to be coupled with oxyalkylated alkylphenols to remain dispersed in the acid. When the acid blend comes in contact with oil, up to 99% of the DDBSA will be extracted during the oil phase, making it an effective anionic anti-sludge agent.

Cationic surfactants are generally not effective at preventing emulsions and can stabilise acid–oil emulsions. Typical cationic surfactants that are capable of preventing solid asphaltene precipitation are shown in the table in Figure 8.34.

Of note is the fact that cationic surfactants with longer-chain lengths are more effective at preventing particle aggregation due to the steric layer around the asphaltene [122].

Other materials, such as acetylenic alcohols and alkyl diphenyl oxide sulphonic acid, can help prevent the formation of emulsions and sludges during matrix acidising [125].

Anionic surfactants	**Cationic surfactants**
Alkyl benzylsulphonic acid (DDBSA)	Alkyl, aryl amines
Alkyl sulphosuccinates	Quaternised ammonium salts

Figure 8.34 Anionic versus cationic surfactants.

In the North Sea, some biodegradable ester quaternary surfactants have been used (H.A. Craddock, unpublished results); see also Section 3.4.2.

8.4.3.4 Retardation of Acids

HCl and HF are two acids reacting quickly with carbonates and silicates. However, the objectives of acid treatments are to increase porosity and permeability as deep as possible into the formation. Methods have been developed to slow the acidising process and include the following [126]:

- Emulsifying the aqueous acid solutions in oil (or solvents such as kerosene or diesel fuel) to produce a mixture, which is slower reacting. Surfactants are used here to aid this process.
- Dissolving the acids in a non-aqueous solvent such as an alcohol.
- The use of non-aqueous solutions of organic chemicals, which release acids only on contact with water.
- The injection of solutions of methyl acetate, which hydrolyses slowly at very high temperatures to produce acetic acid.

In addition to these methods, of which emulsifying the acid is probably the most important, some retardation of the reaction rate can be achieved by gelling the acid or oil wetting the formation solids.

The use of retarders and related agents is more problematic. Retarded acid systems can increase the acid penetration by slowing or blocking the acid reaction with the formation and associated formation damage. This can reduce the rate of acid leak-off, allowing the developing matrix of wormholes to be more extensive and give deeper penetration.

The addition rates of these materials can vary greatly but are usually between 10 and 50 US gallons per thousand gallons of acid.

Gelled acids are mainly used to retard acid reaction rate in acid fracturing treatments [127]. Retardation results from the increased fluid viscosity, reducing the rate of acid transfer to the fracture wall. However, the use of the gelling agents (normally water-soluble polymers) is limited to lower-temperature formations as most gelling agents degrade rapidly in acid solution at temperatures above $55\,°C$.

Some retardants can be added to the mud acid (HCl–HF mixture) to slow the reaction rate of acid with the minerals. The key is to inject a solution not containing HF acid explicitly but a compound able to generate HF acid at greater depth of penetration and longer reaction time for maximum dissolution of fines. This retardant hydrolyses in water when it enters in the reservoir to form HF acid according to the reaction in Eq. (8.1):

Formation of HF acid from bifluoride retardant:

$$HBF_4 + H_2O \rightarrow HBF_3OH + HF \tag{8.1}$$

Another system proposed for moderate-to-deep penetrations uses a phosphonic acid complex (1-hydroxyethane-1, 1-diphosphonic acid (HEDP)) to hydrolyse HF acid. HEDP has five hydrogen atoms available that dissociate at different stoichiometric conditions. A mixture of HEDP acid with $NH_4 HF_2$ produces an ammonium phosphonate salt and HF acid [128].

8.4.3.5 Surfactants

The use of surfactants as anti-sludging agents has already been described in Chapter 3. The formation of emulsified acids, in order to retard the rate of dissolution produced by the acid, involves the use of surfactants as an emulsifier. These types of retarded acid are particularly useful in stimulating formations with low permeability [129].

Surfactants are also added to acid formulations to reduce surface tension. This is necessary as the spent acid in low permeability (tight) reservoirs can be trapped at the pore throats by capillary forces. Nonionic surfactants, particularly fluorocarbons and mutual solvents, are commonly used to lower the surface tension of the acid package. The amount of surfactant added can vary greatly depending on the final surface tension values to be achieved; however it is generally around 0.1–0.5 wt%. It is also necessary to take account of potential surfactant loss due to rock adsorption.

Demulsifier and non-emulsifying surfactants are applied in acid treatments in order to 'break' emulsions that tend to form between crude oil and live or spent acid. Such emulsions can be very viscous and lead to pore throat blockage. These are normally added to a spearhead in a solvent in front of the main acid treatment. Some demulsifiers such as the N-alkylated polyhydroxyetheramines have been specially designed for this type of application [130].

8.4.3.6 Diverting Agents

Surfactants are also used in diversion techniques particularly as foaming agents [131].

There are a number of chemicals and related techniques to aid in place of the acid treatment. The operational mechanics of this is outside the scope of this study. The main categories of diverter are now detailed.

8.4.3.6.1 *Solid Particle Diverters*   Traditionally, solid diverters have been rock salt, benzoic acid flakes, oil-soluble resins and wax beads or sticks. Wax and resin materials cannot be used in low-temperature wells, as they may not melt at the well temperature. A number of polymers have been used, and in the North Sea area polymers, which are hydrolysable, are favoured. Indeed, some can be particularly useful in high-temperature wells as their hydrolysis can be temperature triggered [132].

8.4.3.6.2 *Polymer Gel Diverters*   Polymer gels have a higher viscosity than the acidising solution, which means they will penetrate the lower permeability intervals prior to the acid package. Generally, in wells below 95 °C, a gelled hydroxyethyl cellulose (HEC) pill is injected as a pre-flush. Above this temperature, the viscosity and life of the pill with HEC is greatly reduced and is regarded as ineffective. In such cases, other polymers such as polyacrylamides, and in the North Sea biodegradable polysaccharides, are used [133]. There is much overlap in the technology applied here with general viscosifying of fracturing fluids and also with water shut-off polymers and related materials.

8.4.3.6.3 *Viscoelastic Surfactants*   The main surfactants chemistries used here are [134]

- Anionic surfactants such as sulphosuccinates and methyl ester sulphonates
- Ethoxylated fatty amines
- Zwitterionics and amphoterics
- Cationic surfactants such as alkyl methyl bis(2-hydroxyethyl)ammonium chloride
- Amine oxides and amidoamine oxides (e.g. dimethylaminopropyltallowamine oxide)

A number of other chemical agents can also be added to the acidisation package depending on conditions likely to be encountered:

- *Clay stabilisers* – These are usually polyamines or quaternary salts thereof.
- *Fines fixing agents* – These are organosilanes such as 3-aminopropyltriethoxysilane.
- *Calcium sulphate scale inhibitor* – May be necessary if sulphate concentrations are high.
- *H_2S scavenger* – May be required if acidisation is in contact with sulphate scales.
- *Foaming agents* – These surfactants are used in gas well treatments in conjunction with nitrogen gas to aid in the spent acid removal and lift the well back into production.
- *Drag reducers* – These are used in deep well treatments to provide the fluids with low friction. These are normally high molecular weight polyacrylamides.
- *Low molecular weight alcohols* – These are applied in gas well treatments to lower the surface tension and help release the spent acid for removal.

Acidisation is an established, and reasonably well understood, method of reservoir stimulation. There have been relatively few recent advances in the technology applied, and the preferred methods of stimulation, especially matrix stimulation, involve strong mineral acids or mixtures thereof. Chemical additives are used primarily to minimise the corrosive effects of the acid on metal surfaces, to control dissolved iron, particularly Fe^{3+}, and to retard the action of the acid to allow better penetration into the formation. A number of other chemical additives, as described, can also be added. Fundamentally, however, the acid medium is still hydrochloric or HF acids or combinations of both.

8.4.4 Production Chemicals

The formulation of production chemicals falls into two main categories: those that are formulated in water and those that are formulated in solvents or solvent systems. This has usually, but not always, to do with the intended receiving environment, that is, the water or oil phases of the production fluids. For example, scale inhibitors are most often formulated in water as it is in the reservoir and produced water that the scaling issues occur. On the other hand, wax inhibitors and PPDs are formulated in organic solvents as they are required to act upon the crude oil and therefore must be soluble in it.

Some products can be made to be soluble in both water and oil, and often surfactant-based corrosion inhibitors can be soluble in oil or water and at the same time be dispersable in the other phase.

Very few production chemicals are applied unformulated even if that is only a simple dilution.

In addition to formulation to be compatible with the receiving environment, there are a number of other issues surrounding the compatibility of production chemicals that the formulator needs to be aware of, and these are listed below:

- Compatibility with other production chemicals present or likely to be present in the fluid phase to which the chemical is applied.
- Compatibility with the materials found in the line of production, including flow lines, pumps and associated seals, risers and other hardware.
- Potential to cause problems after export such as in the refinery.
- Ensuring the formulated product meets injection requirements.
- Does the injected production chemical cause associated production chemistry problems, e.g. some film-forming corrosion inhibitors can make emulsion and foam problems worse and counteract any demulsifier and defoamer action?

	Solvent	Corrosion inhibitor	Demulsifier	Antifoam
Solvent	Compatible	Compatible	Compatible	Compatible
Corrosion inhibitor	Compatible	Compatible	Compatible	Not compatible
Demulsifier	Compatible	Compatible	Compatible	Compatible
Antifoam	Compatible	Not compatible	Compatible	Compatible
Wax dispersant	Not compatible	Not compatible	Immiscible–two layers no gunking	Not compatible

Figure 8.35 Chemical compatibility matrix.

Often the formulator will use a matrix (see Figure 8.35) for example to ensure as far as it is practical, a logical approach is taken to formulating compatibility understanding.

Material compatibility is also a critically important issue in the use of chemical additives and inhibitors. As stated previously, ethanol is used as a thermodynamic hydrate inhibitor in certain regions of the world; however it is important that the production chemist is aware of the specific factor that ethanol can have a direct bearing on carbon steels by inducing stress corrosion cracking particularly where oxygen is present even at parts per million levels [135].

This is only one of many examples, and the formulator needs to take account of the potential material incompatibilities of individual components and the main solvent.

8.4.4.1 Production Chemicals for Water

Where production chemicals are formulated in water, then of critical importance is their compatibility with the produced fluids, particularly the produced water and also any seawater or brines they may come into contact with. Generally, production chemicals formulated in water fall into three main categories:

- *Biocides* – Usually a straight dilution of the active material to between 20% and 50% active. Often manufacturers will supply diluted strength materials or have their manufacturing process suited to a particularly commonly used concentration.
- *Corrosion inhibitors.* – These are usually required to be water dispersable so they can act in both the oil and water phases to create a molecular barrier to corrosive species such as dissolved acid gases. These are normally formulated to between 20% and 30% active and can consist of a variety of active materials and a solvent system that is partly water soluble as well as water itself.
- *Scale Inhibitors* – These are generally formulated from a single active or a combination of 2–3 actives. They are generally formulated in water to around 25–35%; however some may require pH adjustment particularly phosphonate materials used for 'squeeze' applications in carbonate reservoirs.

Variation of pH in aqueous formulations can be required to enhance either compatibility or activity. For example, many phosphonate scale inhibitors are used for pre-emptive scale control near the wellbore by placing or squeezing onto the surface of rock formation. Rather than just relying on chemical adsorption and desorption that allows a lot of chemical additive to be quickly returned in an uncontrolled fashion. The precipitation squeeze tries to ensure that a higher proportion of the inhibitor is potentially 'fixed' in the reservoir and is available for slow release back to the production system.

Usually calcium chloride is used as the precipitating agent and may be mixed with the inhibitor solution or introduced in a pre-flush or after-flush solution. By pumping the inhibitor plus calcium chloride solutions in mildly acidic form, it is possible to rely on the increased temperature or higher pH conditions in the reservoir to induce precipitation of the calcium salt of the inhibitor. The two most common inhibitor salts are calcium phosphonate and calcium polycarboxylate. These are sparingly soluble, and their solubility further decreases with increasing temperature and increasing pH [136].

Many other factors have been examined and developed to enhance both adsorption and precipitation squeeze efficacy of action and in extension of lifetime of scale inhibition. Numerous publications and papers have been published on the subject and a useful review is cited [137].

Although often simple formulations or dilutions are used in water-based production chemicals, multicomponent blends have also been developed. Biocides that require to act in environments where sessile bacteria are present and have formed a biofilm are usually formulated with a few per cent of a surfactant to act as a 'biopenetrant' [138]

In formulating corrosion inhibitors, particularly those for acid stimulation work, it is often necessary to incorporate an intensifier [139, 140], and some of these molecules have been discussed in Chapters 5 and 6.

As previously stated, it is often useful to formulate corrosion inhibitors so that they are soluble or dispersable in both water and oil. This is to allow surfactant-based film-forming corrosion inhibitors the best possible means of partitioning at a boundary or surface layer and therefore forming a protective film on metal or similar surfaces. As described in Section 3.1.8, this is about achieving a balance between surfactant characteristics particularly the HLB. In practical terms a stable and useful formulation can be achieved by the use of a water-miscible co-solvent such as MEG [113].

Finally, it is worth remembering that water-soluble materials are more likely to be biodegradable and potentially less environmentally harmful than materials that are water insoluble but oil or solvent soluble.

8.4.4.2 Production Chemicals for Oil

Although many production chemicals are formulated in water, an almost equivalent amount is not. Deoilers, hydrate inhibitors, wax inhibitors and dispersants, PPDs and asphaltene control agents, to name some of the most common, are formulated in solvents as they are required to act in the oil phase of the produced fluids. Demulsifiers are a special case; however they are normally formulated in a locally available solvent to make them oil miscible or soluble.

Demulsifiers are based on surfactant formulations, and the surfactants hydrophobic tails of the surfactants prefer to interact with each other rather than being surrounded by water. This typical behaviour originates from the hydrophobic effect, which is the driving force behind surfactants association; see Section 3.1 [141, 142]. Depending on the strength of interaction between the hydrophilic units of a surfactant and the water, the ability for surfactants to self-assemble is affected. The free energy for both the hydrophilic head and the hydrophobic tail of a surfactant, depending upon the surrounding medium, will determine the ability for micellisation or solvation in either water phase or oil phase. This behaviour can be manipulated so that the demulsifier not only is oil soluble but also acts at the water–oil emulsion interfaces, causing oil droplets to form in the water phase and to aid their aggregation and eventual resolution of the emulsion under treatment.

Of particular value to the formulator when assembling a demulsifier formulation is the RSN, and the RSN value can be defined as the amount of distilled water necessary to be added to a solvent system in order to produce persistent turbidity. It is contrary to the HLB value, which is an empirically determined measure of water solubility and hydrophobic–hydrophilic character of a surfactant. The RSN value provides a practical alternative to the HLB method of assessing HLB of surfactants. Traditionally, RSN has been determined by titration of a surfactant in benzene/dioxan solution with water; however, a method for nonionic surfactants, using a less toxic system of toluene and EGDE, has been developed [143]. For example, the RSN values can be determined for a series of nonionic alcohol ethoxylates where the alkyl chain is dodecanol and EO chain ranges from 2 mol up to 10 mol. The measured RSN values will increase with increasing ethoxylation.

Such sequences allow the formulator to anticipate an action of replacing a certain surfactant with another. However, caution is required as these sequences are not universal, and in the example illustrated, further ethoxylation can decrease the RSN value. It is therefore important for the formulator not to consider RSN values in isolation, and it is also important to consider the surfactant's chemical structure for optimum product selection. The RSN value is a useful indicator within a chemical family as illustrated in the example of non-ionic alcohol ethoxylates described in the previous paragraph; comparing RSN values across different surfactant groups and chemical families usually results in erroneous conclsions when related to demulsifier performance. It is essential in developing a demulsifier formulation that field tests with an experienced technician is performed to evaluate the performance of a number of formulated products and ascertain the best performing demulsifier blend [144].

Of interest and importance is the relationship between RSN values (and HLB characteristics) and a surfactants water/octanol partitioning. This later measurement usually expressed as a log factor is an important classification of a chemical's ability to bioaccumulate [145]. This is discussed further in Section 8.4.5.

In general, oil-soluble products are usually formulated in a convenient, cheap, locally available solvent, unless there is a pressing technical or regulatory requirement. In the North Sea region, for example, most demulsifiers use 2-ethylhexanol as a suitable 'green' solvent.

Where kinetic hydrate inhibitors and corrosion inhibitors are required to be formulated together, alcohols, particularly isopropanol, are the solvents of choice [17]. Isopropanol is a versatile solvent and has a greater solubility in hydrocarbons than most simple alcohols; see Section 8.2.2.

Ethylene glycols are used to formulate heat transfer agents being antifreeze chemicals. Although not as active in freezing point depression as alcohols, they exhibit a lower vapour pressure and therefore a lower rate of evaporation from the cooling system. Overall this makes ethylene glycol the most effective freezing point depressant and heat transfer agent. Ethylene glycol also has good compatibility with common elastomers used at ambient and low temperatures. It does however have poor compatibilities with some elastomers at higher temperatures. As well as exhibiting high rates of biodegradation, glycols such as ethylene glycol are also easily recycled by either simple filtration or redistillation.

A number of non-polymeric asphaltene dispersants have been found to be best solubilised in methanol or ethylene glycol instead of the usual naphtha-based solvent, for example tetrahydroxy-*p*-benzoquinone (Figure 8.36) [146].

Although many scale inhibitors are deployed as solutions in water, there is a drawback with downhole water-based 'squeeze' treatments in that these aqueous solutions may change the wettability of the rock. Once water-wet the rock permeability will have changed, sometimes permanently. This can lead to a water coning effect formed from a water channel expanding to

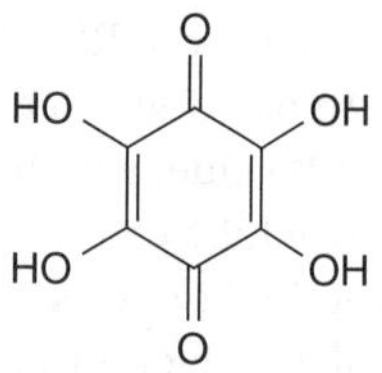

Figure 8.36 Tetrahydroxy-*p*-benzoquinone.

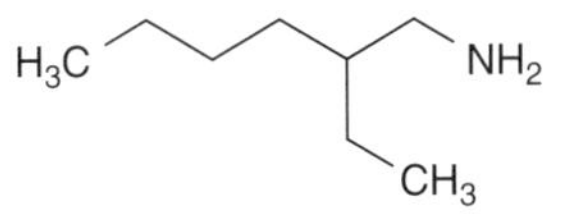

Figure 8.37 2-Ethylhexylamine.

a water pocket and consequently the well is irreversibly damaged. To avoid this from happening, water-based squeeze treatments are not normally used in low water cut wells or water-sensitive formations. Also, high water cut wells can benefit from a non-aqueous-based treatment due to lower densities and therefore less lifting and hydrostatic pressure problems [147].

As a solution, scale inhibitors that are miscible in hydrocarbon solvents have been developed and are often referred to as oil-soluble scale inhibitors. These are usually an alkyl amine and the free acid form of the scale inhibitor, which forms an ion pair, which is oil soluble [148]. It has been found that the preferred amine is 2-ethylhexylamine (Figure 8.37), which is also the most environmentally acceptable [149].

8.4.5 Formulating to Minimise Environmental Impact

Throughout this chapter there have been explanations and suggestions to minimise the environmental impact of solvents by using less harmful variants and substitution to 'green' solvents. The active chemicals have been explained throughout the preceding chapters and their environmental impact detailed. This section attempts to illustrate some methodologies and suggestions that could be adopted by the formulator to further reduce the environmental impact of the overall formulated product. Furthermore, the chemical supplier and formulator working with the end user, oil and gas operator, can, in a number of circumstances, provide chemical solutions to reduce the environmental impact of the overall operation, not just the chemical hazard. A few examples are illustrated in this section. However, this is not a common practice, partly for economic reasons but more often as the regulatory frameworks are not designed to take a wider and more holistic view of chemical use. This dilemma and dichotomy of action will be discussed further in Chapters 9 and 10.

8.4.5.1 The Relative Solubility Number

As described earlier, the RSN can be a useful tool in demulsifier product selection and formulation development. It is also an indicator of potential bioaccumulation, particularly for certain types of surfactant in which chain length extension or a series of esterification or ethoxylation are able to be performed to aid both water solubility and biodegradation, e.g. nonionic surfactants such as alcohol ethoxylates. However, this needs to be used with caution as correlation with other groups is inconclusive [150]. Nonetheless RSN is both a useful tool in product design and formulation when trying to achieve a low bioaccumulation potential for a demulsifier product.

8.4.5.2 Microemulsions

A number of application methodologies have been used to reduce environmental impact, and one of particular relevance to the chemical formulator is the use of microemulsions. A microemulsion is a thermodynamically stable fluid that differs from the normally used kinetic emulsions, which will separate over time. This is due to the particle size of the microemulsion, which covers a range from 10 to 300 nm and because of this is translucent or clear. These materials have an ultralow interfacial tension between the water phase and the oil phase and because of this can deliver a wide variety of oil-soluble oilfield chemicals such as asphaltene inhibitors, corrosion inhibitors and oil-soluble scale inhibitors. There are advantages in that they require to use less organic solvent, particularly useful if toxic aromatic solvents are required, and because of their ability to efficiently disperse the chemical additive or inhibitor into the fluids, the efficiency of the active chemical is increased and less is required [151].

8.4.5.3 Surfactant Compatibility and Switchable Surfactants

As stated earlier in Chapter 3, in general, cationic surfactants are incompatible with anionic surfactants, since they react with one another to produce insoluble cat–anionic compounds. However, it is possible to get around the incompatibility problem by using a nitrogenated surfactant of the amine type, i.e. with no actual positive ion. The environmental advantage of this is that overall water solubility is enhanced by incorporating a second polar group [152]. This can have positive effects in both demulsifier and corrosion inhibitor applications.

In a further twist, it is possible to use the so-called switchable surfactants, where the molecules' action can be reversed by the activation of a trigger. In this action, the molecular structure is changed to another form, and the surfactant properties are transformed from an inactive form to an active one [153]. For example, long-chain alkyl amidine compounds can be reversibly transformed into charged surfactants by exposure to an atmosphere of carbon dioxide, thereby stabilising water/alkane emulsions. Also, the neutral amidines can function as switchable demulsifiers of an aqueous crude oil emulsion.

So far, these switchable surfactants have focused on technical and performance properties, but it should be possible to enhance environmental properties such as biodegradation after switching from a performance mode to an after-use mode.

The other main class of oilfield products is based on polymers, and formulating polymers to a high level of active content can be challenging. Often polymeric materials have high viscosities and do not afford themselves to mobility or pumpability at even reasonably active concentrations of around 20%. Often viscosities are too high at such concentrations, and therefore active contents can be normally less than 5%. This obviously compromises the effectiveness of the product as applied and is compensated for in larger dose rates. It is not uncommon for wax inhibitors and PPDs to be applied in thousands of parts per million or even percentage levels, with the bulk of what is being dosed is a carrier solvent, often an aromatic and toxic marine pollutant. There is an obvious need to address this problem from both technical and environmental perspectives.

Two strategies, so far, seem promising, namely, the use of synergists in the blend and the encapsulation of the active material.

8.4.5.4 The Use of Synergists

Synergism in chemical terms is where the interaction or cooperation of two or more chemical substances produce a combined effect greater than the sum of their separate effects. This is relatively common in the formulation of demulsifiers where surfactants can combine to give enhanced effects in water separation, speed of phase separation and water quality [154].

Other applications are found in corrosion inhibitors. Here there are a number of synergists that are utilised to enhance the film-forming performance, especially with nitrogenous corrosion inhibitors. The most common are thiosulphate salts and mercaptocarboxylic acids. Indeed, thioglycolic acid is commonly used in imidazoline formulations to enhance performance. It is important to note that these products are used synergistically in formulations as they have poor corrosion inhibitor properties if used alone. They are often used to enhance performance in high shear stress environments. They appear to work by oxidation of the thiol(SH) group to a disulphide, which forms a complex with iron ions at the metal surface [155]. Thiosulphate is often added to formulated products to improve corrosion protection; however care is required in certain applications, particularly in the presence of sulphate-reducing bacteria as the addition of thiosulphate can promote certain corrosion problems.

A particular class of synergists has been developed, which are based on thiohydantoin chemistry. These have the benefits of both complexing with the metal surface and having passivating characteristics and are environmentally benign [156].

The amino acids cysteine and cystine or their decarboxylated products cysteamine and cystamine also act as synergists in general formulations with other film-forming corrosion inhibitors [157].

In polymers, particularly in those used in wax control, a number of additives have been investigated to enhance performance and to reduce dosage levels. Small amines as well as polyethyleneimines have been found to be particularly useful (H.A. Craddock, unpublished results). Silicon surfactants have also been claimed as synergists for PPDs and wax inhibitors [158].

Other synergists have been used across the field to enhance and improve technical performance, and a number have been directed at improving environmental properties, particularly biodegradation. More examples are found in Chapters 2 and 4 in particular.

8.4.5.5 The Application of Encapsulated Materials

Encapsulation is generally defined as something being enclosed in a protective shell. In terms of oilfield chemicals, this is the chemical active being enclosed as either a solid particle or a formulated liquid in a protective membrane. This has been applied to a number of products to help in application, delivery and efficiency. It has also allowed the formulator to achieve higher concentrations of actives. Encapsulated products tend to be at a finite size in the millimetre to centimetre range. It is essential to ensure that such materials can be effectively slurried or suspended for pumping for delivery to all the application points. As there are minimum size restraints on encapsulation, microencapsulation of products has been applied to effect greater pumping and delivery efficiencies. Microencapsulation is a process in which tiny particles or droplets are surrounded by a coating to give small capsules. In a relatively simple form, a microcapsule is a small sphere with a uniform wall around it. The material inside the microcapsule is referred to as the core, internal phase or fill, whereas the wall is sometimes called a shell, coating or membrane. Most microcapsules have diameters between a few micrometres and a few millimetres.

For the oil and gas industry, most products need to have the following properties once encapsulated:

- Dry-free flowing powder or granules
- Have a short range of particle size distribution, typically 100–440 μm for microcapsules
- Be extremely stable in water

Encapsulation has been applied to a number of chemical types in the oil and gas sector.

8.4.5.5.1 *Scale Inhibitors* These are by far the most commonly supplied encapsulated or microencapsulated products. These types of scale inhibitor have been designed to allow chemical release over an extended time period. Early products developed in the 1990s encapsulated simple calcium carbonate scale inhibitors such as phosphonates in a permeable polymer membrane. These 'beads' are aggregated and weighted for gravity placement in the sump (rathole) of the completion, and through diffusion the scale inhibitor continuously inhibits the produced water entering the completion and the production stream [159]. They are ideal for applications where a low threshold concentration is required for scale mitigation. They have an overall low treatment cost and a long treatment lifetime, and in many cases, they have substantially replaced the more expensive 'squeeze'-type treatments.

An additional benefit, from the application encapsulated inhibitor method, is realised when treating marginal high water cut wells and wells located in low-pressure areas by ensuring that the wells sustain flow after the treatments without the need for extensive stimulation and workover. This type of treatment has also been carried out for selective offshore, subsea wells where intervention costs are reduced significantly [160].

Polyphosphate inhibitors have also been encapsulated and have been applied in hydraulic fracturing stimulation treatments. A solid inhibitor that can be applied in conjunction with a fracturing treatment allows two treatments to be combined, saving the operator time and cost. It has been shown that calcium–magnesium polyphosphates are useful in these applications and have good compatibility with borate and zirconium cross-linked fracturing fluids [161].

8.4.5.5.2 *Combined Proppant and Scale Inhibitor Products*

Within the last two decades, products that act as proppants in hydraulic fracturing and also have scale inhibitor impregnated in them have been developed and applied [162]. Such products have the advantage in combining functionalities, and therefore only one ingredient has to be handled and delivered to the required area. Particles made from a variety of substances have been examined; however ceramic beads are the most popular. These particles containing scale inhibitor may be made by contacting porous ceramic beads with a solution of scale inhibitor and then drying the beads. The beads may be used as fracture proppants or in a gravel pack so as to suppress scale formation in an oil or gas well.

In placing scale inhibitor into the hydraulic fracture, the well is protected from flowback throughout the lifetime of the scale inhibitor and/or the fractured well, without the need potentially for retreatment. Also, as scale inhibitor is being applied at the production zone, it therefore will have the capability to treat the well system including the production tubing to the wellhead and also perhaps further in the production system, with only the need for a small amount of top-sides scale inhibitor to be considered [163, 164]. This approach has economic, technical and environmental benefits.

8.4.5.5.3 *Corrosion Inhibitors*

The technologies applied in scale inhibitors can also be deployed in other chemical additives such as corrosion, wax and asphaltene additives. To date, outside scale control, there has been little application of the technology because unlike scale control the cost advantages of using a solid or encapsulated product have not, as yet, been sufficient to make its use more prevalent.

Corrosion inhibitors, however, have been encapsulated and are used in certain applications. Indeed, the concept and development of encapsulated corrosion inhibitors predates the work on scale inhibitors by some 20–30 years. Originally these products were developed for enclosed heat exchanger applications. Recently a number of coating additives containing encapsulated corrosion inhibitors have been developed [165], and also nanotechnology for the encapsulation has been investigated [166].

In oil and gas production, the timely delivery of corrosion inhibitor can be critical to the implementation of a proper chemical treatment programme. Corrosion inhibitors are normally applied to producing wells via two primary modes: batch treatment and continuous chemical injection. Both of these methods have advantages and disadvantages. Solid encapsulated time-release corrosion inhibitors are able to provide near-continuous corrosion inhibition in a batch application manner.

In a conventional batch chemical application, a large quantity of corrosion inhibitor is applied at one time, and the retreatment interval is defined primarily by the film persistency of the inhibitor. In other words, batch chemical treatment, while convenient, results in many instances in an inefficient use of the product because much of the inhibitor applied is returned with produced fluids very soon after the well is brought back into production. In effect, only the chemical that is involved in forming a thin film of inhibitor is used; the rest is present in excess. If not retreated at the appropriate interval, the batch mode of chemical application can result in periods of little or no corrosion inhibition in the well.

From a chemical performance standpoint, the application of corrosion inhibitors in a continuous manner is generally preferred because an effective level of corrosion inhibitor can be maintained at all times. However, using continuous chemical injection has its drawbacks. First, the cost associated with the installation and maintenance of chemical tanks and pumps, as well as injection capillaries, can be appreciable or even prohibitively high. Second, if any part of the injection system fails, the well and associated flow lines can quickly enter into a corrosive regime. Further, maintaining the security of chemical injection equipment can be difficult in poorly guarded or remote locations.

Encapsulated time-release corrosion inhibitors have been seen to be highly effective against CO_2 corrosion in onshore oil producing wells. The technology relies upon a product delivery mode in which the corrosion inhibitor is released in a time-controlled manner from the water–hydrocarbon interfacial region in the annulus of the well. This mode of product delivery gives a near-continuous corrosion inhibitor treatment while using a conventional batch-type application technique [167].

Encapsulated inhibitors therefore can allow oil and gas companies to capitalise on the benefits associated with both batch chemical treatment (e.g. no on-site chemical injection equipment) and continuous injection (e.g. constant inhibitive chemical level).

8.4.5.6 Fracturing Fluid and Stimulation Additives

In the last decade or so, there has been a few interesting applications developed for encapsulated products in stimulation. Primarily, these are acid materials that have been mixed with a gelling agent and then encapsulated in an oil or polymer membrane.

Production from carbonate formations is often controlled by the degree of interconnectivity or lack thereof from compartmentalised reservoirs. Effective methods for achieving hydraulic interconnectivity with such compartments are limited in horizontal open-hole and long pay sections in deviated wells. Solid acid capsules that function as both a fluid-diverting agent and a fracture conductivity enhancer have been developed. The degradable sized particulates are incorporated into the acidising fluid, which is designed to enhance inflow from natural fractures and to help enable propagation of hydraulic fractures that can breach or achieve wellbore communication with the divided compartments [168].

The first commercial field trial of a novel acid fracturing technique, using encapsulated citric acid as the etching agent, was performed over a decade ago, and although deemed an operational success, post-treatment results did not show any improvement in production from the test well [169].

8.4.5.6.1 Encapsulated Breakers These materials are used both in conventional drilling operations and in increasing volumes in hydraulic fracturing operations. Here the main purpose of encapsulation is to ensure that the breaker does not come into contact with the polymer or other substrate initiating premature degradation.

The active chemical, such as a peroxide or persulphate oxidising agent, is contained within a membrane that is impermeable or only slightly permeable, preventing the breaker forming any initial contact with the polymer. The active breaker diffuses slowly from the capsule, or the capsule can be destroyed by time or chemical action, releasing the breaker at a desired time and place in the operation to act successfully [170].

Encapsulated gel breakers are now widely applied across the drilling and stimulation and are probably the application methods of choice for both enzymatic and oxidising breakers where delayed action gel breaking is required. The breaker is encapsulated in a water-resistant coating, which effectively shields it from the drilling or fracturing fluids so that rheology is unaffected. In this manner, a high concentration or loading can be added without premature loss of fluid properties [171].

The barrier properties of the coating, release mechanisms and the reactive chemical additives properties are the critical factors in the design of encapsulated breakers. A number of formulated breakers have been encapsulated with a variety of membranes [172].

8.4.5.6.2 Biocides The encapsulation of biocides in applications such as coatings, paints and antifouling agents is now well established [173]. Blends of encapsulated 4,5-dichloro-2-*n*-octyl-3(2*H*)-isothiazolone with free biocides have been developed as antifouling coatings [174].

To date, however, there appears to be no field application or known use of this technology in oilfield exploration, development or production operations. This is a little surprising as some benefits from protection, time delay and placement of targeted application could be gained, as well as a supposed environmental benefit from reduced chemical application and more effective treatment.

8.4.5.7 Active Dispersions

Active dispersions are the basis of a competing technology to solvent-based formulations and encapsulation. They are able to deliver chemical inhibitors and other additives with a heavy active loading while still maintaining flow characteristics and technical performance. To date this technology has been mainly concentrated on high molecular weight polymers, as these are difficult to deliver in conventional solvent systems and therefore have a low active concentration. The key applications focused on are in oil drag reducing agents and paraffin wax control agents [175].

It is believed that this technology could be more widely applied and constant dose dispersion, perhaps involving some encapsulation, could be developed to provide further alternative delivery systems to solvent-based materials.

8.4.5.8 Solid and Weighted Inhibitors

Solid and weighted corrosion inhibitors have been around for some time in the upstream oil and gas sector. The stick inhibitor was probably the earliest form of a high-density corrosion inhibitor [176]. It consists essentially of a conventional inhibitor mixed and/or intermingled with a resinous, or wax, substance and barite. The entire mix was heated above the melting point of the wax and then cast into sticks. When cooled below the melting point of the wax, a stick is obtained in a semi-rigid form. The barite provides the additional weight for the stick. Products made by crushing the inhibitor sticks are also used in treating down the casing annulus of pumping wells. Encapsulated inhibitors have also been examined for use in this application.

There are many other methods and techniques in formulation of oilfield chemicals, and this section has attempted to exemplify a few of them from both technical enhancement and environmental gain perspective.

References

1 Reichardt, C. and Welton, T. (2010). *Solvents and Solvent Effects in Organic Chemistry*, 4th updated and enlarged edition. Wiley-VCH.

2 Miller-Chou, B.A. and Koenig, J.L. (2003). A review of polymer dissolution. *Progress in Polymer Science* **28**: 1223–1270.

3 King, G.E. (2012). Hydraulic fracturing 101: what every representative, environmentalist, regulator, reporter, investor, university researcher, neighbor, and engineer should know about hydraulic fracturing risk. *Journal of Petroleum Technology* **64** (04): 34–42, SPE-0412-0034.

4 Craddock, H.A. (May 2012). Shale gas: the facts about chemical additives. www.knovel.com.

5 Yu, M., Weinthal, E., Patino-Echeverri, D. et al. (2016). Water availability for shale gas development in Sichuan Basin, China. *Environmental Science & Technology* **50** (6): 2837–2845.

6 Kappel, W.M., Williams, J.H., and Szabo, Z. (2013). Water Resources and Shale Gas/Oil Production in the Appalachian Basin–Critical Issues and Evolving Developments. US Geological Survey, US Department of the Interior Open File Report *2013-1137*, August 2013.

7 Caenn, R., Darley, H.C.H., and Gray, G.R. (2016). *Composition and Properties of Drilling and Completion Fluids*, 7the. Gulf Professional Publishing.

8 Yamamoto-Kawai, M., McLaughlin, F.A., Carmak, E.C. et al. (2009). Aragonite undersaturation in the Arctic Ocean: effects of ocean acidification and sea ice melt. *Science* **326** (5956): 1098–1100.

9 Anderson, F. and Prausnitz, J.M. (1986). Inhibition of gas hydrates by methanol. *AIChE Journal* **32** (8): 1321–1333.

10 Saneifar, M., Nasralla, R.A., Nasr-El-Din, H.A. et al. (2011). Surface tension of spent acids at high temperature and pressure. SPE/DGS Saudi Arabia Section Technical Symposium and Exhibition, Al-Khobar, Saudi Arabia (15–18 May 2011), SPE 149109.

11 Palmer, A.C. and King, R.A. (2008). *Subsea Pipeline Engineering*, 2nde. Tulsa, OK: PennWell.

12 Jordan, M.M., Feasey, N.D. and Johnston, C. (2005). Inorganic scale control within MEG/methanol treated produced fluids. SPE International Symposium on Oilfield Scale, Aberdeen, UK (11–12 May 2005), SPE 95034.

13 Fu, D., Panga, M., Kefi, S. and Garcia-Lopez de Victoria, M. (2007). Self-diverting matrix acid. US Patent 7,237,608, assigned to Schlumberger Technology Corporation.

14 Tephly, T.R. (1991). The toxicity of methanol: minireview. *Life Sciences* **48** (11): 1031–1041.

15 Ramirez, A.A., Benard, S., Giroir-Fendler, A. et al. (2008). Kinetics of microbial growth and biodegradation of methanol and toluene in biofilters and an analysis of the energetic indicators. *Journal of Biotechnology* **138** (3–4): 88–95.

16 Budavari, S., O'Neill, M.J., Smith, A., and Heckelman, P.E. ed. (1989). *The Merck Index*, 11ee, 820. Rahway, NJ: Merck & Co.

17 Dahlmann, U. and Feustal, M. (2007). Additives for inhibiting the formation of gas hydrates. US Patent 7,183,240, assigned to Clariant Produkte (Deutschland) Gmbh.

18 Cobb, H.G. (2010). Composition and process for enhanced oil recovery. US Patent 7,691,790, assigned to Coriba Technologies LLC.

19 Lakatos, I., Toth, J., Lakatos-Szabo, J. et al. (2002). Application of silicone microemulsion for restriction of water production in gas wells. European Petroleum Conference, Aberdeen, UK (29–31 October 2002), SPE 78307.

20 Rettinger, K., Burschka, C., Scheeben, P. et al. (1991). Chiral 2-alkylbranched acids, esters and alcohols. Preparation and stereospecific flavour evaluation. *Tetrahedron: Asymmetry* **2** (10): 965–968.

21 Thieu, V., Bakeev, K.N. and Shih, J.S. (2002). Gas hydrate inhibitor. US Patent 6,359,047, assigned to ISP Investments Inc.

22 Huddleston, D.A. (1989). Hydrocarbon geller and method for making the same. US Patent 4,877,894, assigned to Nalco Chemical Company.

23 Richardson, W.C. and Kibodeaux, K.R. (2001). Chemically assisted thermal flood process. US Patent 6,305,472, assigned to Texaco Inc.

24 Yang, J. and Jovancicevic, V. (2009). Microemulsion containing oil field chemicals useful for oil and gas field applications. US Patent 7,615,516, assigned to Baker Hughes Incorporated.

25 Fink, J. (2013). *Petroleum Engineers Guide to Oilfield Chemicals and Fluids.* Elsevier.

26 Delgado, E. and Keown, B. (2009). Low volatile phosphorous gelling agent. US Patent 7,622,054, assigned to Ethox Chemicals LLC.

27 Fisk, J.V. Jr., Krecheville, J.D. and Pober, K.W. (2006). Silicic acid mud lubricants. US Patent 6,989,352, assigned to Halliburton Energy Services Inc.

28 Smith, K., Persinski, L.J. and Wanner, M. (2008). Effervescent biocide compositions for oilfield applications. US Patent Application 20080004189, assigned to Weatherford/Lamb Inc.

29 Rebsdat, S. and Mayer, D. (2005). *Ethylene Glycol in Ullmann's Encyclopedia of Industrial Chemistry.* Weinheim: Wiley-VCH. doi: 10.1002/14356007.a10_101.

30 Kunzi, R.A., Vinson, E.F., Totten, P.L. and Brake, B.G. (1995). Low temperature well cementing compositions and methods. US Patent 5,447,198, assigned to Halliburton Company.

31 Poelker, D.J., McMahon, J. and Schield, J.A. (2009). Polyamine salts as clay stabilizing agents. US Patent 7,601,675, assigned to Baker Hughes Incorporated.

32 Kelly, P.A., Gabrysch, A.D. and Horner, D.N. (2007). Stabilizing crosslinked polymer guars and modified guar derivatives. US Patent 7,195,065, assigned to Baker Hughes Incorporated.

33 Wang, X., Qu, Q., Dawson, J.C. and Satyanarayana Gupta, D.V. (2010). Thermal insulation compositions containing organic solvent and gelling agent and methods of using the same. US Patent 7,713,917, assigned to BJ services Company.

34 Davies, S.N., Meeten, G.H. and Way, P.W. (1997). Water based drilling fluid additive and methods of using fluids containing additives. US Patent 5,652,200, assigned to Schlumberger Technology Corporation.

35 Mueller, H., Herold, C.-P., Bongardt, F. et al. (2004). Lubricants for drilling fluids. US Patent 6,806,23, assigned to Cognis Deutschland Gmbh & Co. Kg

36 Ford, W.G.F. (1991). Reducing sludging during oil well acidizing. US Patent 4,981,601, assigned to Halliburton Company.

37 Fu, B. (2001). The development of advanced kinetic hydrate inhibitors. *Chemistry in the Oil Industry VII* (13–14 November 2001). Manchester, UK: The Royal Society of Chemistry.

38 Hall, B.E. (1975). The effect of mutual solvents on adsorption in sandstone acidizing. *Journal of Petroleum Technology* **27** (12): 1439–1442, SPE-5377.

39 Kelland, M.A. (2014). *Production Chemicals for the Oil and Gas Industry*, 2nde. CRC Press.

40 Collins, I.R., Goodwin, S.P., Morgan, J.C. and Stewart, N.J. (2001). Use of oil and gas field chemicals. US Patent 6,225,263, assigned to BP Chemicals Limited.

41 Dixon, J. (2009). Drilling fluids. US Patent 7,614,462, assigned to Croda International PLC.

42 Wang, J. and Buckley, J.S. (2003). Asphaltene stability in crude oil and aromatic solvents – the influence of oil composition. *Energy Fuels* **17** (6): 1445–1451.

43 Delbianco, A. and Stroppa, F. (1996). Composition effective in removing asphaltenes. EP Patent 0737798, assigned to AGIP SpA.

44 Dubey, S.T. and Waxman, M.H. (1991). Asphaltene adsorption and desorption from mineral surfaces. *SPE Reservoir Engineering* **6** (03): 389–395, SPE 18462.

45 Piro, G., Barberis Canonico, L., Galbariggi, G. et al. (1996). Asphaltene adsorption onto formation rock: an approach to asphaltene formation damage prevention. *SPE Production & Facilities* **11** (03): 156–160, SPE 30109.

46 Trbovich, M. and King, G.E. (1991). Asphaltene deposit removal: long-lasting treatment with a co-solvent. SPE International Symposium on Oilfield Chemistry, Anaheim, CAL (20–22 February 1991), SPE 21038

47 Bailey, J.C. and Allenson, S.J. (2008). Paraffin cleanout in a single subsea flowline environment: glycol to blame? Offshore Technology Conference, Houston, TX (5–8 May 2008), OTC 19566.

48 Barberis Canonico, L., DelBianco, A., Galbariggi, G. et al. (1994). A comprehensive approach for the evaluation of chemicals for asphaltene deposit removal. *Recent Advances in Oilfield*

Chemistry, Chemicals in the Oil Industry V (13–15 April 1994). Ambleside, Cumbria, UK: Royal Society of Chemistry.

49 Lightford, S.C., Pitoni, E., Mauri, L. and Armesi, F. (2006). Development and field use of a novel solvent water emulsion for the removal of asphaltene deposits in fractured carbonate formations. SPE Annual Technical Conference and Exhibition, San Antonio, TX (24–27 September 2006), SPE 101022.

50 Fattah, W.A. and Nasr-El-Din, H.A. (2008). Acid emulsified in xylene: a cost-effective treatment to remove asphalting deposition and enhance well productivity. SPE Eastern Regional/AAPG Eastern Section Joint Meeting, Pittsburgh, PA (11–15 October 2008), SPE 117251.

51 Appicciutoli, D., Maier, R.W., Strippoli, P. et al. (2010). Novel emulsified acid boosts production in a major carbonate oil field with asphaltene problems. SPE Annual Technical Conference and Exhibition, Florence, Italy (19–22 September 2010), SPE 135076.

52 Salgaonkar, L. and Danait, A. (2012). Environmentally acceptable emulsion system: an effective approach for removal of asphaltene deposits. SPE Saudi Arabia Section Technical Symposium and Exhibition, Al-Khobar, Saudi Arabia (8–11 April 2012), SPE 160877.

53 Straub, T.J., Autry, S.W. and King, G.E. (1989). An investigation into practical removal of downhole paraffin by thermal methods and chemical solvents. SPE Production Operations Symposium, Oklahoma City, OK (13–14 March 1989), SPE 18889.

54 Boswood, D.W. and Kreh, K.A. (2011). Fully miscible micellar acidizing solvents vs. xylene, the better paraffin solution. SPE Production and Operations Symposium, Oklahoma City, OK (27–29 March 2011), SPE 140128

55 Jones, T.G.J. and Justin, G.J. (1999). Hydrophobically modified polymers for water control. WO Patent 1999049183, assigned to Sofitech N.V., Dowel Schlumberger S.A. and Schlumberger Canada Ltd.

56 Wang, Y., Wang, L., Li, J. and Zhao, F. (2001). Surfactants oil displacement system in high salinity formations: research and application. SPE Permian Basin Oil and Gas Recovery Conference, Midland, TX (15–17 May 2001), SPE 70047.

57 Hutchins, R.D. and Saunders, D.L. (1993). Tracer chemicals for use in monitoring subterranean fluids. US Patent 5,246,860, assigned to Union Oil Company of California.

58 Trimble, M.I., Fleming, M.A., Andrew, B.L. et al. (2010). Method for removing asphaltene deposits. US Patent 7,754,657, assigned to Ineos USA LLC.

59 Knox, D. and McCosh, K. (2005). Displacement chemicals and environmental compliance- past present and future. *Chemistry in the Oil Industry IX* (31 October – 2 November 2005). Manchester, UK: Royal Society of Chemistry.

60 Neff, M., McKelvie, S., and Ayers, R.C. Jr. (2000). Environmental Impacts of Synthetic Based Drilling Fluids. U.S. Department of the Interior, Minerals Management Service, Gulf of Mexico OCS Region, OCS Study, MMS 2000–064.

61 Peresic, R.L., Burrell, B.R. and Prentice, G.M. (1991). Development and field trial of a biodegradable invert emulsion fluid. paper SPE IADC 21935 presented at the 1991SPE/IADC Drilling Conference, Amsterdam (1–14 March 1991).

62 Williams, C.G. (1859). On isoprene and caoutchine. *Proceedings of the Royal Society of London* **10**: 516–519.

63 Ford, W.G. and Hollenbeck, K.H. (1987). Composition and method for reducing sludging during the acidizing of formations containing sludging crude oils. US Patent 4,663,059, assigned to Halliburton Company.

64 Goncalves, J.T., DeOliveira, M.F. and Aragao, A.F.L. (2007). Compositions of oil-based biodegradable drilling fluids and process for drilling oil and gas wells. US Patent 7,285,515, assigned to Petroleo Brasileiro S.A.

65 Javora, P.H., Beall, B.B., Vorderburggen, M.A. et al. (2009). Method of using water-in-oil emulsion to remove oil base or synthetic oil base filter cake. US Patent 7,481,273, assigned to BJ Services Company.

66 Lightford, S.C. and Armesi, F. (2007). Compositions and methods for removal of asphaltenes from a portion of a wellbore or subterranean formation using water-organic solvent emulsion with non-polar and polar organic solvents. WO Patent Application WO 2007129348, assigned to Halliburton Energy Services Inc.

67 Penna, A., Arias, G. and Rae, P. (2006). Corrosion inhibitor intensifier and method of using the same. US Patent Application 20060264335, assigned to BJ Services Company.

68 Craddock, H.A., Mutch, K., Sowerby, K. et al. (2007). 1A case study in the removal of deposited wax from a major subsea flowline system in the gannet field. International Symposium on Oilfield Chemistry, Houston, TX (28 February–2 March, 2007), SPE 105048.

69 Blunk, J.A. (2001). Composition for paraffin removal from oilfield equipment. US Patent 6,176,243, assigned to Inventor.

70 Craddock, H.A., Campbell, E., Sowerby, K. et al. (2007). The application of wax dissolver in the enhancement of export line cleaning. International Symposium on Oilfield Chemistry, Houston, TX (28 February–2 March 2007), SPE 105049.

71 Jones, V.W. and Perry, C.R. (1973). Fundementals of gas treating. Proceedings of Gas Conditioning Conference, Norman, OK (15–16 March 1973).

72 CCME (2006). *Canadian Environmental Quality Guidelines for Sulfolane: Water and Soil, Scientific Supporting Document.* Canadian Council of Ministers of the Environment.

73 Scovell, E.G., Grainger, N. and Cox, T. (2001). Maintenance of oil production and refining equipment. WO Patent 0174966, assigned to Imperial Chemical Industries PLC.

74 Bacon, D.R. and Trinh, T. (1996). Fabric softener compositions with improved environmental impact. US Patent 5,500,138, assigned to The Proctor & Gamble Company.

75 Paul, J.M. and Fieler, E.R. (1992). A new solvent for oilfield scales. SPE Annual Technical Conference and Exhibition, Washington, DC (4–7 October 1992), SPE 24827.

76 Bradshaw, R.W. and Brosseau, D. (2009). Low-melting point inorganic nitrate salt heat transfer fluid. US Patent 7,588,694, assigned to Sandia Corporation.

77 McFarlane, J., Ridenour, W.B., Luo, H. et al. (2005). Room temperature ionic liquids for separating organics from produced water. *Separation Science and Technology* **40**: 1245–1265.

78 Abai, M., Shariff, S.M., Hassan, A. and Cheun, K.Y. (2015). An ionic liquid for mercury removal from natural gas. Abu Dhabi International Petroleum Exhibition and Conference, Abu Dhabi, UAE (9–12 November 2015), SPE 177799.

79 Nares, R., Schacht-Hernandez, P., Ramirez-Garnica, M.A. and del Carmen Cabrera-Reyes, M. (2007). Upgrading heavy and extraheavy crude oil with ionic liquid. International Oil Conference and Exhibition in Mexico, Veracruz, Mexico (27–30 June 2007), SPE 108676.

80 Sakthivel, S., Velusamy, S., Gardas, R.L. and Sangwai, J.S. (2015). Nature friendly application of ionic liquids for dissolution enhancement of heavy crude oil. SPE Annual Technical Conference and Exhibition, Houston, TX (28–30 September 2015), SPE 178418.

81 Berry, S.L., Boles, J.L., Brannon, H.D. and Beall, B.B. (2008). Performance evaluation of ionic liquids as a clay stabilizer and shale inhibitor. SPE International Symposium and Exhibition on Formation Damage Control, Lafayette, LA (13–15 February 2008), SPE 112540.

82 Yang, D., Rosas, O. and Castaneda, H. (2014). Comparison of the corrosion inhibiting properties of imidazole based ionic liquids on API X52 steel in carbon dioxide saturated NaCl solution. CORROSION 2014, San Antonio, TX (9–13 March 2014), NACE-2014-4357.

83 https://water.usgs.gov/edu/earthhowmuch.html.

84 Beltran, J.M. (1999, 1999). Irrigation with saline water: benefits and environmental impact. *Agricultural Water Management* **40** (2–3): 183–194.

85 Lord, P., Weston, M., Fontenelle, L.K. and Haggstrom, J. (2013). Recycling water: case studies in designing fracturing fluids using Flowback, produced, and nontraditional water sources. SPE Latin-American and Caribbean Heath, Safety, Environment and Social Responsibility Conference, Lima, Peru (26–27 June 2013), SPE 165641.

86 'l-Razi, A.F., Pendashteh, A., Abdullah, L.C. et al. (2009). Review of technologies for oil and gas produced water treatment. *Journal of Hazardous Materials* **170** (2–3): 530–551.

87 Lawerence, S.J. (2006). Description, Properties and degradation of Selected Volatile Organic Compounds Detected in Ground Water—A Review of Selected Literature. Open File Report *2006-1338*. US Department of the Interior and US Geological Survey.

88 http://ozone.unep.org/en/treaties-and-decisions/montreal-protocol-substances-deplete-ozone-layer

89 Capello, C., Fischer, U., and Hengerbuhler, K. (2007). What is a green solvent? A comprehensive framework for the environmental assessment of solvents. *Green Chemistry* **9** (9): 927.

90 Patel, A.D. (1999). Negative alkalinity invert emulsion drilling fluid extends the utility of Ester-based fluids. Offshore Europe Oil and Gas Exhibition and Conference, Aberdeen, UK (7–10 September 1999), SPE 56968.

91 Burrows, K., Evans, J., Hall, J. and Kirsner, J. (2001). New low viscosity ester is suitable for drilling fluids in deepwater applications. SPE/EPA/DOE Exploration and Production Environmental Conference, San Antonio, TX (26–28 February 2001), SPE 66553.

92 Pereira, C.S.M., Silva, V.M.T.M., and Rodrigues, A.E. (2011). Ethyl lactate as a solvent: properties, applications and production processes – a review. *Green Chemistry* **13** (10): 2658.

93 Todd, B., Funkhouser, G.P. and Frost, K. (2005). A chemical "trigger" useful for oilfield applications. SPE International Symposium on Oilfield Chemistry, The Woodlands, TX (2–4 February 2005), SPE 92709.

94 Schriener, K. and Munoz, T. Jr. (2009). Methods of degrading filter cakes in a subterranean formation. US Patent 7,497,278, assigned to Halliburton Energy Services Inc.

95 Denney, D. (2004). Flow assurance in Brazil's deepwater fields. *Journal of Petroleum Technology* **56** (10): 48–49, SPE-1004-0048-JPT.

96 http://www.dccleaningsystem.com/according-the-change-of-drilling-fluid-to-understand-under-well-condition/

97 Yassin, A.A.M., Kamis, A. and Abdullah, M.O. (1991). Formulation of an environmentally safe oil based drilling fluid. SPE Asia-Pacific Conference, Perth, Australia (4–7 November 1991), SPE 23001.

98 Muller, H., Herold, C.-P. and von Tapavicza, S. (1991). Use of hydrated castor oil as a viscosity promoter in oil-based drilling muds. WO Patent 911639, assigned to Henkel Kommanditgesellschaft Auf Aktien.

99 Mueller, H., Herold, C.-P., von Tapavicza, S. et al. (1994). Use of selected ester oils of low carboxylic acids in drilling fluids. US Patent 5,318,954, assigned to Henkel Kommanditgesellschaft Auf Aktien.

100 Hille, M., Wittkus, H., Windhausen, B. et al. (1998). Use of acetals. US Patent 5,759,963, assigned to Hoechst Akteingesellschaft.

101 Lin, K.-F. (1996). Synthetic paraffinic hydrocarbon drilling fluid. US Patent 5,569,642, assigned to Albemarle Corporation.

102 Berg, E., Sedberg, S., Kararigstad, H. et al. (2006). Displacement of drilling fluids and cased-hole cleaning: what is sufficient cleaning? SPE/IADC Drilling Conference, Miami, FL (21–23 February 2006), SPE 99104.

103 Getliff, J.M. and James, S.G. (1996). The replacement of alkyl-phenol ethoxylates to improve the environmental acceptability of drilling fluid additives. SPE Health, Safety and

Environment in Oil and Gas Exploration and Production Conference, New Orleans, LA (9–12 June 1996), SPE 35982.

104 Marshall, D., Jones, T., Quinterop, L. et al. (2007). Improved chemical designs for OBM skin damage removal for production and injection wells. Chemistry in the Oil Industry X, Manchester UK (5–7 November 2007).

105 Kalfayan, L. (2008). *Production Enhancement with Acid Stimulation*. Tulsa, OK: Pennwell Corporation.

106 Taylor, K.C. and Nasr-El-Din, H.A. (1999). A systematic study of iron control chemicals – Part 2. SPE International Symposium on Oilfield Chemistry, Houston, TX (16–19 February 1999), SPE 50772.

107 Coulter, G.R. and Jennings, A.R. Jr. (1997). A contemporary approach to matrix acidizing. SPE Annual Technical Conference and Exhibition, San Antonio, TX (5–8 October 1997), SPE 38594.

108 Buijse, M., de Boer, P., Breukel, B., and Burgos, G. (2004, 2004). Organic acids in carbonate acidizing. *SPE Production and Facilities* **19** (3): 128–134, SPE 82211.

109 Husen, A. Ali, A., Frenier, W.W. et al. (2002). Chelating agent-based fluids for optimal stimulation of high-temperature wells. SPE Annual Technical Conference and Exhibition, San Antonio, TX (29 September-2 October 2002), SPE 77366.

110 Frenier, W.W., Fredd, C.N. and Chang, F. (2001). Hydroxyaminocarboxylic acids produce superior formulations for matrix stimulation of carbonates. SPE European Formation Damage Conference (21–22 May 2001), The Hague, Netherlands, SPE 68924.

111 Mohamoud, M.A., Nasr-El-Din, H.A. and De Wolf, C.A. (2011). Stimulation of sandstone and carbonate reservoirs using an environmentally friendly chelating agent. *Chemistry in the Oil Industry XII* (7–9 November 2011). Manchester, UK: Royal Society of Chemistry.

112 Huang, T., McElfresh, P.M. and Gabrysch, A.D. (2003). Carbonate matrix acidizing fluids at high temperatures: acetic acid, chelating agents or long-chained carboxylic acids? SPE European Formation Damage Conference, The Hague, Netherlands (13–14 May 2003), SPE 82268.

113 Finsgar, M. and Jackson, J. (2014). Application of corrosion inhibitors for steels in acidic media for the oil and gas industry: a review. *Corrosion Science* **86**: 17–41.

114 Khan, G., Newaz, K.M.S., Basirun, W.J. et al. (2015). Application of natural product extracts as green corrosion inhibitors for metals and alloys in acid pickling processes – a review. *International Journal of Electrochemical Science* **10**: 6120–6134.

115 Delorey, J.R., Vician, D.N. and Metcalf, A.S. (2002). Acid stimulation of sour wells. SPE Gas Technology Symposium, Calgary, Alberta, Canada (30 April–2 May 2002), SPE 75697.

116 Nasr-El-Din, H.A., Al-Mutairi, S.H. and Al-Driweesh, S.M. (2002). Lessons learned from acid pickle treatments of deep/sour gas wells. International Symposium and Exhibition on Formation Damage Control, Lafayette, LA (20–21 February 2002), SPE 73706.

117 Crowe, C.W. (1985). Evaluation of agents for preventing precipitation of ferric hydroxide from spent treating acid. *Journal of Petroleum Technology* **37** (04): 691–695.

118 Brezinski, M.M. (1999). Chelating agents in sour well acidizing: methodology or mythology. SPE European Formation Damage Conference, The Hague, Netherlands (31 May–1 June 1999), SPE 54721.

119 Nasr-El-Din, H.A., de Wolf, C.A., Stanitzek, T. et al. (2013). Field treatment to stimulate a deep, sour, tight-gas well using a new, low corrosion and environmentally friendly fluid. *SPE Production & Operations* **28** (3): 277–285, SPE 163332.

120 Jacobs, I.C. and Thompson, N.E.S. (1992). Certain dithiocarbamates and method of use for reducing asphaltene precipitation in asphaltenic reservoirs. US Patent 5,112,505, assigned to Petrolite Corporation.

121 Brezinkski, M.M. and Gdanski, R.D. (1993). Methods of reducing precipitation from acid solutions. US Patent 5,264,141, assigned to Halliburton Company.

122 Lalchan, C.A., O'Neil, B.J. and Maley, D.M. (2013). Prevention of acid induced asphaltene precipitation: a comparison of anionic vs. cationic surfactants. SPE International Symposium on Oilfield Chemistry, The Woodlands, TX (8–10 April 2013), SPE 164087.

123 Fogler, H.S. and Chang, C.-L. (1993). Stabilization of asphaltene in aliphatic solvents using alkylbenzen-derived amphiphiles. 1. Effect of the chemical structure of amphiphiles on asphaltene stabilization. *Langmuir* **10**: 1749–1757.

124 Rietjens, M. (1997). Sense and non-sense about acid-induced sludge. SPE European Formation Damage Conference, The Hague, The Netherlands (2–3 June 1997), SPE 38163.

125 Rietjens, M. and Nieupoort, M. (1999). Acid sludge: how small particles can make a big impact. SPE European Formation Damage Conference, The Hague, The Netherlands (May 31–June 1 1999), SPE 54727.

126 Crowe, G., Masmonteil, J., and Thomas, R. (1992). Trends in matrix acidizing. *Oilfield Review* **4**: 24–40.

127 Nasr-El-Din, H.A., Al-Mohammed, A.M., Al-Aamri, A., and Al-Fuwaires, O.A. (2008, 2008). Reaction of gelled acids with calcite. *SPE Production & Operations* **23** (03), SPE 103979.

128 Malate, R.C.M., Austria, J.J.C., Sarmiento, Z.F. et al. Matrix stimulation treatment of geothermal wells using sandstone acid. Proceedings of the 23rd Workshop on Geothermal Reservoir Engineering, Stanford, CA (26–28 January 1998).

129 Nasr-El-Din, H.A. and Al-Mohammed, A.M. (2006). Reaction of calcite with surfactant-based acids. SPE Annual Technical Conference and Exhibition, San Antonio, TX (24–27 September 2006), SPE 102383.

130 Treybig, D.S., Chang, K.-T. and Williams, D.A. (2009). Demulsifiers, their preparation and use in oil bearing formations. US Patent 7,504,438, assigned to Nalco Company.

131 Zeiler, C.E., Alleman, D.J., and Qu, Q. (2006). Use of viscoelastic surfactant-based diverting agents for acid stimulation: case histories in GOM. *SPE Production & Operations* **21** (04): 448–454, SPE 90062.

132 Solares, J.R., Duenas, J.J., Al-Harbi, M. et al. (2008). Field trial of a new non-damaging degradable fiber-diverting agent achieved full zonal coverage during acid fracturing in a deep gas producer in Saudi Arabia. SPE Annual Technical Conference and Exhibition, Denver, CO (21–24 September 2008), SPE 115525.

133 Southwell, G.P. and Posey, S.M. (1994). Applications and results of acrylamide-polymer/ chromium (III) carboxylate gels. SPE/DOE Improved Oil Recovery Symposium, Tulsa, OK (17–20 April 1994), SPE 27779.

134 Al-Ghamdi, A.H., Mahmoud, M.A., Wang, G. et al. (2014). Acid diversion by use of viscoelastic surfactants: the effects of flow rate and initial permeability contrast. *SPE Journal* **19** (06): 1203–1216, SPE 142564.

135 Sridhar, N., Price, K., Buckingham, J., and Dante, J. (2006). Stress corrosion cracking of carbon steel in ethanol. *CORROSION* **62** (8): 687–702.

136 Rabaioli, M.R. and Lockhart, T.P. (1995). Solubility and phase behaviour of polyacrylate scale inhibitors and their implications for precipitation squeeze treatments. International Symposium on Oilfield Chemistry, San Antonio, TX, SPE 28998.

137 Craddock, H.A. (2012). Scale "Squeezing" www.knovel.com (November 2012).

138 Jones, C.R. and Talbot, R.E. (2004). Biocidal compositions and treatments. US Patent 6,784,168, assigned to Rhodia Consumer Specialties Ltd.

139 Brezinski, M.M. and Desai, B. (1997). Method and composition for acidizing subterranean formations utilizing corrosion inhibitor intensifiers. US Patent 5,697,443, assigned to Halliburton Energy Services.

140 Cassidy, J.M., Kiser, C.E. and Lane, J.L. (2008). Corrosion inhibitor intensifier compositions and associated methods. US Patent Application 20080139414, applied for by Halliburton Energy Services Inc.

141 Ben-Naim, A.Y. (1982). Hydrophobic interactions, an overview. In: *Solution Behavior of Surfactants, Theoretical and Applied Aspects*, vol. **1** (ed. K.L. Mittal and E.J. Fendler), 27–40. Springer.

142 Kronberg, B., Costas, M., and Silveston, R. (1995). Thermodynamics of the hydrophobic effect in surfactant solutions-micellization and adsorption. *Pure and Applied Chemistry* **67** (6): 897–902.

143 Wu, J., Xu, Y., Dabros, T., and Hamza, H. (2004). Development of a method for measurement of relative solubility of nonionic surfactants. *Colloids and Surfaces A: Physicochemical and Engineering Aspects* **232**: 229–237.

144 Poindexter, M.K., Chuai, S., Marble, R.A. and Marsh, S.C. (2003). Classifying crude oil emulsions using chemical demulsifiers and statistical analyses. SPE Annual Technical Conference and Exhibition, Denver, CO (5–8 October 2003), SPE 84610.

145 McWilliams, P. and Payne, G. (2001). Bioaccumulation potential of surfactants: a review. *Chemistry in the Oil Industry VII* (13–14 November 2001). Royal Society of Chemistry: Manchester, UK.

146 Ferrara, M. (1995). Hydrocarbon oil-aqueous fuel and additive compositions. WO Patent 1995020637, assigned to Meg S.N.C. Di Scopelliti Sofia & C.

147 Jenvey, N.J., MacLean, A.F., Miles, A.F. and Montgomerie, H. (2000). The application of oil soluble scale inhibitors into the Texaco galley reservoir. A comparison with traditional squeeze techniques to avoid problems associated with wettability modification in low water-cut wells. International Symposium on Oilfield Scale, Aberdeen, UK (26–27 January 2000), SPE 60197.

148 Watt, R., Montgomerie, H., Hagen, T. et al. (1999). Development of an oil-soluble scale inhibitor for a subsea satellite field. SPE International Symposium on Oilfield Chemistry, Houston, TX (16–19 February 1999), SPE 50706.

149 Reizer, J., Rudel, M., Sitz, C. et al. (2002). Oil-soluble scale inhibitors with formulation for improved environmental classification. US Patent Application 20020150499, assigned to authors.

150 Elder, J. (2011). Relative solubility number RSN – an alternative measurement to log pow for determining the bioaccumulation potential. Master's Thesis. Goteborg, Sweden: Chalmers University of Technology.

151 Yang, J. and Jovanciecevic, V. Microemulsion containing oil field chemicals useful for oil and gas field applications. US Patent 7,615,516, assigned to Baker Hughes Incorporated.

152 Jovancicevic, V., Ramachandran, S., and Prince, P. (1999). Inhibition of carbon dioxide corrosion of mild steel by Imidazolines and their precursors. *Corrosion* **55** (5): 449–455.

153 Lui, Y., Jessop, P.G., Cunningham, M. et al. (2006). Switchable surfactants. *Science* **313** (5789): 958–960.

154 Yang, M., Stuart, M.C. and Davies, G.A. (1996). Interactions between chemical additives and their effects on emulsion separation. SPE Annual Technical Conference and Exhibition, Denver, CO (6–9 October 1996), SPE 36617.

155 Phillips, N.J., Renwick, J.P., Palmer, J.W. and Swift, A.J. (1996). The synergistic effect of sodium thiosulphate on corrosion inhibition. Proceedings of the 7th Oilfield Chemistry Symposium, Geilo.

156 Craddock, H.A. (2002). The use of thiohydation as a corrosion inhibitor and a synergist in corrosion inhibitor formulations. European Patent EP145758.

157 Fan, L.-D.G., Fan, J., Ross, R.J. and Bain, D. (1999). Scale and corrosion inhibition by thermal polyaspartates. Paper 99120, NACE Corrosion 99, San Antonio, TX (25–30 April 1999).

158 Craddock, H.A. (2010). Silicon materials as additives in wax inhibitors. Patent Application 1020439.4, Filing date 2 December 2010.

159 Al-Thuwaini, J.S. and Burr, B.J. (1997). Encapsulated scale inhibitor treatment. Middle East Oil Show and Conference, Bahrain (15–18 March 1997), SPE 37790.

160 Hsu, J.F., Al-Zain, A.K., Raju, K.U. and Henderson, A.P. (2000). Encapsulated scale inhibitor treatments experience in the Ghawar field, Saudi Arabia. International Symposium on Oilfield Scale, Aberdeen, UK (26–27 January 2000), SPE 60209.

161 Powell, R.J., Fischer, A.R., Gdanski, R.D. et al. Encapsulated scale inhibitor for use in fracturing treatments. SPE Annual Technical Conference and Exhibition, Dallas, TX (22–25 October 1995), SPE 30700.

162 Read, P.A. (1999). Oil well treatment. US Patent 5,893,416, assigned to AEA Technology PLC.

163 Szymczak, S., Brown, J.M., Noe, S.L. and Gallup, G. (2000). Long-term scale prevention with the placement of solid inhibitor in the formation via hydraulic fracturing. SPE Annual Technical Conference and Exhibition, San Antonio, TX (24–27 September 2000), SPE 102720.

164 Selle, O.M., Haavind, F., Haukland, M.H. et al. (2010). Downhole scale control on Heidrun field using scale inhibitor impregnated gravel. SPE International Conference on Oilfield Scale, Aberdeen, UK (26–27 May 2010), SPE 130788.

165 Wang, H. and Akid, R. (2008). Encapsulated cerium nitrate inhibitors to provide high-performance anti-corrosion sol–gel coatings on mild steel. *Corrosion Science* **50** (4): 1142–1148.

166 Shchukin, D.G. and Mohwald, H. (2007). Surface-engineered nanocontainers for entrapment of corrosion inhibitors. *Advanced Functional Materials* **17** (9): 1451–1458.

167 Weghorn, S.J., Reese, C.W. and Oliver, B. (2007). Field evaluation of and encapsulated time-release corrosion inhibitor. CORROSION 2007, Nashville, TN (11–15 March 2007), NACE 07321.

168 Blauch, M.E., Cheng, A., Rispler, K. and Kalled, A. (2003). Novel carbonate well production enhancement application for encapsulated acid technology: first-use case history. SPE Annual Technical Conference and Exhibition, Denver, CO (5–8 October 2003), SPE 84131.

169 Burgos, G., Birch, G. and Bulise, M. (2004). Acid fracturing with encapsulated citric acid. SPE International Symposium and Exhibition on Formation Damage Control, Lafayette, LA (18–20 February 2004), SPE 86484.

170 King, M.T., Gulbis, J., Hawkins, G.W. and Brannon, H.D. (1990). Encapsulated breaker for aqueous polymeric fluids. Annual Technical Meeting, Calgary, Alberta (10–13 June 1990), PETSOc-90-89.

171 Watson, W.P., Aften, C.W. and Previs, D.J. (2010). Delayed-release coatings for oxidative breakers. SPE International Symposium and Exhibition on Formation Damage Control, Lafayette, LA (10–12 February 2010), SPE 127895.

172 Gupta, D.V.S. and Cooney, A. (1992). Encapsulations for treating subterranean formations and methods of use thereof. WO Patent 9,210,640.

173 Haslbeck, E. (2004). Extended antifouling coating performance through microencapsulation. 12th International Congress on Marine Corrosion and Fouling, Southampton, UK.

174 Reybuck, S.E. and Schwartz, C. (2008). Blends of encapsulated biocides. US Patent 7,377,968, assigned to Rohm and Haas Company.

175 Oschmann, H.J., Huijgen, M.C. and Grondman, H.F. (2011). Production chemicals based on active dispersions: alternatives to conventional solvent based products. Chemistry in the Oil Industry X11, Manchester, UK (7–9 November 2011).

176 Havlena, Z.G. and Wasmuth, J.F. (1970). Weighted inhibitors solve special oil-well corrosion problems. *Journal of Canadian Petroleum Technology* **9** (3): 206–208.

9

The Regulation and Control of the Use and Discharge of Chemicals in the Oilfield

In previous chapters throughout this book, the environmental impact of chemicals and classes of chemicals has been discussed. Along with many chemicals described, their biodegradation profile, bioavailability, environmental toxicity, environmental persistency and other impacts on the environment have been explored. This has been with particular regard to their use and discharge in oilfield operations. This chapter is concerned with national and international controls that have been placed on this use and discharge, especially those originated for legislative controls imposed by government agencies and related organisations such as the OSPAR Treaty organisation. The fitness for purpose of these controls will be discussed, particularly with regard to environmental protection.

Although most national governments have some form of regulatory control particularly relating to chemical use and their health and safety hazards, only a few have a comprehensive wide coverage of controls relating to the use and discharge of chemicals and their hazard assessment with regard to the environment, particularly the upstream oil and gas industry. In relation to these specific national and regional authorities, the subject matter of this chapter will address specifically the following countries and regions:

1) Europe (the OSPAR Treaty region)
2) United States
3) Canada
4) Australia
5) Other national authorities

A comprehensive documentation and analysis of the regulations governing chemicals used and discharged within these nations with respect to the upstream oil and gas sector is not the purpose of this chapter. Indeed, attempting such a discourse is a somewhat futile exercise as the nature of these regulations is one of constant review, revision and change, much of it directed by political nuance.

It is also necessary to remember that chemical regulation does not control the environment, rather that inputs from many sources both chemical and physical and natural and man-made have overall complex effects on the environment and environmental change is a consequence of sometimes what is perceived as small impacts [1].

Therefore, this chapter will attempt to give the reader an outline of the fundamentals governing the regulations in the specific geographical regions and countries stated. Reference to current and comprehensive information is given in terms of relevant website addresses in the reference section to this chapter.

Oilfield Chemistry and Its Environmental Impact, First Edition. Henry A. Craddock.
© 2018 John Wiley & Sons Ltd. Published 2018 by John Wiley & Sons Ltd.

9.1 Chemical Regulation in Europe: The OSPAR Treaty

Within the European oilfield and in particular the North Sea Region, the OSPAR Treaty organisation is the dominant regulatory policy body; the implementation of the treaty is governed and managed by the OSPAR Commission [2]. Although it is the major source of legislative control with respect to chemical use and discharge in the European oilfield, the OSPAR Treaty is not the primary legislation, particularly with respect to countries that are members of the European Union (EU) or are affiliated to its health, safety and environmental regulations. In the first instance the now well-established REACH requirements on registration of chemicals have to be met.

9.1.1 REACH (Registration, Evaluation, Authorisation and Restriction of CHemical Substances)

These regulations are set out in European community directive EC1907/2006 [3], with the expressed purpose of implementing a high level of human health and environmental protection from the effects and impacts of chemical substances. There are a number of exceptions to the legislative controls, namely, medical, veterinary, alimentary and cosmetic products, polymers and some on-site isolated intermediates. Of note to the upstream oil and gas sector is the exemption for polymers; however this does not apply to the monomers that are required to be REACH registered. The legislation applies also to companies importing materials into the European Free Trade Area, and also as the EU is a signatory to the OSPAR Treaty, the North Sea and North East Atlantic. The legislation requires the evaluation and subsequent registration of all chemical substances (excluding the previous exemptions), which are manufactured or imported at levels over 1 tonne per annum. This process requires the collection of substantive data on the human health and potential environmental impacts of the chemical substance being registered.

This legislation was commenced in 2007; however it has been phased in and continues to be so over a period of over 10 years. This major piece of regulatory control will replace around 40 pieces of legislation and improve and streamline chemical registration and regulation across Europe. It is, however, been designed not to conflict or overlap with other EU chemical legislation [4]. In broad terms the legislation requires the review and, if required and applicable, the substitution of the most dangerous chemicals. The review and data provided can be used to justify the authorisation and continued use of chemical substances that may have been candidates for substitution, and as will be outlined in the Section 9.1.3, it follows a similar approach to that for requirements for authorisation under the OSPAR Treaty. Therefore, it could be supposed that as such REACH could replace the requirements of the Harmonised Mandatory Control System (HMCS) adopted under the OSPAR Treaty; however in practice it has been found that they are run in sequence and that the HMCS particularly applies to what under REACH is known as 'downstream users'.

Under REACH the obligations for assessment, which includes the collection of data and registration of chemical substances, apply primarily to the manufacturer or importer. The subsequent users within the supply chain, the 'downstream users', although having obligations to provide and pass on information does not take part in the actual registration for chemical substances. This registration process is defined for each single chemical substance, and to achieve this, manufacturers and importers are encouraged or required to participate in a Substance Information Exchange Forum (SIEF). Such forums have the purpose of sharing existing data and share costs to acquire other data to complete the registration process. Such data is targeted at consumer and environmental protection and therefore includes toxicology and environmental impact data. It is up to the industry to agree and provide the necessary requirements to meet the registration process and subsequent authorisation for use.

As already mentioned the primary responsibility for data generation lies with manufacturers and importers; however downstream users are under obligations to provide data on the use and exposure of chemicals in the industrial sectors to users further downstream in the supply chain. In the oil and gas industry, the major suppliers of chemicals to the end users (oil and gas companies, drilling companies, other service providers, etc.) are the formulating service companies supplying blended and formulated products across the upstream industry. This has been extensively discussed in Section 8.4. To this end the oil and gas sector has developed a useful tool to provide a series of generic exposure scenarios across all sectors of the upstream industry, providing specific information on how to use the substance or preparation safely, how to protect users and their customers from any harmful effects and how to minimise any risks to the environment. The chemical suppliers in the upstream oil and gas industry through their trade association [5] have also developed a useful tool to assess both human health exposure and environmental risk [6]. These exposure scenarios are not only a consequence of REACH but also the Classification, Labelling and Packaging (CLP) regulations [7], which became law in 2009 and are based on a United Nations (UN) Globally Harmonised System (GHS) for the Classification and Labelling of Chemicals [8]. In addition to the classification and identification of chemical products, these regulations set out standards for the composition of safety data sheets (SDS) and particularly the now mandatory extended SDS, which include exposure scenarios [9].

Most European nations, if not all, have guidance provided by the relevant national authority in respect of REACH requirements and how manufacturers and importers can form an SIEF or join a relevant one. In the United Kingdom, for example, this is provided by the Health and Safety Executive [10]. This is an important difference from the HMCS, as the controlling authority for use and discharge of chemicals in the offshore oil and gas industry for the United Kingdom is the Department for Business, Energy and Industrial Strategy, and this is often the case with respect to other national authorities.

As stated earlier it is not the aim of this chapter to provide the reader or practitioners with a handbook of regulations as these are continuously evolving and can be readily found in appropriate websites. This is to provide them with an overview of the aims of the regulation and make a critic on their effectiveness. In this respect, the REACH process relies on the following:

- The collection of data either from literature and /or generated from prescribed and approved testing
- The provision of a portfolio of data for assessment
- Assessment of the data by ECHA
- Authorisation for use or not

Fundamental to the assessment process is the use of expert opinion and judgement on the data presented and its adequacy to form a coherent and robust opinion. The data is usually presented as single end points, for example, a toxic concentration or a factor for rate of biodegradation in a given time period. Later in this chapter the inadequacies of this approach, from an environmental impact viewpoint, are discussed; see Section 9.6. However, from a regulatory control viewpoint, this makes the process of hazard assessment simply a matter of ranking in a numerical form.

9.1.2 Biocides and the Biocidal Product Regulation (BPR)

As stated, there are a number of notable exceptions to the REACH assessment process and a key product class that is separately regulated in the European area is biocides and related biocidal products such as preservatives.

The use of biocides and biostatic agents in the upstream oil and gas sector in Europe and the North Sea in particular is well established and fairly substantial with several thousand tonnes of formulated products being applied. The dominant non-oxidising biocide classes, as have been discussed in previously, are aldehydes, particularly glutaraldehyde (see Section 6.1.3), which is applied either as an aqueous solution or in combination with quaternary ammonium salts and tetrakis(hydroxymethyl)phosphonium sulphate (THPS) (see Section 4.6), usually as supplied by the manufacturer or diluted with water. These products are applied, as would be obvious, to control populations of microorganisms in the production fluids, which can have both fouling and corrosive influence. This biofouling and microbial-induced corrosion (MIC) can adversely affect production rates [11, 12]. They are also used to prevent sulphate-reducing bacteria (SRBs) from having large populations both in the produced fluids and in the reservoir where the production of hydrogen sulphide is a natural consequence of these organisms' metabolism [13].

On 1 September 2013, the Biocidal Product Directive (BPD) was revoked and replaced by the directly acting EU Biocides Regulation (528/2012). Like REACH, the BPR is an EU directive, which is brought into force through national legislation in each member state. As in the previous directive (BPD), the approval of active substances takes place at union level and the subsequent authorisation of the biocidal products at member state level. This authorisation can be extended to other member states by mutual recognition. However, the new regulation also provides applicants with the possibility of a new type of authorisation at union level (Union authorisation). Currently, each member state has its own authorisation process that can differ, although not usually significantly, from member state to member state. In the United Kingdom, for instance, the process is again controlled and governed by the Health and Safety Executive [14].

In general terms however, this legislation places onerous requirements for registration, authorisation and use of biocides, either derived and manufactured, within the EU or imported into the EU. The main thrust of the legislation is to provide data, in particular human health data, to provide better risk evaluations of chemical substances.

Although all biocidal products require an authorisation before they can be placed on the market, and the active substances contained in that biocidal product must be previously approved, there are, however, certain exceptions to this principle. For example, biocidal products containing active substances in the Review Programme can be made available on the market and used (subject to national laws) pending the final decision on the approval of the active substance (and up to 3 years after). Products containing new active substances that are still under assessment may also be allowed on the market where a provisional authorisation is granted.

The BPR aims to harmonise the market at union level, simplify the approval of active substances and authorisation of biocidal products, and introduce timelines for member state evaluations, opinion-forming and decision-making. It also promotes the reduction of animal testing by introducing mandatory data sharing obligations and encouraging the use of alternative testing methods. A dedicated IT platform, the Register for Biocidal Products (R4BP 3), will be used for submitting applications and exchanging data and information between the applicant, ECHA, member state-competent authorities and the European Commission [15].

As far as the upstream oil and gas sector is concerned, all biocidal products require to be registered and approved through the BPR prior to use. The same compliance is required for all other appropriate chemical substance with regard to REACH with polymers being a notable exception. These two regulatory requirements have to be appropriately met as well as the necessary criteria for registration with the HMCS, whose requirements are discussed in Section 9.1.3.

9.1.3 The OSPAR Treaty and the Harmonised Mandatory Control Scheme (HMCS)

The OSPAR Treaty [1] is implemented in a similar fashion to European directives across signatures to the treaty. The treaty governs a number of issues regarding the use and discharge of materials into the North Sea and North East Atlantic. It has become a cornerstone of how all chemicals used in the upstream oil and gas sector are controlled not only in the treaty area but also with other areas in Europe both onshore and offshore.

The Barcelona Process [16], which controls similar uses and discharges in the Mediterranean, has become somewhat obsolete with respect to chemical approval for use, as the OSPAR Treaty requirements apply; however it is much more complex with regard to discharge as the convention members, which include Middle Eastern and North African countries, appear to have no coherent policy. Consequently, a divergent approach appears to apply in various Mediterranean waters.

In recent years, many multinational operators look to the OSPAR requirements in chemical governance being applied on a global basis or certainly required as a minimum standard for other operations outside of the treaty organisation's jurisdiction.

The stated aim of the OSPAR Treaty is

> "... to prevent and eliminate pollution and to protect the maritime area against the adverse effects of human activities..."

and in particular,

> "... to prevent and eliminate pollution from offshore sources (offshore installations and pipelines from which substances or energy reach the maritime area)."

The control of chemicals is referenced under the statements below:

> "The use and discharge of hazardous substances in the offshore oil and gas industry have been identified as a cause for great concern. To reduce the overall impact of offshore chemicals on the marine environment, OSPAR has adopted a harmonised mandatory control system for use and reduction of discharges of offshore chemicals. This system promotes the shift towards the use of less hazardous or preferably non-hazardous substances."

The implementation of the OSPAR Treaty is through national authorities, which have brought in to force the treaty requirements in different ways although all have adopted, albeit with some differences, the HMCS and in particular the Harmonised Offshore Chemical Notification Format (HOCNF).

Chemical suppliers must provide the national authorities with data and information about chemicals to be used and discharged offshore according to the HOCNF [17]. OSPAR has guidelines [18] on how to complete the format. Based on the information sent by the chemical supplier, the national authorities conduct a prescreening assessment and take the appropriate regulatory action, such as issuing discharge permits.

Chemical suppliers should follow the OSPAR Guidelines for Toxicity Testing of Substances and Preparations Used and Discharged Offshore [19]. OSPAR has also prepared a protocol on methods for the testing of chemicals used in the offshore oil industry [20]. The OSPAR List of Substances/Preparations Used and Discharged Offshore, which is considered to pose little or no risk to the environment (PLONOR) [20], contains substances whose use and discharge offshore are subject to expert judgement by the competent national authorities or do not need to be strongly regulated. References [16]–[20] provide web links to the relevant documents; however practitioners should ensure they have the most up-to-date documents and that it applies to the relevant national authority as differences can occur. For example, Norway has issued a supplementary guideline for completion of the HOCNF [21]. In the prescreening assessment, a number of decision trees are adopted and these have been neatly summarised with regard to testing required per individual substance in Figure 9.1 [22].

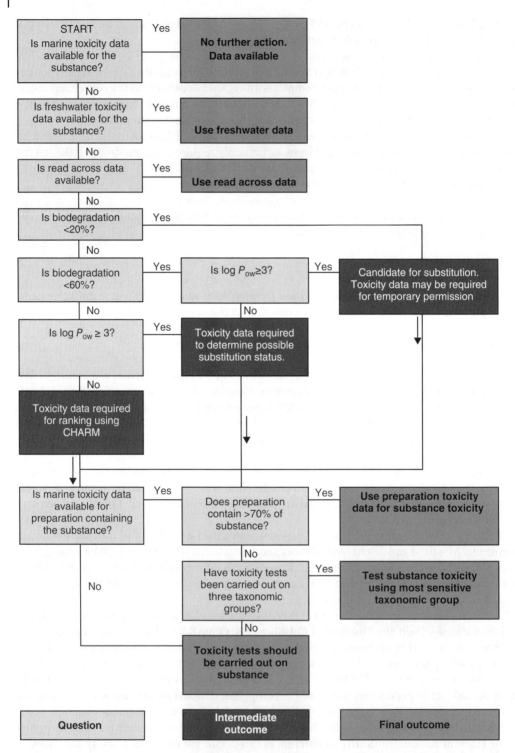

Figure 9.1 Decision tree to determine additional toxicity testing. Source: Reproduced with permission of the European Oilfield Speciality Chemicals Association.

Chemical use and discharge by the offshore oil and gas sector is managed by the implementation of the Offshore Chemical Notification Scheme (OCNS), which is regulated by the Department for Business, Energy and Industrial Strategy in the United Kingdom and by the State Supervision of Mines in the Netherlands, both by following scientific and environmental advice from Marine Scotland and government agencies of the Netherlands, respectively, and from Centre for Environment, Fisheries and Aquaculture Science (Cefas).

The OCNS uses the OSPAR HMCS developed through the OSPAR Decision 2000/2 on an HMCS for the use and discharge of offshore chemicals (as amended by OSPAR Decision 2005/1) and its supporting recommendations. This ranks chemical products according to hazard quotient (HQ), calculated using the Chemical Hazard and Risk Management (CHARM) model. The CHARM model was developed by the contracting parties to the OSPAR Treaty alongside industry and in particular the European Oilfield Speciality Chemicals Association in an attempt to bring a degree of simulation of what occurs in the environment and to afford a measure of risk control [23]. It is not a mandatory requirement that a CHARM-based assessment is conducted; however chemical suppliers need to present data to allow one and may also conduct other similar assessments themselves.

The CHARM model calculates the ratio of predicted effect concentration against no effect concentration (PEC:NEC), which is expressed as an HQ and then used to rank the product. The data used in the CHARM assessment is provided by the chemical supplier in terms of toxicity, biodegradation and bioaccumulation.

The CHARM model is divided into four main algorithms:

- Production
- Completion/workover
- Drilling
- Cementing

The product ranking based on the HQ is published as a colour band as detailed in Figure 9.2.

Products not applicable to the CHARM model (i.e. inorganic substances, hydraulic fluids or chemicals used only in pipelines) are assigned an OCNS grouping, A–E. Group A includes products considered to have the greatest potential environmental hazard, and group E the least.

In this way, each individual substance in an offshore chemical is ranked by applying the OCNS ranking scheme. The overall ranking is determined by that substance having the worst-case OCNS ranking scheme assignment [24].

This type of methodology applies to the United Kingdom and Dutch regulatory areas; however other national authorities have adopted similar but not identical methods and some important differences around interpretation of toxicity and related data often applies, particularly with regard to Norway.

As observed, this previous overview equates to a complex series of regulations for registration and compliance for the use and discharge of chemicals in and into the North Sea basin and

Minimum HQ value	Maximum HQ value	Colour banding	
>0	<1	Gold	Lowest hazard
≥1	<30	Silver	
≥30	<100	White	
≥100	<300	Blue	
≥300	<1000	Orange	
≥1000		Purple	Highest hazard

Figure 9.2 OCNS HQ and colour bands.

North East Atlantic. Considerable resources are employed in the bureaucracy of applying and maintaining the systems and permitting of the chemical use and discharge both in industry and in government agencies. A critique as to the value of the system is presented in Section 9.6.

9.2 Chemical Regulation in the United States

The United States both onshore, on its mainland, offshore in the Gulf of Mexico and in Alaska has different regulations from that of Europe although many similarities prevail, in particular, the use of expert judgement, the use of a permitting system and the reliance in some areas such as the Gulf of Mexico on point toxicity data albeit for different species than in the North Sea. Another difference, although it could also be considered a similarity given differences between European countries in regulation, is that the discharge of chemicals between individual states can have differing regulations. This can have profound effects such as chemicals being used in one state and the 'polluted' water being shipped to another state for discharge.

The key permitting regulation is the requirement for most chemical substances to be registered under the Toxic Substances Control Act (TSCA) [25]. This act was brought into force in 1976 and provides the Environmental Protection Agency (EPA) with authority to require reporting, record keeping and testing requirements and restrictions relating to chemical substances and/or mixtures. Certain substances are generally excluded from TSCA, including, among others, food, drugs, cosmetics and pesticides.

Various sections of TSCA provide authority for the EPA to require

- Pre-manufacture notification for 'new chemical substances' before manufacture.
- Testing of chemicals by manufacturers, importers and processors where risks or exposures of concern are found.
- Issuing of significant new use rules (SNURs) when it identifies a 'significant new use' that could result in exposures to, or releases of, a substance of concern.
- Maintenance of the TSCA Inventory, which contains more than 83 000 chemicals as of the end of 2016. As new chemicals are commercially manufactured or imported, they are placed on the list.
- Those importing or exporting chemicals to comply with certification reporting and/or other requirements.
- Reporting and record keeping by persons who manufacture, import, process and/or distribute chemical substances in commerce.
- That any person who manufactures (including imports), processes or distributes in commerce a chemical substance or mixture and who obtains information that reasonably supports the conclusion that such substance or mixture presents a substantial risk of injury to health or the environment to immediately inform EPA, except where EPA has been adequately informed of such information. EPA screens all TSCA b§8(e) submissions as well as voluntary 'For Your Information' (FYI) submissions. The latter is not required by law, but is submitted by industry and public interest groups for various reasons.

The TSCA has recently been amended by the Frank R. Lautenberg Chemical Safety for the twenty-first century Act (LCSA), which is claimed to bring improvements to TSCA and most importantly an inventory reset [26]. It brings into play a new set of different requirements particularly in terms of chemical testing and risk assessment; however again the same fundamental principles of point data testing and expert judgement apply. Another fundamental difference between US and European regulatory control is encapsulated in the approach in the new act for a fast track process for persistent, bioaccumulative and toxic (PBT) chemicals where the following can apply:

- Risk evaluation not needed, only use and exposure to chemical needed.
- Action to reduce exposure to extent practicable must be proposed no later than 3 years after the new law and finalised 18 months later.
- Additional requirements for PBTs in the prioritisation process for assessments may be required.

In the OSPAR convention treaty, the use of the precautionary principle is paramount and hence no such fast track process would be part of a regulation.

Until now, in the United States once a chemical was registered and listed under the TSCA, it could be used without much restriction unless it was particularly hazardous such as asbestos or mercury. The emphasis was then on, and still is on, individual states to control the discharge of such chemicals.

In the US Gulf of Mexico and US Gulf Coast Outer Continental Shelf, the emphasis is on the toxicity of the produced water rather than on the hazard posed by each specific chemical. This is controlled by the issue of National Pollution Discharge Elimination System (NPDES) permits [27]. Created in 1972 by the Clean Water Act, the NPDES permit programme is authorised to state governments by EPA to perform many permitting, administrative and enforcement aspects of the program. The requirement falls on oil and gas operators to write and obtain the permit, and guidance is given on how to do this [28]. Critically they are required to submit toxicity data on their whole effluent under two categories:

- *Produced water:* 7 days no observed effect concentration (NOEC) using the following species:

- Mysid shrimp (*Mysidopsis bahia*)
- Silverside minnow (*Menidia beryllina*)

- Miscellaneous discharges of seawater or freshwater to which chemicals have been added again with the same species as earlier but in a 48 h NOEC assessment.

The outcome of such testing and assessment is the allocation of a critical dilution factor (CDF), which is assigned to each offshore facility. Effluent samples are then collected at intervals as detailed in the facility permit and the samples tested to show that the effluent is not toxic to the prescribed species at or below the CDF.

In 2017, there were some significant changes to the current produced water discharge requirements being introduced for both offshore and onshore leases. However, the changes presented in the following paragraphs can be broadly characterised as representing a clarification of existing regulations, rather than a substantial overhaul of discharge requirements [29].

It is noted that water generated from mono-ethylene glycol (MEG) reclamation processes, including salt slurries generated from salt centrifuge units, are regulated as produced water. If this salt slurry is discharged without mixing with the bulk-produced water stream, then it is treated as a separate discharge point and must meet any and all requirements associated with produced water:

- Capacity for calculating CDFs for discharge systems with multiple ports is now included. This enables the equivalent diameter of all openings to be calculated using the following formula so that this can be used to calculate CDFs for a single total discharge volume. Equivalent Diameter = $\sqrt{(A_{total}*4/\pi)}$, where A_{total} is the total area of all openings.
- An entirely new requirement for the characterisation of produced water discharges from active leases has been added. This requires that operators conduct an individual study of at

least one produced water from each active lease block or, perhaps preferably, participate in a joint survey that analyses a minimum of ten samples from active leases in each lease area.

As a minimum, the water samples shall be analysed for dissolved arsenic, dissolved cadmium, dissolved chromium (VI), dissolved copper, free cyanide, dissolved lead, dissolved mercury, dissolved nickel, dissolved selenium, dissolved silver and dissolved zinc.

Also changes to the miscellaneous discharge category have been made to specifically include brines used as piping or equipment preservation fluids and to include hydrate control fluids discharged with produced water. A similar distinction is also made in the miscellaneous discharges or chemically treated seawater section for seawater that is used as a piping or preservation fluid.

Major changes in this section of the permit are as follows:

- Leak tracer fluids made from a powder dye are now exempt from the requirement for a 7-day NOEC test value of no less than $50\,\mathrm{mg\,l}^{-1}$ that is applied to all other subsea preservation fluids, control fluids, storage fluids, leak tracer fluids and riser tensioning fluids. However, it is still necessary for the product to pass the 7-day NOEC test at the as-supplied concentration.
- A significant change to prior regulations is that hydrate control fluids are specifically identified as requiring toxicity testing during discharge. The new requirements are as follows:

  - When hydrate control fluids are discharged with produced water, the toxicity limitation established for produced water shall assess the overall impact caused by hydrate control fluids.
  - If the hydrate control fluid is discharged with other miscellaneous discharges, a representative sample shall be used for the toxicity test for the miscellaneous discharge.
  - In situations where a discharge of hydrate control fluids is not monitored by the toxicity testing of either produced water or miscellaneous discharge, then a 7-day chronic toxicity test for that specific hydrate control fluid must be undertaken prior to discharge, and the final concentration in the discharge must not exceed the NOEC at the applicable critical dilution at the edge of 100 m from the point of discharge.

The restrictions above do not apply if the total discharge volume of methanol within a 7-day period is less than 20 bbl or the total discharge volume of ethylene glycol within a 7-day period is less that 200 bbl.

- Brines used for pipeline and equipment preservation must meet three criteria prior to their use as preservation fluids: no free oil, oil and grease content below $29\,\mathrm{mg\,l}^{-1}$ and no content of priority pollutants except in trace amounts. The concentration representing a trace amount is not defined.

Finally, a number of changes have been made to the final NPDES permit and these are summarised as follows:

- A requirement to undertake characterisation studies prior to the use of water-based drilling muds has been included.
- The respective exemption volumes for methanol and ethylene glycol of 20 and 200 bbl in a 7-day period were added.
- Chlorine and bromine were added to the list of chemicals that are excluded from toxicity testing requirements.
- An amendment was made to allow biocides to be added to sump and drain systems, where this was otherwise prohibited in the earlier regulation.

Although the preceding paragraphs do not identify each and every change in the NPDES general permit, as stated earlier they do not amount to a substantial overhaul of discharge requirements. This is important, because although it is claimed there is a substantial effort across the industry to minimise discharges to sea, it has been successfully argued that further restriction of discharge limits for existing assets would compromise their original design intent and, in many cases, would require result in the abandonment of recoverable reserves due to the commercial and physical restrictions of the process and equipment upgrades this would necessarily entail.

9.3 Chemical Regulation in Canada

The Government of Canada controls chemical substances to protect human health and the environment using a variety of tools. These range from providing information about proper use and disposal to regulations that restrict or prohibit use. Scientific understanding, assessment and monitoring, combined with a variety of tools for protection, form a risk-based approach to chemical substances regulation in Canada [30].

For the oil and gas industry, specifically Clause 23 of Canada Oil and Gas Drilling and Production Regulations SOR/2009-315 covers chemical use and discharge to the environment [31].

> Handling of Chemical Substances, Waste Material and Oil
> *23 The operator shall ensure that all chemical substances, including process fluids and diesel fuel, waste material, drilling fluid and drill cuttings generated at an installation, are handled in a way that does not create a hazard to safety or the environment.*

The Canadian Environmental Protection Act of 1999 (CEPA 1999) [32] is one of Canada's most important pieces of legislation with respect to pollution prevention and the protection of the environment and human health. CEPA 1999 supports a 'precautionary approach' and makes pollution prevention the cornerstone of efforts to reduce risks from toxic substances. CEPA 1999 covers a range of activities to address any pollution issues that may not be covered by other federal laws; establishes information-collection authorities; mandates environmental and human health research activities; sets out processes to assess risks posed by substances in commerce; imposes timeframes for managing certain toxic substances; provides a wide range of instruments to manage substances, pollution and wastes; and requires that the most harmful substances are phased out or not released into the environment in any measurable quantity (this in effect equates to virtual elimination). The Act and its administration must be reviewed by Parliament every 5 years, providing an opportunity for Canadians to provide feedback on how well they feel it is protecting the environment and human health.

Other key pieces of legislation governing the use of chemical substances include

- The Hazardous Products Act that establishes standards for chemical classification and hazard communication and the authority to regulate or prohibit consumer products and workplace chemicals that pose a risk to their users
- The Pest Control Products Act that ensures the protection of human health, safety and the environment by regulating products used for the control of pests
- The Transportation of Dangerous Goods Act that promotes public safety in the transportation of dangerous goods
- The Fisheries Act that prohibits the deposit of toxic or harmful substances into fish-frequented waters
- The Canada Labour Code that regulates issues related to occupational safety and health

This legislation along with Canada's Toxic Substances Management Policy has resulted in a national Chemicals Management Plan (CMP) [33], which was launched in 2006, to bring all existing federal programs together into a single strategy. The CMP is a science-based approach that aims to protect human health and the environment through

- Setting priorities and government-imposed administrative timelines for action on chemicals of concern
- Integrating chemicals management activities across federal departments and choosing the best placed federal statute under which to take action
- Enhancing research, monitoring and surveillance
- Increasing industry stewardship and responsibility for substances
- Collaborating internationally on chemicals assessment and management
- Communicating to Canadians the potential risks of chemical substances

Of particular importance to the oil and gas sector is that the CMP introduced a targeted petroleum sector approach for managing approximately 160 high priority petroleum substances with unique and complex characteristics [34].

The main strategic direction of Canada's legislative programme for chemical management is more aligned to Europe than the United States; however other than substances for high priority action that in effect are being eliminated, there is no overall system for testing toxicity and other properties of chemical substances. Rather this is conducted on a case-by-case basis.

9.4 Chemical Regulation in Australia

Australian Government legislation governs the supply of chemicals in Australia, and importers or manufacturers of chemicals or chemical products must comply [35].

Legislation covers assessing and registering chemicals by national chemicals schemes and – to minimise duplication or unnecessary regulatory burdens on industry – the schemes complement each other. They cover

- Industrial chemicals, including chemicals used domestically
- Agricultural and veterinary chemicals
- Medicines and pharmaceuticals
- Chemicals used in (or with) food, including additives, contaminants and natural toxicants

In addition, several chemicals regulation frameworks support chemicals management:

- Poisons scheduling – protecting public health
- Maintaining safety in the workplace
- Transporting of dangerous goods
- Managing chemicals in the environment
- Chemicals of security concern
- Illicit drugs precursor chemicals

Critical for chemists and environmentalists working in the oil and gas sector is the need for chemical substances to be registered on the Australian Inventory of Chemical Substances (AICS) database that identifies chemicals available for industrial use in Australia [36].

Before importing and/or manufacturing an industrial chemical, users must check the AICS to see if it is listed and if there are conditions for using it.

If a chemical is not listed on AICS, or if the intended use is different to the condition of use as described on the database, it is regarded as a new industrial chemical to Australia.

Unless an exemption applies, the new industrial chemical will need to be assessed by the National Industrial Chemicals Notification and Assessment Scheme (NICNAS) for risks to the environment and human health before it can be imported and/or manufactured [37].

There is a lengthy checklist [38] to initiate and aid the registration process that will be followed by the need to provide specific data if not already known, and this can be toxicological and ecological data as required. Once the chemical substance is registered, it is allowed to be used within the conditions of use laid down. In many respects this is something of a hybrid of the US and European systems; however the detail of assessments is not as complex as in Europe.

9.5 Other National Authorities

Most nations have some form of chemical control legislation and associated regulations, some of which will apply to the oil and gas sector. It can, however, be difficult for the practitioners to ascertain the correct legislation and to therefore be able to comply with such. Indeed, many nations have less than adequate or well-defined regulations and in these cases either the oil and gas operating company or the appropriate government agency may ask for an equivalent compliance status to that applied in the North Sea. More frequently operating companies and drilling service companies operating on a global basis have adopted the OSPAR Treaty requirements as their internal standard when it comes to chemical use and discharge. Where national authorities are involved, then local marine species for toxicity testing may be designated and often this means that local testing laboratories have to be used.

9.6 Conclusions and Critique

The examples of regulatory control presented essentially follow some common themes with regard to ecological toxicity testing and involve a registration process of the chemical on some form of national or area database. The collection of data on toxicity and other ecological properties such as biodegradation is normally done by some form of assessment by a modelling system or by expert judgement or both. The aim is to protect the environment from the use of toxic and potential harmful substances and to prevent or minimise their discharge to the environment.

Undoubtedly, in strongly regulated areas such as Europe, there has been a reduction in the amount of chemical products discharged in overall tonnage terms, into the marine environment. In the UK continental shelf across the period 2009–2013, the tonnage of chemicals discharged is reduced from over 129 000 to over 83 000 tonnes; however in this same period the reduction of chemicals used that were identified as candidates for substitution was only 2% [39]. In recent years this percentage has increased, but still a number of chemicals of environmental concern continue to be allowed to be discharged. The dilemma here, as has been illustrated throughout this study and as also further discussed in Chapter 10, is the possible reduction in efficacy coupled with increased unit costs associated with the use of 'greener' products. Countering this argument for the use of green products are the many unexplored possibilities for 'greening' the product portfolio, as has been described throughout this study, and the resultant possibilities of the application of greener alternatives without loss of efficacy. However, again this is driven by the cost of research and development and the prohibitive costs of introducing new chemicals into the marketplace that requires to be regulated through REACH and associated legislation. Similar trends exist in other areas of the globe where monitoring figures are available; however large regions have little or no visible control over the quality and, to some extent, the volume of discharges.

This highlights a common and persistent problem in localised regulation in that ecosystems are globally connected, but attempts at global agreements are fraught with difficult political challenges and even questioning of the science behind the necessity for such agreements. The recent Paris agreement on climate change [40] after a change of executive in the United States was challenged, and at time of writing, the United States has withdrawn from the agreement. Without such agreements, local area initiatives and regulations can only affect local areas to a limited extent, and the greater global systems can still sustain serious ecological damage, in particular the marine system and the atmosphere, which in turn seriously damage local ecosystems.

Although discharges from the oil and gas industry have declined in Europe, this must be put into perspective in terms of other sources entering the marine environment, of which there are many. These include shipping, mariculture, seabed mining, dredging and dumping at sea. Although offshore oil and gas activities still account for almost 40% of the inputs into the European marine environment particularly in the North Sea, the chemical concentrations in these inputs upon discharge to the sea are recognised as being comparatively low [41]. The level of contamination is also further reduced due to natural dispersion, volatilisation and bio-degradation. Consequently, any large-scale negative biological effects, particularly from pro-duced water, at offshore installations are likely to be minor, with the possible exception of bioaccumulation in shellfish [42, 43]. This however does not negate the need for a better under-standing of the produced water's constituents, environmental fate and potential effects, the latter being particularly true for chronic, low-level exposure to a variety of chemicals [41, 44]. Also despite the trend in recent years towards environmental assessment of the overall impact of produce water discharges, it is still considered a major source of marine pollution due to the volumes involved [45].

This leads on to criticisms of the chemical assessment processes and what understanding it presents of the ecosystem under assessment. The focus is to use data collected from testing or that is available in the literature. Such data at best provides a snapshot of toxicity and ecological effects against a particular species in the environment or how a chemical degrades or accumu-lates under certain laboratory conditions. This is especially true for studies related to the aquatic environment where the basis for such assessments is centred around work from GESAMP, Joint Group of Experts on the Scientific Aspects of Marine Environmental Protection [46]. This is an advisory body, established in 1969, that advises the UN system on the scientific aspects of marine environmental protection. At present GESAMP is jointly sponsored by nine UN organisations with responsibilities relating to the marine environment, and they utilise GESAMP as a mechanism for coordination and collaboration among them. GESAMP func-tions are to conduct and support marine environmental assessments; to undertake in-depth studies, analyses and reviews of specific topics; and to identify emerging issues regarding the state of the marine environment. GESAMP itself today consists of 16 experts, drawn from a wide range of relevant disciplines, who act in an independent and individual capacity. Studies and assessments are usually carried out by dedicated working groups, most of whose members are not sitting members of GESAMP but are part of the broader GESAMP network. This means that data is assessed through expert opinion.

The data collected with regard to chemical inputs, as has been described in itself, is not com-prehensive but based on a snapshot of what is agreed to be the most relevant data set with regard to toxicity and other environmental impacts. However, this data in itself carries the risk of flaws particularly with regard to certain molecules that are poorly soluble [47] and as have been described in Chapter 3. Also, selecting the most appropriate tests for certain chemicals such as surfactants and polymers can be difficult as what is packaged as the neat chemical for test may not resemble what is in the formulated product or in the applied solution or other media. The chemical structures that end up in the produced water, drilling fluids and other waste streams may be quite different from those that formed the basis of the applied formulated chemical additive. This would therefore argue in favour of some testing, as is already in use in the United States, on the toxicity and other aspects of the discharge stream.

9.6.1 Expert Opinion

All of the regulatory systems described in operation rely heavily on expert opinion and judgement-based criteria. This can lead to a number of flaws in the assessment process as at the end of the day the modelling data and outputs are assessed by humans and humans make mistakes. However, perhaps more importantly their opinions are formed by a number of factors that are outside the realm of scientific facts, and therefore the data presented can be misinterpreted, particularly in relation to the time and events we live in. This is neatly exampled by a couple of predictions by experts in the past.

> *This 'telephone' has too many shortcomings to be seriously considered as a means of communication.* – Western Union internal memo, 1876.
> *I think there is a world market for maybe five computers.* – Thomas Watson, Chairman of IBM, 1943.

As observed, the art of prediction is fraught with hazard. Expert opinion in terms of regulatory control is further influenced by the government of the day, policies they create and non-governmental organisations exerting pressure and public opinion. All of these can sway the expert's opinion away from a wholly scientific judgement.

Expert judgement or opinion is therefore subject to the following problems:

- It is affected by the process used to gather the opinion.
- It has uncertainty that requires to be analysed and characterised.
- It is conditioned by certain factors such as question phrasing, information considered, assumptions made and inherent bias.
- It can be influenced by other unrelated data and opinion.

There are mathematical methods that can be used to weigh and pool scientific opinion, and these are required to be considered by the regulatory authorities who in the main are making expert judgements. Such methods are usually based on Cooke's principles [48] and where applied can quantify uncertainty therefore removing bias and lead to improved decision-making. In essence Cooke's principles can be condensed to the following bullet points:

- *Scrutability and accountability:* All data including expert judgements must be subject to peer review and the results of review must be open, transparent and reproducible to competent reviewers.
- *Empirical control:* Expert assessments are subjected to empirical quality control.
- *Neutrality:* Methods for combining and evaluating expert opinion should be utilised and encouraged in order to elicit true unbiased opinions.
- *Fairness:* Experts should not prejudge outcomes prior to completion of assessment process.

In its very nature, expert opinion will lead to disagreement, and attempts to impose agreement will lead to confusion between consensus and certainty. This in turn leads to an undermining of the whole process based on the expert judgement. The goal should be to quantify the uncertainty so that a proper understanding of risk can be agreed. This is particularly the case when data is unavailable, not known or unreliable [49]. Although expert elicitation has been recommended as a tool for the IPCC *'to ensure that areas of uncertainty poorly captured by models are better represented, such as whether the Greenland ice cap might melt'* [49], expert opinion can be very valuable and in the right setting, and controls can be a robust form of analysis. However, in the regulation of chemical use and discharge in the oil and gas sector, it seems to be applied in a purely subjective fashion with little transparency, whereas in many industrial settings has become a valuable addition in risk assessment [50].

9.6.2 Modelling Systems

All of the modelling systems employed in chemical hazard assessment rely on data inputs and the well-known adage of garbage in–garbage out is one to be conscious of, as at best the data sets and data points used are only representations of what might be occurring in the receiving environment. Also, the amount of data is still embryonic compared with what is truly required to give an accurate representation of the natural environment.

One of the now most used data assessment models is the CHARM system, as employed in the OSPAR Treaty region. This has been described in overview in Section 9.1.3 and is further detailed here.

9.6.2.1 The CHARM Model

The CHARM model was designed to provide an estimate of the amount of chemicals, which might be present in a 'sacrificial' volume of water surrounding the rig or production platform during a given job, under certain appropriate conditions such as current speed. These 'location' values can be fixed in the model for comparative purposes or varied to provide site specific values – this is termed the predicted environmental concentration (PEC). The PEC is then compared with the predicted no effect concentration (PNEC), which is derived from ecotoxicity data within the HOCNF data set. The result of the PEC divided by the PNEC produces an HQ.

It logically follows that HQ values that exceed '1' represent an unacceptable concentration of product in the given volume of water.

In the HOCNF marine toxicity, data are required for all preparations or their constituent substances (unless they appear on the PLONOR list). However, for many offshore chemicals, only data on the toxicity of the preparation are available, and not on the individual component substances. OSPAR Recommendation 2000/4 [51] allowed the use of the toxicity data of the preparation to estimate the toxicity of a substance contained in it, taking into account the concentration of the substance in the preparation.

The CHARM User Guide [52] describes how the toxicity data of substances, if available, and that of preparations is used to calculate the HQ of the substances also the preparations containing constituent substances. If both data for PEC and PNEC are available at the substance level, then

$$HQ_{substance\,i} = \left[\frac{PEC_{substance\,i}}{PNEC_{substance\,i}} \right],$$

$$HQ_{preparation} = Maximum \left[\frac{PEC_{substance\,i}}{PNEC_{substance\,i}} \right]_{substance\,i\,to\,n}.$$

If data for PEC is available on substance level and data for PNEC is only available on preparation level, then

$$HQ_{preparation} = Maximum \left[\frac{PEC_{substance\ i}}{PNEC_{preparation}} \right]_{substance\ i\ to\ n} .$$

However, the prescreening assessment requires knowledge of the toxicity of the individual substances. The problem has been that for most offshore chemicals, only data on the toxicity of the preparation is available, and little toxicity data exists for the individual component substances. Even to date, the data available is point data on lethal toxicity, biodegradation and bioaccumulation. This data, as has been discussed, is also based on simulated laboratory tests.

The CHARM model uses a number of different defaults for different aspects of defined oilfield operations. These were determined by the working groups through industry surveys and experimentation; at the time the model algorithms were written. This again returns us to expert opinion and judgement. Taking each of the three main oilfield operations.

9.6.2.1.1 Drilling Three different 'typical' hole size sections are recognised within the CHARM model and are used as defaults: 17.5″, 12.25″ and 8.5″. Each of these is seen as being drilled for a different time and resulting in a different amount of mud being discharged and are referenced in Tables 6 and 7, of the Charm Model Guidelines [52].

Since the concentration of chemical released is determined by the amount of discharged mud, it can be seen that if the concentration of chemical is the same in each mud mix, the PEC will be altered by the amount of mud that is ultimately discharged. This is why the values of HQ vary for each hole section.

9.6.2.1.2 Completion/Workover A number of different types of completion and workover operations are recognised within the CHARM model. As with the drilling model, each type varies in the default amounts of chemical that are deemed to be discharged. This is represented in Table 9 in the Charm User Guidelines [52].

This algorithm makes the assumption that 10% of the chemical dosed is discharged. However, is this a correct default given that many chemicals in completion and workover may be wholly discharged and other may have 'zero discharge' status?

9.6.2.1.3 Production In production operations, a number of different types of operation are recognised within the CHARM model. As with the drilling model, each type varies in the default amounts of chemical that are deemed to be discharged. Chemicals are assigned as a 'standard' production chemical, an 'injection' chemical or a surfactant. The fraction released of the chemical is determined by the Log P_{ow} according to Equation 3 in the CHARM model; see the following:

$$C_{o/c} \approx 10^{\log P_{ow}} {}^\circ C_P$$

in which $C_{o/c}$ = concentration of the chemical in oil or concentrate (mg l^{-1}); P_{ow} = partitioning coefficient between octanol and water as determined by laboratory testing; C_{pw} = concentration of chemical in the produced water (mg l^{-1})

Other algorithms, as defined by the CHARM model, provide the drivers for other parts of the PEC water calculations, and standard flows are noted in Table 3 of the Charm Model User Guidelines [52]. It is important to note that the PEC used to drive the production HQ is by the Log P_{ow} partitioning data, affecting the fraction released into a daily fixed amount of water, as the production algorithm does not assume a batch-wise discharge.

As observed, this rapidly become a complex situation to assess and model a chemical's potential environmental impact, and yet, as also stated earlier, the overall impact of the chemical compared with other discharges in the produced water is seen as minimal.

9.6.2.2 Other Models

There are a number of other models of environmental hazard and impact, which can be used to assess chemical hazard and environmental effect; however, in the oil and gas sector, the driving force behind all these models is the relationship between the predicted environmental concentration and the PNEC.

For example, the more sophisticated DREAM [53] (Dose-Related Risk and Effect Assessment Model) is a three-dimensional, time-dependent numerical model that computes transport, exposure, dose and effects in the marine environment. The model can simulate complex mixtures of chemicals. Each chemical component in an effluent mixture is described by a set of physical, chemical and toxicological parameters. The environmental impact factor (EIF), first developed for the water column, has been extended to include ecological stresses in the benthic community. The EIF is a standardised method for marine environmental risk assessment that does not require explicit information on the local biological resources. This makes the methodology relatively easy to apply to new geographical areas.

These models although giving something of a picture of the effects of chemicals and other discharges in the aquatic marine environment have two areas of weakness that need to be borne in mind when interpreting data outputs. Firstly, they operate to a standard model of the environment and by their very nature do not take into account local environmental effects, some of which could be highly significant. Secondly, they rely on point data from laboratory testing, which is also removed from the real environment compounding the difference between data outputs and actuality. These differences reinforce the need for expert judgement; however this is also open to criticism as has been described in Section 9.6.1.

9.6.3 Regulatory Anomalies in Assessing Chemical Environmental Impacts

Throughout the preceding chapters of this book, the environmental behaviour of the classes of chemicals defined has been examined. Due to the uniformity of many regulatory approaches and the compartmentalisation of the ecosystem, many chemical substances and classes of chemical substance do not easily fit into environmental assessment models, and therefore regulation is further reliant on expert opinion and judgement. In this respect, it is necessary for those involved in the process to take a comprehensive view of the chemical substance's physical and chemical properties and its ecotoxicology. Chemical reactivity and physical properties such as vapour pressure, density, water solubility, lipid solubility and, importantly, molecular weight define how and where the substance is likely to be distributed in the environment and what, if any, undesirable ecological consequences are to be anticipated. It is particularly important to realise that a chemical substance does not represent an ecological threat simply because it is toxic or persistent, or lipid soluble or bioconcentratable. Hazard is proportional to the exposure intensity that is defined by the parameters of time and concentration. Therefore, only those materials that are toxic and persistent and bioconcentratable are likely to manifest undesirable ecological consequences.

This state particularly applies to a number of chemical classes and substances, and each of these is now examined, with reference to material in previous chapters.

9.6.3.1 Polymers – Chapter 2

Polymers are exempt from registration as new molecules under both US and European regulatory frameworks, and it is recognised that they are not usually bioavailable. It is highly likely that the substances that are likely to have environmental impact when applying and using polymers and copolymers are the thermal and other breakdown products chemical and otherwise and not either the polymer itself or the constituent monomer. Yet current regulatory controls are driven by direct impact of the polymer and the possibility of residual monomer. Also, despite their lack of bioavailability, chemical manufacturers are driven to improve biodegradation in polymers with inevitable loss of performance and still with the potential to increase the environmental burden. Also, as described in Chapter 2, this means that the economics of application of such green products has become less favourable.

This is not to say that for other environmental reasons (easy of reuse and recycling) it is not desirable to have biodegradable polymers; however, the dilemma is in applying biodegradable polymers is the receiving environment truly protected.

9.6.3.2 Surfactants – Chapter 3

Surfactants cause regulators and environmental scientists considerable problems as they are not neatly characterised and their behaviour is not easily predictable, especially with regard to environmental fate. For example, alkylphenol ethoxylates are key monomers in a surfactant polymer with many uses in the oilfield and in particular separation of oil and gas emulsions and form the basis of many demulsifier formulations [54, 55].

However, they are also under extreme regulatory pressure as nonylphenol ethoxylate is a suspected endocrine disrupter particularly in fish species [56, 57]. The polymeric form, however, is not bioavailable, having a large molecular weight and has little or no impact on the environment. The material is also only poorly biodegradable if at all and the degradation pathways do not reverse to the nonylphenol or nonylphenol ethoxylate monomers [58]. This evidence is largely ignored, and the regulation and other pressure are placed on not using these materials as directed by a precautionary approach. This pressure over time has been applied to all alkylphenol ethoxylate polymers and not just the nonylphenol-based materials. There is little or no application of expert judgement here and no holistic evaluation of risk to the environment. Consequently, this chemistry is no longer applied in the North Sea and is under pressure in other regions.

There is no reasonable scientific evidence for this approach, and many manufacturers can show that the amount of residual monomer is well below any possible toxic effect levels [59]. Furthermore, it could be argued that this material, which is highly effective on many crude oils at very low dosages, has been replaced by other surfactants whose environmental fate is more problematic and at higher dose levels, leading to an increased environmental impact.

This is a real case for open and analytically based expert judgement and to evaluate risk and not hazard.

For surfactants generally, the environmental risk has been little examined and the precautionary approach applied to a great degree. More understanding of the bioavailability of surfactants, their degradation and potential bioconcentration and biomagnification is required. This is not a trivial exercise as the chemistry is complex as are the interactions of the surfactant with marine organisms. At present, there is a reliance on toxicity tests that at best provide minimal data; other test work, and in particular bioaccumulation, is often meaningless [60].

In 2003, the author illustrated an already significant decline in the use of certain types of surfactant in the upstream oil and gas industry over the previous decade due to increasing regulatory pressures [61]. He would argue that a fresh approach is required to the regulation and environmental impact assessment of surfactants, leading to a risk-based approach to their use and discharge.

9.6.3.3 Phosphorous Products – Chapter 4

One of the major, if not the major, environmental impacts of phosphorous products is their potential to accelerate eutrophication in the aquatic environment.

In many areas of the United States, restrictive phosphorus discharge limits are being enacted or proposed where the receiving water streams are particularly sensitive to changes in phosphorus levels. Similar restrictions are considered by a number of oil and gas regulatory bodies responsible for the control and discharge of production waters and associated chemicals [62]. However, in other areas we continue to focus on toxicity and other ecological impacts such as biodegradation rate, whereas many phosphorous compounds are readily biodegradable and relatively non-toxic. However, their fate to compounds and salts of phosphorous that can lead to uptake by algae and related organisms should be of higher concern. Therefore, it is their bioavailability that is probably the overriding EIF, and in the author's opinion it is here that the regulations regarding phosphorous materials should be focused.

In addition, chemicals from oil and gas exploration development and production are only one factor affecting an imbalance of phosphorus in a specific environment; indeed in a risk analysis scenario, it is outweighed by riverine discharges of agricultural pesticides that are a major source of pollution with regard to algae growth stimulation [63]. There is clearly an imbalance in strongly regulating oil and gas discharges of phosphorous and allowing higher levels of discharges from other sources.

This leads to a major criticism of offshore oil and gas regulations for chemical use and discharge that they do not consider the other major discharges such as from freshwater sources into the marine basin in question. This is further discussed in Section 9.7.

9.6.3.4 Inorganic Salts and Metals – Chapter 5

The environmental impacts of metals particularly heavy metals such as zinc are more complex than a straightforward correlation of dose to concentration. Bioconcentration and biomagnification processes can be very important in a particular ecosystem, and this is especially in respect to the aquatic environment.

Bioavailability in the context of environmental science is generally accepted as a molecule being bioavailable when it is available to cross an organism's cellular membrane from the environment where the organism has access to the chemical [64]. It has been generally accepted by the scientific community and regulatory authorities that most organic molecules having a molecular weight greater than 700 will not cross cellular membranes and are therefore not bioavailable. This is a somewhat arbitrary cut-off and is further discussed in Section 9.7.

Most metals, metal salts and other inorganic complexes previously discussed particularly in Chapter 5 have molecular weights below 700. They are also not suitable for environmental impact assessment through models such as CHARM as they do not usually biodegrade. Most regulatory authorities rely on a hazard ranking system based on toxicity and expert judgement. Bioavailability is mostly driven by solubility both in the aquatic environment and in soil [65]. However the partitioning of metals and inorganic salts has complex metabolic pathways in both animals and plants and can lead to not only their removal from biological pathways but also their accumulation. The bioaccumulation of heavy metals and other toxic inorganic compounds in soils and marine sediments can lead to their entrance to the food chain through processing of such sediments by microorganisms and algae [66].

To date, the regulatory authorities have taken a fairly simplistic approach, involving substitution to perceived, less environmentally hazardous materials based on toxicity and mainly toxic properties inherent to the substance. It would seem reasonable to consider the development of models to include bioaccumulation potential; however there are complications in this, and again a simplistic approach can be misleading as to potential ecological impacts. This is particularly the case for heavy metals, which due to their toxicity require to be strongly regulated in many regions, particularly with regard to such materials entering the food chain [67]. The case of zinc is considered in the following subsection as an example.

9.6.3.4.1 *Environmental Impact of Zinc in the OSPAR Area*

European legislation, and in particular EC Council Directive 92/43/EEC on the conservation of natural habitats and of wild fauna and flora, aims to protect some of the most valuable parts of Europe's natural heritage and ensure its wise use. To achieve this, it has identified a set of habitats and species known as Annex I habitats and Annex II species, which require special conservation measures to be taken by member states. In particular, these measures include the designation and management of special areas of conservation (SACs) for the habitats and species.

In response to the challenges presented by this legislative framework, the UK Marine SACs Project was set up to establish management schemes on selected marine SACs in coastal waters. Its activities have focused on a selection of 12 marine SACs around the UK coastline and on the development of specific areas of knowledge needed for the management and monitoring of European marine sites. As part of this work over the last 2–3 decades, this scope of work has included the monitoring of zinc [68].

The work of the UK Marine SACs monitoring programme has concluded that the most sensitive species to the toxic effects of zinc are the marine invertebrates. This work has led to the adoption of an environmental quality standard set for the monitoring of zinc at $40\,\mu g\,l^{-1}$. The monitoring programme has shown a number of sites exceeding this standard. There is currently a proposal to revise this standard down to $10\,\mu g\,l^{-1}$; conversely there is also much scientific argument about the validity of the values of these standards as many factors are arbitrarily applied [69].

There is a reasonable body of work relating to toxicity of aquatic species to zinc, in which much of it was conducted in the 1950s in the United States on freshwater fish and other species. In general, this shows that a variety of species were killed at concentrations of $>15\,mg\,l^{-1}$ for a period of 8 h or greater; however the hardness of the water and its temperature also effected this lethal dose concentration, but the data is widely scattered and somewhat flawed [70]. Nonetheless in a wide-ranging study of marine species along the US Atlantic coast and although slight variations in the level of zinc in various areas of the sea were observed, the amount of zinc found in marine organisms particularly shellfish such as oysters, clams and scallops was thousands of times more than the seawater per unit of weight [70].

Figure 9.3 shows the different bioconcentration factors (BCFs) for various marine fauna in relation to zinc.

Concentration in seawater mg l⁻¹	Algae	Soft invertebrate	Skeletal invertebrate	Soft vertebrates	Skeletal vertebrates
10	100	5000	1000	1000	30 000

Figure 9.3 Bioconcentration zinc in marine species.

This means that biomagnification of the increased rate of bioaccumulation occurs for zinc in marine organisms such as shellfish and fish more than other organisms. The critical regulatory factor here is the impact on commercial fishing as these species will accumulate zinc and have tainting problems.

Work in Australia in the mid 1970s determined concentrations of zinc, cadmium, lead and copper in whole soft parts of the common mussel *Mytilus edulis* (L.) [71]. The mussels were sampled according to procedures suggested by previous studies in order to eliminate the effects of natural environmental variables. Results of the analytical studies were compared with data on the quantities of trace metals known to be discharged by industry into the catchments of each area under study. This has allowed the evaluation of an indicator of the ability of the musse to uptake trace metals without the need for multiple analyses of water samples. The results suggested that the mussel is capable of acting as an efficient time-integrated indicator of zinc, cadmium and lead over a wide variety of environmental conditions.

The work of the UK marine coastal monitoring programme has shown that zinc is an essential element of most marine species and is therefore readily bioaccumulated [68]. Several species of *Crustacea* have been noted as able to regulate the uptake of zinc. It has also found, as in later works, that the uptake of zinc is a complicated process as the organisms can uptake zinc, which is then reflected as a BCF, but the actual concentrations in the animal tissues are of no toxicological significance.

More recent work [72] has examined the theoretical and experimental basis for the use of BCF/BAF in the hazard assessment of Zn, Cd, Cu, Pb, Ni and Ag. The BCF/BAF data for Zn was characterised by extreme variability in mean BCF/BAF values and a clear inverse relationship between BCF/BAF and aqueous exposure. This study illustrates that the BCF/BAF criteria, as currently applied, are inappropriate for the hazard identification and classification of metals. Furthermore, using BCF and BAF data leads to conclusions that are inconsistent with the toxicological data, as values are highest (indicating hazard) at low exposure concentrations and are lowest (indicating no hazard) at high exposure concentrations, where impacts are likely.

Bioconcentration and bioaccumulation factors do not distinguish between essential mineral nutrient, normal background metal bioaccumulation, adaptive capabilities of animals to vary uptake and elimination within the spectrum of exposure regimes, nor the specific ability to sequester, detoxify and store internalised metal from metal uptake that results in adverse effect. An alternative to BCF, the accumulation factor (ACF) for metals, was assessed, and while providing an improvement, it did not provide a complete solution. A bioaccumulation criterion for the hazard identification of metals is required, and work directed at linking chronic toxicity and bioaccumulation may provide some solutions.

Finally returning to the ongoing work of the UK Marine SAC monitoring programme [68] , it has listed the potential effects of zinc in the marine coastal environment as

- Acute toxicity to algae, invertebrates and fish above the proposed EQS of 10 $\mu g\,l^{-1}$ for dissolved zinc.
- Accumulation in the sediments can pose a hazard to sediment dwelling organisms at concentrations above 124 $mg\,l^{-1}$.
- Bioaccumulation in marine organisms poses a potential threat the higher organisms including fish, birds and marine mammals.

As can be seen the metabolic and ecotoxicological pathways for zinc are complex and this is the case for other metals and their salts. It is naive to think that the potential environmental hazard can be evaluated in simplistic toxicity and solubility terms. A more complex model has to be constructed to give a more accurate picture of the EIF of metals and related substances as discharged by upstream the oil and gas industry. Again, where expert judgement is used, an open, transparent and accountable methodology has to operate. The emphasis where substances are not bioavailable should be then on risk of environmental harm rather than assumed.

9.6.3.5 Small Organic Molecules – Chapter 6

Low molecular weight organic molecules are usually under the molecular weight threshold and therefore for regulatory purposes are deemed to be bioavailable. In many cases biodegradation rates are known or can be determined experimentally, and therefore such molecules with environmental impact can be evaluated in model systems such as CHARM. Generally, these molecules give good representation of ecotoxicological outcomes in model systems as accurate data, albeit, from restricted data sets, is available.

Some areas are, however, of question. First is the use of toxicity data in an adverse way when rates of biodegradation show these products to be of low environmental concern. This often leads to products be discounted for use when a more holistic view may reduce the overall chemical environmental burden as these products biodegrade rapidly to non-toxic substances. Second, some small molecules, particularly amines, can biodegrade to more toxic substances and here the opposite is true. Often the substance is considered to be less environmentally harmful than what would be the case if secondary biodegradation products were considered.

Finally, the ultimate fate of many small organic molecules is to become carbon dioxide, carbon monoxide and water. The release of further carbon gases is not considered in the environmental assessment of any of the chemicals used in the upstream oil and gas industry, and perhaps it should be. Although a caveat would be in terms of the additional loading, it would be very small compared with other sectors such as transport. This is further discussed in Chapter 10 with regard to sustainability where chemical use could have a positive environmental effect in terms of carbon capture.

9.6.4 Silicon Chemicals and Polymers – Chapter 7

Silicon products are among the most environmentally acceptable chemical additives used in a number of industrial sectors including the oil and gas industry.

Silica and its associated materials are not bioaccumulating as they are not bioavailable and also do not biodegrade being highly unreactive except under extreme acid conditions. In terms of silica as used in nanotechnology, the evidence is still somewhat circumspect and caution should be taken as to the possible effects of all chemically based nanotechnology [73].

There is a potential for positive environment effects to be utilised from using silica due to its unreactive nature and capability of encapsulating other materials. For example, oil and hydrocarbon drill cuttings have been proposed to be microencapsulated to allow their disposal in an environmentally safe manner [74].

Silicate muds are deemed to be environmental safe and are mainly utilised in environmentally sensitive and highly regulated regions such as the North Sea basin. Such muds are based on potassium and sodium silicate and have the added advantage of being somewhat cheaper than synthetic-based fluids. These materials are preferred in high temperature, high pressure drilling and highly deviated drilling [75].

Silicones, such as polydimethylsiloxane (PDMS), contain an organosilicon linkage that is not found in nature, and it had been assumed that these polymers do not degrade naturally in the environment. However, it has been found that minor degradation takes place by hydrolysis and subsequent oxidations to an ultimate fate of benign silica, water and CO_2 [76–78]. In general silicones are not bioavailable and although often persistent in the environment are not bioaccumulating and are often non-toxic [79]. In earlier studies in order to assess reliably and safely the potential threats posed by the environmental presence of silicones, a number of key parameters were examined. Based on extensive toxicological and environmental fate studies, it was concluded that commercially important organosilicon materials do not appear to present any ecologically significant threat [79].

Therefore, overall silicon products and polymers are not biodegradable, are generally non-toxic and non-polluting and do not bioaccumulate. Indeed, where they are bioavailable they are utilised by a number of microorganisms in building skeletal structures.

Their regulation is through the CHARM model or similar system and/or expert judgement. It could be argued however that silicon products, especially silicon polymers, are a special case as their low toxicity and non-bioavailability is not outweighed by their lack of biodegradation, which in most model studies leads them to receive a relatively high environment hazard factor. However, they are highly effective materials that can in some applications operate at very low dose rates such as 1 ppm or less. This is particularly the case in gas defoaming [80] where due to regulatory requirements they have been replaced by biodegradable products that are only partially effect even then at doses rates in 100s of ppm if not 1000s of ppm. This is placing an increased environmental burden into the aquatic ecosystem and also an increased level of potentially more hazardous and toxic materials.

9.6.4.1 Solvents and Other Substances – Chapter 8

Generally, the solvents' high vapour pressure allows for their easy dispersion in air, promoting contact with living organisms and the pollution of the environment. The high volume, the particular properties of each solvent and the volatility factor in high-risk solvents can have a serious impact on the environment and threaten animal and plant life.

The environmental impact of solvents and related materials has been discussed in Section 8.3, in some detail. As stated, the Montreal Protocol [81] has identified the need to re-evaluate chemical processes with regard to their use of volatile organic compounds (VOCs), which in the main are solvents and the impact these compounds have on the environment. Work in assessing the environmental impact of solvents has shown that simple alcohols (methanol, ethanol) or alkanes (heptane, hexane) are environmentally preferable solvents and the use of solvents such as dioxan, acetonitrile and tetrahydrofuran is not recommended. Results also indicate that methanol–water or ethanol–water mixtures are environmentally favourable compared with pure alcohol or propanol–water mixtures. This study has demonstrated a framework that can be used to select green solvents or environmentally sound solvent mixtures for processes in the chemical industry. The same framework can also be used for a comprehensive assessment of new solvent technologies, provided sufficient data is available [82]. To date this tool has not been applied in the upstream oil and gas sector.

In the main the upstream oil and gas sector has relied again on environmental models or expert judgement to assess the environmental impact of solvents. In Europe, many such products have been placed on a list not requiring further assessment and can be used with the minimum of environmental scrutiny as they are deemed to be included in the PLONOR list [83].

A number international treaties on the subject have generated guidelines to ban the usage of highly volatile chlorinated solvents such as carbon tetrachloride and the more toxic aromatic solvents such as benzene. The oil and gas sector does not use chlorinated solvents and has substantially replaced solvents such as benzene with safer alternatives.

As illustrated in Chapter 8, some solvents have specific environmental impacts, while others are more severe. Many volatile solvents have detrimental effects on atmospheric quality and can disturb the climate; many others are water miscible and can pollute both surface waters and the marine environment. Many of these substances are difficult to remove from waste streams due to the very characteristics that are utilised for their use, namely, their solvency. These characteristics often also make many solvents poorly biodegradable and therefore persistent in the environment [84]. Care needs to be taken when introducing a low level of regulation around such materials.

Oil-based muds offer unrivalled performance characteristics in terms of rate of penetration, shale inhibition, wellbore stability, thermal stability, lubricity and salt tolerance. They are of particular use in highly deviated and horizontal drilling where water-based muds unfortunately do not give the necessary technical performance [85]. For some time now their use and discharge are subject to strict environmental regulations, which now apply almost globally. In effect the contaminated cuttings must be removed from the well site or offshore location and shipped for processing and reuse; this is almost always the case in whatever geographical location [86]. This practice has shifted the environmental burden to a different disposal route, and the use of landfill options does not address the fundamental pollution problem.

Suppliers of mud systems have responded to this problem. Since 1990, several non-toxic, biodegradable, synthetic-based muds (SBMs) with desirable performance and environmental characteristics have entered the market. Synthetic-based drilling fluids have several technological and environmental advantages over water-based and oil-based drilling fluids and can cut total well costs in many cases [87].

Spent drilling fluids and drilling cuttings are among the most significant waste streams from exploration and development activities. They pose a serious and costly disposal problem for offshore operators who must in many cases transport spent mud and cuttings to shore for land disposal. There seems to be no appetite for the use of SBMs as an alternative with disposal as with WBMs at the well site.

9.7 Further Conclusions

In this final section, the author attempts to bring together the critique of a myriad of regulations that govern chemical use and their discharge and disposal in the upstream oil and gas industry. There is an overriding principle to these controls that in the author's view is often lost by both regulators and industry, which is to aid the protection of the environment, often the aquatic environment, and to prevent its pollution.

As in the previous section, there are several of examples where this may not be the case. In the author's view, this is compounded by the positions of the interested parties in framing the control regulations. The regulators come from a somewhat bureaucratic position in following government policy and treaty obligations. Industry has, in general, a compliance position of following the regulations as written. Other parties, such as NGOs, have particular agendas often relating to a narrow or specific issue, e.g. 'Fracing'. None of this lends itself to moving to a more cooperative and meaningful dialogue about the framework of environmental protection. This is not an easy task as it requires cooperation on a global basis and both historically and recently, achieving cooperation and agreement on climate change issues internationally is fraught with national self-interest [88]. Nonetheless it is certainly fundamental that regulatory control of chemical substances is moved to a global basis alongside other pollution prevention measures. The ecosystems that are examined nationally and regionally are part of a much greater whole, and impacts can have serious consequences across the biosphere, as is now recognised by a large body of scientific opinion [1].

In moving towards a more harmonious and cooperative regulatory framework, where many of the issues regarding anomalous criteria in controlling chemical substances and that are also detrimental to the protection of the environment, the author believes two fundamental principles need to be applied: firstly, a transparent and robust basis for expert judgement and opinion is to be used in which the mathematical assessment is based on Cooke's principles and, secondly, there is a movement away from hazard assessment and ranking that allows regulators to rank materials based on criteria that may not be the best environmental option.

The second principle, which leads in concert with the first, is to establish quantifiable risk assessment of chemical use and discharge and the options that could allow supposedly higher hazard materials to be used, which for other factors present a lower environmental impact. The dilemma of materials that are toxic but not bioavailable could therefore be more systematically addressed and reasoned risk assessment for use or non-use presented. At present the overriding principle is to apply the *precautionary principle*, which negates the potential for better environmental options.

9.7.1 The Precautionary Principle

The precautionary principle has become embodied in numerous international environmental treaties and is enshrined in the 1992 EU treaty, 'The Maastricht Treaty' [89]. The precautionary principle has been difficult to define and legal opinion is highly variable; indeed one legal analysis has identified fourteen different definitions of the principle in different declarations [90]. The 1992 Treaty does not define the principle, and despite a body of case law, legal opinion is divided on its meaning and application. The author believes the objective of the precautionary principle is to prevent harm – not progress – and to support a sustainable future for our children and grandchildren. Inventiveness – driven in part by precaution – can encourage competitiveness in a global market that no longer tolerates products that harm public health or the environment. However, in the regulation of the oil and gas sector, it appears to be an option where insufficient information is available or effects are not fully understood, and in such cases activities are halted. Indeed, a declaration in 1990 on the protection of the North Sea calls for action to be taken even if there is scientific evidence to show a no link between emissions and waste streams into marine waters and any effects arising [91]. This is in the author's view contrary to the principle as if this interpretation is applied, then no new technology or chemistry, including those that are trying to enhance the environmental impact of activities, could proceed. Surely the precautionary principle is to proceed with caution not to halt scientific progress. The precautionary principle, if used properly, is based on sound science and is not irrational or emotional. Furthermore, if employed properly, industry will not go bankrupt but could enhance its ability to produce better and safer products. The level of evidence that is needed to trigger action based on the precautionary approach will be less than that needed for a full risk assessment. However, the available scientific evidence should be solid and should be no less rigorously developed than any other scientific information used. On the contrary, use of the precautionary principle should be a challenge to the scientific community to improve methods and procedures needed for studying complex natural ecosystems and for developing sound risk assessment.

9.7.2 Risk Assessment[1]

The author firmly believes that all interested parties in the regulation of chemicals and related substances used and discharged in the upstream oil and gas sector need to move forward to

---

1 At the time of writing, the UK regulatory authorities and their advisors were evaluating a risk-based approach (RBA) to produced water discharges including the chemical content therein. In 2012, OSPAR adopted the recommendation 2012/5 for a 'Risk Based Approach (RBA) to the management of Produced Water (PW) discharges from offshore installations' – OSPAR (2012a) *Recommendation 2012/5 for a risk-based approach to the Management of Produced Water Discharges from Offshore Installations.*

embracing a robust system of risk assessment based on sound scientific principles and evidenced-based criteria.

To an extent, the CHARM model is able to use hazard data to input into a risk management scenario. However, the regulation has encouraged only chemicals whose HQs are low to be examined. In effect, the regulation has driven the discontinued application of a number of chemicals. For example, only two non-oxidising biocides are now used in any volume in the oil and gas sector with a large number not being considered for use. This has been driven both by the Biocidal Product Regulation (BPR) in Europe and the application of the HMCS arising from the OSPAR Treaty. This reduction in biocides has occurred globally and been implemented by default in non-regulated regions due to the major oil and gas companies adopting a policy of only selecting certain accepted products. This has resulted in very few options in biocidal treatment when the main biocides are found to be less effective, leading to increased dose rates and increased environmental burdens. It is presumably not the intention of regulation to move in this direction and a re-addressment of risk rather than hazard is urgently required.

Also, as stated in Section 9.6 under solvents and other substances, these are regulated in the OSPAR region by being on the PLONOR list. It is the author's view that the list has little or no relevance and that all chemical substances and other materials should go through usual channels. In response in UK advisory body, Cefas has published guidance for the addition of new chemical substances to the list [92] in which data in support of inclusion is required. Surely this is an overabundance in regulation, and the list should simply be abandoned in favour of use of the HMCS or other mechanism, leading to risk assessment.

9.7.3   A Holistic View

The author also believes that the systems of assessment are too narrow in their scope and that there is an overabundance of regulations and criteria applying to chemicals whose contribution to the environmental impact globally needs to be placed into significance. This has led in a number of regions to veritable armies of bearcats in both industry and government focused on compliance and adherence to regulations with the creation of new and the development of old regulations seemingly never ending. The system is almost self-perpetuating and now involves a job creation aspect.

This vicious or virtuous circle, depending on your viewpoint, needs to be broken, and a fresh approach assembled in which environmental protection is the key objective. A major criticism of offshore oil and gas regulations for chemical use and discharge is that it does not consider the other major discharges from freshwater sources into the marine basin in question. This then skews the environmental control away from where the real environment burden and critical pollution impact may be.

In the previous chapters and again in the concluding chapter, a number of chemical solutions to reduce the environmental impact of the overall operation have been cited. However, this is not a common practice, partly for economic reasons, but more often as the regulatory frameworks are not designed to take a wider and more holistic view of chemical use. There surely needs to be criteria to show a reduced environmental burden can affect the regulatory control and EIF in such cases. Again, a risk assessment process can incorporate this.

In this chapter, there has been an attempt to describe the key regulatory frameworks and the principles behind them. There have been several of criticisms levelled at the current regulatory controls. The author believes that there needs to be a fundamental shift in regulations away from complex and over-burdensome regulatory compliance to an open and quantifiable approach to environmental protection involving a risk-based approach in a holistic setting.

Many regulatory authorities and regions have bound themselves into the compliance and hazard reduction route by target setting and counting how many chemicals or how much chemical has been discharged, which is not the same as measuring the amount of pollution in the ecosystem. It will, therefore, take a change of mindset in both regulators and industry and will require an open and justifiable methodology to satisfy policy makers, pressure groups and NGOs. Given the pace of change in governmental policy and the, no doubt frustrating, inability of regulators to alter legislation in this area, the author believes that only an industry-led initiative can achieve this. The European Oilfield Speciality Chemicals Association has in some minor ways contributed where it can to some useful changes and has produced some helpful tools to aid in compliance. However, like other trade associations, it is instructed by its member companies who are conscious of the positions of their clients, the oil and gas companies. It is with the latter that the author believes change needs to be initiated and for them to move away from compliance-driven policies by actually doing what they say in many of their mission statements with regard environmental policy.

> *"We are committed to continually improving our environmental performance and reducing the potential impacts of our operations through the implementation of our Operational Excellence Management System."– Chevron* [93]

> *"Our goal of achieving no damage to the environment guides our actions. We consider local conditions when determining which issues would benefit from the greatest focus." – BP* [94]

> *"Statoil's aim is to avoid causing significant harm to the local or regional environment. We take a precautionary approach and apply a combination of corporate requirements and risk-based local solutions to manage our environmental performance. We strive to adhere to high standards of emissions to air, waste management and impact on ecosystems— wherever we work." – Statoil* [95]

> *"Prevent and minimize environmental impacts of projects, processes, and products." –Petrobras* [96]

The examples of policy statements given above are purely arbitrary and any number of similar statements could have been quoted. The author would like to make it clear that he on no way is being critical of the individual companies quoted and is merely using them as examples. The point is that what is stated has to be delivered and time is now of the essence as shall be explored in Chapter 10.

References

1 Lovelock, J. (2006). *The Revenge of Gaia*. London: Penguin Books.
2 https://www.ospar.org (last accessed 7 December 2017).
3 https://osha.europa.eu/en/legislation/directives/regulation-ec-no-1907-2006-of-the-european-parliament-and-of-the-council (last accessed 7 December 2017).
4 https://echa.europa.eu/ (last accessed 7 December 2017).
5 http://eosca.eu/ (last accessed 7 December 2017).
6 Payne, G., Still, I., Robinson, N., and Groome, S. (2009). *The Development of the EOSCA Generic Exposure Scenario Tool (EGEST)- Why we Need it*. Manchester, UK: The Royal Society of Chemistry, Chemistry in the Oil Industry XI, 2–4 November,.

7 http://www.hse.gov.uk/chemical-classification/legal/clp-regulation.htm (last accessed 7 December 2017).

8 United Nations (2011). *Globally Harmonised System of Classification and labelling of Chemicals (GHS)*. New York and Geneva: ST/SG/AC.10/30/Rev.4.

9 https://echa.europa.eu/documents/10162/22786913/sds_es_guide_en.pdf (last accessed 7 December 2017).

10 http://www.hse.gov.uk/reach/ (last accessed 7 December 2017).

11 Sanders, P.F. and Sturman, P.J. (2005). Biofouling in the oil industry. In: *Petroleum Microbiology* (Chapter 9) (ed. B. Ollivier and M. Magot), 171. ASM Press.

12 Crolet, J.L. (2005). Microbial corrosion in the oil industry – a corrosionist's view. In: *Petroleum Microbiology* (Chapter 8) (ed. B. Ollivier and M. Magot), 143. ASM Press.

13 Corduwisch, R., Kleintz, W., and Widdel, F. (1987). Sulphate reducing bacteria and their activities in oil production. *Journal of Petroleum Technology* **39**: 97, SPE 13554.

14 http://www.hse.gov.uk/biocides/eu-bpr/ (last accessed 7 December 2017).

15 https://echa.europa.eu/support/dossier-submission-tools/r4bp (last accessed 7 December 2017).

16 http://www.barcelona.com/barcelona_news/the_barcelona_process_or_euro_mediterranean_partnership (last accessed 7 December 2017).

17 www.ospar.org/documents?d=33025 (last accessed 7 December 2017).

18 www.ospar.org/documents?d=33043 (last accessed 7 December 2017).

19 www.ospar.org/documents?d=32611 (last accessed 7 December 2017).

20 www.ospar.org/documents?d=32652 (last accessed 7 December 2017).

21 http://www.miljodirektoratet.no/Global/dokumenter/tema/olje_og_gass/OSPAR_recommendation2010-13_supplementary_guideline_norway.pdf (last accessed 7 December 2017).

22 Thatcher, M. and Payne, G. (2003). Moving towards substance based toxicity testing to meet new OSPAR requirements. In: *Chemistry in the Oil Industry VIII, The Royal Society of Chemistry*, 3–5 November. Manchester, UK. doi: 10.2466/pms.2003.97.3.995.

23 Still, I. (2001). *The Development and Introduction of Chemical Hazard and Risk Management (CHARM) into the Regulation of Offshore Chemicals into the OSPAR Convention Area: A Good Example of Government/Industry Co-operation or a Warning to Industry for the Future?* Manchester, UK: Chemistry in the Oil Industry VII, The Royal Society of Chemistry, 13–14 November.

24 https://www.cefas.co.uk/cefas-data-hub/offshore-chemical-notification-scheme/ (last accessed 7 December 2017).

25 https://www.epa.gov/laws-regulations/summary-toxic-substances-control-act (last accessed 7 December 2017).

26 https://www.epa.gov/assessing-and-managing-chemicals-under-tsca/highlights-key-provisions-frank-r-lautenberg-chemical (last accessed 7 December 2017).

27 https://www.epa.gov/npdes (last accessed 7 December 2017).

28 https://www.epa.gov/sites/production/files/2016-11/documents/memobestpractices_npdes-pretreatment-r.pdf (last accessed 7 December 2017).

29 https://www.epa.gov/wqc.

30 http://www.chemicalsubstanceschimiques.gc.ca/about-apropos/canada-eng.php (last accessed 7 December 2017).

31 http://laws-lois.justice.gc.ca/eng/regulations/SOR-2009-315/FullText.html (last accessed 7 December 2017).

32 http://www.ec.gc.ca/lcpe-cepa/default.asp?lang=En&n=26a03bfa-1 (last accessed 7 December 2017).

33 https://www.canada.ca/en/health-canada/services/chemical-substances/chemicals-management-plan.html (last accessed 7 December 2017).

34 https://www.canada.ca/en/health-canada/services/chemical-substances/petroleum-sector-stream-approach.html (last accessed 7 December 2017).

35 https://www.nicnas.gov.au/chemical-information/Topics-of-interest/subjects/chemical-regulation-in-australia (last accessed 7 December 2017).

36 https://www.nicnas.gov.au/chemical-inventory-AICS (last accessed 7 December 2017).

37 https://www.nicnas.gov.au/ (last accessed 7 December 2017).

38 https://industry.gov.au/industry/industrysectors/chemicalsandplastics/RelatedLinks/ChemicalsBusinessChecklist/Pages/default.aspx (last accessed 7 December 2017).

39 OSPAR Commission (2015) Assessment of discharges, spills and emissions from offshore oil and gas operations on the United Kingdom Continental Shelf.

40 http://unfccc.int/paris_agreement/items/9485.php (last accessed 7 December 2017).

41 Tornero, V. and Hanke, G. (2016). Chemical contaminants entering the marine environment from sea-based sources: A review with a focus on European seas. *Marine Pollution Bulletin* **112** (1–2): 17–38.

42 Neff, J.M., Lee, K., and DeBlois, E.M. (2011). Produced water: overview of composition, fates, and effects. In: *Produced Water. Environmental Risks and Advances in Mitigation Technologies* (ed. K. Lee and J. Neff), 3–56. Springer.

43 Lourenço, R.A., de Oliveira, F.F., Nudi, A.H. et al. (2015). PAH assessment in the main Brazilian offshore oil and gas production area using semi-permeable membrane devices (SPMD) and transplanted bivalves. *Continental Shelf Research* **101**: 109–116. doi: 10.1016/j.csr.2015.04.010.

44 Bakke, T., Klungsøyr, J., and Sanni, S. (2013). Environmental impacts of produced water and drilling waste discharges from the Norwegian offshore petroleum industry. *Marine Environmental Research* **92** (2013): 154–169.

45 Meier, S., Morton, H.C., Nyhammer, G. et al. (2010). Development of Atlantic cod (Gadus morhua) exposed to produced water during early life stages: effects on embryos, larvae, and juvenile fish. *Marine Environmental Research* **70**: 383–394. doi: 10.1016/j.marenvres.2010.08.002.

46 http://www.gesamp.org/ (last accessed 7 December 2017).

47 Tolls, J., Muller, M., Willing, A., and Steber, J. (2009). A new concept for the environmental risk assessment of poorly water soluble compounds and its application to consumer products. *Integrated Environmental Assessment and Management* **5** (3): 374–378. doi: 10.1897/IEAM_2008-067.1.

48 Cooke, R.M. (1991). *Experts In Uncertainty: Opinion and Subjective probability in Science.* Oxford: Academic Press. doi: 10.1111/j.1432-1033.1991.tb16379.x.

49 Aspinall, W. (2010). A route to more tractable expert advice. *Nature* **463**: 294–295. doi: 10.1038/463294a.

50 Rosqvist, T. and Tuominen, R. (1999). *Expert Judgement Models in Quantitative Risk Assessment.* Vienna, Austria: Atomic Energy Agency.

51 OSPAR Recommendation 2000/5 on a Harmonised Offshore Chemical Notification Format (HOCNF), http://rod.eionet.europa.eu/obligations/484 (last accessed 7 December 2017).

52 http://www.eosca.eu/wp-content/uploads/CHARM-User-Guide-Version-1.4.pdf (last accessed 7 December 2017).

53 Reed, M. and Rye, H. (2011). The DREAM model and the environmental impact factor: decision support for environmental risk management. In: *Produced Water*, 189–203.

54 Berger, P.D., Hsu, C., and Aredell, J.P. (1988). Designing and selecting demulsifiers for optimum filed performance on the basis of production fluid characteristics, SPE 16285. *SPE Production Engineering* **3** (6): 522.

55 Stais, F., Bohm, R., and Kupfer, R. (1991). Improved demulsifier chemistry: a novel approach in the dehydration of crude oil, SPE 18481. *SPE Production Engineering* **6** (3): 334.

56 Tyler, C.R., Jobling, S., and Sumpter, J.P. (1998). Endocrine Disruption in Wildlife: A Critical Review of the Evidence. *Critical reviews in Toxicology* **28** (4): 319–361.

57 Barber, L.B., Loyo-Rosales, J.E., Rice, C.P. et al. (2015). Endocrine disrupting alkylphenolic chemicals and other contaminants in wastewater treatment plant effluents, urban streams, and fish in the Great Lakes and Upper Mississippi River Regions. *Science of the Total Environment* **517**: 195–206. doi: 10.1016/j.scitotenv.2015.02.035.

58 Craddock, H.A. (2002). A review of the degradation and bioavailability of phenol formaldehyde condensation polymers, the so-called resins, as applied in the OSPAR region. *The European Oilfield Specialty Chemicals Association (EOSCA).*

59 Leber, A.P. (2001). Human exposures to monomers resulting from consumer contact with polymers. *Chemico-Biological Interactions* **135-136**: 215–220. doi: 10.1016/S0009-2797(01)00219-8.

60 McWilliams, P. and Payne, G. *Bioaccumulation Potential of Surfactants: A Review*. Manchester, UK: Chemistry in the Oil Industry VII, Royal Society of Chemistry, 13–14 November 2001. doi: 10.1377/hlthaff.2017.0814.

61 Craddock, H.A. (2003). *Environmental Pressures on the Use of Surfactants in the Offshore oil and Gas Industry*. Manchester, UK: Industrial Application of Surfactants V, Royal Society of Chemistry, 16–18 September. doi: 10.1179/cim.2003.4.4.161.

62 https://www.epa.gov/nutrientpollution/problem.

63 Sharpley, A.N., Smith, S.J., and Waney, J.N. (1987). Environmental impact of agricultural nitrogen and phosphorus use. *Journal of Agricultural and Food Chemistry* **35** (5): 812–817. doi: 10.1021/jf00077a043.

64 Semple, K.T., Doick, K.J., Jones, K.C. et al. (2004). Peer reviewed: defining bioavailability and bioaccessibility of contaminated soil and sediment is complicated. *Environmental Science & Technology* **38** (12): 228A–231A. doi: 10.1021/es040548w.

65 McKinney, J. and Rogers, R. (1992). ES&T metal bioavailability. *Environmental Science & Technology* **26** (7): 1298–1299. doi: 10.1021/es00031a603.

66 Haritonidis, S. and Malea, P. (1999). Bioaccumulation of metals by the green alga Ulva rigida from Thermaikos Gulf, Greece. *Environmental Pollution* **104** (3): 365–372. doi: 10.1016/S0269-7491(98)00192-4.

67 COMMISSION REGULATION (EU) No 1275/2013of 6 December 2013 amending Annex I to Directive 2002/32/EC of the European Parliament and of the Council.

68 http://www.ukmarinesac.org.uk (last accessed 7 December 2017).

69 Comments from the International Zinc association on the report (2012). Proposed EQS for water framework directive Annex VIII substances: zinc (for consultation) – WFD UKTAG.

70 Chapman, W.A., Rice, T.R., and Price, T.J. (1958). Uptake and accumulation of radioactive zinc by marine plankton, fish and shellfish. *Fishery Bulletin* **135**: (Fishery Bulletin of the Fish and Wildlife Service, Vol. 58).

71 Phillips, D.J.H. (1976). The common mussel *Mytilus edulis* as an indicator of pollution by zinc, cadmium, lead and copper. II. Relationship of metals in the mussel to those discharged by industry. *Marine Biology* **38** (1): 71–80. doi: 10.1007/BF00391487.

72 McGeer, J.C., Brix, K.V., Skeaff, J.M. et al. (2003). Inverse relationship between bioconcentration factor and exposure concentration for metals: implications for hazard assessment of metals in the aquatic environment. *Environmental Toxicology and Chemistry* **22**: 1017–1037. doi: 10.1002/etc.5620220509.

73 Colvin, V.L. (2003). The potential environmental impact of engineered nanomaterials. *Nature Biotechnology* **21**: 1166–1170. doi: 10.1038/nbt875.

74 Quintero, L., Lima, J.M. and Stocks-Fisher, S. (2000). SPE 59117, Silica micro-encapsulation technology for treatment of oil and/or hydrocarbon-contaminated drill cuttings. *IADC/SPE Drilling Conference*, New Orleans, Louisiana (23–25 February).

75 Alford, S.E. (1991). SPE 21936, North sea field application of an environmentally responsible water-base shale stabilizing system, *SPE/IADC Drilling Conference*, Amsterdam, Netherlands (11–14 March).

76 Lehmann, R.G., Millar, J.R., and Kozerski, G.E. (2000). Degradation of a silicone polymer in a field soil under natural conditions. *Chemosphere* **41**: 743–749. doi: 10.1016/S0045-6535(99)00430-0.

77 Smith, D.M., Lehmann, R.G., Narayan, R. et al. (1998). Fate and effects of silicone polymer during the composting process. *Compost Science and Utilization*, Spring **6** (2): 2–12. doi: 10.1080/1065657X.1998.10701916.

78 Stevens, C. (1998). Environmental degradation pathways for the breakdown of polydimethylsiloxanes. *Journal of Inorganic Biochemistry* **69** (3): 203–207. doi: 10.1016/S0162-0134(97)10019-8.

79 Frye, C.L. (1988). The environmental fate and ecological impact of organosilicon materials: a review. *The Science of the Total Environment* **73**: 17. doi: 10.1016/0048-9697(88)90182-9.

80 Pape, P.G., SPE 10089(1983). Silicones: unique chemicals for petroleum processing. *Journal of Petroleum Technology* **35** (06).

81 http://ozone.unep.org/en/treaties-and-decisions/montreal-protocol-substances-deplete-ozone-layer (last accessed 7 December 2017).

82 Capello, C., Fischer, U., and Hengerbuhler, K. (2007). What is a green solvent? A comprehensive framework for the environmental assessment of solvents. *Green Chemistry* (9): 927–934.

83 http://www.cefas.co.uk/media/1384/13-06e_plonor.pdf (last accessed 7 December 2017).

84 Description, Properties and degradation of Selected Volatile Organic Compounds Detected in Ground Water – A Review of Selected Literature, Open File Report 2006-1338, US Department of the Interior and US Geological Survey.

85 Houssain, M.E. and Al-Majed, A.A. (2015). *Fundamentals of Sustainable Drilling Engineering*, 1ee. Wiley-Scrivener.

86 Hanna, I.S. and Abukhamsin, S.A. (August 1998). Landfills and recycling provide alternatives for OBM disposal. *Oil and Gas Journal* .

87 Burke, C.J. and Veil, J.A. (November 1995). Synthetic-based drilling fluids have many environmental pluses. *Oil and Gas Journal* .

88 http://www.telegraph.co.uk/news/2017/06/01/trump-pull-paris-accord-seek-better-deal/ (last accessed 7 December 2017).

89 https://europa.eu/european-union/sites/europaeu/files/docs/body/treaty_on_european_union_en.pdf (last accessed 7 December 2017).

90 Vanderzwaag, D. (1999). The precautionary principle in environmental law and policy: elusive rhetoric and first embraces. *Journal of Environmental Law & Practice* **8**: 355–375.

91 Ministerial Declaration of the Third International Conference on the Protection of the North Sea, The Hague, 8th March, 1990.

92 http://www.cefas.co.uk/media/1383/adding-new-chems-to-plonor-list-5-dec-2010.pdf (last accessed 7 December 2017).

93 https://www.chevron.com/corporate-responsibility/environment (last accessed 7 December 2017).

94 http://www.bp.com/en/global/corporate/sustainability/environmental-impacts.html (last accessed 7 December 2017).

95 https://www.statoil.com/en/how-and-why/sustainability/environmental-impact.html (last accessed 7 December 2017).

96 http://www.petrobras.com.br/en/society-and-environment/environment/safety-environment-and-health-policies/ (last accessed 7 December 2017).

10

Sustainability and 'Green' Chemistry

This book has been concerned with the use of chemicals in the upstream oil and gas sector and their environmental impact. Until the <u>last</u> few decades there was little or no regulatory controls on their use and discharge and has been exemplified in the previous chapter, this has undergone significant change. In this concluding chapter, the sustainability of the practice of using chemicals in the oil and gas sector will be examined as well as how their impact on the environment may be further reduced, minimised and in some circumstances potentially eliminated. This discourse will also be placed within the context of the sustainability of the oil and gas sector.

10.1   Sustainability and Sustainable Development

Sustainability is something that has been ill defined in recent applications across a variety of industrial sectors and is often driven by regulatory requirements, such as packaging reuse and customer concerns on environmental impacts. The general concept has also been somewhat deviated to apply to overall corporate responsibility, providing the company or large corporation with a good public image. In the oil and gas industry, at a superficial level, how do the concepts of minimising environmental impact and reuse of materials balance with their undoubted impact on sustainability, from the removal and use of large amounts of carbon that have taken millennia to develop? This is an impossible circle to square as the timescales are vastly different.

However, many firms do state their desire to act in a sustainable way, for example, the energy giant Shell:

> Sustainability at Shell is about delivering energy in a responsible way to meet the world's growing needs [1].

Many of the chemical manufacturers that supply the industry have similar statements, for example, BASF, a major chemical corporation, views sustainability as a series of values and commitments to delivering a better future and using chemistry to do so [2]. In this they claim to use the following key processes:

- *Sourcing and producing responsibly*
- *Acting as a fair and reliable partner*
- *Connecting creative minds to find the best solutions for market needs*

Oilfield Chemistry and Its Environmental Impact, First Edition. Henry A. Craddock.
© 2018 John Wiley & Sons Ltd. Published 2018 by John Wiley & Sons Ltd.

The service companies also have similar statements; however in this many are aligned or directed at enhancing their customer's environmental performance, for example, Nalco states the following [3]:

> Growing customer interest in sustainable development is driven by both increased consumer demand for responsible business management and additional regulatory requirements. We are uniquely positioned to help customers improve their environmental performance, with nearly all of our offerings providing social benefits such as:

- Cleaner water and air
- Improved freshwater availability driven by reduction, recycling and reuse of water in production and other processes
- Reductions in greenhouse gases resulting from energy efficiency improvements
- Reduced solid waste generation and hazardous materials exposure.

Undoubtedly these are all laudable aims; however it could be concluded that such statements are more about enhancing corporate image than undertaking the hard road to sustainably managing a business and driving an industrial sector to a sustainable future.

There is much public and industry misunderstanding of what sustainability is and how it fits into the modern industrial world. It is not as some would have us believe a look back to some pre-industrial idyll where in a mythological agrarian society everything was in a balanced and harmonic status with nature. Sustainability sits at the heart of what has been described by some scientists, environmentalists, economists and others as the 'trilemma' [4], that is, seeking ways forward for society, the economy and the environment that are equitable, acceptable and practical and doing so simultaneously. Resolving this does not only involve industry but also reconciling the needs of societies against those of economic development and a rapidly expanding world population. The pressure on resources is becoming ever greater, and the environmental damage caused may already be in some cases irreversible.

Sustainability and sustainable development are often used interchangeably; however, they are different things. Sustainability is an abstract concept of an ideal state of existence in which mankind and nature are in balance. This has been nicely defined by Roland Clift [5] as being

> 'a state of existence, in which humanities techno-economic skills are deployed, within long-term ecological constraints imposed by the planet, to provide resources and absorb emissions and to provide the welfare on which human society relies for an acceptable quality of life'.

Sustainable development, however, is the methods and means we employ to move towards this ideal. There are hundreds of definitions of sustainable development, but none of which is universally agreed – perhaps one of the most famous, robust and objective is in a statement made by Angela Merkel in 1998: [6]

> '...using resources, no faster than they can regenerate themselves and release pollutants to no greater extent than natural resources can assimilate them'.

However, many definitions are at best imprecise and unclear and do little to help policymakers and other decision makers in guiding them towards more sustainable technologies.

In parts of the energy sector, the problem is further compounded by the geological timescales required to renew and replenish hydrocarbon fuel stocks and the polluting effects that their combustion to release their stored energy has had. Of course, it is easily argued that it is best to abandon this energy and feedstock source. Most commentators and others would agree that we must aim to progress society in a sustainable fashion, and some parts of this development are obvious:

- Reduce use of non-renewable resources.
- Reduce emissions of wastes and pollutants.

Both of these criteria are of critical significance to the oil and gas industry, and it should be noted that the technological challenge here is to reduce use rather than abate emissions. The reduction in use of hydrocarbons is of critical importance at least its reduction in use as a fuel as the current known reserves of hydrocarbons represent many times the quantities of carbon dioxide emissions that can be tolerated according to any of the scenarios examined by the Intergovernmental Panel on Climate Change [7]. However, over the last two to three decades, the present economic systems seems to be incapable of dealing with a constraint on the emission side.

The other aspects of sustainable development are even more challenging, such as bringing human expectations into compatibility with the planet's carrying capacity and breaking the connection between material consumption and quality of life. These are more difficult because they are not essentially scientific or technological problems; they are ethical issues. How does the oil and gas industry fit in this trilemma, and significantly how do we reduce consumption of hydrocarbons with suitable renewable resources to ensure *an acceptable quality of life* at least in the short to medium term?

10.2   Sustainable Development in the Oil and Gas Sector

As has been described in the previous section, oil and gas companies and their suppliers are conscious of the need for developing their business in a sustainable fashion, and ultimately as society dictates, this may require a move away from the extraction of hydrocarbons as a non-renewable fuel source. The author would argue that this would mean a greater emphasis on the use of hydrocarbon resources in a beneficial and renewable fashion and that the use of hydrocarbon feedstocks for chemicals, polymers, plastics and other materials will remain into the foreseeable future. This will primarily be driven by two critical factors:

- The economics of using renewable, mainly plant resources as chemical and other hydrocarbon replacement feedstocks
- The competing needs of land use, primarily the growing need for the use of land for food supply with an increasing population

To date the only viable commercial source of biofuel is ethanol from fermentation of sugar crops. A few years ago, increasing fuel prices, concern about climate change and future energy security led to tremendous global interest in the use of liquid biofuels in the transport sector that, in turn, has driven large-scale land acquisitions in developing countries for biofuel feedstock production. However, regardless of the vast nature of reported land deals and widespread concern about their potential negative consequences, implementation of most of the reported biofuel land deals has not yet happened. This was particularly the case for marginal land where the agricultural and economic viability of food plants was questionable. However, a reversion in the price of crude oil and other factors such as the use of untested planting material and conflict with local communities over the land use were important factors that contributed to termination of many biofuel projects [8].

Similarly, in the United States a number of biofuel companies have gone bankrupt due to changing economic conditions. Also, there seems to have been a lack of foresight and care in the investment strategy for biofuel recovery, and this is due to two factors in the attempts of biofuel to replace hydrocarbon-based fuels:

1) It is extremely difficult to achieve the yields necessary to be profitable.
2) Even if the required yield can be achieved, there is not enough feedstock to permit the production of enough biofuel to replace a significant amount of oil.

This latter reason is neatly exemplified by a biofuel company that claimed to achieve a yield of 91 gal per tonne of wood from pine trees. Unfortunately, there are not enough pine trees throughout the United States to replace the jet fuel used by the world's airlines. And jet fuel represents only 6% of the oil used by the world. The result has been millions, if not billions, of wasted investment in this sector [9]. Much of the investment has been funded by public subsidies, of which companies have only paid back a fraction before filing for bankruptcy [10].

This is a salient lesson in the assumption that in the given economic climate, sustainable development can somehow be rapidly implemented and renewable fuels will solve transportation requirements. Undoubtedly renewable energy sources are already making significant contributions to energy requirements, and the use of hydrocarbons, particularly oil as a source of electricity, is in decline in many developed countries and rapidly being examined in many others. The Paris Climate Agreement [11] calls for a global reduction in emissions in order to keep global warming within 2 °C above pre-industrial levels. A key part of achieving this goal will be a shift away from fossil fuels towards renewable energy sources, and in recent years, record levels of investment in green energy have occurred [12]. Indeed, it is entirely possible that many countries will have no dependence on fossil fuels to generate electricity within a generation. Some countries already do, such as Norway, where over 99% of the electricity production on the mainland is from hydropower plants, and others such as Iceland have only a small component from fossil fuels (15%), and this is directed at fuels for transportation. In Iceland geothermal energy provides about 65% of primary energy, and the remaining 20% is from hydropower.

It is significant, however, that transportation is still reliant on hydrocarbon fuels. The alternatives are either an alternative fuel such as biofuel, which has already been described, or fuelled by electric batteries. In writing this chapter electric cars have been the topic of much debate regarding the vehicles of the future. A number of car manufacturers have announced they will be only making electric or hybrid (electric/petrol) cars in the near future [13]. In 2015 the US EPA estimated that 27% of greenhouse gas emissions originated from transportation, primarily from burning fossil fuel for our cars, trucks, ships, trains and planes, and that over 90% of the fuel used for transportation was petroleum based, which includes gasoline and diesel [14]. Clearly the use of electric vehicles where the electricity is generated sustainably can have a significant effect in reducing these emissions. However, as with biofuels, caution is required as there are still many significant challenges in the change from internal combustion engine and its convenience, familiarity and range compared with a battery-powered vehicle needing regular charging, lack of availability of charging facilities and easy of travel across a several hundred mile range.

Of perhaps greater concern is our growing dependence on the use of highly complex and powerful technology fuelled by batteries. Most of these batteries in this technology from mobile phones to portable tablet computers require metals and other valuable minerals. Cobalt is particularly important in the battery that fuels these devices, and more than half of the world's cobalt is extracted in the Democratic republic of Congo (DRC) where it is estimated around 40 000 children, some as young as seven, work in unregulated mines [15]. Mobile phones and other devices have become indispensable; however, in truly sustainable economy, can we dispense with the rights of the men, women and children whose labour powers our phones?

All of this is to state that in the immediate future of the next two decades, the oil and gas sector is unlikely to dramatically change, although there will be pressures on its use in the generation of electricity particularly from crude oil and on its change as a primary fuel for transportation where more hybrid fuelled cars are likely to prevail and diesel-powered vehicles are likely to be phased out [16].

10.2.1 The Case for Shale Gas

Natural gas is an efficient energy source and the cleanest burning fossil fuel. The natural gas extracted from shale rock could become a significant global energy source [17]. In 2015 the US Energy Information Administration (EIA) estimated that the global recoverable reserves of shale gas were over 7500 trillion cubic feet [18]. Although the energy industry has long known about huge gas resources trapped in shale rock formations, particularly in the United States, it is only over the last two decades or so that energy companies have combined two established technologies – hydraulic fracturing and horizontal drilling – to unlock this resource.

As developing and developed nations move from traditional fossil fuel sources to other means of power generation such as renewable and new nuclear plants to replace aging ones, shale gas could provide a significant supply where power demands are stretched and potential power deficiencies could occur through lack of power generation capacity. For some time, commentators, engineers and scientists have warned a number of nations, particularly in Europe, of a significant energy gap arising in the near future [19]. In addition world energy markets continue to be vulnerable to disruptions precipitated by events ranging from geopolitical strife to natural disasters. The Ukraine–Russia gas dispute in January 2009 caused the largest natural gas supply crisis in Europe's history. With increasingly integrated electricity grids, blackouts can cascade and affect multiple economies simultaneously. Therefore, energy security is becoming a strategically important political issue, and here also shale gas has a role to play. Nations with shale gas reserves can be become less reliant on the big energy power brokers such as Saudi Arabia and Russia for supplies.

The United States has led the way in that shale gas extraction and development has allowed it for the first time in many decades to be an exporter of oil [20], much of it to its near neighbour Mexico. It has also allowed it to cut its consumption of oil for power generation, thereby reducing its overall emissions. Some of this may not necessarily be a good thing by just shifting the environmental burden of less efficient and less clean crude oil to another part of the world. Nonetheless the growth of the shale gas industry in the United States has raised expectations that other nations could boost domestic gas production, leading to lower energy prices and improved energy security. However, the degree to which the US experience is transferable to other countries is uncertain. The literature suggests that it is possible to develop shale gas in a sustainable way, but its future will depend on the industry being able to address the environmental concerns, the political will to see the industry through to maturity and public support, with the latter most likely being the biggest determinant [21].

The public's main concern appears to arise over concerns that the chemicals used in the hydraulic fracturing operations will contaminate the water course and water supply from the drilling and hydraulic fracturing process used in the extraction of the shale gas. The main use of chemicals in any volume is in the hydraulic fracturing operation, and it is worth giving more detail on this.

10.2.1.1 Hydraulic Fracturing

Generally, the process of hydraulic fracturing involves pumping a viscous fluid, or pad, into a well faster than the fluid can escape into the formation. This causes the pressure to rise and the rock to break, creating artificial fractures or enlarging existing ones. A hydraulic fracture is a structure that is superimposed on the natural fracture, leaving it undisturbed. As a result, the effective permeability of the reservoir remains unchanged; however, the wellbore radius is increased. This leads to increased productivity because of the larger surface area created between the well and the reservoir.

Hydraulic fracturing is a relatively new technique in terms of petroleum science, having been introduced just over 50 years ago [22]. Classically, the fractures produced are approximately perpendicular to the axis of least stress, and for most deep reservoirs the minimal stresses are horizontal; hence, the fractures occur in the vertical plane.

The actual stresses can be calculated by balancing the vertical (geostatic) stresses with the horizontal stress. Taking into account a number of factors, the horizontal stress can be calculated from the corrected vertical stress. In some circumstances, in shallow reservoirs in particular, horizontal stresses can be created as well as vertical stresses.

Knowledge of the stresses in a reservoir is essential to establish the pressure at which the initiation of a fracture can take place. The upper limit of this pressure usually can be calculated from the formula [23] shown in Eq. (10.1).

Von Terzaghi's consolidation equation:

$$Pb = 3s_{H,min} - s_{H,max} + T - p \tag{10.1}$$

The pressure response during fracturing provides important information about the success of the operation. The efficiency of the fracturing fluid can be estimated from the closure time. The closure pressure is the pressure at which the width of the fracture becomes zero; this is normally the minimal horizontal stress. All this information is important in the design and selection of the fracturing fluid and its chemical components.

10.2.1.2 Fracturing Fluids

The use of fracturing fluids is critical in the stimulation of productivity by hydraulic fracturing. These fluids provide the means to produce the hydrostatic pressure required to create the hydraulic fracturing. In creating a hydraulic fracture, the pumped water has to be treated to increase its viscosity; this is done through the addition of viscosifiers or gelling agents.

After the formation has been fractured, a proppant, mainly sand, is added to the pumping fluid. This forms a slurry that is used to prevent the newly formed fractures from closing when the pumping (hydrostatic) pressure is released. The transportability of this propping agent is dependent on which viscosifying agents have been added to the water or base fluid [24].

The composition of the fracturing fluid therefore consists of a base fluid, usually water; a proppant, usually sand; a viscosifying agent; and some chemical additives. The total number of chemical additives, including the viscosifying agent, used is less than 0.5% of the volume of fluid [25].

The fracturing fluids are injected into the formation for the following critical reasons:

1) To create a conductive path from the wellbore to the formation
2) To carry proppant into the fracture to create a conductive path for the fluids produced

The main fracturing fluid categories are as follows:

- Gelled fluids, including linear or cross-linked gels
- Foamed gels

- Plain water and potassium chloride (KCl) water
- Acids
- Combination treatments (any combination of two or more of the fluids mentioned earlier)

The fluids used most commonly in shale gas formations are the gelled fluids, and it is in the main to these gelled fluids that the chemical additives are formulated.

10.2.1.3 Chemical Additives

In general, the chemical additives in the table shown in Figure 10.1 with the relative amounts used that are added to the fracturing fluid.

The amounts of chemical are relatively small; however, the amounts of fluid are large – 7–14 million litres in a typical fracturing operation. Although a number of chemical additives can be used, any single job would only use a few of these. Of the 12 additives commonly used as shown in the table in Figure 10.1, anything from 3 to all 12 of these could be used in a specific fracturing fluid. The chemistry of all of these types of additives has been considered in the previous chapters and some of these, the more important, and therefore those potentially with the greatest environmental impact, are now discussed.

10.2.1.3.1 *Viscosifiers and Cross-Linking Agents*   The majority of fracturing treatments conducted have used fluids that have been viscosified by guar gums or guar derivatives, such as hydroxypropyl guar (HPG).

Guar gum is a branched polysaccharide composed of the sugars mannose and galactose in the ratio 2 : 1 [26] (see Figure 10.2).

Guar gums and related derivatives are not self-gelling and require a cross-linking agent to be added to cause it to gel in water. In general, these cross-linkers (such as borates) are environmentally unsuitable. Also, they do not impart high enough temperature stability to the gel to provide rheological and fluid loss control and fracture conductivity properties in fluid temperatures higher than 105 °C. However, guar gum is very economical because it has almost eight times the water-thickening potency of similar materials, and, therefore, only a very small quantity is needed for producing sufficient viscosity. Also, guar gum is a direct food additive and is registered for use and discharge in the North East Atlantic, including the North Sea. It is highly biodegradable and is recognised as posing no environmental or toxicological problems [27].

Type of chemical additive	Percentage of fracturing fluid
Gelling agent	0.05–0.06
Cross-linker	<0.01
Surfactant	0.08–0.09
Potassium chloride	<0.08
Scale inhibitor	0.04–0.005
Gel breaker	<0.02
pH adjusting agent	0.01–0.02
Iron control agent	<0.01
Corrosion inhibitor	<0.01
Biocide	<0.001
Acid	0.12–0.13
Friction reducer	0.08–0.09

Figure 10.1 Table of chemical additives.

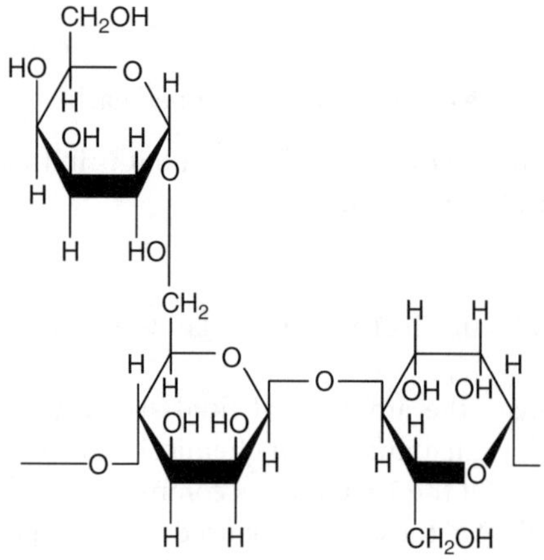

Figure 10.2 The structure of guar gum.

Borates are the most common fracturing fluid cross-linkers and can be formed by boric acid, borax, an alkaline earth metal borate or an alkali metal alkaline earth metal borate. It is essential that the borate source has around 30% boric acid. The boric acid forms a complex with the hydroxyl units of the guar gum polysaccharide, cross-linking the polymer units. This process lowers the pH (hence the need for pH control) [28]. These fluids provide excellent rheological, fluid loss control and fracture conductivity properties in fluid temperatures up to 105 °C [29]. To allow use at higher temperatures, the use of magnesium oxide and magnesium fluoride has been developed. This effectively extends the range of these materials up to 150 °C [30].

As has been described in Chapter 5, the environmental fate and toxicology of boron cross-linkers is more complex than food-grade polysaccharides. Boric acid naturally occurs in air, water (surface water and groundwater), soil and plants, including food crops. It enters the environment through weathering of rocks, volatilisation from seawater and volcanic activity [31]. Most boron compounds convert to boric acid in the environment, and the relatively high water solubility of boric acid results in the chemical reaching aquatic environments. Boric acid is therefore the boron compound of environmental significance [32].

It is assumed that boric acid is adsorbed to soil particles and aluminium and iron minerals and that this adsorption can be either reversible or irreversible, depending on soil characteristics. It is known that boric acid is mobile in soil [33]. The US EPA does not anticipate adverse effects to birds from the use patterns of boric acid products.

10.2.1.3.2 Surfactants The surfactants used in fracturing fluids, such as cetylammonium bromide – a long-chain quaternary ammonium salt – are used to generate viscoelasticity. They congregate into micelles, which interact to form a network imparting viscous and elastic properties to the formulated fluid. Surfactants are included in most aqueous fracturing fluids to improve compatibility with the hydrocarbon reservoir. It is also important that the formation rock is water wet in order to achieve the maximum conductivity of hydrocarbon gas or fluids.

As has been discussed in some detail in Section 3.9, the environmental fate of surfactants can be a complex issue, and some careful consideration is required as to which surfactants to use in order to ensure the best technical and environmental outcome. However, the use of oleochemicals based on naturally derived fatty alcohols is well documented [34]. These molecules being naturally derived materials have good ecotoxicological profiles [35].

10.2.1.3.3 Biocides Where the fracturing fluid is composed of guar gum or other natural polymers, a small amount of biocide is included in the formulation, preventing any undesired degradation and changes in rheological properties. A number of biocides are utilised for this function. All are commonly used in the oil and gas industry and are under strong regulatory control. By their nature, they are highly toxic to all sorts of aquatic organisms; however their concentration in the chemical mix is very small, and many are short lived in action and highly biodegradable.

This all adds up to a case for shale gas to be considered a genuine replacement for higher emissions bearing fuels, and it should not be ignored as a means to both prove an efficient energy stopgap, which reduces discharges to the environment and allows the developed economies to move to a lower carbon, more renewable energy provision, without severely compromising both the environment and standards of living.

The oil and gas sector has yet to fully embrace the reality of a sustainable future and its role in such an economy particularly where more and more countries are driving towards a low carbon, renewable energy-based economy. This being stated in the remainder of this chapter, a number of issues are explored, which, if embraced by the industry and policymakers and supported by the public, will ensure a role for the sector long into the future.

10.2.2   Recovery and Reuse

In operations management, new challenges are being encountered in integrating issues of sustainability with their traditional areas of interest. During the past 30 years, there has been growing pressure on businesses to pay more attention to the environmental and resource consequences of the products and services they offer and the processes they deploy. One symptom of this pressure is the movement towards triple bottom line (3BL) reporting concerning the relationship of profit, people and the planet. The resulting challenges include integrating environmental, health and safety concerns with green product design, lean and green operations and closed-loop supply chains [36]. In achieving a sustainable business model, the issues of recovery and reuse of materials is a considerable challenge, none more so than in the upstream oil and gas sector. Here the practice for many decades has been to drill, dump and move on. However, since the Brent Spar [37] in which Shell changed its mind on 'dumping' at the 11th hour driven by pressure from the NGO Greenpeace, public and political pressure has forced oil and gas companies to re-examine the need for recovery and reuse both from an economic and from environmental perspective.

10.2.2.1   The Brent Spar
The Brent Spar was a North Sea oil storage and tanker loading buoy, 147 m in height and 27 m in diameter, and it displaced 60 000 tonnes of water. It was located in the Brent oilfield in the UK North Sea and was operated by Shell UK. In 1991, it was considered to be of no further value. Shell UK took over 3 years to evaluate disposal options; it complied with national and international regulations, holding consultations with 'representative' environmental and fishing organisations in the United Kingdom. Careful analysis and planning went into developing (as required by UK law) the best practicable environmental option (BPEO), which resulted in scientists and specialists agreeing that disposing of the Brent Spar in the deep ocean would have negligible environmental effects and would thus be the BPEO.

The site selected 'North Feni Ridge' was within UK waters and the UK government issued a licence for disposal. The disposal operation was cleared through the Oslo and Paris Commissions (OSPAR) to secure international agreement.

After its change of mind, Shell towed the Brent Spar to Norway and moored her in a fjord, where she remained until 1998. In July 1998 all the governments of the North East Atlantic region agreed to ban future dumping of steel-built oil installations. In November of the same year, the company began dismantling the structure. The Brent Spar was broken into five sections in an operation costing £43m compared with £4.5m cost of dumping the structure under the sea. The scrap was eventually used to build the foundations of a new ferry terminal.

Whether or not recycling Brent Spar was the right decision both environmentally and economically, it certainly could be argued that it was a more sustainable outcome. Also, the international public outcry led to a change of management and a revamp of Shell's ethical standards with repercussions across the oil and gas industry.

10.2.2.2 Facilities Decommissioning

The decommissioning of facilities and infrastructure at the end of field life particularly in offshore locations has now become part of the life cycle planning in development of an oilfield [38]. Case studies are growing particularly in areas where decommissioning is becoming established as mature oilfields near end of economic life such as in the UK North Sea. Indeed, for almost 20 years, decommissioning projects have been established and concluded [39].

In general, the chemicals used in decommissioning activities are not considered as an important challenge as far as either technical requirements or environmental acceptability is concerned. A majority of products used are selected from the PLONOR List (see Section 9.1.3). However, the application of chemical additives in the decommissioning project is vital to provide safe and environmental acceptable treatment solutions of chemical and hydrocarbon waste products found in storage on the facility [40]. Furthermore, chemicals are often employed in the following end of field life activities, well plugging and abandonment [41], waste treatment, tank cleaning, oil displacement, pigging and pipeline decommissioning [42].

Consequently, important work has to be conducted to obtain safe and efficient products to facilitate cleaning operations and to limit the volumes of polluted water. The main constraint is to determine detergent formulations that are non-toxic and biodegradable. It is also desirable that formulations are compatible with cold seawater and must allow a fast separation of hydrocarbons from the aqueous phase. A methodology for selecting such products has been proposed, which is firstly based on a preselection of safe molecules coming from a green chemistry, e.g. alkyl polyglycosides, and secondly to test formulations based on mixtures of several efficient and safe active materials on different supports (glass, steel, cement, etc.) in order to assess their detergent efficiency against several types of crude oils and weathered crude oils [43]. In addition other testing is conducted on the following:

- The kinetics of separation of the formed emulsion
- The formation of foam and the quality of the separated aqueous phase

Ecotoxicity tests on selected molecules are performed according to OSPARCOM guidelines (aquatic toxicity, biodegradation and bioaccumulation), and complementary toxicity tests are also carried out on the separated water after cleaning in order to estimate the toxicity of the aqueous phase to be discharged.

The main objective is to develop green formulations for cleaning oil-polluted surfaces through a rigorous methodology based on technical and environmental criteria. Therefore, as can be seen the development and selection of a suitable formulated product is far from a trivial matter.

This is particularly the case for biologically contaminated waste materials such as contaminated hydrocarbon waste that often contains hydrogen sulphide and mercaptans, and the decontamination regimen also has to have a treatment step to scavenge or remove the sulphide contaminants [44].

Recent decommissioning projects, particularly in the UK North Sea, have claimed that over 90% of materials originally used are being recovered and reused [45].

10.2.2.3 Oil-Based Muds and Cuttings

Some of the environmental impact issues involving oil-based muds (OBMs), drill cuttings and their replacement alternatives have been described previously in Chapters 8 and 9. Section 8.4.1, in particular, describes synthetic and natural product alternatives to the oil component in OBMs. However, the recycling and reuse of materials in drill cuttings and in particular those contaminated with oil seems to have gained acceptance with both oil and gas operators and regulators [46]. This has meant that the uptake of synthetic-based muds (SBMs) and other environmentally acceptable materials has had a variable uptake by operators. Also, in land-based drill sites, often cuttings are permitted to be removed to an approved land-based site, which is merely shifting the environmental burden.

OBMs are probably the most effective drilling fluids, and although most of the OBM is removed from cuttings on the rig, some adheres to the cuttings and can be discharged. OBM-coated cuttings do not disperse readily when discharged, resulting in accumulations in the offshore environment of cuttings and OBM beneath installations, so-called cuttings piles. Many wells are drilled with water-based mud (WBM), and although some of the WBM is discharged with cuttings, it readily disperses in the aquatic environment and tends not to form cuttings piles. There is, however, the potential for these cuttings to contain oil from the reservoir section of an oil well.

In the North Sea, for example, cuttings piles started accumulating at the feet of installations, and oiled cuttings have identified as having the potential for long-term environmental impact on seabed fauna [47].

Historically, OBM was formulated with diesel oil. As the potential environmental implications of the discharge of OBMs became apparent in Europe, North America and other regions, the industry in collaboration with regulators introduced a voluntary reduction programme that led ultimately to regulations that effectively banned the discharge of OBM contaminated cuttings in 1996. Subsequently, synthetic or non-mineral base oils were developed, but issues with biodegradability were identified, and these products were phased in between January 1997 and January 2001 [48, 49]. Alpha-olefins had a major use in the formulation of synthetic-based drilling fluids. Both linear alpha-olefins and polyalphaolefins were used in the formulation of SBMs [50].

OSPAR Decision 2000/3 came into effect on 16 January 2001 and effectively eliminated the discharge of cuttings contaminated with oil-based fluids (OBFs) (includes OBM and SBM) greater than 1% by weight on dry cuttings [51]. The Offshore Chemical Regulations 2002 implement this decision and require a chemical permit for the use and discharge of chemicals including drilling muds [52]. The Offshore Petroleum Activities (Oil Pollution Prevention and Control) Regulations 2005 also require a permit for the discharge or reinjection of cuttings containing hydrocarbons from the reservoir [52].

OSPAR agreed Recommendation 2006/5 on a Management Regime for Offshore Cuttings Piles, which required all cuttings piles to be assessed against set criteria to determine if any were of immediate environmental concern. This assessment was completed in 2009, showing that existing piles were not of immediate concern and that appropriate management strategies for individual piles could be determined at the time of decommissioning of the installation [53].

In regard to cuttings piles, most findings are in broad agreement, suggesting that the major seabed effects associated with oiled cuttings discharges were smothering and subsequent development of anaerobic conditions due to the microbial degradation of the base oil within a close range of the installation. The impact is therefore expected to be localised and to gradually reduce over time thanks to biodegradation processes occurring in and around cutting piles. The conclusion of the Drill Cuttings Initiative suggests that cuttings piles on the UKCS do not pose an environmental threat that requires immediate remedial action. A range of potential management options for cuttings piles, ranging from removal to leaving in place, were identified, and that the best option would be decided on a case-by-case basis following detailed assessment at the time of decommissioning of the installation [53].

Currently, only water-based cuttings are discharged to sea. These materials are relatively environmentally inert; therefore the main mechanism for seabed effects would be via the physical smothering of seabed organisms in areas of high deposition. The long-term environmental toxicity and/or food chain effects associated with these materials are deemed to be negligible [54].

The overall result of this history is that OBMs are currently used and all cuttings are collected and shipped to shore. The base oil is recovered and reused and the cuttings are reused where not radioactive. Currently in the North Sea, base oils are now traditional enhanced mineral oils and to a lesser extent synthetic paraffins.

Despite a weight of evidence that shows LAOs and PAOs on drill cuttings are non-toxic, biodegrade and do not bioaccummulate [55, 56], the regulated route is to ship cuttings to shore and recover and reuse materials. It is a matter of debate as to whether this is the best environmental option as a full life cycle analysis of the recovery processes needs to be conducted to give a weighted assessment. This would allow a comparison to using a biodegradable synthetic fluid that is discharge to the marine environment.

10.2.2.4   Produced Water Reuse and Reinjection

Produced water is a term used in the oil and gas industry to describe water that is produced as a by-product in the extraction and recovery of oil and gas. Oil and gas reservoirs often have water as well as hydrocarbons, sometimes in a zone that lies under the hydrocarbons and sometimes in the same zone with the oil and gas. Oil wells can produce large volumes of water along with the oil being extracted, while gas wells tend to produce water in smaller proportions.

To achieve maximum oil recovery, water flooding is often implemented, in which water is injected into the reservoirs to help force the oil to the production wells. The injected water eventually reaches the production wells, and so in the later stages of water flooding, the produced water proportion ('cut') of the total production increases.

In recent times produced water recycling and reuse has been critically examined. This is particularly in relation to shale gas extraction, described in Section 10.2.1, where the hydraulic fracturing process uses large volumes of water and the operations are often located in desert or arid locations. The management and reuse of so-called 'frac' water has been discussed in Section 8.3.1 as well as the management of produced water in general. The recycling and reuse of produced water is a major specialist discipline within the upstream oil and gas sector. It is of particular concern in a number of regions where mature fields often produce considerably more water than hydrocarbons. This associated production water from usual production streams can be recycled through reinjection into the reservoir or into a disposal well.

Produced water reinjection (PWRI) has been explored and developed for more than two decades [57]. It is in the nature of many production systems that the costs of produced water handling increase significantly as the oil field matures, which normally coincides with the decline in revenues. To some extent PWRI has been driven by environmental concern for the effects of produced water discharges; however PWRI can also lead to improved performance at a time when profit margins are small. There is often significant potential for making cost, space and weight savings through optimisation of water treatment facilities and PWRI system during the life of a field.

The reinjection of produced water has been carried out on several locations around the world, and in most cases the activities have been concentrated on individual wells and have not included mixing the produced water with seawater prior to injection. In a significant number of cases, some loss in injectivity has been observed, in which in some cases problems have been more severe and accelerated reservoir souring and increased scaling have also been noted [57].

In the last decade, a number of key factors have been established in framing a produced water management strategy including a company's internal and external environments, technology and business drivers [58].

Best practices resulting both from comprehensive assessments of current produced water management tools and from the insights obtained from a decade-long joint industry project (JIP) on PWRI have been established, which have included lessons learned for injector design, operation, monitoring and assessment and intervention that provide the basis for cost minimisation and good environmental practice. Despite using clean water however, field evidence has indicated that injectivity performance is decreased in matrix injection schemes. Alternatively, injectivity maintenance using untreated produced water is feasible [58].

Although there may be various drivers to implement PWRI, reinjecting produced water from the same field cannot replace the voidage created by production, especially early in the life of the field, since most of that voidage is created by hydrocarbon extraction. Thus, other water such as seawater may have to be considered to 'top up' PWRI. This raises the question of what are the implications for scale control of mixing potentially incompatible brines before injection compared with the conventional injection scenario where the mixing takes place in the reservoir. It has been shown that the scale risk at the producers is much lower than if only seawater had been injected and that any residual scale risk at the producer wells could be managed by bullhead squeezing. However, the corollary is that the scale risk at the injectors is much higher. The trigger for scale precipitation in this scenario is brine mixing, but instead of that happening in the reservoir, here it occurs before injection. Thus, the location of greatest scale risk is moved much further upstream in the flow process [59].

Produced water management is already a major concern in many mature reservoirs both from the economics and from the environmental impact; however it is important that regulators, policymakers and other interested parties understand the balance in economics between using PWRI and making field operations uneconomic. This is particularly prevalent when crude oil prices are low as mature fields may be uneconomic, and therefore sustainability as issue is somewhat academic.

Produced water reuse is becoming more imperative and, as has been described, driven by economics and regulation. The demands for the fresh water used in many hydraulic fracturing operations are placing pressure on water sources in some regions of the United States. Because of the high volumes of water needed for fracturing (e.g. in the Marcellus, a typical hydraulic fracturing operation for a horizontal gas well in a tight shale formation requires from 3 to 5 million gallons of water over a 2- to 5-day period), the competing demand driven by industrial, municipal and agricultural users has in some cases decreased the availability of fresh water and increased associated costs. In such areas, the importance of produced water recycling cannot be understated [60].

PWRI offers the intriguing possibility of the potential reuse of some chemical additives and/or reduction in their use by a topping-up mechanism. To date there is no substantial evidence that this is feasible; indeed there is evidence that chemical interactions can cause later production problems. Water injection tests suggest that it is possible for the injector wells to take more water due to the interaction of the corrosion inhibitors. The corrosion inhibitor solution when applied to cores that had been solvent cleaned showed stimulation (increased brine permeability), suggesting that the chemical interacted with mineral surfaces rather than improved brine permeability due to change in residual oil saturation within the pore space of the rock. However, the production well data was found to be in marked contrast to the water injector results, and the change in wettability in both carbonate and sandstone observed with the introduction of the corrosion inhibitors tested would suggest a reduction in oil production and an increase in water production due to the change in relative permeability observed with corrosion inhibitor adsorption onto the mineral substrate (both silicates and carbonate) [61]. Similar observations have been made with respect to kinetic hydrate inhibitors (KHIs) with potential formation damage being observed in carbonate rocks [62].

As has been described in Section 5.9.4, 'Nitric Acid and Nitrate Salts', nitrate and nitrite salts can be effective inhibitors of sulphate-reducing bacteria activity as well as an enhancing agent for biocides. However, it has been observed that increased rates of corrosion may occur [63]. Further work [64] has shown that duplex stainless steel did not show any enhanced corrosion in an inert or CO_2 environment in the presence of nitrite. In the case of carbon steel, nitrite displayed an inhibiting role when exposed to 1 bar CO_2 at 80 °C. Increases of chlorides or sulphates had only a small effect on increasing corrosion rates in the presence of nitrite and CO_2. However, in the presence of an inert gas, increases in nitrite concentration were observed to significantly increase corrosion rates of carbon steel. The addition of small amounts of sulphide in the presence of nitrite resulted in a small increase in corrosion rates of carbon steel as did the impact of 25 Pa of sheer stress. The presence of nitrite interfered with the corrosion inhibiting activity of a commercial inhibitor in the presence of CO_2. These results are supportive of some earlier work [65] where the use of nitrate in PWRI systems has sometimes been associated with increased general and localised corrosion.

These results would certainly suggest that the idea of reuse of the chemical mixture in produced water, particularly corrosion inhibitor, which intuitively would be the best reuse candidate, may not be a viable concept.

It has been proposed that produced water may however be a feedstock for some basic chemical compounds [66]. Produced water is highly saline, and it is postulated that it may be used for soda ash (sodium carbonate) production. This process, in use since 1861, is also known as Solvay process, of which the main component for the industrial manufacture of soda ash is brine. It is proposed that produced water will undergo a pretreatment step designed for removing metals such as calcium, magnesium and iron. After this procedure, produced water will be evaporated until it reaches the brine concentration for Solvay process. This operation will also produce high purity water, suitable for many purposes. To date only some laboratory-scale experiments have been reported.

Considering the environmental benefits obtained from this process, both in wastewater treatment and in carbon sequestration, as well as the production of a valuable commodity, applying Solvay process for treating produced water could be an alternative and promising technology for the future.

The recycling and reuse of produced water, particularly onshore, has been evaluated for other purposes other than PWRI, in particular for other industrial purposes and even drinking water, for some time [67]. However to date little progress has been established except for some pilot studies [68], and it is believed that economic factors, particularly instability in the price of crude oil, are significant factors in promoting the development of this recycling and reuse option.

10.2.2.5 Carbon Capture

Much has been written about the need to capture carbon dioxide (CO_2) from industrial plants, especially coal power plants [69]. The primary driver for this growing attention has been concerns of accelerating accumulation of greenhouse gases in the atmosphere. As this awareness grew, considerable funding and research time has been devoted to the following:

1) Developing the technologies for capture
2) Investigating the subsurface reservoirs in which to permanently place the CO_2

Recently there has been a reassessment of the economics surrounding carbon capture and storage (CCS), and it has been proposed that it will be significantly more expensive than previously thought because previous studies miscalculated the energy required. As such, it is unlikely to provide an economically viable solution to CO_2 pollution from coal-based power generators [70].

The process of carbon capture is inherently simple but challenging to engineer and deploy on an industrial scale and needs economic support from governments. In the oil and gas industry, where this has been economic and therefore implemented, such as in Salah, hidden deep in the Sahara Desert some 700 miles south of Algiers [71]. The field's gas is about 7% CO_2, which must be cut to 2% or less before it can be shipped on to European markets. In Salah, the CO_2 level is cut to 0.3%, but instead of simply venting the removed CO_2, it is pumped into an aquifer below the gas reservoir. Given the scale of the gas flow, it is the environmental equivalent of taking 200 000 cars off the road. This however is a relatively rare example of a wholly sustainable implementation and therefore requires governments to support the technology for the undoubted environmental benefits. Carbon capture is a viable technology already, though without subsidies it has remained too expensive for widespread implementation. About 30 million tonnes of CO_2 is captured annually already. But that number has to be much higher in a zero-carbon society. The International Energy Agency (IEA) has already stated the pressing need for CCS technology implementation, as they believe that without its deployment, climate change goals cannot be met [72, 73]. In many countries, policymakers are faced with a dilemma over its implementation as it admits fossil fuels into a zero-carbon future, and in the United States in particular, it is not supported in a number of quarters because it means admitting climate change is real.

In the oil and gas sector, aside from the demand for carbon dioxide for enhanced oil recovery (EOR) applications (see Section 5.9.4) [74], the fundamental premise of most of the research has been that a governmentally driven imperative(s) would force the application of the technologies and that the CO_2 would be injected into deep saline formations. It had been assumed that the commercial aspects of the work would be offset by what is commonly referred to as a price on carbon, either directly such as a carbon tax or indirectly as in emission trading credits, or similar. Also, there has been an inaccurate assumption that the storage of CO_2 for EOR was not an answer to CO_2 emissions control as it was as follows:

1) Too small a solution to matter in the end
2) Storing only about half of its injected CO_2
3) That it only prolonged a society dependent on carbon-based fuels by adding more oil production to the national and international combustion (emission) profiles already in existence

The problem is that the first two statements can be shown to be inaccurate and the third ignores the reality and magnitude of the role of oil and hydrocarbons in modern society. Nonetheless, the commercial driver for the capture via sales of the 'commodity' CO_2 and the ongoing demonstration in EOR projects of the technology and application are viewed by some as impediments to the long-term strategies for converting to carbon capture. To date, for example, there are no fully deployed carbon capture plants in the United Kingdom, and in November 2015 the UK government cancelled its £1bn competition for CCS technology, just 6 months before it was due to be awarded.

Throughout this book, a number of chemistries have been exemplified, which can be utilised in the use of carbon capture technologies; many are very simple and many are based on amines and amides as described in Section 6.1.7 [69, 75].

Ionic liquids have been proposed as an absorbent in carbon capture. They have various advantages over traditional absorbents, such as the currently dominant amine-based

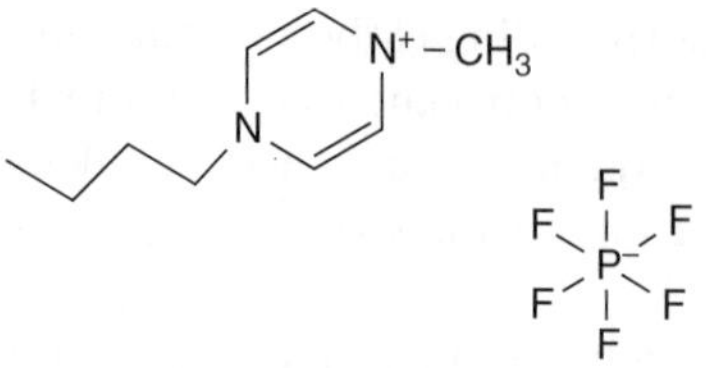

Figure 10.3 1-Butyl-3-methylimidazolium hexafluorophosphate.

technologies. 1-Butyl-3-methylimidazolium hexafluorophosphate (Figure 10.3) is one example of a proposed CO_2 absorbent.

The solubility of CO_2 in ionic liquids is governed primarily by the anion, less so by the cation, and the hexafluorophosphate (PF_6^-) and tetrafluoroborate (BF_4^-) anions have been shown to be especially amenable to CO_2 capture [76].

It would seem reasonable to state that if the world is ever to achieve net zero carbon emissions, societies must capture emissions that cannot be avoided and CCS technology and the chemistry it involves is critical to this goal. In such the oil and gas sector can play a critical and sustainable role but requires support and an economic push from governments around the globe.

10.2.2.6 Recovery and Reuse of Other Materials

As offshore fields reach the end of their economic lives, there is an increasing need to decommission platforms and other structures to promote safety, protect the environment and comply with governmental regulations. Decommissioning is a complex undertaking, requiring a range of equipment and expertise. In decommissioning processes the recovery and potential reuse of materials salvaged in the operation is a major driver and cost operating condition [77].

Modern practices look for more efficient methods of platform decommissioning that integrate multiple services and manages them as a single project, with the objective of completing the decommissioning job safely and efficiently at a fixed cost to the operator [78]. This approach makes the recovery and reuse of as many materials from the steel jacket to the adhesive coatings of paramount importance.

10.3 The Environmental Fate of Chemicals in the Upstream Oil and Gas Sector

In Chapter 9 a number of criticisms of the current regulatory practices in the upstream oil and gas sector were documented. Perhaps the underlying premise of the criticism lies in how the industry and regulatory authorities view the environmental fate of the chemicals in use. In a number of regions, priority is placed on a chemical's inherent capacity to biodegrade to such an extent that those that do not biodegrade but are also not bioavailable are penalised and often barred from use.

Environmental fate is normally defined as the life cycle of a chemical or biological pollutant after its release in the environment [79]. The environmental fate of chemicals entering an ecosystem is often much more complex than examining a single parameter, and in this section, some of the accepted orthodoxy is challenged.

10.3.1 Dilution

Dilution is particularly an important process for reducing the concentration of conservative substances, i.e. those that do not undergo rapid biodegradation, such as metals and other inorganic materials.

Also, as has been described in Section 8.4, most if not all chemicals applied in the upstream oil and gas sector are formulated and therefore are not neat, concentrated substances but mixtures of substances in a carrier solvent. Therefore, dilution of the substance has already occurred. After use, the chemical substances are further diluted in the receiving environment, and this can be a large dilution depending on this environment.

In the marine environment, the dilution capacity of the receiving water can be defined as the effective volume of receiving water available for the dilution of the effluent. The effective volume can vary according to tidal cycles and transient physical phenomena such as stratification. In estuaries, in particular, the effective volume is much greater at high spring tides than at low neap tides. It is important to consider concentrations of substances in worst-case scenarios (usually low neap tides except, for example, when pollutants might be carried further into a sensitive location by spring tides) when examining the potential risk of pollution from such a chemical discharge. Stratification can reduce the effective volume of the receiving water by reducing vertical mixing and constraining the effluent to either the upper or lower layers of the water column.

The process of dilution can be separated into initial dilution and secondary mixing.

For many discharges in the upstream oil and gas sector, the effluent is principally some form of water (seawater, aquifer water and produced water), containing a mixture of pollutants. In offshore operations, the discharge point is generally located such that the effluent is released under seawater. Initial dilution occurs as the buoyant discharge rises to the surface because of the density differential between the saline receiving water and the effluent water. Under certain circumstances of stratification or where the effluent comprises seawater (in a cooling water discharge, for instance), the effluent may not rise to the surface but may be trapped in the lower layers of the water column. Onshore the design of sewage outfalls including the use of diffusers can maximise the initial dilution by entraining as much receiving water into the effluent as possible. Guidelines for the amount of initial dilution expected for the design of discharges are usually set by the competent authorities [80].

For the many buoyant discharges, the effluent rises to the surface where it can form a 'boil'. The plume then forms and spreads and secondary mixing takes place. Eventually, the plume disperses both vertically and horizontally in the water column as the density differential becomes inconsequential and the concentration of pollutants in the water column approaches uniformity. Further dilution occurs as a result of the action of tide, wind and wave-driven currents.

10.3.1.1 The Mixing Zone

In the offshore environment, a mixing zone is an area of sea surface surrounding a surface boil. It comprises an early part of the secondary mixing process and is prescribed to ensure that no significant environmental damage occurs outside its boundaries.

An individual mixing zone has been defined with respect to an established environmental quality standard (EQS) for a particular polluting substance. The mixing zone is the area of sea surface within which the EQS will be exceeded [80]. The relation of the mixing zone to the location of European marine site features will be a key consideration for determining the acceptability of dilution criteria.

Dilution within the mixing zone consists of initial dilution (the dilution received as a plume rises from the discharge point to the water surface) and secondary dilution (a slower rate of dilution occurring between the surface 'boil' and the edge of the mixing zone). It has been recommended that to ensure that the integrity of a marine site is not affected, the minimum size of a discharge (in terms of flow or load) should be assessed on a site-specific basis. This will depend on the substances and/or physico-chemical parameters associated with the discharge, together with the positioning of the discharge in relation to the biotopes(s) or species for which it was designated. The initial dilution of discharges also needs to be considered. For example, an initial dilution of 50 times may be considered appropriate for low toxicity effluents, but highly toxic effluents may require a minimum initial dilution of 100 times [80].

A maximum size of a mixing zone as 100 m around the centre boil in any direction that the plume may travel has been suggested; however, the flow and concentration of pollutants within the discharge are critical to defining the size of the mixing zone.

In terms of environmental impact, it is important to understand not only the ecotoxicological properties of the chemicals in the discharge but also the dilution factors affecting these. Modelling studies have been proposed for a variety of effluent discharges [81], and some models have been developed for the offshore oil and gas industry as described in Section 9.6.2. However, their sophistication in examining dilution factors can be limited and therefore open to interpretation and expert judgement, criticism of which has also been detailed in Chapter 9.

10.3.2 The Fugacity Approach

It has been known for some time that the environmental fate of chemicals can be understood in terms of a conceptual scheme linking the thermodynamic state of a given compound in each environmental compartment and the fugacity approach is one such scheme [82]. This approach is based on the thermodynamic property of mater in that it flows 'downhill' in a thermodynamic sense, in other words mass flows from a high chemical potential to a low chemical potential. From this it can be derived that concentration is proportional to chemical potential in a single environmental compartment, e.g. the aquatic environment, but not across media boundaries such as from water to air. The objective of this approach is to express concentrations in dissimilar media in a consistent, thermodynamically based method so that the downhill flow is apparent. In the fugacity approach the strategy is to express concentration in any medium as the equivalent in a thermodynamic sense to the vapour phase partial pressure or fugacity (f); therefore fugacity can be thought of as a chemical's tendency to escape. A high fugacity is therefore a high tendency to migrate. This leads to the concept that each phase has a fugacity capacity, and therefore for the same contaminant concentration, the fugacity gradient driving escape is less from a medium with a high fugacity capacity than from one with a low fugacity capacity. This can be alternatively stated that an environmental compartment with a higher fugacity capacity can accept a higher concentration of a given compound [82].

This approach has been the basis of a number of models in examining environmental pollution, and work continues in building complex models incorporating the concepts of mass balance and mass transfer [83]. This has led to the development of a general multimedia environmental fate model that is capable of simulating the fate of up to four interconverting chemical species. Multispecies chemical assessments are warranted when a degradation product of a released chemical is either more toxic or more persistent than the parent chemical or where there is cycling between species, as occurs with association, disassociation or ionisation. Such a situation could often occur in the oil and gas sector where many chemical species could be discharged to the environment simultaneously.

There have now been established four levels of multimedia fugacity models applied for prediction of fate and transport of organic chemicals in the multi-compartmental environment as illustrated in Figure 10.4. [84]

Depending on the number of phases and complexity of processes, different level models are applied. Many of the models apply to steady-state conditions and can be reformulated to describe time-varying conditions by using differential equations. The concept has been used to assess the relative propensity for chemicals to transform from temperate zones and 'condense

Level I	Closed system in equilibrium	Equilibrium between compartments according to thermodynamics assumed (partition coefficients such as K_{OW}, K_{AW} or K_S); transformation and active transport not taken into account
Level II	Open system in equilbrium	In addition to level I: continuous emissions and transformation (e.g. biodegradation, photolysis) taken into account
Level III	Open system in steady state	In addition to level II: active transport and compartment-specific emissions taken into account
Level IV	Open system, non-steady state	In addition to level III: dynamics of emissions and resulting temporal concentration course taken into account

Figure 10.4 Fugacity complexity levels.

out' at the polar regions. The multi-compartmental approach has been applied to the 'quantitative water air sediment interaction' (QWASI) model designed to assist in understanding chemical fate in lakes [85]. Another application can be found in POPCYCLING-BALTIC model, which describes the fate of persistent organic pollutants (POPs) in the Baltic region [86].

This type of environmental fate and transport analysis is becoming more common in its application across a variety of industrial sectors where chemicals have the potential to impact the environment. However, it is fair to state that little of this approach has found its way into the assessment of the environmental fate of chemicals used and discharged in the upstream oil and gas sector. Although the models described in Chapter 9 do view the environment, particularly the marine environment, in discrete compartments, the utilisation of the fugacity concept to view the total environment is not implemented, at least to the authors' knowledge.

10.3.3 Risk Assessment

Regulatory requirements in many areas, as described in Chapter 9, has led to the development of hazard evaluation criteria and ranking of chemical as to their potential environmental impact based almost exclusively on their inherent potential hazards. These hazard quotients are almost exclusively based on point data for rates of biodegradation and toxicological data expressed as lethal dosages against prescribed species. This is particularly the case for the marine environment where, as has been detailed in the previous section, dilution factors play a significant role in the presentation of toxic effluent to the environment.

The author would argue that in a 'sustainable' oil and gas sector, the development of more sophisticated risk assessment models is required. Some of this work has already been conducted and should be reconsidered. Some chemical service companies have adopted a global approach to the development of parts of their chemical supply to the upstream oil and gas sector [87]. This however has focused on regulatory compliance particularly within the OSPAR region and places emphasis on the development of more environmentally acceptable products requiring thorough screening of environmental fitness and status of compliance with relevant laws and regulations. The key strategy has been to improve the characteristics of chemical products to reduce risk or damage to marine life and requires changes in previously acceptable products, such as elimination of restricted materials and incorporation of components with improved ecotoxicity values. However, in pursuit of this strategy, a chemical product life cycle management process (C-PLMP) was adopted and covered the 'cradle-to-grave' life cycle of a chemical product where each step of the process offers guidelines for critical product development activities.

Such an approach should be considered as a critical analysis method in achieving a sustainable product life cycle for chemical manufacture, supply, use and disposal (recycling and reuse).

In an attempt to improve chemical product risk profiles, again with view to improve chemical product environmental characteristics and enhance product development for regulatory compliance, a comparative scoring process has been developed [88]. The method uses environmental hazard profiles of developmental products that are compared with current products used for similar purposes. Using the method, hazard profiles of formulations may be improved during the development phase of a project; however the method is reliant on the availability of data for existing and proposed substances, and the data is robust and valid.

A number of models described in Chapter 9 such as CHARM and DREAM have risk assessment elements; however the overwhelming use is as hazard ranking tools, as has been described and critiqued.

10.3.4 Bioavailability

In considering the environmental fate of chemical products, it is paramount to assess their bioavailability. For many regulatory processes, this is considered by molecular weight, and a cut-off of 700 is commonly used by a number of regulatory authorities. This means that chemical molecules and polymers having a higher molecular weight are not considered to be bioavailable. However, there is recent evidence to suggest that the correlation for dissolved organic carbon based on molecular weight may be more complex than previously thought and that higher molecular weight species may be bioavailable [89].

As has been discussed in some detail in Section 3.9, the primary assessment of bioavailability of molecules below the molecular weight cut-off is by assessment of the bioaccumulation potential, which for many oil and gas chemical additives has little or no validity being surface-active chemicals (surfactants) [90]. It seems that the molecular weight cut-off criteria is somewhat arbitrary and also the application of bioaccumulation assessment as a measure of bioavailability is somewhat flawed, at least in adopting a measurement of the partitioning of a chemical between octanol and water.

Researchers and other workers in environmental toxicology have struggled for decades with concepts and definitions of bioavailability. It seems remarkable that such an important report lacks a working definition of the term. Given the legal and regulatory implications of the bioavailability concept as part of hazard and risk assessment framework, the term must be clearly understood.

For example, the UK Contaminated Land Regulations under Part IIA of the Environmental Protection Act of 1990 defines contaminated land as 'land that appears to the local authority to be in such a condition, by reason of substances in, on or under the land, that significant harm is being caused, or there is significant possibility that harm is being caused' [91]. Thus, just the presence of substances of concern is not sufficient; harmful interaction with a receptor must be a possibility. Because toxic effects require that an organism takes up the contaminant, the extent to which substances are bound to soil particles or are available to cause harm needs to be considered.

The following definition of bioavailability is proposed:

> Bioavailability is the inherent ability of a chemical substance to cross a cellular membrane from the medium that an organism inhabits.

Once transfer has occurred the assimilation of the chemical substance can be of critical concern. However, if the substance is not toxic or indeed useful to the organism such as silica in certain cases, (see Section 7.2.7), then such a substance should not be considered of concern. Therefore bioavailability in itself is not necessarily the critical environmental impact component but needs to be linked to toxicity and biodegradation, among other factors.

In addition, in considering bioavailability, it is necessary to consider the potential for bioaccumulation and trophic transfer particularly in the aquatic environment. Organisms that bioaccumulate contaminants including chemicals from water and sediments may transfer these to predators higher in the food chain [92]. The extent to which these contaminants can be transported through the food chain and potentially affect higher organisms in the ecosystem is critical when examining the environmental fate of chemical substances.

It should be noted that while the bioaccumulation of a chemical can still present a problem where exposure levels and uptake rates are sufficiently high in relation to depuration and metabolism rates, a high bioaccumulation potential does not automatically imply the potential for biomagnification. Indeed, for some chemicals, which are readily taken up by organisms near the bottom of the food chain, a capacity for metabolism is more likely in successively higher trophic levels. Biomagnification occurs when the chemical is passed up the food chain to higher trophic levels, such that in predators it exceeds the concentration to be expected where equilibrium prevails between an organism and its environment [93].

Bioaccumulation is considered more fully in the next subsection.

10.3.5 Bioaccumulation and Persistence

Bioaccumulation has been defined as the intake of a chemical and its concentration in the organism by all possible means, including contact, respiration and ingestion, [94] whereas bioconcentration is considered as the intake and retention of a substance in an organism entirely by respiration from water in aquatic ecosystems or from air in terrestrial ones. Thus, the fatty tissues of animals may accumulate residues of heavy metals or organic compounds. These can be passed up the food chain (e.g. through fish, shellfish or birds) and reach greater, possibly harmful, concentrations at high trophic levels among top predators and human beings. Another way to view bioaccumulation is that lipid-loving (lipophilic) contaminants have much lower escaping tendencies (fugacities) from fatty tissues than from water.

The OECD defines persistence as the length of time that a compound is able to remain in the environment after being introduced into it. Some compounds may persist indefinitely [95]. However, in terms of potential environmental impact and harm, such a compound has to be bioavailable and bioaccumulating.

As has been discussed in a number of chapters in this study and in particular Chapter 3 in relation to surfactants, determining bioaccumulation potentials for oilfield chemicals can be fraught with difficulties. However, where it can be shown that a substance is not bioaccumulating, it does not necessarily gain favour for use with many regulatory bodies as the emphasis remains on rates of biodegradation.

The standard method of determining bioaccumulation potential is to measure the log of the partitioning coefficient between octanol and water. There are three different methods most commonly used for the determination of the octanol–water partition coefficient:

- The shake-flask method, OECD 107 [96]. This test guideline describes a method to determine experimentally P_{ow} values in the range $\log P_{ow}$ between -2 and 4 (occasionally up to 5). This method cannot be used with surface-active materials. The partition coefficient is defined as the ratio of the equilibrium concentrations of a dissolved substance in a two-phase system consisting of two largely immiscible solvents.

- OECD 117 [97] uses reverse-phase high-performance liquid chromatography (HPLC) and compares retention time of the specific chemical/surfactant, and the result is compared to a reference sample with known retention time. The method covers $\log P_{ow}$ in the range of 0–6, but it can be expanded to cover the $\log P_{ow}$ range between 6 and 10 in exceptional cases. The partition coefficient of the test substance is obtained by interpolation of the calculated capacity factor on the calibration graph. For very low and very high partition coefficients, extrapolation is necessary.

- OECD 123 [98] is the third used and approved method and is a slow-stir method. This method was mainly developed and used to support the shake-flask method, OECD 107. It permits the determination of the 1-octanol–water partition coefficient (P_{ow}) values up to a $\log P_{ow}$ of 8.2. The partition coefficient between water and 1-octanol (P_{ow}) is defined as the ratio of the equilibrium concentrations of the pure test substance in 1-octanol saturated with water (CO) and water saturated with 1-octanol (CW). In order to determine the partitioning coefficient, water, 1-octanol and the test substance are equilibrated with each other at constant temperature in a thermostatically stirred reactor at 25 °C and protected from daylight. Exchange between the phases is accelerated by stirring. The concentrations of the test substance in the two phases are determined. Each P_{ow} determination has to be performed employing at least three independent slow-stirring experiments with identical conditions. The regression used to demonstrate attainment of equilibrium should be based on the results of at least four determinations of CO/CW at consecutive time points. In this method, it is possible for octanol droplets to form in the water phase.

These methods although relatively inexpensive can be time consuming and have some serious experimental error possibilities, and it is a matter of some debate whether surfactants in particular actually have a true partitioning coefficient. This is because of surfactants' ability to self-associate, to stabilise emulsions, to foam and to concentrate at an oil–water interface. In addition, octanol is a surface-active compound itself and has a $\log P_{ow}$ of 3.

Alternative methods around the derivation of a bioconcentration factor are expensive and also complex and time consuming. In Section 8.4.5 a relatively inexpensive method for determining bioaccumulation potential of surfactants was briefly described to determine $\log P_{ow}$ [99].

During the measurement of $\log P_{ow}$, the surfactant concentration is below the critical micelle concentration (CMC), ensuring that no micelles are formed. It has been debated if the surfactant in the mixture is present at the octanol–water interface or if it is adsorbed to the sample glass surface.

The relative solubility number (RSN) is contrary to the HLB value, an empirically determined value to characterise water solubility and the hydrophobic–hydrophilic character of a surfactant (see Section 3.1.8). The RSN value is commonly used to characterise emulsifiers and is defined as the amount of distilled water necessary to be added to a solvent system in order to produce persistent turbidity.

Commonly a mixture of benzene/dioxane is used as a solvent system. This solvent system can be replaced by a toluene/ethylene glycol dimethyl ether (EGDE) system [100]. The properties of these two phases and the composition of them together give understanding to the mechanism behind the method. When water is started to be titrated into the solvent, consisting of dioxane and toluene, dioxane dissolves both water and toluene. At a certain point where the water percentage becomes high enough, dioxane can no longer hold the toluene and water together, and the water separates out, forming a water-rich phase. This is the point where the solution gets turbid. Addition of surfactants will influence the solvent system. A more hydrophobic surfactant will enhance the hydrophobicity of the solvent system, while the hydrophilic head groups are hydrated with water molecules, i.e. the amount of water that the solvent system can handle is depending on the hydrophobic–hydrophilic characters of the surfactants. Therefore, a high RSN value indicates a more hydrophilic surfactant, and to hydrate the polar head group, more water is needed before the solution becomes turbid. Both RSN and HLB determine similar properties of surfactants, where a higher value indicates higher water solubility, while a lower value refers to a less water-soluble surfactant. Determination of a universal correlation between RSN and HLB for all categories of surfactants has not yet been investigated. This task seems challenging and therefore to generalise RSN across all surfactants is as yet unproven. There however appears to be a good correlation within specific families of surfactants, and this could prove to be a useful tool in the development of environmentally acceptable surfactants without the need to resort to time-consuming partitioning studies or expensive bioconcentration work.

Both bioaccumulation and persistence can be expressed in numerical terms for individual media or environmental compartments. For a multimedia system as illustrated in the earlier figure, the concept of half-life, which is often used in quantifying persistence, cannot be applied because some compartments experience faster reactions than others and overall behaviour is not first order. It is therefore essential to express persistence as an overall residence time attributable to reaction. This leads to a number of complications and complexity in studying these relationships in a model system.

In summary, an individual reaction residence time or persistence can be calculated for each compartment or box for which a mass balance equation applies. Values can also be deduced for the system of boxes as a whole. If the boxes are at the same fugacity (i.e. equilibrium applies, as in level II), mode of entry is not important. If they are at different fugacities (i.e. not at equilibrium, as in levels III and IV), mode of entry influences overall system residence time, but not the individual compartmental residence times [101].

In conclusion, it has been shown that the combination of multimedia and bioaccumulation models can provide a comprehensive assessment of chemical fate, transport and exposure to both humans and wildlife. A logical step would be to incorporate toxicity information to assess the likelihood of risk. Indeed, this capability already exists for many well-studied chemicals, but it could be argued that there is a need to extend this capability to other more challenging chemicals and environmental situations and perhaps to all chemicals of commerce. Finally, the author would argue that to derive the full benefits of these applications relating to environmental fate of chemical substances, risk assessment requires a continuous effort to develop quantitative structure–activity relationships (QSARs) that can predict relevant chemical properties and programmes to validate these models by reconciliation between modelled and monitoring data [102].

10.4 Environmental Pollution in the Oil and Gas Sector and Its Control

In considering the sustainability of the upstream oil and gas sector, it is ultimately essential to consider its huge, and sometimes disastrous, impacts on ecosystems from pollution events, such as the Deepwater Horizon explosion in 2010 [103]. Indeed the ecosystems and wildlife affected by this last large oil pollution event would seem to be only recovering slowly.

One of the few success stories heralded from the Deepwater disaster was the use of a dispersant hosed directly at the source of the well leak. Engineers believed the dispersant prevented the stream of oil particles from coagulating into dense oil patches that would reach land and smother shoreline ecosystems [104]. However other research suggests that the physics of deepwater spills may naturally prevent coagulation and that the dispersant did not play a role [105]. Recent research has claimed that the dispersants may have significantly reduced the amount of harmful gases in the air at the sea surface, diminishing health risks for emergency responders and enabling them to keep working to stop the spill and clean it up sooner. Oil spill dispersants are also considered in more detail in Section 3.8.

As can be seen, responding to an extensive emergency such as a large oil spill requires both a sense of urgency and control. Prevention planning and emergency response contingency are still the best tools we have for primarily preventing such events and responding to them if necessary. It is essential that such large pollution events are seen in context and prevented if at all possible. It is not realistic as some would argue to stop all oil and gas exploration and development; [106] however it is also not justifiable to allow unfettered and uncontrolled access to the exploitation of natural resources. Western societies need to lead the way towards sustainable methods of the extraction and use of hydrocarbons while moving to a lower carbon economy [107]. It is also not justifiable for Western governments and others to blame the growing impacts of hydrocarbon use and emissions on the developing nations.

As has been argued extensively in the previous chapter, there is a need for industry to up its game and face the challenges of creating methodologies and measures for environmental protection from its activities and not to only comply with the regulations. This means rigorous scenario planning and systems implementation for the prevention of oil spill disasters. Rightly many are concerned about the effects of a major oil spill in Arctic waters under the ice sheet, and much work is being undertaken by both government institutions and industry to develop methods of mitigation [108–110]. It is essential that this and other work continues so we are able to firstly have the best possible systems and methods in place to prevent oilfield pollution and secondly in the event of a major incident, our response is the best possible to minimise and mitigate effects from such an event. It is not a sustainable position to rely on principled stance of stopping the continued extraction of oil and gas, but it would be far more productive to use these resources to effect real change in operational methods and to aid what are often developing economics to a better future and eventually to a lower carbon economy. The author agrees that time is running out in terms of pollution reduction and climate change mitigation; however, in a sustainable future with continued reasonable standards of living, and given the current state of technology in other forms of power generation, fuel supply and, probably most importantly, chemical and other material feedstocks, hydrocarbons will need to be exploited for some time into the future.

Also in a sustainable approach to examining pollution of the environment, particularly from a holistic view, other pollutants into the environment from other man-made activities are at least as important as those from hydrocarbon extraction and use. This is particularly the case if the complete life cycle of the supply of the sources of such pollutants is considered. A particularly relevant example is that of plastic pollution, particularly microplastics, in the marine environment [111].

Although plastics are substantially derived from petroleum products, the pollution of the marine environment is almost wholly due to plastic being wastefully dumped into the sea. The first reports of plastic litter in the oceans are from the early 1970s when offshore oil and gas exploration was in its infancy [112]. Even then pellet-shaped plastic particles were being detected about 0.25–0.5 cm in diameter. Since these initial reports, many more ecological damage incidents have been reported with entanglement of species being widely documented [113]. In recent years the priority of the pollution has shifted to the microplastic particles or 'nurdles', and there is now a large body of evidence to support serious levels of marine pollution due to microplastics and their relationship to POPs [114]. These microparticles are defined as having a size range from 5 mm to 1 nm and have arisen from the embrittlement and degradation of plastics in the marine environment over time. The ingestion of these materials by marine species and the incorporation of them by microorganisms are of serious ecological concern. The source of these materials is through mankind's profligate use of plastics and its lack of reuse and recycling, as well as the disposal at sea of these materials. The UN has called on governments globally to introduce measures to reduce plastics entering the world's oceans [115]. It seems increasingly unlikely that the materials can be removed from the environment once in situ and hence efforts must be directed at reduction, reuse and recycling.

10.5 Life Cycle Management

In a sustainable approach to the future of the oil and gas exploration and production sector, a process of life cycle management will be essential. This can be defined as managing the entire life cycle of a product from inception through design and manufacture to service and disposal of manufactured products. Life cycle management integrates people, data, processes and business systems and provides a product information backbone for companies and their extended enterprise [116].

In the oil and gas sector to date, life cycle management has mainly concentrated on the areas of managing and maintaining the complex equipment, particularly in the context of the diverse and demanding working conditions of the oil and gas sector. Efficiently and effectively maintaining the physical infrastructure in good working condition (and limiting the costs and disruptions of doing so) is essential for optimising performance, maintaining a safe environment and ensuring no pollution events. As technologies continue to evolve and necessarily become more sophisticated in response to new operational mandates in a variety of rugged extreme conditions around the world, these challenges become ever more salient.

Simultaneously oil and gas operators are becoming increasingly reliant on the dependable functioning of their physical assets to drive competitive advantage, and new challenges are emerging with respect to the rigorous and highly variable regulatory frameworks intended to ensure worker and environmental safety. As a result, an effective enterprise asset management strategy must be comprehensive, flexible and adaptable. There is a requirement to drive down operating costs as well as increase safety and environmental protection, and an integrated asset life cycle management process is one answer to this paradox. Such a system provides a sophisticated, innovative and holistic approach to asset management that uses a single coordinated system to deliver potentially dramatic efficiencies.

This integrated approach to asset-based life cycle management that can support a company's maintenance management evolution from a corrective mode to a preventive one and further forward to a predictive maintenance systems is a potential game changer for oil and gas companies. Successful implementation provides a company with the ability to capture and utilise common maintenance processes to service, overhaul and, when necessary, replace critical physical assets in a way that can unlock powerful new efficiencies as a company evolves its maintenance practices. The best systems are not only flexible and comprehensive but also integrated into a company's operational framework in a way that facilitates truly predictive, reliability-centred maintenance management. The result is more uptime, lower downtime costs and a more efficient use of company resources. All of this can lead to the ability to reinvest in additional maintenance and reliability technologies to continue to extend the life of oil and gas enterprise assets. In this respect, the benefits towards a sustainable operation are obvious:

- Higher productivity due to more reliable equipment that requires less downtime
- Regulatory and compliance requirements proactively managed particularly in the complex and constantly changing regulatory frameworks across the world
- Demonstrable and verifiable compliance with safety and environmental standards

Preventive and predictive maintenance management not only reduces the chances of whole-sale breakdowns and expensive interruptions but also makes regular servicing a more efficient and effective process. The ability to configure systems with company-specific variables, tracking and analytics and to provide a level of procedural and material standardisation across the frequently disparate components of an operation lowers both the risks in safety and environmental protection and operational costs. Such a system also creates an essential connection between the supply chain and inventory and financial controls.

The resulting operational and strategic efficiencies are that oil and gas companies can ensure they are minimising waste, maximising operational uptime and having critical control of the supply chain including the supply and use of chemical additives.

This overall lower cost base also has an impact on the potential extension of an oilfield. Where this at first sight may not appear to be a sustainable position, it is surely far more environmentally beneficial to maximise the efficient extraction from one asset than to abandon and decommission such a facility to only look for the next prospect [117].

10.5.1 Life Cycle Assessment (LCA)

A critical part of any life cycle management is the introduction into the organisation environmental management and the supply chain of life cycle assessment (LCA). An international standard has been developed, ISO 14040:2006, which sets out the principles and framework for LCA [118].

LCA is a methodology by which organisations can analyse the environmental impacts and effects of their products and services. The duration of this assessment extends across the entire life cycle of the products and services ('cradle to grave'). This process allows for product comparison and strategic decision making with regard to systemic inputs and outputs, as well as the development and incorporation of end-of-life design strategies. Therefore, in the context of chemical supply in the oil and gas industry, it should address the following:

- The origin of the chemistries involved in the chemical supply
- The sustainability of the manufacturing processes and their environmental impact
- The use and discharge of the chemicals supplied
- The ultimate environmental fate of the chemicals used and discharged

As has been described the use and discharge of chemicals has been focused on almost entirely by the industry and has to a large extent been driven by regulatory compliance. In moving the oil and gas sector to a more sustainable position, the use of LCA in the supply chain is critical, and this includes the supply of chemical additives.

LCA examines the environmental aspects and potential impacts in a cradle-to-grave fashion, i.e. from its raw material acquisition, manufacture, supply, use end-of-life treatment and final disposal. For chemical additive use in the upstream oil and gas sector, this is likely to translate to examination of the following aspects in the supply chain as well as use and discharge, i.e.

- The raw material source, petrochemical or naturally derived
- Energy utilisation in manufacture and carbon footprint

- Transport and supply
- Use in processing
- Environmental discharge impacts
- Environmental fate

Typically, it does not address the economic and social impacts of the product and its use.

10.5.2   Societal Life Cycle Assessment (SCLA)

The importance of the social dimension of sustainable development increased significantly during the last decade of the twentieth century. Many industries have experienced a shift in stakeholder pressures from environmental to social-related concerns, where new developments in the form of projects and technologies are undertaken. This has been particularly relevant in the shale gas debate across Europe within the context of the overall energy balance and the shift away from carbon-based power generation to renewable energy. Although a justifiable case can be made on technical, environmental and energy security grounds (see Section 10.2.1), the social and political uncertainties can act against these arguments [119]. This is where a process such as societal life cycle assessment (SLCA) could possibly help in both decision making and support a sustainable energy policy.

The issues around SCLA centre on the quantification of data. There are often many societal indicators, but most are of a qualitative nature, and there is a need to focus on the few quantitative indicators to establish a robust assessment process [120].

The development of SLCA is in its infancy, and the measurement of social impacts and the calculation of suitable indicators are less well developed compared with environmental indicators in order to assess the potential liabilities associated with undertaken projects and technologies. Many important concepts require clarification including the handling of the more than two hundred social indicators. Therefore, any SLCA methodology must explain why it is midpoint- or endpoint-based as well as its reasons to be complementary with, or included within, LCA. As an analogy to economics and cost estimation, SLCA combines, into its statistics, both data and estimates, some of which are correlated to elements of the product LCA and its impact. A critical indicator could, for instance, focus on the work hours required to meet basic needs; however this may be one of many other indicators, many of which are not as quantifiable. Therefore, it could be concluded that quantitative social impact assessment methods cannot be applied for project and technology life cycle management purposes in industry at present; however this does not mean that, under the move towards a more sustainable environment in the future, the oil and gas sector, among other industries, should not make an effort to develop a suitable and verifiable methodology. It could just persuade the argument in their favour.

The author would argue that the process should start now if not already commenced with identification of key indicators and collection of numerical data.

10.6   'Green' Chemistry

Chemical sciences have been at the heart of both environmental impacts and the environmental burden placed on the Earth's ecosystem. Recently, efforts to address a sustainable way forward through green chemistry and initiatives such as carbon capture have been developed. In 1979, James Lovelock published his now famous Gaia theory [121] in which he stated that we were close, or perhaps beyond, the limit of greenhouse gas burden in the planet's atmosphere. There is now an overwhelming consensus among most scientists that accelerated global temperature rise is inevitable and it is a question of limiting the damage. However, despite many words little concrete action is taken to reduce greenhouse gas emissions, and the bomb keeps ticking. Policy action and its implementation are required.

Throughout this study, chemistry and its deployment as chemical additives in the oil and gas industry have been described as well as the use of so-called 'green' chemistry particularly in the development of more environmentally acceptable chemical products.

'Green chemistry' can be a difficult area to define, and in general there are a number of ways to approach the subject. Much is, however, industry and application dependent. A possible accepted definition across the chemical industry is given as follows: [122]

> Green chemistry, also known as sustainable chemistry, is a philosophy of chemical research and engineering that encourages the design of products and processes that minimise the use and generation of hazardous substances.

The International Union of Pure and Applied Chemistry (IUPAC) defines it as [123]

> Design of chemical products and processes that reduce or eliminate the use or generation of substances hazardous to humans, animals, plants, and the environment.

It is also accepted that green chemistry applies the engineering concept of pollution prevention and zero waste, both at laboratory and industrial scales. It encourages the use of economical and eco-compatible techniques that not only improve the yield in production processes but also bring down the cost of disposal of wastes at the end of a chemical process.

'Green' chemistry could be one of the main technological planks to build towards a sustainable future. It takes into account the potential environmental impacts and seeks to lessen these or ideally prevent them through application of certain principles. These 12 principles were set out by Warner and Anastas in 1998 [124], in which they describe the practice of green chemistry in terms of sustainability, efficiency and environmental safety. Most researchers and other practitioners exploring and developing green chemistry do not faithfully deploy all 12 principles, but even adopting a few, such as procuring from sustainable resources, could make a major difference to the environment of the future:

1) Prevent waste – Design chemical syntheses to prevent waste, leaving no waste to treat or clean up.
2) Design safer chemicals and products – Design chemical products to be fully effective, yet have little or no toxicity.
3) Design less hazardous chemical syntheses – Design syntheses to use and generate substances with little or no toxicity to humans and the environment.
4) Use renewable feedstocks – Use raw materials and feedstocks that are renewable rather than depleting. Renewable feedstocks are often made from agricultural products or are the wastes of other processes; depleting feedstocks are made from fossil fuels (petroleum, natural gas or coal) or are mined.
5) Use catalysts, not stoichiometric reagents – Minimise waste by using catalytic reactions. Catalysts are used in small amounts and can carry out a single reaction many times. They are preferable to stoichiometric reagents, which are used in excess and work only once.
6) Avoid chemical derivatives – Avoid using blocking or protecting groups or any temporary modifications if possible. Derivatives use additional reagents and generate waste.
7) Maximise atom economy – Design syntheses so that the final product contains the maximum proportion of the starting materials. There should be few, if any, wasted atoms.

8) Use safer solvents and reaction conditions – Avoid using solvents, separation agents or other auxiliary chemicals. If these chemicals are necessary, use innocuous chemicals.
9) Increase energy efficiency – Run chemical reactions at ambient temperature and pressure whenever possible.
10) Design chemicals and products to degrade after use – Design chemical products to break down to innocuous substances after use so that they do not accumulate in the environment.
11) Analyse in real time to prevent pollution – Include in-process real-time monitoring and control during syntheses to minimise or eliminate the formation of by-products.
12) Minimise the potential for accidents – Design chemicals and their forms (solid, liquid or gas) to minimise the potential for chemical accidents including explosions, fires and releases to the environment.

Under these criteria green products can be defined as those derived from sustainable and bio-derived sources.

In the oil and gas exploration and production industry, however, it is accepted that green chemistry – or more particularly green chemicals – are those that meet certain regulatory requirements, which are often particular to the oil and gas industry. In the North Sea, for example, the 'green' term means being able to satisfy the criteria set out in the Oslo Paris Convention (OSPAR) Treaty [125] and meeting agreed criteria on biodegradation, bioaccumulation and marine species toxicity [126]. Certain contracting parties, in particular Norway, have added additional controls with regard to other properties of solvents, most notably carcinogenic, mutagenic, toxic for reproduction (CMR) status, flammability and other health and safety criteria [127].

In examining the development of 'green' chemicals for application in upstream oil and gas sector, this book has examined at a wide variety of chemistries and examples within classes of effect. It has also illustrated the route taken to develop, in the main, more environmentally acceptable chemicals. In this respect two complementary approaches have been adopted:

1) Derivation of the existing chemical type to provide a more biodegradable and less harmful structure
2) Examination of other chemical structures that are known or presumed to be less harmful to the environment, e.g. use of natural product-type chemistry

As can be seen, this hardly covers the 12 principles described previously and does not examine the sustainability of the approaches adopted, although, in general, one could assume that the use of naturally derived products to be more sustainable.

Earlier in this chapter the use of CCS was exemplified as a method of recovery and reuse of a waste product and an undesirable greenhouse gas. As part of any green chemistry strategy, it will be essential to consider what happens to the carbon dioxide generated as even so-called green chemicals can degrade and biodegrade to carbon dioxide as part of their ultimate environmental fate. The author believes a reassessment of this technology is required by policymakers and others in order to ensure a smooth transition to a lower carbon economy.

As is evident from the 12 principles, green chemistry is as much about how the chemical products are manufactured as the chemical product itself. In order to establish a comparative assessment of manufacturing and synthesis techniques, it is necessary to establish the correct measurement techniques and key measurement criteria. As has been illustrated, the use of metrics to drive business, government and communities is critical in achieving more sustainable practices. A number of metrics have also been proposed to make chemists aware of the need to change the methods used for chemical syntheses and chemical processes; in particular a single metric mass reaction efficiency may prove to be a useful comparator for supply chain assessment.

A paper published in 2002 explores several metrics commonly used by chemists and compares and contrasts these metrics with a new metric known as reaction mass efficiency [128].

Throughout this book, against the individual classes of chemistry, a number of examples of greener alternative products, which are used or could be applied, to various sectors of the upstream oil and gas industry have been exemplified. These products have mainly focused on achieving regulatory compliance (see Chapter 9) and have therefore been designed or derived to achieve greater rates of biodegradation.

10.6.1 Polymers

Many polymers are now being designed to be biodegradable, and many are being derived from bio-based materials and natural products. Polysaccharides for use in commercial polymers have been developed for some time [129] and have been detailed in Chapter 2; many classes of polymer are being designed for improved biodegradation. Poly(2-oxazolines) (see Figure 10.5 for an example) have been intensively examined for applications in the medicinal and pharmaceutical industries due to their numerous properties that can be induced by variation and manipulation of key functional groups [130].

These materials are highly water soluble and therefore have increased bioavailability that also makes them more biodegradable. However, it can be argued that perhaps in terms of improved environmental protection, a more balanced approach involving both biodegradation and bioavailability should be considered.

A number of polymers are not biodegradable, and as has been illustrated in Chapter 2, particularly those of high molecular weight are also not bioavailable [131]. Although this may have some advantages, a variety of molecular weights and functionality are required for polymer use in the upstream oil and gas industry. Also in efforts to become more sustainable, monomer feedstocks are being utilised, which are natural products or derived from natural products. A number of these types of polymer have been described in some detail in Section 2.2. In recent years, much attention has been placed on the so-called cellulosic polymers, many of which can be microbially derived [132], and a number of alternative monomers have been explored, which are based on natural substrates such as lignocellulose [133].

Although a number of polymers may not be biodegradable at least in the test protocols that have been approved to assess such properties, many polymers are subject to photolysis, particularly photo-oxidation. Photo-oxidation is the degradation of a polymer surface in the presence of oxygen or ozone. The effect is facilitated by radiant energy such as UV or artificial light. This process is the most significant factor in weathering of polymers. Photo-oxidation is a chemical change that reduces the polymer's molecular weight. As a consequence of this change, the material becomes more brittle, with a reduction in its tensile, impact and elongation strength. Discoloration and loss of surface smoothness accompany photo-oxidation. High temperature and localised stress concentrations are factors that significantly increase the effect of photo-oxidation [134].

Such degradation can compromise the recycling and reuse of polymers and plastics. Often photo-stabilisers such as zinc oxide are added to prevent and retard this [135]. This treatment has been made more effective by the incorporation of zinc oxide nanoparticles in the polymer

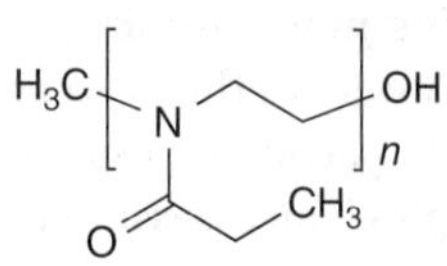

Figure 10.5 Poly(2-ethyl-2-oxazoline).

matrix [136]. Other antioxidants such as flavonoids can also be used to retard the UV light degradation of polymers [137]. Given the inherent toxicity of zinc, such compounds may be more environmentally acceptable and preferred by regulatory authorities when assessing use of polymers, particularly liquid additives in the upstream oil and gas sector.

It is of note that in the marine environment, photolysis is able to occur in the surface waters and the process can be extensive and complex involving surface layers, the water column and sediments. [138] This means that in probably most instances of polymer degradation, photolysis is an initial step. In assessing the suitability of polymers for use, these studies are not conducted. It is the non-degraded material that is assessed for rates of biodegradation. However, it is this process that the most significant factor in weathering of polymeric materials and is of concern as it aids in the generation of microplastics, which are now widely recognised for being responsible in the marine environment for the concentration of POPs [113, 116] (see Section 10.4).

As was detailed in Section 2.6, there are other methods of polymer degradation that also affect the polymers' structural integrity. It is important in assessing the environmental impact of polymers and polymer-derived materials that these pathways are also taken into account as they can affect rates of biodegradation, for instance [139].

As detailed earlier in this chapter, the reinjection of produced water is becoming a more standard practice particularly in certain regions such as the Norwegian sector of the North Sea [140]. It potentially offers the possibility that chemical residues, in particular polymer additives that can be made stable to degradation processes, could be recycled through the production process and reused. This therefore causes a dilemma in environmental assessment of stable polymers that are designed to be reused, in that they could be viewed unfavourably in a normal straight through assessment, being only poorly biodegradable, but offer an overall reduction in environmental impact as dose rates would be significantly reduced due to the recycling process. This exemplifies a dichotomy in the process of environmental impact assessment and the needs of sustainability, a dilemma that will be further illustrated.

10.6.2 Surfactants

As has been extensively described in Chapter 3, surfactants are a major class of oilfield chemical additives. Many are derived from natural products and have excellent environmental properties in terms of low toxicity and good rates of biodegradation. The difficulty in completely assessing the environmental impact of surfactants lies in measuring their bioavailability as described in Section 3.9.

A major class of surfactants that presents a significant problem to both environmentalists and oilfield chemical practitioners are the alkylphenols and their ethoxylates, particularly nonylphenol ethoxylate, as they are known to be endocrine disrupters particularly in fish species [141, 142]. It has been determined that it is the alkyl ethoxylate monomer that is the main concern regarding oestrogenic properties [143]. The mechanism of degradation of these molecules is not a reversible reaction of the polymerisation to the alkyl ethoxylate but rather much more complex process involving oxidation of the ethoxylate functionality and ring-opening processes [144].

It is however possible to design these polymeric surfactants to have ultralow levels of residual monomer much lower than any measurable toxicity level. Under a risk assessment scenario, this may allow consideration of these products where previously they have been discounted; however the precautionary principle is still at play in the European area and will preclude this for at least the present. Again, this contrary juxtaposition of regulatory compliance effects the greater environmental protection possibility, as these non-ionic surfactants are highly efficacious and much lower dose rates are required than most, if not all, of the alternatives.

In terms of a sustainable strategy for surfactant use, it is essential that dose rates are minimised and that the use of biosurfactants is increased. Although many common surfactants are naturally derived, biosurfactants offer a much greater degree of sustainability as they are primarily derived from microorganisms in controlled conditions and are not reliant on a crop-based feedstock such as coconut oil. Biosurfactants are described in more detail in Section 3.6.

10.6.3 Phosphorus

As has been described in Chapter 4, a number of oilfield additives are dependent on phosphorus chemistry; however, from a sustainable viewpoint, this is a difficult position to maintain. Mankind's use has mobilised nearly half a billion tonnes of the element from phosphate rock into the hydrosphere over the past half century. The resultant water pollution concerns are now a main driver for sustainable phosphorus use (including phosphorus recovery).

The emerging global challenge of phosphorus scarcity with serious implications for future food security means phosphorus will also need to be recovered for productive reuse as a fertiliser in food production to replace increasingly scarce and more expensive phosphate rock [145].

There is a need for new sustainable policies, partnerships and strategic frameworks to develop a means of global phosphorus security and in particular renewable phosphorus fertiliser systems for farmers.

There is no single solution to achieving a phosphorus-secure future: in addition to increasing phosphorus use efficiency, phosphorus will need to be recovered and reused from all current waste streams throughout the food production and consumption system including industrial uses (from human and animal excreta to food and crop wastes). It may be that where alternative products exist as in the oil and gas sector, it may be more sustainable and environmentally responsible to utilise these alternatives than to add to the burden of phosphorus use and scarcity.

10.6.4 Microbes and Enzymes

An area that has not been previously referred to in the other chapters is the deliberate use of microorganisms and biological enzymes in upstream oil and gas sector. Historically, enzymes have been used in the oil sector both for improving the characteristics of a range of biopolymers employed in the industry and as breakers for biopolymer gels [146].

The use of enzymes downhole has been limited to gel breaking applications where suitable enzymes are used to break down or degrade a specific gel. In this case, the enzyme is used to remove a chemical that is no longer required, such as biopolymers in filter cakes following drilling or in fracking gels after the frac has occurred [147]. More recently, enzymes have been used to produce useful chemicals in situ. An enzyme-based method has already been reported, which generates organic acids for a variety of acidising applications such as matrix acidising, the stimulation of natural fracture networks, damage removal over long horizontal intervals or gel breaking. The generation of acid in situ following placement of the fluid ensures the even delivery of acid over the whole of the treated zone [148].

The use of enzymes for generating useful oilfield chemicals has now been extended further, and the use of enzymes for the in situ generation of minerals, gels and resins with potential for use in applications such as EOR [149], sand consolidation [150] and low-temperature scale inhibition [151] has been reported.

The direct use of microorganisms has been widely established as a method for enhancing oil recovery [152]. The concept is more than 60 years old; however, early proposals were poorly conceived and in most instances of no practical value. The last 30 years have seen the development of microbial biotechnology to resolve specific production problems such as pressure depletion and sweep inefficiency in a target reservoir. Additionally, microorganisms have been genetically engineered to provide a specific chemical well treatment in which chemicals are modified in situ by thermophilic bacteria [153].

It would be fair to state that both technologies, enzyme and microbial, are in the 'slow lane' of uptake in the application area of the upstream oil and gas sector. However, the use of biotechnology as a means of supplying 'green' chemicals has become well established and is seen as a renewable and sustainable technology [154]. This is particularly the case for solvents where the application of fermentation and other biotechnologies has the capability to deliver a truly sustainable source of chemicals. In the upstream oil and gas sector, very few truly green solvents have been deployed. The 'green' solvents that are used such as methanol have properties that make them environmentally acceptable but are not necessarily derived from a sustainable source.

10.6.5 Other Products

Many of the chemical products and additives described in the previous chapters are manufactured from petrochemical feedstocks derived from crude oil. As is somewhat obvious, the extraction, production and processing of crude oil are not wholly sustainable especially as some of the uses are for power generation and transportation fuel. However, a sustainable future may lie in its use as a raw material feedstock for chemicals and other materials such as plastics that in turn can be reused and recycled.

Nonetheless a more sustainable pathway for chemical manufacture in general is to derive feedstock materials from natural products or to use these directly in a variety of applications as additives in the upstream oil and gas sector. As has been already described, much design and development work has gone into the modification of polymers and surfactants to make them more biodegradable and also to utilise natural products and feedstocks in their manufacture. Other chapters have illustrated how other classes of chemicals can be made more environmentally acceptable, at least from a regulatory compliance viewpoint. There is of course a competing difficulty in these primarily lower molecular weight organic materials or inorganic materials, in that often they are also readily bioavailable and hence if toxic can be environmentally hazardous.

For example, boric acid and its related compounds are among the most common cross-linkers for the polymers, such as guar, used in fracturing fluids [155]. Boric acid, as has been described in Chapter 5, is a compound of environmental significance; however it has to date not been possible to replace this material with technically acceptable alternatives as it has significant performance enhancement at suitable pH ranges and temperatures. Research and development has however provided more efficient boron cross-linkers allowing low polymer loadings [156] and fracturing fluid systems that can be regenerated and reused. A system based on polyvinyl alcohol/organic boron fracturing fluid has the regeneration mechanism based on the reversible cross-linking reaction between $B(OH)_4^-$ ions and the hydroxyl groups of the polyvinyl alcohol, as the pH changes [157]. It is therefore possible to make the use more sustainable and less environmentally damaging.

The author has argued throughout this study for an increased examination of natural products both directly in oilfield applications and as chemical feedstocks in the manufacture of other additives.

For example, an extract of the plant *Aloe vera* can be used directly as a scale inhibitor [158]. It is applicable at low and high calcium concentrations and will not precipitate due to hydrolysis. Indeed, hydrolysis favours the interaction of the scaling ions with the inhibitor and may increase its efficiency. *A. vera* is a crop plant in many parts of the world be used for cosmetic purposes and is therefore a potentially sustainable chemical additive of low environmental impact.

Within Europe and in the OSPAR Treaty region, many low molecular weight organic molecules have been regarded by regulatory authorities as having little or no risk to the environment, and these have been listed and designated [159]. It is stated that these substances do not need to be strongly regulated, as from an assessment of their intrinsic properties, primarily their toxicity, they pose little or no risk to the environment. However, as has been argued in Section 9.7.2, it is the author's view that the list has little or no relevance; indeed it may be that some of the substances on the list do pose an environmental risk. Therefore, all chemical substances should be evaluated as to environmental hazard and risk, but it could also be argued that their sustainability should also be considered.

As has been illustrated throughout this study, a fundamental dilemma often presents itself when examining 'green' chemistries for oil and gas additive applications, particularly in enhancing the biodegradability of chemical products, in that there is often a reduction in efficacy coupled with increased unit costs associated with the use of 'greener' products. Countering this argument for the use of green products is the many unexplored possibilities for 'greening' the product portfolio, as has been described throughout this study and in particular the use of natural products and bio-based materials.

Finally, with regard to low molecular weight organic molecules in particular, their ultimate environmental fate is usually as carbon dioxide, carbon monoxide and water. The release of further carbon gases is not considered in the environmental assessment of any of the chemicals used in the upstream oil and gas industry, and perhaps it should be. This would be another factor in the promotion of CCS as discussed in Section 10.2.2. An additional caveat to this may be in terms of the additional loading, it would be very small compared with other sectors such as power generation and transport.

10.7 The Future of Oilfield Chemicals and Some General Conclusions

Crude oil and natural gas are two highly important commodities in the modern world's economies. Crude oil in particular has a variety of uses with most of it being refined to a variety of petroleum products. The largest use by volume is as gasoline for fuelling motor vehicles, followed by diesel fuel and heating oil for a variety of power uses including power generation. The main use of natural gas is for power generation and domestic heating. Around 15–20% of crude oil feedstock is used for petrochemical manufacture and therefore is the main feedstock to the chemical industry [160].

10.7.1 Petrochemicals

Naphtha and other oils refined from crude oil are used as feedstock for petrochemical crackers that produce the basic building blocks such as alkanes and olefins to make a variety of polymers, plastics and other chemicals for making plastics. The most basic petrochemicals (ethylene, propylene, butadiene, benzene, toluene and xylene) are considered the building blocks for organic chemistry. From this base set of petrochemicals comes a very large number of other chemicals, which are called 'petrochemical derivatives', or simply 'derivatives'. The derivatives are grouped according to how many steps it takes to convert the basic compound into the new derivative. For example, it takes one step to convert ethylene to acetaldehyde; therefore, acetaldehyde can be considered a first derivative of ethylene. If you go one step further and convert the acetaldehyde to acetic anhydride, which is second derivative of ethylene and so on.

Many petrochemicals are produced using extreme temperatures (over 1500 °F) and pressures (over 1000 psi). This process requires large amounts of energy and sophisticated engineering. Because of these extreme operating conditions, energy consumption accounts for a significant portion of the total cost of production. As energy costs rise, the cost of doing business also rises. Access to inexpensive and reliable energy sources (such as natural gas) is therefore essential.

Finding alternative feedstocks that can produce these base materials and also have the flexibility to synthesis derivatives would be a considerable challenge in a carbon-free future. Nonetheless there is progress with a number of routes as described earlier in Section 10.6, 'Green Chemistry'. However, for the foreseeable future, a critical non-renewable area of crude oil and natural gas use will be the generation of petrochemical feedstocks. Having stated this, the reuse and recycle of the materials produced is possible, and some commentators estimate that as much as 90% of plastic could be recycled.

However, in the United Kingdom the government announced that targets for plastic recycling would be reduced from 57% to a 49% for 2016 and then increased by 2% each year until 2020, to a maximum of 57% by 2020 [161].

As a corollary to this the multinational corporation Johnson & Johnson is to stop selling plastic cotton buds – one of the most common item of litter found on Britain's beaches – in half the countries of the world after a campaign to cut marine pollution [162]. The company will instead use paper to make the stick of the buds. This has been driven by campaign groups and environmentalists making the public aware of the impact of plastic pollution in our seas.

This latter example is a useful reminder of how it is unlikely that society will change its way of living but can try to minimise its impact on the environment. It is mankind's behaviour in the end that is the ultimate arbitrator of pollution and its control. In any moves towards a more sustainable way of life, it is certain that this can only come from such changes being acceptable to humanities current living standards and expectations. It is totally unrealistic to expect societies to change to what is perceived as and may well be in fact much lower standards of living.

10.7.2 Reducing the Impact: Towards Sustainability

There is a huge paradox in the use of chemicals to extract oil and gas. Fossil fuel is one of the major inputs into atmospheric emissions, particularly carbon dioxide, and yet we spend a huge amount of time and effort in ensuring minimal impact from using a small proportion of chemicals to aid in what would seem to be a highly polluting feedstock to other activities further up the supply chain, namely, power generation, refining for transport fuel, etc. Most of these uses are not renewable and, therefore at least in part, not sustainable. However, there also is an economic dimension to having a sustainable energy provision. The cost of renewables is still at the time of writing this study, more than the consumption of fossil fuels, particularly natural gas.

The future however probably lies in renewables for primary energy generation as global capacity has grown dramatically in the last few years [163]. At the end of 2016, more than 24% of global electricity was produced by renewables, dominated by hydropower and with wind contributing 4.0% and solar 1.5%. For all energy, renewable energy – excluding traditional wood burning – contributed 10%, overshadowed by the 80% coming from fossil fuels such as oil and gas. This is a considerable advance from a decade previously where renewables contributed a tiny fraction of the overall energy provision. Investment in renewable technologies is now larger than that for all fossil fuels despite subsidies for green energy being much lower than those for coal, oil and gas.

There is an urgent need for a dramatic reduction in carbon emissions as argued throughout this chapter. In Section 10.2.1, a case was made for shale gas development on the basis of energy security and as transition energy feedstock in the move away from energy produced with high carbon footprint. In the United States, it is claimed that the increased availability and use of natural gas from shale has been responsible for emissions falling to 1990 levels [164]. This is due to the fact that natural gas produces around 50% of the carbon dioxide emissions as coal for the same heat output. An additional advantage of gas power plants is their flexibility to change loading to electricity generated by renewables that other fuelled power plants such as coal and nuclear are unable to do [165].

North America leads the worldwide production of shale gas, with the United States and Canada having significant levels [166]. Beyond the United States and Canada, shale gas is so far produced at a commercial scale only in Argentina and China. While the shale gas potential of many nations seems promising, there are several obstacles spanning several economic, environmental, technical and social issues. Nonetheless a switch in China, for example, which is a major user of fossil fuels, coal and oil, to shale gas would have a dramatic impact on global carbon emissions. China also has by far the largest reserves of recoverable shale gas at least according to the EIA [18].

The two key industries pertinent in this study, namely, the upstream oil and gas industry and the chemical industry, are changing, faced with massive challenges as Western societies move from a carbon-based economy to a low carbon-based economy. Often it is industry and consumers working in the age-old customer–supplier relationship that has driven these changes with policymakers somewhat lagging behind.

The current economics in the oil market have been driven by the major OPEC countries happy to see low prices with Saudi Arabia and others content to see oil companies struggle elsewhere in the world as they drive down costs to remain competitive. This has a knock-on effect on the chemical suppliers. In 2016 Brent crude was trading at $30 per barrel, whereas in 2014 it was at $120 per barrel – a spectacular fall from which at the time of writing many oil and gas companies are only coming to terms with. In this new reality sustainability still has to be addressed. In Section 10.1 reference was made to the 'trilemma' in attempting to find solutions to society, the economy and the environment that work for the benefit of all. The World Energy Council has refined the terms to the 'energy trilemma' to express the challenges of maintaining secure, reliable energy that is affordable and accessible for all while ensuring environmental sustainability [167]. The pressure on resources is becoming ever greater, and the environmental damage caused may already be in some cases irreversible. Chemistry and the chemicals provided in achieving this will be critical in ensuring that environmental impacts are minimised and that sustainable technologies are practical and economic.

The pricing economics being practice by OPEC and others in the crude oil market is now seen by many companies as yesterday's fight. Many companies have cut their costs but are also looking to innovate to cut cost further and boost production while also maintaining the highest possible environmental standards. Some commentators are now talking about 'clean oil' in which comparisons are made as to the carbon standard in obtaining the oil. Where prices are equivalent, will consumers, especially those in Western Europe and North America, demand that the oil is sourced from the 'cleaner' suppliers?

The chemical industry is also changing with many companies investing in more sustainable feedstocks and replacing, at least in part, petrochemicals. Many are involved in the production of cellulosics, i.e. basic chemicals made from cellulose. This can provide biofuels such as cellulosic ethanol and can be manufactured from crop plants such as sugar cane (see also Section 8.2). Cellulosic polymers have been described previously in Section 2.2, 'Biopolymers'. Undoubtedly where economic, the chemical industry will continue to look for more sustainable feedstocks and cut its dependence on petrochemicals; however the petrochemical feedstocks offer considerable advantages in flexibility especially when manufacturing new products and polymers.

10.7.3 The Future?

At the recent (September 2017) Offshore Europe Conference in Aberdeen, United Kingdom, both Ben Van Beurden, Shell's chief executive officer, and Bob Dudley, BP's chief executive officer, focused on a changing landscape for the oil and gas industry, in particular that renewable energy is changing the future of the upstream business [168].

Undoubtedly the world needs renewable sources of energy. However, renewables produce electricity, and electricity is less than 20% of the world's energy system. Electric cars will not meet all of the planet's transportation needs. Also with global population continuing to grow, the transition across the globe to a low carbon economy may take generations to be realised.

As stated in the previous section of this chapter, chemical companies are investing heavily in biofuels and bio-feedstocks; however, so are oil and gas companies, BP, for example, are on track to commercialise the manufacture of bio-butanol [169], and many are transitioning themselves from majority oil producers to mostly producing natural gas, which is designated by many as 'lower carbon fuel'. Also, the global oil and gas infrastructure has only recovered around 30% of the reserves under its current operating mandate. The cash potential of trapped assets could represent an injection of revenue into many oil-producing countries that, under the right economic circumstances, is unlikely to be ignored. It can be anticipated that EOR methods including chemical EOR are likely to be adopted and that the chemical industry will be required to aid in these projects with sustainable solutions.

In a different direction, many oil and gas operators, particularly offshore, are investing in the installation of renewable technology, particularly wind-driven turbines, directly. Many of the design practices of structures required are similar as are the work vessels needed to build and maintain them. Much of the offshore oil and gas skills are transferable. The UK sector of the North Sea, in particular activity in the offshore wind sector, has overtaken traditional oil and gas activity [170].

An area that will be required to be given more attention to achieve a sustainable future will be for practitioners and regulators, in a balanced methodology, to assess chemicals not only for their physiochemical and toxicological properties when considering their environmental impact but also for more profound longer-term aspects.

For example, when selecting or regulating the use and discharge of a biocide, the following should be considered:

1) Uncharged species will dominate in the aqueous phase and be subject to degradation and transport, whereas charged species will sorb to soils and be less bioavailable.
2) Many biocides are short lived or degradable through abiotic and biotic processes, but some may transform into more toxic or persistent compounds.
3) The understanding of biocides' fate under downhole conditions (high pressure, temperature and salt and organic matter concentrations) is limited.
4) Several biocidal alternatives exist, but high cost, high energy demands and/or formation of disinfection by-products limits their use.

Similar processes should be adopted for all chemicals being used, and in addition the study of their bioavailability and ultimate fate in the environment should be considered.

Hydrocarbons and fossil fuels look set to play a major role in the near to medium future, and hopefully this can be utilised in a more sustainable and less polluting manner than in previous decades. Chemicals will continue to be a vital component in the extraction processing and production of crude oil and natural gas; probably for several more decades, changes and new innovations will undoubtedly have significant impacts. Considerable and interesting work is being conducted in the use of catalysis to convert atmospheric carbon dioxide directly into methane [171], if commercial this could have a significant economic impact not only on natural gas production but also on carbon dioxide emissions. It is also possible that similar processes can be used to capture waste methane directly and therefore also cut carbon emissions [172]. These are but a couple of examples of work ongoing to create energy from new sources or more efficiently from existing ones.

In our mixed-energy future, it is a highly likely scenario that both hydrocarbons and renewables will coexist for many years, and a sustainable oilfield chemistry discipline will be required to aid and service its delivery. This book has been an attempt to describe the chemistry currently involved, its environmental impact and its potential to move towards a more sustainable future alongside a changing upstream oil and gas sector.

References

1 http://www.shell.com/sustainability.html
2 https://www.basf.com/en/company/sustainability.html
3 http://www.nalco.com/sustainability.htm
4 Winterton, N. (2011). *Chemistry for Sustainable Technologies: A Foundation.* Cambridge: RSC Publishing.
5 Clift, R. (2000). From the forum on sustainability. *Clean Products and Processes* **2** (1): 67–70.
6 Merkel, A. (1998). The Role of Science in Sustainable Development. *Science* **28** (5375): 336–337.
7 http://www.ipcc.ch/
8 Wendimu, M.A. (2016). Jatropha potential on marginal land in Ethiopia: reality or myth? *Energy for Sustainable Development* **30**: 14–20.
9 https://dddusmma.wordpress.com/2016/01/29/false-promise-of-biofuels/
10 https://www.independentsciencenews.org/environment/biofuel-or-biofraud-the-vast-taxpayer-cost-of-failed-cellulosic-and-algal-biofuels/
11 https://www.weforum.org/agenda/2015/12/the-paris-climate-agreement-what-happens-now
12 http://www.un.org/apps/news/story.asp?NewsID=53550#.VwZK5vkrKUk
13 https://www.theguardian.com/business/2017/jul/05/volvo-cars-electric-hybrid-2019
14 https://www.epa.gov/ghgemissions/sources-greenhouse-gas-emissions
15 https://www.amnesty.org/en/latest/campaigns/2016/06/drc-cobalt-child-labour/
16 https://www.theguardian.com/politics/2017/jul/25/britain-to-ban-sale-of-all-diesel-and-petrol-cars-and-vans-from-2040
17 http://www.bbc.co.uk/news/science-environment-11175386
18 https://www.eia.gov/analysis/studies/worldshalegas/
19 https://www.theguardian.com/environment/2016/jan/26/engineers-warn-of-looming-uk-energy-gap
20 https://www.forbes.com/sites/judeclemente/2017/05/21/the-great-u-s-oil-export-boom/#676e12707e5b
21 Cooper, J., Stamford, L., and Azapagic, A. (2016). Shale gas: a review of the economic, environmental, and social sustainability. *Energy Technology* **4** (7): 772–792.
22 Hubbert, M.K. and Willis, D.G. (1957). Mechanics of hydraulic fracturing. *Transactions of the AIME* **210**: 153–166.

23 Von Terzaghi, K. (1923). Die Berechnung der durchlässigkeit des tones aus dem verlauf der hydromechanischen spannungserscheinungen. *Sitzungsbericht der Akademie der Wissenschaften (Wien): Mathematisch-Naturwissenschaftlichen Klasse* **132**: 125–138.

24 Lukocs, B., Mesher, S., Wilson, T.P.J. et al. (2011). Non-volatile phosphorus hydrocarbon gelling agent. US Patent 8,084,401.

25 Chemicals Used in Hydraulic Fracturing. http://www.fracfocus.org/water-protection/drilling-usage (accessed 14 December 2017).

26 Mathur, N.K. (2017). *Industrial Galactomannan Polysaccharides*. CRC Press.

27 Code of Federal Regulations, Food and Drugs, Title 21, Sec. 184.1339. http://www.accessdata.fda.gov/scripts/cdrh/cfdocs/cfcfr/CFRSearch.cfm?fr=184.1339&SearchTerm=guar%20gum (accessed 14 December 2017).

28 Ainley, B.R., and McConnell, S.B. (1993). Delayed borate cross-linked fracturing fluid. EP Patent 5,284,61.

29 Brannon, H.D. and Ault, G.M. (1991). New, delayed borate-crosslinked fluid provides improved fracture conductivity in high-temperature applications. SPE Annual Technical Conference and Exhibition, Dallas, TX (6–9 October 1991), SPE 22838.

30 Nimerick, K.H., Crown, C.W., McConnell, S.B. and Ainley, B. (1993). Method of using borate crosslinked fracturing fluid having increased temperature range. US Patent 5,259,455.

31 World Health Organization (1998). *Boron, Environmental Health Criteria 204*. Geneva, Switzerland: World Health Organization.

32 Eisler, R. (1990). Boron hazards to fish, wildlife, and invertebrates: a synoptic review. Contaminant Hazard Reviews Report 20; Biological Report 85 (1.20). *US Fish and Wildlife Service* **85**: 1–32.

33 Reregistration Eligibility Decision Document: Boric Acid and its Sodium Salts (1993). U.S. Environmental Protection Agency, Office of Pesticide Programs. Washington, DC: U.S. Government Printing Office, EPA 738-R-93-017.

34 Noweck, K. (2011). Production, technologies and applications of fatty alcohols. Lecture at the 4th Workshop on Fats and Oils as Renewable Feedstock for the Chemical Industry, Karlsruhe, Germany (20–22 March 2011).

35 Mudge, S.M. (2005). *Fatty Alcohols – A Review of Their Natural Synthesis and Environmental Distribution For SDA and ERASM*. Soap and Detergent Association 2005.

36 Kleindorfer, P.R., Singhai, K., and Van Wassenhove, L.N. (2005). Sustainable operations management. *Production and Operations Management* **14** (4): 482–492.

37 http://news.bbc.co.uk/onthisday/hi/dates/stories/june/20/newsid_4509000/4509527.stm

38 Twomey, B. (2012). Life cycle of and oil and gas installation. Presented at CCOP and EPPM Workshop on End of Concession & Decommissioning Guidelines, Bangkok (13 June 2012).

39 Kirby, S. (1999). Donan field decommissioning project. Offshore Technology Conference, Houston, TX (3–6 May 1999), OTC 10832.

40 Pebell, U., Kohazy, R. and Dietler, A.. (2000). Experiences in field decommissioning. SPE International Conference on Health, Safety and Environment in Oil and Gas Exploration and Production, Stavanger, Norway (26–28 June 2000), SPE 61477.

41 Desai, P.C., Hekelaar, S. and Abshire, L. (2013). Offshore well plugging and abandonment: challenges and technical solutions. Offshore Technology Conference, Houston, TX (6–9 May 2013), OTC-23906.

42 Craddock, H.A., Campbell, E., Sowerby, K. et al. (2007). The application of wax dissolver in the enhancement of export line cleaning. International Symposium on Oilfield Chemistry, Houston, TX (28 February-2 March 2007), SPE 105049.

43 Dalmazzone, C., Carrausse, M., Tabacchi, G. et al. (2005). Development of green chemicals for cleaning operations in platforms decommissioning. SPE International Symposium on Oilfield Chemistry, The Woodlands, TX (2–4 February 2005), SPE 92826.

44 Craddock, H.A. Selection of appropriate hydrogen sulphide scavengers for decommissioning activities, unpublished results.

45 Oudman, B.L. (2017). Green decommissioning: re-use of North Sea offshore assets in a sustainable energy future. Offshore Mediterranean Conference and Exhibition, Ravenna, Italy (29–31 March 2017), OMC 2017-572.

46 Hanna, I.S. and Abukhamsin, S.A. (1998). Landfills and recycling provide alternatives for OBM disposal. *Oil and Gas Journal* **96** (23). http://www.ogj.com/articles/print/volume-96/issue-23/in-this-issue/drilling/landfills-and-recycling-provide-alternatives-for-obm-disposal.html.

47 Gerrard, S., Grant, A., Marsh, R. and London, C. (1999). Drill Cuttings Piles in the North Sea: Management Options During Platform Decommissioning. Research Report No. *31*. Norwich: Centre for Environmental Risk, School of Environmental Sciences, University of East Anglia.

48 Burke, C. and Veil, J.A. (1995).Synthetic drilling muds: environmental gain deserves regulatory confirmation. SPE/EPA Exploration and Production Environmental Conference, Houston, TX (27–29 March 1995), SPE 29737.

49 Peresic, R.L., Burrell, B.R. and Prentice, G.M. (1991). Development and field trial of a biodegradable invert emulsion fluid. paper SPEIIADC 21935 presented at the 1991SPE/IADC Drilling Conference, Amsterdam (1–14 March 1991), SPE 21935.

50 Gee, J.C., Lawrie, C.J. and Williamson, R.C. (1996). Drilling fluids comprising mostly linear olefins. US Patent 5,589,442, assigned to Chevron Chemical company.

51 OSPAR (2000). OSPAR Decision 2000/3 on the Use of Organic-Phase Drilling Fluids (OPF) and the Discharge of OPF-Contaminated Cuttings, OSPAR Convention for The Protection of The Marine Environment in The North-East Atlantic, Meeting of the OSPAR Commission, Copenhagen 26–30 June 2000).

52 https://www.gov.uk/guidance/oil-and-gas-offshore-environmental-legislation

53 OSPAR Commission (2009). *Implementation Report on Recommendation 2006/5 on a Management Regime for Offshore Cutting Piles.* OSPAR Commission.

54 Bakke, T., Klungsoyr, J., and Sanni, S. (2013). Environmental impacts of produced water and drilling waste discharges from the Norwegian offshore petroleum industry. *Marine Environmental Research* **92**: 154–169.

55 Neff, J.M., McKelvie, S., and Ayers, R.C. Jr. (2000). Environmental Impacts of Synthetic Based Drilling Fluids. U.S. Department of the Interior, Minerals Management Service, Gulf of Mexico OCS Region, OCS Study, MMS 2000-064.

56 Visser, S., Lee, B., Fleece, T. and Sparkes, D. (2004). Degradation and ecotoxicity of C14 linear alpha olefin drill cuttings in the laboratory and in the field. SPE International Conference on Health, Safety and Environment in Oil and Gas Exploration and Production, Calgary Canada (29–31 March 2004), SPE 86698.

57 Hjelmas, T.A., Bakke, S., Hilde, T.. et al. (1996). Produced water reinjection: experiences from performance measurements on Ula in the North Sea. SPE Health, Safety and Environment in Oil and Gas Exploration and Production Conference, New Orleans, LA (9–12 June 1996), SPE 35874.

58 Abou-Sayed, A.S., Zaki, K.S., Wang, G. et al. (2007, 2007). Produced water management strategy and water injection best practices: design, performance, and monitoring. *SPE Production and Operations* **22** (01): 59–68, SPE 108238.

59 Mackay, E.J., Jones, T.J. and Ginty, W.R. (2012). Oilfield scale management in the Siri asset: paradigm shift due to the use of mixed PWRI/seawater injection. SPE Europec/EAGE Annual Conference, Copenhagen, Denmark (4–7 June 2012), SPE 154534.

60 Boschee, P. (2014). Produced and flowback water recycling and reuse: economics, limitations, and technology. *Oil and Gas Facilities* **3** (01): 16–21, SPE 0214–0016 OGF.

61 Johnston, C.J., Sutherland, L. and Jordan, M.M. (2016). The impact of corrosion inhibitors on relative permeability of water injector and production well performance. SPE International

Conference and Exhibition on Formation Damage Control, Lafayette, LA (24–26 February 2016), SPE 178981.

62 Jordan, M.M., Weathers, T.M., Jones, R.A. et al. (2014). The impact of kinetics hydrate inhibitors within produced water on water injection/disposal wells. SPE International Symposium and Exhibition on Formation Damage Control, Lafayette, LA (26–28 February 2014), SPE 168173.

63 Martin, R.L. (2008). Corrosion consequences of nitrate/nitrite additions to oilfield brines. SPE Annual Technical Conference and Exhibition, Denver, CO (21–24 September 2008), SPE 114923.

64 Jenneman, G.E., Achour, M. and Joosten, M.W. (2009). The corrosiveness of nitrite in a produced-water system. SPE International Symposium on Oilfield Chemistry, The Woodlands, TX (20–22 April 2009), SPE 121463.

65 Stott, J.D., Dicken, G., Rizk, T.Y. et al. (2008). Corrosion inhibition in (PWRI) systems that use nitrate treatment to control SRB activity and reservoir souring. CORROSION 2008, New Orleans, LA (16–20 March 2008), NACE 08507.

66 Grimaldi, M.C., Castrisana, W.J., Tolfo, F.C. et al. (2010). Produced water reuse for production of chemicals. SPE International Conference on Health, Safety and Environment in Oil and Gas Exploration and Production, Rio de Janeiro, Brazil (12–14 April 2010), SPE 127174.

67 Doran, G.F., Carini, F.H., Fruth, D.A. et al. (1997). Evaluation of technologies to treat oil field produced water to drinking water or reuse quality. SPE Annual Technical Conference and Exhibition, San Antonio, TX (5–8 October 1997), SPE 38830

68 Doran, G.F., Williams, K.L., Drago, J.A. et al. (1999). Pilot study results to convert oil field produced water to drinking water or reuse quality. International Thermal Operations/Heavy Oil Symposium, Bakersfield, CA (17–19 March 1999), SPE 54110.

69 Leung, D.Y.C., Caramanna, G., and Maroto-Valer, M.M. (2014). An overview of current status of carbon dioxide capture and storage technologies. *Renewable and Sustainable Energy Reviews* **39**: 426–443.

70 Supekar, S.D. and Skerlos, S.J. (2015). Reassessing the efficiency penalty from carbon capture in coal-fired power plants. *Environmental Science and Technology* **49** (20): 12576–12584.

71 https://sequestration.mit.edu/tools/projects/in_salah.html

72 https://www.forbes.com/sites/jeffmcmahon/2017/06/28/the-world-has-to-develop-carbon-capture-iea-warns-and-its-not/#50171b621b51

73 http://www.iea.org/etp2017/summary/

74 Martin, D.F. and Taber, J.J. (1992). Carbon dioxide flooding. *Journal of Petroleum Technology* **44** (4): 396–400, SPE 23564.

75 Booth, A., da Silva, E. and Brakstad, O.G. (2011). Environmental impacts of amines and their degradation products: current status and knowledge gaps. 1st Post Combustion Capture Conference, Abu Dhabi (17–19 May 2011).

76 Ramdin, M., de Loos, T.W., and Vlugt, T.J.H. (2012). State-of-the-art of CO_2 capture with ionic liquids: review. *Industrial and Engineering Chemistry Research* **51** (24): 8149–8177.

77 Elkins, P., Vanner, R., and Firebrace, J. (2006). Decommissioning of offshore oil and gas facilities: a comparative assessment of different scenarios. *Journal of Environmental Management* **79** (4): 420–438.

78 Price, W.R., Ross, B. and Vicknair, B. (2016). Integrated decommissioning – increasing efficiency. Offshore Technology Conference, Houston, TX (2–5 May 2016), OTC – 27152.

79 Rand, G.M. ed. (1995). *Fundamentals of Aquatic Toxicology: Effects, Environmental Fate and Risk Assessment*, 2ee. CRC Press.

80 Cole, S., Codling, I.D., Parr, W. and Zabel, T. (1999). Guidelines for Managing Water Quality Impacts within UK European Marine Sites. Report part of UK SACs Project.

81 SEPA (2013). *Regulatory Method (WAT-RM-28): Modelling for Water Use Activities*. Scottish Environmental Protection Agency (SEPA).

82 Mackay, D. (2001). *Multimedia Environmental Models: The Fugacity Approach*, 2ee. CRC Press.

83 Cahill, T.M., Cousins, I., and Mackay, D. (2003). General fugacity-based model to predict the environmental fate of multiple chemical species. *Environmental Toxicology and Chemistry* **22** (3): 483–493.

84 Mackay, D., Di Guardo, A., Paterson, S. et al. (1996). Assessment of chemical fate in the environment using evaluative, regional and local-scale models: illustrative application to chlorobenzene and linear alkylbenzene sulfonates. *Environmental Toxicology and Chemistry* **15** (9): 1638–1648.

85 http://www.trentu.ca/academic/aminss/envmodel/models/Qwasi.html

86 http://www.trentu.ca/academic/aminss/envmodel/models/Wania.html

87 Hill, D.G., Dismuke, K., Shepherd, W. et al. (2003). Development practices and achievements for reducing the risk of oilfield chemicals. SPE/EPA/DOE Exploration and Production Environmental Conference, San Antonio, TX (10–12 March 2003), SPE 80593.

88 McLean, T.L., Dalrymple, E.D., Muellner, M. and Swofford, S. (2012). A method for improving chemical product risk profiles as part of product development. SPE Annual Technical Conference and Exhibition, San Antonio, TX (8–10 October 2012), SPE 159355.

89 Hull, D. and Ruttenberg, K.C. (2016). Variable molecular weight distribution and bioavailability of DOP from coastal and open ocean waters suggests compositional heterogeneity. Presented to Ocean Sciences Meeting, New Orleans, LA (21–26 February 2016).

90 McWilliams, P. and Payne, G. (2001). Bioaccumulation potential of surfactants: a review. Chemistry in the Oil Industry VII (13–14 November 2001). Manchester, UK: Royal Society of Chemistry.

91 Department for Environment and Rural Affairs (2012). United Kingdom's Contaminated Land Regulations under Part IIA of the Environmental Protection Act of 1990- Statutory Guidance.

92 Suedel, B.C., Boraczek, J.A., Peddicord, R.K. et al. (1994). Trophic transfer and biomagnification potential of contaminants in aquatic ecosystems. In: *Reviews of Environmental Contamination and Toxicology*, vol. **136** (ed. G.W. Ware), 21–39. New York: Springer-Verlag.

93 Neely, W.B. (1980). *Chemicals in the Environment: Distribution, Transport, Fate, Analysis*, 245. New York: Marcel Dekker.

94 Alexander, D.E. (1999). Bioaccumulation, bioconcentration, biomagnification. In: *Environmental Geology, Part of the Series Encyclopedia of Earth Science*, 43–44. Springer.

95 https://stats.oecd.org/glossary/detail.asp?ID=2044

96 OECD (1995). Guidelines for testing of Chemicals Section 1: Physical Chemical Properties, Test No. 107: Partition Coefficient (n-octanol/water): Shake Flask Method.

97 OECD (2004). Guidelines for testing of Chemicals Section 1: Physical Chemical Properties, Test No. 117: Partition Coefficient (n-octanol/water), HPLC Method.

98 OECD (2006). Guidelines for testing of Chemicals Section 1: Physical Chemical Properties, Test No. 123: Partition Coefficient (1-Octanol/Water): Slow-Stirring Method.

99 Elder, J. (2011). Relative solubility number RSN – an alternative measurement to Log P_{ow} for determining the bioaccumulation potential. Master's thesis. Goteborg, Sweden: Chalmers University of Technology.

100 Wu, J., Xu, Y., Dabros, T., and Hamza, H.A. (2003). Development of a method for measurement of relative solubility of nonionic surfactants. *Colloids and Surfaces A: Physicochemical and Engineering Aspects* **232**: 229–237.

101 Mackay, D., Webster, E., Cousins, I. et al. (2001). An Introduction to Multimedia Final Report Prepared as a Background Paper for OECD Workshop Ottawa. July 2001, CEMC Report No. *200102.*

102 Mackay, D., Arnot, J.A., Webster, E., and Reid, L. (2009, 2009). The evolution and future of environmental fugacity models. In: *Ecotoxicology Modeling*, Part of the Emerging Topics in Ecotoxicology Book Series (ETEP), vol. **2**, 355–375. Springer.

103 http://www.nytimes.com/2010/05/28/us/28flow.html

104 Canevari, G.P., Fiocco, R.J., Becker, K.W. and Lessard, R.R. (1997). Chemical dispersant for oil spills. US Patent 5,618,468, assigned to Exxon Research and Engineering Company.

105 Aman, Z.M., Paris, C.B., May, E.F. et al. (2015). High-pressure visual experimental studies of oil-in-water dispersion droplet size. *Chemical Engineering Science* **127**: 392–400.

106 https://earthjustice.org/about

107 DTI (2003). Our Energy Future: Creating a Low Carbon Economy, Department of Trade and Industry, February 2003, UK.

108 Fingas, M.F. and Hollebone, B.P. (2003). Review of behaviour of oil in freezing environments. *Marine Pollution Bulletin* **47**: 333–340.

109 Wilkinson, J.P., Boyd, T., Hagen, B. et al. (2015). Detection and quantification of oil under sea ice: the view from below. *Cold Regions Science and Technology* **109**: 9–17.

110 Arctic Response Technology Oil Spill Preparedness (2017). Arctic Oil Spill Response Technology Joint Industry Programme Synthesis Report, D. Dickins - DF Dickins Associates, LLC.

111 Andrady, A.L. (2011). Microplastics in the marine environment. *Marine Pollution Bulletin* **62** (8): 11596–11605.

112 Carpenter, E.J. and Smith, K.L. Jr. (1972). Plastics on the Sargasso sea surface. *Science* **175** (4027): 1240–1241.

113 Derraik, J.G.B. (2002). The pollution of the marine environment by plastic debris: a review. *Marine Pollution Bulletin* **44** (9): 842–852.

114 Cole, M., Lindeque, P., Halsband, C., and Galloway, T.S. (2011). Microplastics as contaminants in the marine environment: a review. *Marine Pollution Bulletin* **62** (12): 2588–2597.

115 http://www.unep.org/newscentre/un-declares-war-ocean-plastic

116 King, W.R. and Cleland, D.I. (1988). *Life-Cycle Management*. New York: John Wiley & Sons, Inc.

117 Khatib, Z. and Walsh, J.M. (2014). Extending the life of mature assets: how integrating subsurface & surface knowledge and best practices can increase production and maintain integrity. SPE Annual Technical Conference and Exhibition, Amsterdam, The Netherlands (27–29 October 2014), SPE 170804.

118 ISO 14040:2006 (2006). Environmental management - life cycle assessment - principles and framework. International Standards Organisation.

119 https://www.theguardian.com/sustainable-business/2016/sep/29/fracking-shale-gas-europe-opposition-ban

120 Hunkeler, D. (2006). Societal LCA methodology and case study. *International Journal of Life Cycle Assessment* **11** (6): 371–382.

121 Lovelock, J. (2006). *The Revenge of Gaia*. London: Penguin Books.

122 United States Environmental Protection Agency (2006). Green Chemistry.

123 Martel, A.E., Davies, J.A., Olson, W.W., and Abraham, M.A. (2003). Green chemistry and engineering, drivers, metrics and reduction to practice. *Annual Review of Environment and Resources* **28**: 401.

124 Warner, J.C. and Anastas, Green, P. (1998). *Chemistry: Theory and Practice*. New York: Oxford University Press.

125 OPAR Commission. http://www.ospar.org (accessed 14 December 2014).

126 EOSCA OSPAR Regulations on the Use of Chemicals, including PLONOR List. http://www. eosca.com/OReg.htm (accessed 14 December 2014).

127 Thatcher, M. and Payne, G. (2001). Impact of the OSPAR decision on the harmonised mandatory control system on the offshore chemical supply industry. Chemistry in the Oil Industry VII (13–14 November 2001). Manchester, UK: Royal Society of Chemistry.

128 Constable, D.J.C., Curzons, A.D., and Cunningham, V.L. (2002). Metrics to 'green' chemistry—which are the best? *Green Chemistry* **4** (6): 521–527.

129 Mooney, D.J., Buhadir, K.H., Wong, W.H. and Rowley, J.A. (2008). Polymers containing polysaccharides such as alginates or modified alginates. European Patent 09271196.

130 Rossegger, E., Schenk, V., and Wiesbrock, F. (2013). Design strategies for functionalised poly(2-oxazolines) and derived materials. *Polymers* **5**: 956–1011.

131 Hamelink, J. ed. (1994). *Bioavailability: Physical, Chemical, and Biological Interactions*, SETAC Special Publications Series. CRC Press.

132 Kuhad, R.C., Gupta, R., and Singh, A. (2011). Microbial cellulases and their industrial applications. *Enzyme Research* **2011**: 10, Article ID 280696. doi: 10.4061/2011/280696.

133 Delidovich, I., Hasoul, P.J.C., Pfutzenreuter, R. et al. (2016). Alternative monomers based on lignocellulose and their use for polymer production. *Chemical Reviews* **116** (3): 1540–1599.

134 Larche, J.-F., Bussiere, P.-O., Theria, S., and Gardette, J.-L. (2012). Photooxidation of polymers: relating material properties to chemical changes. *Polymer Degradation and Stability* **97** (1): 25–34.

135 Guedri-Knani, L., Gardette, J.L., Jacquet, M., and Rivaton, A. (2004). Photoprotection of poly(ethylene-naphthalate) by zinc oxide coating. *Surface and Coatings Technology* **180– 181**: 71–75.

136 Miyatake, N., Sue, H.-J., Li, Y. and Yamaguchi, K. (2007). Stabilization of polymers with zinc oxide nanoparticles. International Patent Application WO 2007075654, assigned to Texas A & M University and Kaneka Corporation.

137 Tatraalji, D., Foldes, E., and Pukansky, B. (2014). Efficient melt stabilization of polyethylene with quercetin, a flavonoid type natural antioxidant. *Polymer Degradation and Stability* **102**: 41–48.

138 Crosby, D.G. (1994). Photochemical aspects of bioavailability. In: *Bioavailability: Physical, Chemical, and Biological Interactions*, Setac Special Publications Series, Chapter 1 (ed. J. Hamelink), 109. CRC Press.

139 Leja, K. and Lewandowicz, G. (2010). Polymer biodegradation and biodegradable polymers – a review. *Polish Journal of Environmental Studies* **19** (2): 255–266.

140 Norwegian Ministry of the Environment (1996–1997). White Paper No. 58 Environmental Policy for a Sustainable Development – Joint Effort for the Future.

141 Tyler, C.R., Jobling, S., and Sumpter, J.P. (1998). Endocrine disruption in wildlife: a critical review of the evidence. *Critical Reviews in Toxicology* **28** (4): 319–361.

142 Barber, L.B., Loyo-Rosales, J.E., Rice, C.P. et al. (2015). Endocrine disrupting alkylphenolic chemicals and other contaminants in wastewater treatment plant effluents, urban streams, and fish in the Great Lakes and Upper Mississippi River Regions. *Science of the Total Environment* **517**: 195–206.

143 Getliff, J.M. and James, S.G. (1996). The replacement of alkyl-phenol ethoxylates to improve the environmental acceptability of drilling fluid additives. SPE Health, Safety and Environment in Oil and Gas Exploration and Production Conference, New Orleans, LA (9–12 June 1996), SPE 35982.

144 Hawrelak, M., Bennet, E., and Metcalfe, C. (1999). The environmental fate of the primary degradation products of alkylphenol ethoxylate surfactants in recycled paper sludge. *Chemosphere* **39** (5): 745–752.

145 Cordell, D., Rosemarin, A., Schroder, J.J., and Smit, A.L. (2011). Towards global phosphorus security: a systems framework for phosphorus recovery and reuse options. *Chemosphere* **84** (6): 747–758.

146 Harris, R.E. and McKay, I.D. (1998). New applications for enzymes in oil and gas production. European Petroleum Conference, The Hague, Netherlands (20–22 October 1998), SPE 50621.

147 Nasr-El-Din, H.A., Al-Otaibi, M.G.H., Al-Qahtani, A.M. and McKay, I.D. (2005). Lab studies and field application of in-situ generated acid to remove filter cake in gas wells. SPE Annual Technical Conference and Exhibition, Dallas, TX (9–12 October 2005), SPE 96965.

148 Rae, P. and Di Lullo, G. (2003). Matrix acid stimulation - a review of the state-of-the-art. SPE European Formation Damage Conference, The Hague, Netherlands (13–14 May 2003), SPE 82260.

149 Feng, Q., Ma, X., Zhong, L. et al. (2009). EOR pilot tests with modified enzyme--Dagang oilfield, China. *SPE Reservoir Evaluation & Engineering* **12** (01): 79–87, SPE 107128.

150 Larsen, T., Lioliou, M.G., Josang, L.O. and Ostvold, T. (2006). Quasi natural consolidation of poorly consolidated oil field reservoirs. SPE International Oilfield Scale Symposium, Aberdeen, UK (31 May-1 June 2006), SPE 100598.

151 McRae, J.A., Heath, S.M., Strachan, C. et al. (2004). Development of an enzyme activated, low temperature, scale inhibitor precipitation squeeze system. SPE International Symposium on Oilfield Scale, Aberdeen, UK (26–27 May 2004), SPE 87441.

152 Zajic, J.E., Cooper, D.G., Jack, T.R., and Kosaric, N. (1983). *Microbial Enhanced Oil Recovery.* PenWell Books Oklahoma.

153 Andre, H.J. and Kristian, K.H. (2005). Genetically engineered well treatment micro-organisms. UK Patent 2,413,797, assigned to Statoil Asa.

154 Gavrilescu, M. and Christi, Y. (2005). Biotechnology—a sustainable alternative for chemical industry. *Biotechnology Advances* **23** (7–8): 471–499.

155 Dobson, J.W., Hayden, S.L. and Hinojosa, B.E. (2005). Borate cross-linker suspensions with more consistent crosslink times. US Patent 6,936,575, assigned to Texas United Chemical Company Llc.

156 Legemah, M., Guerin, M., Sun, H., and Qu, Q. (2014). Novel high-efficiency boron crosslinkers for low-polymer-loading fracturing fluids. *SPE Journal* **19** (04): 737–743, SPE 164118.

157 Shang, X., Ding, Y., Wang, Y., and Yang, L. (2015). Rheological and performance research on a regenerable polyvinyl alcohol fracturing fluid. *PLoS One* **10** (12): e0144449. Published online 2015 December 7.

158 Viloria, A., Castillo, L., Garcia, J.A. and Biomorgi, J. (2010). Aloe derived scale inhibitor. US Patent Application, 20100072419, assigned to Intevep S.A.

159 http://www.cefas.co.uk/media/1384/13-06e_plonor.pdf

160 US Energy Information Administration. https://www.eia.gov/tools/faqs/faq.php?id=34&t=6

161 http://www.independent.co.uk/news/business/news/government-recycling-targets-cut-pressure-plastics-lobbying-industry-a7585501.html

162 http://www.independent.co.uk/environment/johnson-johnson-cotton-buds-plastic-half-world-marine-pollution-sea-life-a7577556.html

163 https://www.theguardian.com/environment/2017/jun/06/spectacular-drop-in-renewable-energy-costs-leads-to-record-global-boost

164 https://www.wsj.com/articles/u-s-carbon-dioxide-emissions-hit-new-25-year-low-1476298479

165 Heath, G.A., O'Donoughue, P., Arent, D.J., and Bazilian, M. (2014). Harmonization of initial estimates of shale gas life cycle greenhouse gas emissions for electric power generation. *Proceedings of the National Academy of Sciences of the United States of America* **111** (31): E3167–E3176.

166 https://www.eia.gov/todayinenergy/detail.php?id=13491

167 https://www.worldenergy.org/work-programme/strategic-insight/
assessment-of-energy-climate-change-policy/

168 JPT (2017). BP and shell agree: a new energy future is coming. *Journal of Petroleum Technology* .

169 BP Strategic Report 2014.

170 Rassenfoss, S. and Jacobs, T. (2017). A growing range of renewable options for oil and gas. *Journal of Petroleum Technology* **69** (8): 32–33.

171 Rao, H., Schmidt, L.C., Bonin, J., and Robert, M. (2017). Visible-light-driven methane formation from CO_2 with a molecular iron catalyst. *Nature* **548**: 74–77.

172 Sushkevich, V.L., Palagin, D., Ranocchiari, M., and van Bokhoven, J.A. (2017). Selective anaerobic oxidation of methane enables direct synthesis of methanol. *Science* **356** (6337): 523–527.

Index

Oilfield Chemistry and Its Environmental Impact, First Edition. Henry A. Craddock.
© 2018 John Wiley & Sons Ltd. Published 2018 by John Wiley & Sons Ltd.